METHODOLOGY
IN MAMMALIAN GENETICS

Sponsored by

THE ROSCOE B. JACKSON MEMORIAL LABORATORY
THE GENETICS STUDY SECTION
DIVISION OF RESEARCH GRANTS
AND DIVISION OF GENERAL MEDICAL SCIENCES
NATIONAL INSTITUTES OF HEALTH *

* Grant RG 7097

METHODOLOGY *in* MAMMALIAN GENETICS

Edited by

WALTER J. BURDETTE, A.B., A.M., Ph.D., M.D.

*Professor and Head of the Department of Surgery and
Director of the Laboratory of Clinical Biology,
University of Utah College of Medicine;
Surgeon-in-Chief, Salt Lake County Hospital;
Chief Surgical Consultant, Veterans Administration Hospital,
Salt Lake City, Utah*

HOLDEN-DAY, INC., *San Francisco*
1963

200213

PREFACE

Information recently acquired about the biochemistry of heredity has increased the likelihood of determining precisely how the more formal transmission of genetic information is accomplished in higher organisms and has imposed the obligation to renew the task of controlling changes in composition, propagation, and action of the genetic material. The complexity of mammalian genesis and development is regarded no longer as a barrier to investigation at the molecular level, but more as an opportunity to study mechanisms that do not exist in lower organisms in relation to common hereditary units and to choose between alternate explanations for a given process. Possibly the greatest advantage the laboratory mammal offers is not only the possibility of developing strains of animals having remarkably similar partial or total genome but also the opportunity to breed representatives from diverse strains in a manner appropriate for the elucidation of genetic mechanisms. In addition, these uniform lines are available for comparison of the genetic behavior of cells with known properties *in vivo* and *in vitro*. Recent evidence for fusion of mammalian cells *in vitro* suggests that the analytical advantages of sexual reproduction may be extended to studies of the somatic cell as well. Also, the many hereditary diseases known to occur in inbred mammals and the varied response of different strains to bacterial, viral, and parasitic inoculation offer means to determine parameters that may also be operable in similar diseases in man.

A host of methods have been evolved for providing stocks of mammals with uniform genotype suitable for given experimental objectives. The theoretical and practical aspects of insuring this type of control and some indication of how methodology from other scientific disciplines may be used in mammalian genetics are mandatory for an approach to the solution of problems that engage the attention of many contemporary investigators. The intent of the contributors to this volume is to provide a selected array of methods applicable to mammalian genetics that may prove useful in implementing the ideas of those striving to solve the intricate problems encountered in investigations concerned with differentiation, homotransplantation, directed mutation, repair of deleterious mutants, genetic determinants in disease, and the like.

WALTER J. BURDETTE

Chairman,
Genetics Study Section

Salt Lake City, Utah
August, 1962

PARTICIPANTS

Thomas Anderson, Ph.D.*
  The Institute for Cancer Research, Fox Chase, Philadelphia 11, Pennsylvania

H. B. Andervont, Sc.D.
  Laboratory of Biology, National Cancer Institute, National Institutes of Health, Bethesda, Maryland

Louis Baron, Ph.D.*
  Division of Immunology, Walter Reed Army Institute of Research, Walter Reed Army Medical Center, Washington, D.C.

Morris K. Barrett, M.D.
  National Cancer Institute, National Institutes of Health, Bethesda, Maryland

Michael A. Bender, Ph.D.
  Biology Division, Oak Ridge National Laboratory, Oak Ridge, Tennessee

Howard A. Bern, Ph.D.
  Department of Zoology, Cancer Research Genetics Laboratory, University of California, Berkeley, California

S. E. Bernstein, Ph.D.
  Roscoe B. Jackson Memorial Laboratory, Bar Harbor, Maine

Walter J. Burdette, Ph.D., M.D.*
  Department of Surgery, University of Utah, Salt Lake City, Utah
    * Member, Genetics Study Section

C. K. Chai, Ph.D.
Battelle Memorial Institute, Columbus, Ohio

Herman B. Chase, Ph.D.
Department of Biology, Brown University, Providence, Rhode Island

Carl Cohen, Ph.D.
Battelle Memorial Institute, Columbus, Ohio

Douglas L. Coleman, Ph.D.
Roscoe B. Jackson Memorial Laboratory, Bar Harbor, Maine

James Crow, Ph.D.*
Department of Medical Genetics, University of Wisconsin, Madison, Wisconsin

Karl H. Degenhardt, M.D.
Department of Human Genetics and Comparative Pathology, University of Frankfurt, Frankfurt, Germany

Margaret K. Deringer, Ph.D.
National Cancer Institute, Department of Health, Education and Welfare, Bethesda, Maryland

Margaret M. Dickie, Ph.D.
Roscoe B. Jackson Memorial Laboratory, Bar Harbor, Maine

Donald P. Doolittle, Ph.D.
Department of Biostatistics, University of Pittsburgh, Pittsburgh, Pennsylvania

Sheldon Dray, M.D.
Laboratory of Immunology, National Institute of Allergy and Infectious Disease, National Institutes of Health, Bethesda, Maryland

D. S. Falconer, Ph.D.
Institute of Animal Genetics, Edinburgh, Scotland

Morris Foster, Ph.D.
Department of Zoology, Mammalian Genetics Center, University of Michigan, Ann Arbor, Michigan

F. C. Fraser, Ph.D., M.D.*
Department of Genetics, McGill University, Montreal, Canada

John L. Fuller, Ph.D.
Roscoe B. Jackson Memorial Laboratory, Bar Harbor, Maine

Norman Giles, Ph.D.*
Department of Botany, Yale University, New Haven, Connecticut

Benson E. Ginsburg, Ph.D.
Department of Psychology, University of Chicago, Chicago, Illinois

Francis B. Gordon, Ph.D.*
Naval Medical Research Institute, National Naval Medical Center, Bethesda, Maryland

John W. Gowen, Ph.D.
Department of Genetics, Iowa State University, Ames, Iowa

Douglas Grahn, Ph.D.
Division of Biological and Medical Research, Argonne National Laboratory, Argonne, Illinois

Margaret C. Green, Ph.D.
Roscoe B. Jackson Memorial Laboratory, Bar Harbor, Maine

Earl L. Green, Ph.D.*
Roscoe B. Jackson Memorial Laboratory, Bar Harbor, Maine

Leonard Herzenberg, Ph.D.
Department of Genetics, Stanford University, Palo Alto, California

Walter E. Heston, Ph.D.*
National Cancer Institute, National Institutes of Health, Bethesda, Maryland

Warren G. Hoag, D.V.M.
Roscoe B. Jackson Memorial Laboratory, Bar Harbor, Maine

T. C. Hsu, Ph.D.
Section of Cytology, M. D. Anderson Hospital and Tumor Institute, Texas Medical Center, Houston, Texas

George E. Jay, Jr., Ph.D.
Department of Laboratory Animals, Microbiological Associates, Inc., Washington, D.C.

Nathan Kaliss, Ph.D.
    Roscoe B. Jackson Memorial Laboratory, Bar Harbor, Maine

George Klein, M.D.
    Department of Tumor Biology, Karolinska Institute, Stockholm, Sweden

Joshua Lederberg, Ph.D.*
    Department of Genetics, Stanford University, Palo Alto, California

Edwin P. Les, Ph.D.
    Roscoe B. Jackson Memorial Laboratory, Bar Harbor, Maine

Clarence C. Little, M.D.
    Littlehaven, Ellsworth, Maine

Clara J. Lynch, Ph.D.
    Rockefeller Institute for Medical Research, New York, New York

W. B. McIntosh, Ph.D.
    Department of Zoology, Ohio State University, Columbus, Ohio

Andrew V. Nalbandov, Ph.D.
    Department of Animal Science, University of Illinois, Urbana, Illinois

Ray D. Owen, Ph.D.*
    Division of Biological Sciences, California Institute of Technology, Pasadena,
    California

H. Ira Pilgrim, Ph.D.
    Department of Surgery, University of Utah, Salt Lake City, Utah

Raymond A. Popp, Ph.D.
    Biology Division, Oak Ridge National Laboratory, Oak Ridge, Tennessee

T. Edward Reed, Ph.D.
    Departments of Zoology and Pediatrics, University of Toronto, Toronto,
    Canada

Charles M. Rick, Jr., Ph.D.*
    Department of Vegetable Crops, University of California, College of Agricul-
    ture, Davis, California

T. H. Roderick, Ph.D.
   Roscoe B. Jackson Memorial Laboratory, Bar Harbor, Maine

Elizabeth S. Russell, Ph.D.
   Roscoe B. Jackson Memorial Laboratory, Bar Harbor, Maine

Robert H. Schaible, Ph.D.
   The Hall Laboratory of Mammalian Genetics, University of Kansas, Lawrence, Kansas

William J. Schull, Ph.D.*
   Associate Professor of Human Genetics, University of Michigan, Ann Arbor, Michigan

J. P. Scott, Ph.D.
   Roscoe B. Jackson Memorial Laboratory, Bar Harbor, Maine

Willys K. Silvers, Ph.D.
   The Wistar Institute, Philadelphia, Pennsylvania

Herman M. Slatis, Ph.D.
   Division of Biological and Medical Research, Argonne National Laboratory, Argonne, Illinois

George D. Snell, M.S., Sc.D.
   Roscoe B. Jackson Memorial Laboratory, Bar Harbor, Maine

Joan Staats, M.S.
   Roscoe B. Jackson Memorial Laboratory, Bar Harbor, Maine

Arthur G. Steinberg, Ph.D.
   Biological Laboratory, Western Reserve University, Cleveland, Ohio

Gunther Stent, Ph.D.*
   Virus Laboratory, University of California, Berkeley 4, California

Wilson S. Stone, Ph.D.*
   Department of Zoology, University of Texas, Austin, Texas

John A. Weir, Ph.D.
   The Hall Laboratory of Mammalian Genetics, University of Kansas, Lawrence, Kansas

W. K. Whitten, D.Sc.
  National Biological Standards Laboratory, Department of Health, Canberra, Australia

Katherine S. Wilson, Ph.D.*
  Genetics Study Section, Division of Research Grants, National Institutes of Health, Bethesda, Maryland

Henry J. Winn, Ph.D.
  Roscoe B. Jackson Memorial Laboratory, Bar Harbor, Maine

Sewall Wright, Sc.D.
  Department of Genetics, University of Wisconsin, Madison, Wisconsin

George Yerganian, Ph.D.
  Children's Research Foundation, Inc., Boston, Massachusetts

CONTENTS

PREFACE (*Walter J. Burdette, Ph.D., M.D.*) v

Genetic Stocks and Breeding Methods

SYSTEMS OF MATING USED IN MAMMALIAN GENETICS (*E. L. Green, Ph.D., and D. P. Doolittle, Ph.D.*) 3

METHODS FOR TESTING LINKAGE (*Margaret C. Green, Ph.D.*) . . . 56

GENETIC STRAINS AND STOCKS (*George E. Jay, Jr., Ph.D*) 83

Radiation Genetics

MAMMALIAN RADIATION GENETICS (*Douglas Grahn, Ph.D.*) . . . 127

Physiologic Genetics

GENIC INTERACTION (*Sewall Wright, Sc.D.*) 159

QUANTITATIVE INHERITANCE (*D. S. Falconer, Ph.D.*) 193

PROBLEMS AND POTENTIALITIES IN THE STUDY OF GENIC ACTION IN THE MOUSE (*E. S. Russell, Ph.D.*) 217

METHODOLOGY OF EXPERIMENTAL MAMMALIAN TERATOLOGY (*F. Clarke Fraser, Ph.D., M.D.*) 233

GENETICS OF NEOPLASIA (*Walter E. Heston, Ph.D.*) 247

GENETICS OF REPRODUCTIVE PHYSIOLOGY (*A. V. Nalbandov, Ph.D.*) 269

BEHAVIORAL DIFFERENCES (*J. P. Scott, Ph.D., and John L. Fuller, Ph.D.*) 283

Biochemical Genetics

MAMMALIAN HEMOGLOBINS (*Raymond A. Popp, Ph.D.*)   299

TACTICS IN PIGMENT-CELL RESEARCH (*Willys K. Silvers, Ph.D.*)   .   .   323

Immunogenetics

METHODS IN MAMMALIAN IMMUNOGENETICS (*Ray D. Owen, Ph.D.*) 347

Host-Parasite Relationships

GENETICS OF INFECTIOUS DISEASES (*John W. Gowen, Ph.D.*)   .   .   .   383

Genetics of Somatic Cells

GENETICS OF SOMATIC CELLS (*George Klein, M.D.*)   407

CYTOGENETIC ANALYSIS (*George Yerganian, Ph.D.*)   469

Appendices

I. CONTROL OF THE LITERATURE ON GENETICS OF THE
MOUSE (*Joan Staats, M.S.*)   511

II. INTERNATIONAL RULES OF NOMENCLATURE FOR MICE
(*Joan Staats, M.S.*)   517

III. METHODS OF KEEPING RECORDS (*Margaret M. Dickie, Ph.D.*)   .   522

IV. HUSBANDRY, EQUIPMENT, AND PROCUREMENT OF MICE
(*Warren G. Hoag, D.V.M., and Edwin P. Les, Ph.D.*)   538

V. TECHNIQUES FOR THE STUDY OF ANEMIAS IN MICE (*Elizabeth
S. Russell, Ph.D.*)   558

VI. TECHNIQUE FOR THE TRANSFER OF FERTILIZED OVA
(*Margaret K. Deringer, Ph.D.*)   563

VII. CURRENT APPLICATIONS OF A METHOD OF TRANSPLANTA-
TION OF TISSUES INTO GLAND-FREE MAMMARY FAT PADS
OF MICE (*Staff, Cancer Research Genetics Laboratory, University of California,
Berkeley*)   565

BIBLIOGRAPHY   571

AUTHOR INDEX   622

SUBJECT INDEX   627

GENETIC STOCKS AND BREEDING METHODS

E. L. Green, Ph.D., and D. P. Doolittle, Ph.D.

SYSTEMS *of* MATING
USED *in* MAMMALIAN GENETICS†

Mammalian geneticists use a variety of mating systems, each designed to accomplish a specific purpose. To use the systems effectively, it is necessary to know what each system is, when it can be used, and what its theoretical genetic consequences are. This paper describes seven systems of mating which have passed into general use by mouse geneticists. Each system will be described by means of its mating types and their probabilities through successive generations. In some cases reference will be made to the kinds of genotypes and their probabilities, in particular to the probability of heterozygotes.

The theory of systems of matings has been extensively developed by Wright,[1442] Bartlett and Haldane,[56] and Fisher,[375] on whom we have drawn heavily for this exposition. The system later called the "cross-backcross-intercross system" has not been analyzed heretofore; its theoretic consequences are presented here for the first time. We are indebted to Dr. George D. Snell who described the system to us and who has been the first to use it.

The following sections outline the analysis of the mating systems after first defining some necessary symbols and describing the general steps of the analytical method. The last section suggests a few practical rules for the breeders of laboratory animals who desire to improve the genetic quality of mice, rats, rabbits, and other mammals for research.

† The authors gratefully acknowledge the support of the Sagamore Foundation and the Richard Webber Jackson Memorial Fund.

NOTATIONS AND DEFINITIONS

Three autosomal loci of diploid, sexually reproducing organisms such as mice will be designated by the symbols: a-locus, D-locus, and r-locus. The a-locus is any locus whose heterozygosity is in question as a given breeding system advances from generation to generation. The D-locus is the locus of a dominant mutation; the r-locus, that of a recessive mutation. The D and r mutations are called the genes of interest. The alleles at these three loci will be denoted as A/a, D/d, and R/r, and the genotypes by AA, Aa, aa; DD, Dd, dd; and RR, Rr, rr. The relative frequency (i.e., probability) of Aa will be denoted by h.

The mating types are of four kinds:

Incrosses: $AA \times AA$ and $aa \times aa$, matings of like homozygotes,
Crosses: $AA \times aa$, matings of unlike homozygotes,
Backcrosses: $AA \times Aa$ and $aa \times Aa$, matings of homozygote and heterozygote,
Intercrosses: $Aa \times Aa$, matings of heterozygotes.

When the terms incrosses, crosses, backcrosses, and intercrosses appear in lower-case letters, they refer to the locus with questionable heterozygosity. When they appear in small capitals, INCROSSES, CROSSES, BACKCROSSES, INTERCROSSES, they refer to the locus of interest. The last three of these terms are in general use.

The relative frequencies or probabilities of the mating types will be denoted by p, q, r, ..., v with the definition varying slightly from system to system. In general, p will be used to denote the frequency of incrosses ($AA \times AA$ and $aa \times aa$), the maximizing of which is the objective of all of the systems of breeding, except random mating. A subscript n (or m) denotes generation n (or cycle m). G (or C) will stand for generation (or cycle). **G, P, A**, etc. are matrices. P will designate probability.

The probability of crossing over between the a-locus with questionable heterozygosity and the D- or r-locus carrying the mutation of interest will be denoted by c. To avoid a troublesome complication in notation, for any two loci, c will be treated as equal in the two sexes.

The probability of heterozygosity at the a-locus in generation n (or cycle m) will be denoted by h_n (or h_m). In all cases, as p_n increases, h_n decreases. As will be seen, h_n is a function of the probabilities of backcrosses and intercrosses in each system of mating.

SYSTEMS OF BREEDING

Relatively few of the systems developed by breeders of domestic and laboratory mammals are used frequently enough to warrant exposition here. Parent-offspring inbreeding, line breeding, or systems which use first, second, or third cousins will not be described. The systems included are, with the exception of random mating, all regular systems which permit the development of sequence equations to relate the probabilities of incrosses, etc., of one generation to those of the next. Irregular systems, as well as regular systems, may be analyzed by Wright's method of path analysis. The seven systems are of three types: those based upon relationship, those based upon locus control, and those based upon both relationship and locus control.

The systems based on relationship are:

1. RANDOM-MATING SYSTEM,
2. BROTHER-SISTER INBREEDING SYSTEM.

Those based on controlling a locus of interest are:

3. BACKCROSS SYSTEM,
4. CROSS-INTERCROSS SYSTEM,
5. CROSS-BACKCROSS-INTERCROSS SYSTEM.

Those which use a combination of locus control and relationship are:

6. BROTHER-SISTER INBREEDING WITH HETEROZYGOSIS FORCED BY BACKCROSSING,
7. BROTHER-SISTER INBREEDING WITH HETEROZYGOSIS FORCED BY INTERCROSSING.

In the first two systems, no conscious attention is paid to the phenotypes, except for the inevitable selection in favor of vigorous, healthy animals.

In the third system, in each generation, $(G_0, G_1, \ldots)$ the mating is a BACKCROSS with respect to the locus of interest. While the locus of interest is thus being controlled by a BACKCROSS, the locus whose heterozygosity is in question may be undergoing an incross, a cross, or a backcross. In the fourth system the initial generation (G_0) and all subsequent even-numbered generations $(G_2, G_4, \ldots)$ are CROSSES and all odd-numbered generations $(G_1, G_3, \ldots)$ are INTERCROSSES with respect to the locus of interest. In the fifth system, the initial, third, sixth, etc., generations $(G_0, G_3, G_6, \ldots)$ are CROSSES; the first, fourth, seventh, etc., generations $(G_1, G_4, G_7, \ldots)$ are BACK-CROSSES; and the second, fifth, eighth, etc., generations $(G_2, G_5, G_8, \ldots)$ are INTER-CROSSES with respect to the locus of interest.

In the sixth system, each generation, $(G_0, G_1, G_2, \ldots)$ is a BACKCROSS with respect to the locus of interest with the provision that the mates are related as brother and sister. In the seventh system, each generation $(G_0, G_1, G_2, \ldots)$ is an INTERCROSS with respect to the locus of interest, with the same provision about the mates. Table 1 shows the generation sequence relative to the type of mating.

Table 1

GENERATION SEQUENCE RELATIVE TO TYPE OF MATING

System	CROSS	BACKCROSS	INTERCROSS
3. B		0, 1, 2, . . .	
4. C–I	0, 2, 4, . . .		1, 3, 5, . . .
5. C–B–I	0, 3, 6, . . .	1, 4, 7, . . .	2, 5, 8, . . .
6. BS–B		0, 1, 2, . . .	
7. BS–I			0, 1, 2, . . .

The serial number of the generation G_n given in the table shows the sequence of use of a CROSS, BACKCROSS, or INTERCROSS in controlling the locus of interest in five of the systems of mating.

For each system, the breeder should know the probabilities of the mating types and in particular the probability p of incrosses with respect to the a-locus. He will also want to know the probabilities of the genotypes, particularly the probability h of heterozygotes Aa. To evaluate a system he should know the number of generations required to increase p or to decrease h to any desired level. We shall try to answer these questions about the seven systems.

RANDOM MATING

The system of random mating is used when preservation of the genetic (i.e., genotypic) variability of a population without change is desired. Random mating will not, in theory, create genetic variability or diminish it. Variability may be created by outcrossing or by mutations. It may be diminished by any scheme of inbreeding. Random mating means that the frequencies of matings of various types are specifiable by the product and addition rules of probability applied to the genotypes.

Let the probabilities of the three genotypes at the a-locus in G_0 be

$$P(AA) = k_0,$$
$$P(Aa) = 2l_0,$$
$$P(aa) = m_0,$$

where $k_i + 2l_i + m_i = 1$, $i = 0, 1, 2, \ldots, n$, and $P(x)$ means the probability of x. By definition, the gene or allele frequencies are:

$$P(A) = x_0 = k_0 + l_0,$$
$$P(a) = y_0 = l_0 + m_0,$$

where $x_i + y_i = 1$, and $i = 0, 1, 2, \ldots, n$. The mating-type frequencies in G_0 and the genotype frequencies of their progeny in G_1 are shown in table 2. The probability

Table 2

MATING-TYPE FREQUENCIES IN G_0 AND GENOTYPE FREQUENCIES OF THEIR PROGENY

| | | Genotypes in G_1 | | |
Mating types in G_0		AA	Aa	aa
$P(AA \times AA) = p_0' = k_0^2$		k_0^2	----	----
$P(aa \times aa) = p_0'' = m_0^2$		----	----	m_0^2
$P(AA \times aa) = q_0 = 2k_0m_0$		----	$2k_0m_0$	----
$P(AA \times Aa) = r_0 = 4k_0l_0$		$2k_0l_0$	$2k_0l_0$	----
$P(aa \times Aa) = s_0 = 4l_0m_0$		----	$2l_0m_0$	$2l_0m_0$
$P(Aa \times Aa) = v_0 = 4l_0^2$		l_0^2	$2l_0^2$	l_0^2

The genotype frequencies of G_1 are related to the mating-type frequencies of G_0 in the random-mating system by the laws of Mendelian genetics.

p_0' of the mating $AA \times AA$ is $k_0{}^2$ by the product rule of probability. The probability r_0 of the matings $AA \times Aa$ or $Aa \times AA$ is $k_0 \cdot 2l_0$ plus $2l_0 \cdot k_0$, or $4k_0l_0$, by the product and addition rules of probability, and so forth. In the table, $\mathrm{P}(AA \times Aa)$ is the probability of $AA \times Aa$ and its reciprocal $Aa \times AA$.

The relative frequencies of the three genotypes in G_1, obtained by adding the last three columns of table 2, are:

$$
\begin{aligned}
\mathrm{P}(AA) &= k_1 = (k_0 + l_0)^2 &&= x_0{}^2, \\
\mathrm{P}(Aa) &= 2l_1 = 2(k_0 + l_0)(l_0 + m_0) &&= 2x_0y_0, \\
\mathrm{P}(aa) &= m_1 = (l_0 + m_0)^2 &&= y_0{}^2.
\end{aligned}
$$

If random mating ensues, that is, if the probability of any mating type is the product of the genotypic probabilities of the mates, the mating types of G_1 and the genotypes of their progeny in G_2 can be represented as functions of the initial allelic frequencies, x_0 and y_0, as in table 3. The genotypic frequencies in G_2 obtained by adding the last three colums of table 3 are:

$$
\begin{aligned}
\mathrm{P}(AA) &= k_2 = x_0{}^2(x_0{}^2 + 2x_0y_0 + y_0{}^2) &&= x_0{}^2, \\
\mathrm{P}(Aa) &= 2l_2 = 2x_0y_0(x_0{}^2 + 2x_0y_0 + y_0{}^2) &&= 2x_0y_0, \\
\mathrm{P}(aa) &= m_2 = y_0{}^2(x_0{}^2 + 2x_0y_0 + y_0{}^2) &&= y_0{}^2.
\end{aligned}
$$

Table 3

MATING-TYPE FREQUENCIES OF G_1 AND THE GENOTYPE FREQUENCIES OF THEIR PROGENY

Mating types in G_1	Genotypes in G_2		
	AA	Aa	aa
$\mathrm{P}(AA \times AA) = p_1' = k_1{}^2 = x_0{}^4$	$x_0{}^4$	—	—
$\mathrm{P}(aa \times aa) = p_1'' = m_1{}^2 = y_0{}^4$	—	—	$y_0{}^4$
$\mathrm{P}(AA \times aa) = q_1 = 2k_1m_1 = 2x_0{}^2y_0{}^2$	—	$2x_0{}^2y_0{}^2$	—
$\mathrm{P}(AA \times Aa) = r_1 = 4k_1l_1 = 4x_0{}^3y_0$	$2x_0{}^3y_0$	$2x_0{}^3y_0$	—
$\mathrm{P}(aa \times Aa) = s_1 = 4l_1m_1 = 4x_0y_0{}^3$	—	$2x_0y_0{}^3$	$2x_0y_0{}^3$
$\mathrm{P}(Aa \times Aa) = v_1 = 4l_1{}^2 = 4x_0{}^2y_0{}^2$	$x_0{}^2y_0{}^2$	$2x_0{}^2y_0{}^2$	$x_0{}^2y_0{}^2$

The mating-type frequencies of G_1 and the genotype frequencies of G_2 are functions of the allele frequencies of G_0 in the random-mating system.

These are identical with the genotypic frequencies of G_1. Hence the mating-type frequencies of G_2 will also be identical with the mating-type frequencies of G_1. It follows that $p_{n+1}' = p_1'; \ldots; v_{n+1} = v_1$. After one generation of random mating, the probabilities of the six kinds of matings remain constant for any number of generations of random mating. Random mating thus preserves the genetic variability of the population. The restriction—"after one generation"—may be removed if the initial genotypic frequencies, $k_0, 2l_0, m_0$ are related to the allelic frequencies, x_0 and y_0, as

$$
\begin{aligned}
k_0 &= x_0{}^2, \\
2l_0 &= 2x_0y_0, \\
m_0 &= y_0{}^2.
\end{aligned}
$$

If they are not so related initially, they become so after one generation of random mating as shown above.

GENERAL METHOD OF ANALYZING REGULAR MATING SYSTEMS

In the regular mating systems there is a definable probability that matings of any given type will yield matings of the same or any other type (for example, that backcrosses yield incrosses) in the subsequent generation. The probability that a mating of type j yields a mating of type i in the subsequent generation will be called g_{ij}. With s mating types, these probabilities can be arranged in an $s \times s$ matrix so that the leading diagonal of the matrix contains the elements g_{ii}, row i contains all of the elements $g_{ij}, j = 1, \ldots, s$, and column j contains all of the elements $g_{ij}, i = 1, \ldots, s$. Such a matrix will be referred to as the generation matrix $\mathbf{G}$.

If, in generation n, the mating types have frequencies $p_n, q_n, \ldots, v_n$, the frequencies in generation $n + 1$ will be

$$
\begin{aligned}
p_{n+1} &= p_n g_{11} + q_n g_{21} + \cdots + v_n g_{s1} \\
q_{n+1} &= p_n g_{12} + q_n g_{22} + \cdots + v_n g_{s2} \\
&\ \vdots \qquad \vdots \qquad\qquad\quad \vdots \\
v_{n+1} &= p_n g_{1s} + q_n g_{2s} + \cdots + v_n g_{ss}.
\end{aligned} \tag{1}
$$

The probabilities $p_n, q_n, \ldots, v_n$ can be arranged in an $s \times 1$ vector $\mathbf{P}_n$,

$$
\mathbf{P}_n = \begin{pmatrix} p_n \\ q_n \\ \vdots \\ v_n \end{pmatrix}.
$$

If the matrix $\mathbf{G}$ is then postmultiplied by this vector $\mathbf{P}_n$, following the row-by-column rule of matrix multiplication, a new vector is obtained whose elements are the sums (1). We therefore conclude that

$$
\mathbf{P}_{n+1} = \mathbf{G}\mathbf{P}_n. \tag{2}
$$

If, then, the matrix $\mathbf{G}$ and the vector $\mathbf{P}_0$ can be evaluated, when $\mathbf{P}_0$ contains the frequencies of mating types in the initial generation, the vector of mating-type frequencies $\mathbf{P}_n$ can be obtained for generation n by repetitive application of equation (2). However,

$$
\mathbf{P}_n = \mathbf{G}\mathbf{P}_{n-1},
$$

and

$$
\mathbf{P}_{n-1} = \mathbf{G}\mathbf{P}_{n-2};
$$

therefore

$$
\mathbf{P}_n = \mathbf{G}^2 \mathbf{P}_{n-2'},
$$

and by applying this principle repeatedly, we can show that

$$
\mathbf{P}_n = \mathbf{G}^n \mathbf{P}_0. \tag{3}
$$

This means that if the **G** matrix could be raised to the power n, the mating-type frequencies could be obtained immediately in generation n. Raising **G** to the power n is, however, a difficult task. To avoid this difficulty the method of Kempthorne can be used.[700]

A diagonal matrix is one in which some of the elements of the leading diagonal have values other than 0, while all elements off this diagonal are zero. If the matrix Λ is an $s \times s$ diagonal matrix,

$$\Lambda = \begin{pmatrix} \lambda_1 & & & 0 \\ & \lambda_2 & \cdot & \\ & & \cdot & \\ & & & \cdot \\ 0 & & & \lambda_s \end{pmatrix}$$

Λ raised to the power n is

$$\Lambda^n = \begin{pmatrix} \lambda_1{}^n & & & 0 \\ & \lambda_2{}^n & \cdot & \\ & & \cdot & \\ & & & \cdot \\ 0 & & & \lambda_s{}^n \end{pmatrix}.$$

Thus, if **G** were a diagonal matrix, we could easily obtain $\mathbf{G}^n$. **G** will not, however, be diagonal for any system of inbreeding; but we can use a transformation which will give us a diagonal matrix to work with. Let us define an $s \times s$ matrix, **A**, such that

$$\mathbf{V}_k = \mathbf{AP}_k \tag{4}$$

for any generation k, $\mathbf{V}_k$ being an $s \times 1$ vector. Then from equation (2),

$$\mathbf{V}_k = \mathbf{AGP}_{k-1},$$

and since

$$\mathbf{P}_k = \mathbf{A}^{-1}\mathbf{V}_k,$$

it follows that

$$\mathbf{V}_k = \mathbf{AGA}^{-1}\mathbf{V}_{k-1}. \tag{5}$$

But an $s \times s$ matrix **A** exists such that $\mathbf{AGA}^{-1}$ is an $s \times s$ diagonal matrix. Therefore,

$$\mathbf{V}_k = \Lambda\mathbf{V}_{k-1},$$

and, since this has the same form as (2),

$$\mathbf{V}_n = \Lambda^n\mathbf{V}_0. \tag{6}$$

If, then, we can obtain Λ and **A**, we can transform $\mathbf{P}_0$ to give us $\mathbf{V}_0$, raise Λ to the power n, calculate $\mathbf{V}_n$, and transform to $\mathbf{P}_n$.

The problem is to find **A** such that

$$\mathbf{AGA}^{-1} = \boldsymbol{\Lambda}$$

or

$$\mathbf{AG}\quad = \boldsymbol{\Lambda}\mathbf{A}.$$

AG is an $s \times s$ matrix, row i of which is a row vector which can be written $a_i\mathbf{G}$, where a_i is row i of matrix **A**. $\boldsymbol{\Lambda}\mathbf{A}$ is also an $s \times s$ matrix, row i of which can be written $a_i\lambda_i\mathbf{I}$, where λ_i is the non-zero element in row i of $\boldsymbol{\Lambda}$, and **I** is an $s \times s$ identity matrix, with diagonal elements 1 and all elements off the diagonal equal to zero. Then to satisfy (5),

$$a_i\mathbf{G} = a_i\lambda_i\mathbf{I},$$
$$a_i\mathbf{G} - a_i\lambda_i\mathbf{I} = 0$$
$$a_i(\mathbf{G} - \lambda_i\mathbf{I}) = 0. \tag{7}$$

This defines a set of linear equations in the elements of a_i, the solutions of which give the elements of row i of **A**. If the values of λ_i from the matrix $\boldsymbol{\Lambda}$ are substituted into (7), the **A** matrix required to transform $\mathbf{P}_0$ to $\mathbf{V}_0$ and $\mathbf{V}_n$ to $\mathbf{P}_n$ is obtained.

To obtain this solution, we must first have the $\boldsymbol{\Lambda}$ matrix. Equation (7) can be solved if and only if

$$|\mathbf{G} - \lambda_i\mathbf{I}| = 0. \tag{8}$$

If λ is subtracted from each of the diagonal elements of **G** and the determinant of the resulting matrix set equal to zero, s solutions will be obtained for λ, which constitute the λ_i, $i = 1, \ldots, s$, the diagonal elements of the matrix $\boldsymbol{\Lambda}$.

In each of the systems we shall discuss, one of the roots of equation (8) will be 1. Since the roots in $\boldsymbol{\Lambda}$ can be arranged in any order, we will always designate this root as λ_s. The largest numerically of the remaining roots $\lambda_1, \ldots, \lambda_{s-1}$ we will designate λ_1. λ_1 is often called the characteristic root of matrix **G** and has certain properties which make it of special value to us.

Since λ_1 is the largest of the roots $\lambda_1, \ldots, \lambda_{s-1}$, all of which are less than 1, as the matrix $\boldsymbol{\Lambda}$ is raised to higher powers, all of these elements become smaller (except λ_s, which remains 1 for any power). Since λ_1 is the largest of these, it will retain significance even when the rest of the roots have become negligible. Thus, λ_1 becomes an approximate measure of the rate of change of frequency of the various mating types.

We may also wish to know how many generations are required to exceed a given percentage of incross matings. This will occur (ignoring sampling variations) when

$$p_n > \alpha, \tag{9}$$

if p_n is the proportion of incrosses and α is the desired proportion of such matings.

If we assume that the initial generation includes only crosses which maximize the number of generations required to reach a given percentage of incross matings,

$$\mathbf{V}_0 = \mathbf{AP}_0 = \begin{pmatrix} 1 \\ 1 \\ \vdots \\ 1 \end{pmatrix}.$$

Since
$$V_n = \Lambda^n V_0,$$
$$P_n = A^{-1} V_n = A^{-1} \Lambda^n V_0.$$

Since $V_0 = \begin{pmatrix} 1 \\ 1 \\ \vdots \\ 1 \end{pmatrix}$,
$$p_n = b_{s1}\lambda_1{}^n + b_{s2}\lambda_2{}^n + \cdots + b_{ss}\lambda_s{}^n$$

where b_{si} is the element in row s, column i of A^{-1}.

For all inbreeding systems,
$$b_{ss} = 1,$$
$$\lambda_s{}^n = 1,$$
$$\lambda_2{}^n = \lambda_3{}^n = \cdots = \lambda_{s-1}{}^n = 0 \text{ for } n \text{ of any appreciable size,}$$
$$0 < \lambda_1{}^n < 1,$$

and
$$b_{s1} < 0.$$

Therefore
$$p_n \cong 1 + \lambda_1{}^n b_{s1}.$$

From equation (9), it follows that the number of generations n required to exceed the frequency α of incrosses can be approximated by setting
$$\lambda_1{}^n b_{s1} + 1 = \alpha.$$

Then
$$\lambda_1{}^n = \frac{-(1 - \alpha)}{b_{s1}},$$
$$n \log \lambda_1 = \log b_{s1} - \log (1 - \alpha),$$

and
$$n = \frac{\log b_{s1} - \log (1 - \alpha)}{\log \lambda_1}, \tag{10}$$

Using the nearest integral value of n, we can calculate the true p_n from equation (6). The value of n for which equation (9) is satisfied should be within one or two generations of the approximation derived from (10) in all cases.

In summary, then, the frequencies of the various mating types can be obtained in any generation, if the frequencies in the preceding generation are known, by using equation (2). Equation (3) allows the frequencies of mating types to be derived in some advanced generation n from the frequencies in the initial generation. This, however, involves raising the G matrix to the power n, which may be a difficult task. This difficulty may be avoided by deriving a diagonal matrix from the G matrix; it is easy to raise a diagonal matrix to any power. Doing so requires us to obtain the roots of the original G matrix and to derive a new matrix, A, which is used to transform the frequencies into multipliers for the diagonal matrix, **Λ**, of the **G** matrix roots. The roots of the **G** matrix are also useful in that one of them, called the characteristic root, measures the rate of decrease of the frequency of nonincross matings. This root also can be used to calculate the approximate number of generations required to reach a given proportion of incross matings.

In the subsequent discussion, these principles will be applied to the six systems of mating mentioned above.

BROTHER-SISTER INBREEDING SYSTEM

The system of brother-sister inbreeding is used when it is desired to reduce the genetic (i.e., genotypic) variability at all loci. The probability h of heterozygosity

Fig. 1. THE BROTHER-SISTER INBREEDING SYSTEM.

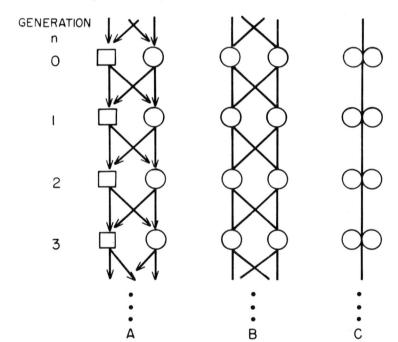

at any locus a is to become as small as possible. Figure 1 shows three ways of representing the system in a diagram. In the following analysis of the system reference is made only to autosomal loci. In the case of sex-linked loci, brother-sister inbreeding also reduces the probability of heterozygosity in the homogametic sex, and, if there is crossing over between the sex chromosomes, in the heterogametic sex. The analysis differs in details from that for autosomal loci.

Assume that A/a are alleles at the a-locus and that the a-locus is autosomal. We desire to know the probability of matings between like homozygotes, i.e., of incrosses in any generation n.   Assume further that the probabilities of various types of matings in the generation $n(G_n)$ are:

Incrosses: $P\begin{pmatrix} AA \times AA \\ aa \times aa \end{pmatrix} = p_n,$

Crosses: $P\,(AA \times aa)\ = q_n,$

Backcrosses: $P\begin{pmatrix} AA \times Aa \\ aa \times Aa \end{pmatrix} = r_n,$

Intercrosses: $P\,(Aa \times Aa)\ = v_n,$

where each mating type also includes the reciprocal mating, if any.   The incrosses each produce one type of progeny, AA or aa, like the parents, and following the system of mating brother by sister, these produce one type of mating, incrosses, in the next generation.   The crosses produce one type of progeny Aa, unlike the parents, and one type of mating, intercrosses, in the next generation.   The backcrosses each produce two types of progeny AA and Aa, or aa and Aa, with probabilities 1/2 and 1/2, and so produce three types of matings, incrosses, backcrosses, and intercrosses, in the ratios $1/4:1/2:1/4$.   Finally, the intercrosses produce three kinds of progeny AA, Aa, and aa in the ratio $1/4:1/2:1/4$ which yield incrosses, crosses, backcrosses, and intercrosses in the ratios $1/8:1/8:1/2:1/4$.

Table 4

GENERATION MATRIX FOR THE BROTHER-SISTER INBREEDING SYSTEM

	p_n	q_n	r_n	v_n
p_{n+1}	1	0	1/4	1/8
q_{n+1}	0	0	0	1/8
r_{n+1}	0	0	1/2	1/2
v_{n+1}	0	1	1/4	1/4

The probabilities of mating types in G_{n+1} may conveniently be shown as functions of the probabilities in G_n in a generation matrix, as in table 4.   The probabilities may be written as equations:

$$p_{n+1} = p_n \quad + (1/4)r_n + (1/8)v_n,$$
$$q_{n+1} = \qquad\qquad\qquad (1/8)v_n,$$
$$r_{n+1} = \qquad (1/2)r_n + (1/2)v_n,$$
$$v_{n+1} = \ q_n + (1/4)r_n + (1/4)v_n.$$

It may be seen at once that the probability of incrosses will steadily rise, since incrosses are produced by incrosses, by backcrosses, and by intercrosses of the preceding generation. Incrosses act as an absorbing barrier. Once a locus reaches an incross, it remains as an incross so long as the brother-sister mating system continues.

If brother-sister mating is started after a cross between unlike strains, so that the initial mating is $AA \times aa$ or the reciprocal, that is, $q_0 = 1$, the probabilities of incrosses in $G_0, G_1, G_2, \ldots$ are:

0, 0, 0.125, 0.281, 0.414, 0.525, 0.616, 0.689,

0.748, 0.796, 0.835, 0.866, 0.892, ...

After 12 generations, 89.2 per cent of the matings will be the desired type, incrosses of $AA \times AA$ or $aa \times aa$.

The probability h of heterozygotes is one-half the probability of backcrosses plus the probability of intercrosses:

$$h_n = (1/2)r_n + v_n.$$

In $G_1, G_2, \ldots$, this probability h takes the successive values:

1, 1/2, 2/4, 3/8, 5/16, 8/32, 13/64, 21/128, ...,

which are the terms of the Fibonacci series. The ratios h_{n+1}/h_n of the probability of heterozygosity in one generation to the probability of heterozygosity in the preceding generation form a series:

0.5, 1.0, 0.75, 0.8333, 0.8, 0.8125, 0.8077, 0.8095,

0.8088, 0.8091, 0.8090, 0.8090, 0.8090, ...,

which shows that, after brother-sister inbreeding has been in progress for several generations, the amount of heterozygosity in G_{n+1} tends to be 80.9 per cent of the heterozygosity in G_n. Or, stated otherwise, the heterozygosity is decreasing at the rate of 19.1 per cent per generation, as first discovered by Jennings[662] and established analytically by Wright.[1442]

Successive values of p_n, h_n, and h_{n+1}/h_n up to $n = 12$ are shown in figure 2.

It may be assumed that, after brother-sister inbreeding has been in progress for several generations, the rate of depletion of the crosses, backcrosses, and intercrosses is constant. That is,

$$q_{n+1} = \lambda q_n,$$
$$r_{n+1} = \lambda r_n,$$
$$v_{n+1} = \lambda v_n,$$

where λ is a factor of proportionality. It follows that

$$-\lambda q_n \qquad\qquad + \qquad (1/8)v_n = 0,$$
$$(1/2 - \lambda)r_n + \qquad (1/2)v_n = 0,$$
$$q_n + \qquad (1/4)r_n + (1/4 - \lambda)v_n = 0.$$

Solving these three simultaneous linear equations for λ yields

$$\lambda = (1/4), \quad (1/4)(1 + \sqrt{5}), \quad (1/4)(1 - \sqrt{5}),$$

of which the largest numerical solution is

$$\lambda_1 = (1/4)(1 + \sqrt{5}) = 0.8090 \quad \text{or} \quad 80.9 \text{ per cent.}$$

λ_1 is known as the characteristic root of the determinant made up of the coefficients in the three simultaneous equations and provides a rapid analytical means of finding the

Fig. 2. THE PROBABILITY OF INCROSSES FOR BROTHER-SISTER INBREEDING SYSTEM.

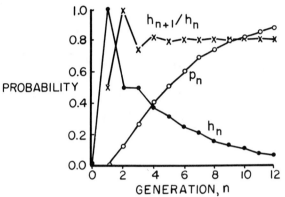

The probability of incrosses p_n and of heterozygosity h_n for the brother-sister inbreeding system when $q_0 = 1$; and the ratios of successive values of h and h_{n+1}/h_n.

expected decrease in heterozygosity after several generations of inbreeding. The largest numerical root is the important one as n gets larger and larger because the probabilities q_{n+1}, r_{n+1}, and v_{n+1} are each functions of $\lambda_1{}^n$, $\lambda_2{}^n$, and $\lambda_3{}^n$ and of the initial probabilities q_0, r_0, and v_0. If λ_2 and λ_3 are smaller than λ_1, $\lambda_2{}^n$ and $\lambda_3{}^n$ will be of diminishing importance, relative to $\lambda_1{}^n$, as n increases in determining the values of q_{n+1}, r_{n+1}, and v_{n+1}. Thus for large n, only the characteristic root need be used for a sufficient approximation in computing the probabilities.

The above results have been arrived at by direct multiplication of the generation matrix by the frequency of each mating type in G_n to obtain the frequencies of the mating types in G_{n+1}, essentially by the repetitive application of formula (1) in the preceding section. The same results can be reached by the more complex algebra of the formulas (3), (5), and so forth.

The generation matrix, G, is given in table 4. The roots λ_i can be obtained by solving equation (8),

$$|G - \lambda_i I| = 0,$$

or

$$\begin{vmatrix} 1 - \lambda & 0 & 1/4 & 1/8 \\ 0 & -\lambda & 0 & 1/8 \\ 0 & 0 & (1/2) - \lambda & 1/2 \\ 0 & 1 & 1/4 & (1/4) - \lambda \end{vmatrix} = 0.$$

Solving the determinant,

$$[1 - \lambda][(1/4) - \lambda][\lambda^2 - (1/2)\lambda - (1/4)] = 0,$$

yielding the roots

$$\lambda_1 = (1 + \sqrt{5})/4,$$
$$\lambda_2 = (1 - \sqrt{5})/4,$$
$$\lambda_3 = 1/4,$$

and

$$\lambda_4 = 1.$$

These are the roots found by solving the three linear equations for a proportionality constant, with the addition of the root $\lambda_4 = 1$. From these the $\mathbf{\Lambda}$ matrix given in table 5 (a) is derived. Substituting the above values of λ_i into formula (7), the matrix $\mathbf{A}$, given in table 5 (b) and its inverse, $\mathbf{A}^{-1}$, table 5 (c) are obtained.

Table 5

THE MATRICES $\mathbf{\Lambda}$, $\mathbf{A}$, AND $\mathbf{A}^{-1}$ FOR BROTHER-SISTER MATING

(a) $\mathbf{\Lambda}$ =

0.8090	0	0	0
0	-0.3090	0	0
0	0	0.2500	0
0	0	0	1.0000

(b) $\mathbf{A}$ =

0	1	0.6545	0.8090
0	1	0.0955	-0.3090
0	1	-0.25	0.25
1	1	1	1

(c) $\mathbf{A}^{-1}$ =

-1.3708	-0.0292	0.4000	1
0.0764	0.5236	0.4000	0
0.8000	0.8000	-1.6000	0
0.4945	-1.2945	0.8000	0

Now,

$$\mathbf{P}_0 = \begin{pmatrix} 0 \\ 1 \\ 0 \\ 0 \end{pmatrix},$$

if all matings in the initial generation are assumed to be cross matings. Then

$$\mathbf{V}_0 = \mathbf{A}\mathbf{P}_0 = \begin{pmatrix} 1 \\ 1 \\ 1 \\ 1 \end{pmatrix},$$

$\mathbf{V}_n = \mathbf{\Lambda}^n \mathbf{V}_0$, and for $n = 12$,

$$\mathbf{\Lambda}^{12} = \begin{pmatrix} 1.835 \times 10^{-1} & 0 & 0 & 0 \\ 0 & 8.311 \times 10^{-5} & 0 & 0 \\ 0 & 0 & 1.526 \times 10^{-5} & 0 \\ 0 & 0 & 0 & 1 \end{pmatrix}$$

$$\mathbf{V}_{12} = \begin{pmatrix} 7.859 \times 10^2 \\ 6.772 \times 10^{-7} \\ 5.960 \times 10^{-8} \\ 1 \end{pmatrix}$$

and

$$\mathbf{P}_{12} = \mathbf{A}^{-1}\mathbf{V}_{12} = \begin{pmatrix} 0.8923 \\ 0.0060 \\ 0.0629 \\ 0.0389 \end{pmatrix}, \quad \text{i.e., } \begin{aligned} p_{12} &= 0.8923 \\ q_{12} &= 0.0060 \\ r_{12} &= 0.0629 \\ v_{12} &= 0.0389. \end{aligned}$$

This corresponds with the value of p_{12} in figure 2, and the other values also agree with results gained by repetitive application of equation (2).

To calculate the generations required to obtain a given percentage of incross matings, formula (10) is used, which gives us an approximation:

$$n \log \lambda_1 \leqslant \log b_{1s} - \log (1 - \alpha_1)$$

for $\alpha_1 = 0.95$,

$$n \simeq \frac{\log 1.3708 - \log 0.05}{\log 0.8090} = 15.6.$$

Therefore it is estimated that 16 generations will be required to give 95 per cent incross matings.   Checking, we find that

$$p_{16} = 0.9539,$$

and

$$p_{15} = 0.9430.$$

Repeating the calculations for $\alpha_2 = 0.99$,

$$n \simeq \frac{\log 1.3708 - \log 0.01}{\log 0.8090} = 23.2,$$

and

$$p_{24} = 0.9915,$$

whereas

$$p_{23} = 0.9895.$$

Thus, 16 generations of brother-sister inbreeding are required to obtain 95 per cent incross matings, and 24 generations to obtain 99 per cent.

Dozens of strains of mice have been inbred by means of brother-sister matings. A few examples are A/J, AKR/J, BALB/cJ, C57BL/6J, DBA/2J, and C3HeB/FeJ. For a complete listing, refer to Committee[220] and Snell and Staats.[1254]

THE BACKCROSS SYSTEM

The backcross system is a means of placing a dominant mutation or a semidominant lethal mutation D on a standard inbred background.   It is assumed that the standard inbred strain is $AAdd$ or $aadd$ and the mutant bearing stock is $--DD$ or $--Dd$.   All matings are BACKCROSSES $dd \times Dd$ with respect to the D-locus and may be incrosses, crosses, or backcrosses, with respect to the a-locus.   The backcross system may also be used to put a recessive lethal or a recessive with low viability or low fertility on an inbred background, by backcrossing heterozygous carriers Rr of the mutation r to a standard inbred strain RR.   In this case it is necessary to be able to distinguish the heterozygotes Rr from the homozygotes RR, each suspected carrier R- being tested by matings with known heterozygotes Rr.

The backcross system is shown in figure 3.

Fig. 3. THE BACKCROSS SYSTEM.

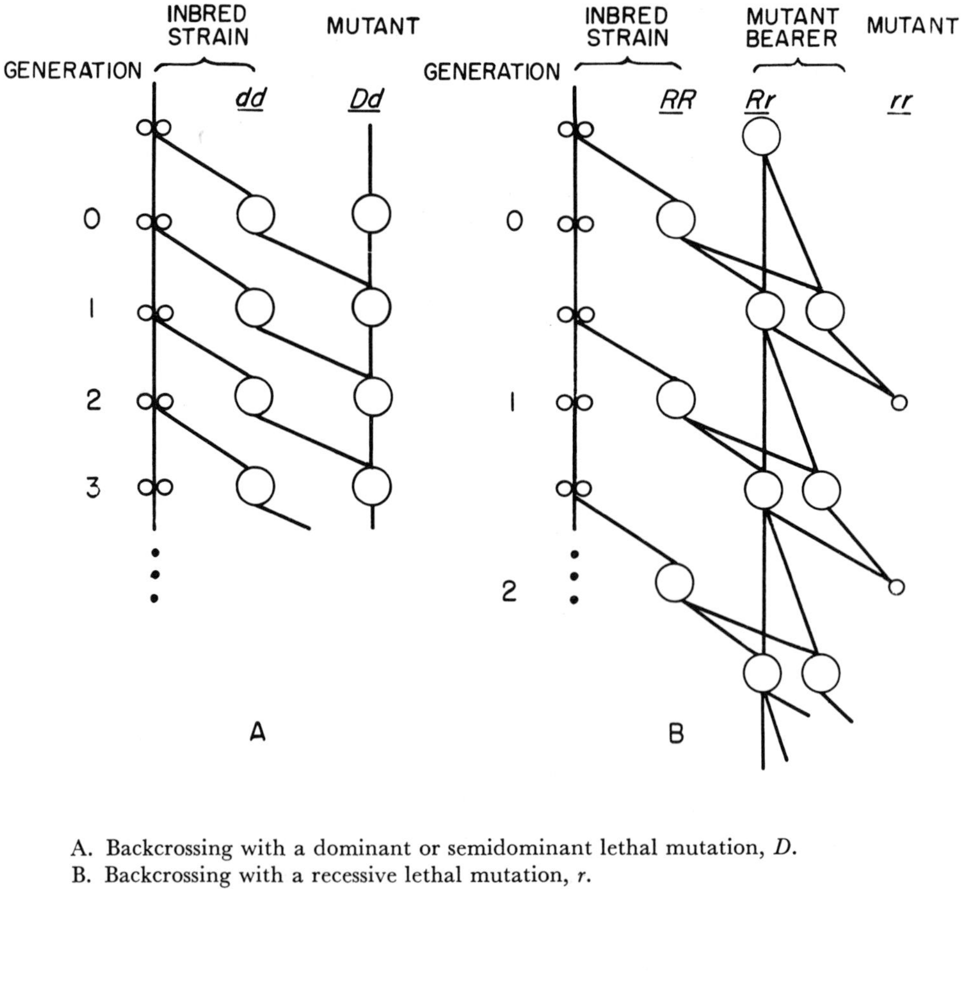

A. Backcrossing with a dominant or semidominant lethal mutation, D.
B. Backcrossing with a recessive lethal mutation, r.

The mating types and their probabilities in G_n are:

$$\text{Incrosses:} \qquad P\begin{pmatrix} Ad/Ad \times AD/Ad \\ ad/ad \times aD/ad \end{pmatrix} = p_n,$$

$$\text{Crosses:} \qquad P\begin{pmatrix} Ad/Ad \times aD/ad \\ ad/ad \times AD/Ad \end{pmatrix} = q_n,$$

$$\text{Backcrosses (first kind):} \qquad P\begin{pmatrix} Ad/Ad \times AD/ad \\ ad/ad \times aD/Ad \end{pmatrix} = r_n,$$

$$\text{Backcrosses (second kind):} \qquad P\begin{pmatrix} Ad/Ad \times aD/Ad \\ ad/ad \times AD/ad \end{pmatrix} = t_n.$$

The standard inbred strain is given on the left, the mutant bearer on the right in each mating type. For the case of a recessive mutation, replace each D by r and each d by R. The probability of crossing over between the a- and the D-locus or the a- and the r-locus is c, where $c = 1/2$ when there is no linkage.

The incrosses each produce one kind of mutant-bearing offspring, which, when mated in turn to the standard strain, yield incrosses in the next generation. The crosses produce doubly heterozygous progeny, considering only the mutant-bearing offspring, which, when mated to the standard inbred strain, yield backcrosses of the second kind. The backcrosses of the first kind produce two kinds of mutant-bearing progeny, AD/Ad and aD/Ad, or aD/ad and AD/ad, in the ratio $(1 - c):c$. The next generation of matings with the standard strain are thus of two types, incrosses and backcrosses of the second kind, in the proportions $(1 - c):c$. The backcrosses of the second kind produce the same two kinds of progeny, but the ratio is $c:(1 - c)$, so the two types of matings in the next generation, incrosses and backcrosses of the second kind, are in the ratio $c:(1 - c)$.

These results may be displayed as a set of linear equations,

$$p_{n+1} = p_n \qquad + (1 - c)r_n + \qquad ct_n,$$
$$q_{n+1} = 0,$$
$$r_{n+1} = 0,$$
$$t_{n+1} = \qquad q_n + \qquad cr_n + (1 - c)t_n,$$

from which it is readily seen that the incrosses will steadily increase as n increases, although very slowly if c is very small. The initial matings may occasionally be crosses or backcrosses of the first kind. Even so they are converted within one generation into other mating types. Therefore, the equations above may usually be written

$$p_{n+1} = p_n + \qquad ct_n,$$
$$t_{n+1} = \qquad (1 - c)t_n.$$

It follows at once that

$$t_n = (1 - c)^{n-1}t_1$$

or

$$t_n = (1 - c)^{n-1}$$

when $t_1 = 1$, as is the case if the initial mating is a cross, that is, if $q_0 = 1$.

The probability of the undesired backcrosses in G_n is t_n and of the desired incrosses is

$$p_n = 1 - t_n = 1 - (1 - c)^{n-1}.$$

If $c = 1/2$, the probability p is successively

$$0, \quad 1/2, \quad 3/4, \quad 7/8, \quad 15/16, \quad 31/32, \ldots,$$

for $n = 1, 2, 3, \ldots$. The probabilities for the desired mating types are shown in figure 4, for selected values of c.

Fig. 4. PROBABILITY OF INCROSSES FOR THE BACKCROSS SYSTEM.

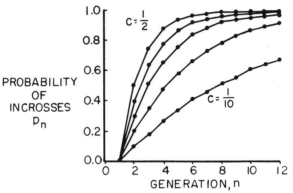

The probability of incrosses p_n for the backcross system starting with $q_0 = 1$. The curves shown are for five selected values of $c = 1/10, 2/10, 3/10, 4/10, 5/10$. The probability of heterozygosity is $h_n = 1 - p_n$. The ratio of successive values of h is constant $h_{n+1}/h_n = 1 - c$, when $n \geq 1$.

The probability h_n of heterozygosity in G_n, when $n \geq 1$, is

$$h_n = r_n + t_n = (1 - c)^{n-1}.$$

The ratio of successive values of h is constant

$$h_{n+1}/h_n = 1 - c,$$

or the probability of heterozygosity in one generation is the fraction $1 - c$ of the probability in the preceding generation. When $c = 1/2$, the probability of heterozygosity is exactly cut in half with each additional generation of backcrossing.

The generation matrix, $\mathbf{G}$, for the backcross system is given in table 6 (a). Because of the fact, mentioned above, that only incrosses or backcrosses of the second type occur after the initial generation, a simplified matrix, $\mathbf{G}^*$, can be used if G_1 is considered as the initial generation. $\mathbf{G}^*$ is given in table 6 (b). Table 6 (c), (d), and (e) give the $\mathbf{\Lambda}$, $\mathbf{A}$, and $\mathbf{A}^{-1}$ matrices for this system.

Table 6

MATRICES FOR THE BACKCROSS SYSTEM

(a) $\mathbf{G}$ = $\begin{pmatrix} 1 & 0 & 1-c & c \\ 0 & 0 & 0 & 0 \\ 0 & 0 & 0 & 0 \\ 0 & 1 & c & 1-c \end{pmatrix}$

(b) $\mathbf{G}^*$ = $\begin{pmatrix} 1 & c \\ 0 & 1-c \end{pmatrix}$

(c) $\boldsymbol{\Lambda}$ = $\begin{pmatrix} 1-c & 0 \\ 0 & 1 \end{pmatrix}$

(d) $\mathbf{A}$ = $\begin{pmatrix} 0 & 1 \\ 1 & 1 \end{pmatrix}$

(e) $\mathbf{A}^{-1}$ = $\begin{pmatrix} -1 & 1 \\ 1 & 0 \end{pmatrix}$

To obtain the frequencies of the various mating types in generation n, $\mathbf{P}_1$, the vector of frequencies in generation 1, is first obtained from equation (2). $\mathbf{P}_1$ will be of the form

$$\mathbf{P}_1 = (p_1 \quad 0 \quad 0 \quad t_1).$$

We then form the vector $\mathbf{P}_1^* = (p_1 \quad t_1)$, transform to $\mathbf{V}_1^*$, obtain $\mathbf{V}_n^*$ and convert to $\mathbf{P}_n^*$. These vectors are the same as the vectors operated with earlier, in equations (2) through (6); the * notation is introduced merely as a reminder that we are operating with only two of the original four mating types.
Note that

$$\mathbf{V}_n^* = \boldsymbol{\Lambda}^{n-1}\mathbf{V}_1^*,$$

the matrix $\boldsymbol{\Lambda}$ being raised to the power $(n-1)$, since one generation has been accounted for already by calculating $\mathbf{P}_1$. $\mathbf{P}_n^*$ gives only the frequencies p_n of incrosses and t_n of backcrosses of the second type; the frequencies of the other two types of matings possible in G_0 are, as has been demonstrated, zero in G_n. The frequencies p_n and t_n calculated by this method agree perfectly with those calculated by repetitive application of formula (2).

The number of generations required to obtain a given percentage of incrosses can also be calculated for the backcross system. When $c = 1/2$, 5 generations are required to obtain a frequency of incrosses greater than 95 per cent and 8 to obtain a frequency greater than 99 per cent. When $c = 1/10$, 29 generations are required to obtain 95 per cent incrosses, and 44 to obtain 99 per cent. It should be noted that if the gene of interest is recessive, test generations will be required, thus increasing the time required to obtain a given percentage of incross matings.

The mean length L of heterozygous chromosome on each side of the locus of interest after n generations of backcrossing is

$$L = \int_0^{1/2} (1 - c)^{n-1} dc$$

$$= \frac{1}{n} [1 - (1/2)^n]$$

$$\cong \frac{1}{n},$$

if n is large and if $q_0 = 1$.   This means, for example, that 40 generations of backcrossing will be expected to reduce the mean length of heterozygous chromosome, on both sides of the locus of interest, to about 5 centimorgans or 2.5 centimorgans on each side.

Several mutations in mice have been put on standard inbred backgrounds by use of the backcross system.   A few examples are: C57BL/6J–A^y, C57BL/6J–$CaSp$, C57BL/6J–Mi^{wh}, C57BL/6J–W^v, BALB/cGnSn–FuC, C3HeB/FeHu–Ds, DBA/1$_o$Hu–Ds. Complete lists of these strains are given by Lane.[751]   Some strains have been propagated by brother-sister matings with forced heterozygosis after 5 or more generations of backcrossing to a standard inbred strain.   Examples are: C57BL/6J–Pt, C57BL/6J–Ra, and BALB/c–Ts.

THE CROSS-INTERCROSS SYSTEM

The cross-intercross system is used to put a recessive, viable gene r on a standard inbred background.   The inbred strain is presumably homozygous, AA or aa, for all loci with heterozygosity in question after the onset of crossing.   The inbred strain is also presumably RR.   Mutant-bearing animals of a stock, inbred or not, are on hand carrying rr at the locus of interest.   The objective is to replace the RR of the inbred strain with rr of the mutant stock, while otherwise preserving the homozygosity of the inbred strain.

The matings proceed in cycles of two generations.   The first mating of each cycle is a CROSS, $RR \times rr$; and the second is an INTERCROSS, $Rr \times Rr$, with respect to the locus of interest (figure 5).

There are three mating types among the CROSSES and three among the INTER-CROSSES.   The mating types and their probabilities in cycle m (C_m) are:

For r-locus: For a-locus:

CROSSES   Incrosses $P \begin{pmatrix} AR/AR \times Ar/Ar \\ aR/aR \times ar/ar \end{pmatrix} = p_m,$

Crosses $P \begin{pmatrix} AR/AR \times ar/ar \\ aR/aR \times Ar/Ar \end{pmatrix} = q_m,$

Backcrosses $P \begin{pmatrix} AR/AR \times Ar/ar \\ aR/aR \times Ar/ar \end{pmatrix} = r_m,$

INTERCROSSES   Incrosses $P\begin{pmatrix}AR/Ar & \times & AR/Ar \\ aR/ar & \times & aR/ar\end{pmatrix}$,

Backcrosses $P\begin{pmatrix}AR/Ar & \times & AR/ar \\ aR/ar & \times & aR/Ar\end{pmatrix}$,

Intercrosses $P\begin{pmatrix}AR/ar & \times & AR/ar \\ aR/Ar & \times & aR/Ar\end{pmatrix}$.

In each mating type of the CROSSES for the r-locus, the standard inbred strain is on the left, the mutant stock on the right.

Fig. 5. THE CROSS-INTERCROSS SYSTEM.

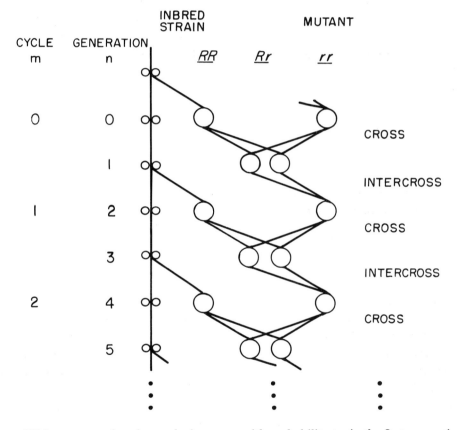

With respect to the a-locus, the incrosses, with probability p_m, in the first generation of C_m, each produce one kind of offspring which yield incrosses in the next generation of the same cycle. These, in turn, contribute exclusively to the incrosses of the first generation of the next cycle, since all progeny other than rr are discarded. The crosses, with probability q_m, also each produce one kind of offspring, all doubly heterozygous. These when intercrossed produce three kinds of rr progeny AA (or aa), Aa, and aa

(or AA) in the ratio $c^2:2c(1-c):(1-c)^2$. These in turn, when crossed to the inbred strain, yield incrosses, crosses, and backcrosses with respect to the a-locus in the ratio $c^2:(1-c)^2:2c(1-c)$. Finally, the backcrosses, with probability r_m, in the first generation of C_m, each produce two kinds of offspring, AR/Ar and AR/ar, or aR/ar and aR/Ar, in the ratio $1/2:1/2$. These yield three kinds of matings in the second generation o C_m, incrosses, backcrosses, and intercrosses, in the proportions $1/4:1/2:1/4$.

The incrosses in the second generation of C_m each produce one kind of rr progeny, with relative frequency $1/4$, and thus yield only incrosses in the first generation of the next cycle, C_{m+1}. The backcrosses each produce two kinds of rr progeny, AA and Aa, or aa and Aa, in the ratio $(1/2)c:(1/2)(1-c)$ and so yield incrosses and backcrosses in the same ratio in the first generation of C_{m+1}. The intercrosses produce three kinds of rr progeny, AA, Aa, and aa, or aa, Aa, and AA, in the ratio $(1/4)c^2:(1/2)c(1-c):(1/4)(1-c)^2$, and so yield incrosses, crosses, and backcrosses in the proportions $(1/4)c^2:(1/4)(1-c)^2:(1/2)c(1-c)$ in the first generation of C_{m+1}. In summary, the backcrosses in the first generation of C_m yield incrosses, crosses, and backcrosses in the first generation of C_{m+1} in the proportions $(1/4)(1+c)^2:(1/4)(1-c)^2:(1/2)(1-c^2)$.

This is so because

$$1/4 + \quad (1/2)c + \quad (1/4)c^2 = (1/4)(1+c)^2,$$
$$(1/4)(1-c)^2 = (1/4)(1-c)^2,$$
$$(1/2)(1-c) + (1/2)c(1-c) = (1/2)(1-c^2).$$

The probabilities of the various mating types in C_{m+1} may then be represented as functions of the probabilities in C_m by three linear equations:

$$p_{m+1} = p_m + \quad c^2 q_m + (1/4)(1+c)^2 r_m,$$
$$q_{m+1} = \quad (1-c)^2 q_m + (1/4)(1-c)^2 r_m,$$
$$r_{m+1} = \quad 2c(1-c)q_m + (1/2)(1-c^2)r_m.$$

Again the incrosses are an absorbing barrier; once matings of like homozygotes are reached, they hold their type in succeeding generations. At the same time fractions of the matings of other types yield more incrosses which are thenceforward fixed, the rate depending upon c. By means of these equations, the probability of the desired type of mating (incrosses) may be computed for any number of cycles and for selected values of c. The results for 6 cycles, or 12 generations, are outlined in figure 6, for the case $q_0 = 1$, that is when breeding starts with a cross $AR/AR \times ar/ar$, or $aR/aR \times Ar/Ar$.

When $c = 1/2$ and $q_0 = 1$, the probabilities p_m of incrosses, the matings of the desired type, are the successive values

$$0, \quad 1/4, \quad 19/32, \quad 203/256, \quad 1835/2048, \ldots$$

for $m = 0, 1, 2, 3, 4, \ldots$.

The probability h_m of heterozygotes is

$$h_m = r_m,$$

which takes the successively smaller values, after the initial generation,

$$0, \quad 1/2, \quad 5/16, \quad 21/128, \quad 85/1024, \ldots$$

for $m = 0, 1, 2, 3, 4, \ldots$, when $c = 1/2$ and $q_0 = 1$.

The ratios of the probabilities of heterozygotes in successive generations tends toward a constant as m increases.   For large m,

$$h_{m+1}/h_m = 1 - c, \text{ approximately.}$$

The probability of heterozygosity in any advanced cycle is, approximately, a fraction, $1 - c$, of the probability of heterozygosity in the preceding cycle.   When $c = 1/2$, the loss tends to be $1/2$.

Fig. 6. PROBABILITY OF INCROSSES FOR THE CROSS-INTERCROSS SYSTEM.

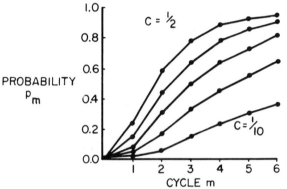

The probability of incrosses p_m for the cross-intercross system, starting with $q_0 = 1$, for five selected values of $c = 1/10, 2/10, 3/10, 4/10, 5/10$.

After the cross-intercross system has been in effect for several cycles, the incrosses will be increasing at a steady rate exactly balanced by the rate of decrease of the crosses and backcrosses.   This rate may be found by finding the characteristic root of the determinant made up of the q and r rows and columns of the cycle matrix (table 7). The desired root is

$$\lambda_1 = 1 - c.$$

Table 7

CYCLE MATRIX OF THE CROSS-INTERCROSS SYSTEM

	p_m	q_m	r_m
p_{m+1}	1	c^2	$(1/4)(1 + c)^2$
q_{m+1}	—	$(1 - c)^2$	$(1/4)(1 - c)^2$
r_{m+1}	—	$2c(1 - c)$	$(1/2)(1 - c)^2$

$$\begin{vmatrix} (1 - c)^2 - \lambda & (1/4)(1 - c)^2 \\ 2c(1 - c) & (1/2)(1 - c)^2 - \lambda \end{vmatrix} = 0$$

$$\lambda = 1 - c, \quad (1/2)(1 - c)^2$$

This result shows that one cycle of the cross-intercross system is genetically equivalent to one generation of the backcross system.

The cycle matrix for the cross-intercross system can be obtained by directly reasoning what proportion of each type of mating a given type will yield in the subsequent cycle, two generations later. It is possible, however, to avoid the confusion attendant on the attempt to reason through two generations. A generation matrix can be constructed for each generation in the cycle. By multiplying these matrices together in order opposite to that in which they occur, the cycle matrix is obtained. Thus, for a cycle of k generations,

$$\mathbf{C} = \mathbf{G}_k \mathbf{G}_{k-1} \ldots \mathbf{G}_1,$$

where $\mathbf{C}$ is the cycle matrix, and $\mathbf{G}_i$ is the generation matrix for generation i in the cycle.

In the case of the cross-intercross system, the matrix for cross generations is

$$\mathbf{G}_1 = \begin{pmatrix} 1 & 0 & 1/4 \\ 0 & 0 & 1/2 \\ 0 & 1 & 1/4 \end{pmatrix}$$

while that for intercross generations is

$$\mathbf{G}_2 = \begin{pmatrix} 1 & c & c^2 \\ 0 & 0 & (1-c)^2 \\ 0 & 1-c & 2c(1-c) \end{pmatrix}$$

and

$$\mathbf{C} = \mathbf{G}_2 \mathbf{G}_1.$$

From the matrices given in tables 7 and 8, $\mathbf{P}_m$ can be calculated for any m, given $\mathbf{P}_0$. Also the number of cycles required to obtain any given percentage of incross matings can be calculated. For both calculations, we must keep in mind that one cycle is the equivalent of two generations.

Table 8

THE $\mathbf{\Lambda}$, $\mathbf{A}$, AND $\mathbf{A}^{-1}$ MATRICES FOR THE CROSS-INTERCROSS SYSTEM

$$\mathbf{\Lambda} = \begin{pmatrix} 1-c & 0 & 0 \\ 0 & (1/2)(1-c)^2 & 0 \\ 0 & 0 & 1 \end{pmatrix}$$

$$\mathbf{A} = \begin{pmatrix} 0 & 1 & 1/2 \\ 0 & 1 & -(1-c)/4c \\ 1 & 1 & 1 \end{pmatrix}$$

$$\mathbf{A}^{-1} = \begin{pmatrix} -(1+3c)/(1+c) & 2c/(1+c) & 1 \\ (1-c)/(1+c) & 2c/(1+c) & 0 \\ 4c/(1+c) & -4c/(1+c) & 0 \end{pmatrix}$$

For $c = 1/2$, 6 cycles (or 12 generations) of the CROSS-INTERCROSS system are required to exceed 95 per cent incross matings, and 8 cycles (16 generations) to exceed 99 per cent incross matings.   For $c = 1/10$, 31 cycles (62 generations) are required to exceed 95 per cent incross matings and 46 cycles (92 generations) to exceed 99 per cent.   This system has been used to put a number of mutations in mice on standard inbred backgrounds.   These include: C57BL/6J–*ja*, C57BL/6J–*le*, C57BL/6J–*ru*, and BALB/cHu–*iv*.   Complete lists of these strains are mentioned by Lane.[751]   Snell[1238] used this system to isolate single genes or small segments of chromosomes which affect histocompatibility.   Examples are AKR–H–2^a (syn. AKR·K), C3H–H–2^b (syn. C3H·SW), and C57BL/10–H–1^d (syn. B1O·BY).

THE CROSS-BACKCROSS-INTERCROSS SYSTEM

The cross-backcross-intercross system, like the cross-intercross system, is used to put a recessive viable gene r on a standard inbred background.   It is especially useful when the mutant phenotype is not visible and easily detectable, but must be determined by a laboratory test, and when the ease of making up another generation of matings (the backcross) outweighs the cost of determining the phenotype.   The system was invented and first used by Snell for developing strains of mice with genes for resistance to tumor grafts on standard inbred backgrounds.

The matings proceed in cycles of three generations.   The first mating of each cycle is a CROSS, $RR \times rr$, when rr is the mutant and RR is the genotype of the inbred strain.   The second mating is a BACKCROSS, $RR \times Rr$, formed by mating the heterozygous progeny of the cross with animals of the inbred strain.   The third mating is an INTERCROSS, $Rr \times Rr$, of the heterozygous progeny of the backcross.   The heterozygotes are discovered to be so by observing which matings produce rr progeny.   All other matings, 3/4 of the total on the average, are wasted.   The rr progeny are then crossed with RR animals of the inbred strain to start another cycle (figure 7).

Among the CROSSES, there are three mating types when any other locus, the a-locus, is considered in addition to the locus of interest.   There are two types of BACKCROSSES and three types of INTERCROSSES.   The mating types and their probabilities in C_m are:

$$\text{CROSSES} \quad \text{Incrosses} \quad P\begin{pmatrix} AR/AR & \times & Ar/Ar \\ aR/aR & \times & ar/ar \end{pmatrix} = p_m,$$

$$\text{Crosses} \quad P\begin{pmatrix} AR/AR & \times & ar/ar \\ aR/aR & \times & Ar/Ar \end{pmatrix} = q_m,$$

$$\text{Backcrosses} \quad P\begin{pmatrix} AR/AR & \times & Ar/ar \\ aR/aR & \times & Ar/ar \end{pmatrix} = r_m,$$

$$\text{BACKCROSSES} \quad \text{Incrosses} \quad P\begin{pmatrix} AR/AR & \times & AR/Ar \\ aR/aR & \times & aR/ar \end{pmatrix},$$

$$\text{Backcrosses} \quad P\begin{pmatrix} AR/AR & \times & AR/ar \\ aR/aR & \times & aR/Ar \end{pmatrix},$$

INTERCROSSES   Incrosses $P\begin{pmatrix} AR/Ar & \times & AR/Ar \\ aR/ar & \times & aR/ar \end{pmatrix}$,

Backcrosses $P\begin{pmatrix} AR/Ar & \times & AR/ar \\ aR/ar & \times & aR/Ar \end{pmatrix}$,

Intercrosses $P\begin{pmatrix} AR/ar & \times & AR/ar \\ aR/Ar & \times & aR/Ar \end{pmatrix}$.

In the CROSSES and BACKCROSSES with respect to the r-locus, the inbred strain is on the left, the mutant-bearing animals on the right.

With respect to the a-locus, the CROSS-incrosses with probability p_m each yield BACKCROSS-incrosses and these yield INTERCROSS-incrosses, which thus lead to CROSS-

Fig. 7. THE CROSS-BACKCROSS-INTERCROSS SYSTEM.

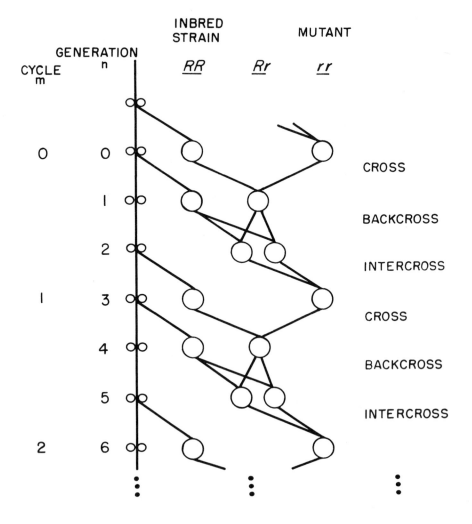

incrosses to begin the next cycle. The CROSS-crosses, probability q_m, produce BACK-CROSS-backcrosses, and these produce INTERCROSSES of three kinds, incrosses, back-crosses, and intercrosses, in the ratio $c^2:2c(1-c):(1-c)^2$. These in turn yield CROSSES of three kinds to begin the next cycle, incrosses, crosses, and backcrosses, in the ratio $c^2(2-c)^2:(1-c)^4:2c(1-c)^2(2-c)$. The CROSS-backcrosses, probability r_m, produce BACKCROSS-incrosses and BACKCROSS-backcrosses in the ratio $1/2:1/2$. The BACKCROSS-incrosses yield INTERCROSS-incrosses and thence CROSS-incrosses to begin the next cycle. The BACKCROSS-backcrosses yield INTERCROSSES of three kinds, incrosses, crosses, and intercrosses, in the ratio $(1/2)c^2:c(1-c):(1/2)(1-c)^2$; and these in turn yield CROSSES to begin a new cycle with incrosses, crosses, and backcrosses, in the ratio $(1/2)c^2(2-c)^2:(1/2)(1-c)^4:c(1-c)^2(2-c)$.

Fig. 8. PROBABILITY OF INCROSSES FOR THE CROSS-BACKCROSS-INTERCROSS SYSTEM.

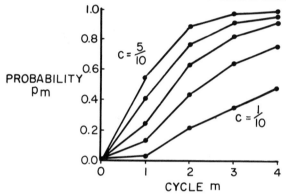

The probability of incrosses p_m for the cross-backcross-intercross system beginning with $q_0 = 1$, for values of $c = 1/10, 2/10, 3/10, 4/10, 5/10$. The ratio of the probabilities of heterozygosity in successive generations is constant, $h_{m+1}/h_m = (1-c)^2$, when $m > 1$.

The probabilities of mating types in C_{m+1} may thus be shown as functions of the probabilities of mating types in C_m:

$$p_{m+1} = p_m + \qquad\qquad c^2(2-c)^2 q_m + \{(1/2) + (1/2)c^2(2-c)^2\}r_m,$$
$$q_{m+1} = \qquad\qquad (1-c)^4 q_m + \qquad\qquad (1/2)(1-c)^4 r_m,$$
$$r_{m+1} = \qquad 2c(1-c)^2(2-c)q_m + \qquad c(1-c)^2(2-c)r_m.$$

Again the familiar result emerges. The incrosses, probability p_m, increase at the expense of the backcrosses, q_m, and the intercrosses, r_m. The rate of increase of p_m depends upon c, the probability of crossing over between the a-locus and the r-locus.

When the initial mating is a cross, that is, $q_0 = 1$, the desired types of matings increase as shown in figure 8, for selected values of c.

When $c = 1/2$ and $q_0 = 1$, the successive values of p_m are

$$0, \quad 9/16, \quad 57/64, \quad 249/256, \quad \ldots,$$

for $m = 0, 1, 2, 3, \ldots$. The probability of heterozygosity is the same as the probability of backcrosses among the CROSSES, i.e., $h_m = r_m$. This probability takes the successive values

$$0, \quad 3/8, \quad 3/32, \quad 3/128, \quad 3/512, \ldots,$$

for $m = 0, 1, 2, \ldots$. It may be seen at once that when $m > 0$,

$$h_{m+1} = (1/4)h_m,$$

when $q_0 = 1$ and $c = 1/2$.

The characteristic root of the determinant formed from the q and r rows and columns of the cycle matrix (table 9) is

$$\lambda_1 = (1 - c)^2.$$

Table 9

CYCLE MATRIX FOR THE CROSS-BACKCROSS-INTERCROSS SYSTEM

	p_m	q_m	r_m
p_{m+1}	1	$c^2(2 - c)^2$	$(1/2) + (1/2)c^2(2 - c)^2$
q_{m+1}	0	$(1 - c)^4$	$(1/2)(1 - c)^4$
r_{m+1}	0	$2c(1 - c)^2(2 - c)$	$c(1 - c)^2(2 - c)$

$$\begin{vmatrix} (1 - c)^4 - \lambda & (1/2)(1 - c)^4 \\ 2c(1 - c)^2(2 - c) & c(1 - c)^2(2 - c) - \lambda \end{vmatrix} = 0$$
$$\lambda = (1 - c)^2, \quad 0$$

This is the factor of proportionality between successive values of q and r,

$$q_{m+1} = \lambda q_m,$$
$$r_{m+1} = \lambda r_m.$$

It also gives the relationship between successive values of h, the probability for heterozygosity, since $h_m = r_m$. When $c = 1/2$

$$h_{m+1} = (1 - c)^2 h_m = (1/4)h_m, \quad m > 0$$

confirming the result of the preceding paragraph.

Again, the cycle matrix may be obtained more easily by multiplying the generation matrices in reverse order to that in which they appear.

The CROSS generation matrix is

$$\mathbf{G}_1 = \begin{pmatrix} 1 & 0 & 1/2 \\ 0 & 1 & 1/2 \end{pmatrix}.$$

The BACKCROSS generation matrix is

$$\mathbf{G_2} = \begin{pmatrix} 1 & c^2 \\ 0 & 2c(1-c) \\ 0 & (1-c)^2 \end{pmatrix}.$$

Finally, the INTERCROSS generation matrix is

$$\mathbf{G_3} = \begin{pmatrix} 1 & c & c^2 \\ 0 & 0 & (1-c)^2 \\ 0 & 1-c & 2c(1-c) \end{pmatrix}.$$

Thus, the cycle matrix is

$$\mathbf{C} = \mathbf{G_3 G_2 G_1} = \begin{pmatrix} 1 & c^2(2-c)^2 & (1/2)+(1/2)c^2(2-c)^2 \\ 0 & (1-c)^4 & (1/2)(1-c)^4 \\ 0 & 2c(1-c)^2(2-c) & c(1-c)^2(2-c) \end{pmatrix}.$$

From the matrices in table 10 the frequencies of various types of matings in any cycle and the number of generations to reach a given percentage of incross matings can be calculated, remembering that a cycle represents three generations.

Table 10

Λ, **A**, AND **A**$^{-1}$ MATRICES FOR THE CROSS-BACKCROSS-INTERCROSS SYSTEM

$$\Lambda = \begin{pmatrix} (1-c)^2 & 0 & 0 \\ 0 & 0 & 0 \\ 0 & 0 & 1 \end{pmatrix}$$

$$\mathbf{A} = \begin{pmatrix} 0 & 1 & 1/2 \\ 0 & 1 & -(1-c)^2/2c(2-c) \\ 1 & 1 & 1 \end{pmatrix}$$

$$\mathbf{A}^{-1} = \begin{pmatrix} c^2-2c-1 & c(2-c) & 1 \\ (1-c)^2 & c(2-c) & 0 \\ 2c(2-c) & -2c(2-c) & 0 \end{pmatrix}$$

Under the cross-backcross-intercross system, with $c = 1/2$, 3 cycles (9 generations) are required to exceed 95 per cent incross matings and 4 cycles (12 generations) to exceed 99 per cent. If $c = 1/10$, 16 cycles (48 generations) and 23 cycles (69 generations) are required to exceed 95 per cent and 99 per cent incross matings.

It should also be noted that approximately four times as many matings as are required to maintain the stock must be made up in the INTERCROSS generation to assure a sufficient supply of the types of matings required. The cost of these extra matings should be borne in mind when the cross-backcross-intercross system is considered for use.

This system is being used by Snell for the isolation of loci affecting histo-compatibility in the mouse. No strains have advanced far enough to have been mentioned in the literature.

BROTHER-SISTER INBREEDING WITH HETEROZYGOSIS FORCED BY BACKCROSSING

The system of brother-sister inbreeding with heterozygosis forced by backcrossing is useful when it is desired to put a recessive mutation and its normal allele or a semi-dominant lethal mutation and its normal allele on a common inbred background. Unlike the three preceding systems, the mutant-bearing animals are not crossed with animals of an existing inbred strain, but are mated as brother × sister. With respect to the locus of interest, all matings are $rr × Rr$, in the case of a recessive viable mutation r, or $dd × Dd$ in the case of a semidominant lethal mutation D. The other homo-zygote is either rejected (RR) or lethal (DD). As before, let A/a be any other locus whose heterozygosity is in question, and let c be the probability of crossing over between the a-locus and the r-locus or D-locus.

There are six types of matings with probabilities denoted as follows:

BACKCROSSES   Incrosses $$P\begin{pmatrix} Ar/Ar & × & AR/Ar \\ ar/ar & × & aR/ar \end{pmatrix} = p_n,$$

Crosses $$P\begin{pmatrix} Ar/Ar & × & aR/ar \\ ar/ar & × & AR/Ar \end{pmatrix} = q_n,$$

Backcrosses (first kind) $$P\begin{pmatrix} Ar/Ar & × & AR/ar \\ ar/ar & × & aR/Ar \end{pmatrix} = r_n,$$

Backcrosses (second kind) $$P\begin{pmatrix} Ar/ar & × & AR/Ar \\ Ar/ar & × & aR/ar \end{pmatrix} = s_n,$$

Backcrosses (third kind) $$P\begin{pmatrix} Ar/Ar & × & aR/Ar \\ ar/ar & × & AR/ar \end{pmatrix} = t_n,$$

Intercrosses $$P\begin{pmatrix} Ar/ar & × & AR/ar \\ Ar/ar & × & aR/Ar \end{pmatrix} = v_n.$$

If each r is replaced by d and each R by D, the matings refer to inbreeding with a semidominant lethal mutation.

The incrosses yield only incrosses in the next generation. The crosses produce two kinds of progeny, aR/Ar and Ar/ar, or AR/ar and Ar/ar, but under the rule of this system of mating brother × sister, which are $rr × Rr$, they yield only one kind of mating in the next generation, intercrosses. The backcrosses of the first kind produce four kinds of progeny which yield incrosses, backcrosses of the second kind, backcrosses of the third kind, and intercrosses in the ratio $c(1 - c):(1 - c)^2:c(1 - c)$ in the next generation. The backcrosses of the second kind also produce four kinds of progeny which yield incrosses, backcrosses of the first kind, backcrosses of the second kind, and intercrosses in the ratio $1/4:1/4:1/4:1/4$ in the next generation.   The backcrosses of the third kind likewise produce four kinds of progeny and these yield incrosses, backcrosses of the second kind, backcrosses of the third kind, and intercrosses in the ratio $c(1-c):$ $c^2:(1-c)^2:c(1-c)$.   Finally, the intercrosses produce seven kinds of progeny which yield all six kinds of matings in the ratio

$$(1/2)c(1-c):(1/4) - (1/2)c(1-c):(1/2)c(1-c):(1/4):(1/4) - (1/2)c(1-c):(1/4).$$

The linear equations which relate the probabilities of G_{n+1} to the probabilities of G_n are:

$$
\begin{aligned}
p_{n+1} &= p_n + & kr_n + (1/4)s_n + & \quad kt_n + & (1/2)\ kv_n, \\
q_{n+1} &= & & & (1/4)(1-2k)v_n, \\
r_{n+1} &= & (1/4)s_n + & & (1/2)kv_n, \\
s_{n+1} &= & (1-c)^2 r_n + (1/4)s_n + & c^2 t_n + & (1/4)v_n, \\
t_{n+1} &= & c^2 r_n + & (1-c)^2 t_n + (1/4)(1-2k)v_n, \\
v_{n+1} &= q_n + & kr_n + (1/4)s_n + & kt_n + & (1/4)v_n,
\end{aligned}
$$

where $k = c(1-c)$.

Fig. 9. PROBABILITY OF INCROSSES WITH HETEROZYGOSIS FORCED BY BACKCROSSING.

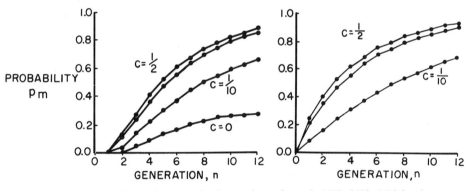

LEFT: The probability of incrosses for four values of $c = 0, 1/10, 3/10, 5/10$ for the system of brother-sister inbreeding with heterozygosis forced by backcrossing, starting with $q_0 = 1$.

RIGHT: The probability of incrosses p_n for three values of $c = 1/10, 3/10, 5/10$, for the system of brother-sister inbreeding with heterozygosis forced by backcrossing, starting with $t_0 = 1$.

Matings in the initial generation may be any of the six types.   Two of these are slightly more likely to occur in laboratory populations.   First, the mutant allele D may be a semidominant lethal in a noninbred line.   When inbreeding is started, the least favorable mating with respect to homozygosity at the a-locus is a cross $Ad/Ad \times aD/ad$, or $ad/ad \times AD/Ad$, i.e., $q_0 = 1$.   The probability p of incrosses increases relatively slowly (figure 9).   Second, the mutant allele r may already exist in one inbred strain and its nonmutant allele R in another. A preparatory cross between the strains will produce double heterozygotes, AR/ar (or aR/Ar). When these are mated with the mutant-bearing strain ar/ar (or Ar/Ar) the initial matings are backcrosses of the third kind, i.e., $t_0 = 1$. The probability p of incrosses increases more rapidly with this starting point (figure 9).

The probability of heterozygosity is less informative than the probability of incrosses in this system of breeding. When it is desired to compute h_n, however, it must be observed that two probabilities are required, one for rr (or dd) and one for Rr (or Dd) animals. They are:

$$P(Aa|rr) = h'_n = s_n + v_n,$$
$$P(Aa|Rr) = h''_n = r_n + t_n + v_n.$$

In general, the probability h'_n of heterozygosity at the a-locus among the homozygous mutants will be less than the probability h''_n of heterozygosity at the a-locus among the heterozygous mutants for all loci linked with the mutant locus. By forcing heterozygosity upon the locus of interest, one also forces heterozygosity upon loci linked with it. The probabilities, h'_n and h''_n, for the first twelve generations, starting with either a cross ($q_0 = 1$) or a backcross ($t_0 = 1$), are shown in figure 10.

If c is near zero, and if $q_0 = 1$, the probability is approximately 2/3 that in a given line the heterozygotes Rr will remain heterozygous Aa at the closely linked a-locus. This is so because the probability t_n of backcrosses of the third kind approaches 2/3 as n increases, while s_n and v_n each approach zero. The probability p_n of incrosses equals $1 - t_n$; as n increases p_n approaches 1/3.

If $t_0 = 1$ and c is near zero, there is practically no chance of getting incrosses, since backcrosses of the third kind with probability t_n yield only backcrosses of the third kind;

$$t_0 = t_1 = t_2 = \cdots = t_n = 1$$

approximately.

The determinant formed from coefficients of $q_n, \ldots, v_n$ in the equations for $q_{n+1}, \ldots, v_{n+1}$ is expressible as an equation of the 5th order:

$$\mu^5 - (3 - 2c - 2k)\mu^4 - c\mu^3 + \{(5 - 7c - 2k)(1 - 2k) + 4(1 - 2c)k^2\}\mu^2$$
$$- (1 - 2c)(1 - 2k)(1 - 4k)\mu - 2(1 - 2c)(1 - 2k)^2 = 0$$

where $\mu = 2\lambda$.

When $c = 1/2$, the characteristic root is, as expected, the same as for brother-sister inbreeding, i.e.,

$$\lambda_1 = (1/4)(1 + \sqrt{5}).$$

The characteristic root rises to one as c decreases to zero. The values of λ_1 for selected values of c are:

$c =$	0.5	0.4	0.3	0.2	0.1	0
$\lambda_1 =$	0.8090	0.8135	0.8270	0.8514	0.8984	1.

It has not been possible to solve the equation

$$|\mathbf{G} - \lambda_i \mathbf{I}| = 0 \tag{8}$$

for this system of brother-sister inbreeding with heterozygosis forced by backcrossing. Values of λ_1 for selected values of c have been estimated from repetitive application of the equation

$$\mathbf{P}_{n+1} = \mathbf{GP}_n \tag{2}$$

for $n = 0, 1, \ldots, 11$. Until a general solution of equation (7) can be derived for these systems, the equation

$$\mathbf{V}_n = \mathbf{\Lambda}^n \mathbf{V}_0 \tag{6}$$

cannot be applied.

Using the above estimate of λ_1, however, the number of generations required to obtain a given percentage of incross matings can be estimated. For $c = 1/2$, 15 generations of brother-sister inbreeding with heterozygous forced by backcrossing are required

Fig. 10. PROBABILITY OF HETEROZYGOSITY WITH HETEROZYGOSIS FORCED BY BACKCROSSING.

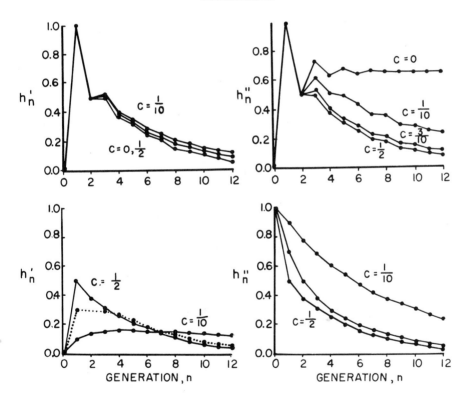

TOP: The probability of heterozygosity, h'_n for rr, h''_n for Rr, for selected values of $c = 0, 1/10, 3/10, 5/10$, for the system of brother-sister inbreeding with heterozygosis forced by backcrossing, starting with $q_0 = 1$.

BOTTOM: The probability of heterozygosity, h'_n for rr, h''_n for Rr, for selected values of $c = 1/10, 3/10, 5/10$ for the system of brother-sister inbreeding with heterozygosis forced by backcrossing, starting with $t_0 = 1$.

to exceed 95 per cent incross matings, starting with $q_0 = 1$; 22 generations are required to exceed 99 per cent incross matings. For $c = 1/10$, 26 and 39 generations, respectively, are required to exceed 95 per cent and 99 per cent incross matings. It should be noted, however, that when c is not zero this system ensures eventual attainment of animals homozygous and heterozygous at the locus of interest, and otherwise coisogenic, aside from recent mutations.

Examples of strains of mice produced by this system of mating are: SEC/1Gn$-se$, FS/Gn$-fs$, and QV/Gn$-qv$. See Lane[751] for a complete list.

BROTHER-SISTER INBREEDING WITH HETEROZYGOSIS FORCED BY INTERCROSSING

This system, brother-sister inbreeding with heterozygosis forced by intercrossing, is useful in putting a recessive lethal mutation on an inbred background. Similarly, any recessive mutation with low viability or fertility when homozygous and its normal allele may be put on a common inbred background. If the new recessive mutation has recently arisen in an already inbred strain, this system provides a way of continued inbreeding. All matings are between heterozygotes, $Rr \times Rr$, with respect to the locus of interest. It will commonly be necessary to determine which are $Rr \times Rr$ matings by observing which matings of $R- \times R-$ produce rr progeny. The system is also useful for inbreeding with a semidominant lethal. In this case, all matings are $Dd \times Dd$, DD being lethal, and dd being rejected.

Let A/a be any other gene pair whose heterozygosity is in question and let c be the probability of crossing over between the a-locus and the r-locus (or D-locus). Five types of matings may occur with probabilities in G_n denoted as follows:

INTERCROSSES	Incrosses	$P \begin{pmatrix} AR/Ar \times AR/Ar \\ aR/ar \times aR/ar \end{pmatrix} = p_n,$
	Crosses	$P\,(AR/Ar \times aR/ar) = q_n,$
	Backcrosses	$P \begin{pmatrix} AR/Ar \times AR/ar \\ aR/ar \times aR/Ar \\ AR/Ar \times aR/Ar \\ aR/ar \times AR/ar \end{pmatrix} = r_n,$
	Intercrosses (first kind)	$P \begin{pmatrix} AR/ar \times AR/ar \\ aR/Ar \times aR/Ar \end{pmatrix} = v_n,$
	Intercrosses (second kind)	$P\,(AR/ar \times aR/Ar) = w_n.$

The incrosses yield incrosses exclusively in the next generation. The crosses yield intercrosses of the first kind and intercrosses of the second kind in the ratio $1/2:1/2$ in the next generation. The backcrosses each produce three kinds of progeny and then yield four kinds of matings—incrosses, backcrosses, intercrosses of the first kind, and intercrosses of the second kind—in the ratio $(1/4):(1/2):(1/4)(1 - 2k):(1/2)k$, where $k = c(1 - c)$, in the next generation. The intercrosses of the first kind produce four kinds of progeny and all five kinds of matings in the ratio $2k^2:2k^2:4k(1 - 2k):$

$1 - 4k + 2k^2 : 2k^2$ in the next generation. Finally, the intercrosses of the second kind produce four kinds of progeny and all five kinds of matings in the ratio $(1/2)$ $(1 - 2k)^2 : (1/2)(1 - 2k)^2 : 4k(1 - 2k) : 2k^2 : 2k^2$ in the next generation.

These results may be shown as a set of five linear equations:

$$p_{n+1} = p_n + \qquad\qquad (1/4)r_n + \qquad\qquad 2k^2 v_n + (1/2)(1 - 2k)^2 w_n,$$

$$q_{n+1} = \qquad\qquad\qquad\qquad\qquad 2k^2 v_n + (1/2)(1 - 2k)^2 w_n,$$

$$r_{n+1} = \qquad\qquad (1/2)r_n + \quad 4k(1 - 2k)v_n + \quad 4k(1 - 2k)w_n,$$

$$v_{n+1} = (1/2)q_n + (1/4)(1.- 2k)r_n + (1 - 4k + 2k^2)v_n + \qquad 2k^2 w_n,$$

$$w_{n+1} = (1/2)q_n + \qquad (1/2)kr_n + \qquad\qquad 2k^2 v_n + \qquad 2k^2 w_n.$$

Fig. 11. PROBABILITY OF INCROSSES WITH HETEROZYGOSIS FORCED BY INTERCROSSING.

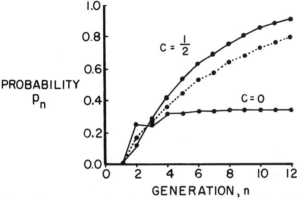

The probability of incrosses p_n for selected values of $c = 0, 1/10, 1/2$ for the system of brother-sister inbreeding with heterozygosis forced by intercrossing, starting with $q_0 = 1$.

The probabilities p_n of incross matings, the desired type, for selected values of c are shown in figure 11, assuming that $q_0 = 1$. This is the case if the mutation of interest arose in a noninbred stock and the initial mating between carriers of the mutation r is $AA \times aa$. This is the least favorable starting point with respect to the value of p_n; other starting points will produce higher values of p_n in fewer generations. The nonlinked loci become fixed at the usual rate for brother-sister inbreeding. Linked loci reach fixation more slowly at rates dependent upon c. When c is near zero, about 2/3 of the matings will remain as intercrosses of the first kind and 1/3 become incrosses of the desired type ($v_n = 2/3$ and $p_n = 1/3$ when n is infinite).

The probability of heterozygosity decreases depending upon c, as seen in figure 12, except when $c = 0$. In general,

$$h_n = (1/2)r_n + v_n + w_n.$$

The determinant formed from the generation matrix by omitting the p_{n+1} row and the p_n column yields the general equation:

$$\mu^4 - (3 - 4l + 2l^2)\mu^3 + (1 - 4l + 4l^2 - 4l^3)\mu^2$$
$$+ (3 - 10l + 14l^2 - 12l^3 + 8l^4)\mu - 2(1 - 5l + 10l^2 - 10l^3 + 4l^4) = 0$$

Fig. 12. PROBABILITY OF HETEROZYGOSITY WITH
HETEROZYGOSIS FORCED BY INTERCROSSING.

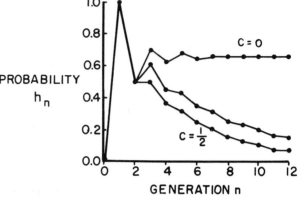

The probability of heterozygosity h_n for selected values of $c = 1/10$, $1/2$ for the system of brother-sister inbreeding with heterozygosis forced by intercrossing, starting with $q_0 = 1$.

where $\mu = 2\lambda$ and $l = 2c(1 - c)$. Solutions for selected values of c give the following characteristic roots:

$$c = 0.5 \qquad 0.4 \qquad 0.3 \qquad 0.2 \qquad 0.1 \qquad 0$$
$$\lambda_1 = 0.8090 \quad 0.8092 \quad 0.8115 \quad 0.8235 \quad 0.8674 \quad 1.$$

Again the equation

$$|\mathbf{G} - \lambda\mathbf{I}| = 0 \tag{8}$$

has not been solved, so that formula

$$\mathbf{V}_n = \mathbf{\Lambda}^n\mathbf{V}_0 \tag{6}$$

cannot be applied. However, the above value of λ_1 can be used to estimate the number of generations required to exceed 95 per cent and 99 per cent incross matings. These are, for $c = 1/2$, 15 and 22 generations, and for $c = 1/10$, 22 and 33 generations. Again it should be noted that all three genotypes at the locus of interest are produced on the same genetic background.

Examples of strains of mice produced by this system are: WB/Re–W, WC/Re–W, WH/Re–W, and WK/Re–W. These four strains in which all matings are $Ww \times Ww$ provide four different genetic backgrounds on which to compare three genotypes, WW, Ww, and ww of the dominant spotting locus.

GENERAL REMARKS

The breeder of laboratory animals for research must face the question of how to produce and propagate animals of the specific types he needs for his specific objective. The choice will depend upon the type of animal, the knowledge of the genetics of the trait of interest, the relative efficiency of the mating systems, the ease or difficulty of determining the phenotype of each animal, and the amount of space in the animal room to be devoted to maintenance of stocks.

The following assertions are intended to help mammalian geneticists in the task of choosing a suitable breeding system.

1. Close inbreeding, such as brother-sister, has been successful on a large scale only with the house mouse. There are in existence, however, lines of rats, rabbits, guinea pigs, and hamsters which satisfy the usual working definition of being inbred, that is, they have survived 20 or more generations of exclusive brother-sister inbreeding. The house mouse does not, however, withstand the depressing effects which accompany inbreeding as well as the existing profusion of inbred strains suggests. The existing strains are the successful survivors of, probably, a twofold or threefold larger number of attempts to establish inbred lines. If one were to start to produce a new inbred strain of mice, he should maintain several lines, say five, to insure that one or two can be propagated to 20 generations at least.

2. Any of the systems of mating (except random mating) described in this chapter will increase the probability of incrosses and will decrease the probability of heterozygosity at all loci except those closely linked with a locus of interest deliberately forced to remain heterozygous. Table 11 shows the probabilities of incrosses after 12 generations for each of the regular systems. For loose linkage (c between 3/10 and 1/2), the methods of crossing with locus-control only are all more efficient than the methods of inbreeding and locus-control, assuming that the inbred strains used in crossing are in fact homozygous. For closer linkage, the methods using forced heterozygosis have greater efficiency. Efficiency may be defined by the number of generations required to achieve a given probability of incrosses or by the probability of incrosses achieved with a fixed number of generations. Table 11 also shows the number of generations required to obtain a 95 per cent or a 99 per cent frequency of incross matings.

Table 11

PROBABILITIES OF INCROSSES AFTER 12 GENERATIONS AND NUMBERS OF GENERATIONS REQUIRED TO ACHIEVE PROBABILITIES OF 95 AND 99 PER CENT FOR INCROSSES FOR SIX MATING SYSTEMS

System	Probabilities of incrosses at G_{12}					Generations required to obtain α per cent of incrosses					
	$c = 0.5$	0.4	0.3	0.2	0.1	α	$c = 0.5$	0.4	0.3	0.2	0.1
2. Brother × sister	0.8922	—	—	—	—	95	16	—	—	—	—
						99	24				
3. Backcross	0.9995	0.9964	0.9802	0.9141	0.6862	95	6	7	10	15	30
						99	8	11	14	22	45
4. Cross-intercross	0.9740	0.9267	0.8282	0.6508	0.3727	95	12	14	20	30	62
						99	16	20	28	44	92
5. Cross-backcross-intercross	0.9932	0.9725	0.9130	0.7718	0.4877	95	9	12	15	24	48
						99	12	15	24	36	69
6. Heterozygosis forced by backcrossing	0.8922	0.8853	0.8618	0.8097	0.6833	95	15	15	16	19	26
						99	22	23	25	29	39
7. Heterozygosis forced by intercrossing	0.8922	0.8920	0.8890	0.8723	0.7953	95	15	15	15	16	22
						99	22	22	23	24	33

3. In the absence of selection, all loci have equal probabilities of becoming homozygous in any one system, except those linked with a locus of interest. Thus alleles with deleterious effects may also become homozygous. It is inescapable that alleles with deleterious effects will be selected against in the simple act of trying to keep the lines in propagation.

4. The backcross, cross-intercross, and cross-backcross-intercross systems are not only efficient means of putting mutant alleles on inbred backgrounds; they also take advantage of prior inbreeding and selection of the inbred line.

5. It is not always possible, and in some instances it may not be desirable, to put a mutant allele on a standard inbred background. That is, the mutant and the available backgrounds may be in some way incompatible. In these cases, the systems of brother-sister inbreeding with forced heterozygosis offer distinct advantages. They allow the background to evolve as inbreeding progresses. The mutant may thus, so to speak, select its own most favorable background from those possible out of the genetic makeup of the mutant generation.

6. A new mutation which arises in an inbred line should be perpetuated within the line of origin in order to keep the new mutant and its nonmutant allele on the same background. Of course, this does not preclude any outcrossing which will be necessary to establish the genetic basis of the new mutation and to explore its interactions with various other alleles and nonalleles.

7. The backcross, cross-intercross, and cross-backcross-intercross systems are effective means of searching for genetic differences between inbred strains. The search is more likely to yield positive results if any two strains having different phenotypes with respect to the trait under study are also different by a relatively small number of loci with discrete effects. However, the true situation cannot be known until after the search has been carried out.

8. The systems described, except random mating and brother-sister inbreeding, all yield mice bearing different alleles on common inbred backgrounds. These strains thus constitute the precision tools for genetic, developmental, physiologic, biochemical, pathologic, behavioral, and immunologic studies. They may be used in either of two ways. First, the alleles caused to segregate (the locus of interest) may be chosen in advance because of their known effect on hair, pigment, blood, skeleton, behavior, and so forth; thus the animals bearing the different alleles are suited for studies of hair or pigment or blood or skeleton or behavior. Second, the animals which differ by one allele only may be studied with a view to seeing if this one allele is concerned with some seemingly unrelated property of the organism. Thus dense (Dd) and dilute (dd) mice may be studied for differences in learning, radiation resistance, or susceptibility to disease. This may be a needle-in-haystack search, but once a difference is found, the genetic explanation is readily at hand.

This second type of study may be contrasted with searching for differences between inbred strains with respect to such characteristics as ability to learn, resistance to irradiation, or susceptibility to disease. Such a search is almost certain to uncover differences between some of the inbred strains.   Yet, once found, there may be great difficulty in accounting for the difference in genetic terms.   If such an analysis is important, a classical analysis of producing first, second, and third hybrid generations and first and second backcross generations may yield the information that two given strains differ by 8 or 10 or 50 pairs of genes.   This information does not, however, easily pave the way for developmental, physiologic, and other studies with precise genetic control.

SUMMARY

Mammalian geneticists use a number of breeding techniques to produce suitable animals for research.   Aside from random mating, the common objective of all of the systems of matings described in this chapter is to reduce the probability of heterozygotes and to increase the probability of matings of like homozygotes with respect to any locus, called the *a*-locus, which is not being specifically controlled in the mating system.

The random mating system is intended to hold constant the probabilities of all genotypes at each locus from generation to generation.

The brother-sister inbreeding system will increase the probability of incrosses, that is, of matings of like homozygotes, with each advancing generation.   For neutral genes, the probability will exceed 95 per cent after 16 generations and 99 per cent after 24 generations.

Three systems of mating (the backcross system, the cross-intercross system, and the cross-backcross-intercross system) are all means of placing mutant genes and their non-mutant alleles on inbred backgrounds.   The backcross system is especially useful with dominant mutations, but it may be used with recessive viable or recessive lethal mutations.   The cross-intercross and cross-backcross-intercross systems were designed for putting recessive viable mutations on inbred backgrounds.   The cross-backcross-intercross system is more efficient than the cross-intercross system in that it requires fewer generations to reach a specified probability of incrosses with respect to any neutral locus (*a*-locus) not being specifically controlled.

Two systems of mating combine inbreeding with locus control.   These are brother-sister inbreeding with heterozygosis forced by backcrossing and brother-sister inbreeding with heterozygosis forced by intercrossing.   The first, or backcrossing system, is useful with a semidominant lethal or recessive viable mutation; the second, or intercrossing system is useful with recessive lethal mutations.   When the neutral *a*-locus is closely linked with the mutant locus $(0 < c \leqslant 0.2)$, these two systems are more efficient than the three systems which require locus control accompanied by crossing with an inbred strain.   For higher values of crossing over, they are less efficient.

DISCUSSION

DR. BURDETTE: One most qualified to discuss this paper is Dr. Sewall Wright, who has consented to review from a different point of view the problems of breeding.

DR. WRIGHT: There are two main questions which arise in connection with such a paper as that presented by Drs. Green and Doolittle: first, the question whether solution of the mathematical problem is useful in genetics and, second, whether the results given are correct.   Dr. Green has brought out so clearly the importance of knowing how many generations are required to bring about reasonable assurance of isogenicity in all neutral genes under diverse systems of mating that no additional attention need be given to this.   With regard to correctness, I have found nothing with which to disagree. I did not attempt to repeat the method of attack used by the authors.   It seemed best to compare their results with those of a wholly different method, path analysis, which I have been applying to such problems since 1921.   This leads to certain extensions.

As applied to mating systems, this method consists in deducing the changes brought about automatically in the correlations between gametes.   As it is a correlation method, it deals only with relations between two things at a time.   Thus, it can deal with the relative amount of heterozygosis expressed as a function of the correlation between uniting gametes but cannot deal with the changes in frequency of types of mating where this involves relations among four varying gametes.   Thus, it does not give as complete a picture as can be obtained from the matrix of mating types (the method of Bartlett and Haldane[56] used by Green and Doolittle), but what it can do, it does more simply. It may be noted that after a sufficient number of generations of a regular system of mating in which the number of individuals in the inbreeding group is constant, all diallelic mating types tend to fall off in frequency at the same limiting rate as does heterozygosis, a rate which is easily obtained by path analysis.   Because of its relative simplicity, generalizations can be made by path analysis which have not been found practicable in terms of mating types.   I will devote most of my discussion to this point.   The method applies only to the effects of accidents of sampling and thus only to neutral alleles.   The joint effects of sampling, selection, recurrent mutation, and occasional outcrossing have been dealt with in other ways.[1417, 1422]

It is necessary to assume that genic frequencies remain constant in a hypothetical, total population consisting of all possible inbreeding lines in order to have a constant basis for comparison of the correlations within lines.   In the case of the two alleles, it is assumed that the frequency array in both ova (o) and spermatozoa (s) always remains $[(1 - q)a + qA]$ in this total population.   The relation between uniting gametes is represented below, in terms of the amount of heterozygosis, h.   The values assigned to a and A obviously make no difference in the correlation between them.   It is convenient to assign O to a and 1 to A.

os	a	A	Total
A	$h/2$	$q - (h/2)$	q
a	$1 - q - (h/2)$	$h/2$	$1 - q$
Total	$1 - q$	q	1

$$\bar{o} = \bar{s} = q$$
$$\sigma_o^2 = \sigma_s^2 = q(1 - q)$$
$$r_{os} = [q - (h/2) - q^2]/[q(1 - q)]$$
$$= 1 - (h/ho)$$

This correlation coefficient is usually designated by F, the inbreeding coefficient (or fixation index). As the proportion of heterozygosis in a random-breeding population is $h_0 = 2q(1 - q)$, the amount of heterozygosis, relative to that in a random-breeding population with the same gene frequencies (the panmictic index P) is equal to $1 - F$.

The frequencies of genotypes can be written in any of the following forms, of which the first is that used above.

	Deviations from array of fixed lines	Deviations from panmixia	Weighted average
AA	$q - Pq(1 - q)$	$q^2 \quad + Fq(1 - q)$	$Pq^2 \quad + Fq$
Aa	$2Pq(1 - q)$	$2q(1 - q) - 2Fq(1 - q)$	$2Pq(1 - q)$
aa	$(1 - q) - Pq(1 - q)$	$(1 - q)^2 + Fq(1 - q)$	$P(1 - q)^2 + F(1 - q)$

It becomes obvious on applying path analysis that the correlation, F, between uniting gametes is a function merely of the breeding system without any assumptions with respect to number or relative frequencies of alleles or of any values which might be assigned them. It is thus a somewhat unusual sort of correlation coefficient, a point that has made some difficulty.[701] It remains to be shown how it is related to heterozygosis in the case of multiple alleles.

Under Mendelian heredity, any set of multiple alleles $a^1, a^2, \ldots a^k$ may be treated formally as a pair of alleles by any sort of dichotomy: a^1 vs. $a^{2, \ldots k}$, $a^{1, 2}$ vs. $a^{3, \ldots k}$, etc. The expressions for the genotypic frequencies in the case of two alleles still hold.

Genotype	Frequency	Genotype	Frequency
a^1a^1	$Pq_1^2 + (1 - P)q_1$	$a^{1, 2}a^{1, 2}$	$P(q_1 + q_2)^2 + (1 - P)(q_1 + q_2)$
$a^1a^{2\ldots k}$	$2Pq_1(1 - q_1)$	$a^{1, 2}a^{3\ldots k}$	$2P(q_1 + q_2)(1 - q_1 - q_2)$
$a^{2\ldots k}a^{2\ldots k}$	$P(1 - q_1)^2 + (1 - P)(1 - q_1)$	$a^{3\ldots k}a^{3\ldots k}$	$P(1 - q_1 - q_2)^2 + (1 - P)(1 - q_1 - q_2)$

Since the value of P for any given pair of neutral alleles is independent of genic frequencies and depends merely on the breeding history, it may be assumed that the value of P is the same for any dichotomy of a neutral, multiple allelic series. Thus, in such a series, the frequency of any homozygote, $a^i a^i$, is of the form $Pq_i^2 + (1 - P)q_i$. By subtracting the frequencies of $a^i a^i$ and $a^j a^j$ from that for $a^{i,j}a^{i,j}$, it may be seen that the frequency of any heterozygote, $a^i a^j$, is of the form $2Pq_iq_j$, and total heterozygosis is given by $h = 2P \sum q_iq_j$. Thus $P = h/h_o$ irrespective of the number of alleles.

If now any arbitrary values V_1, V_2, V_k are assigned to the alleles, the mean and variance for each and the correlation between uniting gametes may be calculated. The symbol, f, is used here for proportional frequency.

$$\bar{o} = \bar{s} = \sum V_i f_i = \sum V_i q_i$$
$$\sigma_o^2 = \sigma_s^2 = \sigma^2 = \sum V_i^2 f_i - (\sum V_i f_i)^2$$
$$r_{os} = [\sum \sum V_i V_j f_{ij} - (\sum V_i f_i)^2]/\sigma^2,$$

but

$$\sum \sum V_i V_j f_{ij} = \sum \sum V_i V_j P q_i q_j + \sum V_i^2 (1 - P) q_i$$
$$= P(\sum V_i q_i)^2 + (1 - P) \sum V_i^2 q_i,$$

thus

$$F = r_{os} = [(1 - P) \sum V_i^2 q_i - (1 - P)(\sum V_i q_i)^2]/\sigma^2$$
$$= 1 - P$$
$$= (h_o - h)/h_o.$$

Thus the inbreeding coefficient, F, can be looked upon either as the correlation between uniting gametes or as a measure of the relative decrease in heterozygosis irrespective of numbers or relative frequencies of alleles or any values that may be

Fig. 13. BROTHER-SISTER MATING, ZYGOTE DIAGRAM.

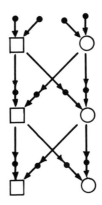

assigned them.   This, of course, includes the assumption that there is no disturbance from assortative mating if there are three or more alleles or from selection.   The correlation between any two gametes with designated rôles in the system of mating is similarly related to the decrease in heterallelism in the designated pairs.

It will be convenient to illustrate the method first by the simple case of brother-sister mating.[1443]   Figure 13 shows the system of mating diagrammatically.   The analysis is reduced to its simplest form, however, by considering only the gametes.   The compound path coefficients relating a gamete produced by an individual to those that united to produce that individual have the value 1/2, since there is a probability of 1/2 that the allele in the former is derived from that in the latter (tending to give a correlation of 1) and a probability of 1/2 that it is not (tending to give a correlation of 0 as far as the path in question is concerned).   Since it is assumed that the variance of the allelic array is constant, the path coefficient is merely the average, 1/2.

In the gametic diagram, figure 14, there are two kinds of correlations in each generation, that between any random gametes from the two parents ($r_{os} = F$) and that between two random gametes from the same individual ($r_{ss} = r_{oo}$ by symmetry).

The corresponding correlations in the preceding generations are symbolized by F', r'_{ss}, and r'_{oo}. Double primes are used for the second preceding generations. Tracing the connecting paths from inspection according to the basic principle of path analysis,

$$F = r_{os} = (1/4)(2F' + r'_{ss} + r'_{oo})$$
$$r_{ss} = r_{oo} = (1/4)(2F' + 2).$$

F can be solved easily in terms of preceding F's by substitution, $F = (1/4)$ $(1 + 2F' + F'')$; but it will be convenient in the more complicated cases considered

Fig. 14. BROTHER-SISTER MATING, GAMETE DIAGRAM.

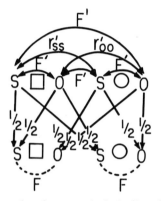

later to replace the correlations by the panmictic indices by substituting $r_{ij} = 1 - P_{ij}$ and using P_{oo} for both P_{oo} and P_{ss}.

$$P_{os} = (1/2)P'_{os} + (1/2)P'_{oo}.$$
$$P_{oo} = (1/2)P'_{os}.$$

Thus

$$P_{os} = (1/2)P'_{os} + (1/4)P''_{os}.$$

This can be written:

$$(h/h_o) = (h'/2h_o) + (h''/4h_o)$$

or merely

$$h = (1/2)h' + (1/4)h''.$$

The changes in heterozygosis can thus be written by inspection for any number of generations. Starting from a cross giving $h = 1$, the well-known series—1/1, 1/2, 2/4, 3/8, 5/16, 8/32, 13/64...—is obtained in which each numerator is the sum of the two preceding numerators (Fibonacci series) if the denominator is doubled in each generation. The ratio of successive terms rapidly approaches constancy. What this ratio is can readily be found by putting $h/h' = h'/h'' = x$.

$$x = (1/2) + (1/4x).$$
$$4x^2 - 2x - 1 = 0.$$
$$x = (1/4)(1 \pm \sqrt{5}).$$

The larger root, $(1/4)(1 + \sqrt{5}) = 0.8090$, is the desired ratio. In more complicated cases it is convenient to derive this from the characteristic equation of the P matrix.

$$\begin{vmatrix} 1/2 - \lambda & 1/2 \\ 1/2 & -\lambda \end{vmatrix} = 0$$

$$\lambda^2 - (1/2)\lambda - 1/4 = 0$$

$$\lambda_1 = (1/4)[1 + \sqrt{5}] = 0.8090 \text{ as the larger root.}$$

This $2 \times 2P$ matrix compares with a 6×6 matrix of mating types assuming multiple alleles in both cases or with the 4×4 matrix of mating types with respect to pairs of alleles as in Green and Doolittle's analysis. The P matrix is usually of lower order than the mating-type matrix, but the root (largest) that gives the limiting ratio, P/P', is always exactly the same.

The case of a population of limited size (N_m males, N_f females in each generation) in which every mating is of the type $Dd \times Dd$ in the locus of interest, will now be considered. The objective is the ratio, P/P' for linked genes, A,a. For the sake of simplicity, it will be assumed that these are related symmetrically to D and d in the total population. Let c_o and c_s be the amounts of recombination in oögenesis and spermatogenesis respectively; figure 15 shows the relation between a sperm carrying D, (D_s) and an ovum carrying d, (d_o). Gamete D_s may derive its A allele from either of the uniting gametes of the preceding generation but with probability $(1 - c_s)$ from the D gamete and probability c_s from the d gamete. These are then the path coefficients in these cases. Similarly the A allele of gamete d_o is related to the d gamete of the preceding generation by a path coefficient with the value $1 - c_o$ and to the D gamete by one with value c_o. The correlation between the A alleles of D_s and d_o is designated $r_{so(N)}$ in which the subscript N indicates *nonidentical* D alleles. The correlation between uniting gametes (F) is of this sort. The correlation between A alleles of sperm and egg that carry identical D alleles $(r_{so(I)})$ and the corresponding correlations between two spermatozoa $(r_{ss(N)}, r_{ss(I)})$ and between two eggs $(r_{oo(N)}, r_{oo(I)})$ are also of concern. Because of the postulated symmetry of the system of mating, it makes no difference whether a parent came from a union of D sperm with d ovum or the reverse. Thus the correlation between A alleles both associated with D or both with d is the average of $r_{ss(I)}, r_{so(I)},$ and $r_{oo(I)}$ with weights of $1/4$, $2/4$, and $1/4$ respectively.

$$r_I = (1/4)[r_{ss(I)} + 2r_{so(I)} + r_{oo(I)}]$$

Similarly

$$r_N = (1/4)[r_{ss(N)} + 2r_{so(N)} + r_{oo(N)}].$$

On tracing the connecting paths in figure 15 it is found that

$$r_{so(N)} = F = [(1 - c_s)(1 - c_o) + c_s c_o]r'_N + [c_s(1 - c_o) + (1 - c_s)c_o]r'_I.$$

If d_o is replaced by D_o, coefficients c_o and $(1 - c_o)$ must be exchanged; and, if D_s is replaced by d_s, c_s and $(1 - c_s)$ must be exchanged. Union of d_s with D_o gives the same result as above.

$$r_{so(I)} = [(1 - c_s)(1 - c_o) + c_s c_o]r'_I + [c_s(1 - c_o) + (1 - c_s)c_o]r'_N.$$

Fig. 15. MATING OF TYPE $Dd \times Dd$.

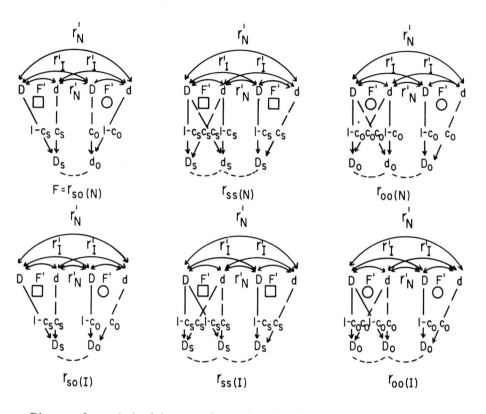

Diagrams for analysis of six types of gametic pairs with respect to linked locus.

In the case of correlations between two spermatozoa or between two eggs, the probabilities that they are produced by the same individual $[(1/N_m)$ in the case of two spermatozoa, $(1/N_f)$ in the case of two eggs] and the probabilities that they are produced by different individuals must be taken into account.

$$r_{ss(I)} = (1/N_m)\{[(1 - c_s)^2 + c_s^2] + 2c_s(1 - c_s)F'\}$$
$$+ [(N_m - 1)/N_m]\{[(1 - c_s)^2 + c_s^2]r_I' + 2c_s(1 - c_s)r_N'\}$$

$$r_{ss(N)} = (1/N_m)\{[(1 - c_s)^2 + c_s^2]F' + 2c_s(1 - c_s)\}$$
$$+ [(N_m - 1)/N_m]\{[(1 - c_s)^2 + c_s^2]r_N' + 2c_s(1 - c_s)r_I'\}.$$

Similarly

$$r_{oo(I)} = (1/N_f)\{[(1 - c_o)^2 + c_o^2] + 2c_o(1 - c_o)F'\}$$
$$+ [(N_f - 1)/N_f]\{[(1 - c_o)^2 + c_o^2]r_I' + 2c_o(1 - c_o)r_N'\}$$

$$r_{oo(N)} = (1/N_f)\{[(1 - c_o)^2 + c_o^2]F' + 2c_o(1 - c_o)\}$$
$$+ [(N_f - 1)/N_f]\{[(1 - c_o)^2 + c_o^2]r_N' + 2c_o(1 - c_o)r_I'\}.$$

It is convenient to let $l_s = 2c_s(1 - c_s)$, $l_o = 2c_o(1 - c_o)$, and $l_m = c_s(1 - c_o)$ $+ (1 - c_s)c_o$ and $a = (N_m - 1)/N_m$, $b = (N_f - 1)/N_f$. The equations for r_N and r_I can be written from the appropriate averages. A further condensation of symbolism is desirable.

$$u = (1 - l_m)$$
$$v = l_m$$
$$w = (1/4)[(1 - l_s)(1 - a) + (1 - l_o)(1 - b)]$$
$$x = (1/4)[l_s(1 - a) + l_o(1 - b)]$$
$$y = (1/4)[2(1 - l_m) + (1 - l_s)a + (1 - l_o)b]$$
$$z = (1/4)[2l_m + l_s a + l_o b]$$

$$F = \qquad ur_N' + vr_I'$$
$$r_N = wF' + yr_N' + zr_I' + x$$
$$r_I = xF' + zr_N' + yr_I' + w.$$

Since the sums of the coefficients are equal to 1 in all cases, the constant terms disappear on substituting $r = 1 - P$ in each case. This leads to the P matrix, the characteristic equation of which follows:

$$\begin{vmatrix} -\lambda & u & v \\ w & y - \lambda & z \\ x & z & y - \lambda \end{vmatrix} = 0$$

$$\lambda^3 - \lambda^2(2y) + \lambda[y^2 - z^2 - uw - vx] + [uwy + vxy - uxz - vwz] = 0.$$

The largest root gives the ratio of heterozygosis in successive generations.

In the case of brother-sister mating ($N_m = N_f = 1$, $a = b = 0$), this reduces to the following which could have been arrived at much more simply if these assumptions had been made in the first place:

$$\lambda^3 - \lambda^2(1 - l_m) - (\lambda/4)[1 - (l_s + l_o)(1 - 2l_m)] + (1/8)(2 - l_s - l_o)(1 - 2l_m) = 0.$$

It is interesting to note that, if there is random assortment in one sex ($l_m = 1/2$), the equation reduces to that for brother-sister mating with random assortment in both sexes.

If there is no crossing over in one sex as in *Drosophila* ($c_s = 0$, $l_s = 0$, $l_m = c_o$), then:

$$\lambda^3 - \lambda^2(1 - c_o) - (\lambda/4)[1 - l_o(1 - 2c_o)] + (1/8)(2 - l_o)(1 - 2c_o) = 0.$$

On substituting $\lambda = \mu/2$, this is an exact divisor of the quartic equation arrived at by Bartlett and Haldane.[56] The conclusions on the rate of decrease of heterozygosis are thus in exact agreement.

If there is equal crossing over in both sexes ($c_s = c_o = c$, $l_s = l_o = l_m = 2c(1 - c)$), then:

$$\lambda^3 - \lambda^2(1 - l) - (\lambda/4)[1 - 2l(1 - 2l)] + (1/4)(1 - l)(1 - 2l) = 0.$$

Bartlett and Haldane[56] and Green and Doolittle arrived at a quartic equation in terms of $\mu = 2\lambda$ from the matrix of mating types in this case. On substituting $\lambda = \mu/2$, the above cubic is found to be an exact divisor, so that again the conclusions on rate of decrease of heterozygosis are in exact agreement.

To hold together a line considerably larger than under brother-sister mating, attention is directed to the limiting case of one male ($a = 0$) and exclusive mating with half sisters ($b = 1$) and also equal crossing over in both sexes. The exact recurrence equation for relative heterozygosis is

$$P = (3/2)(1 - l)P' - (1/16)[5(1 - l)^2 - 13l^2]P'' - (3/16)(1 - l)(1 - 2l)P'''$$

With random assortment,

$$c = 1/2, \qquad l = 1/2:$$

$P = (3/4)P' + (1/8)P''$ as given previously.[1443] The limiting rate with any value of c is again given by the largest root of the characteristic equation

$$\lambda^3 - (3\lambda^2/2)(1 - l) + (\lambda/16)[5(1 - l)^2 - 13l^2] + (3/16)(1 - l)(1 - 2l) = 0.$$

The values of λ for several values of c are given in table 12 in comparison with those

Table 12

THE LIMITING RATIO OF HETEROZYGOSIS IN SUCCESSIVE GENERATIONS (P/P') UNDER VARIOUS SYSTEMS OF MATING AND AMOUNTS OF RECOMBINATION

♂	♀	0	0.02	0.05	0.10	0.20	0.30	0.40	0.50
1 Dd × 1 Dd		1.0000	0.9629	0.9179	0.8674	0.8234	0.8114	0.8092	0.8090
1 Dd† × 1 dd†		1.0000	0.9743	0.9407	0.8984	0.8514	0.8270	0.8135	0.8090
1 dY × 1 Dd		1.0000	0.9900	0.9749	0.9499	0.9013	0.8581	0.8254	0.8090
1 Dd × ∞ Dd		1.0000	0.9657	0.9334	0.9084	0.8940	0.8910	0.8904	0.8904
1 Dd × ∞ dd		1.0000	0.9754	0.9495	0.9273	0.9086	0.8984	0.8924	0.8904
1 dd × ∞ Dd		1.0000	0.9851	0.9657	0.9413	0.9146	0.9017	0.8972	0.8904
1 dY × ∞ Dd		1.0000	0.9949	0.9872	0.9745	0.9497	0.9264	0.9060	0.8904

† Matings of type $dd♂ × Dd♀$ give the same results as those above for the reciprocal type. Equal crossing over in both sexes is assumed.

under brother-sister mating (matings $Dd × Dd$) and various other cases. Whereas mating one male with many half-sisters is only slightly more than half as effective as full brother-sister mating for loci with random assortment, it approaches equal effectiveness for loci closely linked with the gene of interest.

Green and Doolittle (and also Bartlett and Haldane) consider the case of brother-sister matings of the type $Dd × dd$ at a locus of interest. This can also be dealt with by path analysis. First the more general case of a population of N_m males, N_f females will be considered. All males are assumed to be Dd and all females dd, so that only recombination in spermatogenesis (here c) is involved. Letting S and s in subscripts represent spermatozoa carrying D and d respectively, there are six kinds of correlations to consider ($r_{So}, r_{so}, r_{SS}, r_{Ss}, r_{ss}, r_{oo}$). There is no difficulty constructing gametic diagrams in each case from which the formula for the correlations can be written from inspection (figure 16).   In this diagram, $F_1(=r_{So})$ and $F_2(=r_{so})$ are the inbreeding coefficients of males and females respectively.   The coefficients are as follows:

	r'_{So}	r'_{so}	r'_{Ss}	r'_{Ss}	r'_{ss}	r'_{oo}	1
r_{So}	$(1/2)(1-c)$	$(1/2)c$	0	$(1/2)(1-c)$	0	$(1/2)c$	0
r_{so}	$(1/2)c$	$(1/2)(1-c)$	0	$(1/2)c$	0	$(1/2)(1-c)$	0
r_{Ss}	l	0	$(1-c)^2 a$	0	0	$c^2 a$	$(1-l)(1-a)$
r_{Ss}	$1-l$	0	$(1/2)la$	0	0	$(1/2)la$	$l(1-a)$
r_{ss}	l	0	$c^2 a$	0	0	$(1-c)^2 a$	$(1-l)(1-a)$
r_{oo}	0	$1/2$	0	0	$(1/4)b$	$(1/4)b$	$(1/2)(1-b)$

Fig. 16. MATING TYPE $Dd \times dd$.

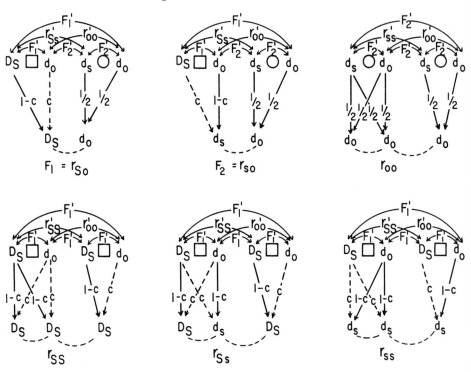

Diagrams for analysis of six types of gametic pairs with respect to linked locus.

The P matrix is the same except for elimination of the constant terms.   The characteristic equation is thus a rather complicated sextic equation.

$$\lambda^6 - \lambda^5/4[4(1-c) + 4(1-c)^2 a + b]$$
$$- (\lambda^4/4)\{(c + 2(1-c)(1-l) - 4(1-c)^3 a - (1-c)b\}$$
$$+ (\lambda^3/16)(1-2c)\{2(3-2l) + 4(1-c)[c + 2(1-c)^2]a$$
$$+ (1-2c)b + 4(1-l)a^2 b\}$$
$$+ (\lambda^2/16)(1-2c)\{2(1-l) - 2(1-c)^2(3-4c)a + 2cla^2$$
$$- (1-l)b + 2clab - 4(1-c)(1-l)a^2 b\}$$
$$- (\lambda/16)(1-2l)\{2(1-c)^2 a - [1-2(1-c)^2]a^2 b\} + (1/16)(1-2l)^2 a^2 b = 0.$$

Under random assortment [$c = (1/2)$, $l = (1/2)$], this reduces to

$$\lambda^2 - (\lambda/2)[1 + (a + b)/2] - 1/4[1 - (a + b)/2] = 0,$$

which agrees with the exact recurrence formula given previously,[1442]

$$P = [1 - (N_m + N_f)/4N_mN_f]P' + [(N_m + N_f)/8N_mN_f]P''.$$

There is considerable simplification in the case of one male ($a = 0$). The limiting case of an indefinitely large number of females ($b = 1$) will be considered.

$$\lambda^4 - (\lambda^3/4)(5 - 4c) - (\lambda^2/4)[1 - 2(1 - c)l]$$
$$+ (\lambda/16)(1 - 2c)(7 - 10c + 8c^2) + (1/16)(1 - 2c)(1 - l) = 0.$$

The case in which all males are *dd* and all females *Dd* is given by interpreting *a* as $(N_f - 1)/N_f$, *b* as $(N_m - 1)/N_m$, *c* as recombination rate in oögenesis, and $l = 2c(1 - c)$ in this sense. The case of one *dd* male and indefinitely many *Dd* females is more complicated than the converse.

$$\lambda^5 - \lambda^4(1 - c)(2 - c) - (\lambda^3/4)[c + 2(1 - c)(1 - l) - 4(1 - c)^3]$$
$$+ (\lambda^2/16)(1 - 2c)[2(3 - 2l) + 4(1 - c)[c + 2(1 - c)^2]$$
$$+ (\lambda/16)(1 - 2c)[2(1 - l) - 2(1 - c)^2(3 - 4c) + 2cl]$$
$$- 1/8(1 - 2l)(1 - c)^2 = 0.$$

Both of these reduce to the equation

$$\lambda^2 - 3/4\lambda - 1/8 = 0,$$

$$\lambda = (1/8)[3 + \sqrt{17}] = 0.8904,$$

as shown previously, if $c = (1/2)$.

In the case of brother-sister mating ($a = 0$, $b = 0$),

$$\lambda^4 - \lambda^3(1 - c) - (\lambda^2/4)[c + 2(1 - c)(1 - l)]$$
$$+ (\lambda/8)(1 - 2c)(3 - 2l) + (1/8)(1 - 2c)(1 - l) = 0.$$

The equation obtained by substituting $\lambda = \mu/2$ is again an exact divisor of the equation (quintic) derived by Bartlett and Haldane[56] and by Green and Doolittle from the matrix of mating types. The exact recurrence equation for *P* in terms of preceding *P*'s is given as usual by replacing the highest power of λ by *P*, the next power by *P'*, and so forth. The results are again in exact agreement.

The limiting values of the ratio of heterozygosis in successive generations, λ, are shown for several values of *c* in table 12 in the cases of male *Dd* × female *dd* (or the reciprocal), male *Dd* and indefinitely many females *dd* and the converse. They are less efficient in reducing heterozygosis than if both parents are always *Dd*. The last is the least effective.

The case in which the gene of interest is sex linked is of some interest in mice. In the case in which the males are dY and the females Dd, there are six types of correlations to consider. The gametic diagrams are shown in figure 17. In the cases of two spermatozoa or two eggs, the alternative possibilities, production by the same individual (probabilities $1/N_m$, $1/N_f$) and by different individuals $(N_m - 1)/N_m$, $(N_f - 1)/N_f$ are both shown.

Fig. 17. Mating type of $dY \times Dd$.

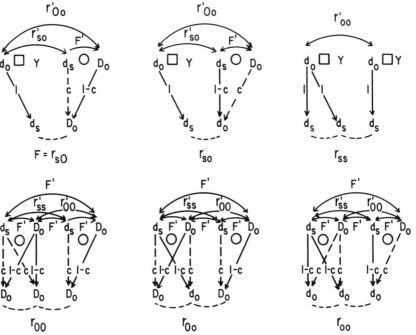

Diagrams for analysis of six types of gametic pairs with respect to sex-linked locus.

The formulae for the correlations can be written from inspection. The coefficients are listed below:

	r'_{sO}	r'_{sO}	r'_{OO}	r'_{OO}	r'_{OO}	r'_{ss}	1
$r_{sO} = F =$	0	c	0	$1 - c$	0	0	0
r_{so}	0	$1 - c$	0	c	0	0	0
r_{OO}	l	0	$(1 - c)^2 b$	0	0	$c^2 b$	$(1 - l)(1 - b)$
r_{Oo}	$1 - l$	0	$c(1 - c)b$	0	0	$c(1 - c)b$	$l(1 - b)$
r_{oo}	l	0	$c^2 b$	0	0	$(1 - c)^2 b$	$(1 - l)(1 - b)$
r_{ss}	0	0	0	0	a	0	$1 - a$

The characteristic equation is thus a rather complicated sextic. In the case of random assortment ($c = 1/2$), r_{so}, and r_{so} become the same and so do r_{Oo}, r_{Oo}, and r_{oo}. The characteristic equation reduces to:

$$
\begin{vmatrix}
(1/2) - \lambda & 1/2 & 0 \\
1/2 & (1/4)b - \lambda & (1/4)b \\
0 & a & -\lambda
\end{vmatrix} = 0
$$

$$\lambda^3 - \lambda^2[(2 + b)/4] - (\lambda/8)(2 - b + 2ab) + (1/8)ab = 0.$$

This agrees with the exact recurrence equation given previously for P as follows:[1426]

$$P = P' - \frac{(N_f + 1)}{8N_f}(2P' - P'') + \frac{(N_f - 1)(N_m - 1)}{8N_m N_f}(2P'' - P''').$$

In the case of a single male ($a = 0$, $r_{ss} = 1$), the last two rows and columns in the P matrix drop out, and the characteristic equation becomes:

$$\lambda^4 - \lambda^3(1 - c)[1 + (1 - c)b] - \lambda^2(1 - c)[(1 - l) - (1 - c)^2 b]$$
$$+ \lambda(1 - 2c)[(1 - l) + (1 - c)^3 b] - (1 - 2c)^2(1 - c)^2 b = 0.$$

The limiting ratios of heterozygosis in successive generations are given for various values of c in table 12. Fixation of sex-linked genes under forced heterozygosis of females is a very slow process.

With brother-sister mating (both $a = 0$, $b = 0$), the characteristic equation reduces to the form given by Bartlett and Haldane.[56]

$$\lambda^3 - \lambda^2(1 - c) - \lambda(1 - c)(1 - l) + (1 - 2c)(1 - l) = 0.$$

Fixation is again interfered with by forced heterozygosis of a gene of interest much more than in the case of autosomal genes.

Details of the various important systems of the backcross type discussed by Green and Doolittle will not be elaborated. The gametic diagram can easily be constructed and coefficients can be assigned to each path under the necessary assumption of a hypothetical, symmetrical, total population in which gene frequencies remain constant. The recurrence formulae for P present no difficulty. Under simple backcrossing $P = (1 - c)P'$. Under the cross-intercross system, the result is the same except that P' here refers to the uniting gametes of the preceding intercross, two generations back. Under the cross-backcross-intercross system, $P = (1 - c)^2 P'$ in a cycle of three generations from P' to P. In these cases the proportions of incrosses in the crosses to the isogenic lines are readily found by path analysis since they involve only two varying gametes. The results all agree with those of Green and Doolittle.

DR. PILGRIM: There is no question concerning the logic of the breeding methods discussed, but the mathematics is based on the assumption of randomness, and these conclusions are only valid in the absence of selection. In the case of selection for the homozygote, homozygosity may be achieved in much less than the predicted time; and, in the case of high selective value for the heterozygotes, it is conceivable that homozygosity may never be obtained. I question whether there are very many genes in mammals which behave in a random fashion with no selection for either homozygosity or heterozygosity.

DR. GOWEN: A factor, generally ignored, is the part chance plays in any quantitative estimates of inbreeding as measured by various coefficients. Sampled numbers of genotypes are almost universally small compared to those of the total population. Chance in "random selection" of parents each generation introduces so much variation in the real rate changes in homozygosis that, individually, the calculated coefficients may be quite misleading even though their average may measure approximately the trends in infinite populations of evolutionary dimensions.

DR. BURDETTE: Dr. Pilgrim, do you wish to clarify your question?

DR. PILGRIM: I wonder how many of these so-called chance elements are really random. How often do we actually select a particular coat color because it happens to be pleasing? There is always selection for breeding performance; otherwise it would be impossible to maintain an animal colony.

DR. GINSBURG: The probability of mutation should also be taken into account in terms of the eventual effect in stabilizing homozygosis in considering the questions about selection.

DR. MCINTOSH: In using the matrix-generation method, can repetitive multiplication be avoided and a final result obtained without going through all the intervening steps?

DR. DOOLITTLE: In answer to Dr. McIntosh's question, yes, one can skip a series of generations in a single step, having the basic matrices.

DR. BURDETTE: Dr. Green, would you close the discussion, please?

DR. GREEN: I am intrigued by Dr. Wright's demonstration that the method of path-coefficient analysis yields the same results as the matrix method in the cases where the matrix method can be used. We should emphasize again that there are irregular breeding systems used by breeders of laboratory animals which cannot be analyzed by the matrix method but which can be analyzed by the method of path-coefficients.

I am glad that the two methods, so different in nature and approach, give identical results where both methods are applicable. Each method has its limitations. Dr. Wright pointed out those of path analysis. The chief limitation of the matrix method is that it cannot be used with irregular systems of mating.

With respect to the other question which was raised, we have to appreciate that selection occurs whenever animals are being propagated. We cannot propagate animals in a vacuum in the absence of selection nor in the absence of mutation nor in the absence of chance. When it comes to dealing with actual animals, these elements are ever present. On paper, a model population can be propagated with any amount of selection or any amount of mutation one chooses. In the method of analysis described, we made deliberate choices, namely, selection zero, mutation zero. Other choices increase the complications. If there were selection for homozygotes, 100 per cent probability of incrosses would be approached more readily.   If there were some selection for heterozygotes, incrosses would be obtained less readily, but could be achieved with brother-sister inbreeding.   We know, in other words, what would be the direction of the change when these other factors are taken into consideration, but we have not worked out the probabilities in detail.

DR. CROW: It is relevant to point out in this connection that, if an organism has something like 10,000 genes, one cannot apply a strong selection to each of them. If the effects of inbreeding on decreasing heterozygosity of any particular locus are, say, 25 per cent per generation, this means that even intense selection cannot counteract the increased homozygosity of more than a small fraction of these in any one population.

DR. WRIGHT: It should be emphasized that these methods have nothing to do with lethals or semilethals.   Dr. Slatis has worked out a different coefficient dealing with lethals in bison.   The average rate of fixation of loci as derived by these methods applies strictly only to neutral alleles but applies approximately when one allele has a selective advantage of lower order than the reciprocal of ten times the effective size of the inbreeding group.   This includes most of the loci, under close inbreeding, for reasons indicated by Dr. Crow.   With stronger selection, the rate for the more favorable allele increases while that for the unfavorable one decreases, with only slight change in the total, unless the selection is rather severe.   With strong selection for the heterozygote over both homozygotes, the total rate is, of course, decidedly lowered just as it is under enforced heterozygosis of a strongly linked gene, the problem investigated by Green and Doolittle.

DR. GOWEN: This approach is really based on an infinite population of gametes, breeding entirely at random.   From an evolutionary point of view perhaps the assumptions may be right, but actual breeding experiments with guinea pigs, mice, *Drosophila*, and so forth never meet the conditions which have been postulated.   Selection continually occurs, however random it may be, with the consequence that individually, inbreeding coefficients may be quite misleading.

Margaret C. Green, Ph.D.

METHODS *for* TESTING LINKAGE†

Methods for detecting and measuring linkage have been summarized and the theory has been explained in an excellent monograph by Mather.[858]   Tables of scores which greatly simplify the calculations for the more common types of data have been published by Carter and Falconer[172] for the detection of linkage and by Finney[364] and Allard[7] for the estimation of recombination frequencies.   Carter and Falconer[172] have also discussed the design of stocks for the detection of linkage and have suggested a particular set of such stocks for the mouse.   It is the intent of this paper to provide only a brief introduction to methods of linkage useful for laboratory mammals, and to direct the reader to the above and other sources for a more detailed discussion of the methods and their rationale.   The linkage testing stocks of the mouse maintained at the Jackson Memorial Laboratory will be described and the latest linkage map of the mouse presented.

Linkage tests fall into two general classes, those in which the main interest is in detecting linkage, and those in which the main interest is in measuring more exactly a linkage already known to exist and determining the order of the two linked genes with respect to other genes in the same linkage group.   Positive tests of the first type, of course, give preliminary information on the closeness of the linkage observed, but usually more extensive tests using different stocks will be desirable.

Whether the purpose of the mating is detection or estimation, only certain types of matings will yield information of any kind about linkage.   The necessary requirement is that one of the mated animals should be heterozygous for each of the two loci with

† Part of the work on which this paper is based was supported by research grants NSF G-6200 and NSF G-7023 from the National Science Foundation and ACS E-162 from the American Cancer Society.

linkage in question.   Consider that two loci *A, a* and *B, b* are to be tested for linkage. The double heterozygote is then either *AB/ab* or *Ab/aB*, depending on whether the recessive alleles, *a* and *b*, came to the heterozygote from the same parent or one from each parent.   In the former case the linkage is said to be in the coupling phase, in the latter in the repulsion phase.

Taking the coupling phase as an example, one can see that four kinds of gametes may be formed by the double heterozygote, *AB* and *ab*, the parental combinations, and *Ab* and *aB*, the recombinations.   In the absence of linkage the parental combinations and recombinations should occur with equal frequency and the proportion of recombination should be 1/2.   In the presence of linkage the parental combinations should occur in excess of the recombinations and the proportion of recombination should be less than 1/2.   The purpose of a linkage mating is to provide a way of estimating the proportion of recombination occurring during gamete formation in the double heterozygote, and of determining whether the proportion so estimated differs significantly from 1/2.

There are three kinds of matings which provide this information, the double backcross, the single backcross or mixed cross, and the intercross.   If there is dominance at both loci, all three kinds of matings produce offspring of the same four phenotypes, *A–B–, aabb, A–bb,* and *aaB–*.   If these types are represented by the numbers 1, 2, 3, and 4 respectively, the outcome of the various matings can be represented as in table 13.

Table 13

$1 = AB, 2 = ab, 3 = Ab, 4 = aB.$

Heterozygous parent		Other parent						
		Double backcross	Single backcross		Intercross			
Genotype		*aabb*	*Aabb*		*AaBb*			
	Gametes	*ab*	*Ab*	*ab*	*AB*	*ab*	*Ab*	*aB*
AaBb	*AB*	1	1	1	1	1	1	1
	ab	2	3	2	1	2	3	4
	Ab	3	3	3	1	3	3	1
	aB	4	1	4	1	4	1	4

The upper two rows represent the parental combinations if the heterozygous parent is in the coupling phase, and the recombinations if the heterozygous parent is in the repulsion phase.   In the double backcross and mixed cross, recombination cannot be detected in the other parent, but in the intercross the other parent is also a double heterozygote and produces detectably different parental and recombinant gametes.

The two left columns of the intercross are the parental combinations for coupling and the recombinations for repulsion.

It can be seen that in the double backcross the parental or recombinant type of the gamete can always be recognized from the phenotype of the offspring.   The proportion of recombination in the gametes can therefore be estimated directly from the proportion of recombinant phenotypes among the offspring.   In the single backcross and intercross, however, some of the phenotypes among the offspring result from both parental and recombinant gametes.   A direct estimate of recombination is therefore impossible and more elaborate methods become necessary.

USE OF CHI-SQUARE FOR TESTING SEGREGATIONS AND DETECTING LINKAGE

Before any effort is expended to detect linkage or estimate recombination, the segregations of the individual genes should be examined to determine whether they are in accordance with expectation, since disturbed segregations may lead to serious errors in estimating recombination.   The most useful method for this purpose is that of χ^2. Since explanations of χ^2 are widely available in textbooks on statistical methods, this discussion will be confined to the special formulas useful with linkage data.

Let a be the observed number of individuals in a class and m be the expected proportion, such that mn is the expected number when n is the total number of individuals.   The general formula for χ^2 is then

$$\chi^2 = \sum \frac{(a - mn)^2}{mn}.$$

An algebraically equivalent form which is easier to compute is

$$\chi^2 = \sum \left(\frac{a^2}{mn} \right) - n.$$

If there are two classes of individuals, denoted by 1 and 2, and if the numbers in the two classes, a_1 and a_2, are expected to occur in the proportion $k:1$, χ^2 can be more easily calculated from the formula

$$\chi^2 = \frac{(a_1 - ka_2)^2}{kn}.$$

An example is the segregation of the gene for ruby eye, *ru*, in the mouse, reported in a cross by Fisher and Snell.[379]   An intercross gave the following segregation:

	+ -	*ruru*	Total
Observed number	117	40	157
Expected proportion	3/4	1/4	

$$\chi^2 = \frac{(117-120)^2}{3 \times 157} = \frac{9}{471} = 0.019.$$

For one degree of freedom, the deviation from expectation is not significant.

In a linkage test, two pairs of genes will commonly be segregating at the same time. Deviations from expectation may be due to faulty segregation of either or both pairs or to failure of random assortment (linkage) between the two pairs. The total χ^2, which for four classes has three degrees of freedom, can be partitioned into three independent parts, each with one degree of freedom, one for segregation at each of the two loci, and one for linkage. Table 14 gives the formulas for partitioning the total χ^2 for the three common kinds of matings.

Table 14

FORMULAS FOR CALCULATION OF χ^2 FOR SEGREGATION AT INDIVIDUAL LOCI, $\chi^2 A$ AND $\chi^2 B$, AND FOR LINKAGE, $\chi^2 L$, FOR THREE KINDS OF MATINGS

Mating	$\dfrac{AB}{m_1}$	$\dfrac{Ab}{m_2}$	$\dfrac{aB}{m_3}$	$\dfrac{ab}{m_4}$	
Double backcross	1/4	1/4	1/4	1/4	$\chi^2 A = \dfrac{(a_1 + a_2 - a_3 - a_4)^2}{n}$
					$\chi^2 B = \dfrac{(a_1 - a_2 + a_3 - a_4)^2}{n}$
					$\chi^2 L = \dfrac{(a_1 - a_2 - a_3 + a_4)^2}{n}$
Intercross	9/16	3/16	3/16	1/16	$\chi^2 A = \dfrac{(a_1 + a_2 - 3a_3 - 3a_4)^2}{3n}$
					$\chi^2 B = \dfrac{(a_1 - 3a_2 + a_3 - 3a_4)^2}{3n}$
					$\chi^2 L = \dfrac{(a_1 - 3a_2 - 3a_3 + 9a_4)^2}{9n}$
Single backcross A intercrossed	3/8	3/8	1/8	1/8	$\chi^2 A = \dfrac{(a_1 + a_2 - 3a_3 - 3a_4)^2}{3n}$
					$\chi^2 B = \dfrac{(a_1 - a_2 + a_3 - a_4)^2}{n}$
					$\chi^2 L = \dfrac{(a_1 - a_2 - 3a_3 + 3a_4)^2}{3n}$

The data of Fisher and Snell[379] on the linkage of ruby, *ru*, and jerker, *je*, in the mouse may be taken as an example again. From an intercross in repulsion the following data were obtained:

	++	+je	ru+	ruje	Total
Observed number	86	31	35	5	157
Expected proportion	9/16	3/16	3/16	1/16	

$$\chi^2 ru = \frac{(86 + 31 - 105 - 15)^2}{471} = \frac{9}{471} = 0.019$$

$$\chi^2 je = \frac{(86 - 93 + 35 - 15)^2}{471} = \frac{169}{471} = 0.359$$

$$\chi^2 L = \frac{(86 - 93 - 105 + 45)^2}{1413} = \frac{4489}{1413} = 3.177$$

The χ^2's for the segregation of ru and je are very small. That for linkage is larger, 3.177, but still below the 5 per cent level of significance for one degree of freedom. The data thus do not give evidence for linkage.

It should be noted that the sum of the three values of χ^2 calculated above, 3.555, is equal to the total χ^2 calculated by the standard formula. This will always be so if the arithmetic has been correctly performed.

If the deviations of the single factor segregations from expectation are significant because of reduced viability, reduced penetrance, or any other disturbing factor, the above formula for the linkage χ^2 does not give a true estimate of the significance of the deviation due to linkage. It will be better in this case to calculate a contingency χ^2 to determine whether the two pairs of genes are recombining at random, regardless of their individual segregation ratios. This can be done by the standard method, calculating the expected frequency in the four classes from the marginal totals. A somewhat simpler method is to use the special formula applicable to two-by-two tables, which eliminates the necessity for calculating the expected frequencies. By this formula,

$$\chi^2 = \frac{n(a_1 a_4 - a_2 a_3)^2}{(a_1 + a_2)(a_3 + a_4)(a_1 + a_3)(a_2 + a_4)},$$

where $a_1, a_2, \ldots$, have the same meaning as in table 14. This χ^2 has one degree of freedom. Its use can be illustrated by the following example.[549]

The segregation of ogligodactyly, ol, and albino, c, in the mouse in an intercross in repulsion was as follows:

	$++$	$+c$	$ol+$	olc	Total
	334	188	107	13	642

The segregation of both ol and c is very abnormal as shown by χ^2 ($\chi^2 ol = 13.6$, $\chi^2 c = 13.6$), there being a deficiency of ol and an excess of c. When the data are arranged in a two-by-two table, it appears that there is an excess in the $+c$ and $ol+$ classes, as would be expected in the case of linkage.

	$+$	c	Total
$+$	334	188	522
ol	107	13	120
Total	441	201	642

To assess the significance of this excess we calculate χ^2 by the above formula.

$$\chi^2 = \frac{642(4342 - 20116)^2}{522 \times 120 \times 441 \times 201} = 28.769.$$

For one degree of freedom this is highly significant.

In cases such as this where segregations are abnormal, the methods of measuring recombination which will be considered below may not be applicable.   For methods dealing with disturbed segregations the reader is referred to Mather,[858] Fisher and Bailey,[376] Bailey,[45] and Carter.[160]

USE OF MAXIMUM-LIKELIHOOD METHODS FOR THE DETECTION AND ESTIMATION OF LINKAGE

The methods most widely in use for the estimation of linkage are those based on Fisher's maximum-likelihood method of estimation.   A simple example will illustrate the principle of the method.

Consider a coupling single backcross $AB/ab \times Ab/ab$.   If p is the probability of recombination, the expected proportions or probabilities of the four types of offspring in terms of p can be determined from table 15.   Combining the like phenotypes, we obtain

$$
\begin{array}{cccc}
AB & Ab & aB & ab \\
\tfrac{1}{4}(2 - p) & \tfrac{1}{4}(1 - p) & \tfrac{1}{4}p & \tfrac{1}{4}(1 - p).
\end{array}
$$

Table 15

EXPECTED PROPORTIONS OF DIFFERENT GENOTYPES AMONG OFFSPRING OF A SINGLE BACKCROSS

Gametes	Expected proportions	Ab 1/2	ab 1/2
AB	$\tfrac{1}{2}(1 - p)$	$AABb$ $\tfrac{1}{4}(1 - p)$	$AaBb$ $\tfrac{1}{4}(1 - p)$
Ab	$\tfrac{1}{2}p$	$AAbb$ $\tfrac{1}{4}p$	$Aabb$ $\tfrac{1}{4}p$
aB	$\tfrac{1}{2}p$	$AaBb$ $\tfrac{1}{4}p$	$aaBb$ $\tfrac{1}{4}p$
ab	$\tfrac{1}{2}(1 - p)$	$Aabb$ $\tfrac{1}{4}(1 - p)$	$aabb$ $\tfrac{1}{4}(1 - p)$

The expected proportions of offspring for the other types of matings are given in table 16.

If we represent the expected proportions or probabilities of the four classes by m_1, m_2, m_3, and m_4, and the observed number of offspring in the corresponding classes by a_1, a_2, a_3, and a_4, then the probability, P, of getting such a set of observed numbers from such a mating is proportional to the product of the probabilities for each individual, or

$$P = Cm_1{}^{a_1}m_2{}^{a_2}m_3{}^{a_3}m_4{}^{a_4}.$$

TABLE 16

EXPECTED PROPORTIONS OF CLASSES OF OFFSPRING IN LINKAGE CROSSES

Mating	Phase	Phenotypes of offspring			
		AB	Ab	aB	ab
Double backcross	C	$\frac{1}{2}(1-p)$	$\frac{1}{2}p$	$\frac{1}{2}p$	$\frac{1}{2}(1-p)$
	R	$\frac{1}{2}p$	$\frac{1}{2}(1-p)$	$\frac{1}{2}(1-p)$	$\frac{1}{2}p$
Single backcross (A intercrossed)	C	$\frac{1}{4}(2-p)$	$\frac{1}{4}(1+p)$	$\frac{1}{4}p$	$\frac{1}{4}(1-p)$
	R	$\frac{1}{4}(1+p)$	$\frac{1}{4}(2-p)$	$\frac{1}{4}(1-p)$	$\frac{1}{4}p$
Intercross	C	$\frac{1}{4}[2+(1-p)^2]$	$\frac{1}{4}[1-(1-p)^2]$	$\frac{1}{4}[1-(1-p)^2]$	$\frac{1}{4}(1-p)^2$
	R	$\frac{1}{4}(2+p^2)$	$\frac{1}{4}(1-p^2)$	$\frac{1}{4}(1-p^2)$	$\frac{1}{4}p^2$

C = Coupling, R = Repulsion

The value of p which maximizes the probability P is the maximum likelihood estimate of p. It can be found by differentiating the above expression with respect to p, setting the derivative equal to zero, and solving for p. The above expression is very difficult to differentiate but the task can be simplified by taking the logarithm L of the expression, differentiating the logarithm with respect to p, and proceeding as before. Since P is maximum when the logarithm of P is maximum the same value of p will be obtained. Thus,

$$L = \log P = C' + a_1 \log m_1 + a_2 \log m_2 + a_3 \log m_3 + a_4 \log m_4$$

and

$$\frac{dL}{dp} = a_1 \frac{d \log m_1}{dp} + a_2 \frac{d \log m_2}{dp} + a_3 \frac{d \log m_3}{dp} + a_4 \frac{d \log m_4}{dp} = 0.$$

For the coupling single backcross,

$$L = C' + a_1 \log \tfrac{1}{4}(2-p) + a_2 \log \tfrac{1}{4}(1+p) + a_3 \log \tfrac{1}{4}p + a_4 \log \tfrac{1}{4}(1-p),$$

and

$$\frac{dL}{dp} = -\frac{a_1}{2-p} + \frac{a_2}{1+p} + \frac{a_3}{p} - \frac{a_4}{1-p} = 0.$$

The variance of p, V_p, can be found from the relationship first shown by Fisher[370] that the second derivative of the likelihood equation is equal to the negative reciprocal of the variance of p, when the expected number of progeny (mn) in each of the four classes is substituted for the observed number (a) and the calculated value of p is used, or

$$\frac{d^2L}{dp^2} = -\frac{1}{V_p} = \sum \left(mn \frac{d^2 \log m}{dp^2} \right).$$

For the coupling single backcross, therefore

$$\frac{1}{V_p} = -\frac{m_1 n}{(2-p)^2} - \frac{m_2 n}{(1+p)^2} - \frac{m_3 n}{p^2} - \frac{m_4 n}{(1-p)^2}$$

$$= -\frac{n}{4}\left(\frac{1}{2-p} + \frac{1}{1+p} + \frac{1}{p} + \frac{1}{1-p}\right)$$

$$V_p = \frac{2p(1-p)(1+p)(2-p)}{n(1+2p-2p^2)}.$$

The solution of maximum-likelihood equations of estimation may be quite tedious in some cases. The use of maximum-likelihood scores first introduced by Fisher[372] and further developed by Fisher,[369] Finney,[364] Carter and Falconer,[172] and others, has greatly simplified the arithmetic. Their use in both detection and estimation can be illustrated with the previous example.

The equation of estimation for the coupling single backcross can be written:

$$\frac{dL}{dp} = \left(-\frac{1}{2-p}\right)a_1 + \left(\frac{1}{1+p}\right)a_2 + \left(\frac{1}{p}\right)a_3 + \left(-\frac{1}{1-p}\right)a_4 = 0.$$

The exact solution of p substituted into the equation will give a value of dL/dp equal to zero. A provisional value of p substituted into the equation will give a value of dL/dp whose deviation from zero, D, is a measure of the deviation of the exact estimate of p from the provisional value. If the provisional value of p is one half, for instance, the calculated value of dL/dp will be greater, the greater the evidence for linkage in the data. For the detection of linkage, then, the values of $-1/(2-p)$, $1/(1+p)$, etc., when $p = 1/2$ can be calculated in advance and used as scores by which a_1, a_2, etc., are multiplied. For the case illustrated the scores are:

AB	Ab	aB	ab
$-2/3$	$2/3$	2	-2

and

$$\frac{dL}{dp} = -\frac{2}{3}a_1 + \frac{2}{3}a_2 + 2a_3 - 2a_4 = D.$$

For the detection of linkage we need to know whether D differs significantly from zero, and for this purpose we need the variance of D. It has been demonstrated[363] that

$$V_D = 1/V_p$$

We could calculate V_p by the method already shown. A more convenient method which makes use of a score calculated in advance is available, however.

The concept, due to Fisher,[370] of the amount of information must now be introduced. The greater the amount of information in a body of data, the greater the precision of the estimate of a parameter calculated from the data, or the smaller the variance of the estimate. It is therefore convenient to speak of the reciprocal of the variance as the amount of information I, or $1/V_p = I_p$.   If I_p is the amount of information about p contributed by a whole body of data, then $I_p/n = i_p$ may be designated as the amount of information contributed by a single individual.

We have already seen that

$$-\frac{1}{V_p} = \frac{d^2L}{dp^2} = \sum \left(mn \frac{d^2 \log m}{dp^2} \right).$$

Therefore

$$I_p = ni_p = -\sum \left(mn \frac{d^2 \log m}{dp^2} \right),$$

$$= -n \sum \left(m \frac{d^2 \log m}{dp^2} \right),$$

and

$$i_p = -\sum \left(m \frac{d^2 \log m}{dp^2} \right),$$

which can be shown to be equivalent algebraically to

$$i_p = \sum \left(\frac{1}{m} \left[\frac{dm}{dp} \right]^2 \right).$$

The quantity i_p is a constant for any particular kind of mating and value of p and can be calculated in advance and used as a weight by which to multiply the total number of progeny to obtain I_p.

The test of significance is therefore

$$t = \frac{D}{\sqrt{V_D}} = \frac{D}{\sqrt{I_p}} = \frac{D}{\sqrt{ni_p}}$$

with probability obtainable from a normal table, or alternatively,

$$\chi^2 = \frac{D^2}{I_p}.$$

D and I_p have the useful property of additivity such that they can be summed for different types of matings to give a test of significance based on all appropriate matings available.

Table 17 (adapted from Carter and Falconer[172]) gives scores and values of i_p for the detection of linkage for a number of different types of linkage matings.   A numerical example will illustrate their use for this purpose.

Table 18 gives the data of Fisher and Snell[379] on the linkage of ruby (ru) and jerker (je) in the mouse.   The calculations of D and I_p are set out in the table.   The χ^2 test for significance of the deviation of p from one half is

$$\chi^2 = \frac{(-140.889)^2}{2655.111} = 7.476.$$

TABLE 17

SCORES OF MAXIMUM LIKELIHOOD AND AMOUNT OF INFORMATION PER INDIVIDUAL CALCULATED FOR $p = 1/2$ FOR DIFFERENT TYPES OF LINKAGE-TESTING MATING‡

Kind of mating	Phenotypic class									i_p
	AABB	AABb	AaBB	AaBb	AAbb	Aabb	aaBB	aaBb	aabb	
A and B semi-dominant										
Intercross	−4	0	0	0	4	0	4	0	−4	4
A semidominant										
Intercross	−4/3		0		4	0	4/3		−4	8/3
Single backcross A†	−2		0		2	0	2		−2	2
AA masks B, b distinction										
Intercross	0		0		0	0	4/3		−4	4/3
Single backcross A†	0		0		0	0	2		−2	1
AA inviable										
Intercross	0				0		4/3		−4	16/9
Single backcross A†	0				0		2		−2	4/3
A and B fully dominant										
Double backcross		−2				2	2		−2	4
Intercross		−4/9				4/3	4/3		−4	16/9
Single backcross A†		−2/3				2/3	2		−2	4/3
aa masks B, b distinction										
Double backcross		−2				2	0			2
Intercross		−4/9				4/3	0			4/9
Single backcross A†		−2/3				2/3	0			1/3
Single backcross B†		−2/3				2	0			2/3
A indistinguishable from B										
Double backcross							2/3		−2	4/3
aa indistinguishable from bb										
Intercross		−4/9					4/7			16/63

† Gene intercrossed

‡ A and B represent the dominant alleles of the two genes in the test, a and b their recessive alleles, irrespective of which are the mutant alleles. The genes are in coupling when A and B enter the test together from the same parent. The signs given are for coupling; they should be reversed for matings in repulsion. Adapted from Carter and Falconer.[172]

Table 18

CALCULATION OF D AND I_p FOR DETECTION OF LINKAGE BETWEEN *ru* AND *je* IN THE MOUSE
(Using data of Fisher and Snell.[379])

Parents	Mating type		Offspring				$D=\sum ns$	Total	I_p
			$++$	$+je$	$ru+$	$ruje$			
$\dfrac{++\ ++}{ruje\ ruje}$	I C	n	151	54	45	20		270	
		s	$-4/9$	$4/3$	$4/3$	-4	i_p 16/9		
		ns	-67.111	72.000	60.000	-80.000	-15.111		480.000
$\dfrac{+je\ +je}{ru+\ ru+}$	I R	n	86	31	35	5		157	
		s	$4/9$	$-4/3$	$-4/3$	4	i_p 16/9		
		ns	38.222	-41.333	-46.667	20.000	-29.778		279.111
$\dfrac{++\ ruje}{ruje\ ruje}$	B C	n	120	. 103	89	122		434	
		s	-2	2	2	-2	i_p 4		
		ns	-240.000	206.000	178.000	-244.000	-100.000		1,736.000
$\dfrac{+je\ ruje}{ru+\ ruje}$	B R	n	9	9	10	12		40	
		s	2	-2	-2	2	i_p 4		
		ns	18.000	-18.000	-20.000	24.000	4.000		160.000
							-140.889		2,655.111

I = intercross, B = double backcross, C = coupling, R = repulsion, n = number, s = score

The deviation is highly significant and linkage must therefore be said to exist.

It is desirable now to estimate p. A fortunate property of D and I_p allows them to be used to obtain a first estimate of p. For small values of D it can be shown that

$$I_p = \frac{D}{p - p_o}$$

and

$$p = p_o + \frac{D}{I_p}, \text{ approximately,}$$

where p_o is the value of p assumed in calculating D, and p is the exact maximum likelihood estimate of p calculated from the same data. In the above example the improved estimate of p is therefore

$$p = \frac{1}{2} + D/I_p = 0.500 - 0.053 = 0.447.$$

The standard error of this estimate is

$$\text{SE}_p = \frac{1}{\sqrt{I_p}} = \frac{1}{51.53} = 0.019.$$

Recombination between *ru* and *je* is thus 44.7 ± 1.9 per cent.

The quantity p estimated in this way is very close to the exact estimate for departures from one half of this magnitude. For large departures from one half it will be necessary to obtain a more exact estimate. For this purpose the first estimate of p could be used to calculate a second set of scores and i_p which could then be used to get a more exact estimate.   It is clear that this labor would be greatly facilitated by the availability of tables of scores for various kinds of matings and values of p.   Several such tables have been published, notably by Allard[7] and by Finney.[364]   Allard's

Table 19

SCORES AND INFORMATION PER INDIVIDUAL FOR $Aa/Bb \times aa/bb$
(Finney[364])

p	i_p	p	i_p
0.01	101.010	0.15	7.843
0.02	51.020	0.20	6.250
0.03	34.364		
0.04	26.042	0.25	5.333
0.05	21.053	0.30	4.762
0.06	17.730	0.35	4.396
0.07	15.361	0.40	4.167
0.08	13.587		
0.09	12.210	0.45	4.040
0.10	11.111	0.50	4.000

If the heterozygous parent is AB/ab, score aB and Ab offspring as i_p, others as zero; for Ab/aB parent, interchange these scores.

Table 20

SCORES AND INFORMATION PER INDIVIDUAL FOR $AB/ab \times Aa/bb$
(Finney[364])

p	AB	aB	Ab	ab	i_p
0.01	−0.2463	100.2563	1.2464	−0.7538	25.626
0.02	−0.2425	50.2625	1.2429	−0.7579	13.126
0.03	−0.2388	33.6022	1.2397	−0.7621	8.961
0.04	−0.2351	25.2751	1.2367	−0.7665	6.878
0.05	−0.2313	20.2815	1.2339	−0.7712	5.629
0.06	−0.2276	16.9545	1.2312	−0.7760	4.797
0.07	−0.2239	14.5799	1.2288	−0.7810	4.203
0.08	−0.2202	12.8007	1.2266	−0.7863	3.758
0.09	−0.2164	11.4183	1.2246	−0.7918	3.413
0.10	−0.2127	10.3137	1.2228	−0.7974	3.137
0.15	−0.1935	7.0137	1.2166	−0.8295	2.313
0.20	−0.1736	5.3819	1.2153	−0.8681	1.910
0.25	−0.1524	4.4190	1.2190	−0.9143	1.676
0.30	−0.1293	3.7923	1.2282	−0.9696	1.530
0.35	−0.1036	3.3596	1.2432	−1.0360	1.436
0.40	−0.0744	3.0506	1.2649	−1.1161	1.376
0.45	−0.0404	2.8269	1.2944	−1.2135	1.344
0.50	0.0000	2.6667	1.3333	−1.3333	1.333

If the heterozygous parent is Ab/aB interchange columns 2 and 4 and also columns 3 and 5 in this table.

Table 21

A. SCORES AND INFORMATION PER INDIVIDUAL FOR $AB/ab \times AB/ab$
(Finney[364])

p	AB	aB, Ab	ab	i_p
0.01	0.3339	100.4958	−1.0219	99.831
0.02	0.3345	50.4915	−1.0442	49.829
0.03	0.3352	33.8205	−1.0670	33.161
0.04	0.3359	25.4828	−1.0903	24.826
0.05	0.3366	20.4784	−1.1141	19.824
0.06	0.3374	17.1405	−1.1383	16.489
0.07	0.3382	14.7550	−1.1631	14.106
0.08	0.3390	12.9646	−1.1885	12.318
0.09	0.3399	11.5710	−1.2143	10.927
0.10	0.3409	10.4551	−1.2408	9.815
0.15	0.3465	7.0970	−1.3821	6.473
0.20	0.3535	5.4040	−1.5404	4.798
0.25	0.3624	4.3763	−1.7189	3.791
0.30	0.3733	3.6806	−1.9216	3.118
0.35	0.3865	3.1742	−2.1538	2.638
0.40	0.4025	2.7860	−2.4223	2.278
0.45	0.4217	2.4765	−2.7369	1.999
0.50	0.4444	2.2222	−3.1111	1.778

B. SCORES AND INFORMATION PER INDIVIDUAL FOR $Ab/aB \times Ab/aB$

p	AB	aB, Ab	ab	i_p
0.01	0.0200	−0.0100	200.0100	1.000
0.02	0.0400	−0.0200	100.0200	1.001
0.03	0.0601	−0.0300	66.6967	1.002
0.04	0.0801	−0.0400	50.0402	1.004
0.05	0.1002	−0.0499	40.0503	1.006
0.06	0.1204	−0.0599	33.3939	1.009
0.07	0.1407	−0.0698	28.6423	1.012
0.08	0.1610	−0.0797	25.0813	1.016
0.09	0.1815	−0.0896	22.3141	1.020
0.10	0.2020	−0.0995	20.1025	1.025
0.15	0.3069	−0.1483	13.4919	1.057
0.20	0.4167	−0.1961	10.2206	1.103
0.25	0.5333	−0.2424	8.2909	1.164
0.30	0.6593	−0.2871	7.0389	1.241
0.35	0.7977	−0.3298	6.1822	1.337
0.40	0.9524	−0.3704	5.5820	1.455
0.45	1.1285	−0.4086	5.1643	1.600
0.50	1.3333	−0.4444	4.8889	1.778

Table 22

SCORES AND INFORMATION PER INDIVIDUAL FOR $AB/ab \times AB/ab$ WITH A SEMIDOMINANT
(Finney[364])

p	AAB	AAb	AaB	Aab	aaB	aab	i_p
0.01	0.9824	201.0024	0.0126	99.9923	100.4999	−1.0178	100.241
0.02	0.9646	101.0046	0.0254	49.9842	50.4996	−1.0362	50.232
0.03	0.9468	67.6735	0.0387	33.3093	33.8326	−1.0550	33.562
0.04	0.9285	51.0087	0.0519	24.9670	25.4985	−1.0747	25.216
0.05	0.9102	41.0104	0.0655	19.9578	20.4976	−1.0948	20.209
0.06	0.8917	34.3455	0.0796	16.6150	17.1634	−1.1155	16.870
0.07	0.8732	29.5853	0.0940	14.2243	14.7815	−1.1367	14.484
0.08	0.8543	26.0153	0.1086	12.4284	12.9945	−1.1586	12.692
0.09	0.8353	23.2389	0.1236	11.0289	11.6043	−1.1811	11.297
0.10	0.8160	21.0180	0.1389	9.9069	10.4917	−1.2042	10.180
0.15	0.7173	14.3576	0.2219	6.5144	7.1504	−1.3287	6.828
0.20	0.6151	11.0317	0.3175	4.7817	5.4762	−1.4683	5.159
0.25	0.5098	9.0431	0.4277	3.7098	4.4717	−1.6236	4.172
0.30	0.4032	7.7292	0.5562	2.9673	3.8077	−1.7946	3.542
0.35	0.2967	6.8087	0.7060	2.4131	3.3455	−1.9825	3.127
0.40	0.1927	6.1451	0.8819	1.9784	3.0201	−2.1883	2.863
0.45	0.0928	5.6658	1.0884	1.6254	2.7984	−2.4150	2.714
0.50	0.0000	5.3333	1.3333	1.3333	2.6667	−2.6667	2.667

For matings $Ab/aB \times Ab/aB$, interchange columns 2 and 6 and also columns 3 and 7 in this table.

Table 23

SCORES AND INFORMATION PER INDIVIDUAL FOR $AB/ab \times AB/ab$, WITH A AND B
SEMIDOMINANT

(Finney[364])

p	$AABB$ aabb	$AABb, aaBb$ $AaBB, Aabb$	$AaBb$	$AAbb$ aaBB	i_p
0.01	−0.0204	100.9897	0.0002	201.9998	199.980
0.02	−0.0416	50.9788	0.0009	101.9992	99.959
0.03	−0.0637	34.3006	0.0020	68.6648	66.605
0.04	−0.0867	25.9550	0.0036	51.9967	49.917
0.05	−0.1105	20.9421	0.0058	41.9948	39.895
0.06	−0.1353	17.5952	0.0086	35.3257	33.207
0.07	−0.1610	15.2000	0.0121	30.5610	28.423
0.08	−0.1876	13.3993	0.0163	26.9863	24.829
0.09	−0.2153	11.9948	0.0213	24.2048	22.028
0.10	−0.2439	10.8672	0.0271	21.9783	19.783
0.15	−0.4027	7.4405	0.0711	15.2836	13.002
0.20	−0.5882	5.6618	0.1471	11.9118	9.559
0.25	−0.8000	4.5333	0.2667	9.8667	7.467
0.30	−1.0345	3.7274	0.4433	8.4893	6.076
0.35	−1.2844	3.1112	0.6916	7.5068	5.121
0.40	−1.5385	2.6282	1.0256	6.7949	4.487
0.45	−1.7822	2.2582	1.4581	6.2986	4.120
0.50	−2.0000	2.0000	2.0000	6.0000	4.000

tables are intended chiefly for use with plant material and do not include single back-crosses, but they do include double backcrosses and intercrosses, with several kinds of epistasis and degrees of dominance. The scores are listed for all useful values of p at intervals of 0.01. Finney's tables, which are reproduced in tables 19 to 23, give scores for five kinds of matings often encountered in animal genetics. Finney's scores are slightly different from those described above. They are based on maximum-likelihood methods, but are calculated so that, instead of giving a correction to be applied to the provisional value of p, they lead directly to the revised estimate, thus eliminating one step in the arithmetic. They are tabled at intervals of 0.01 for values of p from 0.01 to 0.10 and at intervals of 0.05 for values from 0.10 to 0.50.

The use of Finney's scores is shown with data published by Carter[167] on the linkage of luxate (lx) and viable dominant spotting (W^v) (table 24). A recombination value of 0.13 can be calculated directly from the backcross, which suggests the use of 0.15 as a provisional value. Using tables 19 and 20, the appropriate scores (Λ) and values of i_p can be found, proceeding as in the previous example:

$$p = \frac{343.76}{2111.57} = 0.163,$$

$$\text{SE}_p = \frac{1}{\sqrt{I_p}} = \frac{1}{45.95} = 0.0218.$$

This approximation of p will usually differ very little from the exact estimate. In the present case, rescoring at $p = 0.163$ and recalculating gives a value of $p = 0.1624$, which is so small an improvement as not to be worth the effort.

ESTIMATION OF HETEROGENEITY

One may wish to discover whether the several bodies of data which contribute to the estimate of the recombination fraction are homogeneous with respect to the prob-ability of recombination. Fisher[371] has pointed out that the sum of D^2/I calculated separately for each body of data, using the value of p estimated from the total, is distributed as χ^2. Such a χ^2 includes a part derived from the deviation of the total D from zero. When this is subtracted, the resulting χ^2 for $n - 1$ degrees of freedom, where n is the number of separate bodies of data, measures the heterogeneity among the several bodies of data. The data on the linkage of ru and je again may be taken as an example. We would like to calculate D for each of the four kinds of matings for a value of p close to 0.447. Finney's tables do not give D directly, but D can be obtained from them by a simple arithmetic transformation as follows:

$$p = p_o + D/I_p = \Lambda/I_p$$

$$D = \Lambda - p_o I_p$$

Table 24

CALCULATION OF Λ AND I_p FOR ESTIMATION OF RECOMBINATION BETWEEN lx AND W^v IN THE MOUSE

(Using data from Carter[16T])

Parents	Mating type		+W^v	lxW^v	++	$lx+$	Λ	Total	I_p
					Offspring				
$\dfrac{+W^v}{lx+}\ \dfrac{lx+}{lx+}$	B C (Table 7, $p = 0.15$)	n	83	11	13	77		184	1,443.112
		Λ	0	7.843	7.843	0		i_p 7.843	
		$n\Lambda$	0	86.273	101.959	0	188.232		
$\dfrac{+W^v}{lx+}\ \dfrac{++}{lx+}$	SB C (Table 8)	n	35	6	27	16		84	194.292
		Λ	−0.1935	7.0137	1.2166	−0.8295		i_p 2.313	
		$n\Lambda$	−6.7725	42.0822	32.8482	−13.2720	54.8859		
$\dfrac{lxW^v}{++}\ \dfrac{++}{lx+}$	SB R (Table 8)	n	52	42	98	13		205	474.165
		Λ	1.2166	−0.8295	−0.1935	7.0137		i_p 2.313	
		$n\Lambda$	63.2632	−34.8390	−18.9630	91.1781	100.6393		
							343.7572		2,111.569

B = double backcross, SB = single backcross, C = coupling, R = repulsion, n = number, Λ = score.

Table 25

CALCULATION OF Λ AND I_p FOR ESTIMATION OF RECOMBINATION BETWEEN ru AND je IN THE MOUSE

(Using same data as in table 18)

Mating type		Offspring				Λ	$D =$ $A - p_0I_p$	Total	I_p
		$++$	$+je$	$ru+$	$ruje$				
I C	n	151	54	45	20			270	
(Table 9a,	Λ	0.4217	2.4765	2.4765	−2.7369			1.999	
$p = 0.45$)	$n\Lambda$	63.6767	133.7310	111.4425	−54.7380	254.1122	11.2337		539.730
I R	n	86	31	35	5			157	
(Table 9b)	Λ	1.1285	−0.4086	−0.4086	5.1643			1.600	
	$n\Lambda$	97.0510	−12.6666	−14.3010	25.8215	95.9049	−17.1351		251.200
B C	n	120	103	89	122			434	
(Table 7)	Λ	0	4.040	4.040	0			4.040	
	$n\Lambda$	0	416.120	359.560	0	775.680	−13.332		1,753.360
B R	n	9	9	10	12			40	
(Table 7)	Λ	4.040	0	0	4.040			4.040	
	$n\Lambda$	36.360	0	0	48.480	84.840	12.120		161.600
						1,210.5371	−7.1134		2,705.890

I = intercross, B = double backcross, C = coupling, R = repulsion, n = number, Λ = score.

The calculation of D using the tabled values of Λ for $p = 0.45$ is set out in table 25. The heterogeneity χ^2 can then be determined as follows:

Mating		$\chi^2 = D^2/I$	df	P
I	C	0.2338	1	
I	R	1.1683	1	
B	C	0.1014	1	
B	R	0.9090	1	
Sum		2.4130	4	
Total		0.0187	1	
Heterogeneity		2.3943	3	> 0.05

There is thus no significant heterogeneity among the matings.

The estimate of the heterogeneity χ^2 is not accurate unless the total deviation is small. Tables 19 to 23 may not be sufficiently detailed when an accurate value of χ^2 is needed, as when the heterogeneity is on the borderline of significance. Allard's tables may supply the appropriate scores, or it may be necessary to calculate them. Table 26 gives formulas for maximum-likelihood scores and values of i_p for several common kinds of matings.

In cases for which no tables of scores are available, it may sometimes be relatively easy to derive maximum-likelihood scores but may be somewhat more difficult to derive formulas for the amount of information per individual, i_p. Fortunately there exists an arithmetic method which gives an approximate estimate of the total amount of information, I_p, given by the results of any mating or group of matings. Since I_p is equal to the second derivative of the logarithm-likelihood expression, it is also equal to the first derivative of the equation, $dL/dp = 0$. I_p is therefore the rate of change of dL/dp with respect to p. If dL/dp has been evaluated for two values of p (p_1 and p_2) close to the exact estimate, to give two values of D, then $I_p = (D_2 - D_1)/(p_2 - p_1)$. The value of I_p calculated in this way will not be identical to the value calculated from the formula $-\sum \{mn[(d^2 \log m)/dp^2]\}$. The arithmetic method gives the actual amount of information realized in the particular body of data on hand, whereas the formula gives the expected amount of information in bodies of data of this size and value of p.

LINEAR ORDER OF LOCI

Having determined that a mutant is linked with another mutant known to be a member of a particular linkage group, one may wish to determine the linear order of the two mutant loci with respect to another locus in the linkage group. For this purpose it may sometimes be sufficient to determine the distance of the new mutant from the other locus. The recombination for A–B, say, should equal either the sum or difference of the recombinations for A–C and B–C, and the linear order easily follows. If, however, any two of the three loci are closely linked, the errors of estimate

Table 26

FORMULAS FOR MAXIMUM-LIKELIHOOD SCORES AND AMOUNT OF INFORMATION PER INDIVIDUAL, i_p FOR SEVERAL KINDS OF MATINGS†

Kind of mating	Phenotypic class									i_p
	AABB	AABb	AaBB	AaBb	AAbb	Aabb	aaBB	aaBb	aabb	
A and B semidominant Intercross	$\frac{2}{p}$	$\frac{1-2p}{p(1-p)}$	$\frac{1-2p}{p(1-p)}$	$\frac{2(2p-1)}{1-2p+2p^2}$	$\frac{-2}{1-p}$	$\frac{1-2p}{p(1-p)}$	$\frac{-2}{1-p}$	$\frac{1-2p}{p(1-p)}$	$\frac{2}{p}$	$\frac{2(1-3p+3p^2)}{p(1-p)(1-2p+2p^2)}$
A semidominant Intercross	$\frac{2(1-p)}{2p-p^2}$		$\frac{2p-1}{1-p+p^2}$		$\frac{-2}{1-p}$	$\frac{1-2p}{p(1-p)}$	$\frac{-2p}{1-p^2}$		$\frac{2}{p}$	$2 + \frac{(1-p)^2}{p(2-p)} + \frac{(1-2p)^2}{2(1-p+p^2)} + \frac{p^2}{1-p^2} + \frac{(1-2p)^2}{2p(1-p)}$
A and B fully dominant Double backcross				$\frac{1}{p}$		$\frac{-1}{1-p}$		$\frac{-1}{1-p}$	$\frac{1}{p}$	$\frac{1}{p(1-p)}$
Intercross				$\frac{2p}{2+p^2}$		$\frac{-2p}{1-p^2}$		$\frac{-2p}{1-p^2}$	$\frac{2}{p}$	$\frac{2(1+2p^2)}{(1-p^2)(2+p^2)}$
Single backcross (A intercrossed)				$\frac{1}{1+p}$		$\frac{-1}{2-p}$		$\frac{-1}{1-p}$	$\frac{1}{p}$	$\frac{1+2p-2p^2}{2p(1-p)(1+p)(2-p)}$

† From Allard[7]

Formulas are for matings in repulsion; for matings in coupling, substitute $(1-p)$ for p.

of the recombination proportions may be large enough to make the method of sum or difference unreliable. A linkage test using all three loci simultaneously is then advisable. The least frequent class of recombinations obtained from this kind of mating is then assumed to be the double recombinant class. An example will make the reasoning clear.

Falconer and Sobey[345] give data on a cross involving *Rex* (*Re*), *Trembler* (*Tr*), and *shaker-2* (*sh-2*) in the mouse. The cross was *Re Tr*+/+ +*sh-2* × + +*sh-2*/+ +*sh-2*, and the following offspring, grouped in complementary classes, were obtained:

Re Tr +	16		+ *Tr* +	3
+ + *sh-2*	28		*Re* + *sh-2*	8
Re Tr sh-2	1		+ *Tr sh-2*	0
+ + +	1		*Re* + +	0.

The three recombination proportions are therefore:

$$Re - Tr \quad \frac{11 + 0}{57} = 0.193,$$

$$Tr - sh\text{-}2 \quad \frac{2 + 0}{57} = 0.035,$$

$$Re - sh\text{-}2 \quad \frac{11 + 2}{57} = 0.228.$$

Since $0.193 + 0.035 = 0.228$ the order appears to be $Re - Tr - sh\text{-}2$. In this calculation the assumption is made that the two recombination classes which did not occur, $+ Tr\ sh\text{-}2$ and $Re + +$, would require two recombinations each, and that the recombinations in the *Tr–sh-2* interval require only single recombinations. This is a reasonably safe assumption, since, assuming no interference, the probability of double recombinations is $0.193 \times 0.035 = 0.007$. Interference, which undoubtedly exists, would make this probability even smaller. It is therefore very unlikely that the two recombinations between *Tr* and *sh-2* would have been found if the order were $Re - sh\text{-}2 - Tr$ and they were indeed double recombinants.

STOCKS FOR TESTING LINKAGE

For the efficient testing of linkage of new mutations in any organism, it is desirable to have multiple-gene stocks so designed as to cover the greatest length of the known linkage map with the smallest number of crosses. The task will be greater for organisms with greater length of the known linkage map, and will increase as more of the linkage map becomes known. This increase may be partly offset, however, by the discovery of more suitable marker genes (for example, dominants to replace recessives) which increase the efficiency of any one test.

Multiple-gene stocks for testing linkage in the mouse were first developed by Snell (reported by Cooper).[227]   In 1951, Carter and Falconer[172] reported the development of a new set of stocks incorporating mutants discovered in the intervening period, described their use, and discussed the theory of the design of such stocks.   By means of the concept of the swept radius, i.e., the length of linkage map tested by a marker gene in any one cross, they were able to define the principles to be followed in making a choice when two or more linked markers are available.   The concept also allows one to calculate the total length of linkage map tested by a particular set of stocks, a value which is useful for comparing the efficiency of different sets of stocks.

Since 1951, many new mutants and six new autosomal linkage groups have been discovered.   It has thus become possible to construct revised linkage stocks which test a greater length than was possible in 1951.   Such stocks are maintained at the Institute of Animal Genetics in Edinburgh, Scotland, the Radiobiological Research Unit at Harwell, England, and the Roscoe B. Jackson Memorial Laboratory, Bar Harbor, Maine.   Lists of these stocks, as well as stocks at other institutions, are published periodically in the *Mouse News Letter*, a mimeographed bulletin produced and distributed by the International Committee on Laboratory Animals, and Laboratory Animals Centre, M.R.C. Laboratories, Woodmansterne Road, Carshalton, Surrey, England.

The set of stocks for testing linkage now maintained or under construction at the Jackson Memorial Laboratory is as follows:

Stock name	Markers	Linkage groups
EI	Re Sd Va	VII, V, XVI
ROP	Ra Os Pt	V, XVIII, VIII
MWT	a^t Mi^{wh} W^v T	V, XI, III, IX
TCS†	Tw Ca Sl	XV, VI, —
V	a ln fz s v	V, XIII, III, X
PF	a p fr f	V, I, XIV
WSR†	wa-2 se ru je	VII, II, XII

† Still under construction.

Linkage groups IV and XVII are not represented in the stocks, the first because it did not, at the time the stocks were designed, contain a suitable marker, and the second because it contains at present only recessive mutants, which are difficult to incorporate into existing multiple recessive stocks.   No linkage is known for *Sl* but it has been tested with nearly all the other markers and has shown no close linkage with them.

Using the method of Carter and Falconer,[172] the total length of linkage map swept by these stocks can be calculated.   The method requires knowledge of the average map length of mouse chromosomes.   Counts of chiasmata in the male mouse made by Slizynski (quoted by Carter[159]) show the average number per autosome to be 1.9. Earlier counts made by Crew and Koller[229] showed the average number of chiasmata per autosome to be 2.4 in the male and 2.8 in the female.   Since Crew and Koller were unable to count the chiasmata in complete cells, their counts may have been biased upward if they tended to count the longer chromosomes.   The difference between the sexes which they observed, however, is in accord with the known tendency for crossing over in the female to be greater than that in the male. Assuming Slizynski's counts in the male to be accurate but to give an underestimate of the frequency for males and females combined, we may take 2.1 chiasmata per autosome as an approximation of the true value. Since one chiasma corresponds to 50 map units or centimorgans (cM), the average length of autosome can be taken as 105 cM, and the total autosomal map length as 1,995 cM. Using these figures, we can calculate that the total length of linkage map swept by these linkage testing stocks is 1,438 cM if the new mutant is a dominant, and 1,390 cM if the new mutant is a recessive. By raising 100 test progeny from each mating, with these stocks one can test a new dominant against about 72 per cent of the total autosomal map length and a new recessive against 70 per cent. The sex chromosome is not included in these calculations because sex-linkage will be obvious without recourse to these stocks.

LINKAGE MAP OF THE MOUSE

The known linkage map of the mouse is considerably longer than that of any other mammal but it still falls far short of completeness. Nineteen of 20 possible linkage groups are known, of which I, II, III, V, VII, VIII, IX, XI, XIII, and XX have been shown by Slizynski[1218, 1219] to be on different chromosomes. Some of the remaining groups are quite short and may appear to be independent of each other only because they are located far enough apart on the same chromosome. Linkage group IV has not been adequately tested against most of the other groups because the markers in this group were, until recently, highly unsuitable for linkage tests. *Ruby* and *jerker* appeared linked in earlier data but recent tabulation of unpublished data shows them to recombine at random. They may therefore actually represent different linkage groups.

The map in figure 18 summarizes the latest information on linkage in the mouse. Most of it is based on published information but it also makes use of information from personal communications and from the *Mouse News Letter*. The unpublished information is used by permission of the authors. All mutants used in the map have been described either in print or in the *Mouse News Letter* before August, 1960. Some of the information on linkage has been supplied by personal communication as noted in the bibliography. A few mutants for which there was preliminary indication of linkage, but which are now thought to be extinct, have been omitted. *Rodless retina*, r, is probably extinct, but has been included because the evidence for linkage with *silver*, *si*, is reasonably good, and because this was the linkage which first established linkage group IV.

For several reasons the distances between loci on the map are not to be taken too literally. They represent many compromises, and recourse must be had to the references cited for more exact information. The numbers on the map are recombination percentages, but not necessarily those between the two loci on opposite sides of the numbers; it is not possible, without unduly complicating the map, to show which

Fig. 18. LINKAGE MAP OF THE MOUSE.

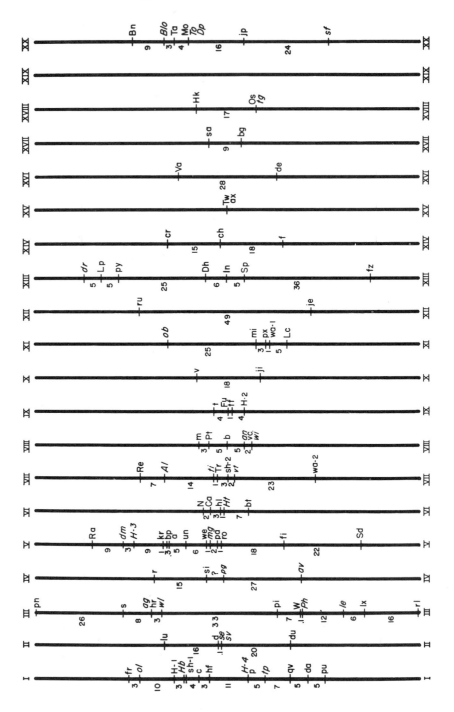

distances are recombinations and which are the result of subtractions.   The linear order of many of the genes is subject to some degree of uncertainty.   Those indicated by symbols in italics have not been critically tested to determine their position relative to all other genes in the linkage group.   The order of the other loci is established with a fairly high probability of correctness.   Many experiments disagree in the estimate of recombination obtained.   In combining estimates from different experiments the usual procedure has been to weight the recombination fraction by the amount of information in the body of data contributing the estimate.   In linkage groups VI and XIII, however, where there are striking differences in amount of recombination between the sexes, weighted averages for each sex separately were calculated, and the unweighted average of the sexes was taken.   There are differences between the sexes in other regions of the linkage map also, but the differences are either small enough or the estimates from the two sexes well enough balanced, so that the unweighted method of averaging would have made very little difference.   In general when sex differences occur recombination is higher in females than in males, but in linkage group VI and between sh-2 and wa-2 in linkage group VII the reverse is true.

Following is a list of the names of the genes used in the map.   The numerous alleles at the histocompatibility loci and at the t locus have not been listed.   For the other multiple allelic series, the mutants which have been given names are listed. The references cited all give information on linkage, with the exception of a few, cited for members of a multiple allelic series, which contain the original description of the allele.   Where no reference is cited, a description has been given by Grüneberg.[507]

Symbol for Gene	Name of Gene	References
	I	
fr	*frizzy*	344
ol	*oligodactyly*	549
H-1	*Histocompatibility*-1	1238, 1244
Hb	*Hemoglobin pattern*	1017
sh-1	*shaker*-1	344, 425, 498, 500
c-series		264, 353, 498, 1017
c^{ch}	*chinchilla*	340, 344, 1267
c^e	*extreme dilution*	
c^h	*himalayan*	487
c	*albino*	143, 180, 183, 263, 297, 425, 500, 518, 549, 1238, 1244, 1466
hf	*hepatic fusion*	143
H-4	*Histocompatibility*-4	1244
p-series		
p^d	*dark pink-eye*	162
p	*pink-eyed dilution*	143, 180, 183, 263, 264, 297, 340, 344, 353, 359, 425, 493, 498, 500, 518, 1112, 1244, 1267, 1466
p^s	*p-sterile*	594

Symbol for Gene	Name of Gene	References
tp	*taupe*	359, 493, 1112
qv	*quivering*	1466
da	*dark*	340
pu	*pudgy*	1267

II

lu	*luxoid*	491
d-series		
d	*dilute*	182, 426, 454, 1231, 1234, 1240
d^l	*dilute-lethal*	1189
se	*short ear*	182, 426, 454, 490, 491, 1231, 1234, 1240
sv	*Snell's waltzer*	490
du	*ducky*	491, 1234

III

pn	*pugnose*	707
s	*piebald*	167, 296, 301, 369, 694, 707, 796, 799, 1240
ag	*agitans*	589
hr-series		
hr	*hairless*	167, 276, 428, 753, 1240
hr^{rh}	*rhino*	589
wl	*wabbler-lethal*	753
pi	*pirouette*	275, 276
W-series		
W	*Dominant spotting*	296, 301, 369, 428, 694, 796, 799
W^v	*Viable dominant spotting*	167, 275, 338, 509, 707
W^j	*Jay's dominant spotting*	1106
W^a	*Ames dominant spotting*	592
Ph	*Patch*	509
le	*light ears*	755
lx	*luxate*	167, 338
rl	*reeler*	338

IV

r	*rodless retina*	691
si	*silver*	343, 691, 1163
pg	*pygmy*	343
av	*Ames waltzer*	1163

V

Ra	*Ragged*	176, 757, 991
dm	*diminutive*	838
H-3	*Histocompatibility-3*	1238, 1251
kr	*kreisler*	549, 755
bp	*brachypodism*	1089
a-series		118, 377, 978, 991
A^y	*Yellow*	169, 1060, 1089
A^w	*White-bellied agouti*	157, 169, 755
a^t	*black and tan*	157, 169, 173, 176, 755, 1355
a	*nonagouti*	157, 173, 176, 336, 368, 549, 755, 757, 838, 1060, 1238, 1251
a^e	*extreme nonagouti*	595

Symbol for Gene	Name of Gene	References
un	*undulated*	157, 336, 368, 377, 978, 1089
we	*wellhaarig*	336, 377, 549, 755, 978, 991
mg	*mahogany*	757
pa	*pallid*	118, 169, 176, 336, 377, 755, 978, 1060
ro	*rough*	336
fi	*fidget*	169, 173, 1355, 1356
Sd	*Danforth's short tail*	118, 176, 336, 1355, 1356

VI

N	*Naked*	227, 845, 926
Ca	*Caracul*	227, 593, 845, 926, 1267
hl	*hair-loss*	593
Ht	*Hightail*	1267
bt	*belted*	593, 845, 926, 1267

VII

Re	*Rex*	171, 177, 266, 337, 345, 755, 870, 871, 936, 1190*a*
Al	*Alopecia*	266, 755
ti	*tipsy*	1190*a*
Tr	*Trembler*	345
sh-2	*shaker-2*	177, 266, 337, 345, 378, 561, 755, 870, 872, 936, 1253, 1412
vt	*vestigial tail*	870, 871, 872, 1190*a*
wa-2	*waved-2*	171, 177, 266, 378, 561, 872, 1190*a*, 1253, 1412

VIII

m	*misty*	753, 1212, 1404
Pt	*Pintail*	596, 753
b-series		
B^l	*Light*	833
b	*brown*	549, 596, 753, 1213, 1212, 1404
b^c	*cordovan*	877
an	*anemia*	549
vc	*vacillans*	1212, 1213
wi	*whirler*	753

IX

t-series		
T	*Brachyury*	12, 302, 303, 305, 816, 819, 1043, 1246
t	*tailless*	302, 305, 819
Fu-series		
Fu	*Fused*	11, 12, 302, 303, 304, 458, 1043, 1246
Fu^{ki}	*Kinky*	12, 298, 302, 303, 304, 305, 1246
tf	*tufted*	298, 816, 819
H-2	*Histocompatibility-2*	11, 12, 458, 1246

X

v	*waltzer*	1241
ji	*jittery*	1241

XI

ob	*obese*	272

Symbol for Gene	Name of Gene	References
mi-series		
Mi^{wh}	*White*	142, 174, 272, 508, 1002
mi	*microphthalmia*	508
px	*postaxial hemimelia*	163
wa-1	*waved-1*	142, 272
Lc	*Lurcher*	174, 1002
XII		
ru	*ruby eye*	268, 339, 379, 1356a
je	*jerker*	268, 339, 379, 1356a
XIII		
dr	*dreher*	817
Lp	*Loop tail*	1252
py	*polydactyly*	374, 990
Dh	*Dominant hemimelia*	161, 817
ln	*leaden*	161, 274, 374, 990, 1252
Sp	*Splotch*	990, 1252
fz	*fuzzy*	274, 990, 1252
XIV		
cr	*crinkled*	709, 1004
ch	*congenital hydrocephalus*	1004
f	*flexed tail*	709, 1004
XV		
Tw	*Twirler*	818
ax	*ataxia*	818
XVI		
Va	*Varitint-waddler*	233
de	*droopy-ear*	233
XVII		
sa	*satin*	1268
bg	*beige*	1268
XVIII		
Hk	*Hook*	489
Os	*Oligosyndactylism*	488, 489
tg	*tottering*	488
XX		
Bn	*Bent*	336, 418, 754, 1001
Blo	*Blotchy*	1112
Ta	*Tabby*	334, 342, 336, 1001, 1112, 1382
Mo-series		
Mo	*Mottled*	336, 342
Mo^{br}	*Brindled*	336, 342
To	*Tortoise*	269, 754
Dp	*Dappled*	1003
jp	*jimpy*	1001
sf	*scurfy*	1382

George E. Jay, Jr., Ph.D.

GENETIC STRAINS *and* STOCKS

INTRODUCTION

In this era of rapidly expanding biomedical research, the need for current information on all aspects of laboratory animal resources is particularly important and will become even more so in the future. From time to time in the past, various listings of laboratory animals have been compiled and made available to the scientific public. The Institute of Laboratory Animal Resources, National Research Council, has prepared a catalogue of commercial sources of laboratory animals in the United States;[639] the *Mouse News Letter*,[903] an informal publication prepared by the Laboratory Animals Centre, Medical Research Council, England, provides a continuing listing of inbred mouse strains from all parts of the world,† and the Committee on Standardized Nomenclature for Inbred Strains of Mice‡ has recently published a revised listing of inbred strains of mice.[219] These listings are of vital importance to the research worker, for they provide information on the location and status of laboratory-animal strains and stocks.

However, the listings indicated above have not adequately covered the extensive genetic resources existing for other mammals such as the rat, guinea pig, hamster,

† Dr. G. D. Snell and Miss Joan Staats of the R. B. Jackson Memorial Laboratory, Bar Harbor, Maine, now prepare a supplement to the *Mouse News Letter*, listing the inbred strains of mice. It is anticipated this supplement will appear every two years.

‡ Committee members: G. D. Snell and Joan Staats, R. B. Jackson Memorial Laboratory, Bar Harbor, Maine; M. F. Lyon, Harwell, Berks, England; L. C. Dunn, Columbia University, New York, New York; H. Grüneberg, University College, London; P. Hertwig, Biolojisches Institut, Halle; and W. E. Heston, National Cancer Institute, Bethesda, Maryland.

rabbit, and so forth.   Consequently, a compilation of this information would be most timely and of considerable interest to all concerned.   The publication of Billingham and Silvers,[99] the *Catalogue of Uniform Strains of Laboratory Animals Maintained in Great Britain*,[742] and the more recent questionnaire of the Committee on Maintenance of Genetic Stocks, Genetic Society of America, provide some information on the whereabouts of such material, but it is believed that many unlisted stocks of genetic interest are being maintained by various investigators.

Thus, the Genetics Study Section, Division of Research Grants, National Institute of Health, thought it most pertinent to develop a composite listing of genetic material for mammalian species most commonly used in laboratory experimentation, and to include such listing in this volume.   This material is the initial effort to bring together all the information alluded to above, plus the results of an extensive inquiry to scores of investigators and organizations around the world.   It is hoped that this compilation is complete, but it is realized that some investigators were no doubt inadvertently overlooked.   I apologize for this oversight and hope their material will be included in the planned revision.

In developing this listing, many sources of information were utilized as well as the talents of many individuals in the United States and abroad.   To all of these sincere thanks and appreciation are extended for permission to use their material and for their efforts and interest.   Thanks are also due to Mrs. Clarice Overath and Mr. Robert Dettman for their untiring help in assembling the data and preparing the manuscript and various others at the National Cancer Institute who have been helpful.

LISTING ARRANGEMENT

The arrangement of this composite listing is by sections according to species as follows: Section I, Mice; Section II, Rats; Section III, Guinea Pigs; Section IV, Hamsters; Section V, Rabbits; Section VI, *Peromyscus sp.*   Each section contains a brief introduction, followed by tables listing (1) established inbred strains, (2) strains in development, and (3) stocks of particular genetic interest.

The distinction between established inbred strains and strains in development was made on the basis of the definition of an inbred strain as recommended in *Standardized Nomenclature for Inbred Strains of Mice*,[219] that is, twenty or more consecutive generations of brother × sister or parent × offspring (provided the mating in each case is to the younger parent).   Two exceptions to this rule were made for rat strains MR and MNR, since the generations of inbreeding presently reported are approximately the theoretic twenty generations.   It is likely that by the time these listings are published twenty generations of inbreeding will have been completed.   Thus the strains in development are those exhibiting various generations of inbreeding from F_1 to something less than F_{20}.   In the case of certain strains of rabbits (table 38), where matings of brother × sister or parent × offspring have not been consistently followed he amount of inbreeding is indicated by an inbreeding coefficient (F), devised by Wright.[1447]

The use of the terms *strain* and *substrain*, again, are used according to the definitions recommended by the Committee on Standardized Nomenclature for Inbred Strains of Mice.   However, some latitude has been exercised in this regard, for the term *strain* has been used to designate those populations in the process of being inbred by brother × sister, parent × offspring, or some other system of inbreeding.   The term, stock, has been used to designate those populations that are maintained for their identified genes or identifiable genetic characteristics of genetic interest, with or without inbreeding.

In the case of mice, only the established strains recently compiled by the Committee on Standardized Nomenclature for Inbred Strains of Mice are included.   Stocks containing identified genes or identifiable genetic characteristics are numerous and with a global distribution.   The Roscoe B. Jackson Memorial Laboratory, Bar Harbor, Maine, is probably the largest repository of murine genes in the world, and along with this the laboratory has developed, and is continuing to develop, isogenic strains and linkage stocks for testing purposes.   Elsewhere in this volume, the linkage stocks at this laboratory are discussed by Dr. Margaret C. Green.   Material of this kind for species other than mice is extremely limited.

Similarly, no attempt was made to ascertain the number or kinds of inbred strains in development for mice, whereas this information has been compiled for the other species.   It is strange that even though some of the earliest mammalian genetics was done in such species as the rat, guinea pig, and rabbit, there is a paucity of such material today.   Fortunately there are indications of renewed interest in these species.

As indicated above, information on the location and status of the various strains and stocks came from previously printed lists, publications, and a questionnaire.   From these, and my own knowledge of the whereabouts of strains and stocks, a mailing list of more than a hundred names was compiled.   At the present time similar surveys are being made in Japan, Canada, Italy, and Germany.   All the information obtained has been arranged in a contributor-strain cross-file.   This file will be kept active by sending out periodic requests for bringing the status of strains and stocks up-to-date.   No information was solicited or obtained (except in a few cases) regarding the environmental conditions under which the animals are maintained, the disease conditions existing (unless of genetic significance), husbandry practices followed, nutritional status, and so forth.   It might be of considerable value for future listings to take into consideration such exogenous factors, since certainly some genetic manifestations are subject to environmental modifications.

MICE

As indicated above, only the established inbred strains of mice compiled by the Committee on Standardized Nomenclature for Inbred Strains of Mice[219] and the list of contributors are included in this section.   The material is presented as published,

with only those editing changes necessary to fit the format of this publication. It i
realized there are probably many strains now in the process of being developed tha
will eventually be listed in future revisions by this Committee, or that will be indicated
in future editions of the *Mouse News Letter*.[903] The importance of this species in mam
malian genetic research is well recognized, and for this reason communication on the
location and status of murine material is much better organized than for the other
rodent species. The far-sighted organization of a committee for the purpose of bring
ing order out of approaching chaos has been a tremendous asset, and consequently
there is little to add to the splendid work already done. Other kinds of useful informa
tion about murine strains, such as composite-gene stocks, linkage stocks, and isogenic
stocks are available through such organizations as the Roscoe B. Jackson Memoria
Laboratory in this country and the various research groups in England that have
done an outstanding job in collecting such material and maintaining it in usable form

Table 27

Established strains of mice†

Name or Symbol	Synonym(s)	Remarks
A		INBR (St): 131. GENET: *aabbcc*. *H-2^a*. ORIGIN: Strong, from cros made in 1921 of albino from Cold Spring Harbor and Bagg albino The majority of sublines trace to a stock which Bittner obtained from Strong in 1927. CHARAC: mammary tumor incidence high in breeders low in virgins. Produces 5–10% young with cleft palate. High incidence of renal disease in old mice. MAINTAINED BY: Bcr, Br, Cam Fa, Fn, Ge, Go, Gr, H, Ha, Icrc, Lab, Mr, Not, Rl, Sn, Sp, Ss, St.
A/Crgl/2		INBR (Crgl): 26. GENET: *aabbcc*. ORIGIN: from a nontumorous ♀ o A/He-Jax. CHARAC: 30–50% mammary tumors in breeders at 12–1 months. Originally very low incidence. MAINTAINED BY: Crgl.
A/Crgl/3		INBR (Crgl): 18 since mutation. ORIGIN: mutation of MTI in 1951 a $F_{84} + 4$. CHARAC: over 90% mammary tumors in breeders at months. Approx. 11% in virgins at 16 months. MAINTAINED BY Crgl.
A/He		INBR (Jax): 111. GENET: *aabbcc*. ORIGIN: see strain A. Strong t Bittner 1927, to Heston 1938, to Jax 1948 at F_{77}. CHARAC: mammary tumors 74% in breeders. Primary pulmonary tumors 30% in ♀♀ 51% in ♂♂. Mast-cell content of spleen very high in old mice especially ♂♂. MAINTAINED BY: Anl, Crgl, Cam, Cbi, Chr, De, Gi Ha, He, Jax, Mas, Mr, Ms, N, Per, Pi, Rd, We.
A/Jax		INBR (Jax): 107. GENET: *aabbcc*. ORIGIN: see strain A. Strong t Cloudman 1928, survivors of Bar Harbor fire to Jax 1947 at F_{73} CHARAC: 28% mammary tumors in breeding ♀♀. Primary pulmonar tumors 41% in ♀♀, 49% in ♂♂. Lower percentage granulocytes than other A sublines. MAINTAINED BY: Anl, Bu, Chr, Fr, Jax, Jb, Ks, Mas.
A/LN	A Lilly	INBR (N): 110. GENET: *aabbcc*. ORIGIN: see strain A. Strong 1921 to Bittner 1927, to W. Murray, to Eli Lilly & Co. 1941, to Jax 1948 to NIH 1951 at F_{84}. CHARAC: resistant to some tumors from othe A sublines. MAINTAINED BY: N.

† A list of abbreviations used is given at the end of this table.

Table 27—*Continued*

Name or Symbol	Synonym(s)	Remarks
AfB		INBR (Gif): 32. GENET: *aabbcc*. ORIGIN: from Netherlands Cancer Inst., a litter of A taken by cesarean section and fostered on C57BL. CHARAC: no mammary tumor, since no MTI. Low percentage of pulmonary tumors. MAINTAINED BY: Gif.
Af/MySp		INBR (Sp): ? + 16. GENET: *aabbcc*. ORIGIN: Murray, from A taken by cesarean section and fostered on strain BD. CHARAC: low mammary tumor, high pulmonary tumor, renal disease. MAINTAINED BY: Sp.
A/Gr$_f$ RIII$_f$B)/Pu		INBR (Mr): F17 since fostering. GENET: *aabbcc*. ORIGIN: one litter of A/Gr mice fostered on RIII$_f$B/Pu, May 1952, without suckling own mother. CHARAC: no mammary tumors, being free of milk factor, otherwise same as strain A. MAINTAINED BY: Mr.
A/Fa-+*c*	ABG	INBR (Fa): N16. GENET: *aabb*+*c*. CHARAC: *c* segregates. MAINTAINED BY: Fa.
A/Be-*CRe*	REB	INBR (Be): 20+. GENET: *aabbReRe*, and either *cc* or +*c*. ORIGIN: from A/Fa *circa* 1948 by addition of *C* and *Re* through five generations of backcrossing to A/Fa; then b × s with forced segregation at the *c* locus, the parents being always *cc* ♂ × +*c* ♀. MAINTAINED BY: Be.
A-*H-2*^{b}	A.BY	INBR (Sn): G12F15. MAINTAINED BY: Kl, Sn.
A-*H-2*^{f}*Fu*	A.CA	INBR (Sn): G12F16. MAINTAINED BY: Kl, Sn.
A-*H-2*^{s}	A.SW	INBR (Sn): G12F18G4F1. MAINTAINED BY: Kl, Sn.
AB	Bluhm-Stamm, Agnes-Bluhm	INBR (Kn): 53. GENET: *cc*. ORIGIN: Dr. Agnes Bluhm, Kaiser Wilhelm Inst., *circa* 1930; b × s until 1943. Breeding history disrupted by war. This stock from Prof. Kaufmann, Univ. Marburg, in 1950. Inbreeding continued. CHARAC: no mammary tumors or sarcomas except after radiation treatment; low rate of pulmonary tumors and lymphomas. MAINTAINED BY: Kn.
AK/n	Ak-n	INBR (Ka): 20+ by Gs+F9. GENET: *aacc*. ORIGIN: Furth, to Gross Nov. 1945; Gross to Kaplan, 1956, and Ontario Cancer Inst., 1957. CHARAC: high leukemia; source of virus that induces leukemia in C3H/BiGs mice. MAINTAINED BY: Gs, Ka, Oci.
AKR	AK, AKm, Afb, R.I.L., Rockefeller Inst. Leukemia	INBR (N): 64. GENET: *aacc*. *H-2*^{k}. ORIGIN: carried by Furth as high-leukemia strain from 1928 to 1936. Then random bred at Rockefeller Inst. for several generations, followed by F$_9$ by Mrs. Rhoads and 30+ by C. Lynch. CHARAC: high leukemia. MAINTAINED BY: Am, Gif, Jax, Jb, Lab, Lhm, Lw, Ms, Mv, N, Pa, Rd, Rho, S, Sp, We.
AKR-*H-2*^{a}	AKR.K	INBR (Sn): G12F11G4F7.
AKR-*H-2*^{m}	AKR.M	INBR (Sn): G12F17.
AU		INBR (Ss): 27. GENET: *aaUU*. ORIGIN: Fisher; Medawar (London) to Silvers in 1957 at F$_{23}$. CHARAC: ♀♀ do not reject ♂♂ isografts. MAINTAINED BY: Ss.
A$_2$G		INBR (Gif): 47. GENET: albino. ORIGIN: cross of A with noninbred albino. CHARAC: useful for endocrine work especially gonadotrophin assay (Lab). Can carry *Salmonella* without showing signs of it (G). Good reproduction (Gif). MAINTAINED BY: G, Gif, Lab, Mr.
BALB/c	BalbC, C	INBR (Rr): 98. GENET: *bbcc*. *H-2*^{d}. ORIGIN: albino stock acquired by Bagg in 1906, to Little, to MacDowell in 1922; b × s inbreeding started by MacDowell in 1923. Prior inbreeding uncertain. Transferred from MacDowell to Snell in 1932 at F$_{26}$ and subsequently

Table 27—*Continued*

Name or Symbol	Synonym(s)	Remarks
		widely distributed especially *via* Andervont. CHARAC: low incidence of mammary tumors, but high incidence when milk agent is introduced. Some ovarian and adrenal tumors. Susceptible to chronic pneumonia. Primary pulmonary tumors, 26% in ♀♀, 29% in ♂♂. MAINTAINED BY: Anl, Crgl, De, Di, Gif, Go, Jax, Jb, Ka, Lab, Mc Ms, N, Pi, Rl, Rr, S, Sp, We.
BALB/Gw	Ba, B alb, Bagg albino	INBR (Gw): ? + 62. GENET: *bbcc*. ORIGIN: see BALB/c. MacDowell to Gowen in 1932 at F_{27}. CHARAC: low resistance to *Salmonella typhimurium*. Radiation susceptible. MAINTAINED BY: Gw.
BALB/MySp	B alb, Bagg albino	INBR (Sp): 47. GENET: *bbcc*. ORIGIN: see BALB/c. Murray to Simpson in 1948 at F_{23}. CHARAC: mammary tumor, pulmonary tumor MAINTAINED BY: Sp.
BALBf	S_1	INBR (Sp): ? + 20. GENET: *bbcc*. ORIGIN: from BALB born by cesarean section and fostered on strain BD. Murray to Simpson in 1948. CHARAC: low mammary tumor, some pulmonary tumor MAINTAINED BY: Sp.
BALB/c_f	C+	INBR (Sp): 47. GENET: *bbcc*. ORIGIN: from BALB/cAn fostered on C_3H. Andervont to Simpson in 1952 at F_{35}. CHARAC: mammary tumor, pulmonary tumor. MAINTAINED BY: Sp.
BALB/c-CFu	CFU	INBR (Rr): N12. GENET: *bb+cFu+*. ORIGIN: from BALB/cSnRr CHARAC: 95% penetrance of *Fu*. MAINTAINED BY: Rr.
BAMA		INBR (Sp): ? + 18. GENET: *cc*. ORIGIN: derivative of A ♀ × MA ♂ CHARAC: low mammary tumor. MAINTAINED BY: Sp.
BD	Bd	INBR (Sp): ? + 16. GENET: *aa*. ORIGIN: Murray, DBA × C57BL N8 to C57BL, then b × s. Warner to Simpson 1950. (In view of the origin of this strain, it should be classed as a subline of C57BL. CHARAC: low mammary tumor. MAINTAINED BY: Sp.
BDP		INBR (Jax): 38+. GENET: *aabbse/dsepprdrd*. ORIGIN: Gates, inbred since 1926. CHARAC: frequent mammary tumors, ovaries hemorrhagic and necrotic, nervous behavior. MAINTAINED BY: Jax.
BL	Bagg L, BALB/R	INBR (De): 68. GENET: *aabbcc*. ORIGIN: Lynch, from Bagg stock via Strong, maintained at Rockefeller Inst. as distinct strain since 1921 Some of original animals obtained from Strong carried agouti. This stock from Lynch in 1951, at F_{46}. CHARAC: low mammary tumor some pulmonary tumors in old mice. MAINTAINED BY: De.
BRS	BrS, Br-S, Br-s	INBR (St): 55. GENET: *aabb*. ORIGIN: Strong, from a branch of NH treated for eight or more generations with methylcholanthrene CHARAC: gastric lesions, adiposity. MAINTAINED BY: St.
BRSUNT	BrSunt	INBR (N): 64. GENET: *aabb*. ORIGIN: Strong, a branch of BRS continued without further methylcholanthrene treatment (UNT = untreated). CHARAC: gastric lesions, adiposity. MAINTAINED BY: N, St
BRVR		INBR (Kp): 57. GENET: *cc*. ORIGIN: from H. A. Schneider, Rockefeller Inst., at F_{57} in 1959. CHARAC: susceptible to bacteria and to Arbor "B" viruses. MAINTAINED BY: Kp.
BSVS		INBR (Kp): 57. GENET: *cc*. ORIGIN: same as BRVR, at F_{56} in 1959 CHARAC: susceptible to *Salmonella* and viruses, particularly hepatitis MAINTAINED BY: Kp.
BUA		INBR (Wi): 36. GENET: albino. ORIGIN: albinos of unknown pedigree at Brown Univ. maintained by random breeding for unknown length

Table 27—*Continued*

Name or Symbol	Synonym(s)	Remarks
		of time. Inbreeding started in 1945. CHARAC: selected for good growth and reproductive performance. No known spontaneous tumors. MAINTAINED BY: Wi.
BUB		INBR (Wi): 44. GENET: *aacc*. *H*-2^{q}. ORIGIN: same as BUA. CHARAC: same as BUA. MAINTAINED BY: Bn, Wi.
BUC		INBR (Wi): 36. GENET: *aacc*. ORIGIN: same as BUA. MAINTAINED BY: Wi.
BUE		INBR (Wi): 19. GENET: albino. ORIGIN: cross betweeen BUD at F_{18} (now discarded) and BUA at F_{14}. CHARAC: selected for circling now 20% penetrance. Onset 14–20 days. Audition and vestibulation present and apparently normal. Circlers vicious fighters; not good mothers. Very high metabolic activity. MAINTAINED BY: Cu, Wi.
C	cinnamon	INBR (St): 86. GENET: *bb*. *H*-2^{q}. ORIGIN: Strong (*see* C3H/St). CHARAC: moderate mammary tumor incidence. Tendency to bifurcation of seminal vesicles. MAINTAINED BY: Ao, St.
CBA	XXXIX	INBR (Jax): 106. GENET: +. *H*-2^{k}. ORIGIN: (*see* C3H/St). CHARAC: mammary tumor incidence variously reported 1.1–22.2%.[742] Some hepatomas. Long-lived. Dietary supplements needed to maintain reproductive efficiency. Absence of lower third molars in about 18%. Few skeletal variants. Moderately resistant to skin cancer induction by chemical carcinogens. MAINTAINED BY: Br, Cbi, Fa, Gr, H, Jax, Lab, Lhm, Ms, No, Rij, Ss, St.
CBAf	ABC, CBAfC3H	INBR (Pa): 28. GENET: +. ORIGIN: Paterson, from CBA mice received in 1949 from Carr and fostered on C3H. CHARAC: spontaneous mammary cancer. MAINTAINED BY: Pa.
CBA-*a*		INBR (Be): 20+. GENET: *aa*. ORIGIN: Carter outcrossed CBA to *aa* and backcrossed to CBA to N6. B × s matings thereafter. MAINTAINED BY: Be.
CBA-a^t		INBR (H): N5 + F24. GENET: $+a^t \times a^ta^t$. ORIGIN: CBA/Fa. MAINTAINED BY: Fa, H.
CBA-*da*		INBR (Fa): ? + 31. GENET: +*da*, *dada*. ORIGIN: Dark (*da*) arose in CBA/Fa *circa* 1955. CHARAC: like CBA. MAINTAINED BY: Fa.
CBA-*H*-2?		INBR: ? + 16. GENET: +. ORIGIN: from Bonser to Mühlbock, to Rijswijk in 1953. CHARAC: not *H*-2^{k}, but an as yet unknown *H*-2 type. MAINTAINED BY: Rij.
CBA-*p*		INBR (Cam): ? + 16. GENET: +*p* × *pp*. ORIGIN: mutation in 1948 or 1949 in CBA/Ca. MAINTAINED BY: Cam.
CBA-*se*		INBR (Gn): 30. GENET: +*se* × *sese*. ORIGIN: mutation in 1948 or 1949 in CBA/Ca stock. CHARAC: high incidence of hydronephrosis in *sese* animals. MAINTAINED BY: Cam, Gn.
CC57BR	BR	INBR (Mv): 41. GENET: brown. ORIGIN: see CC57W. The same cross produced both CC57BR and CC57W. CHARAC: do not develop spontaneous mammary tumors; 55% tumors after administration of milk agent to newborns. 1% pulmonary adenomas in old mice, 15% after urethan treatment. Probably differs from CC57W in leukemia incidence and *H*-2 allele. MAINTAINED BY: Mv.
CC57W	W	INBR (Mv): 40. GENET: albino. ORIGIN: C57BL and BALB/c mice received in Moscow, Sept., 1943, from Nat. Inst. Health (USA).

Table 27—*Continued*

Name or Symbol	Synonym(s)	Remarks
		The one living BALB/c ♀ was mated to a C57BL ♂; selection and inbreeding led to development of the CC57BR and CC57W mice. CHARAC: do not develop spontaneous mammary tumors; 55% tumors after administration of milk factor to newborn, 15% pulmonary adenomas in old mice, some leukemia and skin cancer appear; 100% pulmonary tumors after urethan treatment. Viability and fertility good. MAINTAINED BY: Mv.
CE	Cd, c^e, ce	INBR (Jax): ? + 25. GENET: $A^wA^wc^ec^e$. ORIGIN: wild mutant trapped in 1920 in Illinois by J. E. Knight. Detlefsen studied genetics of the color type. Inbred by Eaton at least 15 generations, some sent to Woolley prior to 1940. Woolley to Speirs, back to Woolley and Jax in 1948. CHARAC: low incidence of mammary tumors, 33% ovarian tumors in old age, some sarcomas, wide range of tumor types, high adrenal cortical carcinoma incidence after neonatal gonadectomy poor breeders, have relatively few litters but eight to ten per litter MAINTAINED BY: Di, Hu, Jax, Lab, Pi.
CFCW		INBR (R1): 29. GENET: $ccCaCa$. ORIGIN: from Carworth Farms in 1948. MAINTAINED BY: R1.
CFW		INBR (R1): 25. GENET: cc. ORIGIN: Webster to Carworth Farms distributed by them. CHARAC: 20% mammary tumors in breeders 9–10 mo.; susceptible to many viruses. MAINTAINED BY: Hd, Ms, Rl
CHI		INBR (St): 93. GENET: +. $H\text{-}2^k$. ORIGIN: Strong (see C3H/St) CHARAC: similar to C3H/St. MAINTAINED BY: St.
CT		INBR (Ch): 24. GENET: $aabbddpp$. ORIGIN: Chase, developed for color testing. MAINTAINED BY: Ch.
C3H/An		INBR (Wi): 82. GENET: +. ORIGIN: see C3H/St. A litter of 4 ♀ and 2 ♂ sent to Andervont in Oct., 1930. Selected in early generations for high and early mammary tumor incidence. CHARAC: over 90% incidence of mammary tumors in breeders and virgins. Higher mammary tumor and hepatoma incidence and lower pulmonary tumor incidence than St subline. Mammary tumors 94% at 8–10 months almost 100% at 14 months. Hepatomas: ♀♀ over 1 year 10%, ♂♂ over 1 year 27%. Pulmonary tumors in ♀♀ 4%, ♂♂ 8%. MAINTAINED BY: Gl, Pa, Sp, Wi.
C3H/Bi Z		INBR (Cr): 56, (Dec. 1957). GENET: +. ORIGIN: see C3H/St. Strong to Bittner, 1931. CHARAC: mammary tumors 90% in breeders 18.4% develop leukemia when given Gross leukemia virus. MAINTAINED BY: Cbi, Cr, Go, Lhm.
C3H/Crgl/2 C3HNT/Crgl		INBR (Crgl): 9, prev. 20+. GENET: +. ORIGIN: from one female of C3H/Crgl which died at 29 mo. without mammary tumor. CHARAC: about 5% mammary tumors at 20 mo. in breeders. MAINTAINED BY Crgl.
C3H/He		INBR (He): ? + 84. GENET: +. $H\text{-}2^k$. ORIGIN: see C3H/An Andervont to Heston in 1941 after 35 b × s generations by An CHARAC: 97% mammary tumors in breeders 7–8 mo., 100% in virgins 10–11 mo., 85% hepatomas in ♂♂ at 14 mo. Low red blood count 1.2% develop leukemia when given Gross leukemia virus. An and H sublines have predominantly six lumbar vertebrae. MAINTAINED BY Am, Chr, Crgl, De, Di, Gif, H, He, Icrc, Jax, Jb, Kp, Lab, Mas, Ms N, Rl, We.

Table 27—*Continued*

Name or Symbol	Synonym(s)	Remarks
C3H/Ks		INBR (Ks): 28. GENET: +. ORIGIN: Strong, 1920. This stock from C3H mice of unknown pedigree which survived the Bar Harbor fire. CHARAC: resistant to some C3H transplantable tumors. This strain appears to fall into Bittner-Strong group of sublines on basis of vertebral counts, but serological tests place the strain in the Andervont-Heston group. MAINTAINED BY: Jax, Ks.
C3H/St	Z	INBR (St): 114. GENET: +. ORIGIN: Strong, 1920, from a cross of ♀ Bagg albino × ♂ DBA. Strains C, CBA and CHI originated from this cross. CHARAC: mammary cancer 70% at 14 months, and almost 100% in those living to 16 months. St and Bi sublines have predominantly 5 lumbar vertebrae. MAINTAINED BY: Cbi, Fn, Ha, Sp, St, Wi. Others maintaining C3H, subline not given: Ber, Ge, Ka, Pe, S.
C3Hf	C3H$_b$, C3H$_f$, C3HfB	INBR (He): 39 since fostering. GENET: +. ORIGIN: Heston, from litter of C3H born by cesarean section and fostered on strain C57BL, 1945. CHARAC: high susceptibility to MTI. Mammary tumor incidence 2% in virgins, 38% in breeders, 20% in males treated with diethylstilbestrol, none in untreated ♂♂. Heptomas 23% in diethylstilbestrol-treated ♂♂. MAINTAINED BY: Anl, Cbi, De, Ge, Ha, He, Ka, N, Pu, Sp. Another line fostered on C57BL by Fekete, maintained by: Hu.
C3Hff		INBR (Oci): 3 since second fostering. GENET: +. ORIGIN: from a litter of C3H/He$_f$ fostered on a C3H/StHa ♀ in March 1958 at Ontario Cancer Inst., to provide a strain with identical genetic background to C3H/He$_f$ but bearing milk factor. MAINTAINED BY: Oci.
C3HeB/De	C3He	INBR (De): 13. GENET: +. ORIGIN: C3H/He ova transferred to C57BL/6 by De. CHARAC: mammary tumor incidence 4% in virgins, 55% in breeders, 74% in forcebred, and 22% in males treated with diethylstilbestrol. Hepatomas 58% in virgins, 30% in breeders, 38% in force-bred, 55% in males treated with diethylstilbestrol, and 90% in breeding males. MAINTAINED BY: De.
C3HeB/Fe	TC3H	INBR (Hu): 31. GENET: +. ORIGIN: C3H/He ova transferred to C57BL/6 by Fekete in 1949, given to Hummel. CHARAC: mammary tumors 10% in breeders. Low sensitivity to FSH. MAINTAINED BY: Gif, Hu, Jax.
C3H-*H*-1^b	C3H.K	INBR (Sn): G14F13. GENET: same as C3H except *H*-1^b substituted for *H*-1^a.
C3H-*H*-2^b	C3H.SW	INBR (Sn): G14F9.
C3H-*H*-2^p	C3H.NB	INBR (Sn): G16F9.
C3H/Ha-*p*		INBR (Ha): 12 since mutation. GENET: +p × pp. ORIGIN: spontaneous mutation at P-locus recovered in F$_{19}$ of C3H/He$_f$Ha. MAINTAINED BY: Ha.
C3H/He-*Sl*	Stock 800, Steel	INBR (Re): 12 since mutation. GENET: *Slsl*. ORIGIN: mutation in C3H/He at Oak Ridge, sent to Re in 1954. MAINTAINED BY: Re.
C3H/N-*W^j*	Stock 93, *W^j*	INBR (Re): 11 since mutation. GENET: *W^{j}w*. ORIGIN: mutation in C3H/N to Russell, May 1955. MAINTAINED BY: Re.
C3HA		INBR: 40. ORIGIN: C3H ♀ × A ♂, subsequent inbreeding of their descendants. CHARAC: 37% mammary tumors in breeders, 45% in virgins; about 25% of these metastasize to lungs. Liver tumors less than 1%. MAINTAINED BY: Lab. Exp. Oncology, AMN, U.S.S.R., Mv.

Table 27—*Continued*

Name or Symbol	Synonym(s)	Remarks
C57BL	C57 black	INBR (Cbi): 60. GENET: *aa*. ORIGIN: Little, 1921, from heterozygous sib pair, ♀ 57 × ♂ 52, from Miss Lathrop's stock. ♂ 52 mated to littermate ♀ 58 gave rise to strain C58. See also C57BL/6, C57BL/10, C57BT/a, C57BR/cd, and C57L. CHARAC: very low mammary tumor incidence; no MTI. Resistant to mammary and ovarian tumor induction by chemical carcinogens. Sporadic juvenile hair loss. Eye abnormalities in about 10% of animals; incidence varies in different sublines. Large numbers of skeletal variants. Various types of internal tumors. Susceptible to lymphoma induction by X ray. MAINTAINED BY: Am, Ber, Cbi, Crgl, De, Fa, Fn, Fr, G, Ge, Gif, Go, Gr, H, Ha, He, Ka, Kn, Lab, Mr, Mv, Not, Oci, Per, Pi, Pu, Rd, Rij, Rl, S, Sp, St.
C57BL/6	B/6, B6, B	INBR (Jax): 61. GENET: *aa*. ORIGIN: see C57BL. Sublines 6 and 10 separated prior to 1957. CHARAC: differs from C57BL/10 in incidence of eye defects. Has 6% lymphocytic leukemia and 9% reticulum-cell sarcoma. MAINTAINED BY: Anl, Chr, Gif, Jax, Lab, Ly, Mas, Ms, N, Sb, Sfd, Ss, We.
C57BL/10Jax	C57 black subline 10, B/10, B10	INBR (Jax): 66. GENET: *aa*. Histocompatibility-2 allele in C57BL/10, and probably all other sublines unless otherwise indicated, H-2^b. ORIGIN: see C57BL. Sublines 6 and 10 separated prior to 1937. CHARAC: differs from C57BL/6 in incidence of eye defects. Comparison of behavior traits of C57BL/10Sc and C57BL/10Jax showed differences. MAINTAINED BY: Ch, Fo, Gn, Jax, Jb, Rl.
C57BL/10Sc	B/10Sc	INBR (Sc): 69. GENET: *aa*. ORIGIN: same as C57BL/10Jax. Some of original stock from Little to Scott. CHARAC: similar to C57BL/10Jax. On basis of behavior tests using both C57BL/10Sc and C57BL/10Jax differences were apparent in the two groups. MAINTAINED BY: N, Sc, Sn.
C57BL-a^t		INBR (H): ? + 13 since mutation. GENET: $a^t a$ × aa. ORIGIN: spontaneous mutation to a^t in C57BL/H in 6th generation at Harwell. MAINTAINED BY: H.
C57BL-a^t2		INBR (H): ? + 12 since mutation. GENET: $a^t a^t$. ORIGIN: derived from C57BL/H-a^t. MAINTAINED BY: H.
C57BL-b		INBR (Bnd): 20+. GENET: *aabb*. ORIGIN: mutation from B to b in 1956 in Barnard Lab. CHARAC: large viable litters; high fertility. MAINTAINED BY: Bnd.
C57BL/Ks		INBR (Ks): 39. GENET: aaH-$2^d H$-2^d. ORIGIN: derived from non-pedigreed mice bred at Jax, sent to Sloan-Kettering Inst., 1947, and returned to Jax 1948. CHARAC: resistant to C57BL/6 tumors E 0771 and C 1498. MAINTAINED BY: Ks.
C57BL/10-H-1^d	B10.BY	INBR (Sn): G18F5. GENET: same as C57BL/10 except for substitution of H-1^d (or H-1^b) for H-1^c.
C57BL/10-H-2^d	B10.D2	INBR (Sn): G12F3G6F3G4F0. GENET: same as C57BL/10 except for substitution of H-2^d for H-2^b.
C57BL/10-H-$2^d N$	B10B	INBR (N): 28. GENET: *aa*. ORIGIN: Borges to NIH 1953 at F_{12}. CHARAC: resistant to C57BL/10 tumors. MAINTAINED BY: N.
C57BL/10-H-2^f	B10.M	INBR (Sn): G12F13G4F1.
C57BL/10-H-2^q	B10.Y	INBR (Sn): G14F10G4F1.

Table 27—*Continued*

Name or Symbol	Synonym(s)	Remarks
C57BL/ 10-*H*-3^b*A*	B10.LP	INBR (Sn): G18F5. GENET: same as C57BL/10 except for substitution of *H*-3^b*A* for *H*-3^a*a*.
C57BL/ 10-H-3^b	B10.LP-a	INBR (Sn): G18F5.
C57BL/ 10-*lu*	luxoid	INBR (Gn): N24. GENET: *aa* + *lu*. ORIGIN: Green, *lu* obtained by mutation in C3H/HeGn in 1950, continuously backcrossed into C57BL/10JaxGn starting about 1950. CHARAC: about 90% penetrance of *lu* in heterozygotes. MAINTAINED BY: Fo, Gn.
C57BL/ 10-*lx*	luxate	INBR (Gn): N21. GENET: *aa* + *lx*. ORIGIN: Green, *lx* obtained from T. C. Carter, backcrossed continuously into C57BL/10JaxGn starting about 1950. CHARAC: about 70% penetrance of *lx* in heterozygotes. MAINTAINED BY: Fo, Gn.
C57BL-*wr*		INBR (Fa): ? + 31. GENET: +*wr*, *wrwr*. ORIGIN: mutation in C57BL/Fa *circa* 1955. MAINTAINED BY: Fa.
C57BR	C57 brown, BR	INBR (Sp): 47. GENET: *aabb*. ORIGIN: see C57BR/a and C57BR/cd; all existing branches probably belong to one or the other of these sublines. MAINTAINED BY: Rl, Sp.
C57BR/a	C57 brown subline a	INBR (Rr): 100. GENET: *aabb*. *H*-2^k. ORIGIN: Little, from same ♀ 57 and ♂ 52 that gave rise to C57BL, C57BR/cd, and C57L. Black and brown lines were separated in first generation. C57BR/a was separated from other brown lines in ninth generation. CHARAC: low mammary tumor incidence. MAINTAINED BY: Ms, Rr.
C57BR/cd	C57 brown subline cd, BR	INBR (Jax): 101. GENET: *aabb*. *H*-2^k. ORIGIN: Little, from same ♀ 57 and ♂ 52 that gave rise to C57BL, C57BR/a, C57L. Black and brown lines were separated in the first generation. Subline cd was established in generation 13 from a cross of two brown branches, one of which had previously given rise to line C57BR/a. CHARAC: low mammary tumor incidence. Hepatomas 14% in ♂♂. MAINTAINED BY: Gif, Jax, Jb, Lab.
C57BR/a-*AFu*		INBR (Rr): N15. GENET: *AabbFufu*. ORIGIN: C57BR/aJRr. CHARAC: 70% penetrance of *Fu*. MAINTAINED BY: Rr.
C57L	C57 leaden, M, L, LN	INBR (Jax): 77. GENET: *aabblnln*. *H*-2^b. ORIGIN: J. Murray, 1933, mutation in generation 22 of a C57BR subline of which the non-leaden (brown) branch is now extinct. Stock maintained by Cloudman, to Heston, 1938, back to Jax, 1947, at F$_{45}$. CHARAC: low mammary tumor incidence. Hematocrit value extremely high. MAINTAINED BY: Anl, De, He, Jax, Jb, Lab, Mc, Ms, N, Rl, S.
C58		INBR (Jax): 111. GENET: *aa*. *H*-2^k. ORIGIN: MacDowell, 1921, from mating of littermates ♀ 58 × ♂ 52 of Miss Lathrop's stock. ♂ 52 mated to littermate ♀ 57 gave rise to C57BL, C57L, C57BR/a, and C57BR/cd. CHARAC: high leukemia. MAINTAINED BY: Cbi, Ha, Hd, Icrc, Jax, Ms.
C58-*wa*		INBR (Ha): 22 since mutation. GENET: *aa* + *wa*. ORIGIN: spontaneous waved mutation in C58/MdHa in 1952 (do not know which "waved"). MAINTAINED BY: Ha.
DA		INBR (Jax): 48. GENET: *cc*. ORIGIN: Hummel, 1948, 1 ♂ and 3 ♀♀ born to nonpedigreed Swiss ♀ with mammary gland tumor. CHARAC: medium mammary tumor incidence, many partitioned vaginas. MAINTAINED BY: Hu, Jax.
DB		INBR (Sp): ? + 18. GENET: *aabbdd*. ORIGIN: Murray, C57BL ♀ × DBA ♂, backcrossed eight generations to DBA ♂, then inbred. To

Table 27—*Continued*

Name or Symbol	Synonym(s)	Remarks
		Simpson in 1948. (In view of the origin of this strain it should be classed as a subline of DRA.) CHARAC: low mammary tumor. MAINTAINED BY: Sp.
DBA	dba, dilute brown, dbr	INBR (Icrc): ? + 46. GENET: *aabbdd*. ORIGIN: Little, 1909, from mice used in color experiments. CHARAC: high mammary tumors in breeders, variable incidence in virgins. MAINTAINED BY: Ch, Ha, Icrc, Ms, Rl, S, Sp.
DBA/1	dilute brown, dba, dbr, D/1	INBR (Jax): 36. GENET: *aabbdd*. *H-2^q*. ORIGIN: Little, 1909. In 1929–30 some crosses were made between sublines and several new sublines were established. Two of these were 12 (now called 1) and 212 (now called 2). DBA/1 to Fekete, 1936; to Jax market, to Hummel, 1947, to Jax, 1948. CHARAC: resistant to DBA/2 transplantable tumors. Mammary tumors 80% in breeding females. MAINTAINED BY: Chr, Gif, Hu, Jax, Jb, Lab, Ma.
DBA/2	D/2, D2, D, dba subline 2, subline 212, dba-2	INBR (Di): 68. GENET: *aabbdd*. *H-2^d*. ORIGIN: see DBA/1 Jax. Some sent to Mider and to Woolley, 1938; Mider to Heston, Heston to Jax, 1948; Woolley to Dickie, 1949. CHARAC: resistance to DBA/1 transplantable tumors. Mammary tumors 43% in breeding ♀♀. High incidence of nodular hyperplasia of adrenal after neonatal gonadectomy; response accompanied by feminizing stimulation of accessory reproductive organs. Audiogenic seizures 100% at 35d. 5% incidence after 55d. Onset of seizures at 14d. Does not develop spontaneous leukemia, but susceptible to leukemia induction; poor reproduction. MAINTAINED BY: Cbi, Crgl, De, Di, Gif, Hn, Jax, Lab, Ly, Mas, N, Oci, Pi, Sp., We.
DBA/2eB		INBR (De): 7 since transfer. GENET: *aabbdd*. ORIGIN: from fertilized ova of DBA/2 transferred to C57BL. MAINTAINED BY: De.
DBAf	dba$_b$	INBR (Icrc): 31. GENET: *aabbdd*. ORIGIN: sublines of DBA fostered on C57BL or BD to remove milk factor. CHARAC: low mammary tumor. MAINTAINED BY: Icrc, Sp.
DBAf-*D*		INBR (Sp): ? + 13. GENET: prob. *aabbDd*. ORIGIN: spontaneous mutation from *d* to *D* in DBAf/Sp in 1951. CHARAC: low mammary tumor. MAINTAINED BY: Sp.
DBA/1$_f$		INBR (Hu): 20 since fostering. GENET: *aabbdd*. ORIGIN: Hummel, before 1950; DBA/1 fostered on C57BL/6. CHARAC: low mammary tumor incidence, otherwise like DBA/1. MAINTAINED BY: Hu.
DBA/1$_0$		INBR (Hu): 10 since transfer. GENET: *aabbdd*. ORIGIN: Hummel, from DBA/1 ovary transplanted to hybrid DBAf ♀ × C3HeB ♂, 1954. CHARAC: low mammary tumor incidence, otherwise like DBA/1. MAINTAINED BY: Hu.
DBA-*D*	DBAbr intense mutant subline	INBR (Ha): 21. GENET: *aabbDd*. ORIGIN: spontaneous reverse mutation from *d* to *D* in 1952. CHARAC: *H-2^d*. MAINTAINED BY: Ha.
DBA/1-H-2^c Dl.C		INBR (Sn): G14F7.
DBA-*p*		INBR (Sp): ? + 6. GENET: prob. *aabbddpp*. ORIGIN: spontaneous mutation in DBA/MySp in 1957. MAINTAINED BY: Sp.
DE	CDE, cdE	INBR (Di): 31. GENET: *c^{e}c^e*. ORIGIN: Eaton, in 1940, from a cross of CE/Wy × E/Gw and selected for *c^e* phenotype. Eaton to Woolley 1948, Woolley to Dickie 1949, to Jax 1954. CHARAC: high incidence (about 100%) of amyloidosis in spleen, liver, and adrenals; poor breeders; no adrenal cortical carcinomas after neonatal gonadectomy. MAINTAINED BY: Di, Jax.

Table 27—*Continued*

Name or Symbol	Synonym(s)	Remarks
DM		INBR (Ms): 27. GENET: *cc*. ORIGIN: Germany; to Institute of Kitasato Med. Res., Tokyo; to Tokyo Univ., to Hokkaido Univ., 1945; to Misima, 1951. MAINTAINED BY: Ms.
D103		INBR (Ms): 27. GENET: *cc*. ORIGIN: from strain DM. CHARAC: high transplantability to MY mouse sarcoma. MAINTAINED BY: Ms.
E		INBR (Gw): 58. GENET: *cc*. ORIGIN: Gowen, from dealer stock in Pennsylvania. CHARAC: medium resistance to *Salmonella typhimurium*. Highly nervous. MAINTAINED BY: Gw.
F		INBR (St): 83. GENET: $aabbc^{ch}c^{ch}ddss$ (white face). *H-2ⁿ*. ORIGIN: Strong, 1926, from a group of unpedigreed Bussey Inst. mice. CHARAC: high leukemia in older animals. MAINTAINED BY: St.
FAKI		INBR: 35 (Nov. '57). GENET: albino. ORIGIN: Furth, 1946. CHARAC: 80% spontaneous leukemia in both sexes. Av. litter size, 4; range, 2–6. (The origin and characteristics of this strain suggest that it is a subline of AKR. If this is the case, the designation FAKI does not conform to standard usage.) MAINTAINED BY: Pollards Wood.
FU		INBR (Rl): 20. GENET: *ccFuFu*. ORIGIN: Russell, from BC2 and F2 of *Fu* on CFW provided by Waelsch in 1947 or 1948. After one or a few backcrosses to CFW, stock mated b × s. MAINTAINED BY: Rl.
GFF		INBR (G): 40 (Nov. '57). GENET: *aabbpp*. ORIGIN: from Agric. Res. Council Field Station, Compton, in March, 1942. CHARAC: can carry *Salmonella* without showing signs of it. Av. litter size, 8.5; age at sexual maturity, less than 10 wks; age at mating, 10 wks. MAINTAINED BY: G.
GM		INBR (Rr): 30. GENET: $a^{t}a^{t}bb + c + ln$. ORIGIN: derived from cross of Large Goodale and Large MacArthur. CHARAC: large body size, become obese. MAINTAINED BY: Rr.
H	High	INBR (St): 43. ORIGIN: Strong, from BRpb and selected for maximal white dorsal spot. CHARAC: low tumor. MAINTAINED BY: St.
HR	hr	INBR (Rl): 25. GENET: $Cp/c^{ch}phrhr$. ORIGIN: Crew, to Carnochan, to Heston, 1947; to Russell, 1948. MAINTAINED BY: Rl.
HR/De	hr	INBR (De): 35. GENET: *hrhr*. ORIGIN: inbred from *hr* stock received from Carnochan in 1947. CHARAC: develops skin papillomas and hemangioendotheliomas. (Strains HR and HR/De seem to have been separated before there was much inbreeding. They may therefore show strain rather than substrain differences.) MAINTAINED BY: De.
I		INBR (St): 68. GENET: *aabbddppss*. *H-2ᵗ*. ORIGIN: Strong, 1926, from a group of unpedigreed mice. CHARAC: resistant to tumors chemically induced. MAINTAINED BY: Ao, Fn, Sp, St.
IF		INBR (Br): 67. GENET: $a^{t}a^{t}$. ORIGIN: Bonser, selection of mice which developed papillomas early after treatment with carcinogens. CHARAC: mammary tissue very sensitive to carcinogenic agents, either local or distant. Very low incidence of spontaneous mammary cancer. MAINTAINED BY: Ber, Br, Rd.
IPBR	IpBr	INBR (St): 25. GENET: *bbddss*. ORIGIN: Strong, PBR × I, progeny inbred. CHARAC: relatively resistant to chemically induced tumors. MAINTAINED BY: St.
IB		INBR (Di): 31. GENET: $aabbc^{e}c^{e}$. ORIGIN: Dickie, from cross of DBA/2Wy ♀ × CE/Wy ♂, b × s thereafter; color was only selective factor. CHARAC: low incidence of mammary tumors. MAINTAINED BY: Di.

Table 27—*Continued*

Name or Symbol	Synonyms(s)	Remarks
JK	short ear	INBR (St): 68. GENET: *aabbppsese*. *H-2ʲ*. ORIGIN: Strong, 1927, from a cross between J and K. CHARAC: low tumor, resistant to chemical induction of tumors. MAINTAINED BY: Ao, St.
JU		INBR (Fa): 25. GENET: *aacc*. ORIGIN: Falconer, from crosses between Goodale's and MacArthur's large strains, Bateman's high lactation strain, and stocks of various mutants with about 50% of C57BL/Fa ancestry. MAINTAINED BY: Fa.
L		INBR (St): 75. GENET: $A^w A^w c^{ch} c^{ch}$. ORIGIN: Strong, about 1926 from a group of unpedigreed mice. CHARAC: low mammary gland tumors, some lymphoblastomas, retinal opacity. MAINTAINED BY: St.
LCH		INBR (We): 31. GENET: *lnln*. ORIGIN: from MacArthur stock (outbred) obtained from Butler, 1949. Selected for high total leucocyte count, with b × s mating. CHARAC: leucocyte count 11,000/mm³. Sex ratio 50% ♂♂. Mean litter size at weaning, 7.4. Leucocyte count rose from 8800 in F_1 to 18,000 in F_{13}, and dropped steadily to present level. MAINTAINED BY: We.
LCL		INBR (We): 31. GENET: *bblnln*. ORIGIN: see LCH. Selected for low count. CHARAC: leukocyte count 3,700/mm³. Sex ratio, 48% ♂♂, mean litter size at weaning, 6.6. No significant change in leucocyte count since F_{11}. MAINTAINED BY: We.
LG		INBR (Rr): 18. GENET: $a^t a^t cc$. ORIGIN: Goodale, to Runner in 1948 (no record of previous inbreeding). CHARAC: large body size, lymphatic leukemia, low incidence of otocephaly. MAINTAINED BY: Rr.
LGW	L	INBR (Gw): 59. GENET: *aasisi*. ORIGIN: "silver" strain from English fancier, to Dunn, to MacDowell, to Iowa State. Inbred at Iowa State. CHARAC: susceptible to *Salmonella typhimurium*. Unreliable breeder. MAINTAINED BY: Gw.
LP	Line 11, Line 11 Pied, AgW(B)	INBR (Jax): 57. GENET: $A^w A^w ss$. *H-2ᵇ*. ORIGIN: Dunn, to Scott, to Dickie, 1947; to Jax, 1949; at F_{33}. CHARAC: some mammary tumors. MAINTAINED BY: Jax, Jb.
LP-*H-2ʳ*	LP.RIII	INBR (Sn): G12F13.
L(P)	L(Paris), 50	INBR (Icrc): 37. GENET: albino. ORIGIN: albino stock from a hybrid (XIX × XLII)F_1 ♀, started by Dobrovolakaia-Zavadskaia in 1940, designated L(C). Breeding stock at F_{17} sent to Tata Institute in Bombay in 1948, to Indian Cancer Research Centre in 1952, designated L(P). Other stock to Busto Arsizio in 1948. CHARAC: no spontaneous mammary tumors. High incidence following methylcholanthrene treatment. No evidence of presence of MTI. MAINTAINED BY: Bax, Icrc.
LT		INBR (Ch): 37. GENET: $B^{lt} B^{lt}$. ORIGIN: dominant allele of *brown* from MacDowell in 1950. MAINTAINED BY: Ch.
MA	Marsh albino, Marsh. Old Buffalo and New Buffalo were branches of strain MA.	INBR (Jax): 47. GENET: *cc*. *H-2ᵏ*. ORIGIN: Marsh's strain 3. Started from pair of mice from Lathrop-Loeb colony; 32 generations of cousin matings by Marsh, to W. S. Murray who started b × s matings, to Warner, to Jax in 1947 at $F_{20}+$. CHARAC: high mammary tumor. MAINTAINED BY: Hu, Jax, Ms, S, Sp.
MA/MY	ND	INBR (Jax): 47. GENET: *cc*. *H-2ᵏ*. ORIGIN: Murray and Warner separated this line from MA after 37 b × s generations. Warner to Jax in 1948 at F_{24}. CHARAC: low incidence of mammary tumor. MAINTAINED BY: Hu, Jax, Sp.

Table 27—*Continued*

Name or Symbol	Synonym(s)	Remarks
MAf		INBR (Sp): ? + 21. GENET: *cc*. ORIGIN: Murray, from MA born by cesarean section and fostered on BD. To Simpson in 1948. CHARAC: high incidence of reticular tissue neoplasm in old age; high pulmonary tumor. MAINTAINED BY: Sp.
MABA		INBR (Sp): ? + 16. GENET: *cc*. ORIGIN: MA ♀ × A ♂ (Murray). To Simpson in 1948. CHARAC: low mammary tumor. MAINTAINED BY: Sp.
N		INBR (St): 67. GENET: *aabbddss*. ORIGIN: Strong, about 1926, from a group of unpedigreed mice. Ancestral to PBR strain. CHARAC: low tumor incidence, resistant to chemical induction of tumors. MAINTAINED BY: Ao, St.
NB		INBR (Mc): 64. GENET: *aabbc^{ch}p/c^{ch}pdse/dse*. ORIGIN: E. L. Green. MAINTAINED BY: Mc, Rl.
NH		INBR (Di): ? 10. GENET: *ddppss*. ORIGIN: Strong, from cross involving strains CBA, N, and JK; to Kirschbaum; to Dickie in 1956. CHARAC: low incidence of tumors, but 30–42% pulmonary tumors in very old mice. Reproductive senility about 8–9 months of age followed by appearance of nodular hyperplasia of adrenals. Animals very sensitive to light, high incidence of blindness. Poor breeders and small litters. MAINTAINED BY: Di, Ms.
NLC		INBR (Rd): 26. GENET: albino. ORIGIN: a pair of mice of uncertain origin in 1949 at Foundation Curie, b × s since then. CHARAC: mammary tumors in 50% of force-bred ♀♀, 18% of virgins; have MTI. Females transmit a nonchromosomal resistance factor against leukemia in NLC × AKR hybrids. Sensitive to leukemogenic action of methylcholanthrene. Vigorous, good breeders, and good mothers. MAINTAINED BY: Rd.
NS		INBR (Fr): 27. GENET: *a^{t}a^t + cN +*. ORIGIN: from heterogeneous *Nn* stock crossed to A from Jax (March, 1945), and inbred with forced heterozygosis for *Nn* and *Cc*. CHARAC: tendency to prolapsus uteri in later life. MAINTAINED BY: Fr.
NZB		INBR (Bl): 42. ORIGIN: from NZO at F_3. CHARAC: hemolytic anemia. MAINTAINED BY: Bl.
NZC		INBR (Bl): 45. CHARAC: about 1% mammary tumors; occasionally cystic kidney; short tail. MAINTAINED BY: Bl.
NZO		INBR (Bl): 43. CHARAC: obese. Mammary tumor less than 1%; high incidence of pulmonary adenoma and of benign tumors of the small intestine. MAINTAINED BY: Bl.
NZY		INBR (Bl): 37. CHARAC: mammary tumors, 60% in breeders, 28% in virgins. Pituitary enlargement in over 90% ♀♀ 12 mo. and older. Since F_{17} about 10% megacolon. MAINTAINED BY: Bl.
NZYf		INBR (Bl): 11 since fostering. ORIGIN: fostered on NZC at F_{26}. CHARAC: marked reduction in mammary tumors. Pituitary tumor incidence not altered. MAINTAINED BY: Bl.
O2O		INBR: 86+. ORIGIN: Korteweg. MAINTAINED BY: not reported.
P		INBR (Jax): 69. GENET: *aabbdse/dsepprdrd*. *H-2^p*. ORIGIN: Snell, extracted following outcross of similar stock developed by Gates. Snell to McNutt at F_{35}, back to Jax in 1948 at F_{39}. MAINTAINED BY: Jax, Jb.

Table 27—*Continued*

Name or Symbol	Synonym(s)	Remarks
PBR	pBr	INBR (St): 56. GENET: *aabbpp*. ORIGIN: Strong, from methyl-cholanthrene-treated NH mice. CHARAC: mammary tumors in breeders. MAINTAINED BY: St.
PHH	pHH, high blood-*p*H	INBR (We): 28. GENET: *aalnln*. ORIGIN: from MacArthur outbred stock, obtained from Butler, 1949. Selected for high blood *p*H, with b × s mating. CHARAC: blood *p*H 7.47; sex ratio 55% ♂♂. Sex ratio now 52% ♂♂. Mean litter size at weaning 6.5. Over-all fertility low. MAINTAINED BY: We.
PHL	pHL, low blood-*p*H	INBR (We): 24. GENET: *aabblnln/a^ta^tbblnln* (leaden brown, and leaden brown-and-tan families). ORIGIN: same as PHH, selected for low blood *p*H. CHARAC: blood *p*H, 7.42; sex ratio 45% ♂♂, now 42% ♂♂. Mean litter size 5.7. Over-all fertility high. MAINTAINED BY: We.
PL	PL(B), Princeton leukemia, LII	INBR (Ly): 43. GENET: *cc*. ORIGIN: from noninbred "Princeton" strain started in 1922 from 200 mice purchased from dealer. Inbred by Lynch, giving rise to a high leukemia line B (=PL) and a second line A (=PLA) with lower incidence. CHARAC: few mammary tumors, 80–90% leukemia. MAINTAINED BY: Jax, Ly, Sp.
PLA		INBR (Ly): 43. GENET: *cc*. ORIGIN: see PL. MAINTAINED BY: Ly.
RF	RFM, Rf, Rfa	INBR (Jax): ? + 18. GENET: *cc*. *H-2^k*. ORIGIN: Furth, 1928, from unknown stock at Rockefeller Institute. He transferred stock to Oak Ridge; to Jax at Oak Ridge's F₁₄. CHARAC: low incidence of leukemia, higher after radiation. MAINTAINED BY: Gif, Jax, Jb.
RI		INBR (Gw): 48. GENET: *aabbcc*. ORIGIN: Gowen, from Webster's resistant stock in 1936. CHARAC: quite resistant to *Salmonella typhimurium*; unreliable to poor breeders. MAINTAINED BY: Gw.
RIET		INBR: 31+. ORIGIN: Korteweg. MAINTAINED BY: not reported.
RUS		INBR (Rl): 31. GENET: *a^ta^truruss*. ORIGIN: from Holman Jan. 1948 (F₄ from Dunn's BT ruby × Holman's black piebald, shaker-1). MAINTAINED BY: Rl.
RIII	Paris, R3	INBR (Crgl): ? + 25. GENET: *cc*. ORIGIN: Dobrovolskaia-Zavadskaia, Paris, 1928. CHARAC: animals have milk agent; mammary tumor in 84% of breeders, 40% of virgins. Nonchromosomal resistance factor against leukoses. Susceptible to pulmonary infection. Good reproducers but bad mothers (cannibalism frequent). MAINTAINED BY: Crgl, Mr, Per, Rd, Rho.
RIII/Jax	Radium Institute (Paris) line III	INBR (Jax): 31. GENET: *cc*. ORIGIN: Dobrovolskaia-Zavadskaia. Sent to Eisen, Eisen to Jax 1948. CHARAC: fairly high incidence of mammary tumors. Resistant to RIII/Wy transplantable tumors. Susceptible to skin ulcerations. MAINTAINED BY: Gif, Jax.
RIII/Wy	R3	INBR (Jax): 48. GENET: *cc*. *H-2^r*. ORIGIN: Dobrovolskaia-Zavadskaia, to Curtin and Dunning, to Woolley before 1947. One pair survived Bar Harbor fire; to Snell in 1950, to Jax in 1957 at F42. CHARAC: fairly high mammary tumor incidence, resistant to skin ulcerations found in RIII/Jax. RIII/Wy transplantable tumors will not grow in RIII/Jax. MAINTAINED BY: Jax.
RIIIfB	RIIIX	INBR (Pu): 58. GENET: *cc*. ORIGIN: RIII fostered on C57BL. CHARAC: does not have milk agent, but very sensitive to it. Mammary tumors in breeders 3%. Fertile but poor mothers. MAINTAINED BY: Pu, Rd.
RIII-*ro*		INBR (Fa): ? + 29. GENET: *cc + ro*. CHARAC: *ro* segregates. MAINTAINED BY: Fa.

Table 27—*Continued*

Name or Symbol	Synonym(s)	Remarks
S		INBR (Gw): 65. GENET: *cc*. ORIGIN: Schott, obtained from dealer (Schwing) in 1927. Subsequently selected for resistance to *Salmonella typhimurium* and inbred by Gowen. CHARAC: resistant to *S. typhimurium* and to radiation. MAINTAINED BY: Gw.
SEA	Ab, Sese-Ab	INBR (Gn): 61. GENET: *bbd* + /*dse*. ORIGIN: Green, from a cross of BALB/c × P. MAINTAINED BY: Gn, Rl.
SEAC		INBR (Gn): 26. GENET: $aabbc^{ch}c^{ch}d$ +/+*se*. ORIGIN: Green, from a cross of SEA × SEC/2. MAINTAINED BY: Gn.
SEAC/c		INBR (Gn): N12 of *se* (Carter) to SEAC. GENET: $aabbc^{ch}c^{ch}d$ +/ +*se*. ORIGIN: Green, by backcross of Carter's *se* from CBA/se into SEAC. MAINTAINED BY: Gn.
SEC	C, Sese-C	INBR (Rl): ? + 27. GENET: $aabbc^{ch}c^{ch}$ + +/*dse*. ORIGIN: E. and M. Green, from cross of NB × BALB/c. Gave rise by crossover at F_{18} to SEC/1 — +*se* and SEC/2 — +*d*. MAINTAINED BY: Rl.
SEC/1	Sese-Cl.	INBR (Gn): 56. GENET: $aabbc^{ch}c^{ch}$ + *se*. ORIGIN: Green. CHARAC: multiple pulmonary cysts, especially in *sese*. MAINTAINED BY: Gn.
SEC/1-*Se*		INBR (Re): 60. GENET: $aabbc^{ch}c^{ch}SeSe$. ORIGIN: E. S. Russell, from SEC/1. MAINTAINED BY: Re.
SEC/2	Sese-C2	INBR (Gn): 55. GENET: $aabbc^{ch}c^{ch}$ + *se*/*dse*. ORIGIN: Green. MAINTAINED BY: Gn.
SHC/a		INBR (Cu): 20. GENET: albino. ORIGIN: outcross of BUD/Wi to CFW at Merck Institute; inbred with continuous selection for circus movement and vigor for 13 generations at Merck, 7 more at Seton Hall College of Medicine. CHARAC: 30% penetrance of circus movement, good fertility and maternal care. MAINTAINED BY: Cu.
SHC/b		INBR (Cu): 20. GENET: albino. ORIGIN: separated from SHC/a at its F_{13}. CHARAC: same as SHC/a. MAINTAINED BY: Cu.
SL	S/L	INBR (Ms): 40. GENET: *cc*. ORIGIN: K. Tutikawa, Nat. Inst. of Genetics, Misima; derived from one of several lines of SMA (q.v.) at F_{10}, selected for high leukemia; to H. Baba, Kyushu University, 1954, at F_{16}. CHARAC: low mammary carcinoma; leukemia 54.5% in ♂♂, 10% in ♀♀; lymphatic leukemia, myeloid leukemia, reticulosarcoma, lymphatic leukemia with a conspicuous myeloid reaction were found in this strain. MAINTAINED BY: Ms, Kyushu Univ., Kyoto Univ.
SM		INBR (Jax): 30. GENET: $A^{w}a$. ORIGIN: MacArthur to Runner in 1948. CHARAC: small body size, renal amyloidosis, leukopenia. MAINTAINED BY: Jax, Rr.
SMA	S, SM	INBR (Ms): 39 (26 at Ms). GENET: *cc*. ORIGIN: S. Makino to Hokkaido Univ., 1944; to Misima, 1951, at F_{15}. CHARAC: good breeder, low incidence of mammary and pulmonary tumor in old age, 15–20% kinked tail variants in each generation. MAINTAINED BY: Ms.
ST	Street	INBR (Jax): 56. GENET: *aabbcc*. *H-2^k*. ORIGIN: Engelbreth-Holm, to Heston at F_{23}, to Jax at F_{25}. CHARAC: 1–2% leukemia including some plasma-cell leukemias. Other tumors, principally pulmonary adenomas and mammary carcinomas, about 3%. MAINTAINED BY: Jax.
STR		INBR (N): 52. GENET: *aabb*. ORIGIN: Strong, to NIH in 1951 at F^{28}. MAINTAINED BY: N.

Table 27—*Continued*

Name or Symbol	Synonym(s)	Remarks
STR/1		INBR (N): 50. GENET: *aabbss*. ORIGIN: piebald branch of STR originating at NIH in 1951 at F_{29}. MAINTAINED BY: N.
SWR	Swiss, Swiss-R, Swiss-8	INBR (Jax): 62. GENET: *cc*. ORIGIN: a strain of Swiss mice from A. de Coulon of Lausanne, Switzerland, inbred by Lynch. CHARAC: 44% pulmonary tumors, 19% mammary tumors in breeding ♀♀. MAINTAINED·BY: De, Gif, Jax, Jb, Ly, Ms.
T		INBR (Sp): 20. GENET: *cc*. ORIGIN: noninbred Tumblebrook stock in 1948, inbred thereafter. Mice were submitted to a single cutaneous application of methylcholanthrene in benzene for seven consecutive generations, F_6 through F_{12}. CHARAC: high incidence of mammary tumors. MAINTAINED BY: Sp.
WB		INBR (Re): GENET: *aaWw*. ORIGIN: heterozygous mice carrying *Ww* were sent by S. Waelsch and S. J. Holman to Russell in 1948. Following several intercrosses and outcrosses to C57BL/6, b × s inbreeding with forced heterozygosis of *Ww* began in 1949. Separate lines were maintained which gave rise to four distinct strains, WB, WC, WH, and WK. CHARAC: homozygotes (WW) are anemic, die at approx. 11 days. MAINTAINED BY: Re.
WC		INBR (Re): 29. GENET: *aaWw*. ORIGIN: same as WB. CHARAC: anemics die at approx. 14 days. MAINTAINED BY: Re.
WH		INBR (Re): 29. GENET: *aaWw*. *H-2^d*. ORIGIN: same as WB. CHARAC: anemics die at approx. 5 days. MAINTAINED BY: Re.
WK		INBR (Re): 27. GENET: *aaWw*. ORIGIN: same as WB. CHARAC: anemics die at approx. 10 days. MAINTAINED BY: Re.
WLL	White Label, Kreyberg, White Label Leeds	INBR (Ge): 53. GENET: *cc*. ORIGIN: Kreyberg, to Bonser at Leeds in 1934. CHARAC: 20–33% mammary tumors in breeders. MAINTAINED BY: Br, Ge.
Y		INBR (Wi): 27. GENET: A^y. ORIGIN: Wilson, from noninbred stock called "a group of mixed ancestry carrying the yellow gene" received from Danforth, Stanford, in 1948. MAINTAINED BY: Wi.
YBR		INBR (He): 70. GENET: $A^y abb$. *H-2$^{d'}$*. ORIGIN: Little, to Andervont, 1936; to Heston, 1946; to Wilson, 1948; to Hauschka, 1951. MAINTAINED BY: Ha, He, Ly, Wi.
YS		INBR (Ch): 40. GENET: $A^y ass$. ORIGIN: original Y from Jax. Spotting introduced, then crossed with C57BL in 1948, then inbred. MAINTAINED BY: Ch.
Z		INBR (Gw): 56. GENET: *cc*. ORIGIN: Gowen, from Swiss mice of the Yellow Fever Lab. of the Rockefeller Inst. CHARAC: medium susceptibility to typhoid. Medium susceptibility to radiation. Capable of having many litters, but raises few. MAINTAINED BY: Gw.
ZBR	Zavadskaia Brachy	INBR (Co): 21. GENET: *ssT+*. ORIGIN: from Dobrovolskaia-Zavadskaia in 1930; outcrossed in succession to several Columbia stocks. MAINTAINED BY: Co.
IV/B		GENET: albino. ORIGIN: from N. Dobrovolskaia-Zavadskaia to Centro per lo studio e la cura dei tumori, Busto Arsizio, in April, 1948. CHARAC: mammary tumor incidence 28%, early breeding and pregnancy increase incidence. Susceptible to multiple tumors. MAINTAINED BY: Baz.
XVII	17	INBR (Rd): 28. GENET: *cc*. ORIGIN: Institut du Radium, 1930. Rigorous inbreeding since 1951. CHARAC: mammary tumors 1–5%.

Table 27—*Continued*

Name or Symbol	Synonym(s)	Remarks
		Sensitive to RIII milk agent (55% tumor incidence after injection of newborns). Spontaneous leukemia less than 2%. Pulmonary tumors 15% in ♀♀ 12–24 mo. Sensitive to pulmonary tumor production by urethan, methylcholanthrene, or X rays. Relative sensitive to Gross tumor agent. Good breeders and good mothers. MAINTAINED BY: Baz, Rd.
101	101BAg	INBR (H): 49 + 17. GENET: A^wA^w. ORIGIN: Dunn. CHARAC: low mammary tumor, slow to breed, fertility moderate. High susceptibility to skin and pulmonary tumor induction by chemical carcinogens, low leukemia, low mammary tumor, high incidence of papillonephritis. MAINTAINED BY: H, Lhm, Rl.
129		INBR (Rr): 33. GENET: $A^wA^{wc^{ch}}p/cp$. *H-2^b*. ORIGIN: Dunn. CHARAC: useful for ovary transplant and ova transfer studies; 1% incidence of testicular teratomas. Low mammary tumor incidence, not susceptible to MTI. High incidence of venous congestion in adrenals and uterus. Highly sensitive to estrogen at all ages. Ultimate source of muscular dystrophy animals. MAINTAINED BY: Ha, Jax, Jb, Lab, Pi, Rl, Rr, Sv.
129-*Fu*		INBR (Rr): N13. GENET: $A^wA^{wc^{ch}}p/cpFu$ +. ORIGIN: *Fu* introduced in 129/Rr. CHARAC: 48–50% penetrance of *Fu*. MAINTAINED BY: Rr.
129/Sv-*CP*		INBR (Sv): N11 + F6. GENET: A^wA^w. ORIGIN: outcross of 129/RrSv to DBA with backcross to 129 to N11, b × s since then. CHARAC: useful for ovarian transplants. MAINTAINED BY: Sv.
129/Sv-*aCP*		INBR (Sv): N11 + F7. GENET: *aa*. ORIGIN: spontaneous mutation from A^w to *a* detected in F_2 of 129/Sv-*Cp*. CHARAC: useful for ovarian transplants. MAINTAINED BY: Sv.

LIST OF SYMBOLS FOR DESIGNATING SUBSTRAINS OF MICE

A   Antoni van Leeuwenhoekhuis, Amsterdam C, Sarphatistraat 108, Netherlands (Dr. O. Mühlbock).

Am   Dr. J. L. Ambrus, Roswell Park Memorial Institute, Buffalo 3, N.Y.

Ao   Dr. D. B. Amos, Roswell Park Memorial Institute, Buffalo 3, N.Y.

An   Dr. H. B. Andervont, National Cancer Institute, Bethesda 14, Md.

Anl   Argonne National Laboratory, Division of Biological and Medical Research, Box 299, Lemont, Ill. (Dr. Robert J. Flynn).

Ba   Instituto de Medicina Experimental, Laboratorio de Genetica, Avda. San Martín 5481, Buenos Aires, Argentina.

Baz   Centro per lo studio e la cura dei tumori, Busto Arsizio, Italy (Dr. G. Ceriotti).

Bcr   University of Birmingham, Cancer Research Laboratories, The Medical School, Birmingham 15, England (Dr. June Marchant).

Be   Dr. R. A. Beatty, Institute of Animal Genetics, West Mains Road, Edinburgh 9, Scotland.

Bi   Dr. J. J. Bittner, University of Minnesota Medical School, Minneapolis 14, Minn. (deceased).

Bl   Drs. M. & F. Bielschowsky, Hugh Adams Cancer Research Department, University of Otago Medical School, Great King Street, Dunedin C.1., New Zealand.

Bn   Dr. S. E. Bernstein, Roscoe B. Jackson Memorial Laboratory, Bar Harbor, Maine.

Bnd   Barnard College, Department of Zoology, Columbia University, New York 27, N.Y. (Dr. Reba M. Goodman).

Bo   Biochemical Department of Cancer Institute, Zlutý kopec 7. Brno, Czechoslovakia.

Br   Dr. G. M. Bonser, Department of Experimental Pathology and Cancer

Research, School of Medicine, Leeds 2, England.

Bs Dr. P. R. F. Borges, Tufts Cancer Control Unit, Tufts College Medical School, 30 Bennet Street, Boston 11, Mass.

Bt Dr. N. Bateman, Field Laboratory, Animal Breeding Research Organization, Dryden Mains, Roslin, Midlothian, Scotland.

Bu Dr. W. J. Burdette, Laboratory of Clinical Biology, Department of Surgery, University of Utah, College of Medicine, Salt Lake City, Utah.

C or University of Cambridge, Department
Cam of Genetics, 44 Storey's Way, Cambridge, England (Dr. M. E. Wallace).

Ca Dr. T. C. Carter, c/o Western Chicken Ltd., London Road, Devizes, Wilts, England.

Cbi Chester Beatty Research Institute, Institute of Cancer Research, Fulham Road, London, S.W.3, England (Dr. P. C. Koller).

Cg Mrs. A. Cohen, Experimental Oncology Laboratory, Radiation Therapy Department, Johannesburg General Hospital, Johannesburg, South Africa.

Ch Dr. H. B. Chase, Department of Biology, Brown University, Providence, R.I.

Chr Children's Hospital Research Foundation, Elland Ave. and Bethesda, Cincinnati 29, Ohio (Dr. Josef Warkany).

Ciu Unidad de Investigaciones Cancerologicas, Avenida Cuauhtemoc 240, Ciudad de México, México.

Cl Mrs. Ruth Clayton, Institute of Animal Genetics, West Mains Road, Edinburgh 9, Scotland.

Co Columbia University, Department of Zoology, New York 27, N.Y. (Dr. L. C. Dunn).

Cr Professor J. Craigie, Imperial Cancer Research Fund, Burtonhole Lane, Mill Hill, London, N.W.7, England.

Crgl Cancer Research Genetics Laboratory, University of California, Berkeley 4, California (Dr. Earl B. Barnawell).

Cu Dr. R. L. Curtis, Seton Hall College of Medicine, Department of Anatomy, Jersey City 4, N.J.

De Dr. Margaret Deringer, National Cancer Institute, Bethesda 14, Md.

Di Dr. M. M. Dickie, Roscoe B. Jackson Memorial Laboratory, Bar Harbor, Maine.

Dm Dr. L. Dmochowski, M. D. Anderson Hospital and Tumor Institute, Virology and Electron Microscopy, Houston 25, Texas.

Eh Dr. J. Engelbreth-Holm, Universitetets patologiskanatomiske Institut, Frederik V's Vej 11, København, Denmark.

Fa Dr. D. S. Falconer, Institute of Animal Genetics, West Mains Road, Edinburgh 9, Scotland.

Fe Dr. Elizabeth Fekete, Roscoe B. Jackson Memorial Laboratory, Bar Harbor, Maine (retired).

Fo Dr. P. Forsthoefel, University of Detroit, Biology Department, McNichols Road at Livernois, Detroit 21, Mich.

Fr Dr. F. Clarke Fraser, Department of Genetics, McGill University, Montreal, Canada.

Fu Dr. J. Furth, Roswell Park Memorial Institute, Buffalo 3, N.Y.

G Glaxo Laboratories, Ltd., Greenford, Middlesex, England.

Ge Dr. H. N. Green, Department of Experimental Pathology and Cancer Research, School of Medicine, Leeds 2, England.

Gi Dr. A. B. Griffen, Roscoe B. Jackson Memorial Laboratory, Bar Harbor, Maine.

Gif Centre de Sélection des Animaux de Laboratoire, 5 rue Gustave Vatonne, Gif-sur-Yvette (S. & O.), France (Dr. M. A. Sabourdy).

Gl Dr. A. Glucksmann, Strangeways Research Laboratory, Wort's Causeway, Cambridge, England.

Glw Dr. Salome G. Waelsch, Albert Einstein College of Medicine, Department of Anatomy, Eastchester Road and Morris Park, New York 61, N.Y.

Gn Drs. M. C. and E. L. Green, Roscoe B. Jackson Memorial Laboratory, Bar Harbor, Maine.

Go Dr. Peter A. Gorer, Guy's Hospital, Department of Pathology, London S.E. 1, England.

Gr Dr. H. Grüneberg, University College London, Department of Zoology, Gower Street, W.C.1, England.

Gs Dr. Ludwik Gross, Cancer Research Laboratory, V. A. Hospital, 130 W. Kingsbridge Road, Bronx 63, N.Y.

Gw Dr. John W. Gowen, Iowa State College, Department of Genetics, Ames, Iowa.

H Radiobiological Research Unit, Harwell, Didcot, Berks., England (Dr. M. F. Lyon).

Ha Dr. T. S. Hauschka, Roswell Park Memorial Institute, Buffalo 3, N.Y.

Hd Mr. L. B. Hardy, Roswell Park Memorial Institute, Biological Station, Springville, N.Y.

He Dr. W. E. Heston, National Cancer Institute, Bethesda 14, Md.

Hg Professor Dr. P. Hertwig, Biologisches Institut, Universitätsplatz 7, Halle (Saale), Germany.

Hn Dr. P. J. Harman, Seton Hall College of Medicine, Department of Anatomy, Jersey City 4, N.J.

Ho Prof. W. F. Hollander, Department of Genetics, Iowa State College, Ames, Iowa.

How Dr. Alma Howard, Research Unit in Radiobiology, Mount Vernon Hospital, Northwood, Middlesex, England.

Hr Dr. G. Hoecker, Catedra de Biologia, Av. Zanartu 1042, Santiago de Chile.

Hu Dr. Katharine P. Hummel, Roscoe B. Jackson Memorial Laboratory, Bar Harbor, Maine.

Icr Institute for Cancer Research, 7701 Burholme Avenue, Fox Chase, Philadelphia, Pa.

Icrc Indian Cancer Research Centre, Parel, Bombay 12, India (Dr. B. K. Batra).

J or Jax Expansion stocks of the Roscoe B. Jackson Memorial Laboratory, Bar Harbor, Maine (under the supervision of M. M. Dickie and E. S. Russell).

Jb Dr. Jan H. Bruell, Psychology Department, Western Reserve University, Cleveland 6, Ohio.

Ka Dr. H. S. Kaplan, Department of Radiology, Stanford University School of Medicine, Stanford, Calif.

Ki Dr. A. Kirschbaum (deceased).

Kl Drs. G. and E. Klein, Department of Cell Research, Karolinska Institutet, Stockholm 60, Sweden.

Kn Prof. Dr. Fr. Kroning, Medizinische Forschungsanstalt, Pharmakologische Abteilung, Bunsenstrasse 10, Gottingen, Germany.

Kp Dr. Hilary Koprowski, The Wistar Institute, 36th at Spruce, Philadelphia, Pa.

Kr Prof. L. Kreyberg, Universitetets Institutt for Patologi, Rikshospitalet, Oslo, Norway.

Ks Dr. N. Kaliss, Roscoe B. Jackson Memorial Laboratory, Bar Harbor, Maine.

L Eli Lilly and Co., Indianapolis 6, Indiana.

La Dr. A. Lacassagne, Institut du Radium, rue d'Ulm, Paris (retired).

Lab Laboratory Animals Centre, M.R.C. Laboratories, Woodmansterne Road, Carshalton, Surrey, England.

Lhm The London Hospital Medical College, Department of Cancer Research, Ashfield Street, Whitechapel E. 1, England (Dr. F. J. C. Roe).

Ln Dr. J. B. Lyon, Department of Biochemistry, Emory University, Atlanta, Georgia.

Lw Dr. L. W. Law, National Cancer Institute, Bethesda 14, Md.

Ly Dr. Clara Lynch, Rockefeller Institute, 66th Street and New York Avenue, New York 21, N.Y.

M Memorial Hospital and Sloan-Kettering Institute for Cancer Research, 410 E. 68th Street, New York 21, N.Y.

Ma Dr. E. A. Mirand, Roswell Park Memorial Institute, Buffalo 3, N. Y.

Mas Department of Zoology, University of Massachusetts, Amherst, Massachusetts.

Mc Dr. W. B. McIntosh, Department of Zoology and Entomology, Ohio State University, 1735 Neil Avenue, Columbus 10, Ohio.

Md Dr. E. C. MacDowell, Department of Genetics, Carnegie Institution of Washington, Cold Spring Harbor, New York (retired).

Me Dr. P. B. Medawar, Department of Zoology, University College London, Gower Street, W.C. 1, England.

Mi Dr. D. Michie, Department of Surgical Science, University New Buildings, Teviot Place, Edinburgh 8, Scotland.

Mo Dr. R. H. Mole, Radiobiological Research Unit, Harwell, Didcot, Berks, England.

Mr Dr. W. E. Miller, Cancer Research Laboratory, Department of Pathology, Royal Victoria Infirmary, Newcastle upon Tyne 1, England.

Ms National Institute of Genetics, Yata IIII, Misima, Sizuoka-ken, Japan (Dr. Tosihide H. Yoshida).

Mv Dr. N. N. Medvedev, Gamaleya Institute of Epidemiology and Microbiology, Schukinskaya 33, Moscow, USSR.

My Dr. W. S. Murray, Roscoe B. Jackson Memorial Laboratory, Bar Harbor, Maine.

N Genetics Research Unit, Laboratory Aids Branch, National Institutes of Health, Bethesda 14, Maryland (Dr. D. W. Bailey).

No Dr. D. J. Nolte, University of the Witwatersrand, Genetics Laboratory, Milner Park, Johannesburg, South Africa.

Not Cancer Research Laboratory, The University, University Park, Nottingham, England (Dr. R. W. Baldwin).

Oci Ontario Cancer Institute, University of Toronto, 500 Sherbourne Street, Toronto 5, Canada (Dr. A. A. Axelrad).

Pa   Dr. Edith Paterson, Christie Hospital and Holt Radium Institute, Manchester 20, England.

Pe   Dr. P. R. Peacock, Royal Beatson Memorial Hospital, 132–138 Hill Street, Glasgow C.3, Scotland.

Per   Division of Cancer Research, Univ. of Perugia, Italy (Prof. Lucio Severi).

Pi   Dr. H. I. Pilgrim, Laboratory of Clinical Biology, Department of Surgery, University of Utah College of Medicine, Salt Lake City, Utah.

Pu   Dr. B. D. Pullinger, Royal Beatson Memorial Hospital, Cancer Research Department, 132–138 Hill Street, Glasgow, C.3, Scotland.

Rb   Dr. R. G. Busnel, Institut National de la Recherche Agronomique, Laboratoire de Physiologie Acoustique, Jouy-en-Josas (S. & O.), France.

Rd   Dr. G. Rudali, Fondation Curie, 26 rue d'Ulm, Paris (V^e), France.

Re   Dr. Elizabeth S. Russell, Roscoe B. Jackson Memorial Laboratory, Bar Harbor, Maine.

Rho   Laboratoire de Recherches de la Société Rhône-Poulenc, 13 quai Jules Guesde, Vitry sur Seine (Seine), France (Prof. R. Paul).

Rij   Medical Biological Laboratory, National Health Research Council TNO, Lange Kleiweg 139, Rijswijk (Z. H.), Netherlands (Dr. D. W. van Bekkum).

Rl   Drs. W. J. and L. B. Russell, Biology Division, Oak Ridge National Laboratory, P.O. Box Y, Oak Ridge, Tenn.

Ro   Dr. U. Roth, Hamburg 36, Jungiusstr. 6, Zoologisches Institut, Germany.

Rr   Dr. M. N. Runner, Roscoe B. Jackson Memorial Laboratory, Bar Harbor, Maine.

S   Mount Sinai Medical Research Foundation, 2755 West 15th Street, Chicago 8, Ill. (Dr. Kurt Stern).

Sb   Dr. Arthur G. Steinberg, Biological Laboratory, Western Reserve University, Cleveland 6, Ohio.

Sc   Dr. J. P. Scott, Roscoe B. Jackson Memorial Laboratory, Bar Harbor, Maine.

Se   Prof. Lucio Severi, Istituto di Anatomia e Istologia Patologica dell'Università degli Studi di Perugia, Italy.

Sid   Stanford University School of Medicine, Department of Anatomy, Stanford, California (Dr. Elizabeth M. Center).

Sk   Dr. H. E. Skipper, Southern Research Institute, Birmingham 5, Ala.

Sn   Dr. G. D. Snell, Roscoe B. Jackson Memorial Laboratory, Bar Harbor, Maine.

Sp   Dr. William L. Simpson, Detroit Institute of Cancer Research, 4811 John R Street, Detroit 1, Mich.

Ss   Dr. W. K. Silvers, The Wistar Institute, 36th at Spruce, Philadelphia, Pa.

St   Dr. L. C. Strong, Roswell Park Memorial Institute, Biological Station, Springville, N.Y.

Sv   Dr. L. C. Stevens, Roscoe B. Jackson Memorial Laboratory, Bar Harbor, Maine.

T   Department of Zoology, University of Toronto, Toronto 5, Canada (Dr. L. Butler).

Tn   Dr. A. Tannenbaum, Medical Research Institute, Michael Reese Hospital, 29th and Ellis Avenue, Chicago 16, Ill.

Tu   Dr. F. C. Turner, Laboratory for research on the Treatment of Cancer, Box 807, Boulder Creek, Calif.

Vi   Miss Emelia Vicari, Roscoe B. Jackson Memorial Laboratory, Bar Harbor, Maine (retired).

Wa   Dr. S. G. Warner (deceased).

We   Dr. J. A. Weir, University of Kansas, Department of Zoology, Lawrence, Kansas.

Wf   Dr. George L. Wolff, Institute for Cancer Research, Fox Chase, Philadelphia 11, Pa.

Wi   Dr. J. W. Wilson, Department of Biology, Brown University, Providence 12, Rhode Island.

Ww   Mrs. E. F. Woodworth, Roscoe B. Jackson Memorial Laboratory, Bar Harbor, Maine.

Wy   Dr. G. W. Woolley, Sloan-Kettering Institute for Cancer Research, 444 East 68th Street, New York 21, N.Y.

II. RATS

Except for the previous publication of Billingham and Silvers,[99] this is the first attempt to obtain a comprehensive listing of established inbred strains of rats, inbred strains in development, or stocks with genic markers. Additional information will doubtless be forthcoming with the passage of time. Since the number of established inbred strains and strains in development is not yet too great, standardization of nomenclature and symbols at the present time should avoid the confusion that once existed for mice.   Consequently, (1) the rules recommended by the Committee on Standardized Genetic Nomenclature for Inbred Strains of Mice should be adopted for inbred rat strains, and (2) the symbols used in the following table should be immediately adopted. The international rules of gene and strain nomenclature are discussed in the chapter by Miss Staats.   Perhaps the Committee will be asked to extend its jurisdiction to rats.

The symbols for strains used in the following listing are based on the rules for mice. Concurrence for the use of these symbols was obtained from most of the contributors maintaining the strains.   The descriptions of the strains are similar to those previously used for mice with the exception of identified genetic characteristics which have been included as part of the characteristics, since the genes are unknown except in the genetic stocks maintained by Dr. E. Dempster.   This list of contributors at the end of this section is in the same format as the list for mice.   The abbreviatiations were checked in order to avoid duplication, and, if a contributor appears on both lists (or all lists), the same abbreviation of the name is used.

Table 28

ESTABLISHED STRAINS OF RATS

Name or Symbol	Synonym(s)	Remarks
ACI	AXC9935, AXC9935 Irish, Irish	INBR: F_{65}. ORIGIN: Curtis and Dunning at the Columbia University Institute for Cancer Research, 1926, to Heston, 1945, at F_{30}, to N 1950 at F_{41}; subsequent sublines from either Du or N colonies. CHARAC: life span 21.7 ± 0.17 mo., susceptible to estrogen-induced tumors, the N subline exhibits a high incidence of kidney abnormalities (from cystic kidney to kidneys absent, unilaterally or bilaterally), spontaneous Leydig cell tumors (25%) and anterior pituitary tumors (15%) in old animals, a high incidence of uterine abnormalities (horns absent, etc.), some resistance to Bartonella infection, susceptible to bronchiectasis (40%) and otitis media (60%), will grow transplantable tumors M-C961, 969, 970, 972, positive for R-1 factor hemagglutinogen *B*, the Br subline shows low defecation response and high activity response in open-field test of emotional behavior, black agouti, irish. MAINTAINED BY: Brh, Du, Ko, N, Seg.
ACH	AXC9935 Piebald	INBR: F_{51} +. ORIGIN: Curtis and Dunning, 1926, at the Columbia University Institute for Cancer Research. CHARAC: high incidence of spontaneous lymphosarcoma of illeocecal mesentery, will grow transplantable tumors R2788, R2572, IRS6820, B-P839, black agouti, hooded. MAINTAINED BY: Du.
ALB	Albany	INBR: F_{21} +. ORIGIN: Wolfe and Wright, Albany Medical College, to N 1950; no inbreeding records available prior to the transfer to N; b × s matings since. CHARAC: exhibits some mammary fibroadenomata, negative hemagglutinogen *B*, large body size, docile, good reproduction, nonagouti, brown. MAINTAINED BY: N.
A990	August 990 I, D (brown hooded).	INBR: F_{61}. ORIGIN: Inbreeding started at Columbia University Institute for Cancer Research by Curtis in 1921. CHARAC: life span 14.03 ± 0.03 mo., resistant to Cysticercus, susceptible to estrogen-

Table 28—*Continued*

Name or Symbol	Synonym(s)	Remarks
		induced mammary and adrenal tumors; will grow transplantable tumor IRC855, good reproduction; Br subline shows low defecation response and high activity response in open-field test of emotional behavior; black agouti, hooded. MAINTAINED BY: Brh, Dem, Du.
A7322	August 7322	INBR: F_{64}. ORIGIN: Curtis, 1925, at the Columbia University Institute for Cancer Research. CHARAC: life span 14.03 ± 0.04 mo., spontaneous mammary tumors frequent, resistant to Cysticercus, will grow transplantable tumors R2857, R2737; positive R-1 factor hemagglutinogen B, agouti, hooded, pink-eye. MAINTAINED BY: Du.
A28807	August 28807	INBR: F_{37}. ORIGIN: Curtis and Dunning, 1936 from a half b × s mating of F_{15} animals of strain A7322. CHARAC: no information submitted. MAINTAINED BY: Du.
A35322	August 35322	INBR: F_{29}. ORIGIN: Curtis and Dunning, 1942, from a mutation originating from an aunt × nephew cross in F_{27} animals of strain A990. CHARAC: vaginal prolapse frequent, will grow transplantable tumor R3280 (bronchiogenic carcinoma), Brh subline shows high defecation response and low activity response in open-field test of emotional behavior, black, nonagouti, hooded. MAINTAINED BY: Brh, Du.
AVO	Avon 34968	INBR: F_{27}. ORIGIN: Curtis and Dunning, 1941. CHARAC: red-eyed, selfed, poor reproduction. MAINTAINED BY: Du.
B		INBR: F_{69}. ORIGIN: Dr. P. Swanson from a Wistar stock; to E. Dempster at F_{43}. CHARAC: large body size (weaning weight 68 grams at 28 days), poor maternal instincts, fertile; albino. MAINTAINED BY: Dem.
BDI		INBR: F_{35}. ORIGIN: H. Druckrey from animals homozygous for certain hair and eye color genes, b × s mating system from the time inbreeding was started. CHARAC: pink-eyed, yellow, selfed. MAINTAINED BY: Dr.
BD II		INBR: F_{20}. ORIGIN: H. Druckrey from animals homozygous for certain hair and eye color genes, b × s mating system from the time inbreeding was started. CHARAC: albino (carries nonagouti, black, hooded). MAINTAINED BY: Dr.
BD III		INBR: F_{35}. ORIGIN: H. Druckrey from animals homozygous for certain hair and eye color genes, b × s mating system from the time inbreeding was started. CHARAC: pink-eyed, yellow, hooded. MAINTAINED BY: Dr, Rot.
BD IV		INBR: F_{30}. ORIGIN: H. Druckrey from animals homozygous for certain hair and eye color genes, b × s mating system from the time inbreeding was started. CHARAC: nonagouti, black, hooded. MAINTAINED BY: Dr.
BUF	Buffalo	INBR: F_{37}. ORIGIN: Heston, 1946, from Buffalo stock of H. Morris, to N 1950 at F_{10}. CHARAC: low incidence of dental caries; negative R-1 factor hemagglutinogen B, will grow Hepatoma 5123 (some enzymes similar to normal liver) and Yoshida ascites sarcoma (16%); spontaneous tumors of the anterior pituitary (30%) and adrenal cortex (25%) in old animals, susceptible to otitis media (70%), bronchiectasis (80%), and myocarditis (30%) in old animals, medium reproduction, albino. MAINTAINED BY: N, Mor.
CAR	Hunt's Caries Resistant, CA/R	INBR: F_{38}. ORIGIN: Hunt, 1937. CHARAC: low incidence of dental caries, negative R-1 factor hemagglutinogen B, fair reproduction, albino. MAINTAINED BY: Hun, N.

Table 28—*Continued*

Name or Symbol	Synonym(s)	Remarks
CAS	Hunt's Caries Susceptible, CA/S	INBR: F_{38}. ORIGIN: Hunt, 1937. CHARAC: high incidence of dental caries, negative R-1 factor hemagglutinogen *B*, poor reproduction, susceptible to respiratory infections, albino. MAINTAINED BY: Hun, N.
C2331	Copenhagen 2331	INBR: F_{53}. ORIGIN: Curtis, at the Columbia University Institute for Cancer Research, 1921. CHARAC: life span 19.08 ± 0.18 mo., spontaneous tumors of thymus, resistant to estrogen-induced mammary tumors, susceptible to estrogen-induced bladder tumors, resistant to Cysticercus, will grow (100%) transplantable tumor IRS4337, positive R-1 factor hemagglutinogen *B*, coat color black, agouti, hooded. MAINTAINED BY: Du.
F344	Fischer 344, Fischer	INBR: F_{75}. ORIGIN: Curtis at the Columbia University for Cancer Research 1920, to Heston 1949, to N 1950 at F_{51}, subsequent sublines from either the Du or N colonies. CHARAC: life span 12.3 ± 0.10 mo., susceptible to Cysticercus, uniform body size (small), susceptible to 2-acetylaminofluorine-induced mammary tumors, will grow transplantable tumors IRC741, IRS1548, L-C18, IRS9802, R3211, R3230, R3251, R3259, N subline exhibits 15% spontaneous tumor incidence of the anterior pituitary gland, uterus and mammary gland, 30–50% incidence of Leydig cell tumors, granulomatosis lesions of the lymph nodes and spleen, old animals manifest otitis media (40–60%), bronchiectasis (61–70%), nephritis (21–30%), myocarditis (21–30%); negative R-1 factor hemagglutinogen *B*, albino (carries nonagouti, black, hooded). MAINTAINED BY: Dem, Du, Fu, Gm, Hi, Ko, N, Nl, Rot, Seg, Sy, Ta.
LEW	Lewis	INBR: F_{36}. ORIGIN: Lewis from Wistar stock, to Aptekman and Bogden, 1954, at F_{20}, to Silvers, 1958, at F_{31}. CHARAC: negative R-1 factor hemagglutinogen *B*; homozygous for red cell antigen *C*, isohistogenic, docile, high fertility, will grow transplantable carcinoma # 10, lymphoma # 8; albino. MAINTAINED BY: Ss, Mai.
M520	Marshall 520, M-520	INBR: F_{76}. ORIGIN: Curtis, 1920, at the Columbia University Institute for Cancer Research, to Heston, 1949, at F_{49}, to N, 1950 at F_{51}. CHARAC: life span 13.5 ± 0.09 mo.; susceptible to Cysticercus; spontaneous tumors rare in Du subline; N subline exhibits 21–25% incidence of adrenal medulla tumors; susceptible to 2-acetylaminofluorine induced tumors; medium susceptible to Bartonella; manifests high incidence of otitis media (71–80%), bronchiectasis (60%), nephritis (60%), periarteritis (20%), cecitis (20%), infections of the Harderian gland (100%); will grow Jensen sarcoma, Yoshida ascites sarcoma (80%), hepatoma 7974 (75%), hepatoma 130 (70%); negative R-1 factor hemagglutinogen *B*; albino (carries black, nonagouti, hooded). MAINTAINED BY: Du, N.
M14	Mi4, L4	INBR: $F_{40}+$. ORIGIN: A. B. Chapman, 1940, from Sprague-Dawley stock, selection for low ovarian response to pregnant mare serum. CHARAC: low ovarian response to pregnant mare serum; albino. MAINTAINED BY: Cp.
M17	Mi7, H7	INBR: $F_{40}+$. ORIGIN: A. B. Chapman, 1940, from Sprague-Dawley stock, selection for high ovarian response to pregnant mare serum. CHARAC: high ovarian response to pregnant mare serum; albino. MAINTAINED BY: Cp.
MR	Maudsley Reactive	INBR: $F_{18}+$. ORIGIN: P. L. Broadhurst, 1954, from a commercial Wistar stock; selection for high defecation response. CHARAC: high defecation response in the open-field test of emotional behavior as well as a heightened susceptibility to certain motivating stimuli; albino. MAINTAINED BY: Brh.

Table 28—*Continued*

Name or Symbol	Synonym(s)	Remarks
MNR	Maudsley Non-reactive	INBR: F_{19} +. ORIGIN: P. L. Broadhurst, 1954, from a commercial Wistar stock; selection for low defecation response. CHARAC: low defecation response in the open-field emotional behavior as well as a reduced susceptibility to certain motivating stimuli; albino. MAINTAINED BY: Brh.
OM	Inbred Osborne Mendel, O-M	INBR: F_{30}. ORIGIN: Heston, 1946, from a noninbred Osborne-Mendel stock obtained from J. White, to N, 1950, at F_{10}. CHARAC: exhibits 70% incidence of adrenal cortex tumors, 21–25% ovarian tumors, 26–30% mammary tumors; susceptible to Bartonella, otitis media (70%), bronchiectasis (80%), myocarditis (30%), positive R-1 factor hemagglutinogen *B*; albino. MAINTAINED BY: N, Mor.
PA	P.A. King Albino	INBR: F_{162}. ORIGIN: King, 1909, from Wistar Institute stock, to Aptekman, 1946, at F_{135}, to Bogden, 1958, at F_{155}. CHARAC: positive R-1 factor hemagglutinogen *B*; isohistogenic, vigorous (and vicious), healthy; good reproduction, will grow ascites tumor 9A, leukemia LK2, carcinoma # 5, lymphoma # 6; albino. MAINTAINED BY: Bg, Nl.
PVC	B (black hooded) (PVG/c) PVG/c, P.V.G./c (piebald)	INBR: F_{30} +. ORIGIN: Glaxo Laboratories, 1947, from stock received from Virol Ltd. CHARAC: isohistogenic, Br subline shows low defecation response and low activity response in open-field emotional behavior, susceptible to otitis media, black, hooded. MAINTAINED BY: Brh, Ct, Gor.
R	R/A	INBR: F_{40}. ORIGIN: O. Mühlbock from a Wistar stock in 1947. CHARAC: albino. MAINTAINED BY: A, Kl.
SEL	Selfed 36670	INBR: F_{28}. ORIGIN: W. Dunning, 1948. CHARAC: positive R-1 factor hemagglutinogen *B*, black, nonagouti, selfed. MAINTAINED BY: Du.
WAB	Wistar Albino, Boots	INBR: F_{81} +. ORIGIN: From the Glaxo WAG stock in 1926 before it became established as a strain, b × s matings since. CHARAC: albino, no other information available. MAINTAINED BY: Ad.
WAG	Wistar Albino, Glaxo, W.A.G., WAG/C, W.A.G./c A(albino (Wag/C))	INBR: F_{70} +. ORIGIN: A. L. Bacharach, 1924, from Wistar Institute stock. CHARAC: susceptible to iron deficiency, Br subline shows high defecation response and high activity response in open-field test of emotional behavior, av. litter size 8.7, albino. MAINTAINED BY: Bk, Brh, Ct, Gor.
WF	Wistar/Furth, W/Fu, WR	INBR: F_{27} +. ORIGIN: J. Furth, 1945, from a commercial Wistar stock. CHARAC: high incidence of mammotropic pituitary tumors and leukemias, tumors 100% transplantable within strains, susceptible to methylcholanthrene-induced mammary tumors, albino. MAINTAINED BY: Fu.
WN	Inbred Wistar	INBR: F_{34}. ORIGIN: W. Heston, 1942, from Wistar stock of Nettleship, to N, 1950, at F_{15}. CHARAC: 30–50% incidence of spontaneous mammary tumors, 21–25% incidence of anterior pituitary tumors in old animals, squamous cell metaplasia and hyperplasia of the thyroid manifested, resistant to Bartonella, poor reproduction, negative R-1 factor hemagglutinogen *B*, albino. MAINTAINED BY: N.
Z61	Zimmerman 61	INBR: F_{70}. ORIGIN: Curtis, 1920, at the Columbia University Institute for Cancer Research. CHARAC: life span 11.9 ± 0.17 mo., susceptible to Cysticercus, susceptible to estrogen-induced tumors and 2-acetylaminofluorine tumors, will grow Jensen sarcoma and R92, albino (carries nonagouti, hooded). MAINTAINED BY: Du.

Table 29

Name or Symbol	Synonym(s)	Remarks
BD V		INBR: F_{10}. ORIGIN: H. Druckrey from animals homozygous for certain hair and eye color genes, b × s mating system from the time inbreeding started. CHARAC: pink-eyed, nonagouti, hooded. MAINTAINED BY: Dr.
BD VI		INBR: F_5. ORIGIN: H. Druckrey from animals homozygous for certain hair and eye color genes, b × s mating system from the time inbreeding started. CHARAC: nonagouti, black, selfed. MAINTAINED BY: Dr.
BD VII		INBR: F_5. ORIGIN: H. Druckrey from animals homozygous for certain hair and eye color genes, b × s mating system from the time inbreeding started. CHARAC: pink-eyed, nonagouti, selfed. MAINTAINED BY: Dr.
BD VIII		INBR: F?. ORIGIN: H. Druckrey from animals homozygous for certain hair and eye color genes, b × s mating system from the time inbreeding started. CHARAC: black, agouti, hooded. MAINTAINED BY: Dr.
BD IX		INBR: F_3. ORIGIN: H. Druckrey from animals homozygous for certain hair and eye color genes, b × s mating system from the time inbreeding started. CHARAC: black, agouti, selfed. MAINTAINED BY: Dr.
BD X		INBR: F_3. ORIGIN: H. Druckrey from animals homozygous for certain hair and eye color genes, b × s mating system from the time inbreeding started. CHARAC: albino (carries nonagouti, pink-eyed, hooded). MAINTAINED BY: Dr.
BN	B.N.	INBR: F_9. ORIGIN: Silvers and Billingham, 1958, from a brown mutation maintained by H. D. King and P. Aptekman in a pen-bred colony, b × s matings with selection for histocompatibility. CHARAC: histocompatible, segregating for red cell antigens C and D, low fertility, not very docile, non-agouti, brown. MAINTAINED BY: Ss.
C		INBR: F_6. ORIGIN: R. Owen, 1958. CHARAC: homozygous for the C blood group allele, albino. MAINTAINED BY: Ow.
D		INBR: F_6. ORIGIN: R. Owen, 1958. CHARAC: homozygous for the D blood group allele, nonagouti, irish. MAINTAINED BY: Ow.
H	hooded	INBR: F_{16}. ORIGIN: H. Newcombe, 1952 (?), b × s matings with selection for large litters. CHARAC: hooded. MAINTAINED BY: Ne.
HB	black mutant of H	INBR: F_9. ORIGIN: H. Newcombe, 1956 (?), b × s matings with *no* selection for litter size. CHARAC: black, hooded. MAINTAINED BY: Ne.
NEDH	Slonaker NEDH	INBR: F_{14}. ORIGIN: W. E. Knox, 1955, from a Wistar stock, b × s matings with selection for moderate body size, good reproduction, longevity, gentleness. CHARAC: moderate body size, good reproduction, longevity, gentle, albino. MAINTAINED BY: Kx.
TEC1	Tec1	INBR: F_5. ORIGIN: W. G. Downs, Jr., 1958, from a Wistar-Sprague Dawley cross, b × s matings with selection for standardized total and differential white blood cell count. CHARAC: albino. MAINTAINED BY: Do.
TEC2	Tec2	INBR: F_3. ORIGIN: W. G. Downs, Jr., 1959, from a Wistar-Sprague Dawley cross, b × s matings with selection for a standardized response to insulin. CHARAC: albino. MAINTAINED BY: Do.
YOS	Yoshida 38366	INBR: F_{14}. ORIGIN: W. Dunning, 1953. CHARAC: will grow Yoshida sarcoma, albino. MAINTAINED BY: Du.

Table 30

Name or Symbol	Synonym(s)	Remarks
M	Microphthal-mic	CHARAC: Microphthalmic or anophthalmic eye to near normal eyes, normal-appearing eyes that are blind due to abberrant pathway of the optic nerve. MAINTAINED BY: Bw.
BLACK SELF	Black	ORIGIN: From Castle's stocks. CHARAC: $aaB–C–H–$. MAINTAINED BY: Dem.
BLUE DILUTE	Blue	ORIGIN: From Castle's stocks. CHARAC: $aa\ B–C–H–dd$. MAINTAINED BY: Dem.
CHOCO-LATE, CURLY, RED EYED, PINK EYED		ORIGIN: From Castle's stocks. CHARAC: $aabb\ C–Cu–rr$ or $aabbc–Cu–pp$. MAINTAINED BY: Dem.
CHOCO-LATE SILVER		ORIGIN: From Castle's stocks. CHARAC: $aabbC–se$. MAINTAINED BY: Dem.
COW-LICK-HOODED NOTCH		ORIGIN: From Castle's stocks. CHARAC: $aaB–C–h^nh^ncwcw$. MAINTAINED BY: Dem.
CURLY 2		ORIGIN: From Castle's stocks. CHARAC: $Cu_2–$. MAINTAINED BY: Dem.
HAIR-LESS		ORIGIN: From Castle's stocks. CHARAC: $hrhr$. MAINTAINED BY: Dem.
JAUNDICE		ORIGIN: From Gunn's stock via Castle. CHARAC: $aaB–hhjj$ or $AaB–hhjj$. MAINTAINED BY: Dem, N.
KINKY		ORIGIN: From Castle's stocks. CHARAC: $aaB–cckk$ or $aaB–C–kk$. MAINTAINED BY: Dem.
RETINI-TIS PIGMEN-TOSA		ORIGIN: Recessive mutation, 1936, in a heterogenous stock maintained by Bemax Laboratories (England), to Dr. D. Campbell, Birmingham Eye Hospital, additional animals added in 1949, gene is maintained by outcrossing and recovering in the backcross, defect ascertained in one eye before breeding. CHARAC: retinitis pigmentosa, stock segregates for various hair colors. MAINTAINED BY: To.
SHAKER		ORIGIN: From Castle's stocks. CHARAC: $srsr$. MAINTAINED BY: Dem.
WOBBLY		ORIGIN: From Castle's stocks. CHARAC: $aaB–h^nh^nwowo$. MAINTAINED BY: Dem.
YELLOWS		ORIGIN: From Castle's stocks. CHARAC: No information available. MAINTAINED BY: Dem.

LIST OF SYMBOLS FOR DESIGNATING SUBSTRAINS AND STOCKS OF RATS

A   Dr. O. Mühlbock, Antoni van Leewen-hoekhuis, Sarphatistraat 108, Amsterdam C, Netherlands.

Ad   Dr. S. S. Adams, Boots Pure Drug Co., Ltd., Pharmacology and Physiology Division, Oaksfield Road, Nottingham, England.

Bg   Dr. Arthur E. Bogden, The Biochemical Research Foundation, Newark, Delaware.

Bk   Dr. D. W. van Bekkum, Radiobiological Laboratory, National Health Research Council T.N.O., Lange Kleiweg 139, Rijswijk (Z.H.), Netherlands.

Brh Dr. P. L. Broadhurst, Institute of Psychiatry, Animal Psychology Laboratory, Bethlem Royal Hospital, Monks Orchard, Beckenham, Kent, England.

Bw Dr. L. G. Browman, Montana State University, Missoula, Montana.

Cp Dr. A. B. Chapman, Department of Genetics, University of Wisconsin, Madison, Wisconsin.

Ct Dr. W. F. J. Cuthbertson, Glaxo Laboratories, Middlesex, England.

Dem Dr. E. Dempster, Department of Genetics, University of California, Berkeley, California.

Do Dr. William G. Downs, Tennessee Polytechnic Institute, Cookeville, Tennessee.

Dr Prof. Dr. Med. Herman Druckrey, Laboratorium D. Chirurg. Univ. Klink., Hugstetterstrasse 55, Freiburg im Breisgau, Germany.

Du Dr. Wilhelmina F. Dunning, Experimental Cancer Research Laboratory, University of Miami, Coral Gables, Florida.

Fu J. Furth, M.D., Roswell Park Memorial Institute, Buffalo, N.Y.

Gm J. P. W. Gilman, D.V.M., Ontario Veterinary College, Guelph, Ontario, Canada.

Gor Dr. W. S. Gordon, Director, Agricultural Research Council, Field Station, Compton, near Newbury, Berkshire, England.

Hi Dr. Russell Hilf, The Squibb Institute for Medical Research, New Brunswick, New Jersey.

Hun Dr. H. R. Hunt, Department of Zoology, Michigan State University, East Lansing, Michigan.

Kl Drs. G. and E. Klein, Department of Cell Research, Karolinsha Institute, Stockholm, Sweden.

Ko Dr. Henry Kohn, Radiological Laboratory, University of California Medical Center, San Francisco 22, California.

Kx Dr. W. Eugene Knox, Harvard Medical School, Harvard University, Cambridge, Massachusetts.

Mai Microbiological Associates, Inc., 4648 Bethesda Avenue, Bethesda 14, Maryland.

Mor Dr. Pablo Mori-Chavez, 779 Sanchez-Carrion, Lima, Peru.

N Genetics Research Unit (Dr. D. Bailey), Laboratory Aids Branch, National Institutes of Health, Bethesda 14, Md.

Ne Dr. Howard Newcombe, Atomic Energy of Canada, Chalk River, Canada.

Nl R. L. Noble, M.D., Department of Medical Research, The Collip Medical Research Laboratory, The University of Western Ontario, London, Canada.

Ow Dr. Ray D. Owen, Division of Biological Sciences, California Institute of Technology, Pasadena 4, California.

Rot Dr. med. W. Rotzsch, Physiologisch-Chemisches, Institut der Universität Leipzig, Liebigstraat 16, Fernruf 311 14, Leipzig, Germany.

Seg Dr. Albert Segaloff, Division of Endocrinology, Alton Ochsner Medical Foundation, 1520 Jefferson Hwy., New Orleans 21, Louisiana.

Ss Dr. W. K. Silvers, The Wistar Institute, Philadelphia 4, Pennsylvania.

Sy Dr. K. L. Sydnor, University of Chicago, Chicago, Illinois.

Ta Dr. Martha J. Taylor, Pathology Division, Fort Detrick, Maryland.

To Miss Eva L. Tonks, Research Department, Birmingham (Dudley Road) Group of Hospitals, Birmingham & Midland Eye Hospital, Church Road, Birmingham 3, Alabama.

III. GUINEA PIGS

The genetic material for guinea pigs is very limited in this country as well as elsewhere. Apparently only three established inbred strains exist; two of the three are the remaining strains from the thirty-five original ones started by the U.S. Department of Agriculture in 1906 and later developed by Wright,[1445, 1447] and the third has been developed by Dr. O. Mühlbock more recently. Several strains are in the process of development, but little is known about them, and, except for the waltzing and silvering stocks, no stocks of genetic significance are maintained. Renewed interest seems to be developing for the use of inbred guinea pigs for immunogenetic work, so the future does hold some promise for the increased use of this species in mammalian genetics.

As in the case of the rats, it is suggested that (1) the rules recommended by the Committee on Standardized Genetic Nomenclature for Inbred Strains of Mice be adopted for inbred guinea pig strains, and that (2) the strain symbols herein used be immediately adopted. Also, possibly the jurisdiction of the Committee will be

Table 31

ESTABLISHED STRAINS OF GUINEA PIGS†

Name or Symbol	Synonym(s)	Remarks
C	C/A	INBR: F_{19}. ORIGIN: O. Mühlbock, 1948. CHARAC: No information submitted. MAINTAINED BY: A.
2	Family 2, Strain 2, STR. 2	INBR: F_{26} +. ORIGIN: U.S. Department of Agriculture, 1906, to S. Wright, 1915, at F_{11}; b × s matings for 33 generations (1933), then within strain random breeding until 1940; to Heston, 1940, at which time b × s mating system restored; to N, 1950, at F_{12}; all subsequent sublines from the N colony. CHARAC: fairly resistant to tuberculosis, N subline has median incidence of deciminated calcification occurring in greater curvature of stomach, colon, kidney, striated muscle of the abdominal wall, lung, and aorta of old animals (34 mo.). Medium reproduction, active sexual behavior, small body size, isohistogenic, tricolor (black, red, white). MAINTAINED BY: Bk, Fn, Ju, N, Ne, Sm, Ss, Wal, Yo.
13	Family 13, Strain 13, STR 13	INBR: F_{26} +. ORIGIN: U.S. Department of Agriculture, 1906, to S. Wright, 1915, at F_{13}; b × s matings for 33 generations (1933), then within strain random breeding until 1940; to Heston, 1940, at which time b × s mating system restored; to N, 1950, at F_{12}; all subsequent sublines from the N colony. CHARAC: less resistant to tuberculosis than Strain 2, medium reproduction, less active sexual behavior than Strain 2, larger body size than Strain 2, isohistogenic, tricolor (black, red, white). MAINTAINED BY: Bk, Fn, Ju, N, Ne, Sm, Ss, Wal, Yo.

† A list of abbreviations follows table 33.

Table 32

STRAINS OF GUINEA PIGS IN DEVELOPMENT

Name or Symbol	Synonym(s)	Remarks
ICRF	ICRF/GP	INBR: F_{12}. ORIGIN: P. C. Williams, 1954, a heterogeneous (?) stock. CHARAC: albino, some polydactylism. MAINTAINED BY: Icrf.
OM 3		INBR: F ? ORIGIN: J. B. Rogers, 1952, from a random-bred commercial stock. CHARAC: no toxemia of pregnancy, no spontaneous tumors, does not reproduce well above 5000 feet. MAINTAINED BY: Rog.
R 7		INBR: F ? ORIGIN: J. B. Rogers, 1941, from a random-bred stock. CHARAC: toxemia of pregnancy (15% incidence in late pregnancy, during parturition, and 1 week post-partum), no spontaneous tumors, life span 3+ years. MAINTAINED BY: Rog.
R 9		INBR: F ? ORIGIN: J. B. Rogers, 1941, from random-bred ancestry. CHARAC: 14% spontaneous tumor incidence in animals over 1095 days of age, life span 4+ years. MAINTAINED BY: Rog.

extended to include guinea pigs. The symbols for strains used are based on the rules for mice. Since few strains are involved, the changes indicated have been made without concurrence from anyone, trusting that the changes will be accepted. The descriptions of the strains are in the same format as for rats and, like the rats, few, if any, of the actual genetic factors involved are known. The list of contributors at the end of the section conforms to that used for mice, and the abbreviations indicated have been checked against both previous lists to avoid duplication.

Table 33

STOCKS OF GUINEA PIGS OF GENETIC INTEREST

Name or Symbol	Synonym(s)	Remarks
Silvering		ORIGIN: This stock was obtained from S. Wright in 1955. CHARAC: homozygous for silvering (*sisi*) and segregating for diminished (*didi* and *Didi*), C, c^d, c^a, P, *p*, *e*, and *ep*. MAINTAINED BY: Re.
Waltzing		ORIGIN: The waltzing manifestation first appeared in 1953 as a probable mutation in a noninbred colony maintained by N. CHARAC: the waltzing condition in typical cases shows a structurally normal hearing apparatus at birth, but shortly thereafter the hair cells of Corti's organ begin to disappear, followed by a gradual atrophy and disappearance of the other cellular elements of the organ; the cochlear neurons atrophy more slowly, with some persisting for more than two years; the trait is dominant, with variable expression. MAINTAINED BY: Mp, N.

LIST OF SYMBOLS FOR DESIGNATING SUBSTRAINS AND STOCKS OF GUINEA PIGS

A   Antoni van Leeuwenhoekhuis, Amsterdam C, Sarphatistraat 108, Netherlands (Dr. O. Mühlbock).

Bk   Dr. D. W. van Bekkum, Radiobiological Laboratory, National Health Research Council T.N.O., Lange Kleiweg 139, Rijswijk (ZH), Netherlands.

Fn   Dr. Frank Fenner, Department of Microbiology, The Australian National University, Canberra, Australia.

Icrf Imperial Cancer Research Fund, Central Laboratories, Burtonhole Lane, The Ridgeway, Mill Hill, N.W. 7, England.

Ju   Dr. Clair W. Jungeblut, Department of Microbiology, College of Physicians and Surgeons, Columbia University, New York, New York.

Mp   Dr. Leo Massopust, Physiological Science Department, Southeast Louisiana Hospital, Mandeville, Louisiana.

N   Dr. D. W. Bailey, Genetics Research Unit, Laboratory Aids Branch, National Institutes of Health, Bethesda, Maryland.

Ne   Dr. W. T. Newton, School of Medicine, Washington University, St. Louis, Missouri.

Re   Dr. Elizabeth S. Russell, R. B. Jackson Memorial Laboratory, Bar Harbor, Maine.

Rog   Dr. J. B. Rogers, Department of Anatomy, School of Medicine, University of Louisville, Louisville, Kentucky.

Sm   Dr. L. H. Smith, Oak Ridge National Laboratory, Oak Ridge, Tennessee.

Ss   Dr. Willys Silvers, The Wistar Institute, Philadelphia 4, Pennsylvania.

Wal   Dr. Roy L. Walford, Department of Pathology, Medical Center, University of California, Los Angeles, California.

Yo   Dr. W. C. Young, Department of Anatomy, University of Kansas, Lawrence, Kansas.

IV. HAMSTERS

During the past several years, there has been considerable interest in the use of hamster species in mammalian genetic research. Consequently, several inbred strains have been established, and a number of additional strains have been developed. The Syrian (golden) hamster, *Mesocricetus auratus*, has been used most, but through the research of Dr. G. Yerganian (see his paper elsewhere in this volume) the Chinese hamster, *Cricetulus griseus*, has proved to be a desirable animal for research. Information on the location and status of genetic stocks of these two species is included in this section.

As recommended for rats and guinea pigs, the adoption for hamsters of (1) the rules recommended by the Committee on Standardized Nomenclature for Inbred Strains of Mice, and (2) the symbols herein used, is urged. Since the basic genetic principles involved in inbreeding hamsters are probably the same as in mice, there is no reason why such definitions and rules cannot be applied equally well. The symbols indicated in the listings conform for the most part to rules for mice. The system of numbers used by Dr. Rae Whitney is not in conformity, but since there has been no opportunity to discuss possible changes with Dr. Whitney, the strains are listed as submitted. Perhaps changes can be made in a revised listing at a later date. Most of the descriptions are abbreviated, probably because not much is known about any of the strains. Future revisions undoubtedly will contain more specific information. The abbreviation of names listed at the end of the section were checked with all the other lists to avoid duplications.

Table 34

ESTABLISHED STRAINS OF SYRIAN HAMSTERS [†]
(*Mesocricetus auratus*)

Name or Symbol	Synonym(s)	Remarks
H		INBR: F_{39}. ORIGIN: O. Mühlbock, 1949, from noninbred stock obtained from Hagedoorn. CHARAC: no information submitted. MAINTAINED BY: A.
HL		INBR: F_{21}. ORIGIN: O. Mühlbock, 1955, from noninbred stock obtained from Horning. CHARAC: no information submitted. MAINTAINED BY: A.
MX	MxBU	INBR: F_{32}. ORIGIN: H. Magalhaes from a noninbred commercial colony. CHARAC: small-medium body size, small litters (4–5 av.), coat color agouti. MAINTAINED BY: Mg.
W	WBU	INBR: F_{27}. ORIGIN: H. Magalhaes from a noninbred commercial colony. CHARAC: small litters (5 + av.), white mottled coat in females only, sex modified lethal (?) trait in males. MAINTAINED BY: Mg.
5.1		INBR: F_{23}. ORIGIN: Department of Biology, Boston University, 1950, from Ingham stock, to R. Whitney at F_{12}, 1956. CHARAC: slow to breed, senile at one year of age, some males show testicular atrophy, some females fail to ovulate, agouti, black eyes. MAINTAINED BY: Wh.

† A list of abbreviations follows table 37.

Table 35

STRAINS OF SYRIAN HAMSTERS IN DEVELOPMENT
(*Mesocricetus auratus*)

Name or Symbol	Synonym(s)	Remarks
CBC	CB	INBR: F_5. ORIGIN: W. H. Hildeman, 1958, from a closed (11 years) random-bred colony maintained at the Chester Beatty Institute. CHARAC: isohistogenic, agouti. MAINTAINED BY: Hi.
CBW	CB	INBR: F_{10}. ORIGIN: W. K. Silvers, 1957, from a closed (11 years) random-bred colony maintained at the Chester Beatty Institute. CHARAC: isohistogenic, agouti. MAINTAINED BY: Ss.
CRB	Cream BU, British Cream	INBR: $F_{10}+$. ORIGIN: H. Magalhaes from stock received from Cook (England). CHARAC: medium-large body size, agouti, black eyes. MAINTAINED BY: Mg.
IGH	ICRF/GH	INBR: F_{17}. ORIGIN: F. C. Chesterman from a heterogenous stock, 1955. CHARAC: agouti. MAINTAINED BY: Icrf.
IWH	ICRF/WH	INBR: F_6. ORIGIN: F. C. Chesterman from a pair born March, 1959, to a golden hamster containing bilateral orthotropic ovarian grafts from a white hamster. CHARAC: white. MAINTAINED BY: Icrf.
LSH		INBR: F_9. ORIGIN: W. K. Silvers received from the London School of Hygiene, 1957. CHARAC: isohistogenic, agouti. MAINTAINED BY: Hi, Ss.
MHC	MHA/c	INBR: F_8. ORIGIN: W. H. Hildeman, 1958, from a closed random-bred colony maintained at the National Institute for Medical Research, Mill Hill, England. CHARAC: isohistogenic, partial albino ($c^d c^d$). MAINTAINED BY: Hi.
MHW	MHA	INBR: F_{10}. ORIGIN: W. K. Silvers, 1957, from a closed random-bred colony maintained at the National Institute of Medical Research, Mill Hill, England. CHARAC: isohistogenic, partial albino ($c^d c^d$). MAINTAINED BY: Ss.
MIT	MIT(IG)	INBR: F_4. ORIGIN: R. Whitney, 1959, from Ingham stock. CHARAC: susceptible to dental caries on special diet. MAINTAINED BY: Wh.
TAB	Tawny BU	INBR: F_{10}. ORIGIN: H. Magalhaes. CHARAC: medium-large body size; pale agouti, black eyes. MAINTAINED BY: Mg.
WHB	White BU	INBR: F_8. ORIGIN: H. Magalhaes. CHARAC: irritable, pink eyes, white with gray ears and pigmented area around tail. MAINTAINED BY: Mg.
1.26		INBR: F_{11}. ORIGIN: R. Whitney, 1956, from a Schwentker × LaCasse hybrid. CHARAC: white, black ears, black eyes. MAINTAINED BY: Wh.
1.50		INBR: F_{12}. ORIGIN: R. Whitney, 1956, from a Schwentker × LaCasse hybrid. CHARAC: white, black ears, black eyes. MAINTAINED BY: Wh.
1.97		INBR: F_7. ORIGIN: R. Whitney, 1958, from Haddow cream × LaCasse white. CHARAC: white and cream, pink eyes, black eyes, black ears. MAINTAINED BY: Wh.
3.19		INBR: F_{14}. ORIGIN: R. Whitney, 1956, from Schwentker stock. CHARAC: agouti. MAINTAINED BY: Wh.
3.20		INBR: F_{12}. ORIGIN: R. Whitney from Schwentker stock, 1956. CHARAC: agouti. MAINTAINED BY: Wh.
4.1		INBR: F_{14}. ORIGIN: R. A. Adams, 1954, from LaCasse noninbred stock, to R. Whitney at F_3, 1956. CHARAC: white, black eyes, black ears. MAINTAINED BY: Wh.

Table 35—*Continued*

Name or Symbol	Synonym(s)	Remarks
4.22		INBR: F_{13}. ORIGIN: R. Whitney from Schwentker stock, 1956. CHARAC: agouti. MAINTAINED BY: Wh.
4.24		INBR: F_{12}. ORIGIN: R. Whitney from Schwentker stock, 1956. CHARAC: agouti. MAINTAINED BY: Wh.
4.39		INBR: F_{12}. ORIGIN: R. Whitney from Schwentker stock, 1956. CHARAC: agouti. MAINTAINED BY: Wh.
7.88		INBR: F_4. ORIGIN: R. Whitney from a Gulf Panda × Haddow Cream hybrid, 1959. CHARAC: white with pink eyes and black ears, or cream (self-spotted) with black eyes and black ears. MAINTAINED BY: Wh.
7.9		INBR: F_1+. ORIGIN: R. Whitney from Ingham stock, 1959. CHARAC: "buff," black eyes, pink ears. MAINTAINED BY: Wh.
8.9		INBR: F_5. ORIGIN: R. Whitney, 1959, from a Gulf Panda × Haddow Cream. CHARAC: white with pink eyes and black ears, or cream (self or spotted) with black eyes and black ears. MAINTAINED BY: Wh.
9.37		INBR: F_{10}. ORIGIN: R. Whitney, 1956, from Schwentker stock. CHARAC: agouti. MAINTAINED BY: Wh.
X.22		INBR: F_{12}. ORIGIN: R. Whitney, 1956, from Schwentker stock. CHARAC: agouti. MAINTAINED BY: Wh.
X.3		INBR: F_{13}. ORIGIN: R. A. Adams, 1954, from LaCasse stock, to Whitney, 1956, at F_3. CHARAC: white, black ears, pink eyes. MAINTAINED BY: Wh.
X.68		INBR: F_9. ORIGIN: R. Whitney, 1957, from a Schwentker stock × LaCasse stock hybrid. CHARAC: white, black ears, black eyes. MAINTAINED BY: Wh.

Table 36

STOCKS OF SYRIAN HAMSTERS OF GENETIC INTEREST
(*Mesocricetus auratus*)

Name or Symbol	Synonym(s)	Remarks
ss	White spotted Panda piebald	CHARAC: urogenital abnormalities (missing kidneys, uteri, seminal vesicles)—10% incidence. MAINTAINED BY: Mg.

Table 37

STRAINS OF CHINESE HAMSTERS IN DEVELOPMENT
(*Cricetulus griseus*)

Name or Symbol	Synonym(s)	Remarks
BUY		INBR: F_{15}. ORIGIN: G. Yerganian from Schwentker stock. CHARAC: fair–good reproduction, 15% incidence of diabetes mellitus, average mature body weight 33–35 g., fairly docile, exhibits brittle bristle, a sex-linked trait that causes hair follicles to fall and break off at 5–6 months of age. MAINTAINED BY: Ye.

Table 37—*Continued*

Name or Symbol	Synonym(s)	Remarks
HGYA		INBR: F_6. ORIGIN: G. Yerganian from a BUY × VSY hybrid; VSY males for this cross were irradiated with 400–500 r localized testicular X irradiation prior to mating. CHARAC: animals are extremely nervous, diabetes mellitus appears spontaneously. MAINTAINED BY: Ye.
HGYB		INBR: F_6. ORIGIN: G. Yerganian from a VSY × BUY hybrid; male BUY animals for this cross were irradiated with 400–500 r localized testicular X irradiation prior to mating. CHARAC: animals are extremely nervous, diabetes mellitus appears spontaneously. MAINTAINED BY: Ye.
JBY		INBR: F_9. ORIGIN: G. Yerganian from a JFY × BUY hybrid. CHARAC: mature body weight large (45–50 g.), past generations (F_{6-9}) manifested 90% incidence of diabetes mellitus, no evidence in current generations. MAINTAINED BY: Ye.
JBYA		INBR: F_3. ORIGIN: G. Yerganian from BUY × JFY hybrids. CHARAC: no information available. MAINTAINED BY: Ye.
JBYB		INBR: F_3. ORIGIN: G. Yerganian from JFY × BUY hybrids. CHARAC: no information available. MAINTAINED BY: Ye.
JFY		INBR: F_{13}. ORIGIN: G. Yerganian from BUY strain. CHARAC: female dominance over male evident, males pugnacious toward each other, mature body weight large (45–50 g.), polyuresis or diabetes insipidis still observed, occasional absence of outer ear in both sexes. MAINTAINED BY: Ye.
ORY		INBR: F_{11}. ORIGIN: G. Yerganian from BUY strain. CHARAC: mature body weight approximately 45 g., medium reproduction mildly susceptible to SE polyoma virus in newborn (tumors of various types appear 7–36 months after inoculation), some diabetes mellitus exhibited in earlier generations. MAINTAINED BY: Ye.
VSY		INBR: F_{13}. ORIGIN: G. Yerganian from BUY strain. CHARAC: late sexual maturity, average body weight 33 g., diabetes mellitus in earlier generations (10% at F_{8-10}). MAINTAINED BY: Ye.

LIST OF SYMBOLS FOR DESIGNATING SUBSTRAINS AND STOCKS OF HAMSTERS

A   Dr. O. Mühlbock, Antoni van Leewenhoekhuis, Sarphatistraat 108, Amsterdam C, Netherlands.

Hi   Dr. W. H. Hildeman, University of California Medical Center, Los Angeles, California.

Icrf   Imperial Cancer Research Fund, Central Laboratories, Burtonhale Lane, The Ridgeway, Mill Hill, N. W. 7, England.

Mg   Dr. Hulda Magalhaes, Department of Biology, Bucknell University, Lewisburg, Pennsylvania.

Ss   Dr. W. K. Silvers, The Wistar Institute, Philadelphia 4, Pennsylvania.

Wh   Dr. Rae Whitney, Bio-Research Consultants, Inc., 9 Commercial Avenue, Cambridge 41, Massachusetts.

Ye   Dr. George Yerganian, Laboratories of Cytogenetics, Harvard University, School of Medicine, Boston, Massachusetts.

V. RABBITS

A formal listing of genetic stocks of rabbits has not been made before, and consequently the information received in this survey is of considerable interest. At least two fairly well inbred strains exist in this country, and several others are well on the way to becoming established strains. The stocks bearing marker genes maintained by Dr. C. Cohen are of considerable value for certain immunogenetic work, and it is likely that more populations of this kind will be developed in the future. There is no doubt that a vast reservoir of genetic material exists in the many stocks maintained by fanciers, but unfortunately they are not readily identifiable at this time. It would be most desirable if genetic material for this species, comparable to the murine material, could be collected in one or more places and properly maintained for mammalian genetic research.

The standardization of strains and stocks of rabbits has not been done in a manner similar to mice, but the adoption of the same rules for designating and defining strains and substrains and the promulgation of symbols is desirable. However, it may be advisable to amend the rules so that inbred strains can be defined in terms of an inbreeding coefficient (F) in addition to the definition based on matings of brother ×

Table 38†

STRAINS OF RABBITS IN DEVELOPMENT

The amount of inbreeding is expressed either as F generations of brother × sister matings or by the inbreeding coefficient (F).

Name or Symbol	Synonym(s)	Remarks
A		INBR: F_{19}. ORIGIN: M. Lurie, 1932, from a heterogeneous stock obtained from E. L. Stubbs, University of Michigan, b × s matings since the start of inbreeding. CHARAC: resistance to tuberculosis variable (highly resistant up to F_8), low fertility, many die of impaction of the large intestine (perhaps faulty peristalsis?). MAINTAINED BY: Lu.
AC		INBR: $F = 36\%$. ORIGIN: P. Sawin from Rockefeller Institute Dutch stock. CHARAC: black, recessive white marking, aa or A-E^dE^ddu-, occasionally segregating for y and ac, mature female body weight 2200 g. MAINTAINED BY: Sa.
ACEP		INBR: $F = 70\%$. ORIGIN: P. Sawin from Rockefeller Institute Dutch stock. CHARAC: aa or A-E^dE^ddu-$epep$ (black, recessive white marking, audiogenic seizures), occasionally segregating for v and agonadia, mature female body weight 2400 g. MAINTAINED BY: Sa.
AD		INBR: F_8. ORIGIN: M. Lurie from a Strain A × Strain D cross (Strain D is now extinct). CHARAC: intermediate resistance to tuberculosis, medium fertility, dutch color. MAINTAINED BY: Lu.
AX		INBR: $F = 36\%$. ORIGIN: P. Sawin from outcross of chinchilla Race V to Races III and X. CHARAC: Ach^dww, $Axas$ forced heterozygosis, occasionally segregating for du, ep, bu, ch^m, c, mature female body weight 3500 g. MAINTAINED BY: Sa.
B		INBR: F_{10}. ORIGIN: C. Cohen from Rockefeller Institute Dutch stock, separated from Strain Y and Strain R at F_2, to C. K. Chai at F_7. CHARAC: segregating for yellow fat y, and angora (1), albino, mature body weight 2000 g. MAINTAINED BY: Ci.

† A list of abbreviations follows table 39.

Name or Symbol	Synonym(s)	Remarks
C		INBR: F_{18}. ORIGIN: M. Lurie, 1932, from Swift stock (originally received from Bull and Webster), b × s matings. CHARAC: highly susceptible to tuberculosis, many adenocarcinoma of the uterus, low fertility, nervous, albino. MAINTAINED BY: Lu.
CAC		INBR: F_8. ORIGIN: M. Lurie from an F_{12} Strain C × Ff Carworth animal, b × s matings since. CHARAC: susceptible to tuberculosis and snuffles, fair fertility, nervous, dutch with occasional albino. MAINTAINED BY: Lu.
DA		INBR: $F = 53\%$. ORIGIN: P. Sawin from a New Zealand white stock (California), 1949. CHARAC: albino (*ce A-E^d*), mature female body weight 4200 g. MAINTAINED BY: Sa.
OS	Os	INBR: $F = 53\%$. ORIGIN: P. Sawin, 1948, from Rockefeller Institute stock. CHARAC: *aa* or E^dE^dOsos (black, minimal recessive white marking), occasionally segregating for hydrocephaly, mature female body weight 3200 g. MAINTAINED BY: Sa.
R		INBR: F_{11}. ORIGIN: C. Cohen from Rockefeller Institute Dutch stock, separated from Strain Y and Strain B at F_2, to C. K. Chai at F_7. CHARAC: segregating for yellow fat, *y*, and angora (1), albino, mature body weight 2000 g. MAINTAINED BY: Ci.
TA	III A	INBR: F_6. ORIGIN: M. Lurie from an F_5 Strain III × F_{14} Strain A, b × s matings since then. CHARAC: medium resistance to tuberculosis, resistant to snuffles, fertile, albino. MAINTAINED BY: Lu.
TTC	III III C	INBR: F_6. ORIGIN: M. Lurie from a Strain III × Strain C cross, backcrossed to Strain III, b × s matings since then. CHARAC: intermediate resistance to tuberculosis, resistant to snuffles, fair fertility, young animals susceptible to infantile diarrhea, albino. MAINTAINED BY: Lu.
T3F	III³F	INBR: F_6. ORIGIN: M. Lurie from a Strain III × Swift stock cross, backcrossed twice to Strain III, b × s matings since then. CHARAC: high resistance to tuberculosis, dutch agouti with occasional albino. MAINTAINED BY: Lu.
Y		INBR: F_{11}. ORIGIN: C. Cohen from Rockefeller Institute Dutch stock, to C. K. Chai at F_7. CHARAC: segregating for yellow fat, *y*, angora (1), albino, mature body weight approximately 2000 g. MAINTAINED BY: Ci.
III	Race III, T	INBR: $F = 70\%$. ORIGIN: P. Sawin from Castle's New Zealand white stock, 1932, probably all subsequent sublines of this strain have come from this colony. CHARAC: $ccAAE^dE^d$ (albino), occasionally segregating for angora, *bu*, *ep*, and scoliosis; Lu subline is now F_9, highly resistant to tuberculosis and snuffles, docile. MAINTAINED BY: Lu, Sa.
III C		INBR: $F = 53\%$. ORIGIN: P. Sawin from Strain III. CHARAC: $aaAAE^dE^d$ (albino), occasionally segregating for *bu*, scolosis. MAINTAINED BY: Sa.
III R		INBR: F_9. ORIGIN: M. Lurie from Strain III animals that survived a virulent bovine tuberculosis infection. CHARAC: similar to Strain III, resistant to tuberculosis and snuffles, docile, albino. MAINTAINED BY: Lu.
X		INBR: $F = 53\%$. ORIGIN: P. Sawin from Castle's small race. CHARAC: $aaeebbC(CH^2)$, sooty yellow, occasionally segregating for r^2, *dk*, *Dw*, scoliosis, *As*, mature female body size 2200 g. MAINTAINED BY: Sa.

sister or parent × offspring, since less stringent systems of mating sometimes have been expedient. Thus, in the case of rabbits, the inbreeding can be expressed either as the number of generations of brother × sister or parent × offspring matings or in terms of the inbreeding coefficient (F). The listings are prepared in this manner. As for the other species already mentioned, it is recommended that the Committee on Standardized Nomenclature extend its coverage to include the rabbits. Some changes have been made in symbols (according to the rules for mice), and concurrence was obtained from the contributor for the acceptance of such changes. The abbreviations in the contributor list, as in the case of previous lists, were checked to avoid duplication.

Table 39

STOCKS OF RABBITS OF GENETIC INTEREST

Name or Symbol	Synonym(s)	Remarks
BLOOD GROUP GENES STOCKS		ORIGIN: C. Cohen, by selection of identified blood group genes. CHARAC: One stock is segregating for Hg^A, the other four stocks are segregating for H_gD, H_gF, Hc, hc, He, Hh, hh. MAINTAINED BY: Cn.

LIST OF SYMBOLS FOR DESIGNATING STRAINS AND STOCKS OF RABBITS.

Ci   Dr. C. K. Chai, R. B. Jackson Memorial Laboratory, Bar Harbor, Maine.

Cn   Dr. C. Cohen, Battelle Memorial Institute, Columbus, Ohio.

Lu   Dr. Max Lurie, The Henry Phipps Institute, Philadelphia, Pennsylvania.

Sa   Dr. Paul B. Sawin, R. B. Jackson Memorial Laboratory, Bar Harbor, Maine.

VI. *PEROMYSCUS SP.*

Peromyscus sp., as a research animal in mammalian genetics and in medicine, has not been widely used, despite the number of identified genetic traits and the possible importance of some of these traits. For this reason, the locations of *Peromyscus sp.* stocks are few. It is believed, however, that additional stocks do exist in the United States, but applicable information had not been submitted by the time this list was closed. It is hoped that future revisions will include such stocks.

The stocks herein listed are given as submitted by the contributors. Since there apparently are no inbred strains as such, but only stocks with identified genes, no changes were made in stock designation. The descriptions given are very brief† but sufficient to illustrate the kind of material available.

Because no inbred strains exist, and, therefore, no problems have yet been created with respect to strain definition, strain symbols, and so forth, it is probably not necessary to establish a standardized system at this time. However, it is likely that such strains

† A more detailed description of these genes can be obtained by writing to Dr. Elizabeth Barto, Mammalian Genetics Center, University of Michigan, Ann Arbor, Michigan.

will eventually be developed, so these problems may occur some day.   If the Committee on Standardized Nomenclature does promulgate rules for rats, guinea pigs, rabbits, and hamsters, possibly they will also consider the potential problem of *Peromyscus sp.*

Table 40

MUTANT STOCKS OF *Peromyscus maniculatus*

Name or Symbol	Synonym(s)	Remarks
Albino		ORIGIN: Wild population, California. GENET: *c.* CHARAC: linkage group I, apparently identical to albino phenotype of the laboratory mouse. MAINTAINED BY: Bar, Mc.
Blaze *		No information available on origin, genetics, or characteristics. MAINTAINED BY: Bar.
Boggler		ORIGIN: No information submitted. GENET: *bg.* CHARAC: no information submitted. MAINTAINED BY: Bar.
Brown	Brown-tip	ORIGIN: Wild population, New Mexico. GENET: *b.* CHARAC: linkage group IV, very similar to the cinnamon phenotype of the laboratory mouse, sepia pigment granules brown, reduced in size, yellow pigment granules not visibly affected. MAINTAINED BY: Bar, Mc.
CNV Convulsive		No information submitted on the origin, genetics, or characteristics. MAINTAINED BY: Bar.
Dilute		ORIGIN: Wild population, New Mexico. GENET: *d.* CHARAC: linkage group IV, similar in appearance to dilute or leaden genes in the laboratory mouse, variable expression. MAINTAINED BY: Bar, Mc.
EP (Sound induced) Convulsive		ORIGIN: No information submitted. GENET: *e.* CHARAC: no information submitted. MAINTAINED BY: Bar.
Flexed Tail		ORIGIN: Wild population, Oregon. GENET: *f.* CHARAC: linkage group I, caudal vertebrae variously malformed, tail usually visibly kinky, knobby at the end, or shortened, highly variable, no evidence of associated belly spotting or anemia. MAINTAINED BY: Bar, Mc.
Gray	Gray-bond	ORIGIN: Wild population, New Mexico, gene frequency high in certain localities. GENET: *g.* CHARAC: superficially resembles chinchilla in the laboratory mouse. MAINTAINED BY: Bar, Mc.
Hairless		ORIGIN: Wild population, California. GENET: *hr.* CHARAC: linkage group III, very similar to hairless in the laboratory mouse, moulting pattern irregular, fertility of females reduced. MAINTAINED BY: Bar, Mc.
Ivory		ORIGIN: Wild population, Oregon. GENET: *i.* CHARAC: similar to albino, but hair retains a moderate amount of sepia pigment in juvenile pelage, slight amount in adults. MAINTAINED BY: Bar, Mc.
Pink Eyed Dilution		ORIGIN: Laboratory stock. GENET: *p.* CHARAC: linkage group I, apparently identical to pink eyed dilution phenotype of the laboratory mouse, dilution and modification of the shape of sepia pigment granules. MAINTAINED BY: Bar, Mc.
Platinum*		ORIGIN: Laboratory stock. GENET: *pt.* CHARAC: similar to dilute and silver, but almost always with high-grade expression, hair bases vary from gray to almost white. MAINTAINED BY: Bar, Mc.
Rotator*		No information available on origin, genotype, or characteristics. MAINTAINED BY: Bar.

Table 40—*Continued*

Name or Symbol	Synonym(s)	Remarks
Silver		ORIGIN: Wild population, Oregon. GENET: *si*. CHARAC: linkage group I, very similar to dilute, equally variable in expression. MAINTAINED BY: Bar, Mc.
Spinner		ORIGIN: Wild population of *P. polionotus rhoodsi*, Florida, now transferred to *P. maniculatus* background. GENET: *sp*. CHARAC: whirling behavior and early deafness, whirling expression variable but almost always detectable. MAINTAINED BY: Bar, Mc.
Spotting	Dominant spot	ORIGIN: Wild population, Illinois. GENET: *S*. CHARAC: high variable amount of white spotting on head, belly and distal portion of tail, homozygotes variable, probably not distinguishable from heterozygotes, some reduction in penetrance. MAINTAINED BY: Bar, Mc.
THF (Sound-induced) Convulsive*		No information submitted on origin, genetics, and characteristics. MAINTAINED BY: Bar.
Whiteside		ORIGIN: No information submitted. GENET: *wh*. CHARAC: No information submitted. MAINTAINED BY: Bar.
Wide Band		ORIGIN: Wild population, Nebraska, of *P. maniculatus nebrassensis*, gene frequency probably high in the Sand Hills region. GENET: *Nb*. CHARAC: linkage group V, longer agouti band and somewhat shorter sepia tip on hair. MAINTAINED BY: Bar, Mc.
Waltzing	Waltzer	ORIGIN: Wild population of *P. maniculatus hairdi*, Iowa. GENET: *v*. CHARAC: linkage group V, similar to Spinner, but deafness not occurring or appearing late in life. MAINTAINED BY: Bar, Mc.
Yellow		ORIGIN: Wild population of *P. maniculatus gambeli*, California. GENET: *y*. CHARAC: linkage group II, longer agouti band, sepia tip of hair much shortened or occasionally absent, white tip of ventral hair lengthened, nonagouti hairs usually absent, monatricks usually with agouti band. MAINTAINED BY: Bar, Mc.

* Tentative designations; not yet described in publications.

Table 41

OTHER STOCKS OF *Peromyscus sp.* OF POSSIBLE GENETIC INTEREST

Name or Symbol	Synonym(s)	Remarks
BW/RX		ORIGIN: Wild population of *Peromyscus maniculatus hairdi*, from Washtenaw County, Michigan. CHARAC: no known mutant genes, wild-type stock, all mutant stocks have been crossed into this stock at least twice, maintained by avoiding full and half sib matings. MAINTAINED BY: Mc.
BW/RA		INBR: F_3. ORIGIN: W. McIntosh from BW/RX, continuing attempts to inbreed b × s have not produced progeny beyond F_5, current attempt now at F_3. CHARAC: wild type. MAINTAINED BY: Mc.
PO		ORIGIN: Wild population of *Peromyscus polionotus* from Ocala, Florida. CHARAC: no known mutant genes although an unidentified whirling trait appears occasionally, small size, will cross reciprocally with *Peromyscus maniculatus* producing fertile hybrids. MAINTAINED BY: Mc.

Bar   Dr. Elizabeth Barto, Mammalian Gene-
      tics Center, University of Michigan,
      Ann Arbor, Michigan.

Mc   Dr. W. B. McIntosh, Department of
      Zoology and Entomoloy, Ohio State
      University, Columbus, Ohio.

SUMMARY

The material on genetic strains and stocks compiled and presented represents an attempt to bring together as much information as possible on the location and status of such strains and stocks. It is hoped that this compilation, although incomplete, will be of value in supplying information on the location and status of laboratory rodent material of medical genetic interest.

The six common species of laboratory rodents (mice, rats, guinea pigs, hamsters, rabbits, *Peromyscus sp.*) are listed. Other species will doubtless be added in the future as their usefulness increases.

Recommendations have been made regarding the adoption of rules for defining strains and substrains, for the designation of symbols, and for the extension of the jurisdiction of the Committee on Standardized Genetic Nomenclature for Mice to include rats, guinea pigs, hamsters, rabbits, and *Peromyscus sp.*

RADIATION GENETICS

Douglas Grahn, Ph.D.

MAMMALIAN RADIATION GENETICS

The best qualification of the mouse for studies on radiation genetics is that it has certain attributes of our best experimental means to the end, *Drosophila*, and the experimental end itself, man.   Although it may be many years before murine genetics will have the esoteric qualities of *Drosophila* genetics, the mouse offers us the opportunity of obtaining quantitatively reliable data on radiation effects that can be checked against *Drosophila* for theoretical consistency and extrapolated to man without zoological inconsistency.   Other mammalian species have been and are presently being employed, such as the rat, the hamster, swine, sheep, bovines, and even the monkey. But, for reasons of economy and general genetic and biological background, the mouse will certainly continue to reign supreme for some time in the field of mammalian radiation genetics.

Prior to World War II, radiation genetics progressed in an orderly fashion as a relatively subsidiary area of interest in the field of genetics.   With the advent of the atomic age and the potentiality of widespread contamination of the biosphere with the by-products of nuclear energy applications, radiation genetics rapidly became a major field of research activity and interest.   Fortunately, many different sources of radiation have become available for critical experimental purposes, but as part of the price of technical advancement, many sources of radiation have also become an ubiquitous part of our general environment.   The quantitative and realistic evaluation of the potential cost of nuclear energy to man's genetic worth has thus become highly important.

Those familiar with modern radiation biology are well aware of the complexities of this cross-bred science.   In many instances, the combined talents of diverse biological and physical scientists are required.   The study of radiation effects is not often suitable for isolated investigators.   Anyone anticipating research activity in the field of mammalian radiation genetics is well advised to enlist the consulting services of a radiological physicist and a radiobiologist.

RADIATION PARAMETERS

As intimated above, there is a wide array of radiation sources available for experimental use; and the source or sources should be carefully chosen for their general applicability and, in some studies, for their use as a simple means of inducing injury. The methodology of exposure is also of extreme importance.   This involves the correct selection of total doses, dose rates, and other temporal factors that will be discussed below.   Before enumerating radiation parameters, it should be stated that accurate and uniform measurement of the absorbed dose is essential.   The techniques and instrumentation of modern radiation dosimetry are sometimes quite specialized and the experimental biologist should not hesitate to ask for outside assistance.   The best biological measures are of little value when the physical parameters are uncertain or inaccurate.   Considerable insight into the questions of dosimetry and radiation sources may be obtained by referring to Hine and Brownell,[582] Glasstone,[438] Fano,[346] and Marinelli and Taylor.[850]

SOURCES OF RADIATION

X and gamma radiations.—These need little description as they have been the principal sources of radiation for most genetic studies.   A wide range of energy levels is available for both radiations, but, for mammalian studies, it is advisable to avoid energies below 120 *kev* in order to avoid the problem of nonuniformity of absorbed dose.[476]

When energies appreciably above 500 *kev* are employed, the specific ionization or linear energy transfer per unit length of the ionization track declines; and a factor of relative biological effectiveness, the RBE, may be needed to provide comparative analysis with data obtained from more conventional deep therapy X-ray units of 200–300 *kev*.   The most economical gamma-ray sources available are in the higher energy class: cesium-137 and cobalt-60 with photon energies of 0.66 *Mev* and 1.25 *Mev*, respectively.

Neutrons.—Neutron energies vary from 0.025 *ev* for thermal neutrons to over 20 *Mev* for fast neutrons.   Linear accelerators can provide reasonably monochromatic energies, whereas the neutron-energy spectrum from the fissioning of uranium-235 is extremely heterogeneous and has an average energy of about 2.5 *Mev*.   Details of the spectrum can be obtained in Lapp and Andrews.[758]   The RBE for neutrons is extremely variable and for most biological effects reaches a maximum value at energies of 2–4 *Mev*.   The RBE itself may vary with the biological endpoint and all too little is known about this factor for induction of mutation in mammalian systems, although considerable work has been done with *Tradescantia* and *Drosophila*.[222, 317] More accurate knowledge of the nonlinear relation between RBE and energy is critical, since the maximum RBE is in the region of the mean energy for fission neutrons.

Unfortunately, neutron sources are generally expensive and require careful monitoring; thus, they are not widely available. On the other hand, the increasing number of training reactors on college campuses may improve neutron source availability for genetic studies. A complete discussion of neutron physics and neutron sources is given in Glasstone[438] and Lapp and Andrews.[758]

Cosmic radiation.—As man prepares to enter the new and rather exciting scientific era of space flight, considerable increase in interest in the estimation of the genetic effects of cosmic radiation can be expected. There are two radiation belts, the Van Allen belts, held in the earth's magnetic field. The outer belt is composed of comparatively low-energy electrons, predominantly in the 20 to 100 *kev* range, which can be easily shielded out with aluminum. However, the production of soft secondary X rays may still present some difficulties. The inner belt, which comes to within 600–700 miles of the earth's surface, contains extremely high-energy protons. The energy spectrum is not fully known, but ranges up to at least 700 *Mev*. Present data suggest that the dose rate can reach 10 *r*/hour at altitudes of about 1,600 miles and 10,000 miles for the inner and outer belts.

In addition to this trapped radiation, there are the primary cosmic particles. These are composed of about 85 per cent protons, 15 per cent helium atoms, and less than 1 per cent heavy nuclei such as carbon, calcium, iron, and oxygen. The cosmic primaries go into the billion-electron-volt energy range and can penetrate deep into tissue with a very dense ionization track. The iron nucleus, for example, has a maximum ionization density of 100,000 ion pairs per micron and can produce a dose to an individual cell of over 1,000 *rep*. A comprehensive recent review of space radiation has been presented by Schaefer.[1161]

Certainly, very few humans will be subjected to these radiations, but their ability to induce severe damage all along the ionizing pathway, their unknown RBE values, and our inability to duplicate the very high-energy heavy nuclei with earthbound machines present an intriguing set of problems. Some progress is being made with high-energy linear-accelerator beams of stripped nuclei and deuterons at the Lawrence Radiation Laboratory, University of California, and at Brookhaven National Laboratory.[201, 234]

Internal emitters or internally deposited radioisotopes.—The vast majority of radionuclides present no unique genetic problem. Those nuclides with an affinity to bone, for example, are of little genetic concern. Gamma-emitting isotopes that seek the soft tissues, such as cesium-137, can be treated as any typical external radiation source. The more difficult problems arise from those isotopes that can become incorporated into the genetic materials. Tritium, carbon-14, and phosphorous-32 are good examples and all emit a beta particle without an associated gamma-photon emission.

Thus, their radiations are highly localized, often to the cell in which they are deposited.

Very little mammalian genetic work has been done with these, but certain techniques appear to be accurate enough to warrant more study. For example, death of spermatogonial cells has been quantitatively measured following the simple intraperitoneal injection of tritiated thymidine into mice.[664] The results compare excellently with those obtained by external gamma-radiation,[952] and offer a straightforward estimate of comparative toxicity.

The importance of carbon-14 to the problem of long-term genetic hazards to man and animals from past nuclear weapons testing has been emphasized by Totter *et al.*[1326] and Pauling,[993] yet only empiric estimates are available on the relative contribution of the three potentially injurious consequences of radioactive decay of C^{14} to N^{14}: transmutation, atomic recoil, and ionization.

Another isotope of considerable interest is deuterium. Though not a radioactive isotope of hydrogen, it is D_2O or heavy water that is commonly used as a moderator and coolant in nuclear reactors. Its introduction into biological systems presents many intriguing problems that have been reviewed by Katz *et al.*[689] and Bennett *et al.*[78] Nearly complete sterility can be induced in mice when they are provided a 30 per cent D_2O concentration in the drinking water. Lesser amounts of D_2O induce partial sterility from which recovery will occur. Since the size of litters remains unaffected, there is no evidence of genetic damage, but deuterium may well be mutagenic when incorporated into the genetic materials.

While the above summary of radiation sources is by no means all-inclusive, the brief descriptions do indicate something of the type and energy of radiations that can be employed and a few of the problems that exist.

TEMPORAL FACTORS IN RADIATION

When one begins to consider the number of permutations and combinations of radiation exposure that can be employed, it soon becomes apparent that it is probably impossible to account for all possibilities experimentally and even operationally important. To start with, the radiation dose can be delivered in the form of a single exposure (often referred to as an acute exposure, which should not be the preferred terminology), multiple exposures or a fractionated exposure, or a continuous exposure (often called a chronic exposure, again not the preferred terminology). Within any one of these, there can be variation in total dose and rate of dose delivered per unit time, or dose rate. Obviously, for continuous exposures, dose rate and total dose are positively related, although termination of the exposure can be varied to hold total dose constant under different dose rates. This, then, varies the duration of exposure or protraction period.

Fractionated exposures may be even more complex. One can vary the number of doses or fractions, the total dose, the dose rate, the interval between doses, the size of the individual fractions, and the total protraction period. Again, a little consideration of these factors points out a number of interdependent relationships which can at times result in difficulties when attempts are made to isolate the effect of some one particular factor.

Variations in the pattern of exposure have been employed for years by geneticists to assist in understanding the kinetics of chromosomal breakage and restitution. The simple use of paired doses usually has been the method of choice. The effects of a given total dose, delivered in a single exposure for determination of base-line, is compared to the effects following the same total dose delivered in two equal parts with a varying interval between halves. The value of this procedure for cytogenetic studies, which have largely been done with plant material and *Drosophila*, needs no documentation here, but apparently the paired-dose method has not been employed for mammalian genetic studies. Weekly and daily fractionations of dose have been used, however, in a number of studies on the induction of sterility.[328, 1085, 1087, 1114] Rather than extend the discussion of exposure pattern, reference will be made to certain applications of these variables at pertinent places in the discussion of genetic tests. The following summary of exposure variables is therefore given.

I. Single exposure; with variation in:
   a. Total dose
   b. Dose rate

II. Fractionated exposure; with variation in:
   a. Total dose
   b. Size of individual dose or fraction
   c. Number of individual fractions
   d. Dose rate/fraction
   e. Interval between fractions
   f. Protraction period or interval between first and last fractions

III. Continuous exposure; with variation in:
   a. Total dose
   b. Dose rate
   c. Protraction period

Of course, these factors are fully applicable to the use of external radiations. When internal emitters are employed, some modification is required because of the lack of discreteness to the exposure period. The duration of exposure will depend upon a combination of the radioactive half-life, metabolic activity, and rate of excretion. Continuous exposure to internal emitters at a constant dose rate can be accomplished by the proper adjustment between input of dose increment and decay and excretion rates.

TECHNIQUES OF EXPOSURE

Although the irradiation of an animal initially may seem a simple, straightforward procedure requiring little more than the placement of the creature under an X-ray tube and turning the machine on and off, contemporary standards require recognition of a few principles of radiological physics and of problems of variation in machine output. It is standard practice to place the animals on a rotating board during X irradiation. This will randomize variations in the dose field due to slight misalignment of the tungsten target of the tube and to unequal absorption of some of the X rays by the target itself. The board, usually a one-half-inch slab of masonite, serves as a back-scattering device to help assure the attainment of electron equilibrium and the elimination of variations in tissue depth dose. Exposure of mice and rats from a single plane, either dorsal or ventral, will usually give a uniform tissue dose, provided that X rays of 120 *kev* or greater are employed with filtration adequate to remove the lowest energy components of the total spectrum.

When photon energies of 1 *Mev* and above are used (Co[60] is an example), the animals should be in a chamber-like device that provides both forward and backward scattering of the secondary electrons. The forward scattering is necessary since high-energy gamma and X rays must penetrate about 3 or 4 millimeters of tissue or tissue-equivalent material to reach secondary electron equilibrium. In the absence of such scattering material, the tissue dose will build up in the animal and lead to an irregular depth-dose curve. Four millimeters of lucite are sufficient to insure equilibrium for the 1.25 *Mev*, Co[60] gamma photon.

When animals the size of guinea pigs, rabbits, monkeys, dogs, and larger are irradiated, it generally becomes necessary to use a bilateral exposure technique. The total dose is delivered in two equal parts, one part to each lateral surface. This prevents sharp changes in depth dosage and accompanying inequalities of exposure of internal organs. Unilateral irradiation of female dogs, for example, could lead to some discrepancy in dose delivered to the two ovaries. Detailed discussions of the pattern of depth dose for different energy radiations in mice, rats, and rabbits are given by Grahn *et al.*[476] and for the domestic animals and man by Bond *et al.*[115]

A common procedure for genetic studies involves the use of partial body exposure. When doses above the midlethal level are required, local irradiation of the gonads must be done. Simple lead hemispheres can be devised for small mammals which permit full exposure of the gonads with only a limited exposure of surrounding tissue. For obvious reasons, the procedure is simplest for males. Sheet lead one-eighth inch thick provides excellent shielding for 250 *kev* radiations and below. Partial body exposure with high-energy X and gamma radiation and neutrons cannot be accomplished through the use of shields alone. Beam collimation is required, although this is not always feasible with all radiation sources. Thus, for some situations, the total dose will be controlled by the survival of the animal rather than by technical manipulations.

RADIATION GENETIC ANALYSIS IN MAMMALS

Apparently, the first effort to study the genetic effects of radiation in mice was reported by Little and Bagg in 1923 and 1924.[801] Although several new mutations were detected, a radiation origin could not be substantiated.   Snell[1250] and Hertwig[551] did report positive evidence for the induction of gene mutations in mice by X irradiation several years after Muller had reported this for *Drosophila*.   All of the early literature has been completely and excellently reviewed by Grüneberg[507] and Russell,[1128] and anyone considering entering the field of mammalian radiation genetics would be well advised to review the pioneering efforts of G. D. Snell, P. Hertwig, H. Brennecke, and their co-workers.

INDUCED STERILITY

Prior to any consideration of genetic analysis, the problem of radiation-induced sterility must be faced.   Interest in this field traces back to the early part of the century[743] and considerable activity continues to the present.   It is, of course, a problem of practical concern to the radiation geneticist, since certain experimental conditions may be precluded by the induction of either temporary or permanent sterility.   A certain amount of trial-and-error methodology is still required for many species, since only the male mouse has been studied in fairly complete detail.[328, 952, 953, 955, 957]   The work of Oakberg can be considered a model for those who wish to study the sterilizing effects of ionizing radiation.   He has pointed out the need for careful timing of post-irradiation intervals for sampling, exact identification of cellular type and maturation stage, and the need for correction for architectural distortion due to shrinkage of tubules.   Spermatogonial cells of the mouse are extremely sensitive to radiation and have an LD_{50} dose of 20 to 24 roentgens.   An increase in cellular death is even detected at single doses as low as 5 *rad* of gamma rays or 2 *rad* of fast neutrons.[1142]

Johnson and Cronkite[664] used Oakberg's techniques in a recent study to evaluate the effect of tritiated thymidine on spermatogonial cells.   Since exact dose rate and total dose from the tritium are not known, a comparison with data from external radiation permits use of the curve for cell-killing as a bioassay.   A dose of 1 μc/gram of body weight, for example, after 60 hours of exposure produced an effect equivalent to something less than 5 *r* of external, Co^{60}, gamma radiation.

The reproductive performance of irradiated males will vary from species to species with respect to dose sensitivity, duration of reduced fertility or complete sterility, and time of recovery to near normal fertility.   As a general rule, there are three distinct periods: the pre-sterile period that immediately follows exposure, the sterile period, and the post-sterile period.   During the pre-sterile period, fecundity gradually declines as post-meiotic germ cells are cleared through.   This period is used for the study of dominant-lethal induction rates, translocation rates, and mutation rates in spermatocytes, spermatids, and spermatozoa.   The post-sterile-period matings are used to study mutations induced in spermatogonial cells.   Thus, the stage of the cell at the time of exposure can be quite accurately defined for genetic analysis.

The sterilizing effects of radiation of the female have not been as thoroughly investigated as in the male, but there is a tremendous species factor in radiosensitivity.

The mature female mouse is permanently sterilized by a single dose of 50 *r* or greater.[1087, 1147] The newborn female, however, is extremely resistant and will show normal fertility even after a single dose of 300 *r*.[1126.] At several weeks of age, however, a period of extreme sensitivity occurs so that even a low dose rate such as continuous exposure to about 8 milliroentgens per minute for a total dose of 85 *r* will cause complete sterility. This is in contrast to the lesser sterilizing effects of a protracted exposure on the mature female in which doses up to nearly 300 *r* of X rays may be accumulated at the rate of 10 *r*/week before complete sterility occurs.[1085, 1087, 1114] Reproduction studies have very clearly indicated that no oögonia are present in the mature ovary and that the bulk of the cells are primary oöcytes.[956] These are most radiosensitive in early stages of development of the follicle. The more mature stages go through to ovulation but are not replaced after a sterilizing exposure.

Females of other mammalian species are not as radiosensitive as the mouse to the sterilizing effect of radiation. This phenomenon is not clearly understood and apparently no working hypothesis has been set forth. Rats and rabbits are only temporarily sterilized by doses above 600 *r*.[743] The female beagle hound that survives an acute radiation syndrome induced by a near midlethal dose of 300 *r* whole-body X irradiation shows excellent reproductive performance in terms of litter size and estrus activity, with even a tendency to improved lactation.[22, 1201] Exposure to fast neutrons is more effective than X irradiation for inducing a form of sterility in dogs, but the data are not adequate enough to determine if an unusually high RBE might be involved.[22] In this instance, the dogs bred but were unable to whelp or lactate.

The above discussion of the sterilizing effects of ionizing radiation was not intended to be complete and a fuller discussion will be found in a report by Oakberg.[954] Additional studies have been reported for dogs, rats, and mice, following single, fractionated, and continuous exposure to X rays, gamma rays, and neutrons.[179, 846, 847, 937, 1194] In summary, the degree and time of sterility varies with species, sex, dose rate, age at exposure, and quality of radiation. However, the pattern of reproductive performance following exposure is employed to control the cellular stage of interest for mutation studies and therefore must be understood by the investigator before genetic analysis can be carried out.

QUALITATIVE GENETIC EFFECTS OF RADIATION

Hereditary partial sterility.—Some of the offspring of irradiated males were observed to produce consistently small litters by Snell[1250] in one of the early radiation genetic studies in mice. Litter size was reduced by about one-half, and one-half of the progeny of these litters expressed the trait in the next generation. The characteristic, which behaves as a dominant trait, is generally classed as an hereditary dominant partial sterility and has been regularly observed in all studies to date with mice.[198, 550, 731, 1132] Snell[1232] demonstrated by the use of marker genes that the characteristic definitely involves a reciprocal translocation. The original interpretation was derived from the genetic behavior of the trait, but cytogenetic proof has also been obtained.[39, 731] Embryonic deaths appear to occur largely at or soon after implantation.[973]

Partial sterility resulting from induced reciprocal translocation is almost entirely restricted to the mature-germ-cell stages, possibly only significantly in spermatozoa. It has been suggested that aberrations of this type would not survive meiosis and therefore the yield in gonial cells would be negligible. This has been challenged by Griffen[497] but his data await complete cytologic confirmation. The simple use of reduced litter size as the detector can be misleading, although it is the normal procedure of first screening for semisterile animals. The frequency in offspring from irradiated females is much less than in offspring from irradiated males.[1127] Since the germ cells of the mature ovary are largely primary oöcytes, this finding indirectly implies that the most mature germ cell is more sensitive to the induction of a transmissible partial sterility; germ-cell death probably culls out this damage in the less mature cells.

The dose-response data are erratic.[1128] Theoretically, one would expect the yield of semisterile mice to be linearly related to the square of the dose since a two-hit aberration is involved. The early work, summarized in the above reference, does not appear to demonstrate any clear-cut dose-response relationship. The data obtained by Charles et al.[198] do, however, fit a (dose)^2 function with a greater reduction of the variance than the simple linear-arithmetic function employed by the authors. In the experiment of Charles, male mice were exposed to four different daily dose levels of X rays but at a constant dose rate. Thus, the physical factors were appropriate for the use of the D^2 function. Matings were continually carried out so that the total dose levels were average values and not carefully separated points. This, along with the fact that there was no sure way of knowing what irradiated cell stage produced the trait, detracts from any quantitative test.†

Dominant lethals.—The induction of dominant lethals in male germ cells can be detected by measuring the reduction in litter size produced by the irradiated sire, although litter size at birth is not usually considered the most accurate measure. A careful evaluation requires sacrifice of the pregnant female at about 15–17 days postconception and the counting of (a) number of live embryos, (b) number of dead embryos, (c) number of corpora lutea, and (d) number of pre-implantation deaths. Most of the losses occur prior to 10–11 days of gestation and pre-implantation losses are closely correlated with dose.[1141] Post-implantation losses rise to about a 20–25 per cent representation of the number of corpora lutea at doses below 100 r and remains at that level. Cytologic studies have demonstrated that the lethal action is predominantly due to aberrant cleavage and chromosomal fragmentation.[128]

---

† The Charles experiment, carried out at the University of Rochester, was nevertheless the first major concerted effort to evaluate the genetic hazards of radiation in a mammal and was set up under the auspices of the Manhattan Engineering District, predecessor of the Atomic Energy Commission. Preparation of a final report was delayed by the untimely death of Dr. Charles, but the report is now being published as a University of Rochester-Atomic Energy Commission Project Report No. UR-565 and should soon become available to interested geneticists.

The frequency of dominant lethals varies with the stage of the cell at the time of irradiation. Mature sperm cells are less sensitive than spermatids but are more sensitive than spermatocytes.[57, 58] The cell-stage factor is relatively easy to control, since it now appears that mice do not store mature sperm and the stage at irradiation can be ascertained by control of the time interval between mating and irradiation.[57] Bateman's report includes a timetable relating post-exposure week with cellular stage.

For matings carried out in the first week after exposure, the dose for inducing 50 per cent dominant lethals is approximately 700 r for 250 kvp X rays, about 100 r for fast neutrons from a nuclear detonation (average energy $\sim$ 2 Mev), about 100 r for 1 Mev cyclotron fast neutrons, and around 300 r for 14 Mev neutrons produced by a Cockcroft-Walton accelerator.[1140, 1142] In all cases the data for survival of embryos fit a simple exponential equation. Bateman[57, 59] has made the most recent and thorough theoretical study of the induction of dominant lethals and the relationship between postulated number of chromosomal breaks and time of death of the conceptus. On the basis of his analysis, the dose dependence of dominant lethals proved to be linear for neutron-induced lethals but nonlinear for X-ray damage. This is consistent with the expectation that neutrons will produce more than one hit per ionization track.

In a series of papers, the Russells have reported on the induction of dominant lethals in female mice.[1123, 1124, 1125] The use of litter size alone is not valid in this sex, since there is a period between 1 and 14 days postexposure when irradiated mice show an excess in ovulation rate. As for studies with males, the pregnant female is sacrificed late in gestation and the number of corpora lutea and living embryos are counted. The ratio of living embryos to corpora lutea is considered an accurate measure of the incidence of dominant lethality. Cellular stage at irradiation is very critical for oöcytes and can be ascertained by the interval between exposure and fertilization. A dose of 400 r will induce more than 98 per cent dominant lethality in oöcytes in first meiotic metaphase, which occurs 8 to 10 hours prior to fertilization. If the interval between irradiation and fertilization is doubled (16 hours), frequency of dominant lethals drops to less than 20 per cent. For X rays, the dose inducing 50 per cent dominant lethals is about 700 r for primary oöcytes, but only about 70 r for the cells in meiotic metaphase.[1120]

Gene mutations.—The greatest interest and concern has been expressed on the following question: what is the radiation-induced mutation rate in a representative laboratory mammal; how does it compare with *Drosophila* and how can it be applied to man? For reasons previously noted, the mouse has been the mammal of choice for this critical area of research in radiation genetics.

The problem can be approached in several ways. The search can be for a total, gametic mutation rate or it can be restricted to specific selected loci. The search can be for dominant visibles, recessive lethals, or recessive visibles. To date, the most successful procedure has been the specific-locus test for the induction of recessive visibles and any associated viability effects, although efforts to make separate estimates of the rates of dominant visible and recessive lethal mutations have also been made.

The specific-locus test procedure is a standard genetic test system that has been employed in *Drosophila* genetics for many years. A wild-type mouse is irradiated and mated to an animal from a multiple recessive tester stock. The immediate progeny are then screened for the appearance of a mutation at any of the loci marked in the parent from the tester stock. Since these F_1 mice are all heterozygous for the markers, a new mutation at any of the loci should appear in this first generation. Mutants with intermediate degrees of expression can also be detected with some degree of precision. Subsequent tests, of course, can check for such homozygous effects as reduced fertility, lethality, reductions in growth rate, and so forth.

The specific-locus procedure is limited by the number of mutants that can be carried in one stock without seriously reducing viability and reproductive performance. Another problem is that of overlapping phenotypes; the array of mutants must be discrete and separable in their expression. The test stock used in both U.S. and British genetics programs using mice at the Oak Ridge National Laboratory (Dr. W. L. Russell and colleagues) and the Atomic Energy Research Establishment, Harwell, England (Dr. A. G. Searle and colleagues, previously Dr. T. C. Carter), respectively, is the seven-locus stock containing the recessive genes: *a* (*nonagouti*), *b* (*brown*), *c^ch* (*chinchilla*), *d* (*dilute*), *p* (*pink-eye*), *s* (*piebald*), and *se* (*short-ear*). This stock was synthesized by Dr. W. L. Russell at Oak Ridge specifically for studies of mutation rate.[1134]

The bulk of information on mutation rates in mammals has been developed from this tester stock, an obvious limitation to current knowledge. It can only be assumed that these seven loci are fully representative of mutability and viability of all genes. A certain degree of doubt is raised by the fact that there is a greater than thirtyfold difference in the spermatogonial mutation rates for several of the loci.[1138]

Alternative methods for detecting the mutation rate for recessive visible genes require three generations of breeding rather than only one for the above procedure. These methods screen the whole genome and require the segregation of the new mutant in homozygous form in the third generation. One method, the backcross method, involves the outcrossing of a son of an irradiated parent and then backcrossing one of his daughters to himself. If the first-generation son carried a mutant, there would be a probability of 0.5 for its transmission to the second-generation daughter. If she carried the mutant, then the third-generation, backcross progeny would have a one-in-four chance of segregating the new mutant in homozygous form. Altogether, the chance of segregation in the third generation is only one in eight, which is not very efficient. The method was employed in the early work of Hertwig[549] and also by Carter and Phillips.[178] The second method for screening the entire genome requires three generations of full-sibling matings. This procedure is only half as efficient as the backcross method and has been used, apparently only briefly, by Carter and Phillips.[175]

Aside from the inherent disadvantages of these two methods of gametic analysis with respect to experimental economy, they will not succeed in detecting mutations that are recessive lethals. Since about three-fourths of the induced gonial mutants reported by Russell are recessive lethals, the yield in the backcross and sibmating procedures may be further reduced by another factor of four. With the exception of the fact that there is no restriction of the number of loci under test, little can be said for these latter methods of genetic analysis. In any of these methods, the mutation rate for either spermatogonial cells or post-gonial cells is isolated by analysis of matings only in poststerile or presterile periods, respectively. Only the rate for oöcytes can be obtained for females.

Recessive lethal mutations.—Of the three methods described above, only the specific-locus test system will permit an estimate of the frequency of recessive lethal mutations. Russell and Russell[1138] report that about 75 per cent of the induced mutations at the seven loci are recessive lethals. This proportion varies among loci, according to present reports, from about 50 to 100 per cent of the number of observed mutants. The locus with the highest gonial mutation rate, the *piebald* locus, also happens to have produced only lethal mutations.

These data pertain to those mutants derived from the single dose tests with the dose delivered at high intensity. Whether or not the same ratio of lethal to viable mutants will occur among those induced by low-intensity continuous exposure remains to be seen. Presumably, there should be no qualitative difference between mutations induced at different rates of radiation dosage, even though the mutation rates themselves may vary (see below). However, the recovery process, acting upon premutational damage to reduce the mutation rate under continuous exposure as compared to single-dose exposure, could conceivably act in a selective manner at the molecular level. On the assumption that variation in genic action is associated with variation in the molecular structure of the gene, it is then conceivable that some forms of genetic damage may be more amenable to spontaneous recovery. Whether or not lethal mutations are selectively acted against cannot be stated, but this would seem to be a potentially important point to have clarified in a mammalian test system.

An additional method that has been employed to determine the recessive lethal mutation rate for autosomal genes can be designated as the linked-lethal procedure. In its simplest form, a single recessive marker gene is carried homozygously in the irradiated parent; the parent is outcrossed and the heterozygous progeny are inbred to produce the normal 3:1 ratio in the F_2. The absence of the marker in the segregating generation is accepted as prima facie evidence of the induction of a lethal closely linked to the marker. The linked-lethal procedure was tried by Snell,[1250] but without success. His test entailed the screening for aberrant segregation ratios in F_3 progeny produced by backcrossing an F_2 mouse heterozygous for a marker with its F_1 parent.

An obvious deficiency of the method is in the limited length of the chromosome under test. Crossing over will naturally occur with greater frequency as the map distance between the marker and the lethal increases. Minor aberrations in the segregation ratio would require extensive testing for proof of presence of a lethal mutant, which could go far beyond its economic value in terms of information yield.

The linked-lethal method can be increased in its efficiency through the use of a number of independent marker genes.[517] The statistical complications of the method have been worked out by Haldane. These include the question of allowance for chance fluctuations in the F_2 segregation ratio and the length of the chromosome scanned on either side of the marker gene. The total swept length will naturally increase with the number of markers employed and with the number of F_2 progeny raised.

The method requires three generations; the first is from crossing the homozygous marker stock *inter se*, one parent of which has been irradiated. The second generation is an outcross of the marker stock which may now carry a lethal in the heterozygous form to a wild strain. The use of what might appear to be an extra generation of breeding, that is, the carrying of the homozygous marker stock one generation beyond the irradiated generation, eliminates sibmating the progeny of the irradiated parents and the accompanying chance of introducing a lethal from both parents independently induced. In other words, the second generation progeny trace back to only a single irradiated gamete in the original parents. The third generation is produced by sibmating the second generation progeny and significant deviations from the expected 3:1 ratios are sought.

Carter[158] tested the system of Haldane with a stock carrying seven recessive, visible genes. These are the same as those noted earlier with the exception that *pink-eye* (*p*) had been replaced by *waved*-1 (*wa*-1). The results suggested that the dose required to induce one autosomal recessive lethal per gamete in spermatogonia is probably no less than about 800 r of X rays delivered as a single dose. Carter concluded that the method is far less efficient than the specific-locus method for the detection of lethal mutations.

An elaboration of the linked lethal procedure has also been described by Carter.[170] This involves the use of linked marker genes and the detection of lethals located between the markers. On a theoretical basis, Carter could not conclude that it would be more efficient than the procedure employing independent markers. Apparently, the method has not been subject to experimental test. Haldane, in an appendix to Carter's paper, noted that the use of linked markers could be reasonably efficient as a means of detecting sublethal recessive mutants, at least for sublethals with a viability between about 5 and 50 per cent. Further exploration of these techniques in the laboratory would appear desirable.

Still another procedure available for the detection of recessive lethals has been described and tested by Carter.[165] This test uses the reduction in litter size produced by the sons of irradiated mice, when these sons are crossed to their daughters. Thus, this is an application of the backcross procedure previously described that can also be used for detection of visible mutations. The probability of homozygous expression of a recessive lethal in the backcross progeny is again one in eight, which now is detected as a one-eighth reduction in litter size in comparison to the control. The regression of litter size on radiation dose was employed for the analysis and led to the conclusion that a dose of about 300 r produced one recessive lethal per gamete in postmeiotic male germ cells.   The exposure in this experiment was protracted over a 5-week period at rates of 1.64, 8.0, and 33.3 r per week.   No exposures of single dose with high dose rate were reported, so no comparison can be made for possible effects of dose rate. This procedure would seem sufficiently straightforward to warrant the development of data for analysis of effects of dose rate on the induction of recessive lethals.

Definitive study of the mutation rate for sex-linked lethals still remains to be done. One attempt was made in Charles' experiment, with equivocal results.   The breeding procedure is simple.   Irradiated males are outcrossed, and their daughters, which may now carry a sex-linked recessive lethal as a heterozygote, are mated to unrelated males. If a lethal segregates, the litters will show a 2:1 sex ratio.   Due to the normal fluctuations in sex ratios, the test system is inefficient.

In recent years, several sex-linked genes have become recorded in the mouse. Recombination percentages vary from 4 to 16 among the known loci,[492] and this is close enough to permit the suggestion that Carter's technique of detecting lethals between linked markers would be worth exploring.   Even the use of a single marker, such as *tortoise* (*To*), which is dominant for its effect on coat color and recessive for its lethal effect, would provide some useful data.   If an induced lethal is closely linked to *tortoise*, test matings would have few or no male offspring, thus simplifying the test system.

Dominant visible mutations.—Data on the induction of dominant visible mutations are exceedingly unsatisfactory and for practical purposes may be considered as virtually nonexistent.   This is not the fault of those investigators who have made valiant attempts to detect these mutations.   The problem is that there is no simple quantitative method available for their detection.   Basically, all that can be done is to screen the offspring of irradiated parents for detectable anomalies and mate these animals to determine the heritability of the trait.   A total gametic rate is therefore observed. Since there is almost no limit to the range of dominant morphologic and physiologic variants that can be measured, it is difficult to determine what the true mutation rate is.

The single greatest effort yet reported was made by Charles et al.[198]   The male offspring of irradiated male parents were sacrificed, dissected, and inspected for morphologic variants.   Obviously, no heritability tests could follow.   The female offspring were mated to unrelated stock; thus externally visible mutants could be studied for transmissibility.   One hundred and twenty possibilities were tested, and fourteen proved to be bona fide mutations.   The data provided an estimated mutation rate of 5.4×10^{-6} per roentgen per gamete.   These mutations were observed among the offspring of males subject to daily X irradiations that were permitted to mate continuously during the course of the experiment.   The mutation rate thus cannot be applied to any single type of germ cell but probably includes contributions from all stages of gametogenesis.

Russell[1134] reported the detection of five X-ray-induced, dominant, visible mutants affecting the coat color, the ears, and the tail.   These were the only morphologic traits screened for dominant changes in the course of early tests on the induction of recessive visibles.   Although no mutation rate could be calculated, Russell noted that 3 of the 5 new mutants affected a trait that was also controlled by one of the specific recessive loci in the main test.   Since 32 recessive visibles were uncovered in the same group of animals, he therefore concluded that the mutation rate to dominant visibles is probably significantly lower per roentgen than the rate for recessive visibles.

Recessive visible mutations.—The determination of the mutation rate for recessive visibles induced by irradiation at specific loci in mice has been one of the major research efforts in radiation genetics by the Atomic Energy Commission for the past decade. The results of this effort, along with the many ancillary findings, now constitute a major contribution to our knowledge of genetics.   This subject has been the topic of so many discussions that the reader should properly address himself to the original reports. The studies do provide some interesting sidelights on the problems of methodology, however.   The basic procedures for the specific-locus test (described above) enjoy the economy of requiring only one generation of breeding to bring the progeny to test. The spermatogonial mutation rate induced by single doses of X rays delivered at a rate of 90 r per minute is approximately $25 \times 10^{-8}/r$/gene for total doses up to 600 r.[1140, 1142] At 1,000 r, the rate falls to about one-half the above figure, which is interpreted to be due to selection against a more radiosensitive class of spermatogonia.[1111, 1131]

The most significant finding, and the one which has received worldwide attention, concerns the effect of the dose rate of radiation on the observed mutation rate.   If the exposure is delivered, essentially continuously, at rates of either 10 r/week or 90 r/week (from 1 to 10 milliroentgens/minute), the mutation rate falls to about one-fourth that observed following the single dose exposure.[1140, 1142]   The difference is significant and has also been confirmed for the female.[1126, 1139, 1143]   The effect had also been noted by Carter,[164] but, unfortunately, his data and experimental conditions could not clearly substantiate the interpretation that an effect of dose rate existed, although it was suggested.   He was forced to compare the mutation rate following single dose exposure in males with the rate induced by continuous exposure in females without the benefit of the reciprocal comparison.

The interesting side effect of this finding in mice was that it produced a great deal of concern and some criticism among geneticists, all of whom had learned as a basic principle that the mutagenic effect of radiation always acted additively, regardless of the manner in which the radiation was delivered.   The original data that suggested an effect of dose rate did contain a few uncertainties to warrant some of the criticism. The protracted exposures were provided by the radioisotope cesium-137, which emits a 0.7 *Mev* gamma ray.   The ionization density is less than that for 250 *kev* X rays, and an RBE of 0.8 or 0.9 might be required to make the physical parameters constant. The standard errors of the initial data were quite large and point-by-point significance testing gave no real assurance of a difference between high- and low-intensity exposure except at the level of the 600 r dose.   It should be recalled, however, that the original concept of dose-rate independence was derived from data obtained on irradiated *Drosophila* sperm.   Analogous data derived from the exposure of mature germ cells of the mouse do not challenge the basic *Drosophila* data. The dose-rate effect is definitely limited to spermatogonia and oöcytes.[1140, 1142] It should also be noted that a significant dose-rate effect has now been demonstrated for *Drosophila* oögonia exposed to Co[60] gamma radiation.[972] The flies were exposed to a total dose of 4,000 *r* over either a period of two weeks or in 31 seconds. The percentage of sex-linked lethals was 1.3 ± 0.5 and 3.4 ± 0.7 for the continuous and single-dose exposures, respectively. The difference is significant.

Studies on microörganisms indicate that some portion of the induced genetic damage falls into a category now labelled as premutational damage. A recovery process, requiring active protein synthesis, acts to prevent the fixation of part of the premutational damage, thus reducing the mutational yield.[514, 708] The reduced mutation rate in mouse gonia and oöcytes following low intensity irradiation is interpreted to be the result of a process of metabolic recovery, which may itself be quite radiosensitive and therefore fail under high-intensity irradiation. The study of mutation at the molecular level has therefore become of increased importance, and new insight into the mechanisms of radiation protection may also result.

Important unanswered questions now concern the effects of fractionated exposures in all their complexities of interminable variables. Is there a limiting low dose rate, above which the mutation rate jumps to the higher level? Will there be all gradations between the high and the low? What will be the effect of intermittent exposure at high intensities but to very small doses? If the answer to this previous question is that the low mutation rate prevails, will it do so only under specific conditions of dose rate, dose per fraction, and fractionation interval? These questions are important for both industrial and medical situations where there is often no regular pattern of exposure.

In addition, the dose-rate effect requires careful evaluation with exposure to the densely ionizing radiations, such as alpha particles and fast neutrons. Because of their greater efficiency in inducing genetic and general cellular damage, is it possible that exposure to neutrons, for example, may produce more concurrent injury to the presumed recovery mechanism and thus a lesser reduction in mutation rate with declining dose rate? The neutron may be of particular value in these genetic studies because of its known peculiar behavior with regard to somatic lethal effects. Let me refer to two studies as an example. Sproul,[1262] using the traditional paired-dose technique to estimate recovery from acute radiation injury, found that the rate of recovery is essentially the same following initial exposure of the whole body to a single dose at high intensity of either Co[60] gamma rays or 14 *Mev* neutrons. On the other hand, Upton *et al.*[1336] have indicated that the RBE for shortening of life induced by neutrons produced by a Po-Be source (∼4 *Mev*) progressively increases as the dose rate declines. The data suggest that the long-term, lethal effects of neutrons are considerably less dependent on dose rates than are those following Co[60] gamma irradiation. It is entirely conceivable that genic mutational damage induced by fast neutrons may also be less dependent on dose rates.

Cytogenetic analysis.—Two subsequent chapters by Drs. Klein and Yerganian amply cover the methodologies of cytology and cytogenetics. The application of tissue-culture techniques to genetic analysis, radiation and otherwise, is also covered by Drs. Ford and Puck in the previous volume in this series, *Methodology in Human Genetics*. The use of mammalian cells in tissue culture for radiation-genetic analysis will assuredly be an area of increasing scientific effort. The opportunity has become available to make elaborate comparative quantitative analyses of radiogenetic sensitivities among all important domestic and laboratory animal species. Besides interspecific comparisons, strain differences in radiosensitivity within species can be explored at the cellular level. To date, the recognized genetic differences in radiosensitivity in mice have been attributed to physiologic genetic factors, rather than to basic differences in the resistance or sensitivity to cellular genetic damage.[473, 474, 475] This point can easily be checked in tissue culture. Strains may differ in their sensitivity to mutagens, at least for chromosomal aberration and restitution rates. If so, new techniques would be provided to study the cellular mechanisms of somatic injury and recovery.

QUANTITATIVE GENETIC EFFECTS OF RADIATION

Although geneticists have been estimating spontaneous and induced mutation rates in a number of species of plants and animals for many years, virtually all of our present estimates of the potential genetic hazards of radiation to man have been made independently of this accumulated experience. There are several reasons for this. First, while fairly good data for spontaneous mutation rate are available for man, the radiation-induced rate is a complete unknown. Second, the bulk of data on detrimental genetic factors in man are in the form of morbidity and mortality statistics. Although many clear-cut mutant phenotypes that severely affect viability are recognized, their frequencies are generally masked by the normal rates of infant and childhood mortality in human populations. Thus, more concern is expressed by geneticists about the total mortality than about the individual syndromes of disease constituting it. An excellent summary of several hundred detrimental genetic qualities in man has been prepared from experience in Northern Ireland.[1282] Mortality rates at birth and among adults are given, although the temporal patterns of mortality for the various characteristics have not been derived. In addition, the genetic basis for many traits is known to be uncertain and to involve considerable environmental interaction.

At present, best approximations of the normal load of detrimental genetic factors in man have been derived from the analysis of the progeny of consanguineous marriages. The analytic technique requires the maximum likelihood estimate of the regression of early mortality on the inbreeding coefficient.[901] The procedure has also been employed by Schull[1169] and Slatis[1216] with considerable success. The basic biological data in nearly all instances include some or all of the following: fetal deaths, neonatal and infant deaths, childhood and juvenile deaths. Such data are comparatively easy to obtain with a high degree of accuracy among most Western nations. Unfortunately, comparable data for laboratory animal populations are not generally available, even though many of the laboratory animals enjoy a standard of living far superior to that of the civilization that maintains them.

The importance to be placed upon these data is emphasized by the fact that the most widely accepted and quoted baseline predictions of the radiation hazard to man are those provided by Crow,[231] involving an expansion of the analysis of Morton, Crow, and Muller, referred to above. These calculations, although ingenious, require the mixed application of parameters from *Drosophila*, mice, and men. As they stand, they are in need of updating and revision, but, basically, the calculations require careful evaluation for populations of laboratory mammals. In addition, the genetic analysis of the progeny of the survivors in Hiroshima and Nagasaki was largely restricted to data on quantitative anthropometry, morbidity, and mortality.[939]

In general, the basic techniques of breeding employed for the detection of qualitative changes can be used for the study of the more subtle quantitative expressions of damage. In many instances, the two types of data can be obtained concurrently. The analytic procedures are more complex, however, and thoughtful statistical design and analysis are required. The key to success is no longer a matter of overcoming the issue of events of low probability by sheer weight of numbers, but rather overcoming the problems of ordinary random fluctuations.

Sex ratio.—Prior to a discussion of the evaluation of mortality statistics, brief mention should be made of studies of sex ratio. This is a semiquantitative trait that does not require any specialized approach. This parameter has become a controversial issue since Schull and Neel[1170] reported a significant shift in sex ratio among the progeny of the survivors at Hiroshima and Nagasaki. Shifts in sex ratio have not been uniformly seen among the progeny of irradiated mice.[1128] A very recent report[729] also indicates that the sex ratio of mice remains substantially unaltered following single and fractionated exposure of the male parent. An earlier preliminary report by Kalmus et al.[676] did indicate a deficiency of female offspring from irradiated male mice for litters sired within 40 days of the exposure. The shift in sex ratio was in the direction of genetic expectation, but this study has apparently not been confirmed. It can be suggested that man may be more sensitive to the induction of sex-linked lethals by radiation. Since the X chromosome of man carries more recognizable detrimental genes than have been noted in the mouse, man may have a qualitatively different genetic potential for induced genetic damage.

Viability.—The general term *viability* is chosen to cover the full range of mortality periods: stillbirths, neonatal deaths, infant or preweaning deaths, and adult mortality. Data on these traits are scattered and incomplete. Russell,[1128] in his review of the early literature, has indicated that an increase in stillbirth rate and mortality to the time of sexual maturity has been observed among the progeny of irradiated male mice for litters sired in both the pre- and poststerile periods. Similar observations have been made with guinea pigs.[1289]

More recently, additional attempts have been made to detect an over-all effect on viability. Boche et al.[113] reported on the life expectancy of the progeny of irradiated male and female mice, as measured from the time of weaning. Their controls consisted of litters sired prior to irradiation and the experimentals were produced at three intervals of time after irradiation: 10 days, 120 days, and 180 days. The study was done in two replications; the first indicated a shorter life expectancy among the experimentals, the second was negative. The use of controls of a different parity than any of the experimental groups is not to be recommended. Litter size and maternal nutritional factors will be confounded with any induced genetic effects of the radiation.

Russell[1133] reported a significant shortening of life among progeny of irradiated males for mice conceived during the presterile period. The offspring had their mean life expectancy from weaning reduced by about 60 days per 100 *rep* exposure of the male parent. This is generally considered to be a maximum effect, since the parents were subject to a prompt neutron exposure from a nuclear weapon detonation and the progeny were conceived from cells in postgonial stages at irradiation. The mutation rate for these stages is twice that for spermatogonia and the n:x RBE may lie anywhere between 2 and 6. Russell and Russell[1137] have indicated that follow-up studies employing X irradiation of spermatogonia confirm the earlier report on neutron effects but the expression is less readily detected.

Preweaning mortality in mice has been measured as a reduction of litter size at three weeks of age.[1137] Litters produced from germ cells exposed in the spermatogonial stage show a 3 to 4 per cent reduction from control values following a 300 *r*, X-ray exposure. No information on the time of death of the animals during the three-week period was given. Total infant mortality, including stillbirths, has recently been studied in rats with the suggestion that some part of the mortality is genetically caused rather than a random function of maternal and litter-size factors.[824] The technique consisted of a comparison of control and experimental litters of similar size for differences in postnatal mortality, rather than a comparison of litter sizes themselves, since progeny were produced during the presterile period and the litter sizes reflected dominant lethal effects occurring prenatally.

Of the studies reported to date, none has really obtained the data required to provide an accurate and quantitative assessment of the full expression of viability mutations. It is imperative that studies be done with the view of encompassing the whole life table; populations must be studied from birth through death as a unit experiment. It is especially important that careful attention be given to the period of neonatal and infant mortality because of the significance of such data for comparison to man.

It is further suggested that the analysis of data for viability be done by standard methods of actuarial statistics. These are amply described by Pearl.[995] Surprisingly, complete life tables have apparently never been obtained for populations of laboratory mammals. The author is presently endeavoring to accumulate a sufficient number of mouse days of experience to generate the tables for about a half-dozen standard inbred strains and hybrids. These are: A/He, A/Jax, BALB/c, C3H$_f$/He, C57BL/6, and F$_1$ and F$_2$ hybrids from the BALB/c × C57BL/6 cross. Rather large numbers of mouse days of experience are required for adequate analysis of the young adulthood portion of the life table. For example, the BALB/c × C57BL/6 F$_1$ hybrid at weaning age has a death rate approaching a low of 1 in 10,000 per day. Prior to this time and beyond 150 to 200 days of age for most mice, the daily rates of death are high enough to permit reasonable data to be derived from samples of only 100 to 200 mice when 50- to 100-day intervals are used. The greatest advantage of using life tables lies in their more direct comparability to human experience. The full form of the life table is remarkably similar for most mammalian species. This similarity has been employed to extrapolate the life-shortening effects of radiation from mice to man.[333, 1144] All analyses to date have concerned only adult populations, however.

Figure 19 shows a comparison of murine and human populations in the neonatal, infant, and childhood periods. The U.S. population data were obtained from the published vital statistics of the United States available from the National Office of Vital Statistics, U.S. Public Health Service. The data for mice are some of those of the writer's. The F$_2$ was derived from the reciprocal crosses of the BALB/c and C57BL/6 inbred strains, and the referred inbred in figure 19 is the BALB/c parental line. Although these data are for the combined sexes, routinely sex is determined at birth and litters are checked daily for deaths. For the sake of ease in final recording and coding of IBM punch cards, the data are assembled in the intervals: 0–5 days, 6–15 days, and 16 days to weaning, which provides a mean of approximately 30 days of age. In time, the data will be analyzed by sex and strain for maternal age, parity, and effects of litter size. Preliminary analysis of the BALB/c strain, for example, shows that litters of 1 to 3 mice have a 2- to 3-fold greater mortality rate during the full preweaning period than do litters of 4 or more.

The point of interest in figure 19 is the nearly complete superposition of the human and mouse infant mortality data. The ratio of time scales is approximately 120:1. This ratio is 2 to 4 times greater than that customarily noted in the comparison of adult populations and undoubtedly reflects the rather attenuated prepubertal developmental period of man. Physiologically, the first 30 days of life for the mouse and the first 10 years of life for man are generally comparable. For example, it is recognized that a small proportion of the females of both species will reach sexual maturity by the end of the indicated age intervals. Many of the major hereditary defects in man express themselves during this early period of life. The same is undoubtedly true for mice and other mammals. The basic similarity in the temporal course of spontaneous mortality for the two species certainly encourages more complete and quantitative study of laboratory animals in order to evaluate the general adherence of the expression of genes for viability to genetic theory and expectation. Of particular note here are the data of Russell and Russell[1138] indicating that most of the recessive lethals detected in the specific-locus tests induce death in the neonatal and preweaning period rather than in prenatal life.

Fig. 19. Neonatal, infant, and childhood mortality rates for inbred and outbred populations of mice compared to United States population.

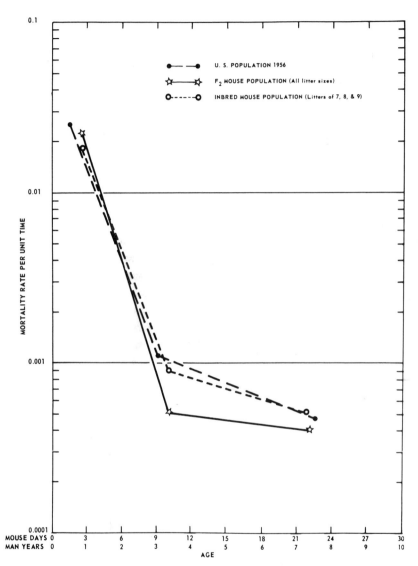

(Age scales adjusted to maximize comparability.)

Growth and maturation.—Although there are no special genetic techniques required for studies in this category, a deficiency of information still remains. Russell[1134] has reported that mutants at the *piebald* locus nearly always show a poor rate of growth and an associated higher mortality. The data of Boche *et al.*[113] did not indicate a growth decrement, however. Possibly most effects on these traits will be due to a few major mutants and thus show up in only a few animals per generation, the detection of which would be statistically difficult.

Reproductive performance.—Here, too, there appears to be a shortage of data, with the exception, of course, of the studies on dominant partial sterility discussed earlier. As noted by Russell and Russell,[1138] there will be a small percentage of the progeny of irradiated parents that will be completely sterile. Minor reductions in reproductive performance most certainly will be difficult to detect and will require good sampling statistics. The author has a small quantity of data for lifetime performance. The experimental group, the progeny of C57BL/6 males subject to 5 r/day of Co[60] gamma radiation for 120 days, shows a 7.5 per cent drop in average size of litter and a slightly more rapid senile decline in productivity. Since only ten breeding females were sampled in each of the two groups, no real significance of the data can be claimed. The study was done on a pilot basis in the period 1956–1958 to test the effects of protracted irradiation on the induction of mutations affecting viability and reproduction and to explore the attendant statistical problems (which are many). It can be noted that the careful initial pairing of the breeding stock by age, for experimentals and controls, will automatically lead to a high degree of comparability of the data, litter by litter, throughout the reproductive lifetime. At the same time, changes in the interval between litters will become apparent, if this is to be one of the manifestations of genetic damage.

Behavioral traits.—There are no data on radiation-induced mutations that affect behavioral characteristics, although attempts to detect genetic damage of this type are being made by Green.[481]

PROTECTION AGAINST INDUCED GENETIC DAMAGE

Basically, there are two principal methods of approaching radiation protection: preventive therapy and supportive therapy. The latter is typified by the successful use of postirradiation injections of bone marrow to improve survival. Bone-marrow therapy replenishes the animal's hematopoietic tissues during a critical period in the acute syndrome. Postirradiation supportive therapy appears to offer nothing to the geneticist interested in reducing the genetic hazards of radiation.

Preventive therapy does hold promise. In this case, treatment is required in the immediate preirradiation period or even during exposure. Various chemicals have been tested along with modifications of oxygen tension. The theory behind the use of chemotherapeutics is that they act to reduce oxygen availability, quench active radical transport, or generally act as a competing target system for the initial events of energy absorption.

While much effort has been spent on radiation-genetic protection studies with microörganisms,[591] very little has been done with mammals. Hypoxia was tried by Russell *et al.*[1136] as a means of reducing the induction of dominant lethals in X-rayed male mice. The animals were exposed to 800 r while in a chamber flushed with

5 per cent O_2 and 95 per cent He.   No evidence for protection was noted.   A similar study with female mice did suggest some minor degree of protection.[1116]   Mature sperm may innately be hypoxic and thus no oxygen effect would be expected.[1140, 1142]

Kaplan and Lyon[686] tried the compound mercaptoethylamine which, like hypoxia, is effective against general somatic damage.   A 4 mg. dose given intraperitoneally about 5 minutes before exposure did not protect against dominant lethal damage in males subject to 500 r of X rays.   The same drug was used in a 10 mg. dosage with rats exposed to 300 r of X rays.[839]   Testicular weight loss was the end point and no effect was noted.   Eldjarn *et al.*[320] point out, however, that only very small amounts of the drugs cysteamine or cystamine are localized in the testes (only 1 to 10 per cent of that observed in other organs).   The same may be true for the compound used by both Kaplan and Maisin, which would suggest that considerably larger dosages would be required.

Rugh and Wolff[1086] succeeded in reducing damage to the ovary of the mouse with a 3 mg. intraperitoneal dosage of either cysteamine or cystamine.   The measures here were the number of litters per mouse-week and the average number of fertile weeks. Animals protected with cysteamine, for example, produced 0.084 litters per week as compared to only 0.030 per week for controls after 50 r of whole-body X irradiation. Protected females were fertile an average of 18 weeks, compared to 6 weeks for the controls.   Recently, Mandl[846, 847] has reported success with male and female rats protected with B-mercaptoethylamine and cysteamine, respectively.   She employed a histologic measure of effectiveness: primordial-oöcyte survival and spermatogonial-cell survival.   Intraperitoneal injections of 20 to 30 mg. were used, and a dose-reduction factor of 1.5 to 3.0 was achieved over the dose range of 100 to 400 r.

No attempts have been made to study protection against specific mutational damage.   As Mandl has pointed out, the earlier failures may not have reflected an inefficacy of the therapeutics but rather the wrong choice of dosage of both drug and radiation, the former too low and the latter too high.   Her success in increasing survival of gonia and oöcytes may encourage a more determined effort to search for an agent that will reduce the yield of genic mutations.

Selection of mammalian species and system(s) of mating.—To a degree, there is really little room for choice when one wishes to carry out studies in mammalian radiation genetics.   The mouse has been and certainly will continue to be the favored species. Some consideration, however, should be given the rat, at least for studies on growth and maturation, since it may offer the opportunity to do more sensitive studies on these traits.   The unfortunate disadvantage for both mice and rats, and to some degree for all litter-bearing animals, is the partialy uncontrollable incidence of neonatal mortality. Since most litters are born during the early morning hours, the stillbirth and immediate neonatal, or perinatal, mortality can be obscured by the cannibalistic tendencies of the dams.   In the writer's experience, it has sometimes been impossible to determine the number born, although positive evidence of stillbirth, or livebirth, or both, in the form of a few scraps of progeny partly eaten, may be available.   An eight- to ten-hour shift in the diurnal cycle could possibly overcome this problem and would seem worth a try. Since most modern animal quarters are windowless and equipped with resettable automatic light switches, no technical difficulties need be anticipated.

For the most part, other species either have too few young per cycle or are too expensive to maintain. Nevertheless, a major Atomic Energy Commission-supported program with swine is under way at the Iowa State University, under the direction of Drs. J. L. Lush and L. N. Hazel, which may provide the controlled early mortality data that are required. However, swine after bearing young also tend to trample and crush part of the litter.

Once the species is selected, the next question concerns the genetic composition of the chosen animals. Should they be inbred, single-cross hybrids, or should they be random-bred or mixed hybrids, such as double crosses? This is not an easy question and the answer any particular investigator gives may not be entirely unbiased. Answering the question raises additional ones on the significance of polymorphism and heterosis in mammalian species. Will inbred animals tend to express heterosis for induced mutation more readily than crossbred animals, or will this expression occur to any significant degree at all? Will genetically heterogenous material have a greater probability of masking the minor detrimental mutations because of an innately greater viability and genetic flexibility?

There may be a way out, however, as in the study Green[481] describes, in which four different levels of inbreeding are employed on two different populations—one genetically homozygous and the other genetically heterozygous. An additional procedure is that used by Chapman.[196] In his study, recurrent incrossing and out crossing is compared with continuous incrossing and outcrossing. In both studies the animals are irradiated every generation in order to check for cumulative genetic injury but under the different selection pressures induced by the varying degree of inbreeding. The only opinion the writer offers is that the outbred system may be the choice for the sake of its greater comparability to man if the investigators cannot afford to check both inbred and outbred genetic systems.

RADIATION AS A TOOL FOR GENETIC RESEARCH

It is an understatement to say that radiation has become an extremely useful tool in experimental biology and medicine. A glance through the contents of this volume should substantiate the fact that many aspects of mammalian genetics have profited as well. No attempt will be made to describe all applications. However, particular contributions to mammalian genetics have been provided through the use of radiation techniques in the areas of physiologic genetics, immunogenetics, developmental genetics, the genetics of disease resistance, the genetics of viability, genetics of cancer and somatic-cell genetics. Pertinent specific references will be found throughout this volume, and the reader should consult the studies of E. S. Russell, H. Chase, R. D. Owen

D. Uphoff, L. B. Russell, J. W. Gowen, D. Grahn, W. E. Heston, K. Atwood, S. Scheinberg, and certainly many others.

Techniques developed by mammalian geneticists, originally for genetic purposes, are now finding usefulness in other areas of radiobiology. For example, a group in the Medical Department of the Brookhaven National Laboratory is making extensive use of measurements of survival of spermatogonial cells for detailed analysis of the relative effectiveness of different energy levels of neutron irradiation.[60, 1075] This judicious mixture of genetic techniques with those of the medical and radiologic physicist promises to offer much valuable data to the science of radiobiology.

The geneticist's adherence to quantitative methodology and to sound concepts of the cellular basis of radiation injury have generally assisted in establishing a high standard of scientific accomplishment in the comparatively young and fast-growing field of radiation biology. Geneticists can take pride in their total contribution, direct and indirect, to this scientifically and politically important field of endeavor. Perhaps many remaining problems in radiobiology will ultimately be solved either by geneticists or through genetic techniques and interpretation.

DISCUSSION

Dr. Degenhardt: I am very pleased to open the discussion of this excellent paper presented by Dr. Grahn. We are all impressed by the recent advances in this specialized field, but we also recognize that our knowledge is fragmentary in certain areas. On the whole, we have advanced little beyond the first steps into the field of radiation genetics, in which methodologic problems still prevent accurate, rapid, and economic progress. The few well-equipped teams of scientists, who are investigating problems in mammalian radiation genetics, are not sufficient even though their individual efforts are excellent. I should emphasize that it is absolutely necessary (1) to encourage international coöperation and international exchanges of ideas and information in the field of mammalian genetics and (2) to encourage coöperation and exchange of ideas between scientists working with experimental mammals and those working with human beings. This would enable us to collect all information in this field within a reasonable space of time and to estimate better than before the hazard of nuclear damage to human beings.

After these general comments I should like to direct attention to a certain problem, mentioned by Dr. Grahn: the induction of dominant lethals in germ cells. This is a critical means for examining the areas of physiologic and developmental genetics in early embryonic stages. W. L. Russell, L. B. Russell, and E. F. Oakberg[1142] have demonstrated that the induction of dominant lethals in male germ cells leads to a high incidence of lost conceptus prior to days 10–11 of gestation. Pre-implantation losses rise to about 7 per cent and post-implantation losses to about 25 per cent of the number of corpora lutea at doses below 100 *r*.

In female germ cells, dominant lethal incidence was shown to be strikingly dependent on the stage of oögenesis during irradiation;[950, 951, 1123] a dose of 70 r will induce about 50 per cent dominant lethality in oöcytes in the first meiotic metaphase. When Bateman[57, 59] computed theoretically the relationship beween the postulated number of chromosomal breaks and time of death of the conceptus due to dominant lethals in the female germ cells, he found that single lethals per egg permit survival to implantation but multiple lethals usually cause loss of the egg before implantation.

These facts should encourage coöperation among geneticists, radiologists, embryologists, cytogeneticists, and biochemists to focus their attention on the direct relationship between chromosomal breaks and abnormal embryonic development.

Regarding post-implantation losses, I wish to draw special attention to seven-day-old embryos in mice. We have information, gained from direct irradiation effects (130 r) upon embryonic differentiation of somatic tissue in successive stages of gestation, that day seven seems to be most critical in early morphogenesis (figure 20). In comparison to embryonic development in human beings, this would be the third week of gestation (day 17 to day 21), a stage of development in which most mothers are not aware of their pregnancy. This conclusion agrees with those given by Russell and Russell,[1121, 1122] and they recommend that, whenever possible, pelvic irradiation of women of child-bearing age should be restricted to the two weeks following the first day of menstruation. Radiation hazards to the embryo are being avoided to a large degree by following this advice.

DR. BURDETTE: Fairly recently we have been able to demonstrate in *Drosophila*, although we have not tried it in mice, that actinomycin D, which is often used in conjunction with irradiation, will reduce the frequency of mutations induced by X irradiation by approximately 50 per cent, although we do not have sufficient data to demonstrate the effect on natural rates of mutation. I am wondering whether you will comment on possible pressures in nature which may reduce the numbers of mutations. Photoreactivation is an example of the type of phenomenon I had in mind.

DR. GRAHN: For mammalian systems or in man, for example, I do not know what particular pressures there would be to reduce the hazard of radiation. Such things as photoreactivation are probably not effective in mammalian systems. Generally it is very much easier to find all the different environmental stresses that will increase the mutation rate rather than to find anything that will act to reduce the potential hazard.

DR. BURDETTE: More attention directed toward reversal and inhibitory effects may be very profitable in view of the current disproportion between this type of study and those dealing with enhancement. Dr. Crow, do you have a comment?

DR. CROW: There was a suggestion a few years ago by Carter and Haldane[170, 517] that one might study the over-all mutation rate in mice by finding lethal mutants linked with known visibles. I wonder how widely this is being applied?

DR. GRAHN: Essentially not at all. Both the use of independent markers and linked markers in the location of a lethal in between the markers should be mentioned. Haldane[517] has worked out most of the statistical problems and has even been able to demonstrate that this procedure can be used to look for sublethals.   Carter[158] tried the linked-lethal procedure and came to the conclusion that it has no improvement in efficiency over the specific-locus, test procedure.   However, the procedure possibly has not been exploited as far as it could be.

Fig. 20. CONGENITAL DEFORMITIES PRODUCED BY ACUTE X IRRADIATION OF PREGNANT MICE OF INBRED STRAIN C57/BL/Ks AT SUCCESSIVE STAGES OF GESTATION.

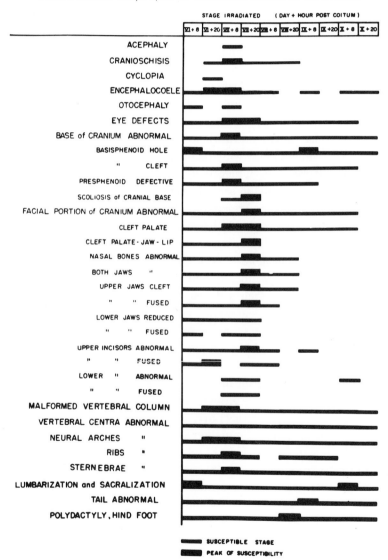

DR. BURDETTE: Would you comment on Dr. Degenhardt's first point?   Is there an effort to correlate research efforts on irradiation globally by the World Health Organization and others?   A summary of these activities should be of general interest.

DR. GRAHN: The United Nations, in conjunction with the World Health Organization, is interested in the problem of effects of low-level radiation in man.[1411] The United National Scientific Committee on the Effects of Atomic Radiation is also interested in improving the collection of vital statistical data so that it will be of some value in human or genetic studies. I know of no specific program or plan to bring together the diverse efforts of different countries other than those agreed upon in an informal way, such as might be arranged between the United States and the United Kingdom and Canada.

DR. GOWEN: In discussing rates of mutation, you placed quite heavy emphasis on the seven loci in mice. Knowing what has been learned about mutations in *Drosophila* and realizing the variations possible when one considers that crossing over occurs within the gene, so to speak, would you put a premium on the seven loci, or would you prefer to have a much more general survey of the mutation frequencies for as many loci as one can keep under reasonable control?

DR. GRAHN: What do you mean when you speak of premium, Dr. Gowen?

DR. GOWEN: Premium in the sense that it would be strange indeed if such a group of loci were representative or even suggested the range of mutation rates or the types of changes which may be expected to occur in, say, mammals or man.

DR. GRAHN: Certainly data on more loci are desirable if for no other reason than to give assurance that these seven loci and the mean or median rates of mutation derived therefrom are reasonable descriptions of the average mutation rate for the total genome. This is neither cheap nor easy, so the solution of the problem is not obvious. Total gametic methods are not too efficient either. In other tester stocks that have been or can be synthesized, their reduced viability, reduced fertility, over-lapping phenotypes, *etc.*, make the problem additionally very difficult to solve.

DR. YERGANIAN: Dr. Grahn's remarks on reverse-lighting patterns of illumination in animal-breeding rooms should be supported wholeheartedly for the benefit of the investigator who wishes to initiate convenient and controlled matings.[1462]

DR. LEDERBERG: It seems to me there has been such a frenzy of interest in radiation-induced mutation that we are very seriously neglecting the other environmental factors which, from a practical point of view, may be much more important to man. I am talking about chemically induced mutation and antimutation effects. As far as I can tell, there has been a studious neglect of the very well-documented observations that mutation rates in bacteria, and very much more recently in *Drosophila*, can be markedly influenced by the presence of various purines in the diet. Furthermore, the compound adenosine will go so far as to reduce the mutation rate in bacteria by the factor of a half. Now if you are concerned about the incidence of mutation in the human population, I think it is reasonably plain that you could exert on a global basis a much more significant effect by alteration of diets which would have some effect on the spontaneous mutation rate than we are now likely to accomplish by small changes in the environmental radiation hazard. I am not saying we are not justified in making strenuous efforts to reduce the radiation hazards to manageable levels; but I feel now that there is a tremendous distortion of interest at the present time in radiation biologic genetics and away from the chemical aspects of mutations, and I think more could be done about this.†

† Subsequent to this symposium, the problem has been reviewed: Genetics, Proc. of the Second Conference, 1960 (Mutation), Josiah Macy, Jr., Foundation, 1962.   See especially the discussion by Dr. A. Goldstein.

PHYSIOLOGIC GENETICS

Sewall Wright, Sc.D.

GENIC INTERACTION

Genic interaction is a subject of great importance in two of the major branches of genetics. It is obviously fundamental in physiologic genetics and is almost as fundamental in population genetics, including the genetic aspects of the theory of evolution. I have been interested in genic interaction for both reasons and in about equal measure since about 1915. In trying to keep up with both, it has sometimes seemed as if I were trying to ride two wild horses bent on going in different directions. I shall try to justify the attempt to keep them together.

INTERACTION IN EARLY GENETIC RESEARCH

The unit characters of the early geneticists probably seemed to most other biologists merely another revival of the ancient doctrine of preformation, discredited by all who, like Aristotle, Harvey, Wolff, and von Baer, had attempted to trace the step-by-step elaboration of complexity in the process of development. The early geneticists did relatively little to disavow this interpretation. They were too busy working out the laws of transmission of these units and the patterns of organzation in the germ cells to have much time for theoretical discussion. In pursuing this task, they were interested in characters mainly as markers for genes. They were interested, for the most part, only in good genes, consistently associated with easily classifiable characters. The catchwords used for convenience in naming genes often seemed to others to imply that geneticists thought of the organism as a mosaic of unit characters.

The extent to which even the earliest Mendelians actually were preformationists can easily be exaggerated. Interaction effects, such as those responsible for the familiar modifications of the $9:3:3:1$ F_2 ratio $(9:3:4, 9:7,$ etc.$)$, were recognized very early[61, 232] and interpreted as due to epigenetic chains of reactions.

ELEMENTARY GENIC ACTION

It is generally accepted now that the physiologic action of genes is wholly by imposition of specific patterns on macromolecules[1450] and even this is thought to be by steps (DNA–RNA–polypeptide, etc.).   The classes of characters that have seemed most

Fig. 21. Diagram of relations of genome and external environment to observed characters at various levels of organization.

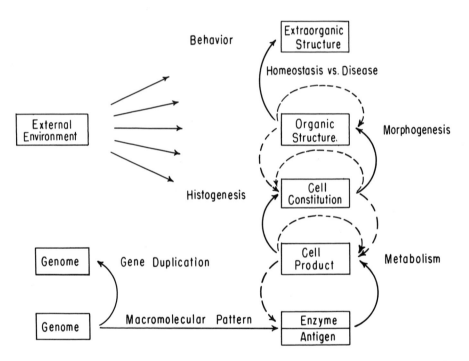

favorable for study of the first links in the control of characters have thus been antigenic and enzymatic specificities and the more elementary metabolic processes.

One of the most significant results from immunogenetics has been the usual occurrence of a one-factor relation between any particular antigenic reaction and a particular gene (or group of alleles) irrespective of the situation at other loci.   Even so, the demonstration of hybrid substances by Irwin shows that there may be interactions among loci even at this level.[641]

Another significant result is the occurrence of a very large number of alleles at many of the loci with antigenic effects.   While attempts at subdivision of the loci according to the pattern of responses to multiple test sera have been demonstrated to be unsound,[985] the extraordinary number of alleles in some cases (more than 250 at the *B* locus in cattle) and the curious intragenic, interaction effects indicate great complexity of genic pattern and the existence of much primary pleiotropy.

The association of blood-group genes in man with fetal hemolytic disease and with various diseases of adults demonstrate much secondary pleiotropy. There is no good reason to suppose that multiple allelism and pleiotropy are any less frequent with loci for which such delicate methods of demonstrating differences are unavailable.

Fig. 22. Diagram of factor interactions in the determination of coat color of the guinea pig.

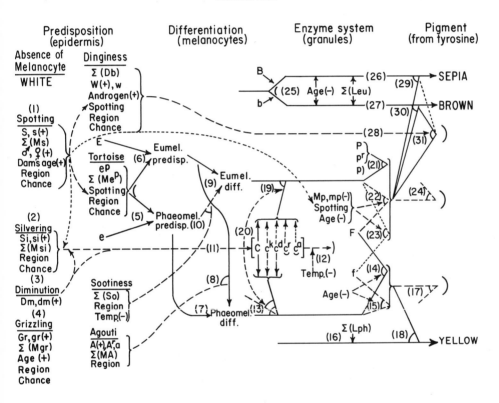

The study of the genetics of metabolic processes in such favorable organisms as *Neurospora, Aspergillus,* and *Escherichia coli* has been characterized by the demonstration of long, branching, reaction chains in which each link is controlled by a particular locus. Nonadditive interaction effects are the rule.[1353] Such processes are merely the first steps leading to the variations observed at the morphologic level in higher animals. Figure 21 is intended to represent the hierarchy of levels at which interaction may occur.[1413, 1424, 1433, 1434, 1441, 1450] The interaction patterns found among the genes concerned with a number of such characters of the guinea pig will be reviewed here.

COAT COLOR OF THE GUINEA PIG

Figure 22 is an interpretation of the interaction pattern with respect to quality, intensity, and pattern of the coat color of the guinea pig. Color depends on the presence of pigment granules (sepia, brown, or yellow) or their absence (white). The pigment granules are produced in special cells (melanocytes) in the hair follicles, basal layer of the epidermis, chorioid coat of the eye and certain other tissues, and in retinal cells. All but the last migrate from the neural crest. The melanocytes may pass the pigment granules to epidermal cells. Absence of pigment may be due to failure of the melanocytes to reach the normal site,[1177a, 1352a] to death in the site, probably in silvering and grizzling (at 1, 2, and 4, respectively, in figure 22), or to failure of melanogenesis without death of the cell (albinism $c^a c^a$).[854, 1204, 1315]

The primary differentiation in quality is between cells that produce eumelanin (sepia, brown) and phaeomelanin (yellow). The most important factor in relation to this differentiation in the guinea pig is the *e* locus (*E* typically self-eumelanic, *ee* typicaly self-phaeomelanic, $e^p e^p$, $e^p e$ a mosaic (tortoise-shell) of the colors found with *E* and *e*). The *e*-alleles cannot be supposed to determine differentiation of melanocytes directly but merely to produce an effect that predisposes toward one or the other type of differentiation (at 5 and 6 in figure 22). Thus with *EA* each hair follicle produces (at 8 in figure 22) the same sort of yellow as with *ee*, during a brief phase in the growth cycle of each hair, in spite of the presence of *E*. This results in a subterminal yellow band in otherwise eumelanic hair. A locus with similar action in the mouse acts on the melanocytes from adjacent epidermal cells.[1203] On the other hand, pigment cells in animals with pure yellow coats at birth because of *ee* may produce much black or brown pigment later (sootiness at 10 in figure 22) in the presence of favorable unanalyzed heredity $\sum (So)$ and low temperature.[1401] Gene e^p is definitely not mutable in the germ line in inbred strains. The nature of the all-or-none process that occurs early in development in tortoise-shells to distinguish different cell lineages (perhaps in the epidermis) is not known.

Cells with phaeomelanic differentiation produce only phaeomelanin but those with eumelanic differentiation produce granules that, while largely eumelanic, may contain a small amount of phaeomelanin, which is revealed under conditions that differentially reduce the eumelanic constituent.

Visual grades, based on standard squares of skin chosen so that each grade is barely distinguished from the preceding, have been assigned each animal at birth and often later. The corresponding amounts of pigment, relative to intense black (grade 21) or intense yellow (grade 11) have been estimated by extraction from weighed samples of hair and colorimetry.[542, 1096, 1453] In the later years,[1415] reflectionmeter readings (*R*) were taken of many genotypes using amber, green, and blue filters. Indices closely paralleling the visual grades were obtained by the equation

$$I_X = 10(\overline{\log R_W} - \log R_X)$$

in which $\overline{\log R_w}$ is the average logarithm of readings of white and R_x in the reading in question. A final intensity index was obtained from the sum of the indices from the three filters. Relative quantities of pigment were estimated from the colorimetric determination by way of the relations to the visual grades. Figures 23 and 24 are based on such estimates for various genotypes; figure 25 is based on the unweighted average of estimates from the indices and earlier estimates from visual grades. An index of quality (Q) was obtained[1438] from the ratio of the index from the blue filter to the total index (I). Average quality ($\bar{Q}$) is plotted against average intensity ($\bar{I}$) for various genotypes in figure 26.

The melanoproteins of the granules consist of melanoid derived by oxidation of

Fig. 23. EUMELANIN AND PHAEOMELANIN IN GUINEA PIGS.

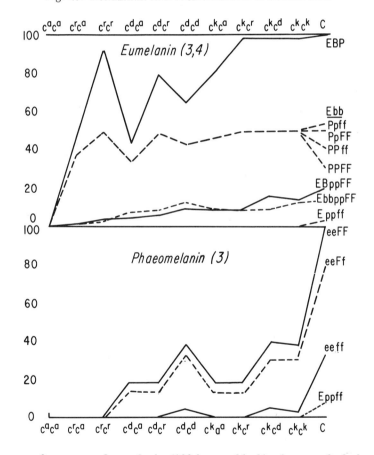

Estimates of amounts of eumelanin (100-intense black), above, and of phaeomelanin (100-intense yellow), below, in guinea pigs with combinations of E, e; B, b; C, c^k, c^d, c^r, c^a; P, p and F, f. The eumelanic estimates are based on animals with three or four plus factors at the loci Si, si; Dm, dm. The phaeomelanic estimates are based on those with three plus factors. The estimates are transformed from reflectionmeter readings.

tyrosine, under control of the enzyme tyrosinase, firmly linked with protein.[1192]   The
specificity of the enzyme seems to depend primarily on the *c* series of alleles which
is interpreted as acting pleiotropically on the phaeomelanic and eumelanic processes
(13 and 19 respectively in figure 22).   The complete or very nearly complete dominance of *C* in all combinations indicates that its product is in excess in both reactions

Fig. 24. EUMELANIN AND PHAEOMELANIN IN GUINEA PIGS.

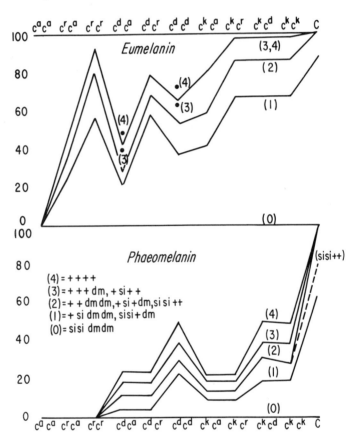

Estimates of amounts of eumelanin (above) in guinea pigs of genotype *EBP* and of
phaeomelanin (below) in ones of genotype *eeFF*, associated in both cases with the various *c*
compounds and with 0 to 4 plus factors at the loci *Si, si; Dm, dm.* Comparable with
figure 23.

while the intermediacy of heterozygotes among the other alleles indicate that their
products are limiting factors in both.   The evidence as a whole indicates the order in
amount of product to be $C > c^k > c^d > c^r > c^a$ but the actual orders of intensity
vary from case to case (figure 23).   The lower alleles are interpreted as competing
(20 in figure 22) for processes 13 and 19 with different degrees of efficiency in order to

account for the qualitative differences shown in figure 26.   Thus c^r and c^a are interpreted as unable to produce any yellow whatever, even in cells with phaeomelanic differentiation, c^d is interpreted as relatively efficient in producing yellow, but less

Fig. 25. Eumelanin in guinea pigs at birth and six months later.

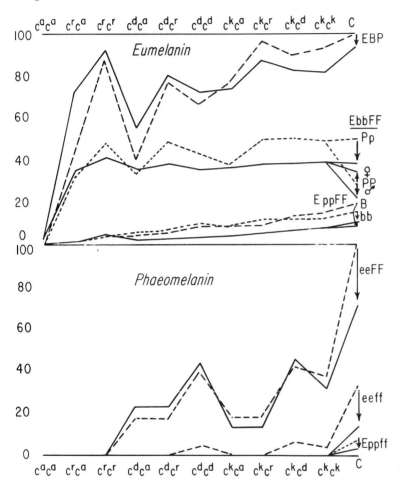

Estimates of amounts of eumelanin (above) in guinea pigs at birth (broken lines B, dotted lines bb) and at about six months of age (solid lines).   Similar estimates for phaeomelanin (below) in guinea pigs at birth (broken) and at about six hours (solid).   These are based on unweighted averages of estimates from transformed visual grades (back as whole) and transformed reflectionmeter readings (darkest spot near midline of back).

efficient than c^r in producing eumelanin because of this competition.   It is more efficient in the eyes (c^dc^d, c^dc^r, c^dc^a black eyed, c^rc^r dark red, c^rc^a light red) in which there is no production of yellow and thus no competition.   Gene c^k is interpreted as somewhat less efficient than c^d in producing yellow but much more efficient in producing eumelanin.

The top lines in the upper and lower parts of figure 23 compare the various *c* compounds in the most favorable combinations with other loci.

Returning to figure 22, the intensity of phaeomelanin is represented as affected by F (strong) and f (weak) at 14 and 15 in figure 22 following reaction 13 and as subject to a threshold (at 17). What is left is represented as interacting (at 18) with a limiting factor $\sum (Lph)$ to give yellow pigment. The averages in figure 23 show a slight reduction brought about by replacing FF by Ff and great reduction by replacing it by ff. The absence of yellow pigment in the heterozygotes represented collectively by $c^{kd}c^{ra}ff$, in contrast with the small amounts in $c^{kd}c^{kd}ff$, is one of the evidences for a threshold. Great differences in the intensity of reds, $eeCF$, among inbred strains (about twice as much pigment in strain 32 as in strain 2) indicate a variable ceiling. Crosses indicated multiple factors, $\sum (Lph)$.

Turning to the eumelanic colors, P (or an allele p^r described by Iljin[626] as similar in effect to P except for lighter eye color) is represented as necessary for the production of

Fig. 26. Mean quality index ($\bar{Q}$) plotted against mean intensity index ($\bar{I}$) for various genotypes.

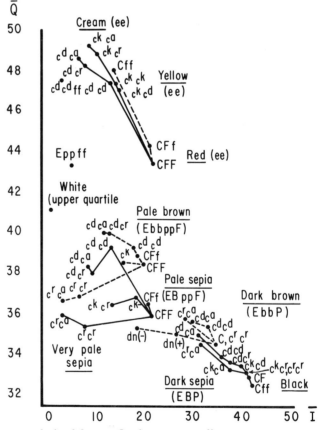

Both indices are derived from reflectionmeter readings.

the full amount possible with each c compound.   Compounds FF, Ff or ff make no detectable difference in the intensities of any of these.   The replacement of P in $EBPF$ by pp (figure 23) results in a great reduction in the amount of pigment in both coat and eyes (pink).   There is disproportionately great reduction in the lower c compounds, especially $c^r c^r$ and $c^r c^a$.   With $c^r c^a$ the color is often indistinguishable from white at birth, indicating a low threshold (at 24 in figure 22).   The fact that even with $c^r c^r$ there is less pigment than with $c^d c^a$, in contrast with the situation in the presence of P, seems to require a carrying through of the effect of the specificity difference between c^r and c^d to the reaction at 22 (in figure 22).

The replacement of F in $EBppF$ by ff reduces pale sepia to a very much paler brownish cream, sometimes indistinguishable from the pure pale yellow of $eec^d c^a ff$. This color is interpreted as due largely to the uncovering of a feeble underlying yellow (at 15) in the absence of both P and F.   The trace of eumelanin, usually present, is attributable to feeble action of f (at 23).   The fact that replacement of F by ff makes no recognizable difference if P is present, but almost complete absence of eumelanin with pp is interpreted as meaning that P, without F, is sufficiently efficient to produce as much eumelanin as the c compound permits but that in the absence of P, F acts as a feeble substitute, and if this also fails, only a trace at most can be produced by f.   It may be noted from figure 23 that with any lower c compound and $Eppff$ the color is pure white, indicating that the eumelanic and phaeomelanic processes fall below the threshold in these cases.

The replacement of B in EBP by bb gives brown in coat, skin, and eyes in place of sepia and approximately halves the apparent intensity by colorimetric determinations of these closely similar pigments.   There is parallelism between sepia and brown among the c compounds (figure 23) to the extent that $c^a c^a$ is white and $c^d c^a$ and $c^r c^a$ definitely more dilute than the others (except certain combinations involving C), but the differences among all the other c compounds are much less in browns not only absolutely but on a percentage basis.   This can be interpreted on the hypothesis that bb imposes a ceiling on the possible amount of pigment at about half that imposed by B and that this ceiling is approached to such an extent even with $c^r c^a$ and $c^d c^a$ that there is no possibility of much further increase with higher compounds.

If P in $EbbPF$ is replaced by pp, there is marked reduction in intensity but not as much proportionately as in the sepias.   The eyes are again pink.   There is a marked qualitative difference from the pale sepias (figure 26) in contrast with the slight apparent difference between dark browns and dark sepias of the same intensity.   Pale browns with C have much more than half as much pigment as the corresponding pale sepias and among the lower c compounds there are no consistent differences in quantity in the somewhat unsatisfactory data on this point.   The smallness of the quantitative difference in comparison with that in the P eumelanics may be interpreted as due to remoteness from the ceiling. $Ebbppff$ is wholly indistinguishable from $EBppff$.

The most remarkable interaction effect is one that affects only browns of genotype $EbbCP$.[1431, 1439]   There is an optimum genotype with respect to the c, p, and f loci, probably Cc^xPpff in which x is any lower c allele. Thus in the presence of Ebb, c^xc^xppff is pure white, $Cppff$ has a trace of color, c^xc^xppF is very pale brown, $CppF$ pale brown, c^xc^aPff slightly dilute brown, c^kc^kPff intense brown, $CPpff$ possibly slightly more intense, but $CPpFF$ slightly dilute (dingy on the head), $CPPff$ slightly more dilute and $CCPPFF$ often with less than half as much pigment as the intense browns. Replacement of CC by Cc^x in these has a slight darkening effect demonstrated only in adults. The type of dilution in the higher combinations is of a peculiar sort (dinginess) which ranges from a mere sprinkling of dark tipped light hairs on the cheeks and nape to uniform dilution of all hairs, except at the extreme tips, to a color as pale as pale sepia (only in $EbbCCPPFF$ with favorable modifiers). This type of dilution could not be produced at all in the presence of c^kc^k or of pp. This effect is represented in figure 22 as the result of a destructive action of CPF product in excess of that necessary to saturate the limited bb product. The failure to observe any such effect in blacks in this colony may be interpreted as due to the higher ceiling provided by the B product. Ibsen and Goertzen[624] described an incompletely dominant modifier of dinginess (W) which completely inhibits brown pigment in a subterminal band in $EbbCCPPFF$. They found a slight effect in blacks in the presence of WW. While W was clearly absent from my colony, dinginess was much affected by other modifiers. It was possible to bring even $EbbCCPPFF$ to full intensity in a few cases by selection.

We will go back to consider modifiers of some of the other processes here. The most important genetic ones are Dm, dm and Si, si (figure 24).[1415, 1432, 1440] The most conspicuous effect of $sisi$ in otherwise intense animals is to cause a sprinkling of white hairs (*silvering*) in the coat which does not progress after birth in contrast with the effect of $grgr$ (*grizzling*)[745, 1431] with no effect at birth but progressive whitening later. We are here concerned with dilution effects of si on colored hairs. Replacement of Dm by dm (*diminution*) has no recognizable effect on intense (C) blacks or yellows but causes dilution in lower c compounds, most conspicuously in c^dc^a (both sepia and yellow) and in c^rc^a sepias. Factors si and dm dilute both colors cumulatively (figure 24). The combination $sisidmdm$ is pure white except for occasional pale spots on the head. Eye color is slightly reduced. There are other effects which suggest that the effect of these genes is on the vitality and metabolic efficiency of certain types of cells rather than on the pigment process in as specific a way as seems to be the case with many of the other loci. Animals with $sisidmdm$ suffer a high mortality after birth and an anemia in which the red blood count is reduced by a third. In males, testicular size is .reduced to 25 per cent of normal and there is complete absence of spermatogenesis and thus sterility. About half the females tested have been sterile and the remainder rather low in productivity. Assuming that the effect of these genes on color is on metabolic efficiency of the pigment cells (if not actually lethal to these as in silvered hairs) it appears that among the specific color factors only the efficiency of the c alleles is affected. The effects of si and dm have been studied intensively only in dark sepias EBP and yellows $eeFF$, but as far as determined they have proportional effects in browns ($EbbP$), pale sepias ($EBppF$), and pale yellows ($eeff$).

There are at least two pairs of modifiers, Mp_1, mp_1, Mp_2, mp_2, that profoundly affect the intensity of pale sepias ($EBppF$), pale browns ($EbbppF$), and probably the trace of eumelanin in pale brownish creams ($Eppff$) but have no recognizable effect in dark sepias (EBP), dark browns ($EbbP$), or yellows (ee).[1439] There are, however, unanalyzed modifiers that affect the height of the ceiling for black or brown [$\sum (Leu)$] as well as those already referred to that affect the ceiling for yellow [$\sum (Lph)$].

ENVIRONMENTAL EFFECTS ON COAT COLOR

There are interaction effects with temperature and age[1401, 1421] (figure 25). Most of the processes become weaker as the animals grow older, independently of temperature. Thus pale sepias fall off some 50 to 60 per cent in intensity by a half year of age (attributable to process 22 in figure 22). In dark browns that are not dingy there is a reduction of about 12 per cent, probably from lowering of the ceiling (27 in figure 22). However, the dingy females darken, indicating a weakening of dingy modifiers (28 in figure 22), but the males become even lighter. Experimental evidence indicated that this lightening was an effect of androgens.[1400] Pale browns diminish in intensity but probably somewhat less than pale sepias. Intense yellows (FF) decrease some 20–30 per cent (14 in figure 22), while fading yellows (ff) show a much greater reduction, some 50–60 per cent (15 in figure 22).

Low temperature has already been referred to as a condition for sootiness of yellow (10 in figure 22). The most striking effects of temperature, however, are those on the lower c alleles, the products of which seem to be markedly thermolabile (12 in figure 22). This effect is not recognizable in the case of C, the product of which is presumably present in great excess. In the case of c^k product, the effect of temperature, if any, does not compensate for the reduction with ageing. Thus yellows of genotypes $c^k c^k F$, $c^k c^r F$, and $c^k c^a F$ become significantly paler after birth when production of pigment occurs at a lower temperature than before birth. This is also true of sepias carrying $c^k P$. On the other hand, yellows carrying $c^d F$ all become more intense. In these, the decreased loss of c^d product with lower temperature more than compensates for the effect of ageing. This is also true of dark sepias with $c^d c^d$, $c^d c^r$, $c^r c^r$ and especially $c^d c^a$ and $c^r c^a$. Albinos ($c^a c^a$) are pure white at birth, although pigment cells are present.[1204] Soon after birth, those with $EBc^a c^a P$ and $Ebbc^a c^a P$ develop much black or brown pigment, respectively, in the skin and in hairs on feet, nose, and ears and traces of eumelanin on the back. In dark browns, the lowering of the ceiling with age and the tendency to darkening, especially of $c^r c^a$ and $c^d c^a$, from lowered temperature almost remove all differences among c compounds except for $c^a c^a$ and $CPPFF$ males.

SPOTTING

We have not yet considered *spotting* and its remarkable interaction effects.[199, 1439, 1454] *Spotting* (colored spots on a white ground) depends primarily on an incompletely recessive gene *s*.   There are unanalyzed modifiers, $\sum (Ms)$, that can shift the median percentage of white in inbred strains with *ss* from about 10 per cent to 98 per cent.   In all cases females average a little whiter than males.   The same female produces whiter young on the average as she becomes older.   The pattern has some orderliness—with the strongest tendency to color near eyes and ears, strongest tendency to white on feet, nose, and midline of belly; but there is always an enormous amount of variability that can only be attributed to accidents of development.   An inbred strain may range from a trace of white to black-eyed self white and yet show by the absence of correlation between parent and offspring that there is no genetic variability.   In such a strain there is indeed no correlation between points one third of the length of the animal apart with respect to presence or absence of white.

Gene *s* is not mutable germinally in inbred strains.   Spotting is probably due to interaction of two modifiable patterns: migration of melanocytes from the neural crest into the skin, and differentiation of the skin.[117a, 1352a]   The points of most interest here are interaction effects with certain other processes indicating that the spotting process is not merely one which leads to presence or absence of pigment cells. It is a process in which the pertinent cells in the areas where color is present fall into different states in a spotting pattern which affects certain color processes but not others. Those affected most conspicuously are the tortoise-shell pattern due to e^p (5, 6 in figure 22) and the pattern of dinginess in browns of genotype *EbbCCPPFF* (28 in figure 22).   The pattern of silvering due to *sisi* (2 in figure 22) and the intensity of pale sepia or pale brown (22 in figure 22) are affected similarly but much less frequently. There are very rare cases of mosaics with respect to other loci which suggest somatic mutation but which are usually related to white spotting in a way that suggests that gene *s* has something to do with them.[1456]

Tortoise shells of genotype SSe^pe^p are predominantly eumelanic but usually show scattered yellow hairs and less frequently more or less yellow in blotches.   With sse^pe^p or even Sse^pe^p, the amount of yellow is increased and there is a strong tendency to segregation of yellow and eumelanin into a few large areas each often with scattering admixture of the other color.[199a, 627, 1419]   These areas are often separated in whole or part by white streaks.   Sometimes a streak between eumelanic areas is white at one end, yellow at the other, indicating that the determination of yellow is related to the process that leads in more extreme cases to white by absence of the pigment cells. Similarly the orderly pattern of dinginess, found in the presence of *SS*, is broken up in the presence of *ss* into a coarse mosaic of light and dark dingy areas, often separated in whole or part by white streaks.   In this case, it seems to be the determination of the darker areas that is most closely related to the determination of white.[1439]

COLOR FACTORS AND ACTIVITY OF TYROSINASE

There has been considerable research on differences in enzymatic activity in pigment cells of different genotypes.   There are definite differences which may be attributed to color factors (E, e; A, a; C, c^k, c^d, c^r, c^a; P, p; F, f), but the correlation with the intensity of pigment production is far from perfect. The dopa reaction has been studied in frozen sections of the skin at birth[735, 1130] and in colorless extracts from the skin at birth.[433] Foster[394] has studied both the oxygen consumption curves and amount of darkening in the Warburg apparatus on adding tyrosine or dopa to homogenates of fetal skin.

There are some apparent inconsistencies in the results, but these seem to trace largely to the stage of the reaction on which observation was focussed. Foster's technique was designed to distinguish differences in moderate to strong reactions and gave less certain results in distinguishing very feeble reactions from the background oxygen consumption of controls. The earlier studies of the dopa reaction distinguished different end results of very feeble reactions from that in controls but gave such saturated results from only moderately strong reactions that these were not always distinguished from very strong ones. The dopa reactions in frozen sections were moreover very seriously obscured where there was much natural pigment (dark sepias, dark browns).

It may be noted first that tyrosinase and dopa oxidase activities were very greatly reduced by replacing E by ee.[394] Foster found similar reduction in the yellow phase of agouti follicles (EA) in comparison with the black phase (Eaa). Much black melanin was, however, ultimately produced from dopa in frozen sections[735, 1130] and extracts[433] from intense yellow skin although both Russell and Ginsburg found significantly less than from even pale sepia.

Foster was unable to demonstrate any tyrosinase or dopa-oxidase activity from lower c compounds including EBc^kc^rPPFF which is intense black. Russell found considerable ultimate darkening of dopa in frozen sections from pale sepias ($EBppFF$) and yellows ($eeFF$) carrying c^kc^k, c^kc^d or c^dc^d though much less than with C, and even some darkening in the corresponding heterozygotes $c^{kd}c^{ra}$. Russell, Ginsburg, and Foster all found no evidence of a reaction from the near blacks of genotype c^rc^r. It would appear that the enzymes produced by the c compounds tend to be lost, after forming the natural pigments, almost as rapidly as they are produced unless produced in great excess by C. We have already noted the thermolability of the product of the lower c alleles which would especially affect Foster's studies of homogenates from fetal skin.

Foster found that homogenates from pale sepias ($EBCppF$) and pale browns ($EbbCppF$) oxidized tyrosine or dopa significantly less rapidly than homogenates from intense blacks or browns (P in place of pp) but the reduction was far from proportional to the great reduction in amount of natural pigment (in contrast with his results from lower c compounds). This agreed with Kröning's results in hair[735] and Russell's results at least in the epidermis. Replacement of F by ff caused no reduction in tyrosine or dopa oxidase activity in intense blacks or browns. With the same replacement in the presence of pp, the moderately strong reactions were reduced to zero (tyrosinase) or very weak (dopa oxidase) according to Foster. These parallel the striking interaction effect *in vivo* (EPF and $EPff$, equally intense; $EppF$, pale; $Eppff$, very pale). The dopa reaction was, however, moderately strong in frozen sections and indistinguishable in pale cream and yellow areas of 14 tortoise shells of genotype $e^p e^p C p p f f$ (W. L. Russell, unpublished results of F. Appel and of L. B. Russell). These reactions were, however, significantly less than in pale sepias (*ECppF*) and reds (*eeCF*). It appears that in the eumelanic series *F* (in *ppF*) and even *f* (in *ppff*) substitute for *P* somewhat more effectively in oxidising added tyrosine or dopa than in producing natural melanin.

Replacement of *B* by *bb* causes no reduction in tyrosinase or dopa oxidase activity by any test. There is, indeed, considerable evidence that the reactions are stronger with *bb*, even in the extreme dingy browns. It appears that *B*, *b* takes no direct part in the enzyme system responsible for oxidising added tyrosine or dopa. Perhaps again greater utilization in blacks than in browns is associated with greater loss.

CONCLUSIONS ON COLOR INTERACTIONS

It may be seen that the products of the color factors enter into a rather complicated pattern of interactions before formation of the final products: sepia, brown, or yellow pigment granules. The pattern resembles somewhat the branching chains of reactions worked out for gene-controlled metabolic reactions in microörganisms. Some of the steps may indeed be of similar nature, since most of the factors probably act in the nuclei of the pigment cells themselves as demonstrated in mice by Reed and Henderson.[1044] Others, however, probably act from adjacent epidermal cells as already noted. The difference between adult male and female dingy browns is probably under endocrine control. Some seem to be concerned directly with the pigment process but others undoubtedly affect pigment merely because of effects on the vitality or metabolic efficiency of the pigment cells.

HAIR DIRECTION

We will consider next a morphologic character, hair direction (figure 27). In wild cavies, the hair is directed away from the snout on the body, with minor qualifications, and towards the toes on the legs. On introducing the dominant gene *R* into a genotype that is otherwise like that of the wild species (*rrMMrereststst*) there is more or less reversal on the feet, especially the hind feet (grade E) and occasionally a slight crest along the back.[1428, 1429] Those with *RMm* are highly variable, but most of them have either a strong dorsal crest (grade D) or a single pair of rosettes, halfway along the back (grade C). Some of those called grade C had only a single dorsal rosette. This was almost always (96 per cent) on the right side, a curious example of regular asymmetry. A few were called grade E and many in certain strains were called grade B, or better, *CA*, with two pairs of dorsal rosettes (as in grade A) but no rosettes on head or belly (as in grade C). With *Rmm*, there are typically two strong pairs of dorsal rosettes, anterior and posterior, radiation from the ears, a strong forehead rosette, eye, check, and groin rosettes, and feathering along the midline of the belly (grade A). The number of dorsal rosettes varied in both directions. It might be at least doubled (grade A^{++}) or it might be reduced to two (grade AC). In cases of asymmetry, rosettes again tended to occur more on the right than on the left side.

Genes M, m are specific modifiers of R with no detectable effect in its absence. Other specific modifiers were clearly present. As indicated above, high grade RMm (grade CA) differed markedly from low grade Rmm (grade AC). It appears that the number of dorsal rosettes was especially subject to modifiers, $\sum (MRd)$, in both of these genotypes. The effects of these modifiers were, however, confused by a very

Fig. 27. DIAGRAM OF FACTOR INTERACTIONS IN THE DETERMINATION OF HAIR DIRECTION AND PLEIOTROPIC EFFECT ON WHITE FOREHEAD SPOT IN THE GUINEA PIG.

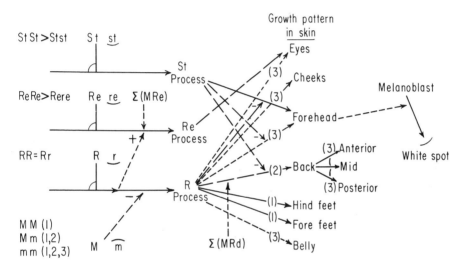

considerable amount of nongenetic variability. Thus the regression of offspring on midparent in matings of $Rmm \times Rmm$, including an excess of the extremes AC and A^{++}, was only 0.48, indicating that only about half of the variability was genetic. Three generations of selection did not suffice to fix grade AC.[1428, 1429]

As indicated above, eye rosettes ordinarily appear only in roughs of high grade (Rmm). Strong eye rosettes were, however, recorded in four individuals that were otherwise of grade E and of genotype RMM, from separate matings that were only remotely related and had no record of the character in their ancestry (other than in grade A). A considerable number (171 born alive, 21 born dead) were derived from one of these lines but a complete analysis of the genetics was not made. There are, however, a number of points of interest with respect to interaction in the unpublished data.

The character proved to be a very troublesome one because of intergradation with the condition in normal, smooth-furred animals (rr) which themselves show a slight divergence of hair direction about the eyes. Among those considered to have the anomalous eye rosette and born alive, 96 were recorded as having a strong pattern and 75 a weak one. There was rather frequent irregular asymmetry, including a few cases in which there was a strong rosette about one eye and none about the other. This indicates that developmental accident played a considerable rôle.

In matings in which *Mm* and *MM* were segregating, there was no significant difference in incidence of eye rosettes among grades E, D, and C contrary to expectation in view of their invariable occurrence in grades AC and A, genotype *Rmm*. There was also no relation to presence or absence of a forehead rosette in matings segregating in *Stst* and *stst*, discussed later. On the other hand, segregation of *R* and *rr* made a great deal of difference. Thus a group of matings that produced 16 per cent with eye rosettes among 208 *RM* young produced only 2 per cent among the 217 *rr* young. A group that produced 55 per cent among 88 *RM* young, produced only 20 per cent among the 46 *rr* segregants. Only liveborn are considered here because of somewhat greater uncertainty in classification of stillborn. Thus *R* greatly favors manifestation of the character but is not necessary. Strong eye rosettes were present in eleven otherwise completely smooth animals (*rr*).

The highly sporadic occurrence at first, including such records as 1 in 15, 16, or 17 *RM* young from scattered matings between normals, of 0 in 27 and 2 in 29 from matings of strong × normal and of 3 in 15 from a mating of strong × strong, suggested either low penetrance or a recombination effect of two or more genes. After all young were examined carefully for weak eye rosettes at both birth and weaning (when they are often more conspicuous), it became apparent that penetrance at this level could be high at least in *RM* young. The pattern of occurrence indicates a single essential gene, *Re* (*rough eye*). The results of outcrosses indicated dominance in some degree but with penetrance far from perfect in heterozygotes carrying *RM* and low but not zero in ones carrying *rr*. There was an approach to 100 per cent penetrance but not expressivity of *RMReRe*, but only moderate penetrance in *rrReRe*.

Another aspect of the full rough pattern of *Rmm*, the forehead rosette, can also appear in the absence of *R*.[1429, 1430] All animals in the colony of this type traced to three obtained from Mr. I. J. Wachtel. This character, *Star* (*St*), behaved in a remarkably different way from *rough eye* (*Re*) in two respects. First, it showed no intergradation with normal and behaved as a simple dominant in eight successive backcrosses to lines in which it had never been present and there was no difference between *rrStSt* and *rrStst*.[1429, 1430] Second, instead of showing enhancement in the presence of *R* there was a tendency toward reciprocal inhibition.[1430] In *rrSt–*, the forehead rosette is almost invariably (99.6 per cent) single and very flat. In about 19 per cent of *RMMSt* it is replaced by two weak rosettes. This forehead roughness is combined with typical grades D or E but the proportion that are grade D is reduced. In *RMmSt–*, this reduction of the rough pattern goes farther. Most show roughness of grades D or even E combined with a forehead rosette that is doubled in 64 per cent. In the 17 per cent called grade C the rosettes were usually weak and lateral. The forehead rosette was weakened in 78 per cent. Finally in *RmmSt* in which an exceptionally strong forehead rosette might have been expected by combination of the effect of *Rmm* and of *St*, there was actually only a slight irregularity of hair direction on the forehead. The anterior dorsal rosettes of grade A were weak and lateral in position, leaving a broad smooth shield anteriorly on the back. This last feature refers to *RmmStSt*. In this case only, *St* ceased to be completely dominant and *RmmStst* showed a great variety of intermediate conditions of the dorsal rosettes. It is interesting that *RMM* and *RMm* tend to inhibit the effect of *St* in a part of the coat, the forehead, in which they have no visible effect in the presence of *stst* and that *St* in turn inhibits the effects of *RMm* and *Rmm* in a region, the anterior back, in which it has no visible effect in the absence of *R*. In view of these effects, it ceases to be surprising that *Rmm* and *St* interact to prevent development of the forehead rosette which each determines by itself.

Gene *St* has an interesting pleiotropic effect in a tendency toward a white spot in front of the center of the forehead rosette. It was shown by Bock[114] that this does not require even heterozygosis in the spotting factor and that it tends to be inhibited by *R*, especially *Rmm*. Table 42 gives a condensed summary of his results for young with no

Table 42

PERCENTAGE OF WHITE FOREHEAD SPOT IN GUINEA PIGS WITH NO WHITE OUTSIDE OF THE FOREHEAD

Grade (excluding forehead)	Typical genotype	*stst*		*St*–	
		No.	Per cent spot	No.	Per cent spot
Smooth	*rr– –*	294	2.4	530	46.6
Rough E	*R–MM*	27	3.7	116	42.2
Rough C, D	*R–Mm*	75	8.0	181	37.0
Rough A, B	*R–mm*	59	5.1	164	15.2
Total		495	3.7		

According to grades of roughness due to *Rr*, *Mm* and to presence (*St*) or absence (*stst*) of forehead rosette from matings that involved *St*.

white outside of the forehead and from matings that involved *St*. The results indicate that the white forehead spot is a secondary consequence of the effect of *St* on the skin of the forehead, manifested in the forehead rosette. The neutralizing effect of *R* on the latter, especially in the combination *R–mm*, also tends to neutralize the white spotting. Another element of the full rough pattern of *Rmm*, the feathering on the belly, has also been observed independently of *R* in a certain somewhat inbred strain.[1416] It followed an irregular course of inheritance, neither a simple dominant nor a simple recessive. Dorsal irregularity of hair direction has also occurred independently of *R* but so sporadically that mere developmental accident seemed indicated. The most important general result of this study of the genetics of hair direction seems to be another revelation of the extraordinary diversity in the relation of genes to characters, and in particular in kinds of interaction, that is to be found in this case in even a rather limited genetic system.

POLYDACTYLY

Attention is now directed to a different sort of morphologic character, the occurrence of atavistic digits. The guinea pig, like all species of *Caviidae*, lacks the thumb, great toe, and small toe. A fourth toe, exactly like the small toe of related rodents, is not, however, uncommon. Among 22 inbred strains there were 11 that were invariably 3-toed on the hind feet (including strains 13 and 32), the small toe appeared sporadically in five (including strain 2, the major branch of which, however, was entirely free of the trait) and in six the incidence was fairly high (including strain 35).[1445, 1446, 1447] A strain, D, with 100 per cent occurrence of a well-developed small toe was produced by Castle[181] by selection. Crosses were made between strain D (4-toed) and the 3-toed strains 2, 13, and 32 and between D and strain 35 with 31 per cent in the branch descended from a single mating in the 12th generation.[1451]

The results from 2 × D and 32 × D gave complete dominance of 3-toed in F_1, except for one individual in 26 in the latter, and passable 3:1 and 1:1 ratios in F_2 and the backcross to strain D respectively. These results suggest segregation of a single, essential, recessive factor for the small toe but tests of the supposed segregants from (2 × D) × D gave results that completely vitiated this hypothesis. The 3-toed segregants produced only 23 per cent 3-toed (in 186), whereas the 4-toed segregants gave nearly the same result (16 per cent 3-toed in 119). The most plausible hypothesis seemed to be cumulative action of multiple factors on the underlying physiology and two thresholds: one for any development of the small toe, below which there is homeostatic control of the normal 3-toed foot and, slightly higher, a threshold (or better, a ceiling) above which development of the small toe is controlled homeostatically, or canalized in Waddington's[1352] terminology. The cross 13 × D gave 67 per cent 3-toed and 33 per cent 4-toed in F_1. Tests of the two F_1 types gave results that did not differ significantly either in F_2 or in the backcross to D. There was no simulation of one-factor heredity, but the results fit well with the interpretation of multiple factors and two thresholds. Most of the young from 35 × D were 4-toed. The results in F_2 and the backcross to 35 again fit the above hypothesis.

The results from these crosses in F_1 and F_2 are represented in figure 28 according to this threshold hypothesis. There was considerable variability in strain 35 in relation to age of mother (high percentage of 4-toed from immature mothers) and season (excess of 4-toed in winter and early spring).[1414] The standard deviation on the physiologic scale can be determined as that of the normal curve that yields the observed trichotomy, taking the interval between the thresholds as the unit. It is approximately 0.80. The same standard deviation was assumed to apply to the other inbred strains and to the three F_1's. The mean of D was put at 2.5σ (2 units) above the upper threshold as a value which might be attained by selection but at which overlap of the threshold would be rare.   Because one 4-toed had been obtained in F_1 (32 × D), a cross which gave results very close to those 2 × D in F_2 and the backcross to D, the mean of (2 × D) was put only 2.5σ below the lower threshold.   The means of the pure strains 2 and 13 were located on the hypothesis that F_1 in each case was exactly half way between the parental strains on the physiologic scale.

Fig. 28. FACTORS DETERMINING DEVELOPMENT OF THE SMALL TOE.

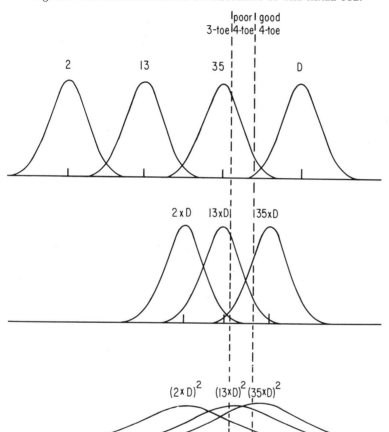

Theoretical distribution of factors determining development of the small toe on an underlying physiologic scale in relation to thresholds for any small toe and for perfect development, in four inbred strains, in F_1 and F_2 of crosses of one (D) with the other three.

The relations of variability in F_2 (2 × D) to the interval between the means of these two most extreme strains permit a minimum estimate of the number of gene differences, assuming equal effects and no dominance.   The number is approximately four.   The hypothesis of one major factor, responsible for half the difference and practically all of the variance plus a great many minor factors, fits equally well.   So does the hypothesis of a great many factors with contributions in geometric series (40 per cent:24 per cent:14.4 per cent:8.6 per cent, and so forth).[1448] Results in third backcrosses best fit an intermediate hypothesis somewhat similar to the last.

It is assumed that the effects are additive on the physiologic scale. However, the factors were not isolated. It may be suspected that, if they had been isolated, a more complex situation would have been found with specific interactions as in the cases of color and hair direction. As indicated above, the hypothesis of two thresholds permits considerable elasticity with respect to the effects on the underlying scale.

Similar small toes have been restored in a wholly different way, in this case often associated with thumbs and great toes resembling those of other rodents.[1413] A single mutant individual showed those characters imperfectly developed. From thousands of descendants it appears that a dominant gene (*Px*) with variable penetrance was responsible. In the stock of origin, about 82 per cent showed one or more atavistic digits. After certain crosses, penetrance fell to 20 per cent or less. Crosses with strain D revealed interesting interaction effects. It was not surprising that penetrance of the small toe in *Pxpx* rose from 62 to 100 per cent by combining these two heredities which favor it. More interesting is the fact that penetrance of the thumb rose from 74 to 100 per cent, although thumbs were wholly absent in strain D. Similarly, the penetrance of the great toe rose from 2 to 18 per cent after one backcross to D and to 55 per cent after two backcrosses.

The homozygote *PxPx* turned out to be lethal and grossly abnormal. It was found by Scott[1184] that about 92 per cent died and were absorbed at about the twenty-eighth day of gestation. Those that reached birth[1185] had short legs, rotated hind legs without tibiae, and broad paddle-shaped feet with 7 to 12 digits each. The animals were always microphthalmic and either hydrocephalic or with brains protruding from the skull, usually harelipped, and grossly abnormal in most of their internal anatomy. The defect was manifest at about 18 days of gestation in limb buds of double width and overgrown midbrains and hindbrains. This constitutes an extreme case of pleiotropy.

The atavistic return of digits, lost in the evolution of the whole family *Caviidae*, raises a question on the nature of homology. Homology is usually treated as an all-or-none matter: particular organs of two species either are or are not homologous. The language used in discussing evolution of organs often seems to imply preformation of an extreme sort, a heredity evolving separately for each part.

It is obvious, however, from the genetics of morphologic characters in all organisms studied, that replicative homologs develop to a large extent under the same systems of genes. Thus genotype *RRMM* has similar effects on hair direction and the underlying pattern of growth of the skin, not only on right and left hind feet but on hind feet and fore feet (with some difference in threshold) and on the separate digits on these feet (a pattern of partings along the sides and across the upper side of these digits shaped like a letter H). The situation is similar with the genes that tend to restore the pentadactyl foot. It is supposed that all of the genes under which any part of the ancestral pentadactyl foot developed were so deeply involved in the development of other parts of the foot and the organism as a whole that most of the genetic system concerned with thumbs, great toes, and small toes is necessarily still present. The loss of these digits is presumably due to some simple, superimposed, inhibitory process that stops formation of the normal number of lobes on the developing limb buds. Any genetic or environmental effect that inhibits this inhibitory process releases the substrate for action of the whole array of genes that shaped these missing digits in the remote ancestors. Homology, whether replicative in the same individual or phylogenetic, must be considered to be a matter of degree, the result of calling into play more or less similar systems of genetic reactions by more or less similar developmental processes.[1434]

The multifactorial polydactyly may conceivably involve ancestral genes that have been carried at low frequencies throughout the history of the *Caviidae* or have been brought back by reverse mutation, but this is hardly likely for *Px*. Yet if *Pxpx* acts merely by inhibiting a relatively simple inhibition acquired in evolution, the thumb, small toe, and great toe that develop in this genotype under the released ancestral heredity may be considered as essentially homologous to these digits in the mammals that have never lost them, in spite of the monstrous characteristics of the foot in *PxPx*.

Thresholds have been shown to play a major rôle in other morphologic deviants of the guinea pig of which otocephaly has been most intensively studied.[1433, 1455, 1458] It may be added that Grüneberg[501] reports that similar "quasicontinuous" variation interrupted by thresholds is the commonest situation in the mouse in the case of morphologic deviants.

TYPES OF INTERACTION

The effects of factor replacement under varied genetic or environmental conditions may be put into four categories.

Constant effects.—There is usually constancy of effect in cases in which the replacement in question is associated with variations in a character that seems obviously unrelated; for example, *R, rr* (rough and smooth fur respectively) and any pair of colors. This also often holds for widely different aspects of a character that is single only in a broad sense; for example, *A, aa* (*agouti* and *nonagouti* respectively) on a *black* (*B*) or *brown* (*bb*) background.

With respect to grades of a single quantitative character, constancy of effect requires somewhat arbitrary definition. One may, for example, choose to treat either a consistent additive effect or a consistent multiplicative effect as a constant effect. Strict examples of either of these types of constancy are, however, probably uncommon. They were not found in the quantitative studies of intensity of the colors of the guinea pig.

Cumulative but not constant effects.—There are many examples of this among the guinea pig colors. Thus on assigning 100 as the measure of quantity of yellow pigment with *CF*, that with $c^d c^d F$ was 38, with *Cff* 33 and with $c^d c^d ff$ 5. Any two-factor case such as this could be made additive arbitrarily by a suitable transformation of scale, just as multiplicative constancy can be converted into additive constancy by the use of logarithms. Such transformations are often useful but usually break down more or less in more complicated systems.

No effect in some combinations.—The frequent reduction of the F_2 ratio $9:3:3:1$ to $9:3:4$ or to $12:3:1$ indicates the frequency of specific modifiers of the effect of a dominant or of a recessive gene, respectively. Thus in guinea pigs carrying *EBF*, *CP* is black, *Cpp* is pale sepia, but either $c^a c^a P$ or $c^a c^a pp$ is white. In those carrying *EBC*, either *PF* or *Pff* is black, *ppF* is pale sepia, and *ppff* is pale brownish cream. The effect of *M, m* in rough (*R*) and smooth (*rr*) guinea pigs illustrates the case of a specific modifier of a dominant that itself lacks dominance (*Rmm*, full rough; *RMm*, intermediate rough; *RMM*, rough only on the toes; but *rrMM*, *rrMm*, and *rrMM* all equally smooth).

There may be mutual instead of one-sided dependence as in the cases of complementary dominants (F_2 ratio $9:7$), complementary recessives (F_2 ratio $15:1$), or complementary dominant and recessive (F_2 ratio $13:3$). In guinea pigs, *CP* gives dark eyes, but *Cpp*, $c^a c^a P$, and $c^a c^a pp$ are all pink-eyed (although *Cpp* can often be distinguished by traces of color). With *sisidmdm*, guinea pigs are anemic, spermatogenesis is absent in the diminutive testes of males, and fecundity is low in females, whereas with either *Si* or *Dm* or both, these are normal. There are, however, cumulative effects of *Si, si, Dm, dm* on intensity of color.

All such two-factor cases can be treated as multiplicative by assigning zero to the effect of one member of one or both pairs of alleles and choice of a suitable scale, but this sort of constancy of effect cannot as a rule be extended to all combinations.

Threshold and ceiling effects of a more complex sort may also be included in this category. All of the colors of the guinea pig are subject to threshold effects. A ceiling effect is most conspicuous in the case of the dark-eyed browns, especially when adult. The irregular penetrance that is characteristic of most mutational effects on morphologic characters implies a threshold. The occurrence and variations in degree of development of the small toe give an example of a case in which there is homeostatic control of development both below a threshold and above a ceiling.

Opposite directions of effect in different combinations.—In guinea pigs with *CPff*, those with *E* (black) have much more pigment than those with *ee* (dilute yellow) whereas with *Cppff*, those with *E* (pale brownish cream) have much less pigment than those with *ee* (same dilute yellow as in the preceding case). The intensifying effects of *C, P,* and *F* in brown (*Ebb*) up to a certain point and their dilution effect beyond this point is another example. In this case, *Pp* always shows overdominance, a phenomenon in which the alleles necessarily reverse their order of effect in different compounds. The increase in intensity after birth of c^d sepias and yellows, in contrast with the decrease in intensity after birth of c^k sepias and yellows, illustrates reversal of effect of a nongenetic condition in different genotypes. In this case c^k sepia is darker than c^d sepia at both ages, and c^k and c^d yellows hardly differ at birth.

In the case of comparison of intense browns (*EbbCPp*) and sepias of genotype

EBc^dc^aP at birth and as adults there is reciprocal reversal of order of effect, the order of intensity being adult sepia (highest), newborn brown, newborn sepia, and adult brown.   The case of *rrSt* and *Rstst*, both with strong forehead rosettes, *RSt* with mere irregularity on the forehead, and *rrstst*, smooth, (*mm* present in all) represents an example in which reciprocal inhibition by two dominants causes reversal of direction of effect in two pairs of loci.

SELECTIVE VALUE

Interaction effects of this fourth type are undoubtedly less common among simple characters than are those of the second or third type.   There is, however, a complex character in which such interaction effect must be almost the rule.   This is selective value, which, in relation to the total environment of a given population, is the character that is all important in the evolution of the latter, with the qualification that an array of genotypes may be even more important than any one genotype.

The reasons for the prevalence of this sort of interaction are partly indicated in the following quotation from an earlier paper[1422] which is cited because a recent account of the history of population genetics[860] attributes to it the assumption of absolute selective values for each gene and thus of no interaction.   The actual assumptions in this respect (and in most others) were the opposite of those stated.

"Selection, whether in mortality, mating or fecundity, applies to the organism as a whole and thus to the effects of the entire genic system rather than to single genes. A gene which is more favorable than its allelomorph in one combination may be less favorable in another.   Even in the case of cumulative effects, there is generally an optimum grade of development of the character and a given gene will be favorably selected in combinations below the optimum but selected against in combinations above the optimum."

The sort of case referred to last is illustrated in the upper part of figure 29 in which it is assumed that a character varying quantitatively is determined by four pairs of alleles with equivalent effects and no dominance.   It is supposed that selective value falls off from the optimum at the midgrade in proportion to the square of the deviation. There are six optimal genotypes in this case, each distinguishable from each of the others by at least two homozygous replacements.   With more loci and multiple alleles at each locus, the number of optimal genotypes in this sort of case may be enormous.[1423]

It is assumed, for simplicity, that the effects of the genes are additive on the underlying physiologic character.   It is probable that, in any real case, the situation would be complicated by specific interactions as in the case of coat and hair direction.   The case of coat color may, indeed, be cited as such an example.   The cream agouti color of the wild ancestor of the guinea pig (*Cavia cutleri*) is presumably the optimum in nature. It is intermediate in intensity and can be simulated more or less in at least four independent ways, none of which are similar to it genetically.[1427]

The quotation above does not refer to a point that was taken up later in the 1931 paper, the probable universality of pleiotropy and its consequences. Pleiotropy makes it unlikely that any allele is the most favorable in all respects even in a given combination with other genes. High selective value in a genotype requires a great deal of compensatory interaction among the genes. The consequence is a tendency toward overdominance and interaction among loci of the fourth type.

Fig. 29. SELECTIVE VALUES OF ALLELES.

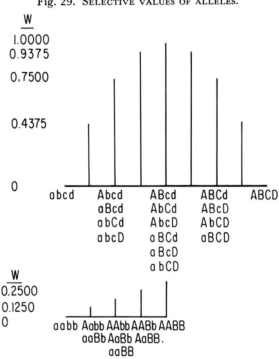

ABOVE: Selective values assigned homozygous combinations of four pairs of alleles, assumed to have equivalent effects on a quantitative character, but intermediate optimum.

BELOW: Selective values assigned to additive pleiotropic effects of two of the pairs of alleles.

In the case of an intermediate optimum illustrated in the upper part of figure 29, the six optimal genotypes all have the same value of W so that it would be a matter of indifference which one is established in evolution. This ceases to be the case if there are small pleiotropic effects. It is assumed in the lower part of figure 29 that genes A and B have additive pleiotropic effects while C and D do not. Figure 30 shows the selective values (W) of all homozygous genotypes except for those at the extremes of the quantitative character (*abcd* with $W = 0$ and *ABCD* with $W = 0.25$). The six peak genotypes are now at three levels: one low, four intermediate, and one high.

In the case of a character that is dependent on cumulative action of multiple factors above a threshold or below a ceiling, even selection directed toward one of these extremes is likely to lead to multiple peak genotypes in the neighborhood of the threshold or ceiling on the underlying scale, on the hypothesis that the pleiotropic effects of the favored genes are largely deleterious.[1427]

Fig. 30. MEAN SELECTIVE VALUES OF 14 HOMALLELIC POPULATIONS.

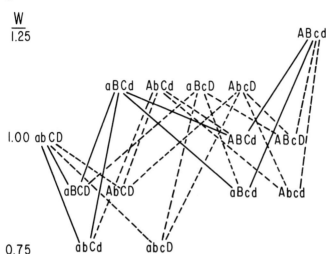

Based on the selective values of figure 29.   Population homallelic in *ABCD* ($W = 0.25$) and *abcd* ($W = 0$) are omitted.

The effects on the underlying scale need not be uniform.   Figure 31 illustrates a case in which a mutation (M') has an effect that offers the promise of a major step in advance but has other effects that make it highly deleterious.   By introducing modifiers A', B', C', D', E', F' that neutralize these other effects of M' and have only very slightly deleterious effects otherwise, the direction of effect of M' may be reversed in a combination that is far superior to the initial one.

THE IMPORTANCE OF INTERACTION IN EVOLUTION

There may be some genes that have such unequivocally injurious effects that these cannot be reversed by any combination with other genes.   In general, however, it seems safe to conclude of any pair of alleles that one is more favorable in some combinations, the other in others.   The primary condition for an effective evolutionary process would seem to be that selection operate to increase the frequencies of favorable multi-factorial genotypes (or even systems of genotypes) rather than of single genes because of their favorable net effect in the population in question.

The least effective evolutionary process is the one that is still, perhaps, most widely accepted: evolution as a succession of rare favorable mutations, each gradually displacing its type allele in what is at each stage essentially a single type genotype.

Morgan[896] stated this viewpoint as follows: "If we had the complete ancestry of any one animal or plant living today, we should expect to find a series of forms differing at each step by a single mutant change in one or another of the genes, and each a better-adapted or a differently-adapted form from the preceding." Under this theory

Fig. 31. MEAN SELECTIVE VALUES OF POPULATIONS HOMALLELIC WITH RESPECT TO A MAJOR PAIR OF ALLELES (*M, M′*) AND SIX MODIFIERS.

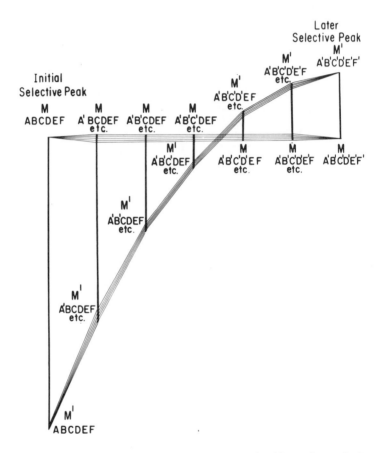

there is no recombination, and interaction effects are significant in evolution only in a restrictive sense.

The other two theories to be considered here are based on the concept that the population is something more than a lot of individuals of the same genotype. Under them a population is characterized genetically by an array of genic frequencies. The elementary evolutionary process becomes a change of genic frequency.[373, 516, 1422] The observations which led to this concept were indeed largely on such obviously variable populations as those of man, cultivated plants, domestic animals, and laboratory rodents rather than on wild species, although some support could be drawn from such studies as those of Sumner[1302, 1303] on the genetics of variability within and between subspecies of the deermouse *Peromyscus maniculatus*. More recently this general viewpoint has received increasingly massive support in species of *Drosophila* under the leadership of Dobzhansky,[283] and we have an enormously richer concept of what the genetics of such wild species is actually like.

We consider first the case of a species (or subspecies) that is essentially homogeneous throughout its range. Because of the conditions of balance that maintain the array of alleles at each locus, this would consist largely of isoalleles. The virtually infinite array of possible recombinations would insure the genetic uniqueness of every individual even though there would be little conspicuous variability. Since, however, the recombinant genotypes are broken up in each generation by the reduction division (with qualifications that are unimportant unless linkage is nearly complete and selection very strong), selection is based only on the net effects of the genes. If the absolute selective values of genotypes are independent of the genic frequencies, Fisher's[373] fundamental theorem of natural selection holds: "The rate of increase of fitness of any organism at any time is equal to its genetic variance in fitness at that time."

Genetic variance was here defined as merely the additive component. Thus favorable interaction effects of genes with unfavorable net effects cannot be utilized any more than under the first theory. There is no way by which a species can work its way from a lower to a higher selective peak with respect to mean selective value. Once the population has arrived at a selective peak, further evolution can occur only by a change of conditions or a wholly favorable mutation, both of which change the whole system of selective values. If, however, such a change occurs, the heterallelic character of the population permits an extensive readjustment on the basis of interaction effects until a new peak is arrived at, a process that cannot occur under the first theory.

In the third theory, it is assumed that there is sufficient isolation of small local populations (demes) in at least some part of the range of the species to permit significant genetic differentiation, but not so much that a successful deme cannot modify the genetic composition of less successful neighboring demes by emigration and crossbreeding.[1422] The introduction of immigration pressure into the conditions of balance at each locus makes for much more strongly heterallelic arrays, locally as well as in the species as a whole, although these conditions are still such that these arrays consist largely of isoalleles and minor modifiers. In such a species, population is continually welling up in some places, falling off in others, but the sites of the population sources and sinks may be changing from time to time. This process of interdemic selection supplements the continuous process of intrademic change of the sort considered in the second theory.

With a suitable balance between selection and immigration on the one hand, and between the resulting tendency toward and away from equilibrium from the cumulative effects of random processes on the other, the array of genic frequencies may occasionally pass from control by one selective peak to control by another. Figure 32 shows the paths which the array of genic frequencies tends to take under selection alone in two small parts of the four-dimensional field of mean selective values ($\bar{W}$) expected from the selective values in figure 30. In addition to selective peaks and pits in the corners, there are shallow selective cols between the lowest peak (*abCD*) and the intermediate one that is shown (*aBCd*), and between the latter and the highest peak (*ABcd*). Figure 33 shows the mean selective values along the most favorable path connecting the lowest peak with the intermediate and this with the highest (as well as a less favorable direct path from lowest to highest through a four-dimensional col not shown in figure 32). The mean selective values of the gene-frequency systems of figures 32 and 33 were calculated from the selective values in figures 29 and 30, using formulas 16 and 19 in a 1935 paper [1444] with respect to interaction effects.

Fig. 32. TRAJECTORIES OF GENE-FREQUENCY SYSTEMS.

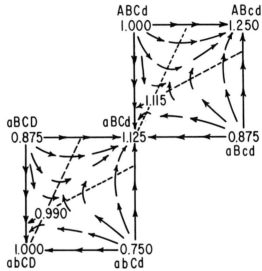

Trajectories of gene-frequency systems on surfaces of selective values on two faces of the four-dimensional field defined in figures 29 and 30.

Random processes need shift mean selective value only 8 per cent as much counter to selection as would be involved in fixation of gene *B*. This figure brings out the reason why I have held that fixation from unbalanced random drift (the Hageodorn effect) is of no importance in progressive evolution. As this is the point on which my position in the 1931 paper has been misinterpreted most frequently, I will give another quotation from it.[1422] The reference is to an extremely small, completely isolated population. "In too small a population, there is nearly complete fixation, little variation, little effect of selection and thus a static condition, modified occasionally by chance fixation of a new mutation, leading inevitably to degeneration and extinction."

It should be said that the conditions of balance are such that passage directly from one selective peak to another is not likely to occur except from arrays of genes with

Fig. 33. MEAN SELECTIVE VALUES OF GENE-FREQUENCY SYSTEMS.

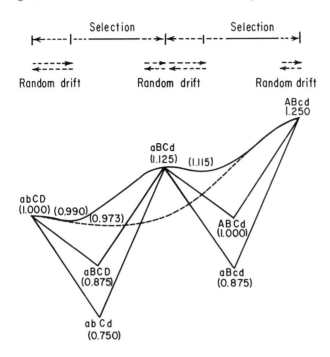

Mean selective values of gene-frequency arrays along the path of least depression from the lowest peak *abCD* through an intermediate peak *aBCd* to the highest peak *ABcd*, of figures 30 and 32, and along the direct path from the lowest to the highest peak through the four-dimensional col.

only slight differences in their net selective values.   But as can be seen from figure 31, the process can apply indirectly to a pair of alleles (M, M') with major differential effects.   Random drift may lead to sufficient frequencies of several of the very slightly deleterious modifiers A', B', C', D', E', and F' at some time in some deme to reverse the selection against M' which, with its associated modifiers, will then start to spread through the species.

Table 43 makes a comparison between the selective process in a homogeneous species[373] and the double process (intrademic and interdemic) in a finely subdivided species.[1422]

The conditions of balance among net selection coefficients, immigration, and random drift may not have held anywhere at any time in the ranges of some species. They almost certainly have held in others (for example in primitive man and many other mammals).   Where they have not held at all, the evolutionary process has been restricted to that of the second theory.   The first theory is merely the limiting form of the second under conditions in which all loci are only very weakly heterallelic. The three are really complementary.

Table 43

COMPARISON OF EVOLUTIONARY PROCESSES IN HOMOGENEOUS AND SUBDIVIDED POPULATIONS

	Homogenous population	Population subdivided into numerous partially isolated demes
Entity selected	Gene, differing from alleles in net selective value	Set of genic frequencies, characterized by harmonious genic effects, controlled by a selective peak
Source of variation	Gene mutation	Shift in controlling peak in a deme (by random drift and selection toward new peak)
Process	Selection among individuals	Selection among demes (by differential growth of population and migration)
(1) Conditions static	Progress restricted to extension of currently controlling peak	Continual shifts in prevailing selective peak
(2) Conditions change	Progress up most available peak in new surface of mean selective values	Interdemic selection relative to all available selective peaks in the new surface of mean selective values

A major evolutionary step under the third theory, as discussed above, simulates fixation of a favorable mutation in the first or second, but in this case is a byproduct of a continuous process of intrademic and interdemic selection that is occurring below a superficial appearance of near-uniformity in the species.   The theory helps account not only for occasional major steps in evolution less miraculously than in the other theories but also for the extraordinary perfection of fine details that we find in the products of evolution.   The essential difference is between a theory in which selection can operate in the virtually infinite field of interaction effects of recombinants instead of almost exclusively on the net effects of each separate gene.

DISCUSSION

DR. BURDETTE: Thank you, Dr. Wright.   The discussion of Dr. Wright's paper will be opened by Dr. Herman Chase.

DR. CHASE: This is an excellent and exhaustive paper, leaving nothing to be taken away and very little to add.   Compared with the guinea-pig work of Dr. Wright, there has been relatively little done with mice in the way of genic interaction.   Since this is a symposium supposedly dealing with methodology, I would like to point out what I consider to be the four main approaches or general methods in physiologic genetics. They are: first, genic substitution; second, genic interaction; third, comparison of genetic constitutions; and fourth, teratogens.   Simple genic substitution with resulting analysis of the varying phenotype is certainly used the most and is basic in work with

Neurospora, bacteria, maize, mice, and so forth. The comparison of genetic constitutions, as when nutritional requirements, sensitivities to drugs, and the like of inbred strains are compared, is another approach often used. The effect of teratogens, the production of phenocopies, is a method which has some merit when cautiously interpreted. Genic interactions are fundamental in the complex of interrelationships involved in the realm of physiologic genetics, the genetics of expression; but genic interaction (the second approach) as a deliberate method of investigation has been used all too little, except for Dr. Wright's long-term studies on guinea-pig coloration and hair direction.

Relating directly to the details of this paper, *pink-eyed dilute* (*p*) in combination with *Light* (*B^{lt}*) prevents basal dilution in the mouse, a situation somewhat analogous to one mentioned by Wright. Recently we have discovered a possible interrelationship between murine polydactyly and anophthalmia. In no case, however, have we made a systematic attempt to combine our mutants for study in the Wrightian manner. Murine genetics during the past 30 years has placed more than enough emphasis on the development and maintenance of inbred strains and on single-gene differences; perhaps now more should be done to explore genic interactions. After all, as Dr. Wright said at the beginning, "genic interaction . . . is obviously fundamental in physiologic genetics and is almost as fundamental in population genetics, including the genetic aspects of the theory of evolution." Whereas he said in his introduction that he was riding the two wild horses of physiologic genetics and population genetics, I would like to make the observation that the one horse of physiologic genetics, tamed by Dr. Wright, has been enough for me.

DR. POPP: Frank Moyer at Johns Hopkins University has been interested in differences among pigmented granules of murine melanoblasts as regards possible differences in tyrosinase in such cells;[914] information on structural differences among tyrosinases of the mouse may be forthcoming in a year or two.

DR. RUSSELL: Dr. Douglas Coleman at the Jackson Laboratory also has under way a splendid program analyzing genetically determined differences in tyrosinases in mouse skin.[213]

DR. YERGANIAN: With respect to outlining the genes involved in diabetes mellitus of the Chinese hamster, *Cricetulus griseus*, we have conducted appropriate hybridizations involving symptomatic animals representing four different strains. To our surprise, instead of the anticipated retention of the high incidence of diabetic animals among the F_1 of single- and double-hybrid crosses, the frequency of symptomatic hamsters was virtually nil. The syndrome[865, 866] reappeared after three and four generations of brother-sister matings. We have estimated the number of genes to be two, in addition to an unknown number of modifiers that control the degree of penetrance and age of onset. These observations have been repeated on numerous occasions, but the rôle of modifiers remains to be fully disclosed.

Since diabetes failed to occur following the hybridization of symptomatic animals, the rôle of modifiers is most perplexing. An interaction among the genes as well as the environment is strongly implicated.   An example of the former or genic rôle is the fact that diabetics have a sixfold increase in alpha-2-serum proteins whereas hybrids, stemming from two or four diabetic parents from different inbred families, exhibit low or normal levels of this diagnostic protein.[494]

DR. DEGENHARDT: Is there any information about a preferred time for genic interaction in early embryonic development?   I am thinking only of polydactyly.

DR. WRIGHT: Dr. Scott worked out the embryology of the polydactylous monster. About the seventeenth or eighteenth day he found broad, paddle-shaped limb buds and that the hindbrain was growing too much for the rest of the body, causing it to bulge out through the top of the head.   Probably action of the gene occurred much earlier than that.

DR. DEGENHARDT: We found in our investigations on the effect of irradiation early in embryonic development that there is a critical stage for inducing polydactyly in the inbred strain C57BL/KsJ, in which polydactyly normally appeared at a 2.2 per cent rate, between day seven to day ten of gestation (table 44).   The table illustrates

Table 44

INCIDENCE OF POLYDACTYLISM IN CONTROLS AND AFTER X IRRADIATION (130 *r*) AT
SUCCESSIVE STAGES OF EARLY GESTATION IN MICE OF THE INBRED STRAIN C57BL/KsJ

CONTROLS Time of X-irrad. in pregnancy		Birth data			Polydactylism				Summary	
Day	hour	Li 39	NB 269	Ø 6.9	right 3	left 3	both sides —	n 6	Per cent of ENB 2.2	269
VI	8 PM	17	108	6.4	4	—	—	4	3.7	107
VII	8 AM	25	106	4.2	2	—	—	2	1.9	103
VII	8 PM	21	112	5.3	17(3)†	5(1)†	—	22	19.6	112
VIII	8 AM	22	112	5.1	13(2)†	2	1	16	14.3	112
VIII	8 PM	17	115	6.8	12	4	1	17	14.9	114
IX	8 AM	20	114	5.7	22(2)†	7(1)†	2	31	27.2	114
IX	8 PM	17	112	6.6	11	4	—	15	13.5	111
X	8 AM	16	105	6.6	13	3	—	16	15.4	104
X	8 PM	18	126	7.0	20	4‡(1)†	1	25	20.0	125
XI	8 AM	15	95	6.3	1	1	—	2	2.1	94
					115	30	5	150	18.2	1096

Li = Litter
NB = Newborn
Ø = Average litter size
ENB = Examined newborn
† = Incidence of 7 toes
‡ = One case of postaxial polydactylism

the incidence rate up to 27.2 per cent, and it may be possible that this is the threshold of genic interaction.   But there seems to be another critical phase.   This work has been done by Dr. Charles P. Dagg.[236]   He gave 5-fluorouracil at a certain stage of embryonic development in the strain 129/J, and he found a second critical phase during the eleventh or twelfth day of inducing polydactyly. Is that the second critical stage for genic interaction?

DR. WRIGHT: These critical stages probably have very little to do with the gene under study. There are certain weaknesses in the developmental process leading to the development of polydactyly, microphthalmia, or anencephaly according to the time of development. Anything environmental or genetic which produces a sufficient disturbance at that critical time would bring out that particular type of defect. The specificity is not so much in the gene, although the timing of genic action may be regarded as specific; the character of the effect depends on the developmental background. Also the same gene may bring out other different developmental effects if it comes into play at different times.

DR. STEINBERG: I hope that in working with mice, investigators will not go through the same series of mistakes that happened when flies were used to detect the time of genic action. When it was found that there is a specific period during development when temperature would affect the size of the eye, it was labelled the temperature-effective period, and it was concluded that this was the time when the gene was acting. However, when the embryology was pursued, it was found that a detectable effect of the gene on the development of the eye occurred some 24 to 36 hours before temperature could affect it.[1273] Furthermore, it was found that oxygen could affect the development of the eye two or three days after temperature could affect the development of the eye.[849] It became, as more and more agents were used, a little more ludicrous to say that the gene was acting at this time, that time, or the other time. The effect of the environment on the development of the organism may be related to the gene, but usually it is not and certainly it need not be. Cause and effect are not so easily related.

DR. BURDETTE: I regret that we do not have more time to devote to the dissection of these complexities. Dr. Wright, perhaps you wish to make a few concluding remarks at this time, particularly with relation to evolution.

DR. WRIGHT: I will not attempt to summarize in any detail what I said in my paper. I wish merely to emphasize again that because of the prevalence of intermediate optima in quantitative variability, and also of pleiotropy, multiple selective peaks are to be expected in any species in the virtually infinite field of possible sets of genic frequencies. Each of these peaks tends to be closely related to an equilibrium point determined in part by the pressures of recurrent mutation and immigration but primarily by mean selective value. These equilibrium points have the property to which Lerner has applied the term genetic homeostasis. After any slight shift, the system tends to return to the same equilibrium.

The selective peaks may be expected to differ greatly in adaptive value. The evolutionary problem is the mechanism by which the species may work its way from lower to higher peaks, necessarily somewhat against the pressure of selection in the first stage. In a sexually reproducing species under given environmental conditions, the only known mechanism is that in which some sort of random drift brings about differences among local populations which can provide material for interdemic selection (by differential population growth and dispersion).

The relation of random drift to selection among demes is precisely analogous to that of random mutation to mass selection within demes, but whereas the latter deals only with the momentary net effects of single gene differences, the former deals with genetic systems as entities. As I have always emphasized, neither random drift nor random mutation is of appreciable evolutionary significance except in conjunction with its appropriate kind of selection.

It is because of the necessary prevalence of types of genic interaction that lead to multiple selective peaks in the genetic system that I have always considered interaction to be as important in evolutionary theory as in physiologic genetics.

D. S. Falconer, Ph.D.

QUANTITATIVE INHERITANCE

A quantitative character is any attribute for which individual differences do not divide the individuals into qualitatively distinct classes. Antigenic differences, for example, separate individuals into qualitatively distinct classes and are therefore not quantitative characters. But an antibody titer, which varies continuously from one individual to another, is a quantitative character. The inheritance of quantitative characters is generally under the control of many genes (that is, it is polygenic), but this is not necessarily so. The essential feature is that the segregation of the genes, whether few or many, is not manifest in phenotypic discontinuity. This feature of quantitative characters precludes the application of the ordinary methods of genetic analysis used for the study of single-gene effects. Special methods are therefore required, and at the same time the questions that can be answered are different in nature. In general, the genetic properties of a quantitative character that can be investigated are those arising from the simultaneous action of many genes. It will not be possible to explain fully the theoretic nature of these properties here; nor will it be possible to describe all the methods available for their study. So this discussion will be confined to a description of the simpler methods which may be of use to those for whom genetics is a secondary rather than a primary interest. Explanations of many of the points which cannot be fully explained here will be found in the treatise by Falconer.[335]

The first and simplest question to be asked about any quantitative character is: are there any genetic differences? To answer this question it is only necessary to compare different strains maintained in the same place and under the same conditions. If strains differ in their mean values of the character this proves the existence of genetic differences. So many characters are known to exhibit strain differences that the existence of genetic differences in any character can almost be taken for granted. The demonstration of strain differences is therefore no more than the starting point of the investigation.

There are, broadly speaking, four sorts of genetic investigations of a quantitative character that may be made. The first two—estimation of the degree of genetic determination and of the heritability—are concerned with the relative importance of genetic and environmental factors as determinants of an individual's phenotypic value of the character. The third is a description of the effect of selection applied to the character. This may be of practical interest, but the additional genetic conclusions that can be drawn are rather limited. The fourth is a description of the effect of inbreeding or of crossing inbred lines, and again the conclusions that can be drawn are rather limited.

None of these investigations can be made without a noninbred, or genetically heterogeneous, strain. The most satisfactory strain for this purpose is one that has been maintained by random mating among a large number of parents over many generations, so that its genetic properties have had time to become stabilized. But to advocate the use of such a strain is obviously a counsel of perfection. In the absence of a random-bred strain the most convenient form of genetically heterogeneous strain is an F_2 of a cross between two inbred lines, or the third generation of a 4-way cross of four inbred lines. These synthetic strains have the advantage over a random-bred strain that they can be reconstituted at will from the original inbreds. Their disadvantages are that the conclusions about their genetic properties have less generality and that their genetic properties are in some ways necessarily unnatural. The lack of generality can be expressed through the inbreeding coefficient. Nobody would claim generality for any biologic property discovered from the study of a single highly inbred line. The F_2 of a 2-way cross is 50 per cent inbred, and the third generation of a 4-way cross is 25 per cent inbred; in other words, a strain derived from a 2-way cross is equivalent to the progeny of a single self-fertilized individual, and a strain derived from a 4-way cross is equivalent to the progeny of a single pair of full sibs. To this extent, therefore, synthetic heterogeneous strains resemble inbred lines in lack of generality. The unnatural features of the genetic properties of synthetic strains are the restricted range of genic frequencies, nonrandom linkage associations, and the absence of lethal and severely deleterious genes. The disadvantages of synthetic strains, however, are only relative. Complete generality is an unattainable ideal; no random-bred strain of mice, however carefully constructed and maintained, is truly representative of all mice, just as no laboratory mammal is truly representative of all mammals.

DEGREE OF GENETIC DETERMINATION

If there are genetic differences, then how important are they? This is the nature-nurture question: what is the relative importance of heredity and environment in determining the value of the character in an individual? (For the sake of convenience all nongenetic differences are referred to as environmental, so that the genotype and the environment of an individual are the only determinants of its phenotypic value.) The question has precise meaning only when framed in terms of the variation between inidividuals: how much of the variation is caused by genetic differences between individuals and how much by nongenetic differences? To answer the question it is necessary to eliminate one cause of variation and to measure the variation remaining, which will be due to the other cause only. If the amount of variation is measured as the variance, then the total variance when both causes are operating is the sum of the variances when each cause is operating separately. It is not possible in practice to eliminate nongenetic causes of variation, because, however careful the experimenter may be in insuring uniformity of external conditions, there always remains a substantial amount of nongenetic variation from intangible causes. The genetic variation, however, can be eliminated by inbreeding, although this is practicable only with fast-breeding laboratory animals. A strain that has been inbred by brother-sister mating for upwards of 20 generations is, for practical purposes, free of genetic variability. The variation within the strain, therefore, is purely environmental in origin. There is, however, a difficulty here, because inbred animals have often been found to be more susceptible than noninbred animals to the normal range of environmental differences. An inbred line therefore does not always provide a reliable measure of the environmental variation for comparison with a noninbred strain. The difficulty can be overcome by taking a cross between two inbred lines. If both lines are highly inbred, then the first generation of the cross (that is, the F_1) is equally free of genetic variation and provides a more reliable estimate of the environmental variation. Better still, however, is to make several different crosses between highly inbred lines and base the estimate of environmental variance on the mean variance within the crosses. This overcomes the possible objection that one cross provides only one particular genotype whose sensitivity to environmental differences may not be typical. The observed variance of the individuals of a genetically variable strain is the sum of the genetic variance in that strain and the environmental variance. Subtraction of the environmental variance, estimated from the cross-bred individuals, therefore gives an estimate of the genetic variance. In this way the total, or phenotypic, variance in the variable strain may be partitioned into two components, genetic and environmental. This partitioning expresses the degree of genetic determination, as the percentage of the total variation that is attributable to genetic differences among individuals of the strain.

An example of the estimation of the degree of genetic determination is provided by the following data on the age of vaginal opening in female mice, from Yoon.[1465] Two inbred strains, BALB/c and C57BL/10, were crossed (Yoon's cross 2), and the age of vaginal opening was recorded from 192 F_1 animals and 163 F_2 animals. The variance among the F_1 animals was 19.1 and the variance among the F_2 animals was 34.9. The difference, which estimates the genetic variance among the F_2 individuals, was 15.8. The degree of genetic determination in the F_2 of this particular cross was therefore $15.8/34.9 = 45 \pm 8$ per cent. (The standard error is calculated from equation 4, as explained below.)

Experimental design.—The only technical problem that needs consideration in connection with the degree of genetic determination is the scale of experiment required to estimate it with a given degree of precision. The standard error to be expected when a given number of animals has been measured can be deduced in the following way.

Let V be the estimate of the variance in the genetically variable group, based on N_v degrees of freedom; and let U be the estimate of the variance in the genetically uniform group, based on N_u degrees of freedom. For the purposes of planning the degrees of freedom may be taken to be the number of animals measured. The sampling variances of these two estimates of variance are given by

$$\sigma^2_v = 2V^2/N_v \quad \text{and} \quad \sigma^2_U = 2U^2/N_u, \tag{1}$$

where σ^2 is the sampling variance. The standard error of each estimate is the square root of its sampling variance. The ratio of the standard error to the variance itself is

$$\sigma_V/V = \sqrt{2/N_v} \quad \text{and} \quad \sigma_U/U = \sqrt{2/N_u}. \tag{2}$$

This ratio is plotted in curve (a) of figure 34.

The degree of genetic determination, g, is estimated as

$$g = (V - U)/V = 1 - U/V \tag{3}$$

The sampling variance of the degree of genetic determination (σ^2_g) can be deduced from the general properties of variances as follows:

$$\sigma_g{}^2 = \sigma^2{}_{(1-U/V)}$$

$$= \sigma^2{}_{(U/V)}$$

$$= \frac{1}{V^4}(V^2\sigma^2_U + U^2\sigma^2_V)$$

$$= \frac{U^2}{V^2}\left(\frac{2}{N_u} + \frac{2}{N_v}\right).$$

Since $U/V = 1 - g$, from (3), the standard error of the estimate reduces to

$$\sigma_g = (1 - g)\sqrt{2\left(\frac{1}{N_u} + \frac{1}{N_v}\right)}. \tag{4}$$

Now, the total number of animals that can be measured is fixed by the amount of space available, or of effort that can be expended. The formula deduced above for the standard error (4) shows that with a fixed total (that is, $N_u + N_v$), the standard error will be minimal if $N_u = N_v$. Therefore the best design for the experiment is to have equal numbers of animals measured from the genetically variable and from the genetically uniform groups. If this number is N (so that a total of $2N$ animals are measured), then we obtain the relationship

$$\frac{\sigma_g}{1 - g} = \frac{2}{\sqrt{N}}. \tag{5}$$

This ratio is plotted in curve (b) of figure 34.

The relationship in equation (4) shows that the precision of an estimate of the degree of genetic determination depends on the value of *g* itself.   The higher the value, the greater the precision with a given number of animals measured.   To plan an experiment it is therefore necessary to guess what the degree of genetic determination may be.   This is not very satisfactory, but it is better than working completely in the

Fig. 34. GRAPHS OF $y = \sqrt{2/x}$ AND $y = 2/\sqrt{x}$.

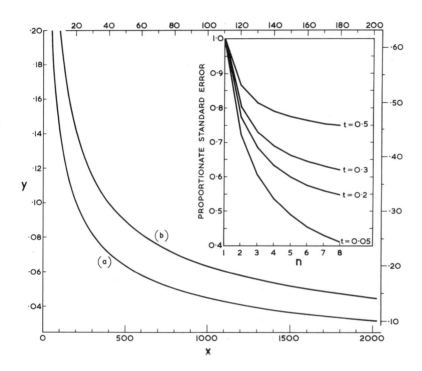

(a) Graph of $y = \sqrt{2/x}$; (b) $y = 2/\sqrt{x}$.   Two scales are given: the left hand scale of *y* is to be read against the bottom scale of *x*, and the right hand against the top.

USES

1. Standard error of estimate of a variance, *V*, (graph a): $\sigma_V = yV$, when *x* is the number of degrees of freedom.

2. Standard error of estimate of the degree of genetic determination, *g*, by comparison of variances of genetically uniform and genetically heterogeneous groups with equal numbers in each group (graph b): $\sigma_g = y(1 - g)$, when *x* is the number of degrees of freedom in each group.

3. Standard error of estimate of heritability by regression of offspring on parents, with one offspring from each parent or pair of parents.   *y* is the standard error when *x* is the number of parents or pairs of parents.   Graph a refers to regression on the mean of both parents, graph b to the regression on one parent.   The inset graphs show the effect of increasing the number of offspring when the number of parents remains the same.   The vertical scale gives the factor by which *y* (from graphs a and b) is to be multiplied to give the standard error.   The four inset graphs refer to different characters with phenotypic correlations (*t*) between offspring of the same parents as shown.

dark. Suppose, for example, that the degree of genetic determination is thought to be 40 per cent, which is a reasonable figure to guess for many quantitative characters; how many animals must be measured to obtain an estimate with a standard error of, say, 10 per cent? Here $\sigma_g = 0.1$, and $1 - g = 0.6$. Entering these values in equation (5) gives $N = 144$. Therefore about 144 animals must be measured in each group, or 288 altogether, to obtain a result that would read $g = 40 \pm 10$ per cent. To take another example showing how the relationship might be used the other way around, suppose a character is difficult to measure and it is decided that the greatest number of animals that could be measured would be 40, twenty of each group. Would it be worth while to try to estimate the degree of genetic determination? Reading graph (b) in figure 34 from the upper and right-hand margins, $\sigma_g/(1 - g)$ is approximately 0.44. So, if g were 40 per cent, its standard error would be $0.44 \times 0.6 = 0.26$. Thus the expected result of the experiment would be $g = 40 \pm 26$ per cent. The estimate would not be significantly different from zero, and therefore even demonstrating the existence of any genetic variation at all could not be expected.

From all this it will be evident that the degree of genetic determination cannot be precisely estimated without a very considerable expenditure of effort, especially if the character is only weakly heritable. But an estimate with a standard error even as high as 20 per cent would not be entirely without interest, and, if suitable data could be accumulated from routine measurements made for the other purposes, information of great interest could be obtained.

HERITABILITY

The degree of genetic determination, although of great intrinsic interest, is of little practical use. It cannot be used to predict the speed of progress to be expected if selection were applied to the character. For this prediction it is necessary to determine the heritability of the character. The determination of the heritability involves the further partitioning of the genetic component of variation into two parts, although the method of partitioning is now entirely different. The need for the additional partitioning arises from the mode of hereditary transmission—from the fact that gametes are haploid and that genes, and not genotypes, are transmitted from parents to offspring. The idea of heritability is therefore of fundamental genetic importance. The genetic variation is of two sorts. Part of it may be thought of as arising from the genes considered singly, instead of paired in the diploid genotype. This component of the genetic variation is called the additive variance. The heritability is the ratio of the additive variance to the total phenotypic variance; or, in other words, the fraction of the total variance made up by additive variance. The other part is the additional variation that arises from the genes coming together in pairs to form genotypes. This component is called the nonadditive variance. The cause of nonadditive variance, in terms of the properties of individual genes, is dominance and interaction (epistasis) between different loci. If there is no dominance and no epistasis there can be no nonaddititive variance;

and, conversely, if no nonadditive variance is found, it can be concluded that there is no dominance or epistasis: the genes are then said to act additively.   Methods are available for separating the different sorts of nonadditive variance.[538]

Because the amount of additive variance reflects the variation that is transmitted from parent to offspring it is responsible for the resemblance between relatives, and its estimation depends on the measurement of the degree of resemblance between relatives.   The most commonly encountered sorts of relationship are between offspring and their parents, and between full or half sibs.   It can be shown from theoretical considerations that the degree of resemblance, measured as a regression or correlation coefficient, in these relationships is related to the heritability as follows (the symbol h^2 stands for the heritability):

Regression of:

$$\text{offspring on one parent} = \tfrac{1}{2}h^2$$
$$\text{offspring on mean of two parents} = h^2$$

Correlation (intraclass) of:

$$\text{half sibs} = \tfrac{1}{4}h^2$$
$$\text{full sibs} > \tfrac{1}{2}h^2.$$

Thus, for example, to estimate the heritability from the resemblance between half sibs it is only necessary to compute the intraclass correlation and multiply this by four; or, from offspring and parents, to compute the regression of the offspring on the mean of their two parents which itself estimates the heritability.   A graphic illustration of the resemblance between offspring and parents is shown in figure 35, with the regression which estimates the heritability.

There are, however, two complications which have an important bearing on the choice of the relationship from which to estimate the heritability.   The first concerns only full sibs.   The genetic cause of resemblance between full sibs is not confined to the additive variance, and the correlation is augmented by part of the non-additive variance if any is present.   Therefore an estimate of the heritability from full sibs is, strictly speaking, valid only in the absence of nonadditive variance.   The error is, however, seldom likely to be serious, and an estimate from full sibs is by no means to be rejected as valueless on these grounds alone.   The second complication is much more important.   It arises from the fact that there are often nongenetic causes for the resemblance between certain sorts of relative, which may increase the regression or correlation and lead to a serious overestimation of the heritability.   These non-genetic causes of resemblance are particularly prevalent in mammals.   They arise from maternal effects and circumstances generally referred to as common environment. For example, large mothers tend to give more milk to their young than small mothers. Therefore the offspring of large mothers tend to be larger than the offspring of small mothers, and so the offspring tend to resemble their mothers in body size for purely nongenetic reasons.   Also, for the same nongenetic reason, offspring of the same mother tend to resemble each other in body size.   It is in the full-sib relationship that resemblance tends most often to be augmented by nongenetic causes. Full sibs always have the same mother and they tend, especially if they are also littermates, to be subjected to similar environmental circumstances. For these reasons the correlation between full sibs is to be avoided as a means of estimating the heritability.

Fig. 35. HERITABILITY OF THE NUMBER OF URETHAN-INDUCED PULMONARY TUMORS IN MICE OF A RANDOM-BRED STRAIN (JC), BY REGRESSION OF OFFSPRING ON THE MEAN OF BOTH PARENTS.

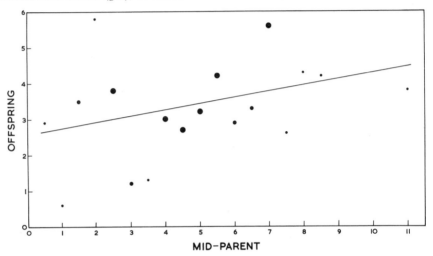

Based on preliminary data from a current experiment by Miss J. L. Bloom, a total of 299 offspring from 46 pairs of parents are represented. Each point shows the mean of all offspring from parents with the given mean tumor-number, the approximate number of off-spring in each point being indicated by the size of the dots: small = fewer than 10, medium = 10–20, large = more than 20. The straight line is the computed regression line, each point being weighted by the number of offspring contributing to it. The regression coefficient, which estimates the heritability, is 0.173 ± 0.043.

The absence of nongenetic causes of resemblance in a relationship is the most important criterion in choosing a method of determining the heritability. After that, the choice of method is a matter of deciding which gives the most precise estimate with a given expenditure of effort. In general, low heritabilities are more efficiently esti-mated from the half-sib correlation, and heritabilities higher than about 20 per cent are more efficiently estimated by offspring-parent regressions. The offspring-parent regression is usually the more convenient method for work with laboratory mammals. It is usually also the more efficient because the total number of measure-ments made is not usually the limiting factor, but rather the number that are made at one time. By the measurement of parents and offspring, the work is spread over two generations. The rest of this section is concerned with the technical problems that arise in the estimation of heritability, first from the offspring-parent relationship and then from sibs.

Offspring-parent regression.—The chief problem of experimental design concerns the number of offspring from each parent or pair of parents that should be measured.   When the most efficient design has been decided, the number of animals that must be measured to attain a given degree of precision can be determined.   The size of an experiment is limited either by the amount of breeding or rearing space available or by the number of individuals that can be measured.   The problem is to decide how to divide the space or effort among the offspring of different parents.   Either few offspring from each of many parents or many offspring from each of few parents can be reared and measured. The solution of this problem comes from a consideration of the expected standard error of the estimate of heritability that will be obtained.   Approximate formulae for the variance of the estimate are as follows:[335]

By regression:

on one parent, $$\sigma_{h^2}^2 = 4\,\frac{1 + (n - 1)t}{nN} \tag{6}$$

on mean of both parents $$\sigma_{h^2}^2 = 2\,\frac{1 + (n - 1)t}{nN}. \tag{7}$$

In these formulae n is the number of offspring measured per parent (equation 6) or pair of parents (equation 7), N is the number of parents or pairs of parents, and t is the phenotypic correlation between the offspring of the same parent or pair of parents.   In work with laboratory animals the offspring are likely to be full sibs.   In this case the phenotypic correlation, t, will be approximately equal to half the heritability, or greater than this if there are also nongenetic causes of resemblance between the offspring of the same parents.   The solution of the problem of design depends on the circumstances that limit the size of the experiment.   The simplest solution is when the limiting factor is the total number of offspring that can be reared or measured.   The total number of offspring is nN, and, if this is fixed, then the denominators of both the above formulae are fixed.   The variance of the estimate is then, in both cases, minimal when $n = 1$.   This means that the most efficient design has only one offspring measured from each parent or pair of parents.   The standard error of the estimate of the heritability then becomes:

By regression:

on one parent $$\sigma_{h^2} = \sqrt{2/N} \tag{8}$$

on mean of both parents $$\sigma_{h^2} = 2/\sqrt{N}. \tag{9}$$

These relationships are also shown by the graphs in figure 34.   Graph (b) refers to the regression on one parent and graph (a) to the regression on the mean of both parents. The horizontal axis shows the total number of offspring measured (that is, N) and the vertical axis the corresponding standard error expected.   It is again evident that precise estimates of the heritability cannot be obtained without the measurement of large numbers of individuals.   For example, to attain a standard error of 0.10 it is necessary to measure 200 offspring if both parents are measured and 400 offspring if only one parent is measured.

The situation discussed above is likely to apply when the labor of measurement is the limiting factor, rather than the amount of breeding or rearing space available. When labor of measurement is not the limiting factor, then it is often possible to rear and measure more than one offspring from each parent or pair of parents without any reduction of the number of parents that can be used. The limiting factor is then the number of parents. If this is fixed, it follows that any increase in the number of offspring measured will improve the precision of the estimate. The question then is how many offspring are worth measuring. The reduction in the expected standard error effected by increasing the number of offspring from 1 to 8 is shown by the inset graphs in figure 34. The figures on the vertical scale give the factors by which the standard error obtained from equations (8) and (9), or from the main graphs in figure 34, are to be multiplied. The improvement in precision depends on the phenotypic correlation between offspring of the same parents, and graphs for four correlations, ranging from 0.05 to 0.5, are given. The cause of the correlation is immaterial in this connection; it may be genetic or nongenetic, or both. When the correlation is high there is little to be gained by increasing the number of offspring measured, but when the correlation is low a substantial improvement of precision results from even a small increase in the number of offspring. If the value of the correlation is not known beforehand, it would seem worth while, as a general rule, to aim at measuring about four offspring from each parent or pair of parents.

Planning on paper is easy, but executing the plan by obtaining and measuring the animals exactly as required is often impossible. Even if it is planned to measure only a few offspring from each parent, some parents will inevitably fail to produce the required number of offspring, and there may be other losses later, so that the plan cannot be strictly followed. This raises two problems. The first is at the stage of the collection of data: should the experimenter discard all families that have failed to provide the right number; and if he does not discard them, should he supplement the total number of offspring by measuring more than was planned from some other families? My answer is that he should include every animal measured and measure any additional animals that he can, because every additional offspring measured adds something to the precision of the estimate. The second problem then arises with the computation of the offspring-parent regression from the data, because the number of offspring will not be the same for all parents. How is the computation to be made? (In what follows I shall use the word family to mean the offspring of one parent or pair of parents, and the word parent to mean equally the single, measured parent or the mean of both parents, according to which regression is to be computed.)

There are two simple courses of action, neither of which makes the best use of the data. One is to take each offspring separately and count each parent as many times as it has offspring. This gives too much weight to the larger families but is reasonably satisfactory when the heritability is low and there is very little resemblance between the offspring of the same parent. The second simple course of action is to take the mean of each family and to regress this on the parental value. This gives too little weight to the larger families, but is reasonably satisfactory when there is a strong resemblance between offspring of the same parents, either for genetic or for nongenetic reasons. A slightly less simple, but much more satisfactory, procedure is as follows. The families are divided into groups according to the number in the family, and the regression of family means on parental values is computed separately within each group. The variance of each estimate is computed in the usual way, the degrees of freedom being based on the number of families in the group. A weighted average of the regression coefficients is then taken, the weight being the reciprocal of the variance of each estimate. The variance of this average regression is the reciprocal of the sum of the weights. The chief disadvantage of this method is that one degree of freedom is lost for each subdivision of the data, a sacrifice to simplicity that may not willingly be made when much effort has gone into the collection of the data. It may therefore seem worthwhile to use a fully satisfactory, but more complicated, method of combining the data from families of different size.

The method that makes the best use of all the data depends on combining the sums of squares and products from families of different size according to a weighting factor appropriate to the family size. The derivation of the weighting factors is explained by Kempthorne and Tandon [702] and Reeve.[1046] The principle is that families of different size are weighted in proportion to the reciprocal of the variance of the estimate of regression that would be obtained from families all of that particular size. It has already been pointed out, in connection with planning, that the proportionate effect of the number in the families on the precision of the estimate is governed by the phenotypic correlation, t, between members of families. But it also depends, to a lesser extent, on the value of the regression being estimated, and, although this factor can be omitted in connection with planning, it must be brought in if the best use is to be made of the data. The first step in computing the weighting factors is therefore to determine the phenotypic correlation, t, between members of families, from an analysis of variance of the offspring, within and between families. Then one must make a rough estimate of the regression, b, which is to be estimated. This need not be very precise, and it can be obtained from a graphical representation of the data in a scatter diagram of the values of parents and the means of their offspring. Approximate values of the correlation, t, and the regression, b, having been obtained, the weighting factors can most easily be calculated in two steps. First compute the quantity T, as follows: when the regression is to be made on a single parent

$$T = (t - b^2)/(1 - t),$$

and when the regression is to be made on the mean of both parents

$$T = (t - \tfrac{1}{2}b^2)/(1 - t).$$

Then the weighting factor, w_n, appropriate to families of n offspring is

$$w_n = (n + nT)/(1 + nT).$$

Table 45

SOME WEIGHTING FACTORS TO BE APPLIED TO THE MEANS OF FULL-SIB FAMILIES OF DIFFERENT SIZES, IN REGRESSIONS OF OFFSPRING ON THE MEAN OF BOTH PARENTS

Listed according to the phenotypic correlation (t) between full sibs, and the heritability (h^2) of the character

$t = 0.05$	0.1		0.2		0.4		0.7	
$h^2 = 0.1$	0.1	0.2	0.2	0.4	0.2	0.6	0.2	0.6
$T = 0.047$	0.106	0.089	0.225	0.150	0.633	0.367	2.267	1.733
n								
1	1.00	1.00	1.00	1.00	1.00	1.00	1.00	1.00	1.00
2	1.91	1.83	1.85	1.69	1.77	1.44	1.58	1.18	1.22
3	2.75	2.52	2.58	2.19	2.38	1.69	1.95	1.26	1.32
4	3.52	3.11	3.21	2.58	2.88	1.85	2.22	1.30	1.38
5	4.23	3.62	3.77	2.88	3.29	1.96	2.41	1.32	1.41
6	4.89	4.06	4.26	3.13	3.63	2.04	2.56	1.34	1.44
7	5.51	4.45	4.70	3.33	3.93	2.10	2.68	1.36	1.46
8	6.08	4.80	5.09	3.50	4.18	2.15	2.78	1.37	1.47

Some examples of weighting factors are given in table 45, the value of T under which the weights are listed being appropriate to the heritabilities shown at the heads of the columns when the regression is to be made on the mean of both parents.   The meaning of the weighting factor, w_n, is that the mean of each family of size n, and the corresponding parental value, are to be counted w_n times in the computation of the sums of squares and products.   [It should be pointed out that the weighting factor given above differs slightly from that of Kempthorne and Tandon in that it gives unit weight to families of $n = 1$, which theirs does not do.   The weight given here is equal to theirs divided by $1/(1 + T)$, which is their weight for $n = 1$.]

The computation of the regression coefficient, b, may be summarized as follows. Let X be any parental value, Y the mean value of the corresponding family of offspring, and w the weighting factor appropriate to the number in that family.   Then the regression of offspring on parents is computed thus, Σ indicating that the quantities are to be summed over all families:

$$b = \frac{\Sigma\,(wxy)}{\Sigma\,(wx^2)}$$

where

$$\Sigma\,(wxy) = \Sigma\,(wXY) - \frac{\Sigma\cdot(wX)\,\Sigma\,(wY)}{\Sigma\,w}$$

and

$$\Sigma\,(wx^2) = \Sigma\,(wX^2) - \frac{[\Sigma\,(wX)]^2}{\Sigma\,w}.$$

Also needed will be

$$\Sigma\,(wy^2) = \Sigma\,(wY^2) - \frac{[\Sigma\,(wY)]^2}{\Sigma\,w}.$$

The variance of the estimate of the regression coefficient should, in my opinion, be computed in the usual way but from the weighted sum of squares. Thus

$$\sigma_b{}^2 = \frac{1}{\sum w - 2}\left[\frac{\sum (wy^2)}{\sum (wx^2)} - b^2\right].$$

The error variance on which this is based is the weighted mean of the squared deviations of family means from the regression line. Kempthorne and Tandon,[702] however, say that the appropriate error variance is the variance (of Y) within families.

Sib analyses.—Although the correlation between sibs is usually less efficient than the offspring-parent regression as a method of estimating heritability in laboratory mammals, there are occasions when it has to be used. If the measurement of the character requires the death of the animals before they can breed, then the regression method is obviously inapplicable, and the correlation between sibs is the only method that can be used.

The correlation of full sibs does not provide a reliable estimate of heritability because it may be augmented by nonadditive genetic variance and often also by non-genetic causes of resemblance. A sib analysis designed to estimate heritability should preferably, therefore, be based on the correlation between paternal half sibs. To provide half-sib data, each of a number of males is mated to several females and the progeny of each male constitutes a half-sib family. Ideally, if the estimation of heritability is the sole object, only one offspring of each female should be measured, so that there are no full sibs within the half-sib families. The problem of design is then a fairly simple one. The total number of progeny that can be measured will be fixed by the amount of space available or by the amount of effort that can be expended on the measuring. The question then is: how large should the families be? Where does the optimum lie between the extremes of having many small families and few large families? It can be demonstrated[1061] that the sampling variance of the correlation coefficient will be minimal when $n = 1/t$ approximately, n being the number of offspring per family and t the phenotypic correlation. So, again, it is found that the optimal design cannot be determined precisely without prior knowledge of the correlation to be estimated. But, by guessing whether the heritability of the character is likely to be high or low, a useful guide to the design can be obtained. Since the correlation between half sibs will be one quarter of the heritability $(t = \frac{1}{4}h^2)$, the optimal design will have $n = 4/h^2$. Therefore if the heritability is 20 per cent, 20 offspring per family should be planned, and if it is 40 per cent, 10 offspring per family should be planned. The precision of the estimate falls off much more rapidly when the families are smaller than the optimum than when they are larger. Therefore it is better to err on the side of having too many offspring per family than too few, and, in the absence of any knowledge of what the heritability is likely to be, it would seem reasonable to plan on having families of about 20 half sibs.

To mate each male to 20 females and to measure only one offspring from each female, as the optimal design requires, is, however, not a convenient design to carry out with laboratory mammals. And, furthermore, the estimation of heritability is not the only interest in a sib analysis. The correlation between full sibs is also of interest, because, when compared with the half-sib correlation, it shows how important nongenetic causes of resemblance between progeny of the same mother are. It would therefore seem desirable to sacrifice some precision in the estimate of heritability in order to have a more convenient design and to obtain the additional information about full sibs. To include full sibs within the half-sib families it is only necessary to measure more than one offspring from each mother. The optimal design is then more complicated and must necessarily be a compromise.[1061] If it is desired to estimate the full-sib and the half-sib correlations with equal precision, and if the full-sib correlation is not augmented by maternal effects, then each male should be mated to three or four females, and between 5 and 10 offspring from each female should be measured. If, however, the full-sib correlation is augmented by maternal effects, then it is better to have more females mated to each male and fewer offspring from each female.

The computation of the heritability from data obtained in this way needs some explanation. The computation consists of an analysis of variance leading to the estimation of three components of variance, attributable to sires, to dams, and to individuals. The correlations are estimated from these components. Sums of squares are computed in the usual way for the following sources of variation:

between sires (that is, between half-sib families);

between dams within sires (that is, between full-sib families within half-sib families);

within dams (that is, among individuals within full-sib families).

The computation of the three components corresponding to these sources of variation is straightforward if all dams have the same number of offspring and all sires are mated to the same number of dams.[1230] But equality of numbers is seldom achieved and a modification of the procedure is required. The procedure will be described without explaining the reasons for it. A full explanation is given by King and Henderson.[710] The sums of squares are composed of the three components in certain proportions which depend on the distribution of numbers within the classes and subclasses. These expected compositions of the sums of squares are given in table 46. It will be seen that each component appears with a certain coefficient which is some function of the numbers. By computing these coefficients and equating the expected composition to the observed value of each sum of squares, three equations containing three unknowns are obtained. (The sum of squares for the total is entered in the table only for the sake of completeness; it is not required.) The coefficient of the within-dam component, W, in each sum of squares is equal to the number of degrees of freedom corresponding to that source of variation. The first step in the solution of the equations is therefore to divide each sum of squares, both expected and observed, by the appropriate degrees of freedom, and so obtain the expected and observed mean squares. The solution of the equation is then straightforward.

Table 46

ESTIMATION OF COMPONENTS OF VARIANCE FROM A SIB ANALYSIS WITH UNEQUAL
NUMBERS IN THE FAMILIES

Source of variation	Degrees of freedom	Expected composition of sums of squares
Total	$N - 1$	$(N - 1)W + (N - K_1)D + (N - K_2)S$
Between sires	$m - 1$	$(m - 1)W + (K_3 - K_1)D + (N - K_2)S$
Between dams (within sires)	$n - m$	$(n - m)W + (N - K_3)D$
Within dams	$N - n$	$(N - n)W$

S = Component of variance attributable to sires.
D = Component of variance attributable to dams.
W = Component of variance attributable to individuals within full-sib families.
N = Total number of offspring.
n = Total number of dams (full-sib families).
m = Number of sires (half-sib families).
k_d = Number of offspring of any one dam.
k_s = Number of offspring of any one sire.

$$K_1 = \frac{\sum k^2{}_d}{N} \qquad K_2 = \frac{\sum k^2{}_s}{N} \qquad K_3 = \sum \frac{\sum k^2{}_d}{k_s}†$$

† This means: compute $\sum k^2{}_d/k_s$ for each sire group and then sum over all sire groups.

Table 47

NUMERICAL EXAMPLE OF THE COMPUTATION OF THE COEFFICIENTS OF THE COMPONENTS OF
VARIANCE IN A SIB ANALYSIS WITH UNEQUAL NUMBERS IN THE FAMILIES

Sire	Dam	Number of offspring per dam k_d	per sire k_s	$\sum k_d{}^2$		
A	a	6			$K_1 = \dfrac{77 + 34 + 120}{43} = 5.372$	
	b	4	15	77		
	c	5			$K_2 = \dfrac{15^2 + 8^2 + 20^2}{43} = 16.023$	
B	d	3	8	34		
	e	5			$K_3 = \dfrac{77}{15} + \dfrac{34}{8} + \dfrac{120}{20} = 15.383$	
C	f	8				
	g	2	20	120		
	h	4				
	i	6				
$m = 3$	$n = 9$	$N = 43$	$N = 43$	231		

Expected composition of:

Source	d.f.	sums of squares	mean squares
Total	42	$42W + 37.63\,D + 26.98\,S$	$W + 0.90\,D + 0.64\,S$
Sires	2	$2W + 10.01\,D + 26.98\,S$	$W + 5.01\,D + 13.49\,S$
Dams	6	$6W + 27.62\,D$	$W + 4.60\,D$
Within dams	34	$34W$	W

Table 48

COMPONENTS OF VARIANCE OF LITTER SIZE IN MICE, ESTIMATED FROM A SIB ANALYSIS
(Data of J. C. Bowman, unpublished)

Source	d.f.	Mean square Expected	Observed
Sires	70	$W + 4.16\,D + 9.75\,S$	$= 17.10$
Dams	118	$W + 3.48\,D$	$= 10.79$
Within dams	527	W	$= 2.19$

Components: $W = 2.19$

$$D = \frac{10.79 - 2.19}{3.48} = 2.47$$

$$S = \frac{17.10 - 2.19 - (4.16 \times 2.47)}{9.75} = 0.48$$

$$S + D + W = 5.14$$

Correlation of half sibs $= 0.48/5.14 \quad = 0.093 \pm 0.064$
Correlation of full sibs $= 2.95/5.14 \quad = 0.57 \pm 0.036$
Heritability: $\qquad h^2 = 4 \times 0.093 = 0.37 \pm 0.258$

The computation of the quantities K_1, K_2, and K_3, which appear in the co-efficients of the components in the sums of squares, need not be as troublesome as the formulae given at the foot of table 46 might suggest. The computation, which is illustrated numerically in table 47, may be done as follows. Tabulate the sires and the dams to which each sire is mated. Against each dam tabulate the number of its off-spring that were measured (k_d). Add these and tabulate the number of offspring measured from each sire (k_s). Now square each k_d and enter the sum of these squares obtained from each sire group as shown under $\sum k_d^2$ in the table. This gives every-thing needed for the computation of the three K's; the computations are shown at the right-hand side of table 47. The coefficients of the components in the sums of squares and mean squares can now be entered, and these are shown at the foot of table 47. An example to show the solution of the equations with real data is given in table 48.

The sib correlations are obtained from the components of variance as follows:

$$\text{Correlation of half sibs} = \frac{S}{S + D + W}$$

$$\text{Correlation of full sibs} = \frac{S + D}{S + D + W}$$

The estimate of the heritability is four times the correlation of half sibs. If the component of variance between dams (D) is not greater than the component between sires (S), then there is no evidence that the full-sib correlation has been augmented by nonadditive variance or by nongenetic causes of resemblance. If this is the case, then the full-sib correlation provides the more reliable estimate of the heritability. The heritability is twice the full-sib correlation.

The standard error of the heritability estimated from a sib correlation is rather laborious to compute.   The following method, based on that of Osborne and Paterson,[971] was devised by Dr. B. Woolf, to whom I am much indebted for explaining it to me and for permission to describe it here.   It simplifies the computation considerably and at the same time makes allowance for unequal numbers in the classes.   The procedure for computation is set out in table 49 and is illustrated numerically from the

Table 49

PROCEDURE FOR COMPUTING THE STANDARD ERROR OF THE HERITABILITY ESTIMATED FROM A
SIB ANALYSIS, ILLUSTRATED NUMERICALLY FROM THE DATA OF TABLE 48

I. Symbols used for the mean squares and the coefficients of the variance components in the mean squares.

| | | | Mean squares | |
| | Degrees of | | | In terms of |
Source	freedom	Observed		components
Sires	$m - 1$	A	$=$	$W + aD + bS$
Dams	$n - m$	B	$=$	$W + cD$
Within dams	$N - n$	C	$=$	W

II. Sampling variances of the observed mean squares.

Sires: $V_A = 2A^2/m - 1 = 2 \times 17.10^2/70 = 8.35$
Dams: $V_B = 2B^2/n - m = 2 \times 10.79^2/118 = 1.97$
Within dams: $V_C = 2C^2/N - n = 2 \times 2.19^2/527 = 0.018$

III. Coefficients of components in mean squares, and other quantities required.

$$a = 4.16 \qquad\qquad bc = 33.93$$
$$b = 9.75 \qquad\qquad x = b - a = 5.59$$
$$c = 3.48 \qquad\qquad y = bc - c - x = 24.86$$
$$T = bc(S + D + W) = 33.93 \times 5.14 = 174.40$$

Half-sib correlation: $P = \dfrac{S}{S + D + W} = 0.093$

Full-sib correlation: $R = \dfrac{S + D}{S + D + W} = 0.574$

IV. Sampling variances of the correlation coefficients.
Half-sib correlation:

$$T^2\sigma_P^2 = (c - cP)^2V_A + (a + xP)^2V_B + (c - a + yP)^2V_C$$
$$9.96 \times 8.35 + 21.90 \times 1.97 + \quad 2.66 \times 0.018 = 126.36$$

Full-sib correlation:

$$T^2\sigma_R^2 = (c - cR)^2V_A + (x - xR)^2V_B + (x - c + yR)^2V_C$$
$$2.20 \times 8.35 + 5.67 \times 1.97 + \quad 544.76 \times 0.018 = 39.35$$

V. Standard errors of correlation coefficients and heritability.

Half sibs: $\sigma_P = \dfrac{\sqrt{126.36}}{T} = 0.064$ $\qquad$ $\sigma_h^2 = 4\sigma_P = 0.258$

Full sibs: $\sigma_R = \dfrac{\sqrt{39.35}}{T} = 0.036$ $\qquad$ $\sigma_h^2 = 2\sigma_R = 0.072$

data of table 48. The computation is divided into five steps in the table and it needs little further explanation. The observed mean squares are denoted by A, B, and C, and the coefficients of the variance components in the mean squares by a, b, and c, as listed in section I of the table, which represents the analysis of variance as already explained and exemplified in table 48. Next (section II), the sampling variances of the observed mean squares are computed; these are twice the square of the mean square, divided by the corresponding degrees of freedom. Under III the values of the coefficients (a, b, and c) of the components are listed and three quantities denoted by x, y, and T are computed for later use. The sib correlations, denoted here by P and R, will have already been obtained from the analysis of variance. Now the sampling variances of the two correlation coefficients (half sibs and full sibs) can be computed fairly simply from the formulae given in section IV. Finally (V), the standard errors of the correlation coefficients are the square roots of their variances. Since the half-sib correlation must be multiplied by four to give an estimate of the heritability, the standard error of the correlation must also be multiplied by four to give the standard error of the heritability. If the heritability is estimated from the full-sib correlation (which, however, is not justified in the data presented), its standard error is twice the standard error of the full-sib correlation.

The heritability estimated from the data of table 48 has a very large standard error. The poor precision of this experiment was due partly to its design. There were 2.6 females per male and 3.7 offspring per female. It would have been better, in view of the high full-sib correlation, to have had more females per male and fewer offspring per female.

SELECTION

Artificial selection, as a method in quantitative genetics, may be used for the utilitarian purpose of producing a strain with a higher, or lower, mean expression of a character; or, alternatively, as a means of investigating the genetics of the character. In both cases the interest centers on the connection between the heritability of the character and the response to selection. The response to be expected can be predicted from a knowledge of the heritability; or, alternatively, the response observed can be used for the estimation of the heritability. However, the words "in principle" should be added to both these statements because experiments with laboratory animals have shown that there is a complication, not yet fully understood, which interferes with the simple theoretical relationship between heritability and response. This complication will be mentioned later; meanwhile it will be ignored.

In principle, the response expected (R) is equal to the product of the heritability (h^2) and the selection differential (S):

$$R = h^2 S,$$

and so the heritability may, in principle, be estimated from the ratio of the response obtained to the selection differential applied:

$$h^2 = R/S.$$

The selection differential, S, is the difference between the mean value of the selected animals and the mean of all the measured animals, including themselves, out of which they were selected. If the individual values of the character are normally distributed and the selected individuals are all those with measurements exceeding a certain value, then the selection differential depends only on the proportion selected and the standard deviation. For example, if the best 25 per cent of individuals are selected their mean superiority, which is the selection differential, will be about 1.3 standard deviations; and if the best 50 per cent are selected the selection differential will be about 0.8 standard deviations. The proportion selected depends, in turn, on the number of offspring produced by the parents. On the average two offspring per pair of parents must be selected if the strain is to maintain itself. So, if each pair produce 8 young, 2 must on the average be selected, and the maximum possible selection differential will be 1.3σ corresponding to 25 per cent selection. The selection differential can be increased by waiting for the parents to produce more young. With laboratory mammals the production of more young means waiting for second or third litters, and this takes time and adds to the interval between generations. The highest rate of progress in time is therefore not necessarily achieved by the highest possible selection differential. The best procedure depends on the number of young per litter and the interval between litters. An examination of the procedure for selection applied to mice,[335] shows that it is best to raise only one litter unless the number per litter is as low as 4, when it becomes worth while to wait for second litters; third litters are worth waiting for only if the number per litter is as low as 2. (The number per litter means, of course, the number of offspring measured and available for selection.)

At the risk of being rash, I shall make the following generalization about how rapidly an investigator may reasonably hope to improve a strain of laboratory mammals, such as mice, rats, and perhaps rabbits, by selection. The selection differential achieved will be about one standard deviation, and the heritabilities of most characters will lie between about 20 and 50 per cent. The rate of improvement will therefore be between about one-fifth and one-half of a standard deviation per generation. If 5 generations can be covered in a year, which is just possible with mice, the investigator can reasonably hope for between 1 and 2.5 standard deviations of improvement in a year.

The prediction of a response to selection is theoretically valid only for one generation, because the selection must be expected to change the genetic properties on which the response depends. But experiments have shown that, in practice, the initial rate of response can be expected to continue for 5 or perhaps even 10 generations. Eventually, selection applied to a closed population leads to a limit beyond which there is no further response. The time required to reach the limit, and the final level reached, cannot be predicted *a priori*. However, experiments suggest, although they are really too few to justify a generalization, that the response may be expected to continue for about 20 generations and that the total improvement one might reasonably hope to achieve by selection is probably of the order of 4 to 10 standard deviations. These conclusions, however, rest on a very tenuous factual basis.

The complicating factor which lays any prediction of the response from a knowledge of the heritability open to doubt is this: several experiments in which selection

Fig. 36. Two-way selection for six-week weight of mice

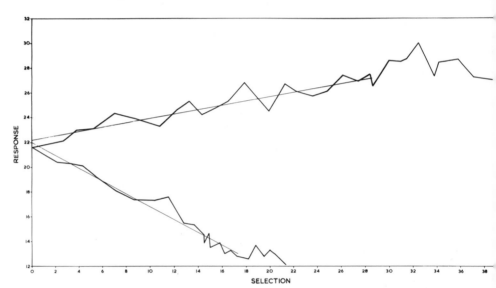

The generation means are plotted against the cumulated selection differentials, and linear regression lines are fitted to the points. The slopes of these lines, which estimate the realized heritabilities, are: upwards, 0.175 ± 0.016; downwards, 0.518 ± 0.023.[341]

was made in both directions, that is, for an increase and also for a decrease of the character, have shown that the rate of response in the two directions was not equal. One such experiment is illustrated in figure 36. The heritability estimated from the response, which may be called the realized heritability to distinguish it from estimates based on the resemblance between relatives, was 17 per cent in the upward direction and 52 per cent in the downward direction. The heritability, if it had been estimated from the resemblance between relatives, would presumably have been about 35 per cent, and, if a prediction had been made for upward selection, the response achieved would have fallen short of expectation. Since the reasons for these asymmetrical responses are not yet fully known, there is no means of predicting when they are likely to occur. Consequently it would be unwise to put much faith in any prediction of a response to selection in one direction.   And, for the same reasons, if selection is to be used as a means of estimating the heritability, it is essential that selection should be made in both directions and the heritability estimated from the divergence between the two selected lines.

There are two details of procedure connected with the estimation of heritability from the response to selection that should be briefly mentioned.   The first concerns the selection differential.   The selected parents may contribute unequally to the off-spring from which the response is measured.   It is therefore necessary to compute the selection differential as the weighted mean of the superiority of each parent, the weight being the number of offspring of that parent that were measured.   It is of interest to compare the weighted with the unweighted selection differential, because if the weighted is less than the unweighted this gives evidence of natural selection opposing the arti-ficial selection.   It shows, in other words, that the best parents have produced fewer offspring than the less good.   The second point of procedure concerns the averaging of the response from successive generations.   The best way to do this is to plot the mean of each successive generation against the cumulated selection differential.   In other words, plot each generation mean against the sum of the previous selection differentials up to that point.   This will give a graph such as that shown in figure 36.   The slope of the line up to any point is the ratio of the response to the selection differential, and this ratio (R/S) provides the estimate of the realized heritability.   The slope may remain constant over many generations, in which case the average slope may be estimated from a linear regression line fitted to the points, as illustrated in figure 36. Or the slope may change, in which case the fitting of a linear regression over the whole experiment would not be justified.   There is always a disconcerting amount of erratic variation of the mean values from one generation to the next.   One cannot, in conse-quence, hope to assess the rate of response with any degree of precision until at least 5, or preferably even 10, generations have been obtained.   Selection is therefore a time-consuming method of study.

Although selection takes a long time before reliable conclusions can be drawn, it does not require a great deal of cage space at any one time.   The amount of space required is inversely related to the rate of inbreeding that can be tolerated.   If the aim is to produce a useful strain, then it is important to keep the rate of inbreeding low, because there is no opportunity for selection between lines and whatever inbreeding depression occurs must be tolerated.   I suggest that for a program intended to last for 10 or 20 generations, the selected individuals should be drawn from not fewer than 10 families.   This can be done with the greatest economy of space by mating 10 pairs in each generation and selecting the best two offspring from each family. Selection within families in this way, however, reduces the selection differential because the variation between families is not utilized.   If, on the other hand, selection is made purely on individual merit, which should usually give a better rate of progress, then it is necessary to mate substantially more than 10 pairs in order that 10 families will be represented among the selected individuals.

INBREEDING AND CROSSING

The effects of inbreeding or of crossing lines on the mean expression of a quantitative character provide some information about the dominance of the genes that affect the character, and may perhaps allow us to draw tentative conclusions about the rôle of the character as a determinant of natural fitness. The effects of inbreeding and of crossing are the same but opposite in direction and they are both the outcome of the same properties of the genes, namely dominance. If inbreeding produces a change in one direction, then the genes on the average should show dominance of one allele over the other and the recessive alleles should affect the character in the direction of the change on inbreeding. To be specific, if the character declines on inbreeding, then the alleles that reduce the character tend to be recessive to their alleles that increase it. If the character does not change on inbreeding it cannot, however, be concluded that the genes do not have dominance. It can only be concluded that they do not have directional dominance; there could be dominance at every locus if some genes were dominant in one direction and some in the other. The presence or absence of directional dominance is the only thing to be learned about the genetic properties of the character from the presence or absence of inbreeding depression or heterosis.

An empiric conclusion based on much evidence is that characters obviously connected with natural fitness exhibit inbreeding depression and heterosis. (By characters obviously connected with natural fitness is meant characters reflecting some aspect of fertility or general viability.) This gives some grounds, therefore, for supposing that if a character shows inbreeding depression and heterosis, then it probably has a close connection with fitness, and natural selection in the past has favored individuals with high rather than low or intermediate values. The presence of heterosis is to be judged, in this connection, as any deviation of the F_1 from the mid-parental value.

THE NUMBER OF GENES

The question of how many genes are concerned in the genetic determination of a quantitative character is one of great interest. But unfortunately the answers that can be obtained have not much meaning, and the question itself is not very meaningful. The number of genes cannot be stated meaningfully without specifying the magnitude of their effects. It is not impossible that all genes segregating in a strain affect every character in some slight degree. However, if the suppositions are made that genes either affect the character or do not and that all those that affect it do so by roughly the same amount, then an estimate of the number that may have some meaning can be obtained.

The procedure for estimating the number of genes is to cross two strains that differ in the mean value of the character and to measure the variance in the F_1 generation and in the F_2 generation. The difference of variance between the F_1 and F_2 gives an estimate of the total genetic variance in this particular F_2 generation.   The estimate of the number of genes that cause this genetic variation then comes from a comparison of the amount of genetic variance with the difference of mean between the parental lines that were crossed.   It is easy to see that if there are very few genes segregating— say one or two—then the parental values will be easily recovered in the F_2; but if there are any more than a few genes the parental values will not be recovered.   Therefore the greater the variance in relation to the parental difference, the fewer are the genes segregating.   The relationship that estimates the number of genes is

$$n = \frac{1}{8} \cdot \frac{R^2}{V_{(F_2)} - V_{(F_1)}}$$

where R is the difference between the parental strains, and $V_{(F_2)}$ and $V_{(F_1)}$ are the variances of the F_2 and F_1 generations respectively.   This relationship, however, is based on a number of assumptions, and if these do not hold it is not valid.   The conditions for its validity are:

1. All the + genes are concentrated in one of the parental strains and all the − genes in the other.

2. The parental strains are highly inbred (or long selected), so that the genes segregating in the F_2 are all at frequencies of one half.

3. There is no dominance.

4. There is no linkage.

5. All the genes have effects of equal magnitude on the character.

The first of these conditions will hardly ever be met unless the parental strains have previously been selected to their limits in opposite directions.   Inbred lines that have not been selected for the character under study are therefore of very little use for this purpose.   The second condition is easily satisfied, and will be automatically satisfied if the first condition is met.   The third condition is probably not very important, because even if all the genes were fully dominant, the error would not be very great compared with the other sources of error.   The fourth condition (absence of linkage) is unlikely to be fulfilled unless there are very few genes.   Linkage sets an upper limit to the estimate that it is possible to obtain for the number of genes.   There cannot be more genes estimated than there are independently segregating segments of chromosome.   The last condition has already been discussed.   It is not so much a condition as a definition of the meaning of the number of genes.   The consequence of any of these four conditions not being fulfilled is that the estimate obtained will be lower than the estimate that would have been obtained if the condition had been fulfilled.   This is the only comforting feature of the situation because it means that we can be sure that the real number of genes is greater than our estimate.

SUMMARY

The two properties of primary interest in the genetics of any quantitative character are the degree of genetic determination and the heritability.

The degree of genetic determination is the proportion of the total phenotypic variance that is attributable to genetic differences between individuals. It is estimated from a comparison of the variances of a genetically uniform group, such as an F_1 of two inbred lines, and a genetically heterogeneous group, such as an F_2 or a random bred strain. The number of animals that must be measured in order to attain a given degree of precision in the estimate can be determined from formulae and graphs presented.

The heritability is the proportion of the total phenotypic variance that is attributable to additive genetic variance. It is estimated from the degree of resemblance between relatives. The regression of offspring on parents is usually the best method for laboratory mammals. The design of the experiment to give maximal precision is discussed. In general it is best to have as many parents as possible at the cost of fewer offspring per parent. The method of computing the regression when the families contain unequal numbers of offspring is explained.

The correlation of half sibs is another method of estimating the heritability. The most efficient design is very inconvenient with laboratory mammals, and as a compromise it is suggested that as many males as possible should each be mated to 3 or 4 females and between 5 and 10 offspring from each female should be measured. The structure of the sib analysis with full and half sibs is described; and the method of computing the components of variance, and from them the correlations when there are unequal subclass numbers, is explained.

The genetic information that can be obtained from artificial selection and from inbreeding or crossing inbred lines is briefly explained. Finally, the estimation of the number of genes contributing to the variation is discussed. Only in very restricted circumstances does the estimate have any useful meaning.

Elizabeth S. Russell, Ph.D.

PROBLEMS *and* POTENTIALITIES
in the STUDY *of* GENIC ACTION
in the MOUSE

Studies of the physiologic genetics of laboratory mammals have a special rôle in the biomedical sciences in that they can and must be the integrating link between basic information from lower organisms and the human problems to which these facts and concepts must ultimately apply. Recent medical advances, by decreasing the prevalence and severity of infectious diseases, have increased the relative importance of constitutional or inherited disease. It is to be hoped that knowledge of the etiology of human and analogous experimental mammalian inherited disease syndromes may be useful both for prophylaxis and for therapy.

During the past quarter century much has been established about the nature of the self-duplicating molecules of deoxyribonucleic acid which transmit genetic information from generation to generation and from cell to cell, along with a beginning insight into the nature of the relationship between these units of genetic transmission and their intracellular products. In experiments with microörganisms many gene-controlled biochemical pathways have been traced. It is very much to be hoped at this particularly stimulating stage in the development of genetics that widespread attention will be turned toward solution of problems of genic action in highly differentiated forms, particularly in mammals. Observed inherited characteristics in them are the raw material of many areas of biology: embryology, anatomy, physiology, biochemistry, endocrinology, immunology, and pathology, for example. The techniques of these disciplines may profitably be applied to analysis of mammalian genic action, and the answers obtained contribute to understanding in these fields. At present the house mouse is the most widely studied mammalian species other than man, because of the large number and variety of named genes and the availability of many inbred strains with genetic homogeneity and well-established physiologic characteristics The purpose of this paper is discussion of the special problems encountered in studies of the physiologic genetics of the mouse.†

The fact of differentiation leads to complications never encountered in micro-organisms. Tracing the pathway between site of original action of the gene and observed character is frequently a major problem. When the site and time of original genic action have been established, it is often difficult to devise methods of analyzing metabolic processes within the affected cells. Examples of profitable approaches surmounting these obstacles have been selected from recent literature and current research projects. It is hoped that presentation of these pertinent examples, and discussion of their relationship to fundamental genetic questions, will lead to some basic generalizations regarding the methodology of mammalian physiologic genetics.

GENES, PROTEINS, AND ENZYMATIC ACTIVITY

It is probable that the primary activity of most, if not all, genes is determination of the specificity of an intracellular macromolecule. Very few of the inherited characteristics recognized in mammals represent these immediate genic products. In the few cases, however, where this may be true, that is, where the observed effect of a genic substitution is change in the structure of a protein molecule, the fact of cellular differentiation may simplify analysis of genic action. The production of this protein may be limited to particular types of cells and may represent a very large part of the total metabolic activity of these cells, greatly facilitating biochemical and physical chemical analysis. Determination of the globin structure of hemoglobin appears to fall in this class,[102, 104, 1181] and experimental genetic analysis of hemoglobin pattern in the mouse, described in this volume by Dr. Popp, shows great promise. Little insight has yet been gained into the developmental processes channeling the metabolism of hematopoietic cells into this limited range of activity, a problem which presents a challenge for future investigations.

The action of the well-known albino series of genes, affecting intensity of hair pigmentation in the mouse and other mammalian species, may very well be determination of the structure of a tyrosinase molecule. Genic substitutions at this locus definitely alter tyrosinase activity,[382] which could mean either alterations in concentration of identical enzymes or qualitative alteration of the enzyme molecule similar to that seen in certain tyrosinase mutants[598] in *Neurospora*. Evidence that this series of genes

† The research reported in this paper, which originates from the Physiological Genetics Group at the Roscoe B. Jackson Memorial Laboratory, has been supported by a contract between the U.S. Atomic Energy Commission and by grants to the laboratory from the U.S. Public Health Service and the American Cancer Society.

may determine enzymatic structure comes from recent studies[213] of tyrosinase activity in mice carrying a new allele of the albino series, *Himalayan* (c^h), discovered recently in the Jackson Laboratory.[487] This mutant almost certainly contains a qualitatively different tyrosinase from those found in mice with any of the other alleles of the albino series. If one gene in an allelic series is responsible for a qualitative enzymatic difference, it is probable that the entire series acts by determining enzymatic structure.

Gene-induced difference in enzymatic activity is not always, of course, attributable to change in enzymatic structure. Recent studies of the mutant allele *dilute* (*d*) and *dilute-lethal* (d^l) have demonstrated a deficiency of the enzyme, phenylalanine hydroxylase, associated with genic substitutions at this locus.[214] Further studies have indicated, however, that the genic action involved is an inhibition of activity of an enzyme formed quite independently of the genes at the *dilute* locus.

FUNCTION OF GENES AND TISSUES

A relatively large proportion of the analyzed effects of substitutions of single genes in the mouse has been traced to differences in the structural differentiation of certain specific tissues only. For these genotypes, the majority of tissues of an affected animal, except in terminal stages of a lethal condition, cannot be distinguished from those of an unaffected littermate. The choreic movements and deafness associated with six different independent mutant genotypes in the mouse have been traced to dedifferentiation of Corti's organ, degeneration of the spiral ganglion, and abnormalities in the stria vascularis.[248, 249, 499] Since effects of these mutants compensate each other in crosses, it is clear that there are at least six different ways in which tissues of the murine vestibular apparatus can fail. The locomotor difficulties, whole-body tremor, muscular spasms, and early mortality found in *wabbler-lethal* mice (*wlwl*) all appear to result from degeneration of the myelin sheath surrounding nerve processes.[273] The retina of a mouse with *retinal degeneration* (*rdrd*) develops normally for the first ten days after birth, then begins to show degeneration of the rods, while those of normal littermates continue differentiation.[1258, 1313]

For each of the mutant types so far described in this section, histologic difference from normal has been described which involves an inability of affected tissue to complete normal differentiation or to maintain normal structure. It is probable that each defect is due either to a metabolic error specific to the abnormal tissue or one imposed upon this tissue by a correlative influence from some other part of the body, that is, either local or distant site of original genic action. For many tissues it is difficult to obtain critical evidence as to this cellular localization, although the limitation of visible difference to one tissue suggests greater probability of local genic action. A somewhat equivocal suggestion as to the site of genic action in retinal degeneration may be derived from an experiment involving organ culture *in vitro* of genotypically normal and *rdrd* eyes explanted at the stage (10 days) when degeneration is first visible.[808]   The degeneration continued in *rdrd* eyes, while normal eyes showed fairly normal differentiation.   This suggests local genic action.   There was, however, some delay in the rate of *rdrd* degeneration *in vitro*, which may have resulted from retardation of autonomous development but which might also indicate that explantation removed the eye from an humoral toxic effect *in vivo*.   Although this experiment provides a suggestion as to a site of original genic action, there is no clue as yet for any of these mutants as to the nature of their metabolic error.

A somewhat different situation exists in the study of hereditary *muscular dystrophy* (*dydy*) in the mouse.   Skeletal muscle is certainly the primarily affected tissue in this disease since the nervous system appears entirely normal.[869]   Detailed histopathology demonstrates great similarity to other hereditary muscular dystrophies, and gives evidence that there is a continuing attempt at muscular regeneration even in individuals severely affected.[1383]   Studies of C^{14}-glycine incorporation and turnover corroborate this finding.[215]   Progressive wastage results from an excess of destruction over synthesis.   Parabiosis experiments, with conjoined normal-dystrophic pairs, indicate clearly that there is no humoral factor responsible for muscular breakdown.[1013] The defect is indigenous to the muscle itself, thus almost certainly a result of local genic action.   All of the differences observed are undoubtedly the result of difference in a single genic pair.   The mutation was first recognized as a deviation within an inbred strain, and the distribution and frequency of dystrophy in pedigrees and in offspring of ovaries transplanted from dystrophic females clearly demonstrated inheritance as a unit autosomal recessive.[869, 1281]   Evidence is now at hand, from outcrosses, repeated backcrosses, and linkage tests, that this substitution of a single genic pair creates essentially the same syndrome of disease in combination with a great variety of genetic backgrounds.[805]   The defect is already apparent histologically at two weeks postnatal, the first age at which the entity can be diagnosed from behavior. At earlier stages study of muscular disease is difficult because normal muscle has not yet assumed its adult form; this finding suggests that in muscular dystrophy, as in previously described syndromes, the tissue defect may involve a failure to complete normal differentiation.   In contrast to the disease entities previously described, however, there is a bewildering array of evidence of deranged metabolism.   K/Na balance is abnormal,[47, 683] creatine/creatinine balance is abnormal,[683, 996] muscle lipid is increased,[1200] distribution of lipoproteins and glycoproteins deranged,[968] myosin concentration decreased,[969] contractibility decreased,[1148] relaxation time lengthened,[1148] and levels of activity of many enzymes altered.[10, 216, 217, 541, 1034, 1035, 1074, 1314, 1372, 1468]   It appears probable that many of these alterations occur whenever muscle is degenerating and thus are results rather than causes of the basic defect. There is no guarantee that the character which is associated with a particular genic substitution is the final stage in a path of genic action; it may be an intermediate step, and the recognized character may have extensive metabolic consequences.   In a situation with multiple metabolic changes, it is difficult to determine which defect is primary.   One helpful approach is retrograde analysis, attempting to find a metabolic deviation present at an especially early stage when other reactions are still normal. This concept underlies some of the especially promising current investigations on murine muscular dystrophy, but the answers are not yet forthcoming.

GENES, HORMONES, AND PATTERNS OF RESPONSE

Localizing the initial site of genic action presents special problems in cases of endocrine defect, because of reciprocal relations between different hormones. This has proved particularly true in recent studies of pituitary dwarfism in the mouse, the first gene whose action was analyzed by substitution therapy.[1222, 1236] The defect is definitely in the anterior pituitary, but the particular type of cell primarily affected is still in doubt to some extent. Gonadotrops are present, although abnormal in appearance, in the dwarf pituitary, and their functional capacity has been proven conclusively in transplantation experiments.[1223] All observers agree dwarfs show a great deficiency in typical murine acidophils, but recent investigations using sophisticated histologic techniques[321, 1000, 1049, 1368] disagree as to whether there are undifferentiated potential acidophils. There is also great reduction in typical thyrotrop cells. Uptake of radioactive iodine suggests but does not prove that the thyroid of dwarfs is especially sensitive to thyrotropin, which favors the interpretation of initial deficiency of thyrotrops and thyrotropin.[1367] Recent experiments in which rate of growth was studied following reciprocal transplants placing pituitaries of littermate normal and dwarf mice 14 to 18 days of age into the sella of hypophysectomized normal and dwarf hosts showed a comparable rate of growth of normals and dwarfs with transplanted normal pituitaries and a reduced rate in normals with dwarf pituitaries.[155] This evidence indicates the defect is in the pituitary rather than the hypothalamus, since the organ functions autonomously following transplantation. It also favors the interpretation of primary deficiency of acidophils and growth hormone. It is to be hoped that in the near future a transplantation approach, combined with histologic study of changing cellular types in the implanted pituitary, may help to identify the types of pituitary cell initially affected in dwarf mice.

Genetically controlled differences between inbred strains have also been reported for androgen level and responses,[189, 191, 192, 194] estrogen level and response,[191] and thyroxin level and response.[190, 193] Since these differences have a polygenic basis, they do not provide good material for retrograde analysis, but the methods used merit description. Secreting glands were extirpated or inactivated, and sensitivity of target organ measured by addition of graded doses of exogenous hormone. Normal circulating level of each hormone was calculated from the dosage producing the end-organ condition normal for the strain.

There is also much evidence of genetically controlled difference in reaction of the adrenal to gonadectomy. The adrenals of some strains fail completely to respond to this stimulus, others show hyperplasia, and still others show cortical carcinoma.[193, 271, 1407, 1408]

GENES IN EMBRYONIC DEVELOPMENT AND DIFFERENTIATION

Many skeletal abnormalities and derangements of organs have been shown to result from genic action during embryonic development. Retrograde analysis, studying the mutant phenotype at progressively earlier stages, is clearly an effective tool for locating place and time of initial genic action but does not always help in determining its nature.

Frequently there is evidence of inductive failure. An abnormality frequently associated with the *Short-Danforth* gene *(Sd)* is reduction or absence of the kidney, which has been demonstrated to result from reduced length of the ureteric outgrowth. In some cases there is no contact with competent tissue, and no induction of metanephros.[440] In addition to the previously mentioned genes causing choreic movements through tissue dedifferentiation, a series of genes with very similar behavioral effects act in a very different pattern by preventing normal induction of the middle ear, each gene acting in a different and specific way.[499, 818]

The most extensively studied series of genes acting through induction is the *Brachyury* or *T* series. In addition to the normal wild-type allele, one dominant *(T)* and numerous recessive *(t^x)* alleles are known; additional instances of mutants have frequently been found in populations of wild mice. Most of the homozygous types are lethal before birth, the time of death varying greatly among alleles, but a few are viable. Only *T* leads to tail reduction in combination with the wild-type allele, and all *Tt^z* combinations lead to short or absent tails. The homozygous lethal alleles of the *T*-series fall into five classes according to time and nature of original genic action.[76] Primary action of *T* seems to be inductive failure at approximately the ninth or tenth day of development resulting from degeneration of the chorda-mesoderm.[204, 502] The *t^{12}t^{12}* homozygote shows the earliest lethal effect, failure of blastocyst formation associated with defective trophectoderm.[1221] In *t^{o}t^o* homozygotes the inner cell mass fails to differentiate into embryonic and extra-embryonic ectoderm.[439] In a group of five relatively early-acting *t^w* alleles (collected from the wild), the first microscopically visible defect is pyknosis in the basal plate of the neural tube around the seventh day of embryonic life. Homozygotes of four relatively late-acting *t^w* alleles show their first abnormality on the ninth or tenth day of development, as pyknosis in the ventral portion of the hindbrain, the notochord and mesoderm being normal.[76] These interesting findings have many embryologic implications for which the reader is referred to the original papers. Several points are, however, worth mentioning in connection with general principles of genic action. Bennett and Dunn[77] point out that although the first microscopically visible defect in all *t* homozygotes is in ectoderm, it is still impossible to decide between local genic action deranging metabolism of the embryonic ectoderm and genic action in the endoderm leading to improper nutrition and consequent death of ectodermal cells. In each case, however, the ectodermal degeneration is associated with failure of induction of mesodermal tissues. The neural tube of both *TT* and *t^{w}t^w* embryos is capable of inducing cartilage *in vitro*.[76] In attempting to bring together the differing effects of this multiple allelic series, Bennett, Badenhausen, and Dunn[76] suggest there are "some grounds for suggesting that at least four different lethal *t*-alleles affect pathways which lead to progressively higher types of neural differentiation. The mutation *T...* appears to affect a different pathway in differentiation, that leading through the chorda-mesoderm. It is not easy to reconcile this diversity in physiological effects with 'unity of action' of a locus that appears on genetical grounds to be a unit not resolved by recombination." In the face of this confusion, it is comforting to remember that tracing the widely differing paths affected by *T* and *t* alleles has nevertheless been very helpful in explaining their complementation in the viability of *Tt* hybrids.

Some genes with shape-determining effects in embryonic life act through mechanisms other than induction. An interesting example is *short ear*, in which a primary effect appears to be reducing growth of cartilage. Since cartilaginous growth is most conspicuous in fetal life, many effects of *short ear* are apparent at this time. However, recent experiments with healing of broken ribs have shown that a metabolic deficiency in cartilage growth persists into adult life in short-eared mice.[486]

MURINE HEREDITARY ANEMIAS

Analysis of genic action in the causation of six different types of inherited anemias in mice is a major concern in current physiologic genetic investigations at the Jackson Laboratory. Genes at three of the loci, *Dominant spotting* (W, W^j, W^v),[503, 504, 1091, 1102, 1106] *Steel* (Sl, Sl^d),[83, 1155] and *Hertwig's anemia* (*anan*)[741] cause varying degrees of macrocytic anemia. Homozygous *flexed*-anemia (*ff*) animals suffer from a transitory macrocytic siderocytic anemia limited to fetal and neonatal life.[505, 506, 885] Animals homozygous for either *jaundice* (*jaja*)[1280] or *hemolytic anemia* (*haha*)[83] suffer from severe hemolytic disease with abnormal nucleated erythrocytes and extensive postnatal jaundice. Animals of five of the available genotypes die shortly after birth, and several other genotypes are semilethal. Each anemia seems to present a different defect in hematopoiesis, and each is present before birth. We are very hopeful that some of these anemia-producing genes may provide especially favorable material for identifying the metabolic error in a tissue primarily affected and possibly may even give clues as to factors limiting genic action to these tissues. The murine anemias named above which we are studying do not exhaust the list of available types. Another dominant-spotting allele has been reported (W^a).[907] A very interesting independently inherited anemia, *diminutive* (*dm*), with associated skeletal defects, has also been described recently.[1279]

It is essential, of course, to have an accurate description of the erythron and the hematopoietic tissue in each mutant genotype. It is also important to measure accurately quantitative differences between the effects of different alleles within the same series and between genes of different series, avoiding difficulties from variations in the genetic backgrounds on which these genes are segregating. As with various entities discussed in the general section on genes and tissue function, it is important to determine the site of initial genic action in each case. It appears probable that most of the defects are indigenous to bloodforming tissue, but this can only be established by transplants between genotypes affected differently. For these reasons, the development of congenic histocompatible strains differing essentially only in the anemia-producing genes is an important part of our work.[1105]

Since animals of several important genotypes are available only in fetal and newborn stages, it is also very important to develop methods for working with extremely small quantities of tissues and for studying metabolism of fetal hematopoiesis. Formation of hemoglobin is, of course, an essential part of all erythropoiesis; and, since many human hereditary anemias have been traced to specific hemoglobinopathies,[619, 938] it seems essential to study the nature of hemoglobin associated with different anemic states. Studies of the reactions of normal and affected animals to particular physiologic stimuli or stresses may be very helpful in identifying metabolic deviations. Several of the genes under study have extensive pleiotropic effects in other tissues. Analyses of these effects will undoubtedly increase knowledge of pathways of genic action, but may also yield clues as to the nature of original genic action in hematopoietic tissue. Attempts are being made to develop each of these approaches in the study of murine anemias, but, except for experiments with the W series, they are still in the stage of potentiality rather than performance.

CHARACTERIZATION OF W-SERIES ANEMIAS

The hematologic phenotype of animals differing in W-series genes ranges from an ostensibly normal picture in ww, Ww, and W^jw animals,[1091, 1102, 1106] to slight macrocytic anemia in W^vw,[1091] and severe macrocytic anemia in animals of all double-dominant genotypes. Order of increase in severity of affliction is ww (normal) $=$ $Ww = W^jw < W^vw < W^vW^v < WW^v = W^jW^v < WW = WW^j = W^jW^j$. (Animals of the last three genotypes are almost invariably lethal in early postnatal life.) The bone marrow of adult W^vW^v and WW^v individuals has almost exactly the same cellularity as that of their normal littermates,[970] although it produces only slightly more than one-half the normal number of erythrocytes, suggesting a delay in maturation of erythroid cells. Careful microscopic study of the marrow confirmed this suggestion, since the ratio of early to late stages in erythropoiesis was significantly higher in the anemic animals.[1110] The first evidence suggesting a biochemical basis for this arrest came from C^{14}-glycine-incorporation experiments using anemic and normal littermates.[16, 1097] There was no difference between genotypes in time of appearance of erythrocytes with labelled globin, but in the anemic mice there was a great delay in appearance of cells with labelled protoporphyrin. Subsequent experiments showed a similar delay in appearance of labelled protoporphyrin in erythrocytes of anemics following injection of radioactive δ-amino-levulinic acid,[15] an intermediate on the path leading to protoporphyrin. This delay in heme synthesis, which is still under investigation, seems a critical feature in action of *W*-series genes. Animals of severely affected genotypes are anemic because of a metabolic defect, which either specifically delays the synthesis of protoporphyrin or nonspecifically arrests erythroid cell maturation at a stage when synthesis of protoporphyrin is an important metabolic activity.

TISSUE LOCALIZATION OF ACTION OF THE *W*-GENE

Genic action leading to the hematopoietic defect of *W*-series anemic animals definitely occurs in the hematopoietic cells themselves, rather than being imposed from another part of the body. This has been demonstrated repeatedly by successful implantation of hematopoietic cells from normal *ww* fetal liver into adult *WW^v* and juvenile *WW^v* and lethally anemic *WW* mice.[84, 1109] The implanted cells function autonomously according to their own genotype, and the blood picture of the host changes gradually but permanently to that of a normal mouse. It is interesting that an initial very small inoculum of rapidly metabolizing cells eventually overwhelms a large body of indigenous defective marrow. These experiments are also an interesting demonstration that hematopoietic cells from the fetal liver at least can implant and function in an adult manner in the marrow spaces of an adult.

RADIATION RESPONSE OF NORMAL AND ANEMIC MICE

The responses to irradiation of normal *ww* and anemic *WW^v* animals are extremely different. A dose of whole-body irradiation (200 *r*) which has very little effect upon the blood picture of normal *ww* mice causes severe and prolonged reduction of the hematocrit level, and, in some cases, death in littermate *WW^v* anemic mice.[85, 1109] The hematocrit level of surviving *WW^v* mice returns to normal suddenly 3 to 4 weeks after radiation treatment. Quantitative microscopic study of the marrow of normal *ww* and anemic *WW^v* mice at successive intervals after 200 *r* whole-body irradiation revealed no difference between genotypes in initial destruction of marrow cells.[1098] The cellularity in both genotypes decreased sharply for the first two days after irradiation. The marrow of *ww* individuals then regenerated rapidly and had returned almost to pretreatment level by the fourth day after irradiation. That of *WW^v* mice, however, regenerated very slowly, and showed no visible increase in cellularity by the eighth day after irradiation. Further evidence that radiation sensitivity depends in this case upon the genotype of bloodforming tissue is found in studies of the reaction of implanted *WW^v* mice, with a normal blood picture, to increasing doses of X irradiation. Their 30-day LD$_{50}$ dose corresponds closely to that of normal *ww* mice.[83] There is even evidence that single doses of *W* or *W^v* (in *Ww* and *W^vw* animals) significantly affect radioresistance.[83]

RESPONSE OF NORMAL AND ANEMIC MICE TO ERYTHROPOIETIC STIMULI

Studies of the responses of *ww* normal and *WWv* anemic mice to known erythropoietic stimuli have yielded valuable information both on the pattern of *W*-gene action and on the nature of hemostatic mechanisms. Twelve moderate-sized daily doses of a purified erythropoietin[807] which has induced extra erythropoiesis in normal animals of six different species,[696] were administered to normal *ww* and anemic *WWv* animals. In all cases the normal *ww* mice responded to the treatment with greatly increased hematocrit level, marked reticulocytosis, and increased total blood volume.[1104] The *WWv* mice, however, were completely unaffected by the injections. This genotypic difference in reaction to a known erythropoietic stimulus depends upon the genetic nature of the bloodforming tissue. The response to erythropoietin of *WWv* mice implanted with *ww* cells is exactly comparable to that of *ww* mice.[697]

There is at least one erythropoietic stimulus to which *ww* and *WWv* mice respond equally. If normal and *W*-anemic mice are subjected to reduced atmospheric pressure[503] or to lowered oxygen tension at normal pressure,[697] animals of both genotypes respond with reticulocytosis and increased hematocrit level. The difference in response of *WWv* mice to different stimuli demonstrates clearly that there must be more than one fundamental erythropoietic stimulus and suggests that the basic genic action in *W*-series anemias may be related to the phase of erythropoiesis affected by treatment with Borsook-type extracted erythropoietin. It is very tempting to speculate that the maturation arrest of *WWv* mice, their special radiation sensitivity through delayed marrow regeneration, and their inability to respond to erythropoietin may all be different aspects of the same basic phenomenon.

ANALYSIS OF *W*-SERIES AND *STEEL* PLEIOTROPISMS

In addition to suffering from the macrocytic anemia already described, animals of all double-dominant *W*-genotypes (*WW*, *WWv*, *WjWv*, etc.) lack pigment in the hair and are almost completely sterile. The sterility results from failure of the primordial germ cells to multiply during their migration to the gonadal ridge, between the eighth and twelfth days of embryonic life.[879, 880] In this analysis the primordial germ cells were visualized in sectioned whole embryos by alkaline-phosphatase staining. The association of reduced germ-cell number with the double-dominant genotypes was determined on a statistical basis rather than by genotypic identification of individual embryos. Only matings between two heterozygous parents produced embryos with defective germ-cell numbers at 9, 10, 11, and 12 days, and the proportion of defective embryos at each age was close to 25 per cent. The germ-cell defect seems to be completely determined by the twelfth day of embryonic life, since 12–16 day fetal gonads transferred to a neutral site with a rich blood supply developed autonomously according to their genotype.[1107] In this study using older fetuses as gonad donors, individual genotypic identification was possible. All double-dominant fetuses were pale at 12–16 days due to their hematopoietic defect. *Luxate*, a third-chromosome gene approximately 18 crossover units from the *W*-locus, provided another check upon fetal genotype. Homozygous *luxate* (*lxlx*) fetuses were identifiable as early as the twelfth day of development by the shape of their abnormal hind feet.[168] This gene was placed in test matings in coupling with W^v, so that more than 96 per cent of all fetuses with abnormal hind feet (*lxlx*) would also be expected to be severely anemic and potentially sterile ($W^v W^v$). Thus, a closely linked gene with early clear-cut expression proved to be a very useful tool in retrograde analysis. An interesting long-term consequence of the paucity of germ cells in $W^v W^v$ mice is that all of the females eventually develop ovarian tumors.[1101] The lack of pigmentation in *WW* and $W^v W^v$ mice is essentially 100 per cent white spotting, involving absence of melanoblasts in the hair follicle.[1207] Melanoblasts, which normally migrate from the neural crest to all parts of the body between the eighth and twelfth days of embryonic life, differentiate from explants of the flank area of normal *ww* 10-day embryos, but do not appear in explants from the same area of *WW* littermates.[100] The pigment defect of homozygous defective *W*-series mice is thus completely determined very early in embryonic life.

Homozygous *SlSl* embryos and neonates[74] and adult $Sl^d Sl^d$ individuals[83] show a triad of pleiotropic effects in bloodforming tissue, germ cells, and hair pigmentation almost identical with that found in *W*-series homozygotes. At least 7 separate instances of mutation at the *W*-locus are known to have occurred spontaneously, in addition to frequent *W*-series mutations in radiation experiments. Four separate instances of *Sl*-locus mutations are known. In each case, all three aspects of the triad were simultaneously altered. It is almost impossible to avoid the conclusion that each of the genes at both of these loci acts as a unit, controlling a single process. The process controlled by the *W* locus and that controlled by the *Sl* locus may be closely related, possibly as different steps in the same synthetic pathway.

In which of the three severely affected tissues is this series of processes initiated? It is known to be important in fetal liver and adult marrow hematopoiesis, but no information is available as to possible yolk-sac hematopoietic defect before the twelfth day of embryonic life. Circulation begins at approximately 9.5 days.[1247] The reduced multiplication of germ cells in *WW* and $W^v W^v$ embryos is apparent at 9 days and very marked at 10 days, the melanoblast defect at 10 days. The very early appearance of these anomalies makes it relatively improbable that they are dependent for expression upon an anemia resulting from defective yolk-sac hematopoiesis. It seems very much more probable that the *W-Sl* series of processes is independently of great metabolic significance in primordial germ cells, in melanoblasts or their embryonic precursors, and in hematopoietic cells. Unity of genic action for these genetic loci may involve processes specifically important in the metabolism of three different tissues.

MATERIALS AND METHODS IN MAMMALIAN PHYSIOLOGIC GENETICS

Review of the examples of genic action in the mouse discussed in this paper may lead to certain generalizations on methodology. The first tool in search for time and place of original genic action is a retrograde analysis. Transplantation and extirpation may be very useful in determining organ localization of the primary defect. Comparison of growth of normal and affected tissues or organs *in vitro* may also prove very useful if culture conditions are sufficiently physiologic to allow deduction as to relation of observed differences to genotype. Linked genes may assist in genotypic identification, particularly in fetal material. Detailed microscopic analysis, sometimes combined with other approaches, helps to distinguish the cells primarily affected. Replacement therapy, isotope incorporation, and tests of functional capacity are useful in determining the nature of intracellular processes affected by particular genic substitutions. Comparison of the effects of different genes in a multiple allelic series and study of dominance relations may provide valuable clues, as may pleiotropic effects of single-gene substitutions.

Studies of genic action are usually limited to unit genes with clear-cut effects, since their effects may most easily be traced. These unit genes tend to affect one or a very few types of cells, probably because the reactions they control are especially important in these cells. (It should be recognized that our methods of detection favor recognition of unit genes with tissue-limited effects; others may either be cell lethal or so diffuse in their effect as to escape detection. Many gene-controlled reactions must alter slightly the metabolism of many kinds of cells.) The characters observed as a result of genic action may be very close to original genic action, as in determination of hemoglobin pattern, or very far removed, as in choreic behavior as a result of defective induction of the middle ear or ovarian tumorigenesis as a result of deficiency of primordial germ cells. It is seldom possible to predict the number of processes between original genic action and observed characteristic; nor is it always possible to be sure when one has reached the end of the road.

Although the examples in this dissertation have tended to be restricted to effects of single-gene substitutions, characteristics of inbred strains form an important part of available material. Comparison of mice from different inbred strains has demonstrated existence of many inherited characteristics depending upon the interaction of polygenic factors the individual effects of which cannot now and may never be identified. Since it is probable that the interacting genes work in different ways, it is unlikely that the paths of genic action lying behind a multigenic character can be traced very far. Provided the limitations are understood, however, it is frequently possible to design excellent experiments identifying terminal stages of the pathway. Chai's work with endocrine level and response, Heston's studies with genetics of neoplasia, and Gowen's studies of disease resistance are excellent examples of such analyses. These cases must be studied if physiologic genetics is to be put to use in the service of man. Any study which traces a segment of a pathway between gene or genotype and observed character, however far this segment may be from original genic action, is a contribution to physiologic genetics.

Although this point has not been stressed in our presentation, it is very clear that for successful analysis of the action of a unit gene it is very desirable to have it segregating against a genetically homogeneous background.   Histocompatibility is essential for transplantation.   Quantitative evaluation of effects of particular genic substitutions depends upon uniformity of the base line used for comparison.   Uniformity of genetic background may come about in three different ways.[483]   With the increasingly wide-spread use of inbred strains, many of the stocks in which deviants are detected may be inbred, so that from its first appearance a new mutant may be segregating against an inbred background.   The best way to maintain high congenicity is by repeated back-cross of the mutant heterozygote to the strain of origin.   If a new mutation is found in a genetically heterogenous stock of animals, a new inbred strain may be produced by successive brother-sister matings with forced heterozygosis for the mutant allele, or the mutant allele may be placed on an existing inbred background by many repeated backcross generations.   It may be worth mentioning that there are at present in the Jackson Laboratory alone 67 stocks designed to place and maintain specified mutant genes on inbred backgrounds.[751]   Twenty-five of these mutants are maintained con-genic with C57BL/6J, which makes for excellent uniformity between experiments and provides very favorable material for new genotypic combinations and comparisons, and for double-genic substitutions in transplantation experiments.   There is evidence, however, that this one inbred strain is not the ideal background for all mutant genes.

As a final plea in methodology of mammalian physiologic genetics, I would like to encourage very widespread use of controlled genetic material by investigators in other biomedical disciplines.   The analyses of genic action cited in this paper include many excellent examples of such utilization.   Physiologic geneticists are forced, by the diversity of the paths of genic action which they encounter, to be jacks of many trades and face the very real possibility of being masters of none.   The studies of *W*-series anemias have involved active participation by pathologists, biochemists, embryo-logists, and physiologists.   Multidisciplinary approach to analysis of the action of a single set of genes has been very useful in this case.   Similar collaboration between geneticists and other types of investigators may prove profitable for many other studies.

DISCUSSION

Dr. Burdette: Thank you, Dr. Russell.   Dr. Russell's paper will first be discussed by Dr. D. L. Coleman.

Dr. Coleman: I would like first to commend Dr. Russell on her stimulating and comprehensive discussion of genic action in the house mouse.   I would like to comment further on two points to which she has alluded in her talk.   First, I would like to clear up a possible misunderstanding.   I do not think that the tyrosinase picture is quite as clear-cut as she has indicated.   However, the best evidence at this time does suggest the situation which she has described.

The other point I would like to discuss relates to my particular studies on the *dilute* mouse[214] and attempts to demonstrate how some of the methods which she has described can be used.   In the study of the *dilute* gene, we were very fortunate in that mutation to the intense color has occurred several years ago in the DBA stocks.   Thus we had genetic homogeneity with an intense animal, *DD*, a heterozygote, *Dd*, and the normal *dilute* animal, *dd*, all available within the same strain.   We were fortunate also here in having a multiple allelic series to work with since there is another allele of this locus, the *dilute lethal*.   This allele, although not available on the DBA background, was in the process of being inbred on another stock.

Also, we made use of pleiotropism.   Both *dilute lethal* and *dilute* mice have diluted pigment, but the *dilute lethal* dies at about 3 weeks and is subject to spontaneous epileptiform seizures.   The DBA strains are also subject to seizures under audiogenic stimuli and it seemed that there should be some relationship between this type of seizure and the pigmentation defect if *dilute* were just a lesser dose of *dilute lethal*.   The diluted pigmentation further suggested a possible abnormality in aromatic amino-acid metabolism.   Thus, one of the first things tried was incorporation studies of radioactive tyrosine (the normal pigment precursor) into the pigment granules of both dilute and intense mice.   It was found that the rate of pigment formation was the same in both genotypes when tyrosine is used as the pigment precursor.   Next the enzymes involved in the formation of tyrosine were examined.   The enzyme, phenylalanine hydroxylase, which forms tyrosine from phenylalanine was found to be deficient in *dilute* strains of mice, having an activity of about 50 per cent in *dilute* mice, *dd*, and about 14 per cent that of the normal in *dilute lethal*.   This, at first glance, provided a ready explanation for the diluted pigmentation.   However, it was pointed out by Dr. Russell that there is not less pigment in the dilute animal but rather that the granules are in a clumped formation.   Also, calculations showed that a leaky enzyme which allowed a 50 per cent production of tyrosine would allow normal pigment formation under most of the conditions that we know, especially when one considers that some tyrosine is derived from the food.   Thus, it appeared that we were working with a secondary effect of the original genic action.   Further studies on the actual amount of the enzymes present indicated that there was no actual difference between the amount found in *dilute* and *nondilute* animals; one function of the *dilute* gene seemed to be the production of an inhibitor of the enzyme, phenylalanine hydroxylase.

In any event, the situation is somewhat analogous to that found in phenylketonuria, diluted pigmentation and inhibition of this enzyme with a subsequent accumulation of phenylacetic acid and other phenylalanine metabolites.   Phenylacetic acid is a compound toxic to the central nervous system which suggests an explanation for the seizure in *dilute* mice.   There are several areas of research which we now are attempting. Could phenylalanine or a metabolite in abnormal concentrations cause an abnormal development of the brain or neural crest which then leads to these seizures?   Could such abnormal concentrations cause an irreversible change in the mode of pigmentation, thus preventing the animal from ever forming normal pigment?   Or, on the other hand, could the decreased levels of tyrosine be critical at an early period of development, thus causing these changes?   The final answers to these problems will only be

obtained by using some of the other methods of which Dr. Russell spoke, more specifically, retrograde analysis and possibly transplantation.

DR. BURDETTE: Do you find very often that the same sort of pathway is present
clinically in patients as in the murine condition?

DR. RUSSELL: In the mouse there is a similar triad of pleiotropic effects[880, 1097,
1155, 1207] in two distinct hereditary anemias. One is the *W*-series anemia, the other
the *Sl*-series anemia. We know the two anemias are not identical since they are
produced by completely nonallelic genes. However, it does seem to me quite probable
that these two genes may very well affect different steps in the same synthetic process.
It is possible that there are also human anemias of maturation arrest very similar to
these conditions in the mouse. These humans might also fail to respond to erythropoietin. One reason for thinking this is that there are many essential anemias in
which affected individuals excrete erythropoietin; and, if they are excreting large
amounts of this substance, it is probable that they cannot respond to it, even though
they are producing it. If histocompatibility could be sufficiently controlled in man,
there might be some purpose to therapy of human anemia by implantation of bloodforming tissue. However, one would have to know a great deal more about human
histocompatibility than is now known before one could possibly determine if this
method would be useful. The answer to Dr. Burdette's question will be more apparent
when more information concerning human histocompatibility is forthcoming.

DR. WRIGHT: The ordinary albino guinea pig has no pigment at birth but later
develops considerable pigment on feet, ears, nose, and even back. This is definitely a
temperature effect, as Dr. George Wolff and others[1401, 1437] showed; there are two
other alleles, c^r and c^d, both of which lead to increased intensity after birth. This
darkening of the c^d and the c^r genotypes is definitely a temperature effect again as Dr.
Wolff has demonstrated. Thus there are three alleles among the five in a guinea pig
that presumably have thermolabile products. They are low in pigment-producing
capacity up to birth. They immediately begin to produce much more afterwards
which can be prevented by high temperature. Most of the other changes in color are
independent of temperature. Sootiness in yellow guinea pigs, however, and also
yellow in the rabbit, as demonstrated many years ago by Walter Shultz,[1176] are
apparently dependent on a temperature effect. The thermolability in these cases is
entirely independent of that of c genotypes.

DR. PILGRIM: Have other heat-labile tyrosinases been found in other Himalayan
albinos or in other species?

DR. RUSSELL: I do not know of any extracted tyrosinase experiment in another
mammal. Of course there is thermostable and thermolabile tyrosinase in *Neurospora*
which is the inspiration for this investigation in the mouse.

DR. GINSBURG: Nachtsheim found that the blue-eyed white rabbit is highly
susceptible to seizures, both sound-induced and spontaneous.[929] However, in a
survey of the races, including various pigment types in Dr. Paul Sawin's very representative rabbit colony, susceptibility to these types of seizures has been found in almost all

of his races of rabbits, including various pigment types. Genetic susceptibility to seizures in the house mouse has been found in a number of pigment types, including black animals.[1343] Inhibition has not been studied very much. However, in the epidermal cells of the skin of the guinea pig there is an easily demonstrable sulfhydril inhibitor which does not correlate quantitatively with differences in pigmentation, but which does definitely inhibit the formation of pigment.[433] This has also been verified by Dr. Rothman in human skin.[1078] I do not know whether it has ever been found for the mouse. It may relate to Dr. Coleman's point on inhibition. I do not think this has ever been studied in relation to genes except in the guinea pig.

DR. COLEMAN: I did not mean to imply that susceptibility to audiogenic seizures and defects in pigmentation are always related. There are many other possibilities here. There is one mutant I have worked on at the Jackson Laboratory which does not have the *dilute* gene, that is, it has the normal intense pigmentation (*DD*) but has a lowered phenylalanine-hydroxylase activity which is about the same level as that found in the DBA. This animal also goes into seizures of an identical pattern to the DBA, suggesting again that the phenylalanine hydroxylase level is important in the induction or seizures in these animals. Although no pigment defect is observed, this also suggests that more than one gene is involved in the control of phenylalanine hydroxylase.

DR. GINSBURG: I did not disagree with the observations on phenylalanine hydroxylase activity, which are extremely interesting and need additional investigation. There did seem to be an impression that this activity was correlated with the pigmentation, and I wish to point out that seizures occur in mice and rabbits of various pigment types and that inhibitory mechanisms of relevance to the problem of pigment formation are known but neglected.

DR. HERZENBERG: Has anyone in this audience any information on the following question? It has been suggested that more genes of physiologic importance may be found in the mouse by applying sensitivity of various inbred strains to drugs or loading them with sugars and seeing what the excretion levels of these sugars might be. Has anyone any information about whether this has been done?

DR. RUSSELL: Water is a fairly good drug, and there certainly are fine examples of fairly simple, genetic differences in reactions to water which have been found recently. There are a number of kinds of polydipsic mice;[267] the imbibition of alcohol varies between strains.[820]

The only thing I would like to say in closing is to express the hope that I have been able to give, through these examples, an idea of how one may go about studying the action in particular genes in mice, why it is a complicated process, and how I believe that differentiation is, in a sense, a tool as well as a problem.

F. Clarke Fraser, Ph.D., M.D.

METHODOLOGY *of* EXPERIMENTAL MAMMALIAN TERATOLOGY†‡

Experimental mammalian teratology deals with anatomical defects arising from errors of development caused by exposing pregnant mammals, or their embryos, to environmental noxious agents, or teratogens. (Mutant genes causing developmental errors may also be considered teratogens, but the analysis of their effects is usually categorized as developmental genetics and is beyond the scope of this paper.)

Why should teratology be included in a volume on mammalian genetics? Because an environmental teratogen must act on a developing organism, and the response that it produces will depend on, among other things, the genotype of the organism. Thus, in studying the effects of an environmental agent on development, it is important to take the genetic constitution into account, and much can be learned about development and its errors by studying the interaction of an environmental agent with a variety of genotypes.

TYPES OF ANIMAL USED FOR TERATOLOGIC STUDIES

A wide variety of mammalian species have been used for teratologic studies. These include rats, mice, rabbits, guinea pigs, hamsters, pigs, and dogs.[681] There is no apparent reason why other species should not also respond to teratogens. Choice of a suitable species for experimentation depends on many factors, not the least of

† Dedicated to Professor L. C. Dunn with admiration and affection.
‡ Financial support received from the National Research Council of Canada and The National Foundation is gratefully acknowledged by the author.

which is the experimenter's purpose. If one intends to analyze the effects of the teratogen in relation to a variety of other factors such as developmental stage, dosage, and maternal physiology, relatively large numbers of embryos will be required for the appropriate statistical analyses. One of the small laboratory rodents may therefore be chosen, since they are relatively economical to house and maintain, and have relatively large litters. Of course, it is necessary to choose a species that responds in the desired way to the teratogen (one cannot study cortisone-induced cleft palate in the rat, for instance). It may also be useful to study the interaction of genotype with teratogen, in which case mice have the advantages of being genetically well known and available in a variety of highly inbred lines.

The use of inbred lines has many advantages. For embryologic studies it may be worth while to search for a strain in which the teratogen to be used produces a high frequency of the malformation to be studied. It also removes the variable of genetic heterogeneity. Some workers[1118] have made use of F_1 hybrids between different inbred strains, thus obtaining genetically uniform embryos with the additional advantages of hybrid vigor. The study of differences between inbred strains and crosses between them in response to a teratogen is a useful way of analyzing the interaction of genotype and environment in producing malformations. The existence of strains in which a particular malformation occurs spontaneously in some animals provides an opportunity to study the interaction of genotype and intra-uterine variables that determine why some animals in a litter are affected and others not.

Workers who are mainly interested in the pathogenesis of malformations may choose a species the embryology of which is well known, or in which the embryos are relatively large, permitting gross anatomic as well as microscopic study. Still others may wish to produce malformations in order to study their diagnosis or treatment, in which case a large species such as dog, sheep, or monkey would be advantageous. It might be useful, for instance, to produce puppies with congenital malformations of the heart and great vessels, to be used for improving the techniques of diagnosis and surgery of such conditions, but this sort of application of experimental teratology has so far not been exploited.

Fowls, amphibia, insects, and other groups have their own special advantages for teratologic work, not the least of which is the fact that embryos can be treated directly, without the complication of a uterine barrier, but consideration of these lies beyond the scope of this article.

CONTROLS

Although it should not be necessary to discuss the necessity of proper controls for any experiment, some of the variables that can confuse the issue in teratologic studies are sometimes overlooked and will therefore be mentioned here.

First, there is the fact that spontaneous malformations do occur in laboratory animals, and these must be distinguished from those resulting from the teratogen being used. Second, there is the possibility that some factor in the experimental procedure other than the teratogen being studied may itself be teratogenic. For instance, some teratogenic procedures may be so stressful to the mother that she will stop eating during the treatment, and maternal fasting is itself teratogenic in some situations. Third, there are a number of variables such as maternal weight,[677] diet,[1190] and season of the year[629] that have been related to the frequency of malformations, and these must be taken into account when comparing the results of different series. Finally, there are genetic differences between strains and substrains that demand caution in comparing results from different laboratories, or even from the same laboratory at different times.

AGENTS USED FOR TERATOLOGIC STUDIES

There are now on record a great number of agents with teratogenic properties. Those that can be classed as metabolic (which includes nearly all of them except irradiation) have been discussed recently by Kalter and Warkany[681] in an exhaustive and useful review. It is difficult to classify teratogens in any logical way, since the modes of action of most of them are not accurately known; for instance, agents usually considered as physical may act through biochemical pathways and *vice versa*. Wilson[1392] has grouped teratogens acting in mammals under the headings of physical agents (X rays, hypothermia, hypoxia, elevated CO_2, puncture of amniotic sac); maternal nutritional deficiencies (lack of vitamin A, riboflavin, folic acid, pantothenic acid, vitamin B12, thiamine, and vitamin D, and fasting); growth inhibitors and specific antagonists (nitrogen mustard, thiadiazole, triazines, other alkylating agents, urethan, azaserine, 6-aminonicotinamide, 8-azaguanine, thioguanine, 6-mercaptopurine, 2-6-diaminopurine, 6-chloropurine), infectious agents (influenza-A virus, attenuated hog-cholera virus, Newcastle virus); hormone excesses and deficiencies (androgens, estrogens, insulin, cortisone, vasopressin, adrenalin, alloxan diabetes); and miscellaneous drugs and chemicals (trypan blue, excess of vitamin A, antibiotics, chelating agents, phenylmercuric acetate, nicotine, salicylates).

Choice of a teratogen depends on the purpose of the investigator. One may wish to know, for instance, whether a particular drug, or other environmental agent, may be teratogenic in humans. In this case the choice of teratogen is decided by the experimenter's question. If the investigator wishes to study the pathogenesis of a particular malformation, he will wish to choose a teratogen which, in the appropriate organism, will produce a high frequency of the malformation to be studied. In this way he can be relatively sure that the embryo examined during early development would have been malformed at birth. Alternatively, he may prefer to study the arrays of malformations produced by exposure to the teratogen at various embryonic stages, on the assumption that the types of malformations produced by teratogens with various metabolic effects may yield information about the nature of the developmental mechanisms involved.

In any case, it will be useful to know the precise stage at which the embryo is exposed to the teratogen. For this reason an agent such as irradiation, that reaches the embryo immediately, may be chosen in preference to one such as cortisone, for which an indeterminate delay occurs between the time the agent is applied to the mother and the time it, or its metabolic consequences, reaches the embryo in effective quantities. Treatment with specific metabolic inhibitors (such as 6-aminonicotinamide) followed shortly afterward by a protective substance (in this case, nicotinamide) may also be used to achieve a short teratogenic episode, precisely timed.[1008]

The use of analogs as teratogens has the further advantage that their metabolic effects can be inferred from their chemical nature, and study of the embryonic effects of specific inhibitions, at particular stages of development, may provide information about the biochemical properties of the embryo. Further factors influencing the choice of teratogens will become apparent later in this discussion.

DEVELOPMENTAL STAGE AT WHICH THE TERATOGEN IS USED

In studying the effects of an agent on development, it is obviously useful to know the gestational stage at which the teratogen is applied. This is usually done by applying the teratogen at a given time after mating, but sometimes by counting back from parturition, assuming gestation length is known from previous observations. The latter is an unreliable method. The time of mating can be established, within limits, by placing the female with a male for a known period of time and observing her at the end of this period for signs of insemination, such as a vaginal plug in mice or rats, or the presence of sperm in the vagina. The frequency of successful matings can be increased by exposing the female to the male when she is in the appropriate stage of estrus, as established by vaginal smear or other signs, depending on the species. It may also be useful (in mice, at least) to maintain the females in a room with a regular light-dark cycle, so that their estrus cycles become synchronized, and they can be exposed to the males at the appropriate time in the cycle.

No matter how precisely the time of mating is known, there is no assurance that all embryos are exposed at the same developmental stage, since within one litter there may be variations in time of ovulation, fertilization, implantation, and post-implantation development, so that no matter how accurately the time from insemination is measured, there is often quite marked variability in developmental stage from one littermate to another. Variation in developmental rate between litters and between strains (or even sublines within strains) must also be taken into account, by using adequate numbers of animals and appropriate controls.

Some confusion exists in the literature as to the terminology of timing gestational stages. Some authors refer to the day on which signs of mating are observed as the first day of pregnancy, or day 1. Others refer to it as day 0. Care should be taken to specify which interpretation is meant in any particular case. The latter system is preferable.

It may be well to point out here the invalidity of the popular notion that a given morphogenetic process is most sensitive to alteration by a teratogen at the time when the process occurs. This is not necessarily true. Each teratogen usually has a period of maximum efficiency in producing a particular malformation (the critical period), but this period may differ from one teratogen to another and often precedes the stage when the morphogenetic process involved takes place. X irradiation, for instance, produces cleft palate in mice with maximum frequency when given about 11 days after insemination,[1118] whereas 6-aminonicotinamide is most effective on day 13, about a day before the palate actually closes.[448] Presumably, normal palatal closure depends on the normal sequence and interaction of a number of previous developmental events, and interference with any one of them will interfere with closure of the palate. Since X irradiation and 6-aminonicotinamide have different critical periods, they presumably interfere at different points in the web of interacting processes leading to closure of the palate.

DOSE

In general, the teratogenic dose of an agent is somewhat below that which causes resorption or abortion of the litter. However, there are quite wide variations in the range between the lethal and the teratogenic[923] doses for different teratogens.

TERATOGENS AND THE ANALYSIS OF MALDEVELOPMENT

Admittedly, some teratological experiments are begun simply because an agent is at hand and an experimenter is curious about what it might do. Others, of course, are conceived to answer specific questions about the nature of malformations and, by inference, about normal development. A number of such approaches will be discussed in the following pages.

1. *Is a given agent capable of producing malformations in humans?*—Since it is usually extremely difficult to demonstrate the existence of an environmental teratogen in human beings, a reasonable first approach to the problem is to see whether the suspected agent is teratogenic in experimental mammals. The work that gave the first great impetus to the field of experimental mammalian teratology—the demonstration by Warkany[1357] and his colleagues that specific maternal nutritional deficiencies caused malformations in embryonic rats—was inspired by the desire to know whether maternal nutritional deficiencies might cause human malformations.

This approach will not, of course, give a definitive answer to the question that (as Warkany has repeatedly emphasized) must rest ultimately upon observations on human beings. If an agent is found to be teratogenic in experimental mammals, this raises the possibility that it may be teratogenic in human beings, but so far very few experimental teratogens have been shown to cause human malformations. On the other hand, if an agent is found not to be teratogenic in experimental animals, even after having been tested in a wide variety of species and strains, the existence of marked species differences in response to teratogens prohibits the conclusion that it is non-teratogenic in human beings.

2. *What are the pathogenetic mechanisms underlying malformations?*—It is often impossible to infer correctly, by observing a malformation at birth, how it got that way. However, if a malformation can be produced with a high frequency by a teratogenic agent, treated embryos at various stages can be compared with untreated controls, and the sequence of events from the first deviation from normality to the full-blown malformation observed at birth can be observed. This approach has been used to elucidate the pathogenesis of a number of malformations. For instance, Monie *et al.*[889] have demonstrated how a variety of urinary tract anomalies produced in rats by a maternal deficiency in pteroylglutamic acid could be explained by retarded development of the urinary tract and vertebral column. Giroud and Martinet[436] have traced the pathogenesis of anencephaly produced in rats, by maternal treatment with large doses of vitamin A, from failure of the encephalic tube to close, through formation of a brain that is in effect turned inside out, and its subsequent degeneration. Warkany *et al.*[1358] have analyzed, by this approach, the origin of myelomeningocoele produced by maternal treatment with trypan blue in rats. Failure of the neural tube to close is followed by overgrowth and eversion of nervous tissue and later by degeneration of the neural plate and formation of a fluid-filled space between the pia of the neural plate and the dura covering the vertebrae. Thus, the myelomeningocoele is essentially a cyst in the subarachnoid space in this case. Many other types of malformations that can be produced at will by teratogenic procedures deserve to have their origins worked out by this approach.

It must not be concluded, however, that the pathogenesis demonstrated for a particular type of malformation produced by one teratogen at one stage of development is the same for all malformations of this type. It has been demonstrated, for instance, that cortisone-induced cleft palate in mice results from a delay in movement of the palatine shelves from their original position on either side of the tongue to their final position above the tongue, and that this delay seems to be due to interference with the mechanism within the shelves that provides the force necessary for this movement.[1354] Cleft palate following amniotic puncture, on the other hand, is probably due to increased resistance of the intervening tongue.[1330] Diminished shelf width, or increased head width, are other possible causes for the failure of the shelves to meet in the midline at the proper time.[405] It is often impossible to tell, from the appearance at birth, which of these mechanisms caused the cleft in a particular case. Here again, extrapolation of experimental findings to human beings should be supported by observations on human embryos, but the experimental observations are useful in illustrating possible pathogenetic mechanisms to be looked for in human malformations.

3. *What are the biochemical properties of an embryo at various stages of development?*—The teratologic approach to this question was first formulated by Warkany and his group, who put female rats on a diet deficient in vitamin A. They showed that the offspring

of these rats had characteristic patterns of malformations, and that the array of malformations could be modified greatly by adding vitamin A to the diet at certain stages of pregnancy.   For instance, when vitamin A was added to the diet before the thirteenth day of gestation, virtually no ocular malformations occurred; but when it was added after the fifteenth day, there was no reduction in the number of ocular defects.   Malformations of the aortic arch, on the other hand, were prevented by addition of vitamin A before the twelfth day, but cardiac defects were not prevented when the vitamin was added as early as the tenth day.[1394]   It could be inferred, therefore, that certain processes concerned with organogenesis required more vitamin A than others (since some organs develop normally in embryos from unsupplemented animals) and that the requirements vary from one developmental stage to another.

The synthesis of analogs to a number of biochemical compounds made it possible to refine this approach, by using an analog of a particular nutritional element to produce a temporary inhibition of the activity of that element.   There is some question as to whether substitution of an inactive analog (which might conceivably have toxic effects in its own right) is strictly analogous to a sudden deficiency of the compound concerned, but the technique does provide a useful analytical tool.   Nelson and her group, for instance, have used a maternal diet deficient in pteroylglutamic acid (PGA) supplemented with a PGA analog and succinyl-sulfathiazole (to inhibit PGA synthesis by the intestinal flora) to produce malformations in rats.[940]   A 36-hour period on the diet followed by high levels of vitamin supplementation to terminate the deficiency was teratogenic, but a 24-hour period was not.   The type and frequency of malformations produced by this transitory deficiency varied with the time of instituting the deficiency and with its duration and severity, providing an opportunity to study the requirements for folic acid at different stages of embryogenesis.   In mice, folic acid analogs have produced malformations in the offspring of treated, pregnant females without the necessity of using a deficient diet.[1327, 1332]

A further refinement to this approach to the analysis of the biochemistry of morphogenesis made use of the nicotinamide antagonist 6-aminonicotinamide (hereafter called 6-AN) which is teratogenic in mice even when a corrective dose of nicotinamide is given as little as two hours after the analog is injected.[1008]   The compound forms a diphosphopyridine nucleotide (DPN) analog that is inactive in some DPN-dependent enzymatic reactions,[227] and it is reasonable to suppose that this is the basis of its teratogenicity.   By varying the dose of 6-AN and the dose of the nicotinamide supplement, it should be possible to obtain useful information about the relative requirements of various organogenetic processes for nicotinamide.   For instance, when a standard dose of 6-AN is given 9.5 days after insemination and a standard dose of nicotinamide is given simultaneously, no malformations occur in the offspring.   When the same dose of nicotinamide is given two hours after the analog many of the offspring have cleft lip.   When twice the amount of nicotinamide is given, again two hours after the analog, the frequency of cleft lip is reduced.   This suggests that a single dose of nicotinamide is not sufficient to correct the metabolic block produced by the analog;

if it were, doubling the dose would not reduce the frequency of defect. At other stages, and for other types of malformation, a double dose of nicotinamide two hours after the standard dose of 6-AN does not produce any fewer malformations than the single dose, showing that in such a case a single dose of nicotinamide is enough to correct the inactivation produced by the analog.[1008] Thus the nicotinamide, or DPN requirements of the maternal-fetal system, appear to vary from stage to stage of embryogenesis, and this approach provides an opportunity to study these variations in a roughly quantitative way.

Further information might be obtained by comparing the arrays of malformations produced by several DPN inhibitors. It is known that 6-AN forms an analog of DPN which is inactive in a variety of DPN-dependent enzymatic reactions but not in all of them.[277] Presumably malformations caused by treatment with 6-AN result from inhibition of one or more of the DPN-dependent reactions, but it is not possible to say which reaction, when inhibited, leads to which malformation. If other DPN analogs that blocked other DPN-dependent reactions were used, presumably a different array of malformations would result. If a given reaction were blocked by both analogs, the same malformation should result from treatment with either one. Thus by using a battery of analogs, and seeing which malformations were produced in common by which analogs, it might be possible to infer which enzymatic system was blocked in order to produce the defect. As the number of analogs increases, and their biochemical effects are better understood, they should provide excellent tools for studying the biochemistry of normal and abnormal morphogenesis.

The immunologic aspects of development, now being energetically studied by experimental embryologists,[316] provide another promising approach to the biochemistry of morphogenesis that has so far been little exploited by mammalian teratologists. Gluecksohn-Waelsch[441] showed that female mice immunized with extracts of brain produced offspring with an increased frequency of central-nervous-system malformations, whereas extracts of heart were ineffective. Wood[1402] confirmed the report of Guyer and Smith[513] that lenticular antiserum injected into pregnant rabbits produced defects in the eyes of the embryos (though the claim of Guyer and Smith that these changes was heritable has not been confirmed). Further studies of this type are needed, particularly in view of the preliminary report by Blizzard et al.[110] that congenital absence of the thyroid in human beings may result from maternal antithyroid antibodies. It would also be interesting to investigate further the suggestion that excessive amounts of specific proteins administered to the embryo would cause specific inhibitions or stimulations of development in mammals as they do in some other organisms.

4. *What are the biochemical effects of teratogens on development?*—As previously suggested, it may be possible to infer, from the biochemical nature of some teratogens, the probable metabolic pathways on which they act to produce their developmental effects, although even with specific analogs the picture may not be entirely clear. With other teratogens, such as cortisone, for instance, the biochemical effects may be so varied and widespread

that it is impossible to foretell which effect is related to the production of a malformation.

A useful approach to this question involves the use of combinations of teratogens. The assumption is made that two teratogens which in combination give the same frequency of a malformation as that produced by the one with the higher effect given singly (a nonadditive effect) act on the same metabolic pathway to produce the malformation, whereas when the combined teratogens produce a frequency of defects which is the sum of the frequencies produced by each one singly, different pathways are involved. This is well demonstrated by the work of Runner and co-workers on the embryonic effects of maternal fasting in mice. Malformations of the vertebrae and ribs were prevented by giving small quantities of glucose or casein during the fasting period, and (to a lesser extent) certain amino acids and acetoacetate. This suggested that protection was provided by supplying substrate for the citric acid cycle. Maternal treatment with insulin, iodoacetate, or a PGA antagonist produced a similar array of malformations, and could reasonably be postulated to interfere with the citric-acid cycle.[1088] Further evidence came from the results of fasting combined with one of a variety of other teratogens.[1090]

When pregnant females were fasted and exposed to hypoxia, the frequency of defects in the offspring was approximately the sum of the frequencies produced by each teratogen separately. This is reasonable on the assumption that the two teratogens decrease two substrates, involving separate metabolic pathways. In Runner's hypothetical scheme, hypoxia affects both the ectodermal and mesodermal components of the inductive system leading to differentiation of the axial skeletal system. Fasting in addition reduces the glucose substrate necessary for the ectodermal component, so the combined treatment should (and does) give a higher frequency of defective offspring than either one alone (fasting 24 per cent, hypoxia 47 per cent, combined 75 per cent). On the other hand, iodoacetate combined with fasting gives almost the same frequency of defects (66 per cent) as iodoacetate alone (62 per cent), as would be expected on the basis that if glycolysis is blocked by the iodoacetate, reducing the glucose substrate by fasting will make no difference, since it involves the same (blocked) pathway.

Synergistic effects (for fasting and cortisone on cleft palate frequency)[678, 878] are more difficult to interpret, especially if the malformation concerned falls into the class of quasicontinuous variations,[501] as postulated for cleft palate.[405] Here, two agents which altered the threshold in an additive manner could produce an apparently synergistic effect as measured by the frequency of induced malformations, by moving the threshold in from the flat tail of the curve to the more steeply sloping portion.

5. *What are the relations of mother and fetus with respect to teratogens?*—The fact that a teratogen must pass through the mother to act on the fetus in most cases is a problem that is not always given sufficient attention. In fact, for many teratogens, it is not clear whether the effects on embryonic development result from direct action of the teratogen on the embryo or from secondary metabolic effects of the teratogen on the mother. Giroud et al.[435] have reported that maternal riboflavin deficiency in rats is accompanied by a decreased concentration of riboflavin on the maternal liver and an even greater

decrease in the embryonic liver, and that a maternal excess of vitamin A results in some increase of vitamin A in the embryonic liver.[434]   The teratogenic effects of radiation are not due to secondary effects of maternal whole-body irradiation.[1391]   Trypan blue has never been observed in the tissues of the embryo, but has been demonstrated in the yolk-sac fluid of rabbit embryos on the seventh, eighth, and ninth days of pregnancy.[357] Wilson *et al.*[1393] have presented evidence to suggest that the yolk-sac epithelium in the rat protects the embryo after the eighth day of gestation by absorbing and immobilizing the dye.   Variation in the protective efficiency of the uterine barrier from stage to stage of gestation is an interesting aspect of teratology that needs further investigation.

The physiology of the mother undoubtedly influences the effect of the teratogen in some cases.   For instance, the frequency of cortisone-induced cleft palate in mice decreases with increasing maternal weight.[677]   This not only means that this variable must be taken into account when designing experimental controls, but raises the question of its biochemical significance.   What aspect of the mother's metabolism is involved, and does it influence the mother's reaction to the teratogen, or have some effect on the embryonic developmental pattern that influences the embryo's response? Or both?

Reciprocal cross differences in response to a teratogen suggest the importance of the maternal environment in determining the embryo's reaction to the teratogen. For instance, the frequency of cortisone-induced cleft palate [404] was much higher in F_1 hybrids from a cross between A/Jax female mice and C57BL males, than from the reciprocal cross (the A/Jax inbred strain being more susceptible than the C57BL). Backcrosses ruled out a permanent cytoplasmically inherited factor as the source of the difference.[679]   Differences in maternal metabolism or placental transmission of the drug were possible explanations, although the latter seems unlikely if the placenta is constituted from fetal tissues, since the two types of hybrids are genetically similar. Another explanation was suggested by studies on the developmental pattern of the embryos.   In crosses that produce embryos with a high frequency of cleft palate following maternal treatment with cortisone, the palate, in untreated animals, tends to close later than it does in crosses that are more resistant to the cortisone treatment.[1329] The palate closes later in A/Jax than in C57BL/6 embryos, and later in A/Jax ♀ × C57BL ♂ than in C57BL/6 ♀ × A/Jax ♂ embryos.   Thus there is no need to invoke a difference in the way the mother handles the cortisone to account for the reciprocal cross difference in cleft palate frequency, although this has not been entirely ruled out. The difference in response to the teratogen can be accounted for simply as the result of a difference in developmental pattern resulting from the interaction of maternal and fetal genotype.

The techniques for transfer of ovum and embryo would be an elegant way to analyze further the interactions of embryonic genotype, maternal, and cytoplasmic factors in determining the embryo's response to environmental teratogens;[485, 826] but such an approach so far does not seem to have been used for this purpose.

Extrinsic variables influencing the probability that a given embryo will have a malformation may be classified as individual factors, affecting littermates differently (local variations in blood supply, implantation, and so forth) and background factors, affecting all embryos within a litter (maternal physiology, variations in size of uterine artery, and so forth). Within inbred strains, it may be possible to detect such background factors by comparing the frequencies with which defective offspring occur within litters with the distribution expected if the probability of being malformed is constant from embryo to embryo. If there is a tendency for clustering, that is, for an excess of litters with many defective offspring and of litters with none, a maternal variable is suggested. A tendency for some mothers to have more defective offspring than others (in the absence of genetic segregation) suggests that the maternal factor is a relatively permanent constitutional factor, whereas fluctuations in the maternal environment may be implicated if the variance is greater between litters than between mothers.

6. *What are the intra-uterine variables related to the production of malformations?*—One of the most difficult problems to attack experimentally in teratology is the fact that, when a teratogen is applied to a genetically homogeneous litter in the same uterus, some embryos may have no malformations and among the malformed ones there may be considerable variation in nature and severity of the defect from one littermate to another. In some cases this can be accounted for on the basis of quasicontinuous variation—a continuous distribution separated by some developmental threshold into discontinuous parts.[501] In the case of cleft palate, for instance, the time of closure of the embryonic palate can be considered a continuously distributed variable. If movement of the shelves from their vertical position on either side of the tongue to the horizontal position above the tongue is sufficiently delayed, the width of the head will have increased so much that the shelves cannot meet in the midline when they do reach their final position. Thus the point at which they can no longer meet separates the continuous distribution of times of palatal closure into a discontinuous one—cleft, or not cleft.[405] However, this interpretation only pushes the problem back one step. What are the factors that determine where on the continuous distribution an individual embryo will lie? Why does shelf movement occur so late in one embryo that the teratogenic procedure pushes it over the threshold, while in its genetically similar littermate shelf movement occurs early enough to keep it on the side of normality? Is some other variable perhaps involved, such as variation in shelf width? So far, no relevant factors have been identified. In addition, there may be variations within a litter in the amount of teratogen that actually reaches the embryo, and this aspect of the question remains almost entirely unexplored.

The problem of intra-uterine variability is perhaps even more important in the case of the spontaneous malformation that appears with a low but characteristic frequency in certain inbred strains, for instance, the cleft lips that occur in about 10 per cent of newborns in the A/Jax strain, or the microphthalmia that occurs in some sublines of the C57BL strain. Here the situation may be more analogous to the situation in humans; a particular grouping of genetic factors increases the probability that an embryo will have a particular malformation, but whether or not a given embryo actually has the defect depends on intra-uterine variables that so far remain an almost complete mystery.   It has been suggested that one of the variables might be segregation of genetic modifiers that may persist in spite of the intense inbreeding that occurs in the development and maintenance of inbred lines.   However, the microphthalmia that occurs in the C57BL/6/Fr subline of mice is just as frequent in the offspring of un-affected mothers as it is in the offspring of their microphthalmic sisters, which argues strongly against genetic segregation in this case (Fraser, unpublished data).

Trasler[1328] has reported that the spontaneous cleft lip that occurs in the A/Jax strain is more likely to appear in the embryo adjacent to the ovary than in embryos at other uterine sites.   This directs attention to the nature of the relevant factors that make the juxta-ovarian site different from other sites, but so far no such factors have been identified.   It would seem that the genetic constitution of an A/Jax embryo (and mother, or both) makes its lip-closing mechanism susceptible to disturbance by rather subtle variations in the uterine environment.   What are the variations and what is there about an A/Jax labial primordium that makes it so sensitive to environ-mental disturbance?   Answers to these questions will eventually require coming to grips with the problem of intralitter variability in biochemical terms.   Perhaps some light could be thrown on the subject by observing the effects of agents with known pharmacologic properties, particularly antimetabolites, on the frequencies of defects occurring spontaneously in inbred lines.   This approach has been well demonstrated by Landauer[748] in chickens.

7. *Interaction of genes and teratogens.*—It should be obvious from the foregoing discussion that the genetic constitutions of the embryo and its mother are influential in determining the response of the embryo to a teratogen.   This fact, which will be no surprise to geneticists, is significant for a number of reasons.   For one thing, it may account for the differences in nature and frequency of defects reported by different workers using the same teratogen but different strains or substrains of animals.   It also emphasizes the fact that even when an environmental teratogen is clearly implicated as a cause of congenital malformations, the genetic constitution may determine whether an embryo exposed to the teratogen will be malformed.   This, for instance, is a possible explanation for the fact that some offspring of mothers having rubella in the early weeks of pregnancy do not have the characteristic malformations shown by others.

As previously mentioned, the study of inbred strains that differ widely in the frequency of a malformation produced by a teratogen, and of crosses between them, can be useful in clarifying the intricate interaction of factors that determine whether or not a given embryo is malformed.   Our studies of cortisone-induced cleft palate in the A/Jax and C57BL strains, F_1 hybrids and backcrosses to the A/Jax strain demonstrated the importance of both fetal and maternal genotype in determining the embryonic response to the teratogen.   Studies of the embryology of cortisone-induced cleft palate in strains with both a high and a low frequency of induced clefts demonstrated, far better than a study of either strain separately would have done, that delay in movement of the palatine shelves from the vertical to the horizontal plane was the cause of the clefts observed at birth.[1354] Furthermore, comparison of the normal time of palatal closure in these strains and crosses showed a correspondence between the normal time of palatal closure and the frequency of induced cleft palate, thus demonstrating how a genetically determined difference in normal developmental pattern was related to the observed differences between strains in response to the teratogen.

Differences between inbred strains in response to teratogens can also be useful in elucidating the metabolic pathways involved.   For instance, the frequency of cleft palate produced by a transitory inactivation of nicotinamide is higher in the A/Jax strain than in the C57BL/6 strain, and in the A/Jax × C57BL/6 than in the C57BL/6 × A/Jax hybrids,[448] just as it is when cortisone is used as a teratogen.   On the other hand, galactoflavin produces more cleft palates in the C57BL/6 strain than in the A/Jax strain.[680]   This shows that the susceptibility of the A/Jax palate to cortisone and 6-AN is not just a nonspecific instability to any environmental insult.   It also suggests that galactoflavin interferes with closure of the palate at a different metabolic point than cortisone and 6-AN, and that perhaps cortisone and 6-AN act on the same pathway. Since 6-AN is known to interfere with DPN synthesis, this implies that cortisone's teratogenicity may reside in its effect on DPN synthesis, or on some related process. Of course we would not want to draw conclusions from these few comparisons, but the approach appears useful.   By observing the effects of a variety of teratogens on a panel of strains and seeing which ones produce the same patterns of strain differences in frequency of malformation, it should be possible to deduce which ones are affecting the same metabolic pathways, and get some idea of which pathways are involved.

SUMMARY

This review of methodology in experimental mammalian teratology has considered the following aspects: choice of experimental animal, choice of teratogen, gestational stage of treatment, dose of teratogen, and contributions of teratologic studies to a better understanding of the causes of malformations.

Possible ways in which teratological methods can make such contributions include:

1. Experimental screening for teratogenicity of agents suspected of causing malformations in human beings.

2. Production of specific malformations with known and controllable frequencies, permitting embryologic studies of the pathogenetic processes leading to the defects observed at birth, and testing of diagnostic and therapeutic procedures on malformed animals.

3. The use of antimetabolites and immunologic preparations to study the biochemistry of morphogenetic processes and the biochemical pathways through which teratogens act.

4. The uses of reciprocal crosses and other methods to investigate the nature of maternal-fetal interactions influencing developmental patterns and their response to teratogens.

5. The use of inbred strains to study the intra-uterine variables that influence the probability that a given embryo, predisposed by its genotype or a teratogen to be malformed, will actually have the malformation.

6. Study of the genetic differences in response to teratogens to clarify the interactions of genes and extrinsic environmental agents in determining normal and abnormal developmental patterns.

Walter E. Heston, Ph.D.

GENETICS *of* NEOPLASIA

The ultimate aim in the study of the genetics of neoplasia in laboratory mammals is the control of cancer in man. In addition, however, such studies contribute much to our general basic knowledge of physiologic genetics. The study of no character, for example, has contributed more information on genes and their relationship to a multitude of nongenetic factors in their paths of action than has the study of mammary tumors in the mouse. It has been demonstrated that the etiology of this neoplasm can include not only the inherited genetic factors but also a virus, exogenous and endogenous hormonal factors, many physical and chemical carcinogens, nutrition, and even temperature, all of which are interrelated in a unified picture of physiologic genetics. It is likely that the etiologic picture of many of the other types of neoplasms will eventually be shown to be equally complex and that all will include genic action.

Anyone interested in undertaking a study of the genetics of neoplasia might well start with a review and digest of Dr. Sewall Wright's papers on otocephaly and poly-dactyly in the guinea pig published nearly 30 years ago.[1414, 1433, 1451] In these studies he established the basic concepts of the inheritance of multiple-factor characters with alternative expression. He established that both genetic and nongenetic factors form a continuous underlying variation, but because of certain thresholds the phenotypic expression becomes discontinuous. The character appears when the combined action of the genetic and nongenetic factors surpasses the threshold. Hence, Grüneberg[501] has more recently referred to this variation as being quasicontinuous, and Falconer[335] has called such characters threshold characters. At least most neoplasms, like oto-cephaly and polydactyly, are inherited as multiple-factor characters with alternative expression, and this early work of Wright provides a firm basis for understanding their

247

inheritance and the relationship of the genetic and nongenetic factors involved in their etiology.

Methods have been worked out for analysis of heritability of threshold characters by Robertson and Lerner,[1062] Dempster and Lerner,[245] and others, and data on tumors as all-or-none characters with continuous underlying genotype could be collected and subjected to such analyses. Furthermore, in certain tumors where a recognizable precancerous lesion occurs, data on the three classes—tumor, precancerous lesion, and nontumor—could be analyzed for number of genes involved much as Green[482] and McLaren and Michie[825] have analyzed the inheritance of number of vertebrae in the mouse.

The usual approach, however, has been by transposing the phenotypic expression of the character from alternative expression to continuous variation. One way of doing this is by measuring susceptibility in terms of latent period of the neoplasm as Anderson[23] did in order to ascertain heritability of bovine ocular carcinoma. The more susceptible the animal, the earlier the tumor appears. Another way is to measure susceptibility by counting the multiple tumors that occur. Multiple tumors occur particularly when the tumors are induced. This manner of measuring response quantitatively was particularly advantageous in estimating the number of genes involved in the inheritance of pulmonary tumors in mice.[557] In this work response was also measured on the basis of latent period [557, 559] and the results from both types of measurement were roughly parallel.

More progress has been made in linkage studies of neoplasms than in any other group of threshold characters. Without the cancer genes being individually identified, however, it has not been possible except in certain situations to distinguish between true linkage and a pleiotropic effect of the gene tested.

Probably the greatest advantage that neoplasms have over many of the other threshold characters is the number of nongenetic factors in this underlying continuous variation that can be identified. These may be any of a host of physical and chemical carcinogens, endocrine factors, nutritional factors, and viruses. Much of our knowledge of the physiologic paths of genic action leading to neoplasia is revealed through the analysis of the interrelationship between these endogenous and exogenous nongenetic factors and the genes.

It is with the methodology employed in these facets of the genetic approach to the etiology of cancer that this paper will be concerned. The genetics of transplanted tumors is considered by others elsewhere in this volume.

INBRED STRAINS FOR CANCER RESEARCH

As early as 1909, Dr. C. C. Little foresaw that in order adequately to analyze the inheritance of neoplasms when the presence or the absence of the tumor was not a full indication of the genotype, it would be necessary to develop inbred strains in which the genotype was uniform. That was the year in which he started inbreeding strain DBA mice and selecting for mammary gland tumors.   Since that time many inbred strains of mice have been developed, but most of the early ones were developed for studies on cancer.   Systems of matings are discussed by Drs. Green and Doolittle and will not be mentioned, other than to point out that brother × sister and parent × offspring matings have been used almost exclusively.   Nor will definitions of an inbred strain, or substrain, be given here since these have been published by the Committee on Standardized Nomenclature.[220]

The introduction of these inbred strains constituted probably the greatest advance in all cancer research.   Genetically suitable material for the study of a wide variety of neoplasms is now available.   Table 50 lists these neoplasms in the mouse along with

Table 50

SOME SUGGESTED INBRED STRAINS FOR STUDIES OF NEOPLASMS

Neoplasm	Strains	Incidence of neoplasm	References
Mammary gland tumors	C3H	Almost 100 per cent in ♀♀	220
	A	High in breeding ♀♀, low in virgin ♀♀	220
	DBA	High in breeding ♀♀	220
	RIII	High in breeding ♀♀	220
	BALB/c	Incidence low, but high following introduction of mammary tumor agent	28
	CC57BR CC57W	Do not develop spontaneous mammary tumors; 55 per cent after introduction of mammary-tumor agent	764, 863
Pulmonary tumors	A	90 per cent in mice living to 18 months	557
	SWR	80 per cent in mice living to 18 months	764
	BALB/c	26 per cent in ♀♀, 29 per cent in ♂♂	1073
	BL	26 per cent in mice of all ages	814
Hepatomas	C3H	85 per cent in males 14 months old	572
	C3Hf	72 per cent in males 14 months old	572
	C3He	78 per cent in males 14 months old	572
		91 per cent in breeding ♂♂; 59 per cent in virgin ♀♀; 30 per cent in breeding ♀♀; and 38 per cent in force-bred ♀♀	257
Leukemia and other reticular-cell neoplasms	C58	High leukemia	767
	AKR	High leukemia	767
	C57L	Hodgkins-like lesion, reticular-cell neoplasm Type B, approximately 25 per cent at 18 months	555
Papilloma and carcinoma of skin	HR	Papillomas occurred in all mice both haired and hairless painted with methylcholanthrene and the carcinomatous transformation occurred in most of the animals	258
	I	Most susceptible of 5 strains tested with methylcholanthrene	31

Subcutaneous sarcomas	C3H	Occurred spontaneously in 57 of 1774 C3H and C3Hf ♀♀	308
		Most susceptible of 8 strains tested with carcinogenic hydrocarbons	30
	CBA	High occurrence after sub-cutaneous injection of methylcholanthrene	146
Stomach lesion	I	Occurs in practically all mice of this strain	25
	BRS BRSUNT }	Occurs following injection with methylcholanthrene and spontaneously	1294
Adrenal cortical tumors	CE	High occurrence following castration	1405
	NH	High occurrence of spontaneous adenoma	711
		High occurrence of carcinoma following castration	419
Teratomas, testicular	129	1 per cent, congenital	1278
Interstitial-cell tumors of testis	A	Readily induced with estrogens (see Gardner)	420
	BALB/c	High incidence following treatment with stilbestrol (Males tolerate stilbestrol better than do those of other strains.)	1197
Pituitary adenoma	C57BL	Occurred in almost all mice treated with estrogens	423
Hemangioendotheliomas	HR	19 to 33 per cent in untreated mice 54 to 76 per cent in mice injected with 4-*o*-tolylazo-*o*-toluidine	259
	BALB/c	High occurrence particularly in interscapular fat pad and lung in mice treated with *o*-aminoazotoluene	27
Ovarian tumors	C3He	47 per cent in virgin ♀♀; 37 per cent in breeding ♀♀; 29 per cent in force-bred ♀♀	257
Myoepitheliomas	A & BALB/c	Occur in region of salivary gland and clitoral gland	790
Harderian gland tumors	C3H	Occur in C3H and hybrids resulting from outcrossing C3H	563, 571

suggestions of strains best suited genetically for the study of each.   Availability of these strains is given in the complete listing of the strains by Dr. George Jay, Jr., to be found in another chapter in this volume.

It should be noted, however, that there are still a number of neoplasms of man, the etiology of which should be studied in the laboratory but for which suitable inbred strains are not available.   Carcinoma of the cervix and uterus are among the more important neoplasms of women but rarely occur in mice.   Such neoplasms occurred in the PM strain, but that strain was unfortunately lost because of sterility factors.

We have observed carcinoma of the uterus in C3H × C57BL hybrids and Allen and Gardner[9] have reported a rather high occurrence of the neoplasm in CBA × C57BL hybrids treated with estrogen. This suggests the value of hybridizing our present strains to obtain a population from which a strain bearing the desired neoplasm might be selected. It also suggests the use of a carcinogen, in this case estrogen, in bringing to expression underlying genetic factors for selection.

Gastrointestinal cancer is the most prevalent cancer of man, but we have no strain of laboratory animal in which it can be studied adequately. Since carcinoma of the glandular stomach of the rat has been produced by injection of the carcinogenic hydrocarbons and carcinoma of the small intestine has been induced in mice by feeding these carcinogens, in this case also a program of selection with the use of the carcinogen to bring out the expression of the genes might result in a strain in which the neoplasm eventually appeared spontaneously. Strong's strain BRS in which the gastric lesion arose following methylcholanthrene was later found to develop the lesion spontaneously.

The pulmonary tumor of the mouse, while a very valuable tumor for basic genetic studies of neoplasia, is not comparable with the bronchiogenic carcinoma of man to which so much attention is directed today. Bronchiogenic carcinomas can be induced, however, in a mash of murine pulmonary tissue implanted subcutaneously along with crystals of carcinogen,[1224] and it is possible that this technique might be of value in selecting genes for this type of neoplasm and fixing them in a strain.

SPECIAL SUBLINES

Sublines free from agents normally passed through the placenta or milk can be of considerable value in separating genetic from nongenetic effects as illustrated by the lines free of the mammary tumor agent, or virus, that is normally transferred through the milk. Those passed through the milk can be satisfactorily eliminated by foster nursing on a strain lacking the agent when one does not want to go to complex germ-free procedures. It is advisable to take the young by cesarean section to insure against their nursing at all upon their own mothers. For survival the fetuses should be removed within 24 hours of term or preferably less. This time can be estimated by observing the vaginal plug when one has some knowledge of the exact gestation period for the strain. However, one experienced in the care of mice can judge the time almost as well by direct observation of the pregnant female without having observed the vaginal plug.

Our own subline of C3Hf mice was started by foster nursing a litter of C3H mice after cesarean birth from their high-mammary-tumor C3H mother with the agent, on a C57BL female without the agent. The subline is now in the forty-fifth inbred generation since the original foster-nursed litter, but mammary tumors have continued to appear at a relatively low incidence ranging in breeding females from 14 per cent in the F_9 generation to 38 per cent in generations F_1 to F_5.[565, 560]   However, the tumors appeared at a much greater age in the C3Hf than in the C3H.   This was an interesting result in that it demonstrated that although the virus could be a very important factor in the etiology of these tumors, it was not essential when there was a strong genetic susceptibility and the females were bred to add hormonal stimulation.

Foster nursing alone, however, does not guarantee freedom from the virus because of the possibility of transmission through the fertilized egg or the placenta, so subsequent testing is essential.   Filtrates of the C3Hf tumors were prepared and tested by injection into agent-free test females and the results were negative, but these tests cannot be considered absolute proof of the absence of the agent because in the preparation of the filtrates viruses can be lost.   Probably the best test for the presence or absence of the agent is in the occurrence of tumors in females of subsequent generations.   Throughout the 45 generations of this subline, the tumors have continued to appear with a relatively low incidence and at an advanced tumor age in females scattered throughout the pedigree chart rather than being segregated in tumor families.   We consider this to be proof that the mammary-tumor agent is not present in this line.   When the agent is introduced into females of the line in even small amounts by injection or transmission by the male or by any other way, it is immediately built up so that the females of subsequent generations appear as a high tumor line comparable with the original C3H.

Sometimes it is desired to have a large number of foster-nursed animals rather than to establish a fostered line.   To compare a mutated strain of the mammary-tumor agent with the original agent, Blair[109] found it necessary to foster nurse a large number of offspring that had not been permitted to nurse upon their mothers.   To do this she devised a screen on which the young were born and immediately dropped through to the foster mother below.

Sublines can readily be developed through transfer of fertilized ova, which eliminates not only possible agents transferred through the milk but also those that might be transferred through the placenta.   This is a delicate procedure, and Dr. Deringer gives a full description of it in the appendix of this volume.   Dr. Deringer's subline C3He produced by transfer of fertilized ova from C3H to C57BL has proved very interesting in that, like the C3Hf, it continues to produce mammary tumors although at a lower incidence and greater age than the C3H with the agent.[257]

A third type of subline can be developed from transplanted ovaries.   Ordinarily, however, transplantation is possible only between histocompatible animals such as those of two isogenic strains, if the gene by which they differ is not an histocompatibility gene, or from animals of a parental inbred strain to its F_1 hybrid.   Recently we have made reciprocal transplants of ovaries between the yellow and the black agouti (C3H $\times$ YBR) F_1 hybrid females.   These F_1 hybrids are alike except for the difference at the *agouti* locus and possibly closely linked loci.   The operation was simple with dorsal incisions and immediate transfer of the ovaries to the capsules of the reciprocal female.   Agouti females bearing yellow ovaries when mated to agouti males gave birth to successive litters containing yellow offspring, proving that the transplanted ovaries were functioning.   This technique is of value in studying the path of action of the lethal yellow gene in the etiology of neoplasms as will be discussed later.

Recently, Krohn[734] succeeded in transferring ovaries homologously.   He obtained young from both C3H and CBA ovaries transplanted into strain A females rendered tolerant by injection of splenic cells from the donor strains.   Such techniques could not be used for developing agent-free substrain because the virus would be transferred by the splenic cells and the ovarian tissue, but it could conceivably be used for certain studies on neoplasia in which one wanted to separate genetic influences from possible intrauterine influences.

USE OF THE F_1 HYBRID

Except for variation between males of reciprocal crosses, F_1 hybrids between two highly inbred strains should be as uniform genetically as the parental strains.   In certain cases, owing to the position of the genetic level in respect to the threshold, the F_1 hybrid may show even greater phenotypic uniformity.   For example, one of the parental strains may have an incidence of a certain neoplasm approximating 50 per cent, or maximum variability, and the incidence in the F_1 hybrid resulting from outcrossing to a very resistant strain may be less than 5 per cent, or minimum variability.   This genetic uniformity together with the hybrid vigor contributing toward long-lived, healthy animals makes F_1 hybrids especially suitable for test animals.   Furthermore, special types of F_1 hybrids can be produced for specific tests.   For example, much of the testing for the mammary-tumor agent has been done in F_1 females from the cross between low-tumor-strain females and high-tumor-strain males.   These hybrids are genetically susceptible to the agent although they do not have it except in rare instances when the agent has been transmitted by the male.

The value of reciprocal F_1 hybrids in revealing extrachromosomal factors in the etiology of cancer is well known.   One of the best known cancer viruses, the mammary-tumor virus, was discovered by geneticists as an extrachromosomal maternal influence revealed through the genetic procedures of making reciprocal crosses between strains having high and low incidences of mammary cancer.[1266]   In contrast, reciprocal crosses in genetic studies of leukemia have revealed an extrachromosomal resistance influence contributed by the low leukemic mother.[763, 837]

SPECIES OTHER THAN THE MOUSE

The value of mammalian species other than the mouse should not be overlooked for genetic studies of certain neoplasms.   The sarcomas of the liver and subcutaneous tissue of the rat induced with *Cysticercus fasciolaris* offer a unique opportunity for a genetic analysis of host-parasite relationship in the etiology of cancer.[309]   This is the only parasite larger than a virus that is now known to cause cancer under laboratory conditions.

The guinea pig is the species to turn to for studies of liposarcomas,[1198] and the rabbit for studies of uterine tumors.[496]   Even animals not yet introduced into the laboratory may offer opportunities for the study of certain neoplasms.   The rodent *Mastomys*, recently brought into the laboratory in South Africa, develops carcinoma of the glandular stomach in the laboratories there.[959]   This neoplasm is rarely found in any other species except man.

ESTIMATION OF NUMBER OF GENES

Incidence data can be used for estimating the number of genes involved in neoplasia provided one goes beyond the classical F_1, F_2, and backcross ratios and obtains breeding tests of the backcross or the F_2 segregants.   In Wright's[1451] analysis of polydactyly, results of crosses between strains 2 and D simulated one-factor Mendelian heredity with dominance of 3-toe over 4-toe in the F_1.   Segregation in the F_2 closely approached a 3:1 ratio, and in the backcross to the 4-toed strain D, a 1:1 ratio.   This interpretation broke down, however, in the tests of the supposed segregants, and at least 3 factors and probably 4 by which the parental strains differed was indicated. He found a close approach to blending inheritance in a character which approached alternative expression because of physiologic thresholds.

In their classical studies of the genetics of leukemia in mice, MacDowell and coworkers[835] were not content to classify their backcross segregants between the high-leukemic strain C58 and the low-leukemic strain StoLi merely on the basis of whether or not each developed leukemia.   Instead, they subjected a series of backcross males to progeny tests by mating them to StoLi females and observing the incidence of leukemia in the family of each.   The incidence in these families varied from 0 to 42.8 per cent with a fairly symmetrical distribution, a result explainable only on the basis of multiple factors.

Some of the early F_2 and backcross segregation ratios obtained by Bittner[107] for spontaneous pulmonary tumors and by Andervont[29] for induced pulmonary tumors suggested single-factor inheritance with susceptibility dominant over resistance.   Such an interpretation was not reconcilable, however, with the data on incidence for the inbred strains.   Not only were there highly susceptible strains such as strain A and highly resistant strains such as C57BL, but other strains such as BALB/c and BL had intermediate incidences.

Estimation of the number of genes involved in pulmonary tumors was made possible by transforming the phenotypic expression of the character from alternative to quantitative expression.   This was done by introducing a chemical carcinogen, in this case dibenz[*a*, *h*]anthracene, and measuring response by counting the resulting multiple tumors.[557]   In crosses between the high tumor strain A and the low tumor strain C57L, the mean of the F_1 was found to be intermediate between the means of the parental strains when age was kept constant.   The mean of the F_2 was likewise intermediate, but the variance of the F_2 was greater than that of the F_1.   The A backcross was intermediate between the F_1 and strain A, and the C57L backcross was intermediate between the F_1 and strain C57L. From the divergence between the parental strains and the excess of variance of the F_2 over that of the F_1, it was estimated that these strains differed by a minimum of 4 pairs of genes for pulmonary tumors.

This accurate quantitative measure of response of induced pulmonary tumors has offered great advantages for genetic and carcinogenic studies. Because of this, presentation of some technical details is justified. A number of chemical carcinogens can be used. Urethan is highly suitable if one wants a water-soluble carcinogen. It should be injected intraperitoneally, however, because when injected intravenously the anesthetic action is quick and there is danger of killing some of the animals. Crystalline dibenz[a,h]anthracene and methylcholanthrene are also highly satisfactory, and with them a water dispersion satisfactory for injection can be prepared easily with a colloidal mill.† Pulmonary-tumor response varies with size of the particles of the carcinogen in the dispersion, and particle size varies from batch to batch, so not more than one batch of dispersion should be used for an experiment. The intravenous injection of these dispersions is most desirable since the crystals are filtered out in the capillaries of the lung, and the pulmonary tumors are the only tumors induced. Mice can easily be injected in the lateral tail vein with a 26 or 27 gauge, $\frac{1}{2}$-inch, short-bevel needle. The vein can be dilated with warm water or xylene, but usually this is not necessary. Adult mice can easily tolerate $\frac{1}{2}$ ml. unless the material is toxic. Air bubbles must be avoided.

Although nitrogen and sulfur mustard are carcinogenic in inducing pulmonary tumors, they are not recommended in genetic studies because of their toxicity and the danger in handling them. Radiation would not be satisfactory since it does not induce enough nodules.

† Preparation of water dispersions of crystalline chemical carcinogens for intravenous injection:

Suggested concentrations: 1 mg./ml. of dibenz[a,h]anthracene, methylcholanthrene, benzypyrene, and so forth, in water plus 0.01 per cent aerosol plus 0.2 per cent Methocel, with NaOH to *p*H 8.

Suggested equipment: Colloidal mill (Eppenbach Processing Equipment, Manufactured by Admiral Tool and Die, Inc., Long Island City, New York, U.S.A., Serial No. QV6–53100).

Preparation of basic solution: Put 2 g. of Methocel 4000 and 0.1 g. aerosol in a large beaker with 1000 ml. water. Heat, stirring constantly to dissolve Methocel. Remove from heat and keep in refrigerator overnight.

Preparation of dispersion: Put about 500 ml. basic solution in mill and start the mill. Then add 1 g. carcinogenic crystals to the solution. Slowly turn the mill down to 0.005 inch and run at this setting for 5 minutes. Then turn mill to 0.003 inch and run for 5 minutes more. Remove carcinogenic dispersion and allow to stand for 5 minutes after which pour off supernatant into a flask (about one half of dispersion). Put remainder of dispersion back in mill using some of original basic solution to rinse flask. Turn mill to 0.002 inch and run for 2 minutes. Remove dispersion to a flask rinsing the mill with the remainder of basic solution. Add supernatant and bring to *p*H 8 with 2N NaOH (about 2 drops).

Throughout procedure stop mill often if necessary to prevent heating.

In view of the fact that we are dealing with a probability that the malignant change will occur and that this probability is increased with increase in time, the temporal factor after the introduction of the carcinogen should be controlled. With susceptible groups, 12 to 16 weeks is about the optimum time for killing the animals, and for most uniform results the mice should be of a uniform age at the time they are injected.

In order to count the nodules accurately, we routinely inject the lungs endo-tracheally with from 1 to 1.5 ml. of Tellyesniczky's fixative before removing them from the chest cavity. This distends the lungs, causing the tumors to stand out as readily discernible spherical nodules. (This, incidentally, gives an excellent fixation of the tissue for later microscopic study.) The counting is then done with the aid of a dissecting microscope.

This manner of transforming the problem from that of a threshold character to that of a quantitative character can also be applied to certain other tumors. When papillomas are induced by painting the carcinogens on the skin, multiple papillomas appear and these can be counted as a quantitative measure of response. Such data on papillomas have not been used for heritability studies, but they have been used for mutation studies[197] as will be described later. Induced hepatomas appear as multiple tumors which offer a quantitative measure of response for genetic studies, and even the spontaneous hepatomas are multiple in the highly susceptible strain C3H and in certain hybrids resulting from outcrossing C3H. Data on number of hepatomas have been utilized to advantage in linkage studies.

Since mammary gland tumors grow rapidly, seldom do more than two or three appear before the female dies, so the number of mammary tumors can hardly be used for a quantitative study. However, as DeOme[251] has clearly shown, mammary tumors are preceded by hyperplastic nodules of the mammary gland which appear in large numbers and can be readily counted. These could be used for a quantitative genetic study.

Latent period is also a quantitative measure of tumor response and has always been considered along with tumor incidence. Latent period can be most readily determined for such tumors as the mammary tumor that can be observed directly or palpated. In studying pulmonary tumors it is necessary to kill and examine sample groups to determine latent period, but this has been done for a genetic analysis, and the results paralleled those obtained by measuring response on the basis of nodule count.[557]

LINKAGE

Although cancers are inherited as threshold characters, considerable progress can be made in linkage studies, particularly in the mouse, which now has a rather extensive chromosomal map. (See Dr. Margaret Green's chapter in this volume.) In 1934, Little[798] demonstrated linkage between lethal yellow at the *agouti* locus and mammary tumors. Since then Bittner[106] has demonstrated linkage of this tumor with *brown*, and we have demonstrated linkage with *agouti*.[562]   Strong[1294] observed linkage between gastric tumors and *brown*.   MacDowell[835] found linkage between *dilution* and leukemia, and Law[768] demonstrated linkage between *flexed tail* and leukemia.   In our laboratory, linkage has been demonstrated between pulmonary tumors and 8 specific genes on 6 different chromosomes.[260, 556, 1347]   These include *hr* in group I, A^y in group V, *vt*, *sh-2*, and *wa-2* in group VII, *Fu* in group IX, *ob* in group XI, and *f* in group XIV.   Burdette[144] also demonstrated linkage between pulmonary tumors and *sh-2* and *wa-2*.   The pulmonary tumor linkages were first demonstrated with induced tumors, the quantitative measure of response in terms of nodule number offering advantages over incidence data in showing the differences to be real although not great.   However, in all cases where the tests were repeated for spontaneous tumors, linkage was confirmed.

What is termed linkage here is an association, and in many cases it is not possible to ascertain whether this association is true linkage or a pleiotropic effect of the gene tested.   *Lethal yellow* offers an advantage in that the test can be made in the F_1 animals which segregate at this locus since the inbred YBR strain has forced heterozygosis here owing to the lethal effect of the gene when homozygous.   There is strong evidence then that any effect noted in the occurrence of tumors is due to this gene itself rather than true linkage.

There is a suggestion in the response of the normal overlaps that the effect of *flexed tail* is also a pleiotropic effect.   The occurrence of pulmonary tumors in phenotypically flexed mice was significantly lower than in the genotypically normals.   However, that in the genotypically flexed but phenotypically normal was comparable with that in the genotypically normal mice.

Identification of effects associated with specific genes offers opportunities for more precise studies on paths of genic action which will be discussed more fully later.   There is evidence that the effects of *lethal yellow* and *flexed tail* on pulmonary tumors may be associated with their effects on normal growth.   A number of studies are suggested here, some of which could carry the problem to the biochemical level.   The association between *flexed tail* and increased occurrence of leukemia may be related to the effect of this gene on the hematopoietic system.   Flexed-tailed animals are born with a transitory anemia.   Other genes associated with anemia should also be tested for linkage with leukemia.

Correlation of these linkage tests with carcinogenic tests can yield interesting information.   *Lethal yellow* increased pulmonary tumors when they arose spontaneously or were induced with urethan, nitrogen mustard, or methylcholanthrene, but had no effect on pulmonary tumors induced with dibenz[*a, h*]anthracene.   The same was observed in the linkage studies on *flexed tail*.   This suggests that the mechanism of carcinogenesis with dibenz[*a, h*]anthracene is in some way unique.   On the other hand some of the other genes were associated with occurrence of pulmonary tumors induced with dibenz[*a, h*]anthracene suggesting that, as would be expected, not all genes are affecting occurrence of pulmonary tumors through the same mechanism.

PATHS OF GENIC ACTION IN NEOPLASIA

No characters provide more interesting complex problems in physiologic genetics than do neoplasms. The late occurrence of most neoplasms allows nearly the whole lifetime of the animal for the primary action of many genes to influence the occurrence of the tumor through a multitude of physiologic pathways. All neoplasia involves physiology of growth and differentiation. Physiology of hormonal production and of response to hormonal stimulation is involved in many. When viruses and other parasites are involved, there is the whole field of host-parasite relationship with not only the action of the genes of the host but also those of the parasite to be considered. When the etiologic factors include exogenous physical and chemical carcinogens, there is the physiology of cellular response to these carcinogens. In all of these areas genic action can be involved.

LOCALIZATION OF GENIC ACTION IN ORGANS AND TISSUES

One of the most fruitful approaches to the localization of the genic action and the paths through which it influences the occurrence of the neoplasm is through the transplantation of tissues and organs of one genotype into hosts of another. This can be done when the donor and the host differ by certain genes but have no conflict in respect to the histocompatibility genes, or when tolerance has been induced in the host. It should be empasized, however, that an animal of one strain made tolerant to tissues from another strain has had its physiology altered.

Often the F_1 hybrid is the more convenient host to use and satisfies the requirements for the experiment. Since the F_1 has all the dominant genes of both parental strains, organs or tissues from both parental strains and thus of different genotypes can be compared while growing in the common F_1 host. Thus, genic action or paths of genic action localized within the tissue or organ can be identified. On the other hand, by outcrossing one parental strain to a number of strains, tissues or organs of the same genotype from this strain can be transplanted into the F_1 hosts of a number of genotypes. In this way genic action can be identified with the host.

Another advantage of using the F_1 as host is that one can readily determine whether or not the neoplasm has arisen from the transplant or from host cells. If it arises from the transplant, it is readily transplantable back to the strain of origin, whereas if it arises from the cells of the F_1 host, it is not ordinarily transplantable to the parental strain.

This method of approach was applied to the problem of pulmonary tumors in mice[567] (figure 37). In earlier studies[557] it had been estimated that the high pulmonary-tumor strain A differed from the low pulmonary-tumor strain C57L by at least four pairs of genes affecting the occurrence of this neoplasm. The question to be answered was, were these genes becoming manifest within the lung or through some other systemic mechanism? Portions of lung from A mice were transplanted with a trocar

Fig. 37. Outline of experiment localizing the action of genes controlling occurrence of pulmonary tumors in the end organ.

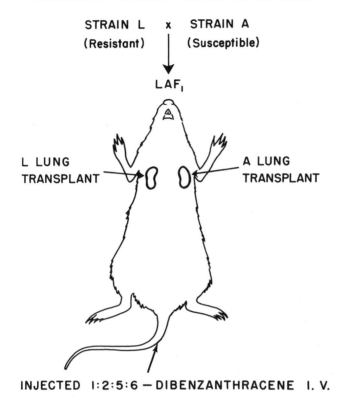

STRAIN L   x   STRAIN A
(Resistant)   (Susceptible)

LAF₁

L LUNG TRANSPLANT   A LUNG TRANSPLANT

INJECTED 1:2:5:6 – DIBENZANTHRACENE I. V.

(Reprinted by permission from the Journal of the National Cancer Institute.)[567]

subcutaneously into one axilla of (A × C57L)F_1 hybrids and portions of lung from C57L mice were transplanted into the other axilla.   To shorten the experiment, the F_1 hosts were injected intravenously with dibenz[a, h]anthracene after the transplants had been given two weeks to establish circulation.   It could be expected that, if the action of the genes were limited to the lung, many tumors would arise in the A transplants and few in the C57L transplants, whereas if the action were manifest through some other systemic mechanism the transplants would lose their susceptibility and resistance identity, and both would develop tumors comparable with the lung of the F_1 host. When the transplants were removed and sectioned four months after the hosts were injected with the carcinogen, many tumors were found in the A transplants and very few in the C57L transplants.   Thus, the actions of the genes controlling occurrence of pulmonary tumors by which these strains differed were being manifest in the end organ.

Shapiro and Kirschbaum[1193] used a variation of this technique in placing the transplants of pulmonary tissue in the ear of the host where they could be observed more directly.   We[569] have used transplants of fetal lung to ascertain that the genetic identity was established early. Fetal lung offers an additional advantage in that the transplanted lobe grows until it forms a mass approximating the size of a lobe of an adult lung, and this provides more tissue in which tumors can develop. The procedure of making a pulmonary tissue mash and treating it *in vitro* with a carcinogen before inoculating it into the host as Smith[1225] and Rogers[1070] have done, could probably also be incorporated into such genetic studies to link mechanisms of action of the carcinogen with mechanisms of the genetic effects.

Similar transplantation procedures have been employed in localizing genic action in the etiology of other types of neoplasms. Trentin and Gardner[1331] transplanted testes from newborn susceptible strain A and resistant strain C3H males into castrated adult (C3H × A) F_1 hybrid males. The transplants were placed at the base of the tail through a subcutaneous tunnel extending from an incision in the scapular region. Two weeks later a 7-mg. pellet containing 25 per cent stilbestrol in cholesterol was implanted subcutaneously in each host to induce the interstitial-cell tumors in the testes. Tumors arose in a much higher incidence of testes from the strain A than in those from the strain C3H, indicating that genic action determining susceptibility to estrogen-induced testicular tumors resides largely in the end organ.

Huseby and Bittner[622] transplanted adrenal glands from strain A, which shows minimal postcastrational adrenal hyperplasia, strain C3H, which shows considerable hyperplasia, and strain CE, which develops carcinoma, into the F_1 hybrids of the various combinations of these three strains. The hosts were then gonadectomized and the adrenal response was observed. The results indicated that the genotype of the end organ determined whether or not this response would be carcinoma, extensive hyperplasia, or very little hyperplasia.

Experiments of transplanting thymuses by Law and Potter[765, 769] and by Kaplan and co-workers[684] from susceptible and resistant parental strains to the F_1 hybrid, indicated genic action within the thymus controlling whether or not leukemia occurred. However, when transplantation of the leukemias was attempted, it was found that some of these occurring late were not transplantable back to the susceptible parental strain, although they were transplantable to others of the F_1 hybrids. Further observations by Law and Potter that the original lymphocytes of the thymic transplant were gradually replaced by lymphocytes from the F_1 host, confirmed that these tumors were of F_1 origin, and indicated that the important genic action was within the stromal cells, somehow controlling whether or not the lymphocytes became neoplastic.

The genetics of mammary tumors is more complex, but certain facets of the problem can be approached through transplantation procedures, particularly those related to hormonal influence. In all mammary-tumor strains of mice, hormonal influences resulting from the females' having litters increase the incidence of mammary tumors, decrease the tumor age, or do both. The extent of the observed effect is governed by the genotype of the strain. In strain A mice the virgins have a very low incidence of mammary tumors, whereas the incidence in the breeders is relatively high. In strain C3H the incidence is high in both the breeders and virgins, but the tumors appear earlier in the breeders than in the virgins.   This strain difference was shown to be a genetic difference, and it was suggested that it could be manifested through the control of hormonal production by the ovary or control of the response of the mammary gland.   Huseby and Bittner[623] transplanted ovaries from strain A and strain C3H subcutaneously into their ovariectomized F_1-hybrid females and observed a higher incidence of tumors among females with C3H ovaries than among those with A ovaries. This result indicated genic action within the ovary that, through the control of hormonal production, eventually influenced the occurrence of mammary tumors.   They also transplanted A ovaries into ovariectomized A females so that an A mammary gland stimulated by an A ovary could be compared with the genetically different F_1 mammary gland also stimulated by an A ovary.   More tumors occurred in the F_1 than in the A glands suggesting that the action of some genes was manifested in the host, controlling the response of the mammary gland.

In testing the effect of the *lethal yellow* gene on mammary tumors in (C3H $\times$ YBR) F_1 females, half of which are yellow (A^y/A) and half are black agouti (A/a), it was observed that the effect of A^y was to cause the tumors to appear in the virgin yellow females with the same incidence and tumor age as was observed in the breeding agouti females.   The tumors also occurred with the same incidence in the virgin agouti females, but at a much later age.   Information on the path of action of this specific gene can now easily be obtained by exchange of ovaries between these yellow and agouti F_1 females.

Endocrine studies suggest that the effect of the ovary on occurrence of mammary tumors is by way of the pituitary gland.   The effect of the pituitary gland also has been noted in that the addition of pituitary transplants greatly increases the occurrence of mammary tumors.[916]   The pituitary glands can be transplanted subcutaneously, although beneath the capsule of the kidney is probably a more effective site.   Proper genetic pituitary-host transplant combinations should reveal whether or not genic action within the pituitary gland is also influential in controlling the occurrence of mammary tumors.

Transplantation of the mammary gland to obtain glands of different genotypes growing in the same host presents some technical problems, particularly in eliminating the host's own glands so they will not interfere with the transplanted glands.   These problems were overcome in part at least by Prehn,[1022] who transplanted rather large portions of the skin with the underlying gland to the backs of females where but little gland normally appears, and he obtained some tumors in the transplanted gland.   In so doing, he demonstrated that genic action within the mammary gland governed its response to either the hormonal stimulation, the mammary-tumor agent, or both.

The most important technique along these lines, however, is that developed by DeOme and co-workers[251] for the transplantation of the hyperplastic nodule of the mammary gland into a cleared fat pad.   (See Appendix VII for a description of this technique.)   The fact that one is dealing here with a rather small clump of cells, from which the neoplasm arises without interference from the host's own glands, makes such studies much more definitive. It also offers opportunities for studying the steps in the development of neoplasia. It is hoped that the possibilities of this technique can be pursued extensively to add to our knowledge of paths of genic action in development of tumors.

HOST-PARASITE RELATIONSHIP

One of the paths through which genes can influence the occurrence of a neoplasm is through their control over the propagation and transmission of an etiologic virus or other parasite. Genetic effects can usually be demonstrated in the development of inbred strains that are susceptible and resistant to the virus or in the testing of existing strains.

Genetic control over the mammary-tumor virus was clearly demonstrated by crossing susceptible strain C3H females that had the agent with resistant strain C57BL males without the agent, and backcrossing the F_1 females all of which had the agent with both susceptible strain C3H males and resistant strain C57BL males.[563] This provided two groups of backcross females which differed genetically because of the difference in their fathers but were alike in that they had received the virus from genetically uniform F_1 females. The occurrence of more tumors in the test females foster nursed by the C3H backcross females than in those foster nursed by the C57BL backcross females indicated that the two backcross groups differed in their ability to transmit the agent because of their differences in genotype. This study was later extended to successive backcrosses to the resistant-strain males. By the third backcross generation the agent was completely eliminated.[564]

In such studies the bioassay of the virus is extremely important. In the mammary-tumor problem the presence or absence of the tumor is not proof of presence or absence of the virus. Some females have the virus and do not develop tumors, and some tumors develop in the absence of the virus. Electron-microscope pictures of particles are not conclusive evidence alone. Even tests of prepared filtrates of tumors or other tissues are not wholly satisfactory since viruses can be lost in the technical procedures. The most reliable test for the presence of the mammary-tumor virus is the development of tumors in females that have nursed upon the female being tested.

One of the difficulties that can be encountered in a detailed genetic analysis is that just as Kappa particles in *Paramecium* can be carried through several generations of cells in the absence of the gene,[1256] so may cancer viruses be transmitted to immediate offspring by females lacking the necessary genes for their propagation. This was found to be true with the mammary-tumor virus, and in testing backcross segregants for their genes controlling the propagation of the agent it was, therefore, necessary to observe tumor occurrence not only in their offspring but also in test females that nursed upon these offspring. The results of this study[571] clearly indicated that multiple genes were involved in the propagation of the agent, and furthermore, that the quantity or quality of the agent could be altered, presumably depending upon the number of these genes present.   It was also evident that the absence of the genes did not result in the elimination of the virus through the production of antibodies.   In the absence of the genes the virus merely did not multiply and was lost through dilution either in the female or in her offspring if they, too, lacked the necessary genes.

BIOCHEMICAL PATHWAYS

Thus far much of the analysis of the links in the chain of events leading from the primary genic action to the final neoplasm has been directed toward those links nearer the neoplasm.   Before these pathways are completely understood, much study must be devoted to the earlier links, many of which are on the biochemical level.   Biochemical genetics in the mouse and other laboratory mammals is lagging behind that in microörganisms and even that in man, but development of the field is now urgent; and, with the background supplied by the studies with microörganisms, rapid progress should be expected.

Our highly inbred strains of mice used in cancer research should be characterized by the enzymatic patterns of their tissues just as they are now characterized by their types and incidences of tumors, or, on a lower level, by their histocompatibilities. Much work has been done on enzymatic patterns of neoplasms *versus* the normal tissue of origin, with emphasis on hepatoma *versus* hepatic parenchyma; but of greater importance in the etiology of cancer is comparison of the enzymatic pattern of one hepatic parenchyma which, because of its genotype, will eventually give rise to neoplasia, with the pattern of another hepatic parenchyma which, because of its different genotype, has very little chance of developing a neoplasm.

The identification of neoplasia with specific genes such as *lethal yellow*, which has a pronounced effect on the occurrence of a number of tumors, offers the possibility of associating specific enzymatic differences affecting neoplasia with specific genes.   The effect of *lethal yellow* on normal growth suggests basic differences in protein synthesis.

GENETICS OF THE NEOPLASTIC CHANGE

The neoplastic change in the cell is a heritable change in that it is continued through successive cellular generations.   With a broad concept of mutation including changes in all hereditary material within the chromosomes, within cytoplasmic bodies, and even within viruses which when introduced into a cell alter its genetic nature, one must accept the hypothesis that a neoplasm arises from a somatic mutation.

Since somatic cells could not be subjected to breeding experiments—the only way in the past of studying mutations—approaches to a genetic analysis of the neoplastic changes have been limited to certain statistical approaches and to studies of correlation between mutagenic capacities of compounds and their carcinogenic capacities.   Within certain limits positive correlations between mutagenesis and carcinogenesis have been obtained,[145] but as a whole this attack has not been too fruitful.   Mutagenic studies can be done best in the lower organisms which, with few exceptions, are not suited to carcinogenic studies, whereas the carcinogenic tests should be made in the mouse or rat which are not well suited to mutagenic tests.

Statistical studies of dose-response curves have been of some interest. Charles and Luce-Clausen[197] found that data on response of mice to repeated painting of a carcinogen on the skin measured in number of induced papillomas gave an exponential curve which they suggested was what would be expected were the neoplastic change due to a recessive mutation. However, our data[568] on number of pulmonary tumors induced in strain A mice with graded doses of dibenz[*a*, *h*]anthracene indicated a straight line response which would suggest a single event, and if this were a genic mutation then it must be dominant.

It would seem that now the way may be opened for an attack on this problem through techniques of tissue culture. With the single-cell cloning techniques now perfected[1026, 1149] and with chemically defined media now in use,[329, 330, 1359] genetic nutritional changes in cells can be identified. By using cells from one of the highly inbred strains so that the point at which they become malignant can be determined by successive transplantation of the cultured cells back into the strain of origin, it may be possible to associate some of these nutritional changes with the neoplastic change.

DISCUSSION

DR. ANDERVONT: Those who have worked in cancer research during recent years recognize the importance of inbred animals that have proved so invaluable. They are used to ascertain the response of different inbred strains to a standard dose of carcinogen or the response of a single strain to graded amounts of a carcinogen. In essence, geneticists have made possible a quantitative approach to the problems of the relationship between host and cancer-inciting factors. This contribution alone is sufficient to justify the use of inbred animals, but, as Dr. Heston has pointed out, cancer is a complex disease.

He used mammary cancer of mice to exemplify this complexity because it is an excellent example of the reactions of the host to a combination of genetic, viral, and environmental influences. Geneticists have given most of their attention to the host, with the result that the most clearly defined factor in the occurrence of the disease is the genetic constitution of the host. Their efforts have yielded important contributions, such as the discovery of the mammary-tumor agent, but they have also uncovered many new problems. The inbred strains display a remarkable difference in susceptibility to the virus as well as to the production of breast cancer by hormonal stimulation in the absence of the virus. It is of some interest that, according to evidence now available, those strains susceptible to hormone-induced breast cancer are also susceptible to the virus. This does not imply that *all* strains susceptible to the virus are also susceptible to hormone-induced tumors, but only that those which respond readily to the administration of estrogenic hormones are also susceptible to the virus.

In contrast, there is no well-defined correlation between susceptibility to the virus and susceptibility to chemically induced tumors.   For example, mice of strain DBA/2 are susceptible to both virus and chemically induced breast tumors, whereas those of strain IF are resistant to the virus but susceptible to a chemical carcinogen.   As Dr. Heston stated, strain C3H mice are susceptible to both virus and hormone-induced tumors but resistant to carcinogen-induced tumors.   These striking variations in susceptibility to induced breast tumors raise the problem of whether we are dealing with a single disease or different diseases.   If mammary cancer in mice is classified according to its etiologic agent, then we are dealing with different diseases.

Hepatoma arising in mice, too, is the result of the interplay of several factors.   A virus has not been implicated but, as Dr. Heston described, heredity is very important. Hormonal influences are involved because the spontaneous tumors occur more frequently in males of all inbred strains susceptible to their development.   Diet exerts a pronounced influence on their development.

Experiments performed with strain C3H mice as test animals reveal the complexity of these tumors.   Males develop more spontaneous tumors than do females, but when the tumors are induced by the administration of an azo dye the females are far more susceptible than males.   However, castrated males are as susceptible as females and can be made resistant again by injection of androgens.   When tumors are induced by the administration of carbon tetrachloride, both sexes appear to be equally susceptible.   However, a single strain of inbred mice exhibits remarkable variation in susceptibility to this tumor.   Such variations suggest to the oncologist that he is dealing with a group of diseases which eventuate in the same final manifestation.   At least the tumor can be studied within a single inbred strain when the genetic factor remains constant.

Any discussion of Dr. Heston's paper would be incomplete without some mention of the contribution the methodology of genetics has made to recent advances in the field of tumor viruses.   During the past seven years at least six viruses have been implicated in the origin of leukemia, or leukemia-like lesions, in laboratory mice.   To those attending this symposium the interesting feature of these viruses is their range of specificity for inbred strains of mice.   Much work remains to be done along these lines, but, to date, strain DBA is susceptible and strain C3H is resistant to one virus, whereas all strains tested thus far are susceptible to another.   These two viruses represent the extremes of the six.

You will recall that the murine mammary-tumor agent varies in concentration in different strains but that a susceptible strain reponds to the virus from other strains. If the specificity of the leukemia virus for certain strains is established firmly by future studies, then the establishment of inbred strains will have enabled the virologist to detect tumor viruses specific for certain genetic constitutions within a species.   Carrying this thought one step further, it is possible that, within a genetically heterogenous species, some tumors are induced by viruses which are so specific that they produce tumors only in certain individuals.   On the other hand, the ability of murine leukemia viruses to elicit a variety of leukemias in their hosts suggests the possibility of a single virus producing a variety of tumors within a species. The logical conclusion to this discussion is that the methodology of genetics has contributed, and will continue to contribute, much to the solution of the cancer problem.

Dr. Nalbandov: A year ago in Russia I learned of a method which impressed me and which may interest you. It involves applying uniform standard stresses (such as electric shock) to inbred strains of mice (C57 and A) and classifying the individuals according to their responses: nervous (unstable nervous system) to nonnervous (stable system). According to V. K. Fedorov (Pavlov Physiol. Inst. Koltushi, Leningrad) mice with an unstable nervous system died very rapidly after inoculation with cancers, whereas mice with a stable nervous system died after a more prolonged period of time and showed many metastases. Mice made neurotic by repeated electrical shocking were more susceptible to cancer (both spontaneous and inoculated) than were their non-neurotic counterparts. In dairy cows responses to light, sound, and electrical shock stimuli were measured and cows classified into four groups from very stable to unstable. The milk yields in 300 days were 6,193 kg. for the most stable group I; 5,516 kg. for group II; 4,918 kg. for group III; and 4,708 kg. for the least stable group IV.[730]

Dr. Pilgrim: I do not know what methods they were using, but most people who work with mice probably consider the C3H one of the calmest and easiest strains to work with and one with the highest incidence of cancer.

Dr. Nalbandov: That is not the point. They subject them to a uniform stress in order to determine the degree of stability of the nervous system.

Dr. Heston: Were the growth curves of the mice followed? Cancer is a disease of the healthy, and almost anything one does to make the mouse less healthy, if such a term is permissible, and which will reduce growth rate will decrease the occurrence of tumors.[570] For example, a deficient diet will decrease the occurrence of tumors and a growth-promoting diet will increase their occurrence.[24, 572, 759, 1311, 1312] Various genes affect susceptibility to pulmonary tumors. The gene which increases growth rate of skeleton and muscle is associated with an increase in these tumors, whereas those that decrease growth rate are associated with a decrease in pulmonary tumors.[556, 566]

Dr. Klein: Dr. Heston, you said that in serial backcrossing the agent was eliminated due to the reactivity of the females by antibodies. What is the evidence for this?

Dr. Heston: No one, to my knowledge, has shown that the mouse can produce antibodies against the mammary-tumor agent. When Dr. Andervont[26] gave the agent to low tumor strain C57BL females these females did not get tumors, but they did not eliminate the agent they received, for they were able to transmit it to foster nursed, susceptible young that later developed tumors. Their own C57BL young, however, could not transmit any agent. In our own segregation studies,[571] among the females resulting from backcrossing F_1 females to C57BL males one would have expected that some would not have genes for susceptibility to the agent; yet, although many of these backcross females did not get tumors, none of them eliminated the agent completely, for all of them transmitted it to susceptible test females that later developed tumors.   In testing their own second-backcross offspring, however, it was found that certain ones of these first-backcross females did not have any offspring that could transmit the agent.   It thus appears the C57BL females and certain backcross segregants lack the genes for propagation of the agent and although they do not eliminate what agent they receive, it does not multiply and is lost completely (presumably by dilution) in the subsequent generation.

DR. KLEIN: I read the paper and was quite convinced the agent was lost, but I was not sure whether the agent was lost because it lacked something in the host or because it was eliminated by a host reaction.   Can you distinguish between these two possibilities?

DR. HESTON: I think it is lost from some process like being diluted out and not by being destroyed by the animal.

DR. YERGANIAN: With respect to Dr. Heston's remarks concerning the rôle of underlying variables, an interesting sequence of events has taken place since 1954, when we provided the Chester Beatty Institute with a subline of partly inbred Chinese hamsters having then undergone four generations of inbreeding.   Shortly thereafter, this new colony provided a subline to the Radiobiological Research Unit at Harwell. I have been informed by Dr. C. E. Ford that the subcolony at Harwell is now in the thirteenth generation of brother-sister matings and has developed an extensive number of pituitary and pancreatic carcinomas among both the control and X-irradiated populations.   In our hands, the parental line (BUY) is now in the seventeenth generation and has a relatively high incidence (7 per cent) of diabetes mellitus by 200 days of age.

I was wondering if genetic drift would have any underlying contributions towards this presumably unrelated dichotomy and, if so, whether or not cancer may then be regarded as a form of metabolic dysfunction which, in the Harwell colony, is expressed in a malignant pattern, whereas a slight alteration in the genetic complex in question within the Boston colony is expressed metabolically as diabetes mellitus.   In both instances, the target cells and organs are identical.

DR. CHAI: I am in agreement with Dr. Heston's concept that tumors are threshold genetic characters.   To this I should like to add (as a speculation) that development of tumors depends not only on genes with major effects but also on the genetic background and interaction between the genes.   There are many published data which can be so interpreted.   For instance, there were differences in incidence of mammary tumors between the F_1 and F_2 hybrids in crosses involving cancerous strains of mice.[793]   These differences can be explained by the threshold concept as the level of threshold fell on certain points in the frequency distributions of the genotypes.   They can also be explained by the differences in genetic background and possibly by different interactions between the genes.   There are many threshold characters, including some hereditary diseases and deformations, for which the incidences are different on different backgrounds, and the genetics is complex and not all clear.   Most of the tumor data in experimental animals come from highly inbred and not from noninbred animals.

This raises the question as to whether the level of homozygosity in an individual affects tumor incidence.

DR. BURDETTE: Dr. Lynch, do you have a comment?

DR. LYNCH: I enjoyed hearing Dr. Heston's survey of this subject to which he himself has contributed a very great deal. The production of strains of mice has entered the modern age. For example, I have been trying to produce a new strain of leukemic mice, one quite different from any used so far. After a good many generations of inbreeding during which with but one exception either one or both parents had the disease, the leukemia was lost rather suddenly. Whether it was due to the loss of a virus or to a genetic change affecting a virus or the host, I do not know. We are trying to introduce a virus into the strain to see whether that will be effective in bringing back the high incidence of leukemia. With respect to the production of some of the tumor strains, we are in a viral age.

DR. BURDETTE: Historically, the two most controversial, opposing ideas of carcinogenesis have been the somatic-mutation and viral hypotheses. In recent years, differences between them have not seemed quite so great as formerly with the advent of more information regarding cellular and viral DNA and RNA. Would you elaborate on how what is known about the similarities between viruses and genes may be related to carcinogenesis at the cellular level?

DR. HESTON: There are new and exciting things in this area that may have some bearing on the problem of cancer. It has been interesting to consider both lysogeny and transduction, although I think that some of the investigators of cancer may have used these terms rather loosely. Lysogeny may be involved in the induction of tumors if it can be shown that the cancer virus does enter the cell and become an integral part of the genetic mechanism of the cell and in so doing causes the cell to become malignant. This would be similar to lysogeny as described for bacteriophage. Such a process would not be outside the limits of the somatic-mutation hypothesis of cancer if we include in this hypothesis any change in the hereditary mechanism of the cell. A situation similar to transduction might be involved in the spread of cancer if, after the virus has entered the cell and caused it to become malignant, some of this information is then carried over to neighboring cells, causing them in turn to become malignant. Such things are interesting to think about, and they are stimulating (and I hope they will continue to stimulate) good research.

A. V. Nalbandov, Ph.D.

GENETICS *of* REPRODUCTIVE PHYSIOLOGY

The scientific literature is replete with examples of heritable differences in reproductive performance between strains or breeds of animals or races of man. Much of this knowledge about hereditary differences in prolificacy had been accumulated before the advent of experimental endocrinology which occurred about 1927. In the infancy of endocrinology the notions concerning the hormonal mechanisms governing reproductive events were vague or incomplete, and it is only relatively recently that a beginning has been made in sorting out the heritable manifestation of hormonal effects and correlate the cause-and-effect relationships between genes and hormones.

BASIC CONCEPTS OF ENDOCRINOLOGY

First, agreement is desirable on some of the basic concepts of endocrinology which define the manner in which hormones affect their target tissues. Some of these interrelations are simple, one-step effects in which a single hormone is known to act directly on its target organ where it produces certain metabolic modifications. Other interrelations are chain reactions which may involve single hormones as the metabolic modifiers between steps, or, in the more complex relationships, may require a plethora of hormones which may be necessary to produce an exceedingly complex metabolic end result. Both the single-step responses and the complex chain reactions are known to be, or can be presumed to be, under genic control. The importance of genic control may vary depending on the step which is being affected and depending, of course, on the importance of this step in the ultimate metabolic response.

It is generally agreed that hormones do not initiate biochemical reactions performed by cells, but that they can govern the rates at which these reactions take place. If, for instance, a hormonal deficiency is induced by surgical or pharmaceutical methods, cells retain a modicum of their inherent ability to perform their highly specialized tasks; but now these tasks are performed slowly, inadequately, and inefficiently. If the level of the missing hormone is raised by exogenous administration of the hormone, the rate of reactions can be raised in proportion to the amount of hormone injected. It is assumed that the genetically determined rates of hormone secretion similarly set the rates at which the inherent cellular reaction take place.

All cells within a body come in contact with all endogenous hormones which are carried by the common blood stream bathing the cells. However, in spite of this general distribution *via* the blood stream, hormones react only with their specific, genetically conditioned end organs. The biochemical reasons for this ability of the different cells to discriminate between hormones and to respond only to their own specific trophic substances remains largely unknown, although it is presumed to be a matter of activation of mechanisms which are permissive of optimal intracellular interaction between enzymatic systems and substrates. Examples of such specialized activation of end organs are plentiful. Gonadotrophic hormone, for instance, is a specialized growth hormone which increases the metabolic activity of the gonads but has no such effect on somatic cells or on the cells of other glands. Conversely, somatotrophic hormone is a growth hormone which stimulates growth of somatic cells but has no effect on the metabolic activity of gonads.

This apparent autonomy of hormonal systems should not lead to the conclusion that the efficiency of any one hormone-end-organ reaction is independent of other endocrine interactions in the body. While there is a specific trophic hormone for almost every reproductive event and almost every end organ, optimal responses are possible only in euhormonal internal environments. For example, only gonadotrophic hormone can stimulate growth of gonads, but optimal growth of gonads can be achieved only in an endocrine environment in which all other glands (thyroids, adrenals, and so forth) are functioning at a satisfactory level. For this reason it is necessary to distinguish between primary, secondary and, occasionally, tertiary deficiencies. For instance, lowered fertility or even sterility of animals may be erroneously attributed to a deficiency of gonadotrophic hormone, while in reality it may be due to the inability of a normal level of gonadotrophic hormone to produce optimal gonadal responses because of a deficiency of hormones other than gonadotrophins.

Since hormones govern rates of reactions it can be presumed, and in some cases demonstrated, that genes or genic complexes determine the rates at which hormones are secreted. Rates of hormonal secretion are frequently the only phenotypic expressions which can be measured and used as a guide to the genotypic differences known to exist. Estimates of rate of hormonal secretion can be obtained in a variety of ways.

In some instances the relationship between genotype and endocrine phenotype is simple, as in cases in which only one hormone is known to be involved and in which this hormone is known to act directly on the end organ.   Size of comb of chickens is an example of such a simple relationship.   More often the pathway is indirect and involves several hormones.   Examples of the latter type include the whole reproductive process, growth, lactation, and, in fact, most of the phenomena governed by hormones. To complicate the picture still further, it is now known that in addition to the endocrine system the nervous system is equally important in the control of phenomena which earlier were thought to be directed by the endocrine system alone.   In spite of these difficulties, valuable information can be obtained by using the various techniques which provide a guide to rates of hormonal secretion.

ESTIMATES OF RATES OF HORMONAL SECRETION

Bioassays of glands yield information on the concentration of hormones in them. This method has the disadvantage that the gland-giving animals must be sacrificed or mutilated.   Furthermore, endocrinologists are not certain whether such bioassays estimate the rate at which hormones are secreted, the rate at which they are stored, or whether they simply indicate the amount of hormone remaining in the gland after its physiologically effective quantity has been released into the blood stream.   Obviously the value of such bioassays hinges on which of these three alternatives is being estimated. While this question has not been resolved to the satisfaction of all endocrinologists, there is much evidence which shows that such assays do indeed estimate the rates at which hormones are being secreted.[930]

It would be most desirable to obtain information on the amounts of the various hormones circulating in the blood, but no good and easy methods are available to make such estimates either by chemical or biological means.   This is because hormones, once they enter the blood, are destroyed or inactivated within minutes or, at most, within a very few hours.   Because of this difficulty, and because the concentration of hormones in the peripheral circulation is extremely low, it becomes necessary to use special techniques which sample the blood as it leaves the more accessible glands or end organs.   These techniques are cumbersome and difficult and have only limited application in the study of populations of the size required for adequate genetic analysis.

In addition to these direct methods of obtaining information on the rates at which certain hormones are secreted, various methods permit indirect estimation of rates of hormone production.   To obtain such estimates, a hormone is allowed to act on its target organ and the degree of stimulation of the target organ is used as an index of the rate of secretion of the hormone.   Differences in rates of ovulation in females belonging to different genetic strains provide good examples of the variable rates of hormonal secretion by pituitary glands.   Rates of ovulation, expressed either in numbers of eggs recovered from the female ductal system, or in numbers of corpora lutea found in the ovaries, have been used to demonstrate pronounced genetic differences between strains of domestic and laboratory animals.   Litter size is an indirect measure of rate of ovulation, but, in defining genetic differences, it must be used cautiously, since litter size is not always determined by rate of ovulation alone.   This will be illustrated in a number of examples later.

The genetic controls of such characteristics as rate of ovulation are probably as complex in mammals as they are known to be in chickens.   Thus, the ability to lay many eggs depends on the combined action of five or more component characteristics each of which is known to be controlled by genes and influenced by hormones.   Chickens which lay the highest number of eggs must show early sexual maturity, the ability to lay many eggs in succession without rest periods (long clutches), and they must be non-broody.[539, 540]   It is highly probable that the original analysis of the numbers of genes involved in the control of the fecundity of fowl is greatly oversimplified and inadequate and that the number of genes governing the expression of each of these characters is larger than originally thought.   This possibility does not alter the fact that each of these five characteristics is known to be affected either by single hormones or interacting hormone complexes which are involved in determining the general level of fecundity in fowl.   No comparable analysis has been made with regard to the genetic control of a similar array of component parts of the reproductive cycle of mammals.

Other indirect measures of rates of hormonal flow can be obtained from observations on the live animals in which rate of growth, rate of lactation, onset of puberty or menopause, and the like reflect endocrine states and are indicative of differences in genotypes.

Still another method of establishing gene-controlled differences between strains of animals involves the measurements of their sensitivity to the injection of exogenous hormones.   In some instances these methods require that the animal be sacrificed following treatment in order to obtain the target organ for analysis or weighing, while in others, the necessary measurements can be obtained from the living animal as in the case of the sensitivity of combs of male chickens to androgens, the sensitivity of the vaginae of rats or mice to estrogen, or the ability of the thyroid to take up radioactive iodine.

Using one or the other of the techniques mentioned, it is possible to obtain reasonably reliable direct or indirect estimates of the rate at which the different hormones are secreted.   Much information has been accumulated in the literature showing conclusively that heritable differences in the rate of hormonal secretion exist between strains of animals.   It is not the purpose of this paper to make a complete survey of the literature illustrating these points and only a few of the more unusual or interesting examples will be presented.   These examples have been selected because they illustrate the problems and principles which confront the scientist interested in this general area of physiologic genetics.   Neither is it possible to present an exhaustive review of the gamut of chemical or methods of biological assay which can be used to obtain estimates of the various hormones contained in glands, tissues, or fluids of animals.   The interested reader is referred to the current endocrine literature as well as the standard texts on endocrinology.   *Hormone Assay*,[323] which is somewhat out of date, and *The*

Hormones[1007] contain methods for biological assay as well as chemical determination of hormones in tissues and fluids.

CORRELATED CHANGES IN RATES OF HORMONAL SECRETION

It appears to be generally true that genetic selection for the increased or decreased rates of secretion of one hormone results in increased or decreased rates of secretion of other hormones produced by the same gland.   There are many examples of this principle in the literature.   Body size is generally determined by the coöperative actions of somatotrophic, thyrotrophic, and perhaps adrenotrophic hormones, the rate of somatotrophin secretion probably being the primary determining factor responsible for the differences in the body size between two breeds of rabbits.   Similarly, rate of ovulation is a reflection of the rate of secretion of gonadotrophic hormone complex. In one breed (table 51) there is a close association between small body size and small

Table 51

COMPARISON OF VARIOUS HORMONE-CONTROLLED CHARACTERISTICS OF LARGE AND SMALL BREEDS OF RABBITS

Breed	Body weight (kg.)	Rate of ovulation	Litter size	Gestation length (days)
Polish	1.66	5.52	4.48	31.1
Large race	4.30	10.73	8.13	31.5

Data from Venge.[1341]

litter size, the former presumably due to a low rate of secretion of somatotrophic hormone, the latter presumably caused by a deficiency of gonadotrophic hormone.   In the large race, the opposite association of the presumed rates of hormone secretion is noted.

If one compares the rates of twinning, lactation, and growth in cattle, one finds that the highest rates of these characteristics (all of which are due to the highest rates of hormonal secretion) appear to be closely associated.   Thus, the Holsteins are highest in all of these characteristics, while the Jerseys are lowest (table 52).   The fact that this association does not seem to hold for body weight and lactation in beef cattle in comparison to Holsteins can be explained on the basis that large body weight in Holsteins was achieved by selection for a high rate of secretion of somatotrophic hormone in euthyroid, and perhaps even hyperthyroid animals, whereas selection in beef cattle was in favor of hypothyroidism, that is, a low basal metabolic rate and a phlegmatic, easily fattened animal.

One of the most interesting examples of this type is provided by MacArthur's[831] experiment.   Using mass selection and avoiding inbreeding, he selected mice for only

Table 52

RANK ORDER OF VARIOUS BREEDS OF CATTLE WITH REGARD TO THREE HORMONE-
CONTROLLED PHYSIOLOGIC CHARACTERISTICS

Breed	Rank order in		
	Per cent twins	Lactation	Body weight
Holstein	1	1	1
Ayrshire	3	2	2
Guernsey	2	3	3
Jersey	4	4	4
Beef cattle	5	5	1

one criterion—body size. His foundation stock weighed about 23 grams, and, after selecting for 21 generations, his small race weighed about 12 grams while the body weight of the large race was 40 grams. Keeping in mind that the only criterion for selection in this experiment was body weight, it is interesting to note that at about the twelfth generation, the small race had a rate of ovulation of 7.2 eggs and a litter size of 5.3 young, while the large race had a rate of ovulation of 14.1 eggs and a litter size of 10.5 young. Although these data suggest that selection for one hormone-controlled characteristic usually leads to an increase in the rate of secretion of other hormones secreted by the same gland, data proving this by actual assay of glands or the determination of hormonal levels are not readily available.

In another study[46, 1066] a comparison was made of the hormonal contents of the anterior pituitary glands of two strains of swine one of which was selected for a slow rate of gain, the other for a rapid rate of gain. A comparison of the concentration of

Table 53

COMPARISON OF THE HORMONAL CONTENT OF THE ADENOHYPOPHYSES OF TWO GENETICALLY
DIFFERENT STRAINS OF SWINE

Line selected for	No. of corpora lutea	No. of embryos	Testis weight of chicks (g.)†	Body weight of pigs 156 days old (kg.)	Width of epiphyses of rats (micra)‡
Rapid growth	14.5	11.62	15.7	48.7	243.4 ± 15.6
Slow growth	10.5	8.53	25.2	24.9	198.6 ± 8.1

† Indicates gonadotrophic hormone content and
‡ content of growth hormone of adenohypophyses.

gonadotrophic hormone and growth hormone in a standard quantity of pituitary tissue was made at comparable ages in both strains (table 53). There was a significant and clear-cut difference with regard to the concentration of growth hormone. Because it was found that the anterior pituitary glands of the rapidly gaining strain contained significantly more somatotrophin per unit pituitary tissue than did the glands from the slowly gaining strain, it was possible to account for the differences in the rates of gain between the two strains.   However, even though there was also a clear-cut difference between the fertility (that is, litter size) of the two strains (table 53), the assay of the hypophyses for concentration of gonadotrophic hormone showed an inverse relationship to rate of ovulation and to size of litter.   Since the assay method used measured the total gonadotrophic hormone content rather than the two component parts of the gonadotrophic complex (consisting of the follicle-stimulating hormone FSH, and the luteinizing hormone LH), it is conceivable that the poorer reproductive performance of the slow strain was due to an abnormal ratio between FSH and LH which was perhaps different from the ratio usually found in normal animals and which was incompatible with optimum reproductive performance.   From the genetic point of view it is clear that the two hormones under discussion were present in the pituitary glands of the slow and fast strains in significantly different concentrations; but the endocrine interpretation of the inverse relation between gonadotrophin and fecundity of these two strains is not obvious.   The slow strain showed a significantly higher incidence of cystic ovaries and generally impaired fertility in comparison to the fast strain.   In this case, then, the higher gonadotrophic hormone potency of hypophyses of the slow strain is indicative of abnormal reproductive performance rather than of greater fecundity.   This example illustrates the fact that estimates of hormonal concentrations are not always reliable indicators of the efficiency with which the animal utilizes its hormones.   Such estimates are useful only when considered in conjunction with the overall performance of the animal.

It is interesting to speculate on the possibility of producing strains of animals in which a high rate of secretion of one hypophyseal hormone is associated with a low rate of secretion of another trophic hormone.   This would create genetic strains in which, for example, large body size is combined with small litter size, or small body size is associated with a litter size significantly larger than that observed in the larger strain.   As a general rule, breeds of chickens with large body size are poorer egg producers than are breeds of chickens with a smaller body size.   Although this suggests an inverse relationship between the hormones concerned with the control of these physiologic differences, it is possible that the principle cited earlier to explain differences between beef and dairy breeds may apply.

There are examples which show that beneficial effects of genic complexes controlling hormonal action at one step of the reproductive process may be partly or completely offset by adverse genic effects acting at another step.   The work of Fekete[350] shows that between two strains of mice ovulation rate is inversely correlated with reproductive efficiency and litter size.   Despite a significantly higher rate of ovulation, the litter size of mice of the DBA strain is significantly lower than the litter size of mice of the C57BL strain (table 54).   Analysis of this situation has shown that DBA ova are as viable as the ova of the other strain, but that the higher loss of ova in DBA mice is

Table 54

COMPARISON OF RATE OF OVULATION AND REPRODUCTIVE EFFICIENCY OF TWO INBRED
STRAINS OF MICE

Strain	Average no. of eggs per ovulation	Average no. of young per litter	Percentage of eggs developing into young
DBA	8.2	4.8	58.3
C57BL	6.7	5.6	83.9

due to an inhospitable uterine environment in which DBA eggs find themselves. A similar situation which results in lowered fertility in cows, and which could be ascribed to an unfavorable uterine environment, has been described by Hawk *et al.*[535]

The comparative rates of the efficiency of reproduction of the two strains of mice present a good example of the necessity of analyzing each step of the reproductive process separately before deciding which stage is adversely affected by genetic mechanisms. Obviously, the genetically determined low size of litters of DBA mice must not be ascribed to a genetically low rate of secretion of gonadotrophic hormones. Several other examples of genic action at secondary or tertiary steps are known in reproduction, lactation, and growth.

The nature of the adverse uterine environment on implanting ova is not known, but it is known that definite differences exist among strains of animals in the intensity with which uteri respond to their trophic hormones and in the types of responses produced. Drasher[288] demonstrated significant differences in sensitivity of uteri of mice of five different inbred strains to a standard dose of estrogen (table 55). She also made comparisons between the immediate effect of estrogen and the residual

Table 55

EFFECT OF 3 GAMMA OF ESTROGEN INJECTED INTO 5 STRAINS OF INBRED, CASTRATED MICE
KILLED 13 DAYS OR 73 DAYS AFTER CASTRATION

Days killed after castration	Strain	129	C57BL/6	C3H/He	C3HeJH	DBA/1
13	Uterine Wt. (mg.)	127.5	63.0	58.5	74.7	80.6
	Total RNA	592	88	35	281	264
	Total DNA	893	378	682	522	668
	Total N	1903	888	835	1012	1060
73	Uterine Wt. (mg.)	+13.9	+0.8	−26.8	−40.7	−64.9
	Total RNA	−81	+273	−5	−95	−108
	Total DNA	−381	−80	−357	−325	−536
	Total N	+594	+242	−307	−477	−855

effect which could still be detected about 60 days after initial treatment.   Drasher found that uteri of DBA mice were much more sensitive to estrogen than were the uteri of mice of the C57BL strain.   However, when the effects of estrogen on the uteri of these two strains were compared some 70 days after the initial treatment, it was noted that uteri of females of the C57BL strain still showed a carryover effect which manifested itself in a smaller loss in uterine weight, as well as in a smaller loss in the number and size of cells than those noted in the DBA strain.   Whether the significantly different rate at which the injected estrogen was metabolized by the two strains is related to the significantly different hospitality of the uteri of these strains noted by Fekete is unknown.

It seems redundant to cite extensive evidence to show that highly significant differences have been found with regard to the sensitivity of different strains of animals to exogenous hormones.   Apparently, degree of inbreeding has little to do with either the sensitivity or with the variability of response.[189]   Some strains may be less sensitive to one hormone but more sensitive to another.   Munro *et al.*[920] compared the sensitivity of various end organs of five different breeds of chicks to the various hormones present in a crude anterior pituitary extract (table 56).   This preparation contained gonado-

Table 56

ORDER OF BREED SENSITIVITY TO THE INJECTION OF EXOGENOUS HORMONES (ORGAN
WEIGHTS EXPRESSED AS PERCENTAGE OF BODY WEIGHT)

Breed	Comb ♂	Thyroid ♂	Thyroid ♀	Gonad ♂
White Leghorn	1	5	2	4
White Wyandotte	5	1	1	2
Light Sussex	4	2	4	3
New Hampshire	2	4	3	5
White Rocks	3	3	5	1

trophic hormones which cause an increase in the weight of testes due to their ability to stimulate growth of the seminiferous tubules as well as of the interstitial tissue.   The latter is known to secrete the male sex hormone, androgen, which causes comb growth. In response to gonadotrophic hormone White Rock chicks showed the greatest increase in weight of testes, while White Leghorn chicks were near the bottom in sensitivity. The androgen secreted by the stimulated testes caused the greatest increase in comb size in White Leghorns and an intermediate increase in White Rocks.   It is probable that the combs of Leghorns were larger because they were more sensitive to smaller doses of androgen rather than because their testes, in spite of their lower sensitivity to gonadotrophin, secreted more androgen.   This illustrates the importance of selecting the proper test animal for each hormone to be assayed.   If one wanted to use chicks as assay animals one would obviously select White Rock chicks for the assay of gonado-trophic hormones and Leghorn chicks for the assay of androgen. The decision whether to choose an inbred or outbred strain of animals for certain assays would depend on many factors such as the hormone to be assayed and the availability of strains. Some F_1 hybrids are more sensitive to hormonal stimulus and less variable in their response than either or both of the inbred parental strains while with other hormones and other strains the opposite may be true.

The relationship of the genotype to sensitivity to stimuli is shown in an analysis of the endocrinology of the broody instinct in chickens.[931]   Normally, cocks do not show maternal instincts toward chicks and when confined in close quarters with chicks they actually become aggressive toward them.   Broodiness is thought to be controlled by complementary autosomal genes and by a sex-linked gene which appears to be present only in chickens of the Cornish breed.   Although it is probable that this genetic scheme is an oversimplification of the actual genetic control of the highly complex physiologic characteristic of broodiness, it will serve to illustrate a principle.   In the experiments in question it was found that the injection of the hormone prolactin can induce cocks to become broody and to exhibit complete maternal behavior toward chicks.   This includes the whole complicated ritual of clucking, tidbitting, hovering, and defending the chicks.   It was further noted that genetically nonbroody breeds of cocks required much larger doses of prolactin and a longer period of injection before exhibiting the full maternal responses (table 57).   The dose of prolactin required to

Table 57

COMPARISON OF PRESUMED PHENOTYPES AND DOSES OF PROLACTIN REQUIRED TO INDUCE
MATERNAL BEHAVIOR IN COCKS

Breed	No. of males	Total dose of prolactin (units)	Phenotype of breed
Cornish	3	300	Intensely broody
W. Plymouth Rock	2	400	Broody or
W. Plymouth Rock	2	500	occasionally broody
S.C. White Leghorn	1	500	Not broody
S.C. White Leghorn	3	700	Not broody

induce broodiness in cocks correlated rather well with the observed phenotypes of hens of the three breeds studied, as well as with the presumed genotypes of these breeds. In some White Leghorn cocks it was difficult to induce the full broody responses even with very large doses of prolactin and after prolonged periods of injection.   It is not known whether the greater sensitivity to prolactin of genotypically "broody" cocks is due to the possibility that their pituitary glands secrete endogenous prolactin which, added to the exogenous prolactin, raises the hormonal level enough to permit expression of maternal behavior in males.

Finally, brief mention should be made of cases of genetic sterility which may be due to a variety of causes, some of them apparently based on endocrine malfunction, although the etiology is not always clear.   In the slow-gaining strain of pigs mentioned earlier there was a high incidence of cystic ovaries which, in this particular instance, appears to be associated with an imbalance of the components of the gonadotrophic hormone.   The propensity to cystic ovaries also has an hereditary basis in cattle. In the Swedish Highland breed of cattle, ovarian hypoplasia is inherited, being most pronounced in animals carrying two pairs of double recessive genes.   The left ovary is most commonly affected but both ovaries may be hypoplastic.   In extreme cases of hypoplasia follicles are completely lacking, but in partial hypoplasia only the primary follicles are absent.   It is interesting to note that bulls of this genetic strain of cattle show testicular hypoplasia which affects the left testicle first and more severely than it does the right.[323]

Genetically sterile mice affected by the obese-sterility syndrome have ovaries which normally remain infantile throughout life but which are capable of responding to gonadotrophic hormones.   With appropriate hormonal therapy, mice of this strain can be made to produce litters.[1226]   It appears in this case that the infantilism of the mice is due to a deficiency of gonadotrophic hormones in genetically obese mice of the *obob* genotypes.   The remainder of the reproductive apparatus is unaffected by this genotype and, as far as its ability to provide fertilizable ova and proper uterine environment for normal pregnancy is considered, it retains its ability to respond normally to the proper exogeneous hormones.

In a recent experiment the site of genic action on hypophyseal malfunction was analyzed.[155]   The pituitary glands from genetically dwarfed mice were implanted into the sellae turcicae of hypophysectomized normal mice, and pituitaries from the latter were implanted into sellae of dwarf mice.   It was found that the genotype of the hypophysis determined the growth rate of the recipients, allowing dwarf recipients of normal pituitaries to grow normally, while normal recipients of dwarf pituitaries failed to grow.   This experiment shows that the primary genic action is on the pituitary gland itself and not on the hypothalamus.   Snell[1235] showed that this type of dwarfing is due to a single, autosomal gene which causes deficiency of the growth hormone secreting acidophils in the hypophysis.[403]

CONCLUDING REMARKS

Data selected from the literature have been presented to show that genes or genic complexes determine (in a general way) the rates at which hormones are secreted by glands as well as the sensitivity of the target organs to their specific trophic hormones. Different sets of genes may affect the chain of reproductive events at several different levels.   Some genic effects may be primary, for example, those acting on the regulator of the rate of reproductive events, the adenohypophysis.   Other genic effects may be secondary in that they influence reproductive events during the intrauterine life of embryos.   The end results of either primary or secondary genic effects may be identical, that is, impairment of fertility, but the immediate endocrine causes of these effects are completely different.

The majority of the crucial reproductive events are governed by several interacting hormones or by overlapping series of hormones.   The genetic mechanism controlling the endocrine steps in the chain of reproductive events are probably as complex as the interactions between the hormones.   Furthermore, the action of target-specific trophic hormones may be modified by other hormones which are nonspecific for a particular target organ but which, nevertheless, contribute to the euhormonal state of the organism as a whole and thus govern the sensitivity of the response of the target organ to its specific trophic hormone.

In spite of these complications, several methods can be used in evaluating the rôle of genes in the regulation of rates at which glands secrete hormones.   These methods include: the direct estimation of the concentration of hormones in the glands producing them; the biological or chemical estimation of hormones in body fluids; the indirect estimation of rates of endogenous hormonal secretion by measuring the changes they produce in their specific target organs (that is, comb size, rates of ovulation, litter size, and so forth); and finally, the sensitivity of end organs to injected hormones.

DISCUSSION

Dr. Burdette: Mr. W. K. Whitten of the Jackson Laboratory will first discuss Dr. Nalbandov's paper.

Mr. Whitten: First of all I would like to mention a small item in terminology. As an Australian, part of my national heritage is the world's largest population of rabbits, and young rabbits are called kittens.   Dr. Nalbandov has given us a critical review of the method to be used in the study of genetics of reproduction, and he has illustrated this with frequent examples from his own research.   I am sure that by application of these techniques, much useful knowledge will be obtained.

The next step in the problem is to identify the primary action of the genes involved, and to do this we will need much more biochemical knowledge.   It will be necessary to know the pathways of hormonal synthesis, catabolism, and also the metabolic processes which they catalyze.   It will also be necessary to identify the hormonal receptors which probably occur on the surface of the target cells, because I feel that it is at this level that the genes will have the effect.

Some progress has been made with the metabolism of the steroid hormones, but the picture is so complex that it reminds me of some of the slides presented at this meeting.   Possibly the hormones of the anterior pituitary are themselves the primary product of the genes.   This is further suggested by the fact that gonadotrophin resembles rather closely the blood-group substance of humans.   The correlation of growth and gonadotrophic action which Dr. Nalbandov brought out is not a surprise to me. In fact, I have always believed that these hormones were in some way linked.   It is possible that the primary genetic unit is a hormonal complex and that the hormones as we know them are units of secretion or perhaps chemists' artifacts.   In this regard, it would be interesting to know the gonadotrophin content of the pigs of the slow-growing strain, the ones that were not cystic.   Were you able to determine this, Dr. Nalbandov?

Dr. Nalbandov has rightly emphasized the need to study each step of the reproductive process separately, and I think this is particularly important because there are many mutually inhibitory reactions in the various steps.   I think it was by the elimination of the reproductive phases of brooding and incubation that it was possible to progress so far with the genetics of egg production of chickens.   With improved techniques of the husbandry of the newborn, we may be able to study reproduction in mammals uncomplicated by the process of lactation.

DR. WRIGHT: I was interested in Dr. Nalbandov's correlation of large size and large litters.   Many years ago, I was working with 23 different inbred strains of guinea pigs.   They came to differ characteristically in each element of vigor: percentage born alive, percentage raised of those born alive, regularity in producing litters, size of litter, weight at birth, and gain to weaning.   The weights and rates of gain were strongly correlated, but for the most part the averages in these different elements of vigor showed no correlation among the strains.   There was consistency in the particular combination of characters of each strain, year after year.   One family would be very high in regularity and low in size of litter, and we encountered almost every possible combination with but one exception.   There was a strong correlation between weight and size of litter such as Dr. Nalbandov has found.

DR. NALBANDOV: I find this very interesting, Dr. Wright.   It is possible that only these two hormones are responsible for body size and for litter size and that they are linked.

DR. WRIGHT: Yes.

DR. NALBANDOV: Mr. Whitten asked about the hormonal content of pituitary glands of noncystic animals.   We had so few animals that the assays are questionable. What data we have show that the content of gonadotrophic hormone of the hypophyses in the small, slowly growing subrace was lower than it was in the rapidly growing strain.

DR. GOWEN: Have you examined mice of the strain Goodale established, as well as MacArthur's?   It has been our experience that the Goodale strain is difficult to maintain in reproduction.   Also, strains of mice differ widely in their reproductivity over their life span.   Some strains have large litters at first and soon play out.   Other lines have relatively uniform litters over long periods of time.   Others show a curvilinear increase or decrease in litter size with aging of the parents.   Others are so fertile they breed every estrus.   Hormonal variations probably contribute to these differences.   Do you have any data covering these possible interactions?

DR. HESTON: We have that line of Goodale's, and my impression is that they have large-sized litters when they do have them.   However, many of them do not raise their litters, and many of them are sterile.

Dr. Nalbandov: The examples given by Drs. Gowen and Heston emphasize the fact that breakdowns in the reproductive processes may occur at different levels. The genetic ability to produce large litters does not guarantee the ability to nurse a large litter or that the young may die of neglect.

J. P. Scott, Ph.D., and John L. Fuller, Ph.D.

BEHAVIORAL DIFFERENCES

Rather than repeat material already familiar to geneticists, we shall discuss in this paper special methods and problems peculiar to behavioral genetics. These special problems arise from the fact that any study relating genotype to behavior requires some extension of the concept of phenotype. Ordinarily the phenotype of an organism is expressed in terms of its structure or color, and these characteristics are relatively similar under a variety of conditions. A behavioral phenotype, however, is transient and can be evoked only under a specified set of conditions. The choice of these conditions and the nature of the phenotypic measurement are the major concern of this paper.

Behavior, while it has a physiologic basis, is usually defined as the activity of an entire organism rather than in terms of the activity of a single organ system. Thus behavior is one additional step removed from primary genic action, and the genetics of behavior occupies a position somewhere between physiologic and population genetics.

Because of the relative remoteness of the behavioral phenotype from the gene, emphasis [410] has been placed on the noncongruence between genetic organization and behavioral organization. Noncongruence might at first thought imply that genetics has very little to do with behavior, but the empiric evidence is quite the contrary. Almost any measure of behavior can be shown to be affected by genotype, and, conversely, almost any major genotypic difference can be shown to affect behavior. Thus the genetic analysis of behavorial variation is, and will continue to be, a lively field of investigation.

BASIC CONCEPTS

The unit of behavior.—The fundamental unit of behavior is the behavioral pattern, which may be defined as a regular combination or series of actions having a particular adaptive function. Behavior can, of course, be almost infinitely subdivided into minutely perceptible movements, but such units themselves have no special function and any one may be part of many different patterns of behavior. To take a familiar example, the crowing behavior of a rooster is a behavioral pattern. Analyzed in terms of bodily movements, components are found which are parts of eating, drinking, visual investigation, and the like.

Behavioral patterns having a similar function can be combined into a loose behavioral system. For example, fighting mice show several patterns of behavior: attack, defensive posture, running away, and complete passivity, and all these may be considered as parts of the agonistic behavioral system. For purposes of genetic analysis it is usually more rewarding to study specific patterns rather than to search for a general trait such as aggressiveness.

Behavioral analysis becomes even more difficult when behavior is measured in terms of performance or adaptation. When behavior is studied on the level of the individual organism, its primary function is adaptation to the environment. Consequently, an important dimension is the degree of adaptation achieved. The vast majority of scientific tests of intelligence and performance are based on a scale of this sort. Obviously this is not a scale found in the physical sciences, nor is it directly related to the action of single genes. The person who is interested in performance is measuring the results of behavior, not behavior itself. When such results can be achieved in only one way, genetic analysis should be reasonably simple. However, in the majority of practical situations, adaptation can be achieved through a variety of behavioral patterns or combinations of patterns, and the test measures only the end result. In short, measures of this sort usually involve extremely complex interaction, and the genetic analysis of differences in performance may therefore be extremely difficult and unclear. For example, we devised a test for dogs in which food had to be pulled out from under a box with a string. Some breeds of dogs would do this with their teeth and others with their paws. The end result was the same, but the motor aptitudes involved were quite different. Most other methods of measuring performance, from the currently popular bar-pressing technique for measuring behavior of rats to human intelligence tests, are subject to the same difficulty.

In a recent paper, Ginsburg[432] has argued that the best natural unit for analysis of behavior is the effect of a single gene. This definition may be of great practical importance in such syndromes as phenylpyruvic oligophrenia. However, the natural units of behavior proposed above do not, for the most part, directly correspond to genes or other natural units on the genetic level of organization. Actually, genes and behavioral patterns are separated by complex pathways of physiologic events. One may start at either end of the path and try to find the way through. In either case, there are apt to be many branchings and turnings so that a single gene may connect with several patterns, and the paths from a single pattern may lead to a variety of genes.

The relationship between genetic and behavioral variation.—Behavior is one of the ways by which organisms maintain homeostasis in the face of environmental changes. In its simplest sense, behavior is an attempt to adapt to some sort of change. The adaptiveness of responses is ordinarily considered to be the outcome of natural selection. Organisms with responses not adaptive are less fit and have fewer survivors. Because of the great variety of possible changes in the external environment, there is little possibility of evolving ready-made behaviors which will adapt an organism to every situation. The course of evolution has been toward an increased capacity for behavioral variability. As Jennings[661] pointed out long ago, some variability of behavior is found even in protozoa. Behavioral variability is, however, most characteristic of the vertebrates and particularly of mammals. Referring again to the fighting behavior of the mouse, we have noted that a caged male attacked by another will at first react by fighting back. If this is unsuccessful, he runs away. If he cannot escape, he assumes a defense posture and as a final resort may become completely passive. A primary characteristic of adaptation through behavior is variation according to the conditions of stimulus. An animal tries first one behavioral pattern, then another, and finally adopts the one which is most successful. (In this account, we are deliberately omitting abnormal or maladaptive behavior which may result from unsuccessful adaptation.)

The science of genetics is concerned with the study of variation. When the effect of heredity on behavior is analyzed, an apparent paradox is discovered: that one of the characters affected by genes is behavioral variability itself. Consequently, a large proportion of behavioral variation independent of genetic variation must always be expected. Homeostasis of the individual may be best accomplished by changes in behavioral patterns rather than constancy.[894]

Another factor which must be included in the study of behavior of higher animals is the process of learning. Repetition of a particular situation or stimulus tends to reduce behavior to learned habits which are relatively nonvarying. Such increased constancy of individual behavior, however, does not necessarily reduce the proportion of variance assignable to heredity.[410]

The concept of threshold.—For the most part, animals do not respond uniformly over a wide range of stimulus intensity. No reaction may be given until a certain threshold is reached, and beyond this point increased stimulation sometimes has little effect. A behavioral character to which the threshold concept may be applied is the syndrome of audiogenic seizures in mice. Sounds below a certain intensity do not evoke seizures. Once threshold is reached, however, the intensity of the convulsive response may be just as great when the response is made to jingling keys as when it is made to a powerful motor-driven bell. The threshold concept may be applied to the genotype as well as to the stimulus. Temporarily it will be assumed that susceptibility to audiogenic seizure is determined by a single pair of alleles A and a. The AA genotype in the model is highly susceptible to sounds and convulses very easily. The *aa* genotype is so resistant that it will not convulse under ordinary circumstances, since compensatory reactions occur within the nervous system before a convulsive level of activity is reached. It is now further assumed that the *Aa* genotype is poised upon a genetic threshold such that the occurrence or nonoccurrence of a convulsion is dependent upon environmental circumstances. This particular model has been employed to account for the pattern of inheritance of convulsions in crosses between DBA/2 and C57BL/6 mice[409] with the added postulation that the threshold was dependent upon multiple loci rather than a single locus. The evidence for multiple factors was derived from breeding experiments. There are good reasons to believe that threshold situations of this sort may be extremely frequent in the field of behavior.[1187] Certain features of this model should be carefully noted. In the first place, a heterozygote theoretically might be expected to be particularly adaptable since its behavior is a function of the immediate environment; but this very fact implies greater variation in behavior, hence lower reliability of measurement. A second feature of the model is that the heritability of a threshold character decreases as the incidence of the character approaches 50 per cent.[245] This fact creates problems in selection for such characters and in the interpretation of Mendelian experiments.

GENERAL METHODS

The developmental method.—Learning itself may be considered as a process of functional differentiation of behavior.[1186] This implies that the nervous system of a mammal is differentiating throughout life in the way that most other organ systems differentiate during the embryonic period. Consequently, the developmental method becomes extremely important in studying the effect of heredity upon behavior.

It is almost a truism to state that any phenotypic character is not inherited but developed. The importance of this concept in the genetic study of behavior rests on the fact that a very large proportion of the behavioral development of a mammal takes place after birth and proceeds far into adult life. A given type of behavior may be completely absent early in development, may appear in one form a little later, and in still another form at a later age. This may enormously complicate the usual method of analyzing genic mechanisms from a single phenotype taken at one age.

An example of changes in a trait with age is our study of the inheritance of barklessness *versus* barking in dogs (figure 38). At 5 weeks of age, barking was extremely uncommon in the test situation, and there was little difference between the breeds. At 11 weeks of age, barking had risen to a maximum, and there was a wide gap between the barkless basenji and the other breeds. By 15 weeks the rate of barking was declining, so that it was again difficult to separate breeds on the basis of this character. In such cases, the necessity of the developmental method is obvious, as is the problem of determining just what is a barkless dog. We are still not sure whether barking should be treated as a unit character or as a quantitative character.

Another example of the change of behavioral phenotype with age is the syndrome of audiogenic seizures.   A number of investigators[407, 1344] have demonstrated that the pattern of age-susceptibility to convulsions differs among strains.   Frings and Frings have pointed out that one could obtain almost any genetic ratio desired by conducting tests at an appropriate age.   Classification of individuals into reactors and nonreactors is thus often a function of some underlying developmental process.   One must be cautious in interpreting such a classification in terms of a dichotomy of genotype.

Fig. 38. AVERAGE NUMBER OF BARKS GIVEN BY FIVE DIFFERENT BREEDS OF DOGS IN COMPETITION FOR A BONE AT DIFFERENT AGES.

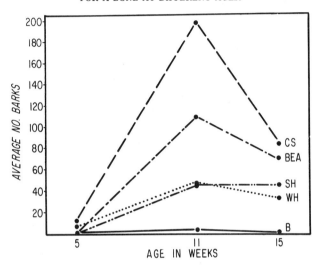

From top to bottom, the breeds are the cocker spaniel, beagle, Shetland sheepdog, wire-haired fox terrier, and "barkless" basenji.

Controlling environmental contributions to variation.—In all genetic experiments it is essential to separate those differences between individuals produced by biological inheritance from those produced by environmental differences.   In the case of behavior, the range of environmental effects is much greater than for such characters as coat color or blood type.   Some characteristics of an organism such as body size are, of course, determined in large measure by environmental factors.   However, the investigator of physical characters need not usually pay as much attention to the previous history and to the conditions of measurement as the behavioral geneticist.   Maternal effects upon behavior, for example, extend beyond the uterine period through lactation and, in some species, far beyond weaning.   Maternal effects may be transmitted by means as diverse as chemicals diffusing through the placental barrier, composition and amount of milk, amount of sheltering of the young, and age of weaning.   Thus, in a recent study of inheritance of behavior in the dog, F_1 males were backcrossed to the parental stock females in order to compare backcross and F_1 generations reared by the same mothers.   Such a design helps to separate environmental and genetic sources of variance. The F_2 generation in our experiment was reared by F_1 mothers whose hybrid vigor created a quite different maternal environment. It would have been desirable in this case to have fostered F_2 puppies upon mothers of the parental strain. With the dog this is not always easy to accomplish, since cross fostering can be successfully carried out only between females in approximately the same stage of lactation.

Another means of obtaining estimates of the effects of random and environmental factors is comparison of successive litters from the same mating. Such litters will, on the average, have similar heredity, although random variation will occur if the parents are heterozygous.

Another solution of the problem of equalizing or eliminating maternal effects is hand rearing. Under such circumstances, of course, it is possible that the behavioral patterns usual in the species will not appear, since the eliciting stimuli may be omitted in the hand-rearing situation. Thus, the greater control of environmental variation may detrimentally affect the appearance of the behavior of interest. Isolation of subjects to insure equality of environmental conditions suffers from the same drawback. It is important in studying the development of behavior to place the subject in an expressive environment.[410] Such an environment is one designed to favor the elicitation of species-specific behavioral patterns. It is conceivable that the technique of semi-isolation[408] may prove to have some value in behavioral genetics. In this method, animals are isolated at weaning and held in cages which permit minimum stimulation. They are removed, however, at regular intervals and placed in a highly stimulating environment. This method insures control of development while providing essential conditions for psychologic development.

The measurement of behavior.—As stated earlier, behavior may be defined as the activity of the intact organism. Behavior is complex and obviously can vary in more than one dimension. Behavioral analysis of a character such as fighting in the mouse begins with a set of descriptions of the various behavioral patterns involved. When two mice are placed together, the measurements consist of the latency, frequency, and intensity of the fighting pattern. Latency and frequency can be measured with a stopwatch and counter. An enormous amount of behavioral work can be done with no more elaborate instruments than an arena, timer, and counter. The intensity of fighting behavior is more difficult to measure, as one cannot usually apply direct measures of force. This is a technical and not a theoretical difficulty. Presumably each blow and each bite involves the expenditure of a definite amount of energy, but the application of a measuring instrument would interfere with the behavior. Rating scales for intensity have been used, but these create problems of reliability and of equality of intervals on the rating scale.

Latency, frequency, and intensity of behavioral patterns are often not independent measures. A pair of mice that fight often are apt to fight quickly and vigorously; but since the correlations between these measures are less than unity, one might expect to arrive at somewhat different genetic conclusions depending on the particular measure employed.

Another difficulty is encountered when several behavioral patterns are used to identify a particular trait. For example, sniffing of the female genitals is part of the male sexual behavior of guinea pigs.[510] This behavioral pattern, however, actually appears more frequently in low-sex-drive males than high-drive males. The latter proceed directly to more active courtship and copulation. As indicated above, the measurement of behavior tends to result in a quantitative analysis of the effects of genetics. However, behavior may also be analyzed in terms of the presence or absence of specific patterns, and the possibility of unitary characters is not excluded.

Choice of conditions for testing.—Conditions of stimulation are critical for the extension of group differences. This is expressed diagrammatically in figure 39, which shows

Fig. 39. HYPOTHETICAL RELATIONSHIP BETWEEN STIMULUS AND INTENSITY OF RESPONSE IN TWO POPULATIONS, A AND B.

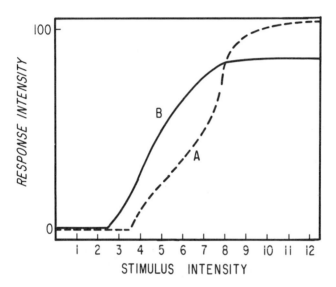

an hypothetical relationship between stimulus and response in two populations, A and B. Note that below a stimulus threshold, 3 for B and 4 for A, no response is elicited in either population. On this measurement, therefore, B would be considered the more responsive. As intensity of stimulus increases up to 6, the differences between populations also increase. Above 6, they converge until they are equal at a stimulus intensity of 8. At still higher levels of stimulation, responses are on a plateau and a physiologic limit has been reached.

Obviously, measurements of strain differences which follow this model are dependent upon the particular level of stimulus in the situation for testing. Measurements of threshold stimuli have not been widely used in behavioral genetics, although any test which involves the presence or absence of a behavioral pattern probably involves such a threshold. Measures taken at a physiologic limit have the advantage that stimulus intensity can vary over a rather wide range without changing the difference between the strains. It is clear, however, that such a measure tells only part of the story and that a more dynamic concept of behavior is obtained when we consider the quantitative stimulus-response relationship over a wide range of intensities. That is, A and B may be compared with respect to the slope of a curve of learning or extinction or in terms of a psychophysical function. Such measures may have complex genetic elements, but they are apt to be more meaningful in relation to the adaptive function of behavior.

Genotypic versus phenotypic orientation.—The preceding remarks point up a major difference in the methodology of behavioral genetics, genotypic *versus* phenotypic orientation. In the former, an investigator starts with a known genotypic difference and studies its effects upon behavior. In such a situation a gene, a chromosome, or a whole genotype is analogous to a treatment applied to an organism. The genotypically oriented investigator does not usually stop with demonstrating a correlation between genotype and behavior. He is also interested in tracing the path between gene and character through intervening physiologic mechanisms.

In using mammals he has available two general methods. The technique of strain comparison makes use of populations with differences at a large number of loci, some of which are known but the majority are unknown. The experimenter using this method has an excellent opportunity to find behavioral differences, but genetic analysis is apt to prove difficult except in terms of a coefficient of heritability. Some success has been obtained in analyzing results of crosses between strains by standard biometric techniques using transformed scores,[140, 410] but these methods sometimes yield ambiguous results.

More precise genetic analysis is possible when a single genic difference can be established between the experimental populations. One of us[1182] used the method of repeated backcrosses to eliminate genes linked with brown and white eye color in *Drosophila*. The gene *bw* appeared to have no important effect upon phototaxis. The white-eye gene did affect phototaxis. Earlier work on the effect of genes on the behavior of *Drosophila* often failed to take into account the numerous loci in addition to the one with conspicuous morphologic effect.

The preparation of stocks of this type in mammals is laborious, but the behavioral geneticist can find a number of inbred strains of mice maintained in forced heterozygosis at a particular locus. The difficulty of such methods is that the genes in question may have no behavioral effects, or, at least, no effects detectable on a background of environmentally induced variability. One can, of course, force the issue by using genes which produce major defects in sensory and motor organs, but the behavioral effects in this situation are not of great psychologic interest.

Here it may be pointed out that behavioral techniques potentially provide a sensitive method for the detection of cryptic mutations—those without readily detectable morphologic defects. When heritable differences in behavior appear between recently separated sublines of established inbred strains, it is conceivable that a single locus may be involved.   A possible instance of this sort has been reported by Denenberg[246] who found that different sublines of C57BL/10 mice differed in rate of avoidance learning even when reared under identical conditions.   Genetic analysis of this phenomenon would be most interesting.   We have no means of knowing how frequently such cryptic mutations may occur, and how successful a systematic search with a number of reliable behavioral measures may be.

The problems of behavioral differences may also be attacked from a phenotypic orientation.   The behavioral patterns which can be chosen for investigation are of course almost infinite, and the choice will depend upon personal interests.   Behavioral patterns which are important in adaptation have especial interest.   Many biologists and psychologists are much more interested in the heritability of important adaptive patterns than in determining the effects of a rare mutant gene upon a standard genetic background.   One can be quite sure that findings on the inheritance of choice of mate, dominance and submission, or problem-solving ability will not be trivial insofar as the welfare of the species is concerned; but the very fact that these behaviors are so important makes it unlikely that they will have a simple genetic base.   Thus the investigator in this area must sacrifice genetic clarity in order to work with characters of major importance for survival.

The distinction between the phenotypic and genotypic orientation is, of course, not absolute, but a matter of degree.   Both approaches have advantages and disadvantages and both will contribute to the development of behavioral genetics.

Statistical analysis.—One of the special problems of the analysis of the inheritance of behavioral patterns is the assigment of variance to genetic and environmental components.   It is relatively easy to determine the heritability of a trait in any particular environmental situation, but such a figure has little general significance because of the responsiveness of behavior to environmental change.   The method may, however, be highly important in determining whether or not there is a major genetic component in a behavioral trait, and Broadhurst[136] has demonstrated how it may be applied to a simplified cross-breeding experiment.

On the other hand, the detailed analysis of Mendelian mechanisms through analysis of variance frequently fails because of complicated interaction between genetic and environmental factors.   Scott, Fuller, and King[1188] have demonstrated how a relatively simple genetic mechanism may interact with environmental factors to produce a highly complex manifestation of a phenotype.   The annual seasonal breeding cycle of the African basenji dog appears to be determined by a single recessive gene, but the expression of the character is also controlled by decreasing diurnal length. F_1 hybrids with cocker spaniels show a tendency to run six-month nonseasonal cycles like the cocker, but also to respond in part to the seasonal changes in light, so that their breeding cycles are much more variable than those of either parental strain.

With more complex genetic mechanisms, the interactions become so complex as to defy analysis in most cases.   Bruell[140] has suggested that it is possible through the use of means and medians to determine whether or not a behavioral trait is consistent with Mendelian inheritance, without attempting to measure the effect of segregation on variance.

Scott[1183] has suggested a method of analysis of segregation independent of variance, based on the point of maximum separation between the two parental strains and a comparison of ratios between the two backcrosses. The method assumes the existence of a threshold, plus the additive effect of genetic and environmental factors which lead to crossing the threshold, and it can be used where such assumptions are justified. It will give an estimate of the number of genetic factors involved and can be used to predict an expected ratio in which correspondence with actual data can be tested by the usual statistical methods.

FUTURE DEVELOPMENTS IN BEHAVIORAL GENETICS

A major problem in behavioral genetics is the investment in manpower needed to determine a behavioral phenotype. A long period of testing under carefully controlled conditions may be required to obtain a score which can become part of a genetic analysis. Mechanization of tests may in the future make possible more elaborate experiments. Hirsch and Boudreau[584] devised a geotactic maze for *Drosophila* which sorted out the population automatically. Bruell[140] has automated various pieces of apparatus for the study of murine behavior. It might be hoped that the operant conditioning procedure (which is almost the limit of automation) would prove useful for behavior genetics, but in actuality the long period of time required for shaping behavior in the operant apparatus seems to preclude such applications, and procedures so far used have been designed to minimize genetic variations. It does, in fact, seem doubtful that the more complex functions of learning can be automated to the extent that large numbers of subjects can be run through tests with little involvement of humans. Many behavioral patterns are more accurately and reliably recorded by the human eye than by elaborate apparatus. Nevertheless, ingenuity and the desire for new tests may well help the investigator with his problems of manpower.

The demonstration of differences in behavior between strains should continue, but it is to be hoped that such demonstrations will merely be the prelude to more elaborate genetic analyses. The modern tools of biometric genetics have actually been applied only sporadically to behavioral characters. Experiments by Bruell[140] and by Broadhurst[136] indicate a trend toward more sophisticated designs.

Possibly the most significant advances will be made in the relation of genes to behavior through biochemical pathways. Ginsburg[431] has outlined the general scheme for such studies. The relationship between investigations of this sort and the search for a biochemical factor in mental disorder in man[704, 705] is obvious. Behavioral phenotypes are more difficult to measure than ordinary phenotypes, but a large number of investigators have demonstrated that the problems are not insuperable. In fact, the difficulties in studying the genetics of behavior are fairly similar to those encountered in studying the inheritance of complex physiologic characters. The methods, too, are similar in that emphasis may be placed alternatively upon the effects of a well-defined genetic entity or upon the genetic contribution to the variance of a character continuously distributed.   The concept of thresholds which appear so important to behavioral genetics is familiar to students of growth and was formally described by Wright[1451] in his discussion of polydactyly in guinea pigs.   Thus the unique nature of behavioral genetics lies predominantly in the need for more extensive and detailed control of the life history of the subjects and of the use of special methods developed by psychologists and animal behaviorists for the objective measurement of behavioral patterns.

DISCUSSION

DR. BURDETTE: Dr. Benson Ginsburg of the University of Chicago will open the discussion of the paper of Dr. Scott and Dr. Fuller.

DR. GINSBURG: One section of the paper by Drs. Scott and Fuller should be emphasized by repeating it.   "The preceding remarks point up a major difference in the methodology of behavioral genetics, genotypic *versus* phenotypic orientation.   In the former, an investigator starts with a known genotypic difference and studies its effects upon behavior.   In such a situation a gene, a chromosome, or a whole genotype is analogous to a treatment applied to an organism.   The genotypically oriented investigator does not usually stop with demonstrating a correlation between genotype and behavior.   He is also interested in tracing the path between gene and character through intervening physiologic mechanisms."   I wholeheartedly agree with this point of view and cite it as the reason that the Scott-Fuller paper really does belong in a section devoted to physiologic genetics.

My own view toward so-called behavioral genetics follows from this position: behavior is a biologic aspect of organisms under genetic control and with an evolutionary history.   I do not think that I am any more or less of a geneticist since turning from pigment studies to behavior.   It is simply that one is studying another phenotype belonging to a different and more fundamental aspect of organisms.   I believe that genetics occupies a central synthesizing position in the biological sciences and that it contains the organizing principles for thinking about evolution; for thinking about ontogeny; and for thinking about methods of investigating the ways in which organisms come to be the way they are, both phylogenetically and ontogenetically, including having the capacities to interact with each other.   Physiologic genetics is thus involved in the study of the capacities of the nervous system, the endocrines, and the way in which all the capacities of the organism behave, including those making possible the complexities of group organization, development, and interaction.

At the level of the gene, genetics occupies the same central position in a broader context.   The new laws of physics and chemistry will probably come from a study of these biomolecules, the genes, and the ways in which they are able to duplicate themselves and control cellular activities.   In this manner, genetics occupies a very central and important position on the threshold of the future of all sciences, from the physical to the psychologic.

Turning to the psychologic, I would like to make a few points in relation to the position that to a psychologist, a rat is a rat and a mouse is a mouse. This should not be so. One can, biologically speaking, parse the creature, mouse, into genera, species, races, sexes, phenotypes, strains, and genotypes. If one takes almost any aspect of the psychologic literature, for example, the effect of early experience, one finds that general principles are developed that purportedly apply to all rats or all mice, if not to all infrahuman mammals, so that such effects as those of early manipulation on adult emotionality can be described and predicted.[1162] If, however, one goes to the inbred mouse and samples different strains, what one finds is something quite different. A majority vote of a population of mixed strains or of random-bred mice may follow the general rules if the minority votes are discarded. Even with genetically controlled materials, one can find situations (genotypes) in which the rules are followed. One can also find genotypes in which, with exactly the same experimental techniques, the behavior, measured in the same way, is exactly the opposite to that predicted by the general rules. In still other genotypes, the same experimental manipulations will make not a particle of difference to the later behavior.[430] Environment or nurture has been nicely classified, controlled, and manipulated by psychologists and sociologists; but nature remains to them merely a rat, a mouse, or a mammal.

Another example approaches more closely the kind of thing that Dr. Russell mentions and which, as Dr. Heston pointed out, is still not as close to the gene as we would like to get, but, nevertheless, much closer. If one considers various genotypes within the house mouse, either on the background of an inbred strain as in Dr. Coleman's work or by comparisons between strains, one finds differences in the ability of the nervous system to handle common substrates. One of the things that we discovered in our group through collaboration with investigators at the Roscoe B. Jackson Laboratory years ago was that the controversy over the rôle of glutamic acid in behavior was a situation of precisely the kind that I have been describing. The ability of this simple substance to affect the nervous system and thereby behavior depended on the genotype within a behavioral phenotype. Within that phenotype some strains existed in which glutamic acid would give all of the positive effects claimed in the literature, but for other strains this was not true. The ability of this substrate to affect the nervous system and through this, the behavior, was a function of the genotype.[432] Moreover, the way in which it affected the nervous system was also a function of the genotype. In mice of the DBA/1 strain, the threshold of audiogenic seizures can be depressed by administered glutamic acid. In other high-seizure strains, one cannot necessarily do this. Along with alleviation of the seizures, one can improve learning performance on mazes which are located on the thresholds of its ability and, therefore, represent a mild stress. If this effect is compared to the effect of the same agent on another high-seizure strain, DBA/2, it is found that, whereas the effect increases with dose and then remains elevated in DBA/1 mice, other ways of handling the glutamic acid metabolically are present in 'DBA/2 mice and the strain very quickly returns to its own behavioral equilibrium.   In this case the return is maladaptive because glutamic acid helps the organism.   The initial doses produce the effect, but an escape occurs with subsequent doses.   The difference between the effect of glutamic acid in the two strains is clearly due to a difference in genetic capacity.

Another important point concerns the definition of a natural unit of behavior or phenotype.   Is gross bodily activity, which is the sum of various behaviors, any more a natural unit than fighting under confined conditions?   The ethologists have attacked the problem of the natural unit by making the behavior misfire.   Imprinting to an inappropriate object identifies an entire behavioral complex and provides one method for extracting a natural biological unit, and this, in a sense, is a kind of natural phenotype.

Another way of defining a natural unit is to grasp whatever genetic difference exists, whether it be a strain difference or a genic difference, whether it involves climbing up on a pole, turning an activity wheel, agressiveness under confinement, or whatever, and consider that this is that part of the iceberg that is above water and that only by analyzing what genetic complex is behind it and then working back to the causal mechanism as well as forward again to produce the phenocopy, can the part of the iceberg that is below water be found.   A behavioral phenotype is the result of research and not something that is given at the beginning.

This, to me, is behavioral genetics.   It involves a physiologic-genetic approach to behavioral phenotypes.   It is in no way a new science.   If one reads the earlier genetic literature, one finds that when Mendelian traits were described, many of them were, broadly speaking, behavioral.   They ranged all the way from mutations that affected the sense organs, such as *rodless retina*,[692] to position preference in coach dogs.[695] That behavior is under some genetic control has been known for a long time, and this so-called phenotypic approach has not advanced our knowledge of it very far.   The physiologic-genetic approach is needed to bring about such an advance.   Darwin had a prototype of a strain difference in behavior that he refers to in *The Expression of the Emotions of Man and Animals*.   He describes a little hybrid girl who was half French and half English.   When she was still very young, she shrugged her shoulders in a completely un-English way, although reared in an English environment.   On the phenotypic side, we have not progressed much beyond this in so-called behavioral genetics to date.   We shall not progress unless a physiologic-genetic approach is followed as outlined in these comments.

DR. FULLER: The two previous speakers have been discussing two different viewpoints.   Dr. Scott emphasized the phenotypic orientation towards behavioral genetics, in which one starts with behavioral patterns which are defined as a phenotype and works backward to find out how they are inherited.   The advantage of this approach is that it deals with behavior important to the organism and significant in evolution. However, one may have to sacrifice genetic clarity for the sake of behavioral significance. In the genotypic approach one uses genes as a form of treatment.   The effects of the treatments may be assayed by measuring behavior which is comparatively trivial, although it serves well for analysis of the genetic effects. The two methods are not really contradictory but are, rather, supplementary.

Dr. Scott illustrates examples of genetic populations which diverged and converged in behavior as they developed. Such convergence or divergence of means does not necessarily mean that genetic effects are becoming less or greater as the animals grow older. In some instances we have been able to show that the contribution of heredity to differences as measured by interclass correlation stays relatively constant as animals age or as they practice a skill. Looking only at means may be deceiving.

Finally, I would like to close by stating that behavioral phenotypes are more difficult to measure than ordinary phenotypes, but a large number of investigators have shown that the problems of behavioral genetics are not insuperable. In fact, the difficulties of studying the genetics of behavior are very similar to those encountered with complex physiologic characters. Great progress may be expected from studies of behavioral effects of well-defined genetic entities. Another important area is the determination of the heritability of continuously distributed traits. The concepts of a developmental threshold is familiar to students of growth and also appears important in behavioral genetics. The unique factor of behavioral genetics is its need of extensive and detailed control of the life history of its subjects and its use of special methods for the reliable objective measurements of transitory phenomena.

Dr. Scott: Dr. Ginsburg argues that behavioral genetics is no different from physiologic genetics, and I agree that one of the basic scientific questions in this field is how heredity affects behavior. However, I think that behavorial genetics goes beyond physiologic genetics in that a higher level of organization is being investigated and consequently phenomena are encountered which are not apparent on the lower level. Dr. Wright has already illustrated the complicated types of interaction involved in the inheritance and production of pigment. Adding another level of organization makes Dr. Wright's system appear relatively simple.

BIOCHEMICAL GENETICS

Raymond A. Popp, Ph.D.

MAMMALIAN HEMOGLOBINS†

Variability among mammalian hemoglobins was recognized many years ago,[33, 51, 285, 732, 1047] but recent advances in methods for their identification, separation, and more complete characterization have enabled biologists to utilize differences between hemoglobin to probe more deeply into the challenging and tantalizing problem of the mechanisms of genic function. Physical methods for distinguishing mammalian hemoglobins have been the subject of many papers since 1940. Technologic advances in protein chemistry are now being applied to analyze the chemistry of hemoglobins as well as other molecules of biological importance. Chemical analyses of the content and sequence of amino acids in some peptide chains have been reported for a variety of mammalian hemoglobins within recent years. The chief objective of this review is to cite methods that have been applied in studies of mammalian hemoglobins and to compare and evaluate the usefulness of each method for survey, preparative, and/or analytical procedures.

GENERAL COMMENTS

Attempts to elucidate the nature of genic action have been based on the premise that the specific physical and chemical properties of macromolecules are reflections of the function and specificity of genes. The molecules chosen for such studies should, as far as can be determined, be primary products of genic action, have similar biological functions and similar physical and chemical properties, yet occur as intra- or interspecies

† Some of the analyses in this paper were made possible through the cooperation of Dr. Norman G. Anderson and his group.

variants that can be easily isolated from other biological contaminants without denaturation in quantities sufficiently large for comparative physical and chemical analyses.   Among other molecules, mammalian hemoglobins seem to satisfy these criteria.

Hemoglobin is a conjugated protein composed of heme prosthetic groups, which are united to the polypeptide chains of a globin.   The physical structure, four heme units associated with four globin coils, appears to be similar for all mammalian hemoglobins.[998]   Heme is compsed of four pyrole groups with one molecule of iron and is similar in all mammals.   Each of the four polypeptide chains of the globin molecule is composed of approximately 150 amino acids.   In contrast to heme, the globin moiety differs among species and frequently occurs in more than one form within a species.

Establishment of homogeneity or heterogeneity of hemoglobin may require the use of many physical and chemical methods, since small interspecies variations frequently cannot be revealed by use of a single method.   Intraspecies heterogeneity is often more difficult to demonstrate; however, the genetic implications that might be derived only through complete characterization of intraspecies differences become the reward for the added effort required to elucidate such variations.   Ultimately, only a complete analysis of the amino-acid sequence will establish unequivocally the homogeneity or heterogeneity of some intraspecies hemoglobin variants.   Use of pooled samples should be avoided if at all possible, since properties of individual samples may be obscured.   However, pooling of samples may be permitted if strains of laboratory animals are used in which homogeneity of individuals has been established.

The objectives of the investigator should be decided at the outset, since the information being sought usually dictates the analytical procedures.   Simplicity is desirable, but simple methods frequently yield equivocal results.   Every method has its limitations and sources of error, and the investigator should be acquainted with them.   Moreover, hemoglobin is a relatively labile protein and may therefore be altered during some preparative or analytical procedures.   Hence, it may be necessary on occasion to establish that minor hemoglobin variations observed with use of a particular technique reflect actual hemoglobin differences *per se* and are not modified forms or complexes of hemoglobin with other molecules, such as haptoglobins.   The validity of the results is strengthened, however, if similar results are obtained by more than one method.

For the purposes of presentation, the methods are classified according to whether they are used to study a physical or a chemical property of the molecule; some methods could be placed under either category.   An evaluation and discussion of each method is presented, and several references are cited for the application of each method to the analysis of mammalian hemoglobins.

PHYSICAL METHODS

Electrophoretic analysis

Electrophoresis is commonly used to determine the isoelectric point and electrophoretic mobility of biological materials.   Knowledge of the isoelectric point is of value not only in characterization of proteins, but also aids in choosing physical conditions necessary for the isolation and purification of proteins. The properties of proteins that enable them to be separated in an electrical field have been discussed.[4] When electrophoretic analyses indicate a single component in the system, the fraction may nevertheless not be pure by other criteria.

Moving-boundary electrophoresis.—The moving-boundary apparatus developed by Tiselius[1318] has been modified many times to increase its resolving power and thereby its usefulness as a tool for research. An important modification has been the development of an apparatus with four electrophoretic cells that can be used simultaneously.[1010] The essential components of the instrument, the U-shaped electrophoretic cells containing two electrode compartments for application of the electric field[804] and an improved optical system for hemoglobin[648] to permit localization and visualization of the shape of the moving solutes, have been described in the articles cited. The methodology for moving-boundary electrophoresis is thoroughly described by Longsworth.[803]

Electrophoretic mobility, as experimentally defined, represents the distance traversed by the boundary or particle under observation per unit time per unit volt. The electrophoretic mobility of mammalian hemoglobins is in the order of 10^{-5}cm^2/sec/volt at pH 7.0. The migration of molecules is affected by the hydrogen-ion concentration, ionic strength of the buffer, chemical composition of the buffer with regard to valence of the ions and viscosity of the buffer, and the surface charge density of the molecule under investigation. Thus, the physical conditions should always be stated precisely. Ideally, as many different moving boundaries are formed as there are different types of molecules in the solution. Observations on many proteins reveal that interactions between different molecular species that affect their electrophoretic mobilities are relatively rare. However, components in the buffer may interact with proteins, such as hemoglobins and enzymes, producing a shift in their isoelectric points.[994, 1195]

Buffers used for studies on hemoglobin by the moving-boundary technique are sodium phosphate with dithionite, 0.1 ionic strength, pH 5.7–8.0,[994] and cacodylate–NaCl, 0.1 ionic strength, pH 6.5.[994, 1211, 1381] More rapid separations can be achieved by using buffers of lower ionic strength; Beaven *et al.*[67] used sodium phosphate, ionic strength 0.05, pH 8.0, and Itano and Robinson[649] used potassium phosphate, ionic strength 0.01, pH 6.85. The lower ionic strength is satisfactory for qualitative analyses, but quantitative analyses may not be highly accurate under such conditions. Differences in the moving-boundary electrophoretic patterns of murine carbonmonoxyhemoglobins are illustrated in figure 40.

Moving-boundary electrophoresis is not a practical method with which to survey for differences among hemoglobins. The required instrument is expensive, only a few samples can be analyzed in one day, large volumes of hemoglobin solutions are needed, and considerable time is required to prepare the samples. Nevertheless, moving-boundary electrophoresis, along with techniques of X-ray diffraction[997, 998] and chemical analysis[1053, 1054] has been a valuable analytical tool for studying problems of theoretical interest to hemoglobin chemists. The structural arrangement of the four globin units[524, 650, 651, 652, 1065, 1210] has been approached by subjecting mixtures of electrophoretically distinguishable hemoglobins to conditions of low or high *p*H to dissociate hemoglobin into half-molecules and subsequently to allow the half-molecules to reassociate at neutral *p*H. The mixtures of reassociated molecules are then analyzed by the moving-boundary technique or other methods, that is, paper, starch-gel, and agar-gel electrophoresis or column chromatography can be used, for presence of fractions other than the two original components. Presumably, half-molecules of structurally similar hemoglobins can hybridize upon molecular reassociation at neutral *p*H. Hybridization of human and canine hemoglobin is reported in the latter two

Fig. 40. Comparison of murine carbonmonoxyhemoglobins analyzed by moving-boundary electrophoresis.

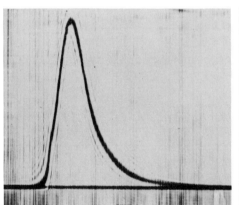 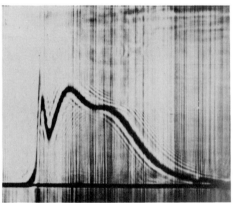

Patterns were photographed after eight hours of electrophoresis; phosphate buffer, ionic strength 0.1, *p*H 7.5, 158 volts, and 20 milliamps.
Left (A). Single hemoglobin of strain C57BL mice.
Right (B). Diffuse hemoglobin of strain 101 mice.

studies cited above, suggesting that canine hemoglobin contains configurational analogs of the *alpha* and *beta* chains of human hemoglobins. If the ionic change on the polypeptide chains of the hemoglobin differ, it may be possible to establish the plane of cleavage of the dissociated half-molecules. The results obtained with human hemoglobin indicate that the hemoglobin molecule dissociates into non-identical half-molecules at low *p*H,[1210] the two *alpha* chains separating as a unit from the two *beta* chains; however, at high *p*H the hemoglobin molecule dissociates into identical half-molecules.[524] Each of the half-molecules of hemoglobin will dissociate further into two subunits, but the reaction is irreversible since such units will not reassociate at neutral *p*H as do the half-molecules.[524, 616]

Zone electrophoresis.—A more compact and less expensive instrument than that required for moving-boundary electrophoresis was necessary before electrophoretic analysis could become a routine laboratory procedure. Zone electrophoresis may be carried out using many types of supporting media. In reviewing zone electrophoresis, the methods are classified by the kind of supporting medium that is employed. Theoretical aspects of electrophoresis discussed above are also applicable to zone electrophoresis. Comparative studies on the electrophoretic mobility of mammalian hemoglobins are commonly carried out with zone electrophoresis, but exact determinations of electrophoretic mobilities are best determined by the moving-boundary techniques. Frequently the latter method also offers higher resolution. Because of its simplicity, zone electrophoresis is generally preferred as a system when surveying populations of animals for hemoglobin differences.

1. Paper electrophoresis: Methods using paper as a supporting medium, first described by Wieland and Fischer[1387] and Haugaard and Kroner,[525] developed more rapidly than other methods of zone electrophoresis. Designs of several types of equipment have been illustrated elsewhere[111] and the methodology has been outlined very well.[203, 648, 760]

For optimal results and reproducibility, the factors that influence the migration of hemoglobin during electrophoresis should be constant throughout the paper, that is, the paper porosity must be uniform and vapor pressure, quantity of hemoglobin applied, electrical current, temperature, and electroösmosis should be constant.[1337] The filter papers commonly used as supporting media are Whatman no. 1, 3 cm. wide, 3MM and 531 or Schleicher and Schuell 2043A for hanging strip methods, and Whatman 3MM, 10 to 20 cm. wide for the sandwich technique. More hemoglobin can be applied to thicker papers and the presence of minor components can therefore be detected more readily.[50] Moreover, thicker paper gives better reproducibility among runs, but thinner paper is more often used when quantitative determinations with recording instruments are to follow. Vapor-pressure equilibrium should be reached before electrophoresis is begun, the case used for hanging strips should be air tight to maintain the vapor pressure, and the glass plates used for the sandwich technique should be sealed with silicone. Simpler instruments without cooling devices may overheat at high currents, and the resulting evaporation of buffers from the paper may produce distortions of electrophoretic patterns and inconsistent results. The addition of inert substances, such as glycerol,[822] inhibits evaporation of buffer from the paper. Pretreatment of the paper with the buffer may possibly reduce the adsorption of proteins and reduce electroösmosis by neutralizing the negatively charged carboxyl ions of the cellulose paper. The buffers commonly used for paper electrophoresis of hemoglobins are barbital, pH 8.6 and ionic strength of 0.025–0.06, and phosphate, pH 6.5 and ionic strength of 0.1.[446]

After electrophoresis, the proteins are fixed to the paper by heat denaturation at 110–120° C. for 15–30 minutes. For quantitative analysis, the temperature and period of time for heat denaturation should be controlled, since the capacity of proteins to bind dyes is affected by different conditions of heat denaturation.[111] The hemoglobin on the paper strips can be stained by one of the following methods: bromophenol blue,[311]

Amido black 10B,[360] or light green.[48]   If desired, either the stained or the heat-denatured, unstained paper strips can be placed in a recording device to determine the mobility and relative amount of hemoglobin present.   Several sources of error are inherent in any type of device chosen; these factors have been reviewed elsewhere.[712] Hemoglobin solutions studied are not pure; for example, they generally contain carbonic anhydrase and methemoglobin reductase.   Staining procedures are not specific for hemoglobin but rather stain proteins.   It is necessary to prove by other means that a stained peak, especially a minor one, is hemoglobin.   A few additional reports, not mentioned above, concerning the use of paper electrophoresis for the investigation of hemoglobin variants in many mammals, are easily available.[153, 331, 392, 443, 656, 1039, 1340, 1384]   Population studies are reported in these papers; the inheritance of some of the variant forms has been established.

Conventional paper-strip electrophoresis is not an efficient technique for obtaining large quantities of a pure fraction of hemoglobin from blood that contains more than one type of hemoglobin.   Continuous flow, paper-curtain electrophoresis[111] has been successfully used to separate serum fractions, but reports were not found of its use for isolation of hemoglobins.   In our laboratory we found that hemoglobins are tightly adsorbed to the curtain and that good resolution of the fractions is difficult unless they are electrophoretically quite different.

2. Starch-block electrophoresis: Starch-block electrophoresis is often used for preparative procedures when larger quantities of an electrophoretically pure fraction are needed than can be obtained by paper-strip electrophoresis.   The technique was first described by Kunkel and Slater.[739]   It is perhaps less suitable for routine analysis than is paper electrophoresis, although comparable electrophoretic patterns are obtained with either technique.   It has the advantage that large quantities, up to 1,000 mg., of hemoglobin can be applied; moreover, adsorption of proteins on starch is less pronounced than on paper, which facilitates removal of individual components by elution. It should be mentioned that electrophoresis on sponge rubber, from which materials can easily be removed by squeezing, has also been investigated,[882] but its usefulness for isolation of hemoglobins has not been reported.   The procedures for starch-block electrophoresis have been set forth previously.[737]   The starch block is usually prepared by pouring a barely liquid paste of washed potato starch and buffer, usually barbital buffer, 0.05–0.10 ionic strength, pH 8.6, into a mold (38 × 10 × 1.5 cm.).   Blotting paper is placed at each end of the tray to remove excess liquid and is retained until no more buffer remains on the surface of the starch.   A slot is made in the starch block and the sample, as either a solution or a starch paste, is added with a pipette.

Minor components (A_2 and A_3) of normal human hemoglobin were first isolated by use of starch-block electrophoresis.[736, 740, 856]   The technique has been used clinically in the diagnosis of the Thalassemia trait, an inherited condition in which the quantity of A_2 hemoglobin is elevated from normal values of 1.2–3.5 per cent to abnormal values of 4.2–6.8 per cent of the total hemoglobin.[429, 738]   An increase in the amount of A_2 hemoglobin above 3.5 per cent is usually diagnostic for Thalassemia, although the quantity of A_2 hemoglobin may increase under the influence of other types of anemia.[671] Cepellini,[187] through the use of starch-block electrophoresis, has recently identified a new human-hemoglobin fraction (B_2) which he suggests may be the product of an allele of the locus that controls the synthesis of A_2 hemoglobin in humans.

3. Starch-gel electrophoresis: Resolution on starch gel is greater than that on paper. The adsorption of proteins on starch gels is low, similar to that for starch-block electrophoresis; the gels are easy to prepare and can be preserved for permanent records. Another advantage of the starch gels over paper is that the rate of migration is less dependent on concentration of protein. Furthermore, samples being compared can be run adjacent to one another on the same starch-gel block without appreciable interference by lateral diffusion of the proteins during electrophoresis. The design of the apparatus, preparation of gels, application of material, staining, and other details of the method for starch-gel electrophoresis have been described by Smithies.[1228, 1229] The technique is not so simple as paper electrophoresis, but the increased resolution achieved more than compensates for its increased complexity. Prehydrolyzed starch has recently become commercially available,[1227] thus removing a major source of variability. Twelve to 16 g. of starch per 100 ml. of buffer is heated until the opaque mixture becomes transparent and thick. The solution should not be removed at this point but is heated longer and shaken vigorously until it begins to become less viscous. The air bubbles are moved by placing the heated solution under reduced pressure for few seconds *using a flask with heavy walls* before the liquid starch is poured into plastic trays. A cover is placed on each tray removing excess starch and producing a uniform starch-gel block. Optimal gelling is achieved by placing the preparations in a refrigerator at 4–6° C. overnight.

Boric acid-NaOH buffers, 0.015–0.04 ionic strength, pH 8.0–8.6, are most commonly used, but other buffers, such as formate, 0.02 ionic strength pH 1.9,[918] phosphate,[1228] and acetic acid-sodium acetate (unpublished) have been used with hemoglobin. The gelling property of the starch is influenced by the type of buffer and the amount of starch used; thus, the percentage of starch may need to be changed when a new buffer is tried.

Several methods have been used to apply samples to the starch-gel preparation. Samples to be inserted may be mixed with starch to make a thin paste that is pipetted into a slot made in the gel, or samples may be applied to filter paper that is placed into a transverse cut made in the gel. The slot method of application is recommended when larger samples are required; however, resolution under such conditions may be somewhat reduced. The quality of paper, when the technique of paper insertion is used, is important; if the paper adsorbs hemoglobin tightly, the bands are spread more. A thin plastic sheet is placed over the gel to reduce evaporation during electrophoresis.

The electrophoretic patterns are better near the center of the gel; therefore, the starch-gel slabs are sliced horizontally and the hemoglobin patterns at the inner surfaces are compared. It may not be necessary to stain hemoglobin owing to its red color, but the minor components of multiple hemoglobins become more readily visible following staining (figure 41) with a saturated solution of Amido black 10B in methanol, H_2O, and acetic acid (5:5:1). It is more difficult to estimate the quantity of different fractions of hemoglobin in starch-gels than on paper strips. Photoelectric devices have been used with preparations of starch gels made temporarily transparent by boiling them for 30 seconds in a 10 per cent solution of acetic acid.[1342] Permanent preparations

Fig. 41. COMPARISON OF MURINE OXYHEMOGLOBINS ANALYZED BY STARCH-GEL ELECTROPHORESIS.

Red-cell lysates were electrophoresed for four hours; borate buffer; ionic strength 0.025, pH 8.5, 6 volts/cm., 6 milliamps. Hemoglobin was stained with Amido black 10B.
ABOVE (A). Single hemoglobin of strain SeC mice.
BELOW (B). Diffuse hemoglobin of strain BALB/c mice.

for photometric analyses can be made[361] by mounting stained, transparent starch-gel blocks in agar. Quantitative analysis of eluted hemoglobin can also be made after freezing the gel, which makes the gel spongelike, and removing the pigment by applying pressure.

The electrophoretic mobility of proteins is not identical in starch gel and on paper,[1020] presumably due to greater adsorption of proteins on paper. Although two-dimensional electrophoresis—paper in one direction and starch gel in another at right angles to it—can be used for some protein separations, this system is perhaps not useful for hemoglobin. However, because of the greater resolution two-dimensional electrophoresis affords, it might be useful for distinguishing hemoglobin from hemoglobin-haptoglobin complexes through the use of the differential peroxidase activities that these substances exhibit.[979, 1228] Application of starch-gel electrophoresis to the study of hemoglobin differences in mammals appears in several reports.[49, 1016, 1017, 1072, 1103]

4. Agar-gel electrophoresis: The use of agar gel in electrophoresis, like starch gel, developed more slowly than paper, but agar is being used in place of starch by some investigators because it makes excellent preparations for quantitative analyses. The electrophoretic patterns in agar are very similar to those obtained by moving-boundary electrophoresis, as 1 per cent agar gel is essentially an aqueous medium, yet agar gel has sufficient structure to reduce free diffusion of macromolecules during and after electrophoresis. Most agars commercially available should be purified[151] before use for best reproducibility of results.

The systems used for agar-gel and starch-gel electrophoresis are very similar with one exception—liquid agar is usually layered about 3–6 mm. deep on photographic plates rather than poured into plastic molds. Slits for application of filter papers containing the samples are cut in the sheet of agar after it gels. Several slits can be made in an agar plate and as many samples of hemoglobin can be analyzed simultaneously. The pH range in which agar gel can be used satisfactorily lies between 6 and 9, which is more restricted than that for starch-gel electrophoresis. Citric acid-sodium citrate buffer, ionic strength 0.025, pH 6.5,[1064] and barbital buffer, ionic strength 0.025, pH 8.2,[471] are commonly used for hemoglobins.

For quantitative determinations, hemoglobin can be eluted from agar gel after freezing in a manner similar to that described for starch gels. The agar can also be dried following electrophoresis and hemoglobin stained with a solution of Amido black 10B dissolved in a 0.1M sodium acetate–1.2M acetic acid buffer containing 15 per cent glycerol, pH 3.5.[360] The stained, washed, and dried agar film can be separated from the plate; the film is transparent and ideal for quantitative analyses in densitometric devices.

Techniques of immune electrophoresis[470] in agar gel have not been reported for the study of hemoglobin, but such methods could be applied in the study of hemoglobins.

Column chromatography

The theory of ion-exchange chromatography has been discussed by Boardman and Partridge.[112] Pure ion-exchange adsorption is a function of the conditions of equilibrium between (1) the protein and the buffer, (2) the protein and the resin, and (3) the resin and the buffer.

(1) $\text{Protein–NH}_2 + \text{H}^{\oplus} \rightleftharpoons \text{Protein–}^{\oplus}\text{NH}_3$

(2) $\text{Protein–}^{\oplus}\text{NH}_3 + \text{Resin-COONa} \rightleftharpoons \text{Resin–COONH}_3\text{–Protein} + \text{Na}^{\oplus}$

(3) $\text{Resin–COONa} + \text{H}^{\oplus} \rightleftharpoons \text{Resin–COOH} + \text{Na}^{\oplus}$

Exclusive of differential filtration, which may also influence separation, it can be deduced from the above formulae that optimal conditions for ion-exchange chromatography should occur in a resin that is in equilibrium between its acid and salt forms near the isoelectric point of the protein.

Column chromatography has been used for identification of hemoglobin differences and also for isolation of hemoglobin fractions. Boardman and Partridge[112] first demonstrated the separation of sheep and bovine hemoglobins on ion-exchange resins, and Huisman and Prins[612] developed the method for distinguishing several abnormal human hemoglobins. The reviews of Hirs[583] and Moore and Stein[891] should be consulted for a complete description of how to assemble a column for chromatographic separations.

In brief, the components of the system are a reservoir for buffer and a column containing the ion-exchange resin through which the buffer and dissolved protein pass under gravitational force. Use of an automatic fraction collector permits continuous recovery of isolated fractions over periods during which the column need not be attended. The quantity of hemoglobin in the effluent fractions is determined by photometric assay, adsorption at 415, 540, or 575 mμ being commonly used. If the hemoglobins collected occur in different forms, that is, HbO_2, HbCO, HbCN, or $Hb^\oplus$, a wave length should be chosen that gives equivalent extinctions for equal amounts of each form of hemoglobin in the mixture.

Selection of a system of ion-exchange resin, buffer, pH, ionic strength, and elution rate is in part obtained by trial and error. Previous knowledge of the differences between the electrophoretic mobilities or isoelectric points of the hemoglobins to be separated is usually helpful, since the binding or adsorption of the molecule on a resin depends on the existence on the protein of a sufficient charge of sign opposite to that on the adsorbent. In general, there is a relationship between electrophoretic mobility and chromatographic behavior, but exceptions are not uncommon.[610] Amberlite IRC–50 (XE–64 or 97), a carboxylic cation-exchange resin, is most commonly used for separating mammalian hemoglobins. More recently, Huisman et al.[611] have investigated the use of carboxymethyl-cellulose, which has a higher capacity for the adsorption of protein, for separating some human and animal hemoglobins. Many ion-exhange resins have a tendency to bind proteins so tightly that denaturation or alterations of the molecule may occur during chromatography; an appreciable amount of methemoglobin is formed from carbonmonoxyhemoglobin unless chromatography is carried out in the cold at temperatures below 5° C. If proper ventilation is available, KCN can be added to the buffer to convert the methemoglobin formed to cyanmethemoglobin, which has chromatographic properties similar to those of carbonmonoxyhemoglobin. The buffer for ion-exchange chromatography should have the same ionic charge as the adsorbent, to avoid formation of pH variables at the point of interaction of buffer and adsorbent. The pH of the buffer should be chosen so that the protein is actually adsorbed or only a filtration process is accomplished. In general, adsorption of hemoglobin varies inversely with the pH of the buffer. As experimentally determined,[112] adsorption of hemoglobin below pH 7 is not due solely to electrostatic forces but is influenced by hydrogen bonding between the undissociated carboxyl groups of the resin and polar groups of the protein; that is, the positive charge of hemoglobin realized above pH 7 becomes less important below pH 7 for specific adsorption on the resin.   The influence of secondary forces, such as hydrogen bonding, at low pH may be helpful in aiding separation of components chromatographically similar at neutral pH.   As indicated by the formulae, adsorption can be reduced by increasing the sodium ion concentration of the buffer.   The salt concentration is generally increased stepwise as the fractions separate, such that elution is accomplished more rapidly and with smaller volumes of effluent.   The elution rate used by most investigators is 4–6 ml./hr. for columns about 1 cm. in diameter.   Specific conditions of buffers, pH, ionic strength, and flow rate for isolation of different components of human hemoglobins can be found in various articles.[8, 207]   Column chromatography has also been used as a preparative method to purify hemoglobin fractions used for further chemical analyses[8, 616, 1395] and to detect and isolate hybrid hemoglobin molecules formed during hybridization.[1345]   Column chromatography through molecular sieves, such as Sephadox, is being adapted for the separation and analysis of peptides of enzymatic digests.[578]

Ultracentrifugation

The ultracentrifuge is used to enhance the migration of molecules through a solution by increasing the intensity of the field of force.   The sedimentation of particles in the ultracentrifuge depends upon volume, shape, and density of the particle; viscosity and density of the medium; and the intensity of the gravitational field.   The sedimentation rate, described in terms of the sedimentation constant, s, can be used to calculate molecular weight by application of the formulae derived by Svedberg.[1304]   The molecular weights of hemoglobins (approximately 68,000) determined by this method agree with those obtained by diffusion methods.[512]   A description of the apparatus and methodology has been presented by Pickels,[1005] and a monograph on ultracentrifugation has recently been published by Schachman.[1159]

In general, the ultracentrifuge is not useful for distinguishing between mammalian hemoglobin variants since their molecular weights are usually near 68,000.   A determination of the molecular weight of a molecule is necessary for interpretation of some chemical analyses, such as amino-acid composition, peptide analysis, sulphydryl analyses, or end group analyses.   Popp and St. Amand[1017] and Gluecksohn-Waelsch[442] have reported that the diffuse-type hemoglobin of the mouse contains a component of heavy molecular weight (figure 42), but additional investigation is required to ascertain whether the heavy hemoglobin is synthesized as such or is formed *in vivo* by dimerization, including disulfide bonding.   In support of the possibility that the component of heavy molecular weight of certain mouse hemoglobins may be formed secondarily by chemical bonding after synthesis, Gluecksohn-Waelsch[442] found that increased quantities of a heavy molecular component were formed during storage at 4° C., although this heavy component might not be the same hemoglobin as that present in fresh preparations.

The physical structure of the hemoglobin molecule under various chemical conditions has also been studied by ultracentrifugation. Field and O'Brien[358] showed that human hemoglobin undergoes dissociation into half-molecules at pH 3.5–5.5, and recently Hasserodt and Vinograd[524] described a similar phenomenon at pH 11.

Fig. 42. COMPARISON OF MURINE OXYHEMOGLOBINS ANALYZED BY ULTRACENTRIFUGATION.

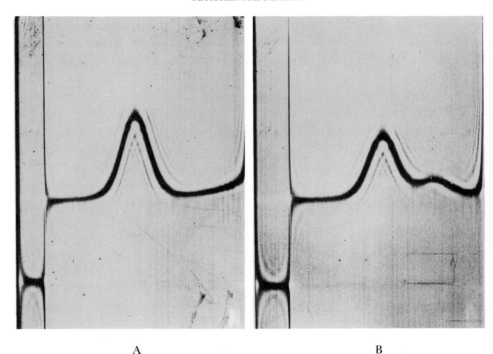

A   B

Patterns were photographed after 80 minutes of ultracentrifugation; phosphate buffer, ionic strength 0.1, pH 7.5, 5° C.

LEFT (A). Single hemoglobin of strain C57BL mice, S_{20} value, ~4.7.

RIGHT (B). Diffuse hemoglobin of strain 101 mice, S_{20} values, ~4.6 and 7.1.

The dissociation is more complete at pH 11, and less denaturation occurs. A disadvantage of dissociation at high pH, however, is that the heme-labelling methods described by Singer and Itano[1210] cannot be used to establish transfer of hemoglobin subunits during hybridization experiments, since methemoglobin of most species is rapidly denatured by alkaline hydrolysis. However, Vinograd and Hutchinson[1345] have used C^{14}-labeled hemoglobin to demonstrate molecular hybridization of carbonmonoxyhemoglobins dissociated at pH 11.

Alkali denaturation

Alkali denatures hemoglobin, producing a brown, insoluble, alkaline-globin hemochromogen. Körber[732] demonstrated by this method that fetal hemoglobin is more resistant than adult hemoglobin to alkaline decomposition. Although fetal hemoglobins of other mammals also differ in alkali resistance from their adult counterparts, greater resistance to decomposition does not apply to all fetal hemoglobins; that is, the fetal hemoglobins of sheep, goat, and cow are less resistant to alkaline denaturation.[134, 669] The number and percentages of different types of hemoglobin in a sample can often be determined from the rate of denaturation, since the decomposition of each hemoglobin in a mixture proceeds according to first-order kinetics.[526] This approach has been used in studying the hemoglobin of monkey,[1191] mouse,[1379] rat,[843] and sheep,[1338] as well as human hemoglobins.[202, 1385]

Crystallography and solubility

The classical monograph of Reichart and Brown[1047] should be consulted for illustrations of the various types of hemoglobin crystals that occur in different taxonomic groups of mammals. Drabkin's method[286, 287] of crystallization of hemoglobin has been used to obtain pure crystalline fractions of hemoglobins for analytical procedures, on the assumption that crystalline hemoglobin was homogeneous. However, Allen *et al.*[8] and Clegg and Schroeder[207] have shown by column chromatography that crystalline human hemoglobin is, indeed, heterogeneous.

Crystallography has also been used to differentiate between fetal and adult hemoglobins[670, 703] and more recently has been developed as an auxiliary method for the identification of different types of murine hemoglobins (figure 43) in conjunction with electrophoretic and solubility properties.[1015] These techniques have also been used in studies on the mode of inheritance of some hemoglobins of the mouse, as is mentioned later.

Crystallographic and solubility studies are usually carried out in parallel; because of the sensitivity of the hemoglobin molecule to physical variations, the methods used separately may not be highly reliable. Determination of hemoglobin solubility is based upon the determination of the amount of protein remaining in solution at a series of salt concentrations. The results are influenced by the pH, ionic species, and temperature of the solution and by the concentration and form of the protein.[212] The comparative solubilities of different hemoglobins are determined by establishing salting-out curves at constant pH, temperature, and hemoglobin concentration.[261, 262] A number of hemoglobins have different solubilities under identical conditions and may be distinguished on this basis.

In choosing a system for salting-out, the following factors should be considered: salts with a high buffering capacity are desirable, since hemoglobin solubility is greatly affected by change of pH; buffers containing multivalent anions, that is, phosphate, sulfate, and citrate, are more efficient for salting-out than those with univalent anions;

the solubility of the salt must be sufficient to allow a precipitating concentration to be reached; and the *p*H of the buffer should be near the isoelectric point of the protein, at which point solubility is generally minimal. The most commonly used buffer for hemoglobin studies is potassium phosphate. The hemoglobin is usually converted to carbonmonoxyhemoglobin, since the latter is more stable than oxyhemoglobin. The amount of carbonmonoxyhemoglobin left in solution is determined photometrically. To compensate for daily variations in the physical conditions, standards are usually run along with each group of assays. Popp[1014] has modified this technique by establishment of nomographs to enable quantitative determinations of mixtures of several murine hemoglobins within ± 5 per cent.

Fig. 43. COMPARISON OF MURINE CARBONMONOXYHEMOGLOBINS ANALYZED BY CRYSTALLOGRAPHIC METHODS.

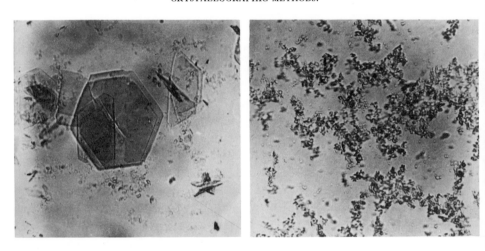

A   B

Precipitates were formed within 21 hours; 3.08 M phosphate buffer, *p*H 6.5, 24° C.
LEFT (A).   Crystals of single hemoglobin of strain C57BL mice.
RIGHT (B).   Amorphous precipitate of diffuse hemoglobin of strain 101 mice.

Literature on the solubility of hemoglobin is not extensive[65, 66, 478, 670, 687, 1068] in view of the fact that a solubility study may be performed with little effort and expense in conjunction with other types of assay. A relationship between electrophoretic and solubility characteristics may exist, as was found for the sheep[1338] and mouse.[1015] Solubility may also be helpful in distinguishing hemoglobin types that are not electrophoretically different. Itano[645] showed that although human hemoglobins *D* and *S* are electrophoretically similar, the reduced form of hemoglobin *D* is more soluble in phosphate buffer than the reduced form of hemoglobin *S*. More recent studies have indicated other differences between these hemoglobins which will be discussed in the next section. In studies on the inheritance of single and diffuse hemoglobins among inbred strains of mice, Popp (unpublished data) has observed that several strains possess single hemoglobins that are electrophoretically similar in starch gels, but three of these single hemoglobins have different solubilities in phosphate buffer (figure 44). The solubility of a 50/50 mixture of SeC and C57BL hemoglobins is intermediate between that of SeC and C57BL. The quantity and form of crystals that

Fig. 44. COMPARISON OF MURINE CARBONMONOXYHEMOGLOBINS ANALYZED BY SALTING-OUT METHODS.

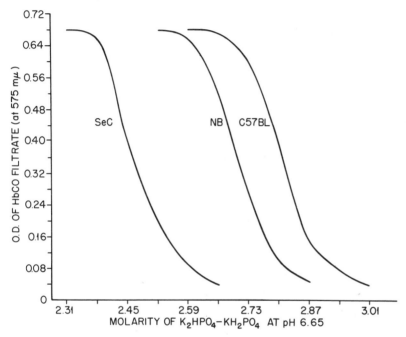

Optical density of carbonmonoxyhemoglobin filtrates obtained in phosphate buffers of variable molarities were read at 575 mμ; physical conditions were 0.3 per cent solution of hemoglobin, 21 hours of incubation, pH 6.65, 30° C.

develop in 21 hours at 30° C. in a 2.73 M phosphate buffer, pH 6.7, can also be used to distinguish the C57BL, (C57BL × SeC)F$_1$, and SeC hemoglobins. Data are presented in table 58 on the inheritance of solubility characteristics of the single hemoglobins of strain C57BL and SeC mice. Previous studies[1017] indicated that the hemoglobin locus that governs the molecular characteristics which determine the electrophoretic behavior of single and diffuse hemoglobins is closely linked to the *chinchilla* locus of linkage group I. Data in table 58 show that the *chinchilla* locus and the locus that controls the molecular characteristics which determine the solubility properties of C57BL and SeC hemoglobins segregate independently. Thus, at least two loci appear to be involved in directing the physical and chemical characteristics of hemoglobins in the mouse. Data on solubility, crystallography, and electrophoretic characteristics

Table 58

PROGENY RESULTING FROM F_1 INTERCROSSES AND F_1 BACKCROSSES OF C57BL AND
SeC CLASSIFIED FOR CHINCHILLA AND HEMOGLOBIN SOLUBILITY

Matings	Phenotype of Progeny†					
	$\dfrac{c^{ch} Hb^{SeC}}{c^{ch} Hb^{SeC}}$	$\dfrac{c^{ch} Hb^{SeC}}{c^{ch} Hb^{C57BL}}$	$\dfrac{c^{ch} Hb^{C57BL}}{c^{ch} Hb^{C57BL}}$	$\dfrac{C Hb^{SeC}}{-Hb^{SeC}}$	$\dfrac{C Hb^{C57BL}}{-Hb^{SeC}}$	$\dfrac{C Hb^{C57BL}}{-Hb^{C57BL}}$
	Number					
$\dfrac{C Hb^{C57BL}}{c^{ch} Hb^{SeC}} \times \dfrac{C Hb^{C57BL}}{c^{ch} Hb^{SeC}}$	4	6	6	18	25	13
$\dfrac{C Hb^{C57BL}}{c^{ch} Hb^{SeC}} \times \dfrac{c^{ch} Hb^{SeC}}{c^{ch} Hb^{SeC}}$ reciprocal crosses	36	28	—	34	38	—

† Symbols C and c^{ch} indicate full color and *chinchilla*, respectively; symbols Hb^{C57BL} and Hb^{SeC} indicate hemoglobin loci in strains C57BL and SeC, respectively.

of hemoglobins obtained from test progeny also indicate that the diffuse-type hemoglobins differ among some strains of mice (Popp, unpublished data).

CHEMICAL ANALYSES

The physical methods described in the foregoing section are used to identify different hemoglobins in mammals and to study their mode of inheritance. Such methods in themselves, however, do not reveal the details of molecular structure which distinguish one hemoglobin from another. Analyses of the amino-acid composition and sequence are required to elucidate these differences. Differences in the protein chemistry of hemoglobin molecules, resulting from genic action, are of concern to biochemical geneticists. Some of the equipment and methods described above are also used for analysis of the composition of the hemoglobin molecule, with the principal difference that the submolecular components are being analyzed and compared rather than the native hemoglobin molecule.

Amino-acid analysis

Column chromatography, used for recognition of different hemoglobins, as described earlier, is also used to separate the amino acids of hemoglobin hydrolysates. The procedures of amino-acid determination on ion-exchange resins, developed by Moore and Stein,[890, 892] has been applied by several investigators to assay the amino-acid composition of hemoglobins.[647] Automatic recording apparatus[1209, 1259] provides complete quantitative analyses within 24–72 hours.

In a comparative analysis of human hemoglobins *A*, *C*, *E*, and *F*, Stein et al.[1272] observed that the amino-acid composition of these hemoglobins is similar except that fetal hemoglobin contains an additional amino acid, isoleucine. Allen et al.[8] noted a similar isoleucine difference between adult and fetal hemoglobin. Since different adult hemoglobins may have similar amino-acid compositions, total amino-acid analyses fail to disclose the nature of the differences between closely related hemoglobins. However, the analyses are necessary for estimating the number of peptides that can be expected from enzymatic digestion for peptide analyses.

Sulphydryl analysis

Attempts have also been made to detect differences in the number of available sulphydryl groups in hemoglobin by amperometric titration.[630, 637]   As described by Ingram, the number of available –SH groups is determined by measuring the number of silver atoms per molecule bound as Ag–S–protein.   A weak ammoniacal silver nitrate solution is added to the protein until Ag ions are no longer bound.   The assumptions are made that the available –SH groups in hemoglobin will react with metal ions such as Ag or Hg in a 1:1 ratio and that the metal ions will not react with non-thiol groups in the protein.   Comprehensive studies have revealed minor differences in the number of thiol groups among hemoglobins of man, horse, ox, dog, cow, and sheep,[637, 921] and comparative studies indicate a similar number of –SH groups for human hemoglobins *A*, *S*, and *C*.   Although earlier reports using silver ions suggest that eight –SH groups are present,[597] recent studies using mercury ions report only six –SH groups in human hemoglobins.[13]   Further application of Hg titration in the study of mammalian hemoglobins can be found in a recent report by Riggs.[1056]

Thus far, configurational differences in hemoglobins that could be produced through disulfide linkage have not been observed.   Nevertheless, formation of intermolecular disulfide linkages could produce hemoglobins of high molecular weight, such as are encountered in some strains of mice.   Comparative results of sulfhydryl content of the native versus the denatured, heavy hemoglobin might be used to support or reject the possibility of intermolecular disulfide bonding.   Results obtained, however, should be interpreted with caution; Murayama[921] has reported that not all the –SH groups in native hemoglobin may be titratable and that the end point determined by amperometric methods is temperature dependent.

End-group analysis

Efforts to establish the amino acids that occupy the terminal position of the polypeptide chains of the globin moiety of hemoglobin have been more fruitful than sulfhydryl studies.   A free *alpha*-amino (N-terminal) group remains at one end of a polypeptide and a free carboxyl (C-terminal) group at the other end.   These end groups are free to undergo a number of organic reactions.

N-terminal analysis.—Several methods for N-terminal analysis of peptides and proteins have appeared in the literature.   The method developed by Sanger[1153] is based on the fact that 2–4-dinitrofluorobenzene will react with the free amino group forming a bond which is more stable than the peptide linkage of the amino acids in the protein.   After addition of the dinitrophenyl group (DNP) the protein can be hydrolyzed and a fraction of the DNP remains attached to the N-terminal amino acid.

DNP confers a yellow color upon the DNP-amino acid complex, which is useful in following the fractionation and subsequently in identifying the N-terminal amino acid by methods of paper or column chromatography.[111, 777] In addition to the terminal α-amino group, the δ- and ε-amino groups of arginine and lysine, which are not involved in the peptide linkage, will also react to form DNP complexes. This must be considered in determining which labeled amino acid is actually in the terminal position. Keil[698] has described a modification of Sanger's procedure for analysis of microgram quantities. The dinitrophenylation, extraction, and hydrolysis can be carried out in a single vessel, eliminating the possibility of loss of the peptide during extraction of the DNP complex from reagents and degradation products formed during the procedure.

The free α-amino acid will also react with thioisocyanates (for example, phenyl-thioisocyanate) to form a phenylcarbamyl-peptide, a principle applied by Edman[318] for the N-terminal analysis of proteins. In this complex, the terminal peptide bond adjacent to that upon which the phenylcarbamyl substitution takes place is more susceptible to acid cleavage than the other peptide bonds of the protein. Controlled acid hydrolysis splits off the N-terminal amino acid in a phenylthiohydantoin form soluble in 5 per cent $NaHCO_3$, while the other peptide bonds of the insoluble peptide are left untouched. The amino acid is freed from the phenylthiohydantoin complex by alkaline hydrolysis and can be identified by paper or column chromatography.

An important adaptation of N-terminal analysis has been its use in analyzing the sequence of amino acids in polypeptide chains. The method of Edman is suited for analysis of amino acid sequence through stepwise degradation of the polypeptide. Ingram[638] has adapted the DNP-labelling method for stepwise degradation through catalytic reduction of the DNP peptide. The sequence of the amino acids in the peptides is established through identification of the amino acid released after each stepwise degradation procedure. Hunt and Ingram[618] and Hill and Schwartz[579] have analyzed the use of the two methods for studying the sequence of amino acids in peptide number 4 of human hemoglobins.

With the use of the methods described above, the N-terminal amino acids and, in some cases, the sequence of adjacent amino acids have been established for hemoglobins of man,[581, 616, 667, 1053, 1054, 1168, 1196] horse, dog, cow, pig, goat, sheep, rabbit, and guinea pig.[785, 987, 988, 1156, 1157]

Enzymatic methods for determining N-terminal residues with the use of leucine aminopeptidase are under development.[580] Because of the rapid rate of the reaction, it is difficult to obtain information on the sequence of the amino acids released. Furthermore, leucine aminopeptidase may not react on all native or even denatured proteins; for example, performic acid oxidation was required before the N-terminal residue of albumin could be hydrolyzed by the enzyme.

C-terminal analysis.—Procedures for the chemical analysis of C-terminal amino acids have developed more slowly than those for N-terminal analyses. The method developed by Akabori[3] and Ohno[960, 961] is performed by hydrazinolysis of the protein which produces amino-acid hydrazides and amino acids; the latter are derived only from amino-acid residues possessing the free carboxyl group. Upon dinitrophenylation the hydrazide forms di-DNP-amino acid hydrazide and the amino acid forms DNP-amino acid. The C-terminal, DNP-amino acids are then separated by fractional extraction and can be characterized chromatographically and estimated colorimetrically. In the analysis the dicarboxylic amino acids, aspartic and glutamic, may be contaminants in the extract of the C-terminal amino acids. The application of the method has been widely demonstrated, and a review of satisfactory results of analyses of C-terminal residues of a variety of proteins is presented by Haruna and Akabori.[523]

The method as described is not useful for sequence analysis since all the peptide bonds of the molecule are destroyed during the hydrazinolysis. Ohno[962] and Niu and Fraenkel-Conrat[945, 946] have modified the hydrazinolysis method for sequential analysis of the amino acids from the C-terminal end. The principle consists of partial, rather than complete, hydrazinolysis of the protein followed by dinitrophenylation of the peptides and extraction of the C-terminal, DNP-peptides from the other di-DNP-peptide hydrazides. The C-terminal peptides are then completely hydrazinized and the sequence inferred from the relative amounts of the hydrazinized amino acids present in the mixture. The method is not highly accurate since some amino acids are partially destroyed and the DNP-derivatives of dicarboxylic amino acid, that is, aspartic and glutamic, are difficult to distinguish from the true C-terminal amino acid.

Enzymatic methods for obtaining C-terminal analysis appear to be as satisfactory as chemical methods. Use of carboxypeptidase-A and -B,[437] protaminase,[1370] and a *Streptomyces griseus* proteinase[1157] have been reported.

Another method (unpublished) can be used to identify the C-terminal peptide. Prior to removing the sample of hemoglobin, lysine can be tagged through injection of C^{14}-labeled lysine. Autoradiography and specific chemical tests can be used to identify the peptides that contain lysine and arginine, respectively. Peptides that possess neither arginine nor lysine are presumably in the C-terminal position.

Peptide analysis

Knowledge of the chemistry of protein molecules expanded rapidly following development of quantitative methods for analyzing the amino-acid content of proteins. Amino-acid analysis fails, however, to disclose the exact nature of the chemical differences that exist between closely related hemoglobins, such as *A* and *S*, and a search was made for another approach to enable investigators to elucidate any structural variations hemoglobins may possess. Scheinberg *et al.*[1164, 1165] reported that the differences observed for the electrophoretic mobilities of *A* and *S* hemoglobins at neutral pH[994] disappeared when these hemoglobins were subjected to electrophoresis in buffers at low pH. From their studies, they deduced that the electrophoretic difference between *A* and *S* hemoglobins could be ascribed to a difference in the number of ionized carboxyl groups each possesses. Ingram[633] reasoned that, if the difference involved even a single charged group, such a specific chemical difference might be more easily identified in a mixture of a smaller number of polypeptides released through enzymatic digestion than in a mixture of a large number of amino acids produced by acid hydrolysis.   Enzymatic digestion with trypsin and chymotrypsin, followed by two-dimensional paper chromatography, was used by Sanger and Thompson[1154] to establish the amino-acid sequence of the glycyl chain of insulin.   This system was adapted to the study of hemoglobin peptides[633] with the incorporation of electrophoresis into the procedure.   The peptides of the trypsin digest were separated by paper electrophoresis in one direction and further characterized by chromatography at right angles to the direction of the electrophoresis.   This procedure is called "finger-printing."   Detailed methodology, along with an improved method for trypsin diges-tion, is described by Ingram.[632]   The improved method reduces the time required for trypsin hydrolysis from 40 to approximately 2 hours.   The trypsin-resistant core that remains after digestion of hemoglobin can be hydrolyzed by chymotrypsin.[617] If necessary, a drop of caprylic alcohol can be added to the solution without ill effects to reduce foaming created by nitrogen stirring.   A thick chromatography paper, usually Whatman no. 3 or 3MM, is used as a supporting medium for electrophoretic and chromatographic separations of the soluble peptides of enzymatic digests.   Care should be taken that the chromatography and electrophoresis are always run in the same manner relative to the machine direction of the chromatography paper.   Follow-ing the method described by Ingram,[632] electrophoresis is performed first, using pyridine:glacial acetic acid:water (10:0.4:90) at pH 6.4, and ascending chromato-graphy second, using n-butanol:glacial acetic acid:water (3:1:1).   However, equally good results are obtained by reversing the procedures;[688] the peptides are separated by descending chromatography first, using n-butanol:acetic acid:water (4:1:5) and electrophoresis second, using pyridine:acetic acid:water (1:10:289) at pH 3.7 (figure 45).

The various peptides of human hemoglobins have arbitrarily been assigned numbers for ease of reference.[632]   An illustration of the chemical differences in no. 4 peptide of *A*, *S*, and *C* hemoglobins has been presented elsewhere.[634, 636]   (It should be noted, however, that the amino-acid sequences are incorrect in the references cited above; the corrected sequences have recently been reported by Hunt and Ingram[618] and Hill and Schwartz.[579])

Comparative studies on peptides of hemoglobins can also be done with instruments used for paper-strip electrophoresis.   Resolution of the peptides is not so good as with two-dimensional separation; however, the application of specific biochemical tests on the separated peptides increases the sensitivity of the system.   Such a system has been used by Benzer *et al.*[79] to establish that there are 3 varieties of human hemo-globin *D*, although the native hemoglobin of the 3 varieties are electrophoretically and chromatographically similar.   That chemical alterations occur in peptides no. 23 and no. 26 for hemoglobins D_α and D_β, respectively, illustrates that chemical alterations in the hemoglobin molecule occur in places other than peptide no. 4, which contains the chemical alterations of hemoglobins *S* and *C*.   It has been found that hemoglobin *I* also has a chemical alteration in tryptic peptide no. 23.[922]

Fig. 45. Comparison of murine hemoglobins analyzed by "fingerprinting."

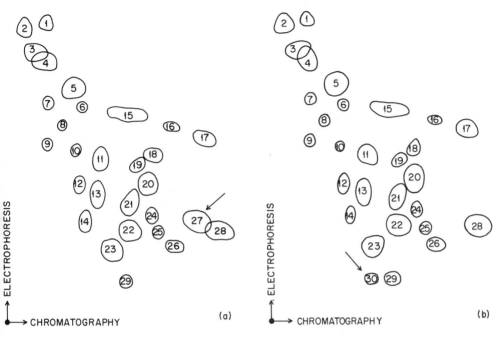

Tryptic digests were chromatographed with n-butanol:acetic acid:water (4:1:5) for 16 hours and electrophoresed in pyridine:acetic acid:water (1:10:289) buffer, *p*H 3.7 developed with ninhydrin.

LEFT.   Single hemoglobin of strain C57BL mice.

RIGHT.   Diffuse hemoglobin of strain BALB/c mice.

In 1957 Rhinesmith *et al.*[1054] established that the hemoglobin molecule is composed of four polypeptide chains of two different kinds, referred to as *alpha* and *beta* chains. Through the use of reversible dissociation and reassociation of mixtures of *A* and *S* hemoglobins Vinograd *et al.*[1346] showed that the *alpha* chains of the half-molecules of *A* and *S* hemoglobins were similar, whereas the *beta* chains were not.   Thus, they argued that the chemical difference between *A* and *S* hemoglobins occurred in the *beta* chain. In view of the basis of the template theory of protein synthesis, it became of interest to know whether the other peptides in which alterations had been observed (that is, no. 23 and no. 26) were in the same chain as peptide no. 4.   The *alpha* and *beta* chains can be separated electrophoretically in a detergent (sodium dodecylsulphate in veronal buffer), ionic strength 0.12, *p*H 8.6, or by urea chromatography on ion-exchange resins (Amberlite IRC-50/XE64 or "CG50") using 2–8 M urea, *p*H adjusted to 1.9 with HCl.[635, 1395]   "Fingerprints" of peptide hydrolysates of each half-molecule showed that approximately one-half of the peptides of the hemoglobin molecule were present in the *alpha* chain, and the remaining peptides were in the *beta* chain.   Peptides no. 4 and no. 26 were found in the *beta* chain, and the *alpha* chain contained peptide no. 23.

(It should be noted that some confusion occurs in the earlier literature regarding which peptides are in the *alpha* and *beta* chains of human hemoglobins; see Hunt and Ingram[619] for schematic representation of correct assigment for *alpha* and *beta* chains of human hemoglobins.)

The observations that chemical alterations occur in both the *alpha* and *beta* chains support earlier suggestions by Schwartz *et al.*[1180] and Smith and Torbert[1220] that perhaps there are two hemoglobin loci in man, one controlling the sequence of amino acids in the *alpha* chain and the other controlling the sequence of amino acids in the *beta* chain. A mutation may affect one or the other of these loci. Hunt[616] and Jones *et al.*[667] established that the *alpha* chain of *F* and *A* hemoglobins are similar but *F* hemoglobin has a modified *beta* chain, called *gamma*, suggesting that separate loci exist for the synthesis of *alpha*, *beta*, and *gamma* chains of adult and fetal hemoglobins. A_2 hemoglobin apparently differs from *A* in several peptides.[1291]

More recently, it has been found by Jones *et al.*[668] that hemoglobin *H* lacks *alpha* chains. A fetal counterpart of hemoglobin *H*, Bart's hemoglobin, that also lacks *alpha* chains has been reported,[620] and further investigations have shown that it is composed of four *gamma chains*.[699] Thus, there is suggestive evidence, at least, for the presence of many different loci which control hemoglobin synthesis in man

SUMMARY

Methods commonly used in studies of mammalian hemoglobins are surveyed briefly. They include moving-boundary and zone electrophoresis, ultracentrifugation, alkali denaturation, crystallography and solubility determinations, and amino-acid, sulphydryl, end-group, and peptide analyses. Immunologic and spectrophotometric techniques, and oxygen- and carbon monoxide-combining capacity, which generally have more special applications, are not discussed in this review. The methods are described in such a way as to give the reader an idea of the facilities required for each analysis, and detailed descriptions are cited. It is hoped that the material presented will help the reader to choose the methods most suited to his own investigation and facilities.

DISCUSSION

DR. BURDETTE: The discussion of Dr. Popp's paper will be opened by Dr. Elizabeth Russell who also is engaged in some interesting studies on murine hemoglobins.

DR. RUSSELL: First let me congratulate Dr. Popp for making clear a great deal of complexity. I hasten to add that I do not pose as one who can tell a great deal about murine hemoglobins, because I am still very much a student in this area. I agree completely with Dr. Popp that multiple approaches from many different angles will be essential to understand how genes are acting in producing the varying hemoglobins found in mice.

It is extremely important to study murine hemoglobins, because there are so many potentialities in mice for experimental analysis.   I hope studies in mice will help us to understand genetic differences in human hemoglobins for which experimental analysis is not possible.   The kind of breeding tests which can be performed with mice are not possible in human populations at any rate.   In connection with that I especially congratulate you on the finding of a difference of the segregation pattern of differences between the hemoglobin of SEC mice, which we have known formerly as an electrophoretically single hemoglobin, and that of C57BL/6, which is another single hemoglobin. These two segregate independently of the *albino* locus, as I understand it.   Differences between single and diffuse hemoglobins previously observed appear to be very closely linked with albinism.   Now we are beginning to progress in using the potentialities of mouse genetics to help in the understanding of how these hemoglobin differences in mice actually arise through genic action.

Recently I have been learning about column chromatography as a method for studying differences in murine hemoglobins.   There are two reasons for using this method.   One is that Dr. Richard Schweet of the University of Kentucky Medical School has used column chromatography very effectively in the study of hemoglobin synthesis.[103]   Since he has become interested in murine hemoglobins and wishes to do this kind of work with them, chromatography is an obvious method of choice. After observing his work with column chromatography, I would comment that the patience required for this method is considerable.

As for the degree of resolution, it appears to be excellent in the separation of particular components of hemoglobin from different strains.   A great advantage in column chromatography is that one can prepare mixtures of different hemoglobins and tell whether the same hemoglobin component is present in both strains.   Perhaps "fingerprinting" could do the same, but I am reasonably sure that we will find out more by using both methods than by utilizing either exclusively.   So far we have found that there are two, different, diffuse types of hemoglobin in the two inbred strains which we have studied extensively; that is, samples from the inbred *flexed* mouse and the AKR mouse give different hemoglobin patterns on column chromatography.   The two single ones with which we were dealing are alike.   One is the C57BL/6 hemoglobin which is like that of SWR/J.   This is fortunate in that one strain is albino, the other full color.   This difference in a gene linked to a locus affecting hemoglobin will help us to use the same tool in different linkage experiments.

The AKR/J hemoglobins have two major components and a minor.   C57BL/6J and SWR/J have a single, major component and a minor.   Extensive experiments involving mixtures and F_1 hybrids must now be carried out if inheritance of murine hemoglobins is to be more clearly understood.   I wish that there were a quick way of doing the hard job, but I really do not think there is any means of avoiding the extensive, difficult work required.

Dr. Popp, have you studied N-teminal amino acids? Something on this is known from Dr. Schweet's work also.   In the C57BL/6 major component, there is an *alpha* chain and a *beta* chain.   In both of the chains the terminal amino acid is valine as in human hemoglobin; the second amino acid in one chain is leucine, but in the other chain is not leucine.   We know there are two chains, but we do not know any more than that about them.

Commenting on the similarity of heme in all hemoglobins, let us call this a similarity, or perhaps identity, of the chemical structure of hemes.   We are talking in theoretical terms about the synthesis of a protein molecule.   We already know from humans of the possibility of independent *alpha*-chain synthesis and *beta*-chain synthesis. Let us remember further that in mice there are at least two independent genes which affect the rate of synthesis of heme.   Thus, before the total, complex, hemoglobin molecule is formed, many kinds of genic action must occur.

DR. BERNSTEIN: Dr. Popp, a number of methods that you elaborated on today require the assumption that hemoglobins are quite stable.   Inasmuch as changes in oxygen concentration (oxygenation or partial oxygenation) of the molecule result in changes in charge and solubility characteristics, I wonder if you would comment on the relative stability of oxyhemoglobin, carboxyhemoglobin, and so forth, with particular reference to species differences.   Of course I am concerned with the mouse.

DR. POPP: Cyanmethemoglobin seems to be the most stable form of murine hemoglobin.   Although methemoglobin is quite stable, for many studies it is a less desirable state in that the electrophoretic mobility and chromatographic behavior of methemoglobin is slower than that of cyanmethemoglobin or carbonmonoxyhemoglobin.

Murine carbonmonoxyhemoglobin is about twice as stable as oxyhemoglobin; however, studies of Douglas *et al.*[285] and Anson *et al.*[33] suggest that the conversion of murine hemoglobin to carbonmonoxyhemoglobin may be incomplete.   Their data show that 30 to 40 per cent of the hemoglobin remained in the oxyhemoglobin state. This in itself is interesting, since hemoglobins of most mammals are readily converted to carbonmonoxyhemoglobin by simply bubbling carbon monoxide into the solution of hemoglobin.

DR. RUSSELL: My impression, based on starch gels rather than on resin papers, is that one sees dense spots but resolution is not good enough to test for identity of hemoglobins in a mixture.   Resolution may be better on resins; I would be glad to try this method, but its suitability may depend upon the question being asked.

DR. BURDETTE: Is the wave length used 415 mμ?

DR. RUSSELL: Yes.

DR. POPP: We have run an absorption spectrum from 220 through 900 mμ using a Beckman automatic recording instrument.   Areas of the spectra that looked slightly different were examined more carefully, using a Beckman spectrophotometer.   The absorption spectra were indistinguishable even for hemoglobins of different electrophoretic character, for example, single and diffuse hemoglobins of strains C57BL and 101, respectively.

Willys K. Silvers, Ph.D.

TACTICS *in* PIGMENT-CELL RESEARCH†

The phenomena of genic action and interaction are so clearly presented by the permanent record which results from the elaboration of melanin granules in the hair matrices and their incorporation in the hair shaft, that the elucidation of the genetic aspects of melanogenesis greatly preceded a true understanding of the anatomical basis and biochemistry of pigment formation. Indeed, it was not until about 1940 that the extraepidermal origin of the pigment cell was unequivocally demonstrated in the classic experiments of Rawles.[1040, 1041] We now are aware that pigment formation in mammalian skin and hair is restricted to specialized branched or dendritic cells—melanocytes—of neural crest origin, usually located in the basal layer of the superficial epidermis, in the case of skin pigmentation, and in the hair bulbs for hair pigmentation.

Once it had been established that the many coat-color patterns found in mammals resulted from genic mutations and genic interactions, geneticists became interested in studying particular coat-color phenotypes, since it was evident that such phenotypes reflected or expressed, rather directly, the end results of specific gene-controlled processes. Interest in coat color was intensified when it was realized that only a single genic change was necessary to alter the phenotype drastically. Thus any study of mammalian coat color could not help but be concerned with the final results of a dynamic interplay of genic actions and interactions originating within the boundaries of a single type of cell, the melanocyte, and its milieu. Moreover, the fact that the entire genetically controlled sequence of chemical reactions in the synthesis of melanin from precursor tyrosine was restricted to the cytoplasm of the melanocyte, and could

† Some of the work referred to in this paper was supported by a grant (C-3577) from the National Institutes of Health, U.S. Public Health Service.

apparently only be greatly influenced through the local environment in which this cell occurred, also commended these cells as attractive subjects for experimental studies in a wide variety of biological fields.

Although there is a great array of pigmentation patterns among different species of mammals, all of these patterns owe their origin to factors fundamentally similar. They are based on regional or local variations in the distributions of two basic pigment types, eumelanin (black, brown) and phaeomelanin (yellow, red). They include phenotypes in which pigment is completely or partially absent. Familiar examples are white spotting involving localized regions of the body and silvering or mottling due to variations in the pigmentation of individual follicles. Furthermore, it is probably true to say that almost all the diverse patterns of pigmentation found in mammals are phenotypically represented within certain domesticated species, particularly in dogs. Despite its presentation of a wide assortment of coat-color patterns which have, to a certain extent, been analyzed genetically,[797] the dog is an unfavorable species for critical studies of genic action because of the difficulty evident in attempts to breed it and inbreed it on a large scale. Attention was, therefore, devoted to the small laboratory rodents as convenient subjects with which to investigate these problems.

Since the effect(s) of a particular genetic locus can only be recognized on the basis of variations (alleles) from wild type produced by mutations, the development of suitable laboratory animals has depended on the recognition and description of coat-color mutants and their preservation.

DEVELOPMENT OF INBRED COLOR STOCKS

A thorough, systematic investigation of the genetics of mammalian pigmentation, however, requires more than maintaining coat-color mutants. It requires the production of different stocks of animals each of which can be defined with respect to its coat-color genotype. Furthermore, since it has been shown that the expression of many genes concerned with melanin formation, or suppression of its production, may be drastically modified by other genes (modifiers), it is obligatory to produce inbred color stocks in order to control genetically the phenotypic variations that are almost unavoidable in noninbred material. For example, in the mouse the genetics of white spotting differs from the inheritance of most color patterns in that it is much more quantitative in character. Success in studying this quantitative character, therefore, depends on developing genetically homogeneous lines.

The relative amounts of eumelanin and phaeomelanin present in the hair shafts, determined by the agouti series of alleles of the mouse, can also be greatly influenced by the cumulative expression of other genes with which these alleles are associated. For this reason there exist certain modified strains of genotypically yellow mice, usually described as being of sable color, which contain a variable amount of black pigmentation down the middle of the back or even the entire dorsum. Some of these animals contain so much dark pigment that, although genetically yellow, phenotypically they resemble the black-and-tan pattern resulting from the agouti allele, a^t–.   The darkening of the fur of these animals is attributable to the cumulative action of a number of factors which are independent of the agouti locus.[300, 1067]   Because of this extreme variation in the amount of eumelanin and phaeomelanin which can and does occur in heterogeneous stocks of agouti-locus genotypes, it should be recognized that in such studies as those concerned with possible enzymatic differences between these genotypes, it is extremely important to control the background on which each of these agouti alleles expresses itself.

Studies on nongenetic aspects of mammalian pigmentation.—Inbred strains are almost obligatory for determining the significance of nongenetic factors in producing coat-color variations.   In the guinea pig Wright and Chase[1454] showed that white spotting depends primarily on a recessive factor, *s*, with many minor genetic factors having additive effects.   Because·of the availability of highly inbred strains of white spotted guinea pigs Wright[1420] was able to demonstrate that, in addition, nongenetic factors also play a part in determining this character.   Variation in the amount of white spotting even within highly inbred lines of these animals was found to be related to the age of the dam, the amount of white spotting of her offspring increasing with her age. A similar situation is also found in mice.   In the C57BR/cd strain a white patch is present on the ventral surface in a majority of the animals.   The extent of this non-pigmented area is extremely variable, ranging in size up to 10 per cent of the ventrum. Although, as Murray and Green[925] demonstrated, the occurrence and extent of these markings appear to have a hereditary basis; the total effect of heredity is relatively slight, most of the variation being nongenetic in origin.   Here, unlike the guinea pig, however, the age of the female apparently has only a very slight influence on the amount of white spotting in her offspring so that most of the nongenetic variation has still to be accounted for.

DEVELOPMENT OF COISOGENIC COLOR LINES

While the establishment of inbred strains of color stocks made available an unlimited supply of genetically uniform animals of specific coat-color phenotypes, it still had to be recognized that the differences between the coat-color patterns of these stocks might be attributable not only to the major color factors with which these stocks differed but also to the different genetic backgrounds on which these major factors were incorporated in each of the established strains.   Although with some of the major genes concerned with pigmentation these variations are almost certainly negligible, with others, as noted above, they are certainly not.   For this reason the only genetic material in which the effects of different alleles and different genetic loci can effectively be compared are coisogenic strains—strains which differ from each other in respect of a single locus concerned with melanin formation or, as in the case of white spotting, with its absence.

Production of coisogenic lines.—The production of such color stocks is, indeed, a very tedious process.   The incorporation of a dominant or semidominant color factor on a uniform genetic background requires repeated backcrossing of the mutant phenotype into the desired isogenic background.   To replace a dominant gene with a recessive gene on the same background it is necessary to alternate backcrossing with inbreeding in order to recover the desired recessive character.   This system of mating is described by Snell[1238] for the production of isogenic resistant lines.   For example, if one wishes to establish albino (*cc*) animals coisogenic with C57 *Black* (*CC*) this is accomplished as follows: First of all, the albino is mated with a C57 *Black*.   All the offspring from such a mating will be black, of course, although all will be heterozygous at the *C* locus (*Cc*). The albino phenotype is then recovered by inbreeding these heterozygous animals (one-fourth will be albino) which are then backcrossed to C57 *Black* and so forth.   With each backcross generation the number of foreign factors (that is, not C57 *Black*) is reduced by half, so that after about ten such backcrosses, albino animals are produced which are essentially identical or isogenic with strain C57 *Black* except with respect to the *C* locus.

It must be emphasized that the rigorous breeding procedures essential for the production of coisogenic stocks cannot be relaxed once the goal of isogenicity with the desired strain has been attained.   In order to maintain these stocks coisogenic with each other and to eliminate all risk of the occurrence of genetic heterogeneity, it is necessary to continue the same procedures employed in the establishment of these stocks initially.

Utilization of coisogenic lines in studying genic interactions.—In the mouse the incorporation of specific genes concerned with melanin formation or white spotting on a C57BL/6 background was initiated some years ago by Dr. Elizabeth Russell at the Roscoe B. Jackson Memorial Laboratory, Bar Harbor, Maine.   From foundation stocks of animals differing from each other with respect to only a single locus concerned with pigmentation, animals can be produced in which the interactions of two or more loci can be studied on a uniform genetic constitution.   These are especially useful in studies to elucidate the effects of alleles at the agouti locus (described in the next section) on the amount of white spotting.†

Little[799] observed a reduction in the amount of white spotting in the coats of yellow mice from that in nonyellow animals when genes responsible for the production of white spotting were incorporated into the same yellow ($A^y a$) and nonyellow (*aa*) genotypes.   Since A^y is lethal when homozygous, all inbred strains of yellow animals *must* be maintained heterozygous at this locus and are, therefore, coisogenic in nature. Subsequently Dunn, Macdowell, and Lebedeff[307] demonstrated that yellow acts as a modifier of one form of spotting, *Ww* (*dominant spotting*,[507]) but has no corresponding effect on other forms.   Although it has been shown that the agouti allelomorphs yellow-bellied agouti (A^w–) and black-and-tan (a^t–), have no effect on piebald spotting (*ss*) or its modifiers,[307] their effects on the modifiers of *dominant spotting* has yet to be

† A good method of measuring quantitatively the amount of white spotting is described by Russell, Lawson, and Schabtach.[1106]

determined. The availability of coisogenic $a^t a$, Ww, $A^y a$, $Mi^{wh} Mi^+$, and aa stocks now makes such an investigation possible, since $A^y a Ww Mi^{wh} Mi^+$ genotypes† exhibit more pigmentation than $aa Ww Mi^{wh} Mi^+$ animals on the C57BL/6 isogenic background (Russell, unpublished). It would be very interesting to determine whether $a^t a Ww Mi^{wh} Mi^+$ genotypes phenotypically resemble the corresponding aa genotype on their dorsum and the corresponding $A^y a$ genotype on their ventrum, or whether there is no effect or a general effect of this allele on the amount of white spotting normally present in $aa Ww Mi^{wh} Mi^+$ animals. Although the former result is not anticipated, its possible occurrence would certainly have to be taken into consideration in explaining two of the most perplexing problems of mammalian pigmentation—the etiology of white spotting and how eumelanin formation differs from phaeomelanin formation.

APPLICATIONS OF TISSUE TRANSPLANTATION IN STUDYING MAMMALIAN PIGMENTATION

Since the acceptance or rejection of a vascularized graft depends upon the presence or absence in the donor tissue of genetically determined transplantation antigens that are foreign to the host, it was not until inbred and coisogenic color stocks became available that the technique of graft transfer could be added to the armamentarium of the mammalian pigment-cell worker. Because of the accessibility of avian embryos and their almost uniform acceptance of grafts of homologous or even heterologous origin, at least until hatching, embryologists, ably led by Willier and his associates, had been making important discoveries bearing upon the nature and origin of melanoblasts, their differentiation and the physiologic factors which influence it. The work did not depend upon the use of inbred animals, which were certainly not available. In retrospect, in the light of the work of Billingham, Brent, and Medawar[92] and others, it now seems certain that the success of these early homologous transfer experiments leading to the production of transient, or even permanent, chimeras depended upon the fact that early exposure of the embryos to the foreign antigens induced a state of partial or complete specific nonreactivity usually referred to as immunologic tolerance. These embryos never developed the ability to reject their foreign grafts.

Application of immunologic tolerance.—Techniques are now available[91] to render adult mice and other mammals tolerant of homologous tissue grafts by inoculating them either directly as fetuses or intravenously at birth or shortly thereafter (that is, before their immunologic mechanism of defense has become functionally mature) with suspensions of living homologous cells. Such animals, when adult, may then accept tissue transplants having the same genetic constitution as the perinatally inoculated cells.

† Mi^{wh}, white, is a semidominant associated with white spotting, and inasmuch as spotting genes tend to interact synergistically with each other, $Ww Mi^{wh} Mi^+$ genotypes may be predominantly white.

Although such immunologically tolerant mammals may have some utility in studies concerned with mammalian pigmentation, it is doubtful whether they will ever prove to be as useful as the tolerant avian embryos. This appears evident for a number of reasons. Mammalian embryos cannot be manipulated experimentally as can avian embryos, so that the transplantation of embryonic tissues must of necessity be made to newborn or older hosts. In mammals the tolerance-responsive period, determined by the genetic disparity between donor and host, usually requires exposure of the fetus to homologous cells. This is certainly the case in the guinea pig and appears to apply to many strain combinations in the mouse. Furthermore, in most species fetal inoculations are attended by a high mortality rate.

The availability of coisogenic color stocks and inbred lines of mice has made it unnecessary to resort to the production of immunologically tolerant animals since these animals offer genetically tolerant combinations for transplantation studies. The existence of these stocks, therefore, not only makes it possible to perform similar transplantation studies in mammals hitherto restricted to avian embryos but, in addition, offers the experimentalist genetically uniform animals which have no parallel in any avian species.

The facility with which these animals lend themselves to transplantation studies has already been utilized in a number of investigations to determine the mode of action of some of the genes concerned with pigmentation or its absence. An account of some of these studies, dealing with the action of genes at the agouti locus and with those concerned with white spotting and albinism, follows.

Analysis of action of the genes at the agouti locus.—The six alleles at the agouti locus of the mouse determine the nature of the melanin produced by the melanocytes of the hair bulbs, that is, whether it is eumelanin (black, brown), phaeomelanin (yellow), or both. By substituting different alleles at this one locus the phenotype can be made to vary from all black (extreme nonagouti, $a^e a^e$) to all yellow (A^y-). Between these extremes are the agouti ($A-$), yellow-bellied agouti (A^w-), black-and-tan (a^t-) and nonagouti (aa) types which exhibit varying proportions of eumelanin and phaeomelanin. The agouti phenotype is characterized by a yellow banding of the otherwise black or brown hairs over most of the body, that is, the main eumelanin coloration is interrupted by a phaeomelanotic band; the yellow-bellied agouti is identical with the above on its dorsum, but like the black-and-tan phenotype, has a yellowish ventrum which, depending upon its genetic background and regional variations, may contain either completely yellow hairs or yellow hairs with dark bases. The dorsum of the nonagouti animal is similar to that of the black-and-tan, containing only eumelanin pigment except in the region of the ear where some yellow-containing hairs are also found. Although the ventrum of the nonagouti animal is also predominantly eumelanotic, yellow pigment is found in some of the hairs of certain regions, for example, around the mammae and perineum. The extreme nonagouti, the last mutant described in the series,[595] is completely black.

The dominance relationships between these various alleles differ between the dorsal and ventral sides of the animal, ventrality appearing to favor yellow pigment formation.   On the dorsum the order of dominance is $A^y > A^w = A > a^t = a > a^e$, whereas on the ventrum A^w and a^t are dominant to A.   Thus the phenotype in Aa^t animals is indistinguishable from that produced in the yellow-bellied agouti, A^w-.

There are two possible mechanisms by means of which the alleles at the agouti locus produce their effect: (1) by acting autonomously within the hair-bulb melanoblasts themselves, in which case each agouti-series genotype would differ from the others at the level of its melanoblasts, or (2) by producing their effects by altering the follicular environment in some manner which, in turn, affects the expression of what are essentially equally susceptible and similar melanoblasts common to all genotypes.

Availability of the appropriate isogenic color stocks and their F_1 hybrids has made it possible to discriminate between these two alternative mechanisms.   This was done by transplanting histocompatible skin from near-term or newborn donors to neonatal recipients which differed in respect to the nature (agouti-locus constitution) and intensity (other loci) of their future pigmentation.   Thus the analysis depended on the mode of expression of potentially intensely pigmented melanoblasts of one agouti-locus constitution when incorporated, following their migration, into the developing follicles of a graft which was not only of a different agouti-locus genotype or of a different tract origin (for example, ventral to dorsal), but with hairs (when pigmented by the melanoblasts indigenous in the graft) either light in color or white.

For example, an experiment was conducted with an inbred stock of yellow animals consisting of two genotypes, $A^y ac^e c^e$ and $aac^e c^e$.   Animals of the $A^y ac^e c^e$ genotype appear as black-eyed whites, and $aac^e c^e$ animals are light gray in color.   Newborn animals from this light-colored stock provided skin grafts, while the hosts for these grafts were intensely pigmented, histocompatible yellow ($A^y aCc^e$) and black ($aaCc^e$) animals obtained by crossing $A^y ac^e c^e$ animals with those of the C57BL/6 ($aaCC$) inbred strain.   Under these conditions it was found that when genotypically black melanoblasts, $aaCc^e$, migrated into the genetically yellow but phenotypically non-pigmented $A^y ac^e c^e$ hair bulbs of the grafted skin, intensely pigmented yellow hairs were produced.   Conversely, when yellow melanoblasts, $A^y aCc^e$, migrated into normally lightly pigmented $aac^e c^e$ hair bulbs, black hairs resulted.

Similar studies have been conducted with other host-donor combinations at the agouti locus.[1202, 1203, 1208]   All the results are consistent with the hypothesis that the agouti-locus genotype of the receiving hair follicle determines whether eumelanin or phaeomelanin or both (the agouti pattern) will be produced by its melanocytes. Furthermore, the results also indicate that this expression of genic activity is dependent not only upon the genotype of the follicular environment but also upon the location of this environment on the integument, for example, whether it is on the dorsum or ventrum of the animal.   These experiments, made possible only by the availability of the proper isogenic strains, show that at least some of the genes affecting hair color act indirectly upon the melanocytes *via* the milieu of the hair follicle.   It is hoped that the mode of expression of other genes concerned with melanin formation will be investigated along similar lines.[1044]

Transplantation to mammalian eye or spleen and to chick-embryo celom.—Other methods for discriminating between those factors which act intrinsically within the melanocyte and those which act *via* the follicular environment involves the transplantation of embryonic melanoblast-containing tissues to the adult spleen,[1205, 1206] the anterior chamber of the eye,[853] or to the celom of the chick embryo.[284, 881] The granular and morphologic attributes of melanocytes which have differentiated in the foreign environment following their migration out of the grafts as melanoblasts, are compared with those of the melanocytes which occur either within the implant or in adult animals of the donor genotype. For example, it has been found that neural-crest cells from genotypically yellow embryos of the mouse not only differentiate into eumelanin-secreting melanocytes in the celom of the chick embryo or in the spleens of phenotypically different murine hosts,[1205] but also in those regions of yellow mice where melanin pigment is normally found outside of the hair, for example, in the Harderian gland or eye.[854] These results established that the intrinsically determined capacity of melanoblasts of all genotypes is to produce eumelanin, and only in the local milieu of physiologically appropriate hair bulbs do they produce phaeomelanin.

Melanoblasts appear to migrate more extensively from grafts placed in the anterior chamber of the eye or in the celom of the chick embryo than from those placed in the spleen.[1205] In addition, the two former sites have the advantage in that grafts residing within them do not stimulate the host immunologically and procure their own destruction. In the case of heterografts in chick embryos their acceptance is, of course, due to the immaturity of the recipient. Homografts in the anterior chamber of the eye are exempted from the consequences of any reaction because they usually remain without vascularization and, therefore, cannot succumb to an immune reaction even if they could elicit one. Grafts in the anterior chamber are rejected if they do become vascularized. The conditions of an unvascularized graft in the anterior chamber of the eye are similar to those within cell-impermeable diffusion chambers which make use of membrane filters to separate grafted tissue from host cells.[5, 6]

It must be stressed that it is of utmost importance not to overlook nongenetic influences in these transplantation studies, stemming, for example, from the purely mechanical or other properties of the unnatural environment into which the donor melanoblasts are introduced, which may affect the differentiation of the latter. That these must sometimes be taken into consideration in interpreting the results has been demonstrated in transplants to the anterior chamber of the eye[853] and in transplants to the chick celom.[944]

White spotting and albinism.—Using appropriate genotypes these transplantation methods have been employed to investigate problems of white spotting and albinism, characters which have been recorded in a great variety of mammals[792] including man. While it is beyond the scope of this paper to discuss these investigations in detail[100] some information is required in order to illustrate the methodology.

Whereas in all species so far investigated (for example, mouse, rat, guinea pig, and rabbit) the hair bulbs of white spots are characterized by matrices consisting of regularly arranged cells of equal size, albino hair follicles contain, in addition, many large clear cells in their upper bulb region.   Histologic observations suggested that these hyaline cells are amelanotic melanocytes, that is, melanocytes which are in every respect normal except for their inability to synthesize melanin, since lightly pigmented genotypes exhibit the same type of cell, except that there are pigment granules present in the cytoplasm.   Furthermore, Quevedo[1033] found that the melanocytes of intensely pigmented mice, artificially depigmented by biotin deficiency, are similar in morphology to the follicular clear cells of the albino.   To establish, beyond doubt, the identity of these cells a determination of their embryologic origin was required.

In the mouse, melanoblasts originate from the neural crest and migrate to their definitive positions in the skin during the eighth to twelfth day of embryonic development.   This was established by Rawles[1040, 1041] who transplanted various portions of potentially pigmented embryos to the celomic cavity of embryonic chicks and observed the development of melanocytes from them.   By this procedure she demonstrated that there was an anterior to posterior mediolateral spread of neural-crest cells.   For example, melanocyte differentiation from 8.5- to 9-day-old embryonic transplants to the chick celom depended on the presence of the neural tube, whereas with 10.5-day-old embryos, skin ectoderm and adhering mesoderm from any level of the trunk, but not from the limb bud, gave rise to pigmented hairs.   The limb buds receive their neural crest component during the eleventh day of development.

Using an essentially similar procedure with the exception that the spleen of the adult mouse instead of the chick embryo celom was used as the site for explants, it was possible to prove conclusively that clear cells, like melanocytes, arise from the neural crest.   The absence of neural crest cells in the transplanted embryonic tissue was associated with the failure of clear cells to appear in the hair bulbs of the developing grafts.   In the design of these experiments care was taken to utilize histocompatible donor-host combinations which were readily available.[1206]

Whereas albinism appears to stem from an inherited metabolic defect—almost certainly a failure of tyrosinase synthesis—in melanocytes which are otherwise present in normal numbers and distribution, the basis of white spotting is still unknown.   It is obvious that white spotting could not have the same etiology as albinism, since in white-spotted animals the ability to synthesize tyrosinase is present in some of the cells of the body.   This becomes even more apparent from the fact that hair bulbs of white-spotted areas appear to lack clear cells and are indistinguishable from those which develop in skin experimentally deprived of neural-crest cells.   Consequently, it can be assumed that a white spot results either from the absence of melanoblasts or from their failure to differentiate in a particular area.   If white spotting does result from a localized environmental effect which inhibits the differentiation of melanoblasts present in the spotted area, one would expect to obtain functional melanocytes from these arrested cells if they are artificially introduced into an environment known to be favorable for melanocyte differentiation. It was with this idea in mind that the experiments described below were undertaken by Markert[851] and Markert and Silvers.[100]

In the mouse there are at least 14 different loci associated with white spotting.[499] Some of these genes, for example, *W*, *W^v*, *W^j* (*dominant spotting*) and *Mi^wh* (*white*), when homozygous, produce animals which are completely white with melanin pigment occurring only in the cells of the retina. Indeed, the fact that the retina is pigmented in these otherwise all white mice indicates that the genetic capacity for tyrosinase synthesis certainly must exist. It is pertinent that retinal melanocytes originate from the outer wall of the optic cup instead of from the neural crest. Since the hair follicles of these white animals contain no demonstrable clear cells, they are essentially one *large* spot and are, therefore, extremely useful subjects for investigating the etiology of white spotting.

To test the hypothesis that these animals are white because of an environmental effect which prevents a step in the differentiation of neural-crest cells into melanocytes, various embryonic tissues containing neural-crest derivatives, from both potentially pigmented and potentially black-eyed white animals, were implanted intraocularly into albino hosts. While such implants from potentially pigmented genotypes usually gave rise to numerous melanocytes that migrated over the inner surface of the host eye,[853] in no case did this occur with implants derived from potentially black-eyed white animals. Thus, if these animals are white because of an arrest in the differentiation of melanoblasts, then this genetic suppression may be the result of a locus which acts autonomously within the melanoblasts themselves rather than through the environment in which these cells occur.

While such a basis for white spotting is attractive in explaining the all-white phenotypes which are produced when such genes as those of the *W* series are homozygous, it does not adequately explain the localized white spotting produced in the otherwise pigmented animals heterozygous for these factors. Further investigations are required to elucidate the etiology of white spotting, which may very well result from a multiplicity of causes inasmuch as there are so many different loci associated with this character.

OTHER INVESTIGATIONS UTILIZING INBRED AND COISOGENIC COLOR STOCKS OF MICE

Other studies concerned with genetic aspects of pigment function in which inbred and coisogenic strains of color stocks have proved invaluable include: those concerned with the capacity of hair bulbs of different genotypes to bring about the oxidative blackening of dopa[1119]; those concerned with the ability of homogenates prepared from skin of different genotypes to oxidize tyrosine, tryptophane, or other amino acids;[395] autoradiographic studies of the incorporation of C^{14}-labeled tyrosine into melanin;[381, 852] and microscopic analyses of the size, shape, number, and arrangement of the pigment granules present in the hair septules in various genic substitutions to determine the effects of individual loci on these attributes.[1092, 1093, 1094, 1095] Studies utilizing tissue-culture methods such as those described by Cohen[210] and Reams, Nichols, and Hager[1042] in the chick might also be profitable with these coisogenic strains.

These stocks also provide genetically uniform material required for investigating the pleiotropic effects which many of the genes associated with pigmentation have in other tissues.   Among these effects are those which involve hematopoietic tissues, absorption of bone, eye size, formation and function of the nervous system, and number of germ cells.[499]

PROBLEMS OF PIGMENTATION IN SPECIES OTHER THAN MICE

Despite the availability of large numbers of genetically uniform coat-color types of mice and their obvious experimental advantages for pigment-cell genetics, there are certain pigmentary phenomena which can only be investigated in the particular species in which they occur.   Among these is partial albinism found in Himalayan rabbits and certain other mammals of similar phenotype.   In this condition the fur is predominantly white but pigmented at the extremities of the body.   However, it has long been known that pigmented hair appears in other normally nonpigmented regions of the body after shaving the skin and maintaining the animal in a cold environment.[1177]

The intriguing tortoise-shell (e^pe^p) pattern in the guinea pig, which is characterized by an irregular intermingling of yellow with black hairs, and the effect which spotting has on this pattern is also without a counterpart in the mouse and certainly deserves investigation.   In the presence of spotting factors, especially ss, not only is there an increase in the amount of yellow in tortoise-shell genotypes but, in addition, in e^pe^pss animals there is a tendency toward segregation of yellow and black so that a tricolored yellow, black, and white spotted phenotype results.[199, 1419, 1449]

Pigment spread.—Finally the provocative phenomenon of pigment spread as it occurs in guinea pigs, Friesan cattle, and other mammals[100] is of interest to those concerned with the potentialities of the melanoblast and melanocyte.   If black skin from a black-and-white-spotted guinea pig is grafted to a white area on the same animal, there is a slow centrifugal encroachment of pigmentation into the white skin from the graft margins.   Conversely, if white skin is grafted into a black area, it gradually blackens from its periphery inward.   This spread of pigment occurs from black into yellow (red) or from yellow into white skin, and it also occurs naturally where skins of two different colors are juxtaposed on the same animal.

Although this phenomenon has been investigated repeatedly, no completely satisfactory or generally acceptable explanation of the underlying mechanism has yet been forthcoming.   The two most attractive hypotheses which have been advanced to explain this phenomenon are: (1) it results from a transformation of amelanotic melanocytes into melanin-producing cells or of phaeomelanin-producing into eumelanin-producing cells by a postulated infective agent derived from pigmented melanocytes,[94, 95, 96, 97, 98] and (2) pigment spread is simply due to the actual migration of melanoblasts or melanocytes.

The fact that pigment spread can be initiated by seeding a white spot with cellular homografts of dissociated epidermal cells including melanocytes,[96] but may subsequently bleach out spontaneously after a variable period or may be caused to bleach out promptly following an immunizing skin homograft from the donor of the melanocyte-containing suspension, has some important implications.   The fact that albino mice will permanently accept coisogenic pigmented skin constitutes strong evidence that

Fig. 46. Principle of method of determining origin of melanocytes in annulus of spread surrounding a black-in-white graft in the guinea pig.

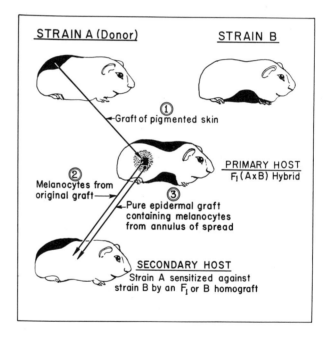

melanin *per se* is not an isoantigen.   Thus if the pigment spread initiated by homologous melanocytes is the result of a progressive infection, it follows that homologous melanocytes must infect the postulated amelanotic melanocytes with transplantation antigen(s) as well as with a capacity to synthesize melanin, unless part of the melanogenic machinery happens to contain a transplantation antigen.

The general biological implication of the infection theory suggests that differences between the various melanocytes of a single guinea pig are due to cytoplasmic rather than to nuclear differences.   Furthermore, if proven, it would represent the first example of an autonomous cytoplasmic component that gives cytoplasmic inheritance in a mammal at the somatic-cell level.

To establish whether the melanocytes in the annulus of pigment spread around a black-in-white graft are of graft origin or not, a crucial point for the migration hypothesis, Billingham and Silvers are currently utilizing two isogenic strains (as confirmed by skin-grafting procedures[99]) of guinea pigs and their F_1 hybrid. These strains 2 and 13 are of the $e^p e^p ss$ genotype (described above) and, therefore, can each provide and receive black grafts to initiate pigment spread in initially white areas.

The experimental procedure represented diagrammatically in figure 46 is as follows. Pigment spread is initiated in white spotted areas of F_1 hybrid hosts by means of black grafts originating from histocompatible parental strain animals. The specific techniques for carrying out these experiments have been described by Billingham and Medawar.[95, 96] After a period of not less than 100 days, when there is a sufficient annulus of pigment spread available, an attempt is made to determine whether the genotype of the melanocytes present in this annulus is of the donor strain or of the F_1 hybrid. This is achieved by isolating pigmented epidermis from the annuli of spread and determining the ability of the melanocytes contained therein to survive in specifically sensitized animals (in respect to F_1 hybrid homografts) of the donor strain.

If pigment spread is simply the outcome of melanocytic or melanoblastic migration, then the melanocytes of the graft and those that surround it in the annulus of spread must be of the same genotype: that of the donor. They should, therefore, survive when transplanted back to the donor strain. If, however, an infection process is involved, then only the melanocytes from the graft should survive, since those from the annulus of spread would be of hybrid genotype and consequently should be destroyed very soon after they are placed on the presensitized, secondary host.

CONCLUSION

An attempt has been made to discuss only some of the many methods which can be employed in studying the genetic aspects of mammalian pigmentation. Most of these methods have involved tissue transplantation, since this technique has proved to be of utmost importance in approaching experimentally many of the problems of pigment-cell workers, particularly those concerned with the physiologic genetics of pigmentation.

It is obvious that the progress which has been made in understanding how genes control such processes as melanoblastic differentiation and synthesis of melanin has rested, first, upon the availability of color stocks uniform with respect to at least those hereditary factors concerned with pigmentation, second on the production of inbred color stocks, and finally on the existence of coisogenic color lines. Although almost all of these have been established in either the guinea pig or the mouse, mainly through the pioneering efforts of Drs. Sewall Wright and C. C. Little, they include such a wide range of different phenotypes that among these two species can be found suitable subjects with which to investigate almost all aspects of pigment-cell biology. There is no doubt that a systematic analysis of mammalian pigmentation still provides one of the best methods in elucidating the many diverse ways that genic action can influence the behavior of a single cell.

DISCUSSION

DR. BURDETTE: Dr. Morris Foster will discuss Dr. Silvers' paper.

DR. FOSTER: I find myself in the quandary of finding something to say when Dr. Silvers' work is so clear that it drastically reduces the number of possible questions any fertile imagination could produce. Therefore I should like to defer any questions directed to Dr. Silvers in favor of anyone in the audience who has perhaps less favorable bias toward his work and to provide instead a very brief supplement describing some recent advances in methodology relating to the genetically controlled biochemistry of mammalian melanin pigmentation.

At the subcellular level, we are confronted with processes which occur in at least three distinct echelons of organization in the sequence from tyrosine to melanin. At the first, small molecular level within differentiated and functioning melanocytes, we deal with substrates such as tyrosine. The precursor undergoes a series of transformations, and we then proceed to a second organizational level, that of the macromolecule, when the late intermediates being to polymerize. We finally arrive at a third level of organization, that of a fully formed cytoplasmic organelle, when a polymerizing product is conjugated with an already formed protein matrix to produce a mature melanin granule.

It should be emphasized at the outset that it is quite naïve to equate tyrosinase activity with the whole process of melanin formation; it is asking too much of mammalian material to expect that assayed tyrosinase activity and the amount of melanin would be modified either to the same degree or, indeed, in the same direction by given allelic substitutions. Thus, before trying to relate a given genotype to a given amino-acid sequence of the tyrosinase molecule, it might be useful first to find out which genic substitution affects which part or parts of the pigment-building system, and at least to obtain, for biochemical purposes, a tyrosinase-rich tissue. At least one such genotype was unexpectedly found, and it seems to offer now the greatest hope for extracting and solubilizing the enzyme and, ultimately, for determining its amino-acid sequence (perhaps also permitting analysis of secondary and tertiary structural features).

Our own recent approach to objective and quantitative assays for genetically controlled, melanogenic attributes involves the use of lyophilized individual skins, from each of which we obtain four, equal, dry-weight subsamples. These subsamples are assayed respirometrically for tyrosinase and dopa-oxidase activity, and subsequently they are also assayed for melanin content by means of turbidimetric measurements of alkaline thioglycolate suspensions of unincubated tissue or of tissue which had been incubated in the Warburg vessels and removed after a standard period of incubation. The genetic materials we used involved separate and combined substitutions at the pink-eyed dilution (*p*), maltese dilution (*d*), and brown (*b*) loci; giving us a total of eight color genotypes. We also used albinos as controls to provide baseline measurements.[396]

Fig. 47. SAMPLE OXYGEN-CONSUMPTION CURVES DURING THE INCUBATION *in vitro* OF SKIN SUBSAMPLES OBTAINED FROM TWO DIFFERENT COLOR GENOTYPES.

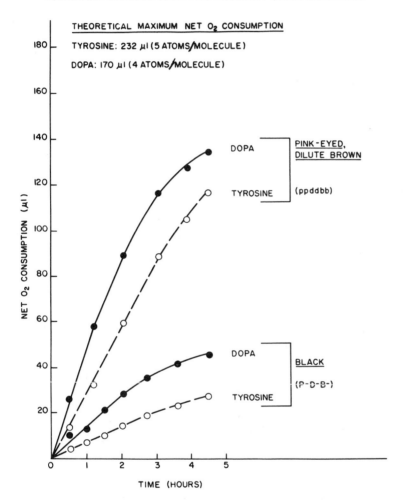

Our first surprise was that a very pale animal (that is, pink-eyed, dilute, brown; genotype *aappddbb*) could have far more assayed oxidative enzymatic activity toward added substrate than a very dark one (that is, intense black, genotype *aa*). (See figure 47.) Our second, and most pleasant, surprise concerns the use of the thioglycolate dermal suspensions which are easily prepared. I should like to compare our own data, obtained directly as readings of the Klett colorimeter scale, with those computed from the histologic data of Elizabeth S. Russell.[1092, 1093] (See figure 48.)

Except for possible differences of scale (Dr. Russell's scale, using an arbitrary unit called pigment volume, is based on the number and average dimensions of melanin granules at a given developmental stage), there is a very striking and gratifying paral-

Fig. 48. Comparison of turbidimetric estimates of natural melanin content ("un-incubated" subsamples) with estimates based on E. S. Russell's histologic data [1092, 1093] ("pigment volume").

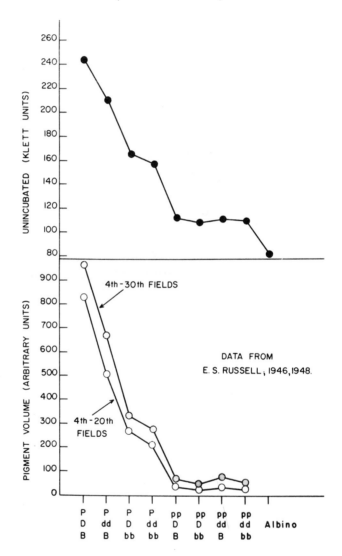

lelism as to order of effect, with respect to natural melanin content (subsamples not incubated), as a function of genotype. This parallelism gives us a reasonable measure of confidence in the use of the thioglycolate suspensions for obtaining measurements of melanin content.   One final point should be made: on Dr. Russell's pigment-volume scale, albino skin would have a zero value.   In our case, however, suspensions of albino skin give a residual turbidity of about 80 Klett units.   (Since making these remarks originally, we have learned that there are important physiologic differences between the intense black homozygote (*aaBB*) and the slightly less pigmented, but enzymically heterotic, heterozygote (*aaBb*).[398]   These recently discovered differences do not, however, contradict our less detailed data summarized here.)

When the data are subjected to an analysis of variance, it turns out that the greatest source of significant biological variation is due to differences in color genotype. The next major source of biological variation is attributable to differences between litters within genotypes.   (See sample analysis in table 59 for tyrosinase activity.)

Table 59

SAMPLE ANALYSIS OF VARIANCE FOR TYROSINASE ACTIVITY

(Maximum rate of net oxygen consumption in microliters per hour)

Source of Variation	D.F.	Sum of Squares	Mean Square	F	P	F'	P'
Between genotypes	8	7,079.85	884.98	173.19	<0.001	58.84	<0.001
Within genotypes	216	3,248.29	15.04			1.00	
1. Between litters	50	2,400.72	48.01	9.40	<0.001		
2. Within litters	166	847.57	5.11	1.00			
Total	224	10,328.14					

F, P stand for variance ratios and corresponding probabilities when "within litters" mean square is used as error mean square.   *F', P'* stand for variance ratio and corresponding probability when "within genotypes" mean square is used as error mean square.   The data analyzed here are more extensive than those previously summarized.[396]

Significant variance ratios of between:within genotypes permit ranking of genotypic means for all measured melanogenic attributes according to the methods developed by Duncan,[293, 294] who also described an admirably compact way of summarizing the results of all possible tests of significant differences between means.   (See examples by Foster.[396])

While many statements are implied in such compact summaries of numerical ranking, I wish to make only a few additional points.   First, black skin, which is obviously much more heavily pigmented than albino skin, is hardly distinguishable from albino skin with respect to rate of respiration upon incubation with tyrosine. Second, the skins of pink-eyed, dilute-brown mice react very strongly with added substrates, despite their very low level of natural melanin content.   (See figures 47 and 48.)   Thus there is no simple correspondence between assayed enzymatic activity and natural content of melanin.   Third, detailed comparisons regarding allelic substitutions on different genetic backgrounds indicate some regular pattern of effects.   For example, in black *versus* brown contrasts, brown skin is more active than black toward added substrates. Black skin, however, is best able to form melanin when relying upon its own endogenous resources, that is, when incubated in the absence of substrates. We therefore conclude that the inherited pigmentary defect in brown skin is in part due to some form of endogenous limitation of substrate. This conclusion was "recently" independently "confirmed" by Sewall Wright in 1916 and 1942.[1416, 1449] (Addendum: Recently we have noted an additional feature of the genetic affliction in brown skin; it cannot form as much total tissue-bound melanin as black skin upon incubation *in vitro*. Thus the genetic damage in the brown genotype may also involve decreased number or effectiveness of the melanin binding sites of the protein matrix.)

Finally, if we make systematic comparisons to determine the effects of single and combined allelic substitutions at the p and d loci, we find that the amount of tissue-bound melanin formed upon incubation *in vitro* does not run parallel to the amount of oxidase activity. We are therefore led to suspect that the limiting factor in these color mutants is not one of oxidase activity but rather one or more defects in the terminal phases of the pigment-building process, phases which are not readily measured respirometrically. A useful by-product of this study is the discovery of an unsuspected, exceptionally rich source for enzymatic extraction procedures, namely, the pink-eyed, dilute, brown genotype.

While the methodologic improvements discussed are encouraging, they have not yet helped us solve two important problems. The first concerns the nature of the process for forming yellow pigment (phaeomelanin). At the present time no evidence conclusively demonstrates tyrosine or any other likely chromogen as the natural precursor of phaeomelanin. (See, however, Foster[395] and also Fitzpatrick and Kukita.[382]) A second apparent paradox is the striking albino-like behavior of deeply pigmented skins obtained from animals bearing intermediate alleles at the albinism locus. As long as allele C is lacking, that is, the chinchilla mouse ($aac^{ch}c^{ch}$), the skin behaves like the completely unpigmented albino skin toward added tyrosine or dopa. A similar situation in the guinea pig was previously reported.[394] No evidence has yet been obtained to explain this apparent paradox. One possible approach involves the use of paper chromatographic methods for identifying diffusible endogenous melanogenic precursors and intermediates. This method was already proved successful in identifying the unstable, diffusible, intermediate dopa, in an incubation medium containing skin with high tyrosinase activity (Foster and Brown,[397] work supported in part by NIH Grant C–4305 and NSF Grant G–6480).

I regret that Dr. Silvers failed to mention the excellent studies performed by himself, in conjunction with Dr. Russell, concerning the rôle of the melanocyte environment in instructing the cell to synthesize either eumelanin or phaeomelanin. At this point I should like at least to draw attention to the pertinent publications.[1202, 1203, 1205, 1208]

DR. BARRETT: I have a question that is not too profound, but I have usually wondered about it when thinking of this work. Regardless of which theoretical explanation applies, why do we not see in nature some explanation for the fact that in spotted, banded, hooded, and otherwise multicolored animals the old animals are not all pigmented?   Why in a multicolored animal is not the old animal characterized by all, or almost all, black pigment?

DR. SILVERS: Part of the answer may be due to the fact that in the general integument of pigmented rats and mice, melanogenesis only occurs in the melanocytes of the hair bulb.   The other melanocytes in the skin do not synthesize melanin.   In the guinea pig there is a population of active melanocytes in the basal layer of the epidermis as well as in the hair bulbs.   In this animal at the region of transition where black and white areas are juxtaposed, one does find a very narrow band of pigmented skin from which white hairs originate.   This area has been secondarily pigmented by a migration of melanocytes from the black area.   However, why this process does not continue throughout the life of the animal, I do not know.   It may at an imperceptible rate. The rate of spread around a black graft to a white area also falls off asymptotically.

DR. NALBANDOV: I have two questions.   First, what is the origin of the pigment cells in such structures as the testes in chickens, which may be completely black? Second, what is the origin of pigment cells in the uterine region?   Neither the testes nor the uterus are of ectodermal origin.   Could chickens which may have one black and one yellow testicle or in which each testicle may be two-toned serve as experimental animals for the study of questions in which you are interested, that is, the origin of the yellow and black pigments?

DR. SILVERS: If the cells you refer to are dendritic in form, then it is almost certain that they originate from the neural crest.   Melanocytes may occur in many regions of the body where one might not expect them, for example, in the spleen, parathyroid, and ovary of the mouse.   It is possible that during that period of development when melanoblasts are making their way from the neural crest to their definitive positions, they migrate to and through many areas of the animal, but only persist, differentiate, and produce melanin in suitable environments.

The existence of both black and yellow pigment in the testis of the chicken is very interesting.   Unfortunately we do not have any inbred strains of appropriately pigmented chickens to investigate this, although it might be possible to make appropriate grafts to chick embryos.   In the mouse there are two types of melanin, pheomelanin and eumelanin.   Although it has not been proved, I am of the opinion that in the agouti mouse a melanocyte can alternate its pigment production, producing eumelanin at one time and pheomelanin at another.   In the chicken, however, all the evidence is consistent with the hypothesis that, although each melanocyte has the ability to synthesize either black or yellow pigment, once it begins to synthesize one type it cannot change and produce the other.   This is a problem which deserves further investigation.

DR. SLATIS: Dr. Silvers described a very good experiment for distinguishing between infective and migration theories, because a positive result would support the migration theory and rule out the infective theory.   The quality that made the experiments on transplanting eye discs in *Drosophila* look so good was that they distinguished between effects due to the genotype of the host and the genotype of the transplanted material. If the infective theory is correct, then one might expect that, if one transplants into a host that cannot produce pigment of certain colors, infected cells would be obtained that would give pigment on a white background, but the pigment would be of the wrong color. It appears to be a more discriminating method than the one that you tried in which a positive result is required in order to give you something on which to base a conclusion.

DR. OWEN: Have you followed this reaction colorimetrically as well as by oxygen consumption?

DR. FOSTER: No, but I think we can now by dealing with material incubated inside dialysis membranes. Although we have not done it because of other work, it is certainly a very worthwhile thing to do in order to have some information about intermediate steps along the way.

DR. COLEMAN: We have used a different assay, in conjunction with those already described, in studying this problem.[213] This assay involves the measurement of the amount of incorporation of radioactive tyrosine into the pigment granules. This assay has been found sensitive enough to distinguish clearly all the alleles of the C locus, a distinction not possible with the manometric method. Otherwise our results are essentially the same as those of Dr. Foster. Our studies with the pink eyed black mouse ($BBCCpp$) suggest that pink eye decreases the amount of tyrosine incorporated into actual pigment in contrast to the increased tyrosine oxidation associated with the pink eye gene as measured manometrically. These observations suggest that pink eye causes tyrosine to be oxidized faster but to some product which is not melanin. This demonstrates the value of using more than one assay method as suggested by Dr. Owen.

As for the yellow mouse our work indicates that less tyrosine is incorporated into pigment in this genotype than either of the normal brown or black genotypes. Both studies *in vitro* and *in vivo* indicate that tyrosine is the normal precursor to yellow pigment. No evidence has been found which would implicate tryptophan or any other amino acid as a precursor to this kind of pigment.

DR. HERZENBERG: The kind of experiments that Dr. Silvers carried out in animals which are inbred could be carried out in other animals as well, particularly chickens and other birds by making them tolerant to the donor of the skin with which they are willing to work. If Dr. Nalbandov is interested in carrying this out, I am sure he could do it.

DR. HESTON: There are strains of chickens at the Regional Poultry Laboratory at East Lansing that may be inbred enough for your experiments.

DR. SILVERS: Dr. Billingham and I have tested some of the inbred lines of chickens maintained at East Lansing, Michigan by skin graftings and have not found any histocompatible strains as yet.[99] However, we are now testing some more of these lines.

DR. SCHAIBLE: With regard to transplantation experiments involving pigmented and unpigmented regions, the chicken has another drawback in that there are no mutants which produce adult spotting patterns like those found in mammals. In chickens, melanocytes seem to be able to proliferate until they occupy all feathered regions.   The only types known to have restricted pigment areas are those carrying E in their genotypes; then only the extremities (wing tips, ventral regions, and feet) are white in the newly hatched chick.   Even when migration of pigment cells appears to be retarded by teratogenic agents until only a small pigmented area remains on the back in the down pattern of E types, the birds eventually become fully pigmented.[747]

IMMUNOGENETICS

Ray D. Owen, Ph.D.

METHODS *in* MAMMALIAN IMMUNOGENETICS

Antigen-antibody reactions depend on mutual complementarity of structure between the antigen, which induced the antibody to be formed, and the antibody, which reacts only with the antigen that induced its formation and with other substances having a sufficient degree of configurational similarity to it. Antibody reactions therefore provide ways of studying individual similarities and differences on a molecular level, less precise but easier than detailed structural-chemical analysis. Because the molecular differences so defined are generally found to relate in rather direct ways to particular genic differences, in almost any species investigated immunogenetics taps a reservoir of genetic characteristics amenable to straightforward genetic analysis. Thus, many of the best genetic markers for studies on either natural populations or laboratory mammals are those identified by serologic methods. This aspect of genetics is one of only a few to which mammals as experimental material have contributed in unique and basic ways. Even in studies of other kinds of animals and plants, it is usually the antibody response of a mammal, most often a rabbit, that is used in providing tools for immunogenetic study. And as a source of genetically regulated antigens, easily obtained and tested, probably nothing at present rivals the mammalian red blood cell. As a demonstration of individuality, probably nothing is so impressive as the regular incompatibility encountered when tissues are exchanged between individuals, to be rejected by the immunologic mechanisms of the host.

Because the genetic qualities dealt with in immunogenetics are generally simple ones, the methodology of the genetics side of this kind of study is mainly simple, straightforward, and traditional; for the most part, these methods need not be exploited here. It will be assumed that it is the methodology of immunology that will be unfamiliar to the users of this paper, and it will be toward that part of our subject that attention will mainly be directed.

A few of the general terms and concepts of immunology should be introduced at the start:

Antigens are molecules with two properties: first, they will induce an antibody response when injected into an animal, and second, they will react specifically with the antibodies they have induced. Often, the term antigen is not defined on a molecular basis; for example, a red-cell suspension injected into an animal may be described as the "antigen" injected. In general under such circumstances, one should recognize that this gross "antigen" is composed of a number of discrete molecular antigens, each involved in its own antigen-antibody reaction systems. Sometimes, a molecular antigen is found to be unable to induce antibody formation although it is able to react with antibodies evoked by a related material. For example, a fragment of a protein may be able to react with the antibody induced by the intact protein, although the fragment itself when injected may not give rise to a detectable antibody response. In general, in order to be a complete antigen in the sense of both inducing and reacting with antibodies, a molecule needs to be relatively large; most of them are in excess of 40,000 molecular weight, and many are in the millions. A small molecule, however, such as a substituted benzene ring, may react specifically with antibodies; the reactive sites on antibody molecules are themselves only small parts of the molecules.[672] Thus, a complete, macromolecular antigen is conceived as being composed of numerous individually small sites on sections of its surface, each site capable of reacting with antibodies specific for it. These sites are spoken of as valence sites; complete antigens are polyvalent and complete antibodies have two valence sites per antibody molecule. The two valence sites on an antibody molecule seem to be identical with each other. Very small reactive compounds may appear to act as antigens, as in idiosyncrasies for particular drugs like aspirin, or dermal hypersensitivities to small compounds such as picryl chloride. In such cases, it is generally assumed that the small molecule complexes with larger ones in the recipient's system; it is the macromolecular complex that acts as a complete antigen, the strange small added grouping being responsible for the foreignness of the total molecule and directing the specificity of the antibody response and reactions.

Antibodies are modified serum globulins, secreted into the circulation by cells of the antibody-forming series as a consequence of the injection of a foreign material and specifically reactive with the material that caused their formation and release. The basis for this remarkable adaptive response on the part of the injected animal is a fascinating problem in somatic-cell genetics, currently under scrutiny and debate. The easiest antibodies to discern are those that result in a visible reaction with the antigen in ordinary media in test tubes, such as the agglutination of red cells in saline or the precipitation of a soluble antigen from saline solution. Such reactions can be formally conceived as involving two processes: first, the specific combination of antibody molecules with sites on the antigen, and second, a derivative reaction in which a visible change occurs in the system. The discrimination of two phases came about largely as a result of observations of systems in which the first step occurred without the second, and in which the second could be promoted by making some nonspecific addition to or change in the system. For example, in some systems, although antigen-antibody unions occur at low salt concentrations or in the cold, visible changes do not follow until salt is added or the system is warmed. Red cells combine specifically with their corresponding antibodies at low temperatures, but they fix complement and lyse in relatively short times in the presence of complement only when the system is warmed. Such observations tend to distinguish between a first stage of antigen-antibody combinations and a derivative second phase which may or may not follow depending on the conditions of the test; nevertheless, in many systems the ultimate effect is generally believed to be the result of the continued operation of the same mechanisms as are involved in the initial antigen-antibody unions. For example, a visible precipitate is generally believed to form as bivalent antibodies combined with polyvalent antigens in a continuous process, the antibody molecules providing bridges in a lattice forming the developing aggregate. The degree to which nonspecific secondary interactions may enter into serologic reactions is often debatable and probably depends on the particular system under test. For example, in some systems of cellular agglutination there is reason to believe that changes in cellular membranes consequent upon the initial antibody attachments at isolated sites may reduce the tendency of cells to repel each other, and therefore lead secondarily to aggregation. Hemolysis clearly depends on attacks on the antibody-sensitized cell by nonspecific components of normal serum, leading only secondarily to cell lysis. Many reactions *in vivo*, such as anaphylaxis, also depend on secondary consequences of the initial antigen-antibody combinations occurring in tissues.

When a system has been defined for a particular test (such as saline agglutination), it is often found that some of the antibodies capable of uniting specifically with the antigen are unable to promote the second, visible stage of the test reaction. Such antibodies are often described as incomplete antibodies. A number of methods are available for their detection; for example, they may block the antigen by combining with it, so that access to the relevant sites on the antigen is prevented for complete antibodies later added to the system. Or they may promote aggregation in other media, though they fail to do so in saline. They may sensitize cells for hemolysis but not promote agglutination, or they may combine with cells and render them subject to aggregation by other antibodies directed against the antibody globulin molecules themselves. Or they may provide a vehicle for passive transfer of tissue sensitization, so that, for example, if they are injected into the skin of an insensitive individual they confer a transient local sensitivity to their corresponding antigen at that skin site. In precipitating systems, multivalence of both antigen and antibody has been mentioned as a necessary condition for the appearance of a visible aggregate. Failure of an antibody preparation to produce an aggregate is sometimes the occasion for the use of the term "univalent" with reference to the antibody. Often the term is an unfortunate one;

antibodies may fail to produce a visible reaction for numerous reasons other than univalence, and the careless use of this term suggests an explanation for their behavior which should be applied in any particular case only when this explanation has been established in fact. Univalence in a proper sense can derive from several different causes: some antibody may be univalent when it is made, an originally bivalent antibody may be split into two univalent fractions, or one of the valence sites may be covered by aggregation, sometimes with quite different materials such as serum albumin.

Aside from their heterogeneity in respect to visible effect produced on the antigen under test, antibodies are heterogeneous in numerous other respects as well. Of primary concern is their heterogeneity with regard to specificity; some antibodies react essentially only with the antigen that induced their formation, whereas others cross-react with related antigens as well. The directions of these deviations in specificity vary within the heterogeneous antibody population, so that within the limits imposed by the general concept of serologic specificity there is a smear of subspecificities in any given antiserum. Recognition of this kind of heterogeneity is important to interpretation of antiserum reaction in immunogenetics.[981, 985] Perhaps a related type of antibody heterogenity involves the avidity with which an antibody combines with its corresponding antigen. Variations in avidity can be conceived as variations in the closeness of fit and in the extent of mutual complementarity between the combining sites of the antigen and antibody. Operationally, avidity is generally expressed in the behavior of the system upon dilution. The equilibrium,

$$\text{Antigen} + \text{Antibody} \rightleftharpoons (\text{Antigen-antibody Complex})$$

lies to the right with avid antibody, to the left with antibody of lesser avidity. Antibody is also heterogeneous on physical-chemical grounds, as measured by solubility, electrophoretic behavior, chromatography, and ultracentrifugation. Changes in the character of antibody, with regard to specificity, avidity, and other characteristics, commonly occur during the course of an immune response.[663]

Some types of immune responses have not been shown to involve the participation of classical, plasma antibody. Among these are reactions that appear to be strictly cell borne and cell mediated, such as the delayed hypersensitivity reactions,[770] and at least some types of tissue-transplant rejection.[861] Except for some reference to tissue-transplant rejection, however, consideration here will be limited to plasma antibody responses and genetic test systems based on such antibody.

IMMUNOGENETICS OF RED BLOOD CELLS

Procurement of antisera.—In general, it is the immune response resulting from exposure to a foreign antigen that provides the most useful tools in immunogenetics. In many important instances, however, antibody-like materials which react specifically with the red blood cells of other individuals are normally present in the plasma of particular individuals.   In such cases no injections are necessary, although injections may increase the titer or the reactivity or change somewhat the character of the antibody normally present.   Materials that will aggregate the red cells of mammals, with some degree of specificity, are not restricted to mammalian sera; lower vertebrates and even many invertebrates have such hemagglutinating materials in their plasma.   Even some plant-seed extracts provide useful reagents for distinguishing individual differences within mammalian species; practically all of the useful extracts of plant seed are from legumes.[101]   The plant-seed extracts have been used mainly for studies of human red cells, and to a lesser degree those of birds; they have as yet found relatively little application in the immunogenetics of mammals other than man.   The preparation of the plant-seed extracts is generally simple; the pulverized seed is extracted with saline, and the extract clarified by filtration or centrifugation before it is used as a test fluid on red-cell suspensions.

The classical and best-studied example of normal antibodies that make individual distinctions within a species is, of course, concerned with the *ABO* blood groups of man. Similar normal antibodies are useful in studies of cattle and sheep and occur in a number of other species as well.   In rats, normal antibodies reacting with the cells of other individuals are rarely encountered; there is, however, at least one simply inherited difference displayed by normal antibodies in the plasma of certain rather rare rats, which agglutinate in saline the red cells of almost all other rats.[147, 980]   Normal antibodies for murine cells also occur in certain sera of the mouse; they are, however, irregular and weak.   Normal antibodies for the red cells of other species are very common.

The techniques used to obtain test sera, either normal or immune, vary with the kind of animal being used.   In such animals as cattle or sheep, blood can be collected in quantity by needle puncture of a large vein at the side of the neck.   In rabbits, small quantities of blood are conveniently obtained by nicking the marginal ear vein with a razor blade, and permitting the blood to drop into dry test tubes.   Warming the ear or rubbing a drop of xylol on its tip distends the vein and promotes more rapid bleeding.   It is useful also to rub a light film of vaseline on the ear at the site of the incision to reduce blood clotting during bleeding.   Bleeding can be stopped by pressing a bit of dry absorbent cotton over the incision and compressing the vein with the fingers distal to the incision for a minute or two.   Larger quantities of blood are obtained from the rabbit by cardiac puncture, a procedure that in experienced hands is of little danger to the animal.   The rabbit is tied down on a board, without anesthesia; anesthesia is a greater threat to the survival of the rabbit than is the operation itself. Bleeding is generally done with an 18-gauge 1½-inch needle on a 50-ml. syringe. Quantities of 50 or 60 ml. can easily be taken at a time without killing the rabbit; he may be bled in this manner on at least two successive days.   Several different approaches to cardiac puncture are in use; in one, the V formed by the lowest attached ribs and sternum provides the point of orientation.   The needle, on the syringe, is then inserted through the second costal space, close to the sternum on the animal's left side, aimed slightly toward the animal's head and toward the midline. It is useful to wet the syringe before use with an isotonic citrate solution (2 per cent sodium citrate, 0.5 per cent NaCl).

In mice and rats, small quantities of blood may be obtained by incision of a vein in the tail; particularly in mice, it is preferable to make a shallow cut with a razor blade across the ventral surface of the tail, rather than to use one of the lateral tail veins. Rapidity of bleeding can be promoted by warming animals under an ordinary study lamp, before bleeding. Hemolysis of the serum can be avoided if the dry tubes into which the blood is allowed to drip have been coated with silicone. Many believe that a more convenient technique than tail-vein incision or cardiac puncture in the mouse or rat is to obtain the blood by Halpern's method from the orbital sinus.[1283] Hematocrit determinations can be made directly in the tubes used for bleeding, without transfer, if heparinized tubes are prepared. Cardiac puncture is also convenient in mice and rats; mice are bled under Nembutal anesthesia and rats under ether or urethan anesthesia. The approach to cardiac puncture described for the rabbit also works for the mouse and the rat; shorter and smaller needles are used (for the mouse $\frac{3}{8}$ inch, 26 gauge; for the rat, $\frac{1}{2}$ inch, 24 gauge). Many find it easier to bleed a mouse by cardiac puncture, using an approach immediately under the tip of the sternum and directing the needle straight forward and almost horizontally. With this approach a 22-gauge, 1-inch needle is proper. Guinea pigs and other small mammals are also conveniently bled by cardiac puncture; we have bled, for example, opossums, monkeys, and raccoons easily this way. When considerable quantities of blood are to be obtained from the mouse, particularly when blood from homogeneous populations is to be pooled, mice are often sacrificed, and blood is collected in a beaker after decapitation.

If the plasma is to be used, or if test red-cell suspensions are the main objective of the bleeding, the blood is collected in an anticoagulant. Several of these are used: for example, heparin, versene (0.1 M, pH 6 to 7, in 0.85 per cent saline), oxalate, or citrate solutions. When, however, it is a test serum that is required, the blood is collected in a dry (preferably silicone-coated) tube, and is allowed to clot. The clot should be freed from the wall of the test tube after it has formed; for most samples, it can simply be shaken free after it has been permitted to stand undisturbed for an hour after bleeding, or the clot may be rimmed with a wooden applicator or glass rod. In most bloods the clot then shrinks, squeezing out the serum; after an hour or two the serum is poured from the clot, the free cells are centrifuged out, and the clear serum is decanted or pipetted into storage bottles.

Better yields of serum are obtained from mice if the blood is collected into a tilted petri dish, so that the clot forms as a film over half of the plate. The clot can then be freed from the glass surface over most of the area, and the plate tilted so that the serum as it is squeezed out of the clot runs down to the clear side of the plate. With the small quantity of blood obtained from an individual mouse it is very desirable to keep the plate covered during the process, to avoid drying. For most practical laboratory purposes, sterile technique is not used in processing sera; the fresh serum itself is effectively bactericidal, and the serum is generally stored frozen between periods of its use. Preservatives are frequently added; Merthiolate (0.05 per cent) or sodium azide (0.1 per cent) are most common. Mouse serum to be used for blood-typing tests sometimes changes in its specificity or loses its activity upon ordinary storage; routinely, mouse-typing reagents are therefore subdivided into small quantities in ampoules, and are then lyophilized and stored in sealed ampoules, if possible after evacuation of the ampoules. Most serologic typing reagents, however, are much less critical in their storage requirements than are those prepared from murine serum.

Immunization.—Numerous routes of injection are used; the amount and the character of the antibody response is often influenced by the route adopted. Rabbits are generally injected intravenously into the marginal ear vein. For immunization with red-cell suspensions, we commonly inject 0.5 ml. of a 20 per cent suspension of washed cells intravenously three times a week for three weeks. The rabbits are then rested for 7–10 days before the antiserum is collected. The injection is made with a $\frac{5}{8}$-inch, 25-gauge needle on a 2-ml. syringe; the use of inexpensive disposable syringes and needles is becoming common and has several advantages. The amount injected, the frequency of injection, and the duration of the immunizing period are arbitrary, and other schedules are as effective as that described. Frequently, rabbits are kept after their first bleeding, and after several weeks are injected again to produce more antiserum. Very often the results of the later immunizations are better than those of the first series. Because of the possibility of an anaphylactic response in the immunized rabbit when he is injected after an interval of rest, we usually make the first reimmunizing injection intraperitoneally rather than intravenously, injecting about four times as much material as is used in a single intravenous dose; this has the effect of desensitizing the rabbit, and two additional intravenous injections can then be made as usual at two-day intervals.

For some types of antigens, especially relatively small protein molecules of low antigenicity, it has been found desirable to use adjuvants in connection with the injections. Alum precipitation of the antigen before injection is often effective. Another type of adjuvant very frequently used is based on the studies of Dr. Jules Freund, and in its several modifications is generally referred to as Freund's adjuvant. Procedures are described in connection with antiglobulin techniques in a later section of this paper.

Material in Freund's adjuvant is generally injected intramuscularly (in rabbits, into the large muscle of a rear leg or into the loin), or subcutaneously (under the skin of the back). Intense local reactivity is induced at the site of injection, and repulsive necrotic lesions may appear. If the outer surface of the needle is kept free of Freund's material, dry and sterile, such lesions are less frequent. An adjuvant often makes the difference between getting a useful antiserum and getting none at all. Adjuvant, however, in our hands generally has had little effect on circulating antibody responses to red cells and similar large materials; it would appear that such preparations generally have sufficient "adjuvant" activity of their own.   In some systems, such as the auto-sensitization of mice to red cells of their own type, Freund's adjuvant may play an irreplaceable rôle.

Avoid young animals, or animals in poor physical condition, for the preparation of antisera.   We commonly use a market strain of New Zealand white rabbits and require that each animal weigh at least 5 lb. before injections are begun.   Mice achieve sufficient serologic maturity after they are about 6 weeks of age, but they are commonly used at 12 weeks.

Mice are commonly immunized by way of intraperitoneal or subcutaneous injections, although intravenous injections into the lateral tail veins are not difficult.   Rats are somewhat more difficult to inject intravenously than are mice; intravenous injections of rats are possible, but intracardiac injections or injections by cannulating a femoral or carotid vein are often easier.[237]   Mice do not produce good antisera to murine red cells in response to the injection of blood; instead, macerated tissues (particularly spleen) or tumor transplants are commonly used to sensitize them.   We have evidence that better hemolytic antisera are likely to be produced in mice if they first reject a skin transplant from the donor line; shortly after the rejection of the sensitizing transplant we give the first of two intraperitoneal injections of splenic cell suspensions (1 donor spleen per 3 recipients) at weekly intervals, then bleed the recipients for test serum about ten days after the second injection.

Preparation of antibody reagents.—Under the simplest and most fortunate circumstances an antiserum, either normal or immune, may contain an antibody population recognizing only a single difference among the individuals tested—those having the specificity recognized by the antibodies, and those lacking it.   Serologically, this simple situation is revealed when absorptions are conducted with the red cells of a panel of test individuals.   Packed, washed, red cells of each test individual are mixed with a saline dilution of the test serum and thrown down in a centrifuge.   The supernatant, absorbed serum is then used as a test reagent; in the simplest situation all positive cells will absorb the antibody for all other reactive cells, whereas negative cells have no effect on the antibody population.   Genetically, this simple serologic situation is almost always found to reflect a single genetic alternative; positive individuals are found to be either homozygous or heterozygous for an allele producing the test specificity, and individuals lacking it are homozygous for an allele of the gene so defined.

Much more commonly, an antiserum will be found to contain two or more antibody populations of discrete specificity.   Serologically, this situation is revealed by absorption analyses; not all reactive cells will remove the antibody for all others.   Table 60 shows an absorption analysis leading to the recognition of two specificities, designated *A* and *B*, by means of anti-*A* and anti-*B* antibody populations in a particular antiserum.   An analysis like that illustrated in table 60 should be followed by further absorptions on the reagents postulated from this set of absorptions to be of single specificities; the test fluid remaining after absorption with cell no. 2, for example, should now be further absorbed with each of the reactive cells separately, and it should

Table 60

ABSORPTION ANALYSIS LEADING TO THE RECOGNITION OF TWO
SPECIFICITIES

	Antiserum									
	Unabsorbed	Absorbed with cells of individual:								
		1	2	3	4	5	6	7	8	9
Test cells from individual: 1 (*A*)	+	0	+	0	+	0	0	+	+	+
2 (*B*)	+	+	0	0	+	+	0	+	0	+
3 (*AB*)	+	+	+	0	+	+	0	+	+	+
4 (–)	0	0	0	0	0	0	0	0	0	0
5 (*A*)	+	0	+	0	+	0	0	+	+	+
6 (*AB*)	+	+	+	0	+	+	0	+	+	+
7 (–)	0	0	0	0	0	0	0	0	0	0
8 (*B*)	+	+	0	0	+	+	0	+	0	+
9 (–)	0	0	0	0	0	0	0	0	0	0

The unabsorbed antiserum evidently contains two distinct antibody fractions discernible with these test cells, recognizing four cell types, *A*, *B*, *AB*, and (–).   Absorbing with type *A* leaves anti-*B*; with *B* leaves anti-*A*.

be shown that each of these cells removes all of the antibody reactive with each of the others.   Very frequently, an antiserum will require three or four or more sets of symbols for its consistent analysis (for example, anti-*A*, anti-*B*, anti-*C*, anti-*D*).   The analysis of such an antiserum provides an interesting exercise.   The behavior of each cell in both the absorption columns and the test rows of the table, and relative to the behavior of all other cells in the test, must be consistent with the symbolic designations. In such complex situations, unit reagents are often achieved only after further absorptions with pools of two or more cells of different symbolic types.   Realistically, we must observe here that frequently deviations from ideal behavior are observed in these complex systems, such that, for example, a given cell may appear to be removing antibody in absorption with which it does not appear to react in tests.   The published work in this field is generally based on selections of reagents that behave in logical and consistent fashion, and authors and teachers in this field often ignore a tangled hinterland of poor reagents set aside, unused and unexplained, that violate the principles of straightforward symbolic absorption analyses.

Genetically, reactions that require two sets of symbols (such as *A versus* non-*A*; *B versus* non-*B*) are often found to depend on two independent pairs of alleles.   Frequently, however, multiple allelic series are discerned; some of the longest multiple allelic series in all genetics are noted in this field.   When sets of specificities are found to depend on a series of multiple alleles, there are very frequently cross reactions among the products of related alleles.   This situation, dependent mainly on the lack of complete specificity of serologic reactions, leads to uncertainties regarding the significance of the symbolic designation of antigens, in terms of their antibody reactions, and of genes, in terms of the antigens they control.   This point of principle will not be discussed further here; it has been considered in detail elsewhere.[981, 985, 1286]

As to the methodology of absorptions themselves, the details will be found to depend to some degree on the particular system to be investigated.   In many systems, such as rabbit antisera to cattle blood, or an antiserum produced by one cow against red cells of another, one simply mixes the packed, washed, absorbing cells with a dilution of the antiserum to be absorbed in nearly equal quantities, allows the well-mixed mixture to stand for a few minutes, then spins the cells down and repeats this procedure on the supernatant with fresh absorbing cells until all of the antibody reactive with the absorbing cells has been removed.   This process may be complete in two successive absorptions (as with most cow-anti-cow sera), or may require four or five successive absorptions (as with rabbit-anti-cow serum).   The objective here is get out all of the antibody with which the absorbing cells can react; further absorptions with these cells or with negative cells then have essentially no effect on the reagent.   In many other systems, however, the amount of packed red cells used in an absorption, and the number of absorptions, must be cautiously controlled; a procedure described by Race and Sanger[1036] as that in use at the Medical Research Council for the preparation of anti-human M and N sera is a useful reference.

The reason generally given for the necessity of proceeding with caution in situations like that for anti-human M and N is that the anti-M and anti-N antibodies are not fully specific for the corresponding antigen, but will in fact unite with low avidity to the other antigen as well.   Even the relatively rare cases in which human beings develop anti-N prove to be specific for N at only certain temperatures; they react with M cells at lower temperatures, or after enzymatic treatment of the cells.[585]   The plant anti-N can be adsorbed by M cells, although it is much more easily dissociated from them than it is from N cells.[784]   Similar situations are encountered with reagents for red-cell differences among mammals other than man.   They lead, of course, to difficulties; for example, in such a situation a questionable test result can be checked only with difficulty by an absorption test because even a negative cell may remove or reduce the antibody in question.   They also make it difficult to judge the specificities of cells by tests on eluted antibody after absorption.   Techniques for elution of antibody will not be discussed in detail in this chapter; one of the easiest involves the principle of adsorbing antibody to the surface of the cell at a low temperature, washing the cells gently in the cold, and then warming and shaking the system in a saline medium. The antibody may then elute from the cell surface into the saline, from which the cells can be removed by quick centrifugation at the higher temperature.

Another basis for deviations from ideal behavior in absorptions has been suggested by the work of Jacquot-Armand et al.[659]   It appears that the red cells of certain animals will remove a fraction of human anti-B only when the cells are used in large excess, and that this adsorption is promoted by nonspecific factors in serum, such as added serum from persons of type AB, which of course lacks the anti-B antibody itself. More work should probably be done along this line; if it is true that nonspecific co-factors may sometimes be involved in the fixation of antibody in absorption, part of the logic upon which absorption analyses are based becomes doubtful. Most immunogeneticists generally choose to believe that absorption tests are the last compelling resort for ascertaining the specificity of questionable test cells; it is now clear, however, that even this resort is not unfailingly reliable. Nevertheless, in addition to their utility in providing simpler test fluids as reagents, antibody-absorption procedures are a useful and often irreplaceable source of detailed information about cell and antibody specificities. If one were to give a value estimate, he would probably conclude that more misleading conclusions have entered the literature through failure to conduct adequate absorption analyses than through the overzealous application of this technique.

When one is working with so small a mammal as the mouse, and particularly in segregating generations when recourse cannot be had to pooled bloods from numerous representatives of an homogeneous group, shortage of absorbing cells and of antisera may become a limiting consideration. Under these conditions, *in vivo* absorptions, as conducted by Amos,[19] may prove useful. For example, when 0.1 ml. of an antiserum produced in BALB/c mice against the C57 *black* leukosis *EL4* was injected intravenously into C57 *black* mice and the recipient animals were bled at intervals, the titer of the serum of the recipient animal against *A*-cells had fallen below detectable levels after 30 minutes. Injected into the BALB/c mice, the same amount of serum remained at a titer of 1/256. The passively administered antibodies persisted in the serum of the nonreactive mouse with a half-life of about two days or longer. Intraperitoneal rather than intravenous injections are now commonly used for this technique. Given an antiserum that can withstand the dilution factor observed (about 1/20), this technique therefore provides a convenient method of absorption analyses in murine test systems.

Red-cell test systems.—A variety of test systems have been elaborated for immunogenetic studies of red cells. These have been described in detail in research papers and books on technique in the literature,[241, 1290] and only a brief survey will be undertaken here.

1. Saline agglutination. Tests for saline agglutination of red cells are commonly performed in small test tubes (10 × 78 mm.), into which one or two drops of the diluted test reagent are dropped from a 1-ml. pipette equipped with a 1-ml. rubber bulb. A drop of a suspension of the washed test cells is then added in the same manner —usually, about a 2 per cent suspension. When a number of samples of cells are to be tested against a number of reagents, usually the reagents are arranged in rows in a series of racks, and the test cells are added in columns, to provide a complete test pattern. For some reagents, it may be necessary to allow the test to stand for an hour or two or more after shaking the cells into smooth suspension, allowing the cells to settle by gravity and then reading for agglutination either macroscopically or microscopically or both. Frequently, however, the test can be read only five minutes or so after the cells have been added, sedimenting the cells by brief centrifugation. The

macroscopic reading is performed by shaking the pellet of sedimented cells gently back into suspension, meanwhile checking for agglutination by examining the suspension for clumps against a light background. For these and similar tests, a special small centrifuge called the *Sero-fuge* (Clay-Adams, Inc., N.Y.) provides a significant saving in time.

Variants of this test system are numerous. For example, for particular reagents the cellular suspensions may need to be heavier than 2 per cent; the temperature may be optimally either higher or lower than room temperature; a slanting capillary tube technique, with examination for the sedimentation pattern on the side of the capillary tube, is a sensitive technique for some systems and conserves valuable test sera. For some tests, such as *Rh* tests with particular reagents, slide agglutination is used. Directions for special techniques are provided with commercial reagents if these are to be purchased.

2. Other systems for agglutination. Antisera that fail to agglutinate red cells in saline, or that do so only weakly or unreliably, may produce strong and reliable agglutination if a medium other than saline is employed, or if the cells are treated in particular ways. One of the most common modifications is to use 20 per cent bovine albumin as a diluting fluid for the test serum and as a medium for the suspension for test red cells. Frequently, a somewhat heavier red-cell suspension is used in such systems—often of the order of 4 per cent. These tests are often incubated for an interval of two hours or so, at 37° C, before being read. Normal serum is sometime used as a diluting fluid; if the normal serum contains antibody, this can be absorbed out with washed red cells before it is used. Dextran, gelatin, and polyvinylpyrrolidone are also sometimes used.[380, 665, 829]

Murine antibodies generally produce only irregular and uncertain agglutination of murine red blood cells in saline media. A major methodologic contribution, which made it possible to bring the mouse into productive use for investigations of red-cell immunogenetics, was the development by Gorer and Mikulska[459] of the Dextran method for hemagglutination. In this method, the test red cells are suspended in a 1/2 dilution with saline of normal human serum, which has been absorbed with washed, pooled, murine red blood cells in order to remove any normal, human-anti-mouse red-cell antibody. We ordinarily make up the test cell suspension in saline to proper concentration and just in the amount needed, then centrifuge down the cells, aspirate off the supernatant saline, and replace it with the absorbed and diluted human serum. The test reagents, mouse-anti-mouse antisera, are diluted with a saline dilution of Dextran, the latter at a concentration of 2 per cent. Unfortunately, different Dextran preparations vary in their desirability for this purpose; some of them agglutinate cells themselves without the addition of antibody in nonspecific fashion, while others fail to promote the agglutination of antibody-sensitized cells. The Dextran of choice is *Intradex*, produced by Glaxo Laboratories, Ltd. (Greenford, Middlesex, England) but this is difficult to obtain.

After adding a drop of the normal, serum-suspended, test cells to a drop of the

Dextran-diluted test reagent, the cells are allowed to settle for a period of an hour or longer, and most of the supernatant fluid is then pipetted off with a capillary pipette. The sedimented cells and a small amount of the remaining fluid are then streaked across a microscope slide, and the slide is rocked back and forth manually five or six times. In negative tests the cells may at first appear to be aggregated, but they spread smoothly during the rocking procedure. In positive tests they remain conspicuously clumped. The technique is very sensitive to handling; negative tests may be made to appear positive if they are insufficiently rocked, or positive tests may be made to appear negative if the fragile aggregates are broken up during the pipetting procedure or through overmanipulation. With experience, however, this test becomes entirely reliable and is still the technique of choice for routine typing and testing of red cell antigens in the mouse.† Two other techniques for red-cell typing in the mouse, agglutination in saline medium after enzyme treatment of the red cells and hemolysis, will be mentioned below.

Treatment of red-cell preparations with particular enzymes sometimes renders them subject to agglutination by antibody that does not cause the reliable agglutination of untreated cells. The technique for trypsin treatment recommended by Morton and Pickles[900] is essentially as follows: A stock solution of 0.1 gram of crystalline Armour trypsin is prepared in 10 ml. of N/20 HCl; this can be kept for several months at 4° C. One part of this stock solution is diluted with nine parts of M/10 phosphate buffer (pH 7.7) on the day it is to be used. Four volumes of the diluted trypsin are added to one volume of well-washed packed cells; the mixture is held at 37° C. for $\frac{1}{2}$ to 1 hour. After a single additional washing the cells are made up to 5 per cent concentration in saline. In *Rh* testing, the test sera as well as the suspensions of cells are equilibrated to 37° C., and the test is incubated at that temperature in order to avoid false positive reactions caused by cold agglutinins. In applying enzymatic treatments to other mammalian red cells, adjustments frequently have to be made for the particular system under test, in order to avoid false positive reactions.

Numerous other enzymes have also been used to treat red cells; in addition to trypsin, papain and ficin are most frequently used.[1388] A rapid and simplified enzyme test has come into common use in human red-cell typing. Löw's technique[806] is as follows: a papain solution is made up by grinding 2 g. of papain (papayotin Merck 1:350) in a mortar with 100 ml. of M/15 phosphate buffer, pH 5.4. This is filtered, and 10 ml. of 0.5 M cystein is added to activate the enzyme. The solution is diluted to 200 ml. with the buffer, incubated for 1 hour at 37° C., and stored at −20° C. It retains its activity over long periods under these circumstances. Three parts of this enzyme solution are mixed with one part of the test serum. Then equal parts of the serum-enzyme mixture and a 3 per cent red-cell suspension are mixed in a test tube and incubated for 2 hours at 37° C. before being read for agglutination. Hekker *et*

---

†See, however, Stimpfling, J. H., Transplant. Bull. **27**:109, 1961 for a technique using polyvinylpyrrolidone.

al.[543] add a drop of a papain solution directly to a drop of test antiserum on a slide and then add a drop of the red-cell suspension. This is reported to give an accurate and rapid test system. These and other test systems are ably described by Race and Sanger.[1036]

Another method of setting up and reading tests is to examine the pattern of the sedimented cells after they have settled out by gravity. This may be done in ordinary round-bottomed test tubes, or through the use of plates containing round- or sloping-bottomed depressions. Antibody-sensitized cells often behave in a nonspecifically sticky fashion so that they do not sediment to the lowest point of the depression but remain attached to the sides and bottom of the tube or depression. Frequently, agglutination is easily read under low-power magnification or even macroscopically. In such receptacles, and in cases in which the agglutinates are very fragile, positive reactions can often be identified in the undisturbed plates much more easily than they can be read by ordinary tube and microscopic methods of examination. Hildemann (personal communication) reports good results through the application of such a system to murine red-cell typing.

3. Hemolysis. Agglutinating tests are sometimes convertible to hemolytic ones through the addition of complement to the system. The most common source of complement is fresh normal guinea-pig serum, a drop of which, usually at a dilution of 1/8, added to the saline tube-agglutinating system described in the preceding section, may cause sensitized red cells to lyse. Another common source of complement is fresh normal rabbit serum, which is generally used at a somewhat lower dilution, usually 1/2. The optimal source of complement in any new system is unpredictable; for example, certain antibody reagents will lyse positive cattle red cells far better with rabbit complement than with guinea pig, whereas others work much better with guinea pig complement. Rabbit complement is best for testing mouse cells with murine antibody.[574]

Rabbit, guinea-pig, or other normal sera sometimes contain antibodies against the test cells and may cause the cells to lyse without the addition of reagent antibody. In such a circumstance it is often advisable to screen individual rabbits or guinea pigs in advance and to select as complement sources those whose sera lack normal antibody against the red cells to be tested. Alternatively, the normal antibody can be removed from a serum by a quick absorption at 0° C.; we generally use a volume of packed red cells for this absorption about equal to the volume of undiluted normal serum to be absorbed, and conduct a 5-minute absorption with prechilled cells and serum and with the tubes in an ice bath, followed by centrifugation in a refrigerated centrifuge. The antibody is removed by the absorbing cells, but the complement activity is little affected. Complement is a relatively labile material; it should never be permitted to sit out at room temperature for any extended interval before its use in the tests. It is inactivated by heating at 56° C. for 20 minutes.

Hemolytic systems can often be rendered more sensitive by the use of an isotonic buffer containing magnesium and calcium for the diluting and suspending medium.

The buffer favored for such purposes was reported by Pillemer[1006] in his studies of complement and properdin.   The recipe is as follows:

> 85.0 g. NaCl
> 5.75 g. 5,5-diethylbarbituric acid
> 3.75 g. 5,5-diethylbarbiturate
> 5.0 ml. M $MgCl_2$
> 1.5 ml. M $CaCl_2$

Dissolve in about 1500 ml. hot distilled water.   Cool, add distilled water to final volume of 2 liters.   Store as stock at 1° C.; add one part stock buffer to four parts distilled water before use; discard diluted portions after 12 hours (pH 7.4).

One laboratory reports that time can be saved in setting up extensive red-cell hemolytic tests if measured amounts of diluted complement are mixed with the test serum and dropped into the tubes in a single operation, before the cells are added. Another laboratory, however, reports that this technique in some systems leads to some inactivation of the complement, and that it is better to add the complement to the sensitized cells as a final step.   Hemolytic tests are often read at an interval of one-half hour after the test has been set up and shaken, in order to detect the quick hemolysis that is often characteristic of cells homozygous for the test antigen, in contrast to the slower hemolysis detected with the cells of heterozygotes.   In any case, the cells are then allowed to settle for a further hour and a half before a definitive reading is taken; often an additional reading is taken after two more hours.   Readings are generally recorded on a scale from 0 to 4, 0 representing no hemolysis, in which the supernatant is free of hemoglobin deriving from the test cells.   With increasing degrees of hemolysis deeper color appears in the supernatant, and the size of the pellet of sedimented cells decreases.   Complete hemolysis (4) leaves no unlysed cells; the shaken tube remains a sparkling clear red.

Hemolytic tests are especially well adapted to evaluating the proportions of positive and negative cells in a mixture, such as the cellular population found in a chimera.   The supernatant of the hemolytic test can be read for hemoglobin colorimetrically; and after subtraction of an appropriate blank prepared to measure the contribution in color, if any, from the non-red-cell components of the test (that is, the complement and the test serum), the degree of color in the test tube can be expressed as a fraction of the color in the tube containing the same amount of red cells osmotically lysed.   This technique has proved useful in following the course of repopulation of the erythropoietic tissues of irradiated mice after homologous bone-marrow transplants.[452, 981, 985]   In other systems, it is reported that more reliable values may be obtained by washing the residual cells after the specific lysis of positive cells in a mixture, then lysing these residual cells osmotically and reading the hemoglobin colorimetrically.[848]

4. Antiglobulin test systems.   A sensitive and versatile technique for the detection of incomplete antibodies was described by Coombs, Mourant, and Race in 1945.[225]

The principle of the test is as follows: Antibody globulin combines firmly with its corresponding antigenic sites on cell surfaces and remains attached during washing of the positive cells. Unbound globulin will be removed from the system during the washing. The washed cells, having the antibody on their surface if they are positive, are then exposed to antibody to the globulin itself. Reaction between the cell-bound globulin and the antiglobulin results in the aggregation of positive test cells.

To illustrate the technique, our procedure (entirely derivative from the experiences of others with other systems) for the detection of individual differences among rhesus monkeys may be used as an example. The tests are set up in small agglutination tubes, as for saline agglutination. The test fluids are rabbit-anti-rhesus red-cell antisera fractionated by antibody absorption to leave reagents that react with the red cells of some monkeys and not others. Two drops of the reagent diluted in buffered saline (130 g. NaCl, 12.3 g. Na_2HPO_4, 3.6 g. KH_2PO_4, 17 liters distilled water, pH 7.08) are first placed in the appropriate tubes. One drop of a 2 per cent suspension of washed test red cells is then added to each proper tube. The tubes are shaken and allowed to stand at room temperature for at least one-half hour, after which they are shaken, filled with buffered saline, and centrifuged; the supernatant is poured off and the cells are washed again with saline. To the pellet of cells remaining after pouring off the last supernatant, one drop of a 1/10 dilution of goat-anti-rabbit-globulin serum, which has been absorbed with rhesus red cells in order to remove anti-rhesus red-cell antibody, is added. The procedure for preparing this anti-rabbit-globulin serum will be described below. The test is shaken again, allowed to stand for another hour, then centrifuged briefly and read for agglutination as the pellet of centrifuged cells is shaken back into suspension. With this test system, antibody reagents that produce little or no reliable agglutination of positive test cells in saline become very sharply discriminating.

The type of antiglobulin serum used in this kind of test depends, of course, on the source of the cell-sensitizing antibody. In the rhesus test described above, since the test antibody was rabbit, the Coombs antiserum was anti-rabbit globulin, in this case prepared in a goat. For test systems in which the cell-sensitizing antibody is other than rabbit, it is commonly the rabbit that is used as the source of the Coombs reagent. For example, sheep cells and rat-anti-sheep red-cell antibody can be used in a Coombs system if rabbit-anti-rat globulin is used as the developing reagent. Similarly, in human tests, where the Coombs system has become a very important test procedure, it is rabbit-anti-human globulin that is generally used as a Coombs reagent.

In preparing the Coombs reagent, first a predominantly globulin fraction is salt-precipitated from an immune serum—for example, to prepare an anti-rabbit-globulin serum in a goat, we first take a rabbit antiserum to bovine serum albumin which we know to contain a good deal of antibody and precipitate this antibody by the addition, at room temperature, of an equal volume of saturated ammonium sulfate solution to a 1/2 dilution of the serum in saline. After adjustment to pH 7.8 with 10 M NaOH, the precipitated globulin is centrifuged out and redissolved in saline. For this purpose, we do not ordinarily purify it more than one further precipitation and resolution, and we recognize that we have more than γ-globulin in this preparation. We then alum-precipitate the protein for injection according to the following procedure.

Dissolve the globulin prepared as above from 40 ml. of rabbit serum, in 25 ml. buffered saline. Add 80 ml. distilled water, then 90 ml. of 10 per cent potash alum in water. Adjust with NaOH to pH 6.5. Let stand ten minutes, centrifuge and wash once with saline. Resuspend in 15 ml. buffered saline, and divide into three equal aliquots. Inject one (intramuscularly); freeze the other two aliquots to be injected (i.m.) respectively one and two weeks later. Bleed after an additional two-weeks' rest. The goat can be restimulated by injecting globulin in solution, or even by injecting whole rabbit serum.

Alternatively, the globulin or whole serum can be injected in Freund's adjuvant. The procedure we use is essentially that described by Munoz.[919]   To 2 parts of paraffin oil (Drakeol 6-VR, obtained from the Pennsylvania Refining Company, Butler, Pennsylvania), containing 2 mg. of heat-killed and lyophilized *M. tuberculosis* (we use an avirulent strain H37Ra, obtained from Dr. Sidney Raffel of Stanford University), is added one part of Falba (obtained from Pfaltz and Bauer, Incorporated, Empire State Building, New York 1, New York). After autoclaving, this mixture is taken at 3 parts of the sterile adjuvant to 2 parts of the salt-precipitated globulin. If this is, for example, mouse globulin to be injected into a rabbit, we prepare three ampoules, each with 1 ml. of the adjuvant-globulin mixture, and inject the contents of one ampoule subcutaneously once each week for three weeks, then bleed after a rest of two weeks.

There are other methods for making antiglobulin sera and conducting anti-globulin tests, some of them better for particular systems than others. References, and a quick and clear survey of the field, can be found in Race and Sanger's book.[1036] Coombs and Roberts[226] have published a brief review, including further applications and adaptations of the method.

5. *Elimination rates of labeled cells.* A useful serologic technique is based on the accelerated clearance of labeled antigen from the circulation of immune animals. With soluble antigens, this is one of the more sensitive and convenient indicators of the immune state.[205, 278]   The method has found little application in immunogenetics, although it offers promise. Some use has been made, however, of the clearance of labeled red cells, and the observations at hand should stimulate further study. The label of choice appears to be Cr^{51}. Mollison,[888] who has contributed actively to this area, has reviewed the methodology recently and has called attention to rather frequent evidence of incompatibility based on reduced survival of Cr^{51}-labeled human cells injected into normal human recipients in whose serum no incompatible antibodies could be found. An example of fruitful application of a similar technique to problems of immunogenetic incompatibilities in mice is a paper by Goodman and Smith.[453]

6. *Tests on mixtures of cells.* As mentioned earlier, the saline-agglutination system, hemolysis, and the antiglobulin system (in the case of rhesus red-cell mixtures)

have all been used for the quantitative identification of cellular mixtures in chimeras deriving from either natural or experimental transplants of erythropoietic tissues. In some mixtures, however, only minute quantities of positive cells may be present in an overwhelming predominance of negative cells, or, in the reverse situation, very small numbers of nonreactive cells may be present among very large numbers of positive ones. Several techniques have been developed to evaluate these minute populations. No detailed description of techniques will be attempted here, but reference should be made at least to an isotope-dilution method together with use of plant lectins for the selective removal of positive cells;[36] to the "mixed agglutination" found with detector cells added to a suspension;[666] and to the potential utility of antibody marked with a radioactive tracer. For a general discussion of this subject, see Cotterman.[228]

Preserving red cells.—Contamination of blood samples can be controlled by adding certain antibiotics. In a procedure described by Cahan,[154] 200,000 units of penicillin G (crystalline-potassium) is dissolved in 100 ml. of a citrate solution (6.0 g. sodium citrate, 7.0 g. sodium chloride, 1 liter distilled water). One gram of Streptomycin sulfate (Squibb) is dissolved in 100 ml. of a phosphate buffer (16.4 g. $Na_2PO_4 \cdot 7H_2O$, 5.36 g. $NaH_2PO_4 \cdot H_2O$, 1 liter distilled water). Stone and Beckstrom[1284] report that, by using 0.25 ml. of the penicillin solution and 0.1 ml. of the streptomycin solution in the anticoagulant provided for 10 ml. of whole cattle blood, the blood can be shipped for long distances without refrigeration and can be stored under refrigeration for extended periods.

Blood cells can also be stored frozen for years in proper media. A technique effective for cattle erythrocytes has been described by Stone *et al.*[1285] Whole blood warmed to 37° C. is centrifuged, and the plasma is removed and replaced with 40 per cent ethylene glycol at 37° C. made up in a 6 per cent sodium citrate solution. After mixing, the blood is put into plastic tubes, stoppered and stored at −20° C. When needed, the blood is thawed at room temperature or 37° C. A minimum of 4 ml. is centrifuged and the supernatant discarded. The cells are resuspended in an excess of 20 per cent ethylene glycol, centrifuged, and the procedure is repeated with solutions of 10, 5, and 2 per cent ethylene glycol in sequence. The cells are then suspended in 0.9 per cent saline. Although many of the cells may lyse during this procedure, enough remain for typing, and they type normally after periods of storage in excess of two years.

For some species, glycerine or glycerine and plasma provide excellent media for freezing and preserving red cells and other tissues.[195] For others, such as cattle, the ethylene-glycol method described above seems to work much better. Several papers on red-cell preservation were included at a meeting of the American Association of Blood Banks in San Francisco in August, 1960, and abstracts appear in the published proceedings of that meeting.

Tests on contaminated blood samples should be regarded with caution, because particular types of bacterial contamination may change the test reactions with particular reagents.[1287]

OTHER CELLS IN SUSPENSION

Although suspensions of red blood cells have provided the most convenient material for mammalian immunogenetics in the past, increasing attention is being given to the antigenic characterization of other types of cells as well. Some of the red-cell specificities seem to be limited to the erythrocytes; others are found in or on diverse types of cells and tissues. Particularly in man, a variety of white-cell antigens has become available for immunogenetic analysis. In some instances, these are of significance in systems of maternal-fetal incompatibility, comparable to those involving the *Rh* and *ABO* red-cell antigens.[152, 586, 744, 1071]

Terasaki[1317] has described a procedure for obtaining rather pure lymphocyte suspensions from chicken blood, and for conducting agglutination tests with lymphocyte suspensions. Agglutinating test systems have also been applied to mammalian leucocyte suspensions; in the past, however, such tests have often given inconsistent results.[242] Absorption tests with *A* and *B* human white cells have given straightforward results, as have such tests with spermatozoa, though sperm are not agglutinated directly by anti-*A* or anit-*B*.[750] Race and Sanger[1036] report that the ability of spermatozoa to absorb anti-*A* and anti-*B* antibodies is not removed by as many as nine successive washings of the sperm, indicating that this property of the spermatozoa is probably not simply adsorbed from the seminal fluid. Kiddy and his colleagues[706] have reported rather extensive tests on the antigenic qualities of rabbit sperm.

The sensitive and versatile technique of mixed agglutination was reported by Coombs and Bedford in 1955.[223] Essentially, it involves sensitizing test cells with antibody, then adding red cells of known type. If the test cells adsorb the antibody to their surface, red cells of the corresponding type cluster on the surfaces of the sensitized cells, whereas failure of the test cells to combine with the antibody results in their failure to accumulate a shell of the corresponding red cells. This technique, with relatively minor modifications, has been used to type human epidermal cells,[224] cells in tissue culture, and other cells. A basically similar procedure was reported by Gullbring[511] to fractionate sperm-cell suspensions from *AB* men; anti-*A* sensitized *A* red cells combined with part but not all of the sperm, leaving an unreactive fraction, and anti-*B*-sensitized red cells had a similar effect. The question of whether a sperm cell may express its own antigenic genotype, however, remains open at present.

Another useful and versatile technique depends upon the cytotoxic effects of antibody on cells, especially in the presence of complement.[456] Several procedures are applied; the one we use, based entirely on the experiences of others and most directly from Vos *et al.*[1351] is as follows: Murine spleen-cell suspensions are prepared by pressing the spleen through a fine stainless steel screen, and suspending in Tyrode's solution. Small clumps are suspended by flushing the suspension in and out of a syringe with a 22-gauge needle. One drop of the cellular suspension is mixed with one drop of murine antiserum or normal serum. One drop of rabbit complement, which has been absorbed with red cells of the mouse in the cold in order to remove normal rabbit-anti-mouse antibody, is then added. The mixture is incubated at 37° C. for 30 minutes. After incubation the tubes are kept at 4–6° C. until the ratio of viable to nonviable cells is determined. The suspension may be stored at this temperature for as long as six hours. Immediately before viability is determined, the cells are resuspended with a capillary pipette and the reaction mixture is allowed to come to room temperature. To one drop of the mixture is added one drop of 0.4 per cent eosin in Tyrode's solution and in a hemocytometer stained nucleated cells are counted as nonviable. Non-stained nucleated cells are counted as viable. About 100–200 cells are counted from each tube; the degree of cytotoxicity of the antiserum is expressed in terms of the fraction of nonviable cells observed, in comparison with controls treated only with normal serum and complement. Standardization of the procedure, particularly in terms of the numbers of cells in the suspension, the various dilution factors, and the number of units of hemolytic complement added, is essential for quantitatively repeatable results. (Dr. Winn comments on this subject in more detail at the end of this chapter.)

An interesting point has been made by Amos and Wakefield:[21] although mouse complement is ineffective for cytotoxic action of murine antibody on murine cells *in vitro*, similar cells are lysed in diffusion chambers *in vivo* without the necessity of added complement.

In some instances, the effects of antibodies on mobile cells provide a good index of serologic reaction. Unfortunately, we have not yet encountered in mammals a system as sensitive and productive in this regard as the antibody-immobilizing systems of the Ciliates. Mammalian sperm-cell suspensions, for example, do not generally appear to be sensitive to antibody immobilization. The ameboid motions of some of the leucocytes, however, are affected by antibody, and this has provided a test system of some utility.

SOLUBLE ANTIGENS

Molecules in free suspension or solution in the body fluids, or obtainable in extracts, should provide a rich source of material for the immunogeneticist. Until recently, however, techniques for working with the serology of soluble antigens could cope only with extreme difficulty and uncertainty with complex antigenic preparations. A rather high degree of purification by chemical or physical methods was a usual prerequisite. A number of powerful methods are now available for working with mixtures, and we can sample this active area only inadequately here.

Inhibition systems.—Materials having a specificity related to one of the red-cell antigens can easily be tested, taking advantage of the red-cell reaction as an indicator. The general procedure is as follows: A determined amount of antiserum, say anti-*A*, is mixed with a measured amount of the solution under test, such as saliva or a saline extract. After an interval, *A* cells are added to the mixture. If the test material had *A* specificity, it will have combined with the anti-*A* antibodies, and thus will inhibit them from reacting with the *A* test cells. If, on the other hand, the test material lacked *A* specificity, the antibodies will not be inhibited and will therefore promote the agglutination of the test cells. This is the kind of system used for the definition of the blood-group substances in animal plasmas and other fluids. The techniques have been discussed in detail recently by Boyd,[121] and will not be further elaborated here.

Enzymatic inhibition.—Enzymes under genetic control are often excellent antigens and the effects of antibodies on enzymatic preparations are sometimes useful in distinguishing genetic differences. This approach has been used more productively with microörganisms than with mammalian material in the past, but there is no reason why it should not be useful in mammalian genetics as well. An antiserum is prepared by injecting an enzymatically active material into an animal, generally a rabbit. The antibodies that are formed may precipitate the enzyme from solution; the enzymatic activity provides a sensitive and convenient tag for the removal of the enzyme from the supernatant by such precipitation. Precipitin tests will be discussed below; we will only note here that the complications of complexity in mixtures are to a considerable degree avoided in enzyme serology, because of the easy identification of the particular antigen, the enzyme, in the mixture by means of its activity. In many but by no means all enzyme-anti-enzyme systems, the antibody may inactivate the enzyme without precipitating it. Under such circumstances, tests for enzymatic activity in mixtures can be made immediately after the addition of antiserum and substrate to an enzyme preparation; the results are quickly and easily read if convenient measures of enzymatic activity are at hand and if there is no interference by nonantibody serum factors in the assay.

Coupling antigens to red cells.—Erythrocytes to which soluble antigens have been coupled are often endowed with the property of agglutinating or giving other visible reactions with antibody to the test antigens. A common technique is the use of "tanned" red cells; the procedure we use is based on a report by Stavitski,[1270] and is derived from the original report by Boyden.[123] Sheep blood in Alsever's solution is washed three times with saline, and 1 ml. of the packed cells is diluted with about 40 ml. of pH 7.2 buffered saline, so that 1 ml. of this diluted cell suspension plus 5 ml. of distilled water gives a reading of 400 with a no. 54 filter in the Klett colorimeter. The buffered saline, pH 7.2, is prepared by diluting 100 ml. of a buffer containing 23.9 ml. of 0.15 M KH_2PO_4 and 76.0 ml. of 0.15 M Na_2HPO_4, with 100 ml. saline. A stock solution of tannic acid (Merck or Mallinckrodt reagent grade) is diluted with saline, 1/100. A further dilution, to 1/20,000 of the acid, is prepared daily.

One ml. of the cell suspension is incubated in a water bath at 37° C. for 10 minutes with 1 ml. of the 1/20,000 dilution of tannic acid. The cells are then centrifuged gently and washed with 1 ml. of pH 7.2 buffered saline, resuspended in 1 ml. saline. The treated cells cannot be kept more than 18 hours before use.

The antigen is prepared by mixing 4 ml. of pH 6.4 buffered saline with 1 ml. of the antigen solution in saline and 1 ml. of the tannic-acid-treated suspension of cells, in this order, and allowing the mixture to stand at room temperature for ten minutes.

The cells are then centrifuged, washed once with 2 ml. 1/100 normal rabbit serum, and then resuspended in 1 ml. of 1/100 normal rabbit serum. The saline at pH 6.4 is prepared by adding 100 ml. of saline to 100 ml. of a buffer composed of 32.2 ml. of 0.15 M Na_2HPO_4 and 67.7 ml. of 0.15 M KH_2PO_4. The pH should be checked on a pH meter and adjusted if necessary with either phosphate solution, 0.15 M.

Many antigens work in this preparation, in mixtures as well as in pure solution. Optimal amounts for sensitizing cells vary around 0.25 mg. of protein or other antigen per ml. treated red-cell suspension.

Compounds other than tannic acid are used as coupling agents. For formaldehyde methods, see Ingraham[631] and Czimas.[235] The hemagglutination of antigen-coupled red cells is a very sensitive technique, acceptably specific if proper controls are run.

Precipitation system.—Classical precipitation systems have been described for single, relatively pure, proteins in solution, in terms of reactions with their corresponding antibodies. Typically, a constant amount of antiserum is placed in each of a series of tubes, and increasing amounts of antigen are added to the tubes in sequence. The classical precipitin curve rises to a maximum at some intermediate antigen quantity, then decreases in antigen excess. Optimal relative concentrations are expressed in various ways—in terms of the rapidity of appearance of a visible precipitate, in terms of a maximum quantity of precipitate formed per mg. antibody nitrogen, and in terms of an equivalence zone within which all of the detectable antigen and antibody are included in the precipitate, none remaining in the supernate. These measures of central tendency do not usually coincide; on either side of a rather broad central zone, however, flocculation times increase, the quantity of precipitate decreases, and either antigen or antibody begins to be detectable in the supernatant fluid after precipitation is complete.

Precipitin tests of this sort are subject to quantitative treatment through measurement of the amount of precipitate by nitrogen determinations or other methods. A chromatographic technique for the quantitative study of the precipitin reaction, especially adaptable to very small amounts of serum, has recently been described by Miquel *et al.*[875] Absorptions can be conducted, in simple systems, by reacting the test antigen with a given antiserum at relative concentrations within the equivalence zone, removing the precipitate and using the partially absorbed supernatant as a further test reagent. Somewhat different results are sometimes obtained if, instead, the antigen is added in small increments, the precipitate being removed after each addition, until precipitation no longer occurs. In the past, there has been a tendency on the part of some geneticists familiar with only the elements of immunology to assume the properties of a simple antigen-antibody system for complex mixtures of indeterminate numbers of related and unrelated antigens, and correspondingly complex antisera. Unfortunately, the system becomes very uncertain under such circumstances; two or more systems, precipitating in a single tube, are not often at equivalence within the same zone, so that antibodies to one may remain in the supernatant after the other has passed into a region of antigen excess. Soluble complexes for the second system are therefore not removed from the absorbed "reagent." The uncertainties of the system probably increase exponentially with the number of components. Unless one is able to purify his antigenic preparation to an essentially single component basis, therefore, or unless he is working with an antigen that is certainly tagged, by something like enzymatic activity, a unique absorption spectrum, or a tracer label that it does not share with other molecules in the solution, or serologic specificities detectable by red-cell tests, there is nowadays little justification for conducting immunogenetic work on soluble antigens in mixtures by means of ordinary saline precipitation methods. The simple precipitin system, however, has probably contributed more to our basic knowledge of serology than has any other. Readers are referred to discussions of this system from a methodologic viewpoint, for example, by Kabat and Mayer[673] and Boyd.[122]

Gel-diffusion serology.—A new era in the immunogenetics of soluble antigens was ushered in by the development of gel diffusion methods by Oudin and Ouchterlony. These methods make use of the diffusion of antigen or antibody or both into a gel medium, usually agar. Since the different components of an antigenic mixture generally diffuse at different rates, this technique permits the separation of components of a mixture in a system comparable in some ways to chromatography. In the Oudin method[974, 975] agar containing a known serum is allowed to solidify in a small tube, and then an aqueous solution of the antigen is poured over the solid agar. As the antigen diffuses into the underlying antibody gel, it forms a zone of precipitate for each antigen-antibody pair present in adequate quantity. The Ouchterlony method is a system of double diffusion; the antigens diffuse through an intervening agar zone, to meet a front of diffusing antibody. Where corresponding antibody and antigen meet, a narrow line of precipitate is formed; and, when the antigenic preparation is a mixture, a number of bands form, each corresponding to a particular antibody-antigen system at positions related to the diffusion rates of the antigenic components and to the relative concentrations of the antigens and antibodies. This system is subject to quantitative analysis.[14, 349]

The method we use, adopted directly from G. J. Ridgway of the Bureau of Commercial Fisheries, Seattle, is as follows. An agar base is prepared by mixing 1.5 per cent Difco agar, 0.72 per cent NaCl, 0.6 per cent sodium citrate, and 0.01 per cent Merthiolate, in distilled water. This solution is adjusted to pH 6.7 with HCl before the addition of trypan blue (to 0.01 per cent) while the agar is still hot. It is then poured into test tubes, 8 ml. per tube.

In setting up the tests, one tube of the hot agar base (kept in a boiling water bath) is poured into a flat-bottomed petri plate. The plate should not be swirled to distribute the agar. When the plate is cool, penicylinders (obtained from Fisher Scientific Co.) are arranged in a pattern on the plate. We commonly use a porcelain penicylinder at the center and stainless steel penicylinders arranged at the points of a regular hexagon so that each is equidistant from the center. The distance from the center of the central penicylinder to the center of each peripheral one is 2 cm. Shreffler, in our laboratory, has prepared lucite guides which fit on the petri plate, with holes bored in the proper position just large enough to drop the penicylinders through to the agar surface. After the penicylinders are in position, another tube of 8 ml. of the hot agar base is poured around them. Again, the plates should not be swirled.

The antiserum is then placed in the central penicylinder, filling it, and the test antigens are placed in sequence in the peripheral ones. Plates are sealed with rubber tape and placed at 37° C. for about a week (the time necessary will vary with the antigens and the serum). The precipitin lines that form are then photographed; a modified dark-field type of illuminator producing good photographs has been described by Klontz *et al.*[728] The trypan blue in the agar, and the porcelain central penicylinder, improve the quality of photographs.

Numerous variations of both the Oudin and Ouchterlony procedures have been described. Preer[1021] has described a method adaptable to the use of very small quantities of material. Many investigators prefer to use molds of various design and to pour the agar medium for gel diffusion tests around the molds. When the molds are removed, wells remain in the solidified agar. Sets of agar-gel cutters for various purposes, together with descriptions of their uses and references, can be obtained from Shandon Scientific Co. Ltd., 6 Cromwell Place, London.

Gel-diffusion methods have made possible the first extensive and definitive work with the immunogenetics of mammalian soluble antigens in precipitating systems, for example, genetically controlled variations in the specificity of rabbit serum globulins[974, 975] and the sharp inherited differences in the quantity of a serum protein distinguishing particular inbred lines of mice.

Immunoelectrophoresis.—Another powerful technique for dealing with the serology of mixtures has been developed mainly by Grabar[472] and extended by many others. Essentially, the procedure first subjects the antigenic mixture to electrophoresis. After the electrical field has separated the components of the mixture along a linear axis as a function of their charge at the particular pH and ionic strength of the medium, the components of the mixture are permitted to diffuse through the agar to meet a front of diffusing antibody, so that a band of precipitate forms where antigens and their corresponding antibodies meet. The preliminary electrophoretic step provides an additional dimension of separation of the antigenic components; the technique has been proved especially effective for the discernment of many components in very complex mixtures, such as whole serum or plasma. Bussard[150] has described a useful modification, in which the precipitation occurs during the course of electrophoretic migration.

A combination of starch-gel electrophoresis with gel-diffusion serology has been reported by Schwartz.[1178] Starch gels, after the method of Smithies,[1229] generally make cleaner electrophoretic separations than do agar gels, and the addition of antibody precipitation to this system is a powerful technique indeed.

Complement fixation.—A classical serologic procedure of high sensitivity and specificity is the system of complement fixation. In principle, the system is simple: an antigen is allowed to react with its corresponding antibody in the presence of complement. In many instances, such reactions "fix" complement if it is present, even though the complement may not be necessary for the reaction to occur. After this phase of the test, in which the antigen-antibody reactions occur, a second indicator system is added in order to determine whether complement has been fixed. The indicator system is usually sheep cells which have been sensitized with rabbit-anti-sheep antibody. If complement is present, the cells will hemolyze; if it has been fixed by a preceding antigen-antibody reaction system, it will not be available to the hemolytic system and the cells will fail to lyse. The sensitized sheep cells, therefore, provide a measure of whether or not an antigen-antibody reaction has occurred in the test system.

Techniques of complement fixation are highly sensitive, and are subject to modification by many external factors. Careful controls must be run; the proper amount of complement must be added; the indicator cells must be properly sensitized, and so on. Descriptions of complement fixation procedures are to be found in Kabat and Mayer[673] and Boyd.[122] In general, although this system is a useful one in serology, it has found relatively little application in immunogenetics, particularly of mammalian materials.

TISSUE TRANSPLANTATION

Consideration of tissue transplantation will be limited to studies of normal rather than neoplastic tissues and to only a small selection from this active area. The methodology of tumor-transplantation is considered elsewhere in this volume.

Skin grafting.—The standard method for skin grafting, especially in mice, was developed by Medawar and his group.[98a] The following technique, which we use in our laboratory, is essentially Medawar's.

Donor mice are usually killed by the intraperitoneal injection of 0.1 ml. of Nembutal sodium, 50 mg. per ml., which is commercially available. Alternatively, if not too many grafts are to be removed, the donor mice may be anesthetized by the injection of 0.1 ml. per 10 g. body weight of a 1/10 dilution of the above Nembutal preparation. Different lines of mice vary somewhat in their sensitivity to the anesthetic, and the first mice to be injected should be carefully observed, with the idea that slightly less or more anesthetic may be desirable for the line in use. After the donor mouse has recovered from anesthesia, the sites from which the grafts are taken may be left open; they will heal rapidly.

If the donors have been killed they can be pinned out on a board; if they have not been killed they can be tied or taped down. The back of the animal, from the ears to the tail, is clipped closely; we use an Oster small-animal clipper, model 2, with a no. 40 head. The surface of the clipped skin is then swabbed thoroughly with Zephiran (1/1000 in 50 per cent alcohol).

An area of the skin of the donor is elevated in a small tent, with a small, very fine pointed forceps, the tips of which have been curved toward each other to meet at a single sharp point. As the skin is held under tension, it is cut from beneath with a concave (no. 12) Bard-Parker scalpel, pulling the scalpel against the firmest skin attachment and curving the cut back up to create a circular graft about 1 cm. in diameter and with even margins. About a dozen grafts can be taken from a single donor, if he has been sacrificed. Each graft is then placed with its dermal side downward on sterile filter paper, moistened with sterile physiological saline in a sterile petri dish. Only the dorsal skin is used, but this can extend to the base of the tail and the ears and along the sides of the animal.

After the grafts are all in the petri dish, spread face downward, the donor is discarded or placed aside to recover, and the grafts are one by one turned dermal side upward and the panniculus carnosus is carefully scraped away with a dull, straight scalpel. To do this, a pointed forceps is used to press down the graft at a point near its margin, and the panniculus can be pulled loose in a single piece. Care must be taken not to injure the underlying dermis, and the last evidences of fat should be scraped away, leaving a thin, even graft. This is again turned dermal side down.

The recipient mouse is anesthetized, and an area of his dorsal surface on the right side in the chest region is clipped and disinfected with Zephiran solution. Use the lower margin of the dorsal rib cage and the scapula as markers. It is convenient to graft on the right side of the midline, reserving the left side for an autograft or a later graft.

Using a fine, sharp, curved scissors carefully pinch off an area of the skin and clip it with the scissors. If this is done properly, the skin through the dermis will be removed in a narrow slit, exposing the recipient's panniculus carnosus. This will be evident by its glistening surface with intact blood vessels running through it. The recipient area is then cleared by cutting away the skin with the scissors at the margins until an area somewhat larger than the prospective graft is exposed. There should be a margin of free surface (1 or 2 mm.) around the graft when it is in place. A selected graft is then put in place, being careful that the dermal side is down, and it is then covered with a piece of vaseline-impregnated, rather fine gauze, about $\frac{1}{2}$ inch by $\frac{3}{4}$ inch. The gauze should have been prepared in quantity and sterilized earlier. It is convenient to sterilize the gauze in large rectangles and then to cut it as required in the petri dish. All of these manipulations are with sterile forceps and scissors. Care must be taken to cover the entire recipient surface with the gauze and not to disturb the graft from this point forward. A piece of plaster-of-paris bandage (we use Gypsona) about 7 inches in length and $\frac{5}{8}$ inch wide is moistened and the excess moisture shaken free. These pieces have been rolled on small lengths of plastic tubing about an inch long, secured with a small rubber band and stored in jars. There is no need to sterilize them. The moist strip is first unrolled and placed over the graft area so that an inch or more projects to the animal's right. The main roll is then passed under the animal's body and returned to the right side ventrally, and the dorsal projecting strip is grasped along its full width with a forceps. Tension is exerted on the lower strip while the forceps holds the upper strip firmly in place so that it does not slide over the vaseline gauze. After the first complete loop, the forceps is carefully released and pulled out gently.   Then the bandaging is continued under tension, so that the animal is tightly bandaged.   At the end, the bandage is moistened with a finger and rubbed to give a smooth and finished surface.   The animal can then be set aside for the bandage to dry and if necessary the bandages can be marked with dots of dye to distinguish individuals or experiments.   It is helpful to keep the mouse warm until his recovery from the anesthetic is complete.

The bandage should not be removed for six days, and unless one is looking for established immunity (a second-set response) it is better to leave the bandage in place for eight or ten days.   If the bandage is removed at six days, it is often wise to re-bandage until the eighth day to prevent damage to the skin by scratching.   After that time the grafted area can be left open and examined daily for the end point of complete graft destruction.   Early evidences of reaction include an easy peeling of the epidermis to reveal the glistening surface of the dermis below.[93]   Total excisional biopsy and microscopic examination is desirable to confirm end points and immune processes.

Several short techniques for skin grafting in mice have been described, such as the use of a tissue punch to take the skin from the donors and to prepare the graft bed.   In our hands, and those of others, however, these techniques have not always proved successful, perhaps because of the loss of the panniculus as a graft bed, and in our opinion it is preferable to follow the established and fully successful methodology described above.

Transplantation of other tissues.—A number of tissues and organs other than skin have been successfully transplanted in small laboratory mammals.   Procedures for ovarian transplants were described a number of years ago by Robertson.[1063]   Thyroid and adrenal transplants in the rat have been adapted to convenient experimental study by Woodruff and Sparrow[1403] using the localization of injected radioactive iodine as an index of thyroid graft survival and function.   Techniques for the transplantation of other tissues will not be referred to here, except for a later consideration of the transplantation of hematopoietic tissues.   The *Transplantation Bulletin* is a rich source of reference materials.

Parabiosis.—Several important compatibility studies and techniques depend on the use of parabiotic animals, generally mice or rats.   A successful technique for parabiosis of mice has been described by Eichwald *et al.*[319]   Rubin[1079] has suggested that somewhat different results may be observed depending on the nature of the surgical union obtained through different techniques of parabiosis.   Finerty[362] gives extensive references to earlier work with parabiosis, especially in the rat.

Tolerance, paralysis, and enhancement.—Various methods of suppressing immune responses have come to attention in recent years.   These include the tolerance to homologous tissue transplants induced in an animal through the injection of viable homologous cells when he is newborn or an embryo; the suppression of immune responsiveness in an adult by injection of large amounts of the antigen, especially effective with polysaccharides (paralysis); and the diversion of the immune response in the direction of a type of antibody that suppresses rather than promotes graft rejection (enhancement).   Techniques for these essentially immunologic systems, of considerable genetic interest, will not be elaborated here; references to most of the current procedures will be found in an earlier review by Owen.[986]

Bone-marrow transplantation.—There is a great deal of activity currently in the area of hematopoietic transplants into irradiated adult mammals, especially mice.[393]   The genetically controlled incompatibilities between host and graft in this system are of particular interest and importance.   Brief mention will be made here of the essential and most-used techniques.

Donor mice are generally killed by bleeding them to death, often by decapitation; there is reason to believe[451] that exsanguination of the donor may reduce the number of a type of cell in peripheral blood that may have an unfavorable effect on the host into which the bone marrow is to be injected.   The mice are pinned or taped securely on a board and each femur is exposed, dissected out, and transferred to a petri dish where the femurs are scraped free of adhering tissue and rinsed with Locke's solution. Generally, the dosage of bone marrow to be used will approximate one or two femurs per recipient.   Both ends of the femur are cut off with a scissors, and the marrow plug is forced out, using Locke's or Tyrode's solution in small quantity to flush out the marrow canal with a 22-gauge needle and a 2 ml. syringe.   The marrow is then suspended evenly by filling and emptying the syringe, forcing the suspension through the needle several times, and the nucleated cell count is taken in a hemocytometer using crystal violet in 1 per cent acetic acid as a stain.   The suspension is adjusted so that it will contain the desired count of nucleated cells per $\frac{1}{2}$ ml., the quantity injected intravenously into each recipient.

Radiation conditions for the recipients are specified in publications in the field.[393] Several techniques have been applied to follow the course of repopulation of the erythropoietic tissues.[983]

Embryonal hematopoietic tissue, especially embryonic liver, is often substituted for adult bone marrow.   It seems to have some advantages, in terms of its lesser tendency to produce a delayed incompatibility reaction and in terms of its ability to transplant, at least in genetically anemic recipient mice, even without irradiation of the host unless strong histocompatibility barriers are present.[1099]

The main genetic tools for studies of transplant incompatibility have been the inbred lines of mice, their hybrids and segregating F_2 and backcross generations.[983] The development of coisogenic lines differing only at specified loci affecting histocompatibility, by George Snell at the Jackson Laboratory, has provided the most precise information about immunogenetics in this important area, and the most useful materials for its advancement.

DISCUSSION

DR. BURDETTE: Dr. Henry J. Winn of the Jackson Laboratory will open the discussion of Dr. Owen's paper.

DR. WINN: Dr. Owen's paper is a very impressive manuscript in which he has skillfully interwoven the complex terminology and methodology of immunology with what I consider the equally complex terminology and conceptions of genetics.

As most of you know, the methodology of immunogenetics would have been a great deal simpler to discuss about ten to fifteen years ago. At that time it consisted largely of the study of the interactions of serum antibodies with intact red blood cells. Now the very rapid growth and development of the field of tissue transplantation has changed all this, and we find ourselves studying a very large variety of reactions involving not only serum antibodies but cellular or cell-bound antibodies. For the most part the work that has been done with the cell-bound or cellular antibodies does not really lend itself to any particular experiments that I know of in immunogenetics. Generally, this work has been directed to some understanding of the mechanism of graft rejection or the relationship, metabolic or otherwise, between soluble and cell-bound antibodies.

Our interest in the cytotoxic technique came about because of the limitations of the red-cell agglutination test. In the mouse one finds that the red-cell agglutination technique devised by Gorer and colleagues makes a really fine tool for the study of antigens controlled by the genes at the H-2 locus, but there are some fourteen or fifteen or maybe twice that number of loci which control the acceptance or rejection of grafts. We had hoped that by studying the effects of antibodies on white cells we might be able to analyze the antigens controlled by the genes at these other loci. The technique is relatively simple. One mixes antiserum with white cells from either lymph nodes or spleen (for some reason we do not completely understand, thymus is not suitable). To this mixture is added measured amounts of complement, and after a suitable period of incubation some vital dye to determine how many of the cells are dead.

Gorer has done some work on this and Schrek has also. For the most part they have added enormous quantities of complement to mixtures of undiluted or only slightly diluted serum and cells. We wanted to use the technique in a more quantitative fashion, and we decided to standardize the requirement for complement and antibodies. The longer we worked at standardizing the technique, the more difficult we made it, and we soon reached the point where in one day we could analyze only a single serum. (With the red-cell agglutination technique I estimate something over a hundred serums could be analyzed in one day.)

It occurred to us that one of the problems we had was the fact that this system required an enormous amount of complement in terms of the amounts that are normally used to lyse red blood cells, described by Dr. Owen. This suggested that instead of going through this very laborious procedure of incubating the cells and obtaining differential counts for each tube, we might actually add a very large excess of complement and measure the amount that was used up. This has worked out very nicely and we are now using a test which is patterned after that used by Osler and his colleagues for the study of soluble proteins and carbohydrates.

I would like to mention just one of the things we have been able to do with this test. Earlier I had mentioned that thymic cells are not suitable for cytotoxic tests, because the cells are not killed when one adds antibody and complement. Based on some preliminary absorption tests, we had postulated that this was because there were not enough antigen sites on the thymus cells and they were not fixing enough complement. Using the quantitative complement-fixation test, we have shown that this is indeed the case; the thymic cells mixed with antibodies (even when the antibodies were made against the thymic cells) fixed far less complement than preparations of cells taken from either bone marrow, lymph nodes, or spleen.

The cytotoxic technique and the complement fixation test may actually be applied to peripheral blood, and if this could be done with humans, it could conceivably open up the possibility, brought out by Dr. Russell, of typing human blood for histocompatibility factors.

DR. SNELL: Dr. Owen and Dr. Winn have given us a very adequate summary of some of the methods of immunogenetics. There is little I could add. I am not an immunologist, but I have acquired some knowledge secondhand from other people working here.

It has interested me to see, over the past ten to fifteen years, how the mouse has finally come to be used as a tool in immunology. Twenty-five years ago about all one could find in the literature were a few unsuccessful attempts to find blood groups in mice. The results were always negative. Now the mouse is very much in business. Of course one presumed the defect of the mouse originally was its small size. Here at the Jackson Laboratory we circumvent that now by using large numbers of mice.

I mention one additional technique to illustrate some of the tricks of the trade. Dr. Owen mentioned absorption as a method of obtaining an antiserum with a single specificity. That is a somewhat messy and time-consuming technique, and rather particularly so in mice, in which the available amounts of serum and tissue are usually small. Also the resulting antiserum may be heavily contaminated with tissue proteins. A very simple and satisfactory alternative is absorption *in vivo*. One simply injects the mice intraperitoneally with antiserum and then bleeds them perhaps an hour later; the antibodies reactive with the recipient tissues are absorbed *in vivo*. The antiserum may be diluted about one in ten, reducing the titer from about one in a thousand to one in a hundred, but the antiserum is still usable for most purposes.

DR. DRAY: Dr. Owen referred to some of the newer methods of immunochemistry pertinent to mammalian genetics which I believe merit further emphasis. During the past fifteen years, two very highly significant advances, agar-gel immunochemical analysis and allotypy, have developed from the work of Dr. Jacques Oudin at the Pasteur Institute. In 1946, Oudin discovered that when a mixture of antigens in solution diffuses into a gel containing a mixture of precipitating antibodies, the multiple bands of precipitate which result may be explained as due to the different antigen-antibody systems present rather than to the Liesegang phenomenon as thought previously.[976] Through his work and others, agar-gel immunochemical analysis has evolved rapidly into a variety of techniques suitable for the analysis of complex mixtures of soluble antigens; for their identification and interrelationships and for the determination of the relative and absolute concentrations of the antigens, their diffusion coefficients, molecular weights, and electrophoretic mobilities.

In 1956, Oudin immunized rabbits with immune precipitates (ovalbumin-rabbit-anti-ovalbumin precipitate) plus Freund's adjuvants and demonstrated production of isoantibodies which would precipitate serum components in some rabbits but not in others, thus resulting in serum groups analogous to red blood cell groups.[977]   The term "allotypy" was proposed to designate the variation of the antigenic specificity of these serum antigens.   This work has opened a new approach to the genetic study of proteins, one which we also have undertaken.[290]   It has been demonstrated that these allotypes may have electrophoretic mobilities of α-, β-, and γ-globulins; and that for the γ-globulins there are at least two genetic systems with at least three alleles at each locus.   In contrast to the polysaccharide antigens attached to red blood cells, which have been studied by agglutination reactions combined with absorption methods as cited by Dr. Owen, these soluble protein allotypes may also be studied by means of the precipitin reaction utilizing the agar-gel immunochemical methods.   Since the antibodies are produced in the same species, specific antibodies are obtained without the necessity of absorption techniques.   It would not be unreasonable to expect that other animal species would show the phenomenon of allotypy.   Just as with the blood groups based on differences of red blood cells, serum groups based on antigenic differences of proteins should open new opportunities for work in mammalian genetics.

I would like to make three additional comments on methodology in immunogenetics.   The introduction of paraffin-oil adjuvants with *Mycobacteria* by Dr. Jules Freund provided the most powerful immunization method known, at least for protein antigens.[406]   Such adjuvants have made it easier to produce antibodies to weak antigens.   One must keep in mind that this method of immunization introduces the complication that very small amounts of antigen or contaminants will produce antibody whereas ordinarily they might not.   Also, the antibodies produced with the use of adjuvants may be more heterogeneous.

When a mixture of several antigens are used for immunization, antibodies to each antigen are not necessarily produced.   In fact, competition may develop among the antigens so that antibody production to one of the antigens, perhaps the one of interest, may be suppressed.[2]   Therefore, purified antigens should be used whenever possible. Cellulose ion-exchange chromatography devised by Peterson and Sober is one of the best methods now available for the fractionation of protein antigens.[999]

Finally, I would like to comment on the choice of species for antibody production. Most commonly the rabbits or horses are used as recipients for preparation of antibodies.   However, a closely related species might be more revealing and perhaps more sensitive to minor differences of antigenic properties.[289]   This principle of immunologic perspective was pointed out many years ago by Landsteiner.[749]

DR. KLEIN: I would like to ask Dr. Winn whether the cytotoxic method is applicable to antigens other than H-2, and, as far as H-2 is concerned, whether he has any way of visualizing the cell-to-cell reaction between the lymph node cells and the target cells.

DR. WINN: I would like to answer the last question first.   No, we do not have any techniques at all for visualizing the action between the specifically activated cells and the target cells.   As a matter of fact, we have considerable evidence that the cells normally taken from draining nodes or spleen actually are immature in the sense that they have not yet developed the capacity to react with the target cell.   This is because these cells are much more active on a cellular basis if they are transferred several days before the test graft is applied.   In a series of studies in which we mixed the target cells with the immune, lymph-node cells and incubated them for very long periods of time and then selectively destroyed the lymph-node cells with antiserum and complement, the tumor cells grew as if they had never been in contact with the tissues at all.   They grew just as if they had been incubated with serum alone.   I think that if you want to visualize this reaction you have to find a source of the more mature cells, possibly in the peripheral blood, or alternatively, one might provide some system of incubating these cells from draining nodes and spleens *in vivo* in chambers or in tissue cultures and testing them several days later.

I think I would have to ask Dr. Snell to comment on the use of cytotoxic tests. There is one system involving coisogenic strains in which we do have excellent complement fixation.   So far as I know, all the evidence indicates that a non-*H-2* difference is involved there.   We also obtained a complement fixation with C3HK and C3H serum which I believe has an *H-*1 difference.   As far as using other cells is concerned, lymph-node cells give far better reactions than any other tumor cells and any other normal tissue cells.   The cytotoxic test, I think you may be aware, is not equally applicable to cells other than those from the lymph node.   When the cellular preparation is made, a high percentage of the cells are already dead; but this is the advantage of complement fixation tests as they can be used on cells whether they are living or dead.

DR. SNELL: I might merely comment as to whether two of the systems which Dr. Winn has employed are established as non-*H-2* systems.   The *H-*1 system is very definitely, there is no question about it; but Dr. Winn did not get quite as clear results with it as with the other system.

DR. WINN: It does not fix nearly as much complement.

DR. SNELL: The other system, which does fix complement well, turned out to be a new locus, *H-*5, but there is one additional test to run.

DR. HERZENBERG: Dr. Owen, would you comment on the uses of labeled antibodies?

DR. OWEN: There have been many uses of labeled antibodies, for example in localizing antigen, but only scattered applications of these techniques that could be described as specifically immunogenetic.   One example is the work of Masouredis,[857] who was able to distinguish a dosage effect in the *Rh* complex by the use of I^{131}-labeled antibody.

DR. COHEN: I have tried the techniques of Masouredis[857] in studying gene-dosage effects on rabbit red-cell antigen and had no success.   There appeared to be too much nonspecific absorption of the tagged antibody so that the small differences were difficult to distinguish.   I used, as in my test system, tagged antibody which had been absorbed on the specific cell and then eluted.   The tagged eluate was then used on cells of known genotypes.   Masouredis did it this way in his human work and was successful.   It did not work in the rabbit red-cell system.

DR. REED: Fluorescent antibodies are, of course, the other major type of labeled antibody.   Also, several tritium-labeled antibodies have been prepared to date.[1037,][1045]   Tritium-labeling is easily done by the Wilzbach method,[1274] by sending the lyophilized antiserum to one of several firms to be exposed to tritium gas.

DR. COHEN: The fluorescent antibody technique has been used for the identification of blood-group antigens by Cohen and her co-workers[209] as a means of investigating problems in immunologic genetics and hematology.   They used human anti-*A*, anti-*B*, anti-*D*, and anti-*C* to identify minor cell populations occurring in a mother which might be derived from the transplacental transfer of cells from the fetus to the mother. This technique could also detect minor cell populations arising through mutations or through induction of chimeras.

DR. GOWEN: It has not been mentioned, but there are real possibilities for isolating animal strains that will be good producers of immune serum.   In my own experience there are great differences in serum titers in random-bred animals, indicating differences in their genetic capacity to make immune antibodies.   The isolation of strains with high production of immune sera could be of both practical importance and contributory to better understanding of immune phenomena.

DR. BARRETT: It has been said by several, on what I believe to be uncontrolled observations of race III rabbits produced by Dr. Sawin, that they are better-than-average antibody producers.   I have used some of these rabbits myself.   I would not say whether they are better or not; they certainly are quite good.

DR. DRAY: Dr. Barrett raises an interesting problem concerning the individual variation of rabbits in their capacity to produce antibodies.   Gamma-globulin groups of rabbits may offer an approach to this problem, since at least 36 genotypes based on three alleles at two loci are now known.[290]   Since antibodies are found among the γ-globulins, the possibility exists that the genetic control of γ-globulin allotypes may be correlated with the genetic control of the capacity to produce antibodies.   Such a result would also have interesting implications concerning theories of antibody formation.

HOST-PARASITE RELATIONSHIPS

John W. Gowen, Ph.D.

GENETICS *of* INFECTIOUS DISEASES

For every disease there is a host species in which the disease is recognized. Since individuals within species vary genetically in many ways, they, as expected, also vary in expressing the syndrome for any particular disease. Similarly if the disease is infectious, another species, pathogenic to the host species, is generally responsible for initiating and carrying on the disease. The organisms within the pathogenic species likewise vary genetically in virulence and other characteristics and, in consequence, also lead to variations in their invasive power to individuals within the host species. Some diseases further complicate the results by requiring other species, as mosquitoes for malaria, for vectors to maintain and carry the disease to the host, thus adding further genetic variables to the already complex situation. Different environmental agents of many kinds likewise strongly affect the expression of disease in numerous ways. These factors tend to confuse the data on causation of disease by inducing variation in disease expression so that individuals may vary all the way from no expression through various grades of morbidity to those that die. So far as is known, there are no problems of genetics that offer more complications in their solutions than those pertaining to disease resistance. The problems are, however, of interest because of the complexity, the genetic implications and discussions brought about thereby, and the methodology developed for their solution.

Discussion of these problems will be limited largely to mice and only to certain aspects of the host-pathogen relations as our investigations fall almost entirely within this field. The research considered is largely that of associates and students who have made extensive contributions to the problems. Papers giving a broader coverage of the subject are listed in the bibliography.

Because of the multifactorial nature of the causation of infectious disease, investigators without exception have limited their quantitative studies to specific pathogens in fixed amounts and under pre-established environmental conditions. For new diseases the limits within which these variables were controlled have been determined by trial and error. Search then has been made for ways and means by which the genotypic variability of the host could be broken up into pure breeding strains so that the factors could be studied within a more limited range. Most workers have limited their attention to one or two of these strains generally derived by repeated inbreeding. Such limitations seriously restrict the generality of any disease studies as they may be applied to the species or to the more important transfer of the information gained to man. Our own viewpoint has been that the host strains for study should be specific samples, each covering a limited range of the whole species' susceptibility to the disease but that when all inbred strains are considered they should cover the full range of susceptibility or resistance to the disease as represented by the species genotypes. The same considerations have been given to the genotypes of the lines which may be developed from the pathogenic species, although in much of the work herein reported only one line of the organism (one that is remarkably consistent so far as virulence is concerned) has been utilized. Other observations have not been reported since this program is designed to cover methodology within mammalian hosts.

Initial attention may be turned to the variations in resistance to murine typhoid due to *Salmonella typhimurium* 11C that may be established for mice as a whole and fixed in inbred lines. The procedures by which these strains were established were diverse, but all have been inbred brother × sister over many generations with selection for the desired disease resistance in fixing the levels of mortality. The results are shown in table 61.

Table 61

RESISTANCE OF INBRED STRAINS OF MICE TO MURINE TYPHOID
Salmonella typhimurium, 200,000 11C

Strains of mice	Per cent survived
S	83.9 ± 0.6
RI	67.7 ± 1.6
K	50.6 ± 1.3
Z	34.2 ± 0.8
C	29.6 ± 1.4
E	23.5 ± 1.0
N	16.9 ± 1.2
L	9.6 ± 0.9
Q	1.7 ± 0.4
Ba	0.2 ± 0.1

These strains have characteristic reactions to this disease. They range from the most resistant to the most susceptible. Binomial standard deviations of the means are calculated for each strain. These data are for the period 1944 to 1956. The representation of the expected performance of each strain is on the whole good, but the data themselves do not tell the full story. The S mice are now more resistant to the invasion and growth of this pathogen than the figures show. For this strain there is a pronounced difference in the reaction of the sexes, the males having less resistance than the females, 74 to 95 per cent, respectively. No other strain shows such a difference between the sexes. Tests on the males show a much higher frequency of deaths, 1 to 4 days following inoculation, than those for any other strain or even for the females of the S strain. It is further found that the males are quite susceptible to large doses of killed S. *typhimurium* 11C. The deaths also occur within the period of 1 to 4 days following inoculation. These facts are interpreted as pointing to two causes of deaths for the males of this particular strain. The deaths which come early in the disease are attributed to rapid release of endotoxin from the digested bacteria. The deaths which occur later are attributed to the common cause of death for all strains, the growth of the organisms in the host.

The genetic techniques which have been used to separate the differences in resistance to disease have varied, but in each case have included inbreeding to fix the particular genetic resistance observed within the strain. Throughout, the breeding stock which has been the source of animals for test has been free from S. *typhimurium* for a period of more than twenty years. Tests for resistance of some twenty different strains (genotypes) have been made over this period. From the years 1944 to 1955, 21,669 mice were observed in these tests. These animals had their first contact with the disease at ages from 45 to 700 days. All tests were made with a single line of S. *typhimurium* 11C at a dose of 200,000 organisms injected intraperitoneally. This line of the pathogen has shown a consistent virulence for over twenty-five years. In tests with this organism, survival ratios include nearly completely resistant to nearly completely susceptible, S and Ba strains respectively; the other strains show intermediate values for resistance. Although these values have fluctuated during the 11-year period included in these data, they have kept their relative positions on the whole. The observed stratifications may be interpreted on the basis of a number of genes affecting different physiologic functions of the host responsible for the resistances or susceptibilities to the bacteria. These resistances or susceptibilities could scarcely be analyzed in a random-breeding population. In any such population, successive mice, when utilized for disease tests, would show hit-or-miss resistance or susceptibility to the disease and thus vitiate results coming from such experiments.

Mortality represents the accumulated dysfunction of the various organ systems which may be attacked by the invading organism. Different organs may be susceptible in different strains. The temporal sequences in the progress of the disease may also differ between inbred strains. The severity of the sickness (morbidity) expresses some of the differences in the patterns of resistance. The S mice are so resistant that they show almost no effects of the pathogenic organisms. The Ba mice are extremely sick almost from the first contact with the disease. The other strains have characteristic patterns of morbidity which are apparent at different stages in the disease cycle. The pattern of morbidity for each strain is characteristic, indicating the specificity of expression of the disease.

TRANSMISSION OF DISEASE RESISTANCE IN CROSSES

Convincing evidence for the dependence of resistance and suceptibility to infectious disease on the strain genotypes comes in repetition of the test results for the different strains generation after generation despite varying environments characteristic of changing seasons and years. However, this is not all the evidence desired. To collect the evidence properly would require an experiment comprising $1,600 \pm$ cells and possibly 100,000 mice. The time, energy, and space required for this analysis together with the three generations of breeding required have prevented us from doing the experiment up to this time. However, valid data covering critical crosses are available through the work of Hetzer,[573] Zelle,[1467] Weir,[1376] and Gowen and Stadler (unpublished). The combined data for the first generation are given in table 62. The S strain was under intense selection for resistance at that time.

Table 62

RESISTANCE OF PARENTS AND CROSSES OF INBRED STRAINS OF MICE TO MURINE TYPHOID
Salmonella typhimurium 11C, 200,000 organisms

Parents Male Female	Progeny tested	Per cent survived	Mating types
S × S	728	89	(Res × Res) P_1
RI × RI	42	80	(Res × Res) P_1
BrR × BrR	851	78	(Res × Res) P_1
Ba × Ba	346	2	(Sus × Sus) P_1
L × L	207	2	(Sus × Sus) P_1
Hybrids			
S × RI	54	78	(Res × Res) F_1
RI × S	35	73	(Res × Res) F_1
BR × S	756	75	(Res × Res) F_1
S × L	88	89	(Res × Sus) F_1
L × S	133	79	(Sus × Res) F_1
S × Ba	53	85	(Res × Sus) F_1
Ba × S	40	80	(Sus × Res) F_1
L × Ba	198	18	(Sus × Sus) F_1
Ba × L	68	37	(Sus × Sus) F_1

Res = resistant. Sus = susceptible. P_1 = parental purebreds. F_1 = hybrids.

Table 62 gives data on 3 resistant inbred lines and 2 that are susceptible. Comparison with table 61 shows that the S mice of this period in the genetic development of the strain were 5 per cent more resistant than in the 1944–1956 period. The same was true for the mice of the RI strain, 12 per cent, and for the Ba mice, 1.8 per cent.

The L mice on the other hand were less resistant, 7.6 per cent. This difference is seemingly important, although numbers of mice are not large and other factors besides binomial variations contribute to the real variance of data dealing with problems of disease.

The differences between the resistant and susceptible strains for either period were pronounced. Confining attention to the first three hybrids, S × RI, RI × S and BR × S, the results showed the hybrid resistances to be closely similar to those of their inbred parents, although they were, in each case, a little less resistant, 9 per cent. This was convincing evidence for the similarity of the genes making up the relatively resistant strains. The results did not indicate heterosis. Rather they suggested partial dominance for some of the genes controlling resistance.

The hybrids S × L, L × S, S × Ba, and Ba × S represent extreme crosses of resistant × susceptible. They were only 6 per cent less resistant than the resistant parental strain. Dominance of most genes for resistance was indicated. There was no significant evidence for the reciprocal crosses being different from each other. In mice the chromosomal arrangement is XY for the male and XX for the female. The male progeny received all X-linked genes from the mother, whereas the females had an X chromosome from each parent. Any major factors for resistance on the X chromosomes should make corresponding changes in the survival ratios of the sexes in the reciprocal crosses. As Hetzer showed this was not the case. However, there was a tendency for the females to have a higher survival than the males; but the sex differences in survival extended to all matings and not just those which were diagnostic of sex linkage. The equality of the resistances of the hybrids further emphasized that resistance or susceptibility to this disease was not due to a maternal effect but to inheritance being transmitted by both sexes. Passive immunity carried through the egg cytoplasm, the fetal circulation, the colostrum, or other maternal environmental effects was ruled out as a factor materially influencing resistance.

The crosses of mice of the L strain mated with those of the Ba strain represented the crosses of two highly susceptible strains. The resulting hybrids' resistances were greater than either parent's and by a noticeable amount. This result indicated that inbred strains L and Ba possessed genes for resistance even though they themselves were quite susceptible. These genes complement each other to give increased resistance to the F_1 progeny. This is in contrast to the results for the crosses of the resistant strains. These crosses dropped slightly in resistance. The comparison between the two classes of hybrids could be used against the argument that the results of the L × Ba crosses were due to heterosis. However, that reasoning may not hold, for data have been collected which indicated that heterosis may be genotype specific.

The backcross data gathered by Hetzer[573] while in this laboratory further supported these views. The main feature of these tested backcross animals was the further regression of their *S. typhimurium* resistances toward those of the susceptible strains. This places their sensitivity to typhoid as intermediate between those of their resistant and susceptible grandparents. Segregation of factors for resistance and susceptibility probably took place, several genes in different loci being important to resistance.

The average mortalities for the crosses with L as the inbred parent were nearly the same as those with Ba as the parent, 53 and 48 per cent, respectively. Survival among the backcrosses ranged from 67 to 40 per cent. Examination of these crosses showed no established reason for the variability. Higher susceptibility came in the progeny of the F_1 (S♀ × L♂) for the L backcross, whereas the higher susceptibility was evident in the offspring of the F_1 (L♀ × S♂) for the Ba backcross. The resistances of these two crosses separately contradict hypotheses often used to explain resistance to disease other than on inheritance grounds. In the first case the data paralleled those of the two hybrid generations in suggesting differences in the reciprocal hybrids. This hypothesis is refuted, however, by the second cross. The second cross suggested that the Ba female parent was responsible for lowering the survival of her progeny, a type of maternal effect; but comparison with the other backcrosses and all the crosses in general denies this conclusion.

Tests on the inbred parents, hybrids between them, and backcrosses to the susceptible strains consequently support the dependence of resistance to disease on genetic factors transmitted along customary paths. The mode of action of this inheritance is capable of wide modification through changes in numbers of the infecting organisms of different lines of the bacterial species. Results of this type are generally sufficient to establish the genetic dependence of disease susceptibility. However, there is another hypothesis, contraindicated by many features of the above data, which often has been cited in the past. Evidence on the validity of these hypotheses, consisting of even more critical data on the problem, may be had by adopting the following approach.

GENETIC ASPECTS OF DISEASE AS INDICATED BY PROGENY OF DOUBLE MATINGS

The other hypothesis is that the progeny acquire an immunity either by active contact with the low-grade disease or passive immunization and thus develop transmitted resistance to the challenge tests. This hypothesis is based on the fact that animals which survive a disease develop an active immunity of irregular duration. Some survivors retain a latent infection which is thought to protect them by keeping up an active immunity. These carriers may spread subclinical infection to their progeny, thereby inducing active immunity in them. Evidence indicates that passive transfer of immunity from dam to progeny may take place, but males are not able to transmit passive immunity to their offspring. A method for evaluating the weight to be given the factors of genetic *versus* acquired immunity in resistance to disease was designed by Gowen and Schott.[467] This was applied to studies on typhoid in the mouse but has general applicability. The method consists in mating two males of different strains to a single female in the same estrus period so that litters will contain young of two known separable types. In our case S mice differ from L mice in their genetics of coat color, so that it is possible to distinguish between S mice as albinos, L mice as silver black, and hybrids as black mice. An L female mated in the same estrus period to two males, one L and one S, may have pure L and hybrid progeny in the same litter. Environmental conditions are common to all animals of a litter, each having the same opportunity to receive immune bodies from the mother and to receive latent infection from the parents. According to the genetic hypothesis, the L progeny should be nearly 100 per cent susceptible and the hybrids should show resistance as may be seen in table 62. According to the hypothesis of acquired immunity, the L and hybrid littermates should be equally susceptible.

The two types of progeny tested showed that the L mice of the litter were all susceptible, whereas the hybrid mice of the same litter were resistant as expected of this hybrid cross. The differences were highly significant. This technique offered a means of showing that the concentration of genetic factors for resistance in the selected strain, S, were truly significant whereas any factors for acquired immunity were without effect when the disease was murine typhoid (*S. typhimurium*). The results further substantiate the earlier results as well as indicate a desirable technique for separating the genetic basis for natural resistance to a disease from that due to factors for immunity harbored by the mothers.

EFFECT OF NUMBERS OF PATHOGENS ON SURVIVAL

The effects of sheer numbers of organisms on the characteristics of a disease are of interest particularly as such data help to understand the pathogenesis whereby genotypes alter the phenotypic expressions of the disease. Schott[1167] utilized a host population composed of a random selection of albino *Mus musculus*. They were infected by intraperitoneal injections of different numbers of the same line of bacteria. The survival for each dosage is shown in figure 49.

Figure 49 shows that the survivals of the mice were greatly influenced by the numbers of organisms which they received. Animals which received the low dosage (1×10^4 organisms) displayed a long incubation period followed by increased and then decreased mortality. An appreciable number of animals survived the disease attack. At the high exposures (1×10^7 organisms), the organisms rapidly overwhelmed the hosts. There was practically no incubation period. The course of the disease ended days earlier than those for the smaller dosages. Between these two extremes intermediate effects were observed. The pathogen was *S. typhimurium* in mice but the data are illustrative of what may be observed for all types of pathogens—chemical poisons, viruses, bacteria, protozoa, and multicellular pathogens. The scales of dosage will change with the material used, that is, with ricin a dose less than 0.001 mg. caused sickness but few deaths, whereas a dose of 0.004 mg. resulted in deaths of nearly all the treated mice, but the principle of mortality dependent on dose was maintained throughout this range. In genetic studies of host-disease syndromes, the investigator must stay within the limits of dosage set by the test curves for the particular material. Experience shows that the preferable limits for resistance studies are those set by

Fig. 49. Survival of mice following acute exposure to different quantities of the pathogen, *Salmonella typhimurium*, and during the course of the disease.

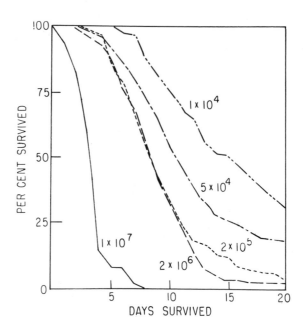

10 to 90 per cent survival of the treated population, as it is within these limits that most information on disease processes may be obtained.

Possibly the most satisfying evidence for progress in selecting for resistance to disease, beginning with an unselected population, comes in the changes which appear in days survived in survival curves for different dosages. Figure 50 illustrates these curves as obtained by Schott[1167] and Hetzer.[573]

The selected mice were unquestionably more resistant to the *S. typhimurium* when tested by each of these three different dosages than the unselected population from which they came. These tests were made in 1936. The S strain was further selected to the nineteenth generation. Selections were discontinued at that time and the strain maintained by sibling matings. Today the mice are as resistant as those shown by the data of figure 50. The lapsed time over which the S mice have retained their resistance is 21 years. This period represents a turnover of 63 ± generations, or in terms of generation time in man about 1,900 years.

The syndrome by which the disease was first described also changed during the 11 generations of selection. The relative dosage effects of the different quantities of the bacteria have remained in similar positions in both the unselected and selected mice, but fewer mice from the eleventh-generation population die. In the unselected mice the lower dose causes somewhat less mortality than the intermediate dose. The high dose is much more lethal than the two lower dosages. The selected mice show greatly increased resistance. The low-dose treatment now has 87 per cent survivors, whereas

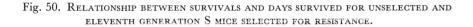

Fig. 50. Relationship between survivals and days survived for unselected and eleventh generation S mice selected for resistance.

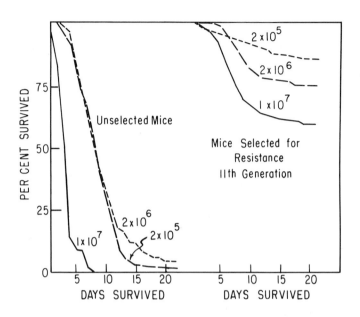

The mice were treated with the pathogen, *S. typhimurium*, in three dosages, 2×10^5, 2×10^6, and 1×10^7 organisms.

only 4 per cent of the unselected mice survived. The median dose definitely caused median mortality. In the eleventh generation, 76 per cent of selected mice survived; whereas only 2 per cent survived in the unselected population. The high dose of ten million organisms was rapidly and completely lethal to the unselected mice, but 60 per cent survived during the eleventh generation of selection.

Not only did the selected mice have fewer deaths than those of the unselected group from which they came, but also the distributions of the deaths took different forms. The frequency curves in per cent of total deaths for those mice which died in each group are given in figure 51. At the highest dose (1×10^7), the deaths occurred early in the unselected group, the modal value being between 4 and 5 days, whereas in the selected group the modal deaths were one or two days later. The mode for the unselected group was more than twice as high as that for the resistant cohort. The range over which deaths in the unselected group occurred was reduced to half that for the mice of the eleventh generation. The syndrome obviously changed. The intermediate dose acted on the selected and unselected mice in a similar manner. The modal day for death was 9 days after inoculation. The proportions of deaths on that day were about equal. Deaths occurred later in the unselected mice. It is difficult to compare the results for the lowest dose. The main effect observed was an increase in the

Fig. 51. MORTALITY OF MICE FROM *Salmonella*

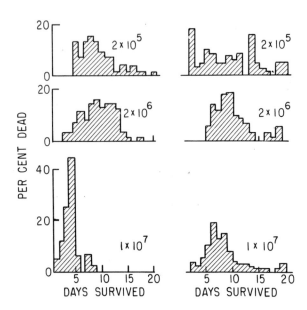

Mortality (percentage frequency) from *S. typhimurium* for mice of the unselected population (left) and mice selected eleventh generations for resistance (right).

numbers of survivors.   In fact the numbers of mice dying in the selected cohort are so few as to give them the appearance of random deaths.

Analysis of pathogenesis in the resistant mice of the eleventh generation would require a notably higher number of invading organisms than when mice of the original population were under study.   In fact it is doubtful if the investigator could obtain the same syndrome by only adjusting the initial numbers of the invading *S. typhimurium* 11C organisms.

Heavy reliance must be placed on data of this kind in the methodologic design of all infectious disease experiments.   The information, however, has greater significance in that it is indicative of how at least some of the different pathogen genotypes may induce their effects.   Increase in numbers of inoculated organisms of the same line would increase the dosages of any substance generated by these organisms that may be inimical to the hosts.   Increased dosages lead to more severe morbid reactions.   Pure lines of the pathogen, when selected, often as apparent mutations, occurring within the same original parental line, are often characterized by differences in virulence of entirely similar patterns to those discussed above, even though a fixed number of organisms are inoculated into each host.   This suggests that, at least in some instances, the genotypic differences in the pathogen act as quantity controls on the same substance or capacity within the given mutant lines rather than in necessarily creating entirely new products detrimental to the host genotypes.

ROUTE OF INFECTION AND SEVERITY OF DISEASE

Between 1920 and 1940 great emphasis was laid on studying disease through utilizing only natural routes of infection. The advocates produced little evidence for their contention that use of other routes could not lead to comparable resistances in the hosts. Attempts to cause natural infections often produced trauma as great as that of experimenters who openly avowed that the routes of infection they were using were not natural. Further difficulties arose in that attempts to introduce the pathogens by natural routes also introduced dosage variations to increase variance in the results. However, the real problems involved, while often overlooked, were of interest and have been pursued by several investigators.

The methodologic problem really turns on whether natural resistance or susceptibility to a disease is dependent on localized differences in resistance of tissue at some customary portal of entry for the disease organism, or is a property of all cells of the body. Roberts and Card,[1059] Irwin,[642] Lambert and Knox,[746] Schott,[1167] Gowen and Schott,[467] Hetzer,[573] and others collected data showing that the normal portal of entry could be bypassed and genetic resistance to any one of several diseases could be established in the host. Webster[1363] collected data for different routes of entry on mice which he had previously selected for resistance and susceptibility to *Salmonella enteritidis* introduced through the so-called natural route, the stomach. The selections were based on intrastomachal instillation of 5,000,000 organisms through a silver tube inserted into the stomach. The relative resistances of these susceptible and resistant strains of mice are indicated in figure 52.

The most striking result of these tests is that each route of instillation of the *S.*

Fig. 52. RELATIONSHIP BETWEEN ROUTE OF ENTRY AND MORTALITY OF SELECTED MICE DUE TO BACTERIAL INFECTION.

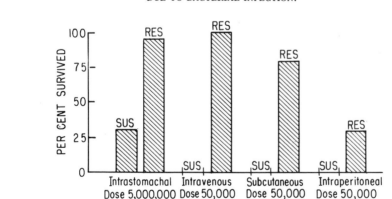

Relative resistance of mice selected for susceptibility (Sus) and resistance (Res) to *Salmonella enteritidis* when the bacteria are introduced into the stomach through the esophagus by silver tube (dose of 5,000,000 organisms), or injected intraperitoneally, subcutaneously, or intravenously (dose of 50,000 organisms).

enteritidis picks out the mice of the susceptible strain from those of the resistant. In this sense, each of the tests is accomplishing its purpose. The second conclusion is that the intrastomachal instillation of the bacteria requires 100 times the dose of the pathogen to cause the disease than the intraperitoneal, subcutaneous, or intravenous routes. Conclusions from this fact may be opposed. (1) It may be supposed that the intrastomachal route is the natural route of infection and as such has had most of the natural resistance mechanisms built up around it. If this were true, the numbers of bacteria immobilized would be large and a higher initial dose be required to cause the typhoid disease. (2) On the other hand, the stomach and contents furnish much material in which large numbers of bacteria may be lost through chance and thus never get an opportunity to reach the vital centers and cause death. The inoculation routes are closer to these centers so there is less chance of loss and fewer organisms are required. Of all the inoculation routes, it is a little surprising perhaps that the intravenous is least efficient, the subcutaneous is next, and the intraperitoneal is the most effective. *S. enteritidis* parasitizes the liver and spleen. In that sense, intraperitoneal inoculation places the bacteria near these organs and so could facilitate immediate infection possibly at the vital organ. On the other hand, at one stage the disease is a septicemia with large numbers of bacteria in the bloodstream. Should the septicemia be a fundamental part of the disease, it would also seem that the intravenous injection should be more helpful in reaching the vital organs than subcutaneous inoculations which distribute the bacteria more widely and also give them more chance for loss before their effects may reach the vital center. However, the data show that this is not the result. In any case, the large dose required by the intrastomachal routes is subject to at least two interpretations having opposite significance.

Another test of this matter came in our own research some 30 years ago while studying genetic resistance to poisons as distinct from reproducing pathogens. The dose of ricin for the mice was 0.002 mg. The mice were of a single inbred strain and were 60 days of age. The routes of instillation were subcutaneous, intraperitoneal, and intravenous (into the tail vein). One hundred and sixteen mice were used in the tests, figure 53.

The data of figure 53 are in order of the susceptibilities of the mice, the subcutaneous route being least toxic, the intraperitoneal route noticeably more lethal, and the intravenous being completely lethal. The order of effect is that which our preconceptions might lead us to expect. There should be more wastage of poison when introduced subcutaneously than when placed near the vital organs in the peritoneal cavity. Introduction into the circulation would seem to give the means to carry the toxic chemical most directly to the organs vulnerable to its effects.

The order of effectiveness of the routes of entry into the body is not the same as that for disease due to *S. enteritidis*. It appears quite likely it will not be the same for other pathogens. In any case the evidence is equivocal on whether or not a given route for introducing a pathogen in testing for natural resistance of a host is better than another.

Fig. 53. RELATIONSHIP BETWEEN ROUTE OF ENTRY AND SURVIVAL OF MICE INOCULATED WITH
0.002 MG. OF RICIN.

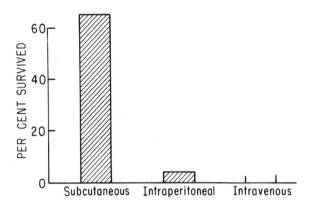

Direct comparisons of the products of the tests furnishes a better comparison. Data on this point are not entirely satisfactory since they were collected by different observers working with different populations of albino mice. Schott[1167] and later Hetzer[573] selected for natural resistance to *S. typhimurium* by selecting for resistance on the basis of survival to test dosages of 50,000 organisms introduced into the peritoneal cavity. The foundation population was rather susceptible since only 17.7 per cent survived the test. The first- and second-generation tests showed the families quite heterogeneous for resistance. Selection for resistance was continued for six successive generations when resistance of these animals gave them 75.3 per cent survival. Hetzer continued the selection for two more generations with the same test dose. The survivors of that dose were 84 per cent. The data show that selection for resistance is an efficient method for establishing natural resistance to *S. typhimurium* when the disease is induced by intraperitoneal injection of the organisms.

Webster[1362] selected for a susceptible and a resistant strain of mice by testing previous siblings by intrastomachal inoculation of 5,000,000 organisms and then later unexposed siblings for breeding on the basis of these susceptibility tests. By this test, the initial population of mice was not as susceptible as that of Schott's as judged by their respective survivals—62.6 per cent on Webster's test *versus* 17.7 per cent on Schott's. In four generations of selection Webster's selected, susceptible mice were susceptible to the point where only 15 per cent survived. The selection dose was then reduced to 100,000 organisms. On this lower dose the survival percentage became 17.6 per cent. Selection toward resistance for two generations gave a population of mice which, when tested with 5,000,000 organisms, survived 88.7 per cent. The test dose was then increased to 50,000,000 organisms. The survival of the fourth-generation mice treated with this dose was 82.5 per cent, but the degree of resistance had reached at least a temporary plateau for this method. In conclusion it may be said that the intrastomachal route proves to be a successful but slower route for selection of mice for resistance and susceptibility.

A test for how successful selections by the different routes have been is made by comparing the resistances of the resulting strains. As Webster showed, the strains of mice resistant to *S. enteritidis* administered by the gastric route were likewise resistant to the organisms introduced subcutaneously, intraperitoneally, and intravenously. Derived susceptible strains by the intrastomachal test route were likewise even more susceptible to the organisms introduced by the other three routes. The degree of progress toward resistance, although apparently less than that for susceptibility, was in fact more significant since resistant, unselected strains of mice were difficult to find in natural populations, whereas those which were as susceptible as or more so than the selected strains were relatively easy to find.

Sample mice of the strains derived by Webster have been bred in our laboratory for some years. They may be compared directly with our S and RI lines developed here at Ames. The S mice are the most resistant and were established by selections based on the results of intraperitoneal inoculations of the test organism (see table 61). The RI mice take their origin from the Webster resistant mice but have had their resistance greatly increased through our selections based upon the intraperitoneal route of inoculation. The K and C mice are two branches of the Webster selections for resistance which were split from the main line on the basis of their resistance and suceptibility to St. Louis encephalitis and louping ill. Similarly the N and Q lines represent strains similarly separated from this established susceptible strain. There is some variation between his two resistant lines and between his two susceptible lines, but the resistant group are definitely more resistant than the susceptible group on the basis of the intraperitoneal tests as they were on the intrastomachal tests. But these resistant strains are not so highly resistant as those selected by the intraperitoneal route in the first place.

These results are important from the methodologic viewpoint in pointing the way toward developing resistant and susceptible hosts to different diseases so that these diseases may be analyzed for the host, gene-based characters which are significant to the given diseases. They are also significant in indicating that, until a natural barrier can be differentiated from the diluting effect of distance from the vital region attacked by the particular disease, it is proper to assume that every cell in the body is endowed with properties which are contributory to making the animal resistant or susceptible to a disease.

Further support for the significant part played by the cells of the whole body in resistance to a specific disease comes in a study of X-ray irradiation effects on resistance mechanisms of genetically differentiated strains of mice exposed to *S. typhimurium*. The experimental design, although straightforward, is at the same time a little complex. Details are found in three papers of Stadler and Gowen.[1264, 1265, 1265a] X rays were chosen as a means of altering the resistance of the cells in different regions of the body. The experiment was designed as a factorial having four elements. Five inbred strains of mice having known differences in X-ray and typhoid sensitivities were utilized in comparable numbers. The numbers for the two sexes were balanced for each strain.

There were four X-ray exposure doses: 0, 320, 480 and 640 roentgens.   The levels of X-ray dosages were chosen to span the range from no effect to nearly complete lethality when the mice were exposed to whole-body irradiation.

There were eight combinations of exposures to the X-ray effects: none, head, mid, rear, head-mid, head-rear, mid-rear, and whole-body exposure.   The head region or anterior third of the body extended into the thorax.   The middle third of the body (mid region) included the lower thorax and abdominal cavity containing stomach and upper intestinal tract, liver, spleen, adrenals, ovaries, and kidneys.   The posterior third of the body (rear region) included the lower intestinal tract, bladder, and urinary system and testes of the males.

The eight groups of different regional exposures were treated with 320 r, 480 r, and 640 r, making a total of 24 different X-ray treatment groups.   In addition, the mice of one group were put in tubes and completely lead-covered for a time comparable to that of the longest dose, 640 r, but were not exposed.   This group acted as a control on handling as well as for the unirradiated, 0 r group.

There were 25 treatment groups with 5 strains and 2 sexes making up 250 cells in the experiment.   Each cell represented a different strain, sex, and treatment.   A minimum of 25 mice were treated in each cell.   Some cells contained a few extra animals.   The completed experiment involved a total of 6,904 mice.

The immediate effects of X ray were largely completed 12 days following irradiation.   Although there were marked differences in survival times between the different strains, few deaths occurred after the twelfth day.   A 15-day interval following irradiation was allowed to cover the direct effects of exposure.

The mice were then inoculated to test for their resistance to murine typhoid. Different lines of the bacterial species $S.$ $typhimurium$ cover the full range in virulence or pathogenicity due to differences in their genetic constitutions.[468, 1467]   This factor has been controlled in this work by limiting all the disease tests to one of our lines, 11C, of $S.$ $typhimurium$.

The data showed that the sexes were not significantly different in this experiment in their reactions to this disease.   The physiologic sex differences introduced by the chromosomal differences played little part in the outcome of disease.   Strain differences were obvious.   The high innate resistance of the S strain prevailed over the latent effects from the X-ray exposures except in the severest treatments, those of whole-body exposure to 480 r and 640 r.   The S strain was more than twice as resistant to the disease as were the Z and K strains.   The survival curves of the S mice showed less marked reductions in survival due to X-ray treatments than those of the other strains.   The differences in the disease-resistance levels of the Z and K strains and the level of the Q strain was even greater.

Because of our previous knowledge of the susceptibilities of the Q and Ba strains, the Q strain received only one-hundredth the dose and the Ba only one-thousandth the dose of $S.$ $typhimurium$ 11C which was given to the Z, K, and S mice.   The extreme susceptibility of the Ba mice under normal conditions allows no opportunity to observe effects like those of irradiation unless very small dosages are administered.

When analyzed for X irradiation alone or typhoid resistance alone, the strain effects were highly significant. For typhoid resistance, differences between strains were more marked in those groups where the irradiation had been more detrimental to survival. In only a few of the exposure groups did the radiation lower the resistance of the S strain to murine typhoid. For Z mice, the previous irradiations lowered survival to typhoid infection in all treatment groups. The Q strain, with its greater ability to withstand radiation, showed somewhat more reduction in survival from typhoid following the exposures of 320 *r* than of 480 *r* or 640 *r*, except in the whole-body treatment groups. Murine typhoid disease following 640-*r*, whole-body exposures was lethal to all the strains. Similarly typhoid following 480-*r*, whole-body treatments was lethal to all strains except S. The 480-*r*, regional exposures were effective in lowering the resistance levels of the Z and K mice. The average values of the regional effects on the four strains show that resistance to *S. typhimurium* was decreased as the exposure doses were increased. These observations support those of Gowen and Zelle[469] who found that X irradiation 15 days before contact with typhoid resulted in reduced survival for the 6 inbred strains of mice tested.

The high resistance level of the S strain to murine typhoid made the S strain the better indicator for differences between the treatment groups. The full range of survival from 100 per cent at 0-*r* to 0 per cent at the 640-*r*, whole-body exposure was covered by the S strain after typhoid infection. The range of survival was limited for the other strains by their greater susceptibility to typhoid even without X irradiation. The natural resistance of the S mice was most severely affected by the whole-body exposures followed by the head-mid exposures. The mid-rear exposure treatments at the three dosage levels were next in order of decreasing survival following typhoid infection. The other regional combinations were, in order of decreasing effects: the mid, head-rear, rear, and head exposures. The latter two exposure treatments had little effect on the S mice. The three other strains, Z, K, and Q, reacted in a similar manner for the different regional irradiations. The Ba strain was so susceptible even to 200 *S. typhimurium* organisms that it contributed no useful information on the effects of previous X-ray treatments.

The typhoid-resistance response was affected most severely by the previous irradiations to the whole body. Of the three body regions taken separately, resistance was lowered most by irradiation to the mid region. Irradiation to the head region reduced resistance somewhat more than irradiation to the rear region. The typhoid response following the X-ray treatments showed the head-mid exposure as somewhat more detrimental to natural resistance than the mid-rear exposure.

Full expression of the effects of body cells in the different regions is exhibited by the S mice. Survival decreased with exposure of any part of the body to X rays at the 640-*r* dose. Exposure of the head and rear regions decreased survival by 4 per cent; exposure of the mid region decreased survival by 13 per cent. Treatment of any two of the regions still further decreased survival proportionate to the regions exposed.

The really big decrease in survival came when all the cells of the body were exposed; but 20 per cent of this decrease was due to direct effects on the mid, head, or rear regions. The 80 per cent was accounted for by the absence of any unexposed cells. It mattered little which of the three body regions was left unexposed. Each third contributed much more than its share to typhoid resistance if it escaped exposure to the irradiation. These results are taken to mean that, when exposed to a disease of this type, all cells of the body retain much of their initial genetically controlled ability to resist disease invasion even when they differentiate into highly specialized tissues. From the methodologic viewpoint, high-energy irradiation offers a valuable means of specifically depressing the functions of individual organs so far as their properties of resistance to disease are concerned. Greater changes may be accomplished by whole-body irradiation when all cells are irradiated. The effects are genotype specific to some degree but are of similar kind for all genotypes.

Methodologic problems pertaining to the basis of genetic resistance to disease such as differences in leucocytes and serum-proteins; growth of heart, kidney, spleen, and other organs; active- and passive-immunity phenomena; mutation and other similar changes in hosts or pathogen; and diverse environmental effects are all part of this complex pattern of disease. Techniques for discriminating between various parameters are available, and introductory references to this material may be found in the bibliography.

SUMMARY

The expression of infectious disease is dependent on complex multivariate forces of host and pathogen genotypes as well as on a wide variety of environmental influences. Quantitative studies of disease have largely depended on restricting the range of variation exhibited by two of these three variables. The methodology by which these restrictions have been imposed and some of their consequences have been discussed for host genotypes and breeding behavior. These include the numbers of pathogens and their route of invasion during initiation of the disease as well as radiation as a means of altering qualities of disease resistance of the host.

DISCUSSION

Dr. Burdette: Dr. J. A. Weir of the University of Kansas will open the discussion of Dr. Gowen's paper.

Dr. Weir: Mice are expensive to maintain and require considerable space; consequently, workers employing mice exclusively are committed in a way that leads to specialization and narrowness in research interests. The rôle of the discussant, as I see it, is to relate the material presented to the larger picture. I find this a difficult task because Dr. Gowen paints with a broad brush, and he follows the axiom of genetics that problems are important and materials only secondary. Although he has limited his discussion largely to mice, he has also worked with large animals, chickens, viruses, *Drosophila*, bees, plants, and other living forms. At this point, I might be tempted to cite some obscure references from the literature, but, if I were to do so, I would likely use one of Dr. Gowen's papers as a general source. His reviews have been both thorough and penetrating.[462, 463, 464, 465]

Before proceeding with my own observations, it is appropriate to point out the similarity between Dr. Gowen's researches on infectious disease and the cancer-research program of the Jackson Laboratory. In both cases, the procedure is to investigate thoroughly a limited number of forms of the disease. On the one hand, several inbred strains differentiated on a basis of genetic resistance to murine typhoid are employed along with a specific, pure strain of the pathogen, *Salmonella typhimurium*. On the other hand, a few, specific, neoplastic diseases have been selected, along with special inbred lines of mice; for example, C3H as an object for study of mammary tumors and hepatomas.

It also seems appropriate to give a short historical sketch. The year 1919 is of particular significance as a starting point (work with plants started somewhat earlier, with Biffen's demonstration in 1905[90] that resistance of certain varieties of wheat to striped stem rust fungus depended upon a single gene). In 1919 Dr. Sewall Wright began sending his surplus guinea pigs from Washington to Philadelphia, there to be inoculated intraperitoneally or subcutaneously with human-type, tubercle bacillus. The importance of host resistance was clearly demonstrated. Over 30 per cent of the variation in survival time after inoculation in crossbreds was determined by relationship to the best inbred family.[1457] The year 1919 was also the time of the great influenza epidemic and this stimulated, on both sides of the Atlantic, work in experimental epidemiology. W. W. C. Topley and associates in England, supported by the Medical Research Council and L. T. Webster and associates at the Rockefeller Institute in the United States, studied the natural history of typhoid in populations of mice.[1365] Host resistance, or susceptibility, was not emphasized at first but it was frequently mentioned.[1325, 1364] This factor obviously could not be ignored and it became a subject of major emphasis.

The scene now shifts to the midwest where Professor W. V. Lambert, in 1924, started his work on the inheritance of resistance to fowl typhoid in chickens.[746] Under Lambert's direction similar work was initiated, using mouse strains obtained from E. C. MacDowell, L. C. Dunn, L. C. Strong, and M. R. Irwin. A culture of the organism, then named *Salmonella aertrycke*, was supplied in 1926 by W. W. C. Topley, Public Health Laboratory, Manchester, England. R. G. Schott and H. O. Hetzer completed their Ph.D. dissertations in 1931 and 1936 respectively.[573, 1167] When Dr. Lambert was called to administrative duties elsewhere, Dr. Lindstrom, with characteristic wisdom and foresight, induced Dr. Gowen to join the department at Ames as Professor of Genetics. In Ames, Dr. Gowen, already a pioneer in the field in his own right, picked up Dr. Lambert's work and also brought his own special strains of mice from Princeton. Later he added the surviving lines from L. T.

Webster's selection experiments. With this great wealth of materials he initiated the intensive investigations that are still in progress.

The type of selection experiment used to produce resistant and susceptible lines of mice bears comment. Parenthetically, a number of strains have been produced by inbreeding without selection, much as the cancer strains have been developed. Proponents of selection theory choose their characters for ease of measurement, and this introduces elements of artificiality. Even selection for DDT resistance falls in a special category, because the host is subjected to an artificial insult unlike anything it has faced during its long evolutionary history. Selection for resistance to infection, on the other hand, duplicates many of the features encountered in naturally occurring epizootics. The disease produced by intraperitoneal inoculation with known numbers of organisms of murine typhoid is identical in all its essential features to the disease as it occurs in an epizootic, because in mice the acute form of the disease is a septicemia and not primarily a gastrointestinal infection.

As in work with malignant neoplasms, inbred lines provide the investigator with a tool of inestimable value for examining the nature of the syndrome, but there are certain pitfalls in a study of the mechanisms of genetic resistance. To look for strain differences is an important part of the methodology, but the number of degrees of freedom available for correlation studies depends on the number of inbred strains and is independent of the numbers of mice. Furthermore, the degrees of freedom may be depleted in the search for clues. It is noteworthy that Dr. Gowen has not been content merely to isolate mechanisms of resistance, but he has attempted to establish physiologic relationships and to apply a variety of techniques. It should also be emphasized that he and his associates have taken elaborate precautions to separate the rôle of natural resistance from that of acquired immunity, to control dosage, and to apply genetic techniques of crossbreeding.

The nature of some components of natural resistance should be commented on briefly. Gowen and Calhoun [466] found that level of resistance and total leucocyte count were related. Subsequent studies using X irradiation have added strength to the conclusion that high leucocyte count and a high level of resistance are causally related. This stimulated some work of my own. I started with what now seems to have been a naïve assumption, namely that resistance could be synthesized by selecting for single components. I selected for total leucocyte count and obtained a high line with total count of more than 15,000 cells per mm^3 and a low line with less than 5,000 cells per mm^3. The outbred line has a mean count of 9,000 cells per mm^3. When mice were inoculated with standard doses of several strains of *S. typhimurium*, there were significant and consistent differences between strains in mortality and days to death. But the resistance levels were opposite to expectation on the assumption that the character selected is an additive component of resistance.[1377] Selection for blood *p*H gave similar results.[1377] Hill *et al.*[577] also obtained similar results in that mice selected for resistance to a partially purified, toxic fraction isolated from *S. typhimurium* proved more susceptible than the controls when they administered live organisms either *per os* or intraperitoneally. Returning to the rôle of leucocytes, the mere presence of large numbers may be a detriment unless the cells are effective in digesting bacteria. Otherwise they serve merely as a means of disseminating the pathogen throughout the host. These studies illustrate the wisdom of Dr. Gowen's approach, namely to deal with resistance as an entity. Work in his laboratory revealed that there are genetic differences in ability of leucocytes[69] and macrophages[949] to digest phagocytized bacteria. The Camp Detrick work with anthrax[1375] has also demonstrated the importance of qualitative differences in leucocytes.

The technique of double matings was mentioned by Dr. Gowen. This tool for investigation has not been fully exploited. We have found in our laboratory that about one litter out of 10 to 15 matings is of mixed parentage, when females are placed with pairs of males. All that is necessary to determine parentage is to have suitable color markers. When the seminal vesicles of one male are removed, to prevent plug formation, the number of mixed litters is not increased as might be expected. We have had fewer mixed litters by this procedure. Direct observation has shown that mixed litters are usually the result of a rapid round robin, with one male providing sperm and no plug or only a partially formed plug. Dominance (in the behavioral sense) of one male over the other has no effect on the percentage of mixed litters.

The experimental work with murine typhoid has a practical aspect. It is noteworthy that in Dr. Gowen's laboratory, even though the precautions are not elaborate, there has never been an outbreak of typhoid in the breeding colony. Mice and *Salmonella* do not necessarily go together, like pie and cheese. In our own laboratory murine typhoid was introduced by mice from elsewhere. Until efforts to eliminate the pathogen were successful, the disease remained confined to the imports and their descendants. Except in the few cases in which crosses were made, our own laboratory stocks remained free from disease.

At present there is a trend toward team research involving more or less intimate coöperation between workers in different disciplines. This is desirable and will no doubt increase. However, there will always be a place for an energetic, imaginative, and well-rounded independent investigator. May Dr. Gowen continue for many years to produce the type of research that he has described.

DR. GORDON: In considering methodology in this field, I would like to take just a minute or two to mention work of one of my colleagues at the Naval Medical Research Institute, Bethesda.[621] Dr. Herbert S. Hurlbut's interest is in the transmission of arthropod-borne viruses, and he has determined the susceptibility of other arthropods than the ones naturally involved in transmission, using representative viruses of this arthropod-borne group. That is, he took arthropods such as house flies, a species of *Lepidoptera*, beetles, *Hemiptera* (all non-blood-sucking), and some hemophagous arthropods, and determined their susceptibility to the selected viruses by inoculation through the body wall with a capillary pipette. He found that many of these species were susceptible in the sense that they would propagate the virus, it could be recovered in fairly high titer in some cases, and could be transmitted from one individual to another and maintained in such hosts indefinitely.   At the present time we have seen no adverse effect on these hosts, and the demonstration of the virus is by subinoculation into mice.

We have discussed these findings among ourselves as a method for investigating genetic factors in the ability of arthropods to propagate or transmit the viruses that infect vertebrates.   With the system Dr. Hurlbut has used, the amount of virus going into the arthropod can be accurately controlled, and the amount present after a suitable incubation period can be satisfactorily measured.   By choosing insect species in which genetic studies have already been made and in which various genetically described strains are available, suitable combinations of insect and virus could probably be found for a fruitful attack on this problem.

Dr. Roderick: Dr. Gowen, did you find any difference between the selected and unselected lines with reference to qualitative manifestations of the diseases?   Differences in the symptoms between the lines might indicate that the selected lines were resistant because they had in some way changed their mode of reaction to certain specifications of the pathogen.

Dr. Gowen: Symbiotic relations of an organism to one host, for which this host may supply a reservoir for nutrients, maturation, or multiplication, as well as acting as a vector for transmission of the organism and pathogenic relations of the same organism to a later ultimate host, constitute real challenges to genetic research.   The two-step relations hold through a range of pathogenic organisms—viruses, bacteria, protozoa, helminths, fungi, and the like.   The genetic controls for these diverse cases would seem to have been developed independently.   Damage to the vector host may range from none observable, through moderate, to that which acts relatively slowly compared with the time necessary for infecting the ultimate host.   The specificity of the host-parasite relations is so exact that basically the relations must be largely under genic control.   For instance, some malarias and viruses seem to damage the mosquito but little, yet the mosquitoes offer conditions suitable for their multiplication.   In retrospect it seems not unlikely that this long-standing relationship has an evolutionary history such that the virus or protozoan was initially pathogenic but that, with time and possibly through genetic modifications of both pathogen and host, they were able to live together in a symbiotic relationship.   That genic changes are basic to this pattern is evident from studies on a number of related species displaying differences in reaction patterns.   Huff[608] some years ago found that *Culex pipiens*, to become a carrier of avian malaria, required the presence of the dominant gene of a pair of alleles, resistance being recessive.

The second question concerns changes in virulence within particular lines of the pathogen.   In our work *Salmonella typhimurium* 11C rarely changed its virulence.   Zelle,[1467] in a series of experiments covering three years, obtained convincing mutants in virulence only four times.   The parental organisms were passed through successive transfer in mice of a resistant line or mice of a susceptible line.   The mutants were always directional to greater virulence.   Very rarely an avirulent mutant may also occur when planted on culture media.   *Salmonella gallinarum*, on the other hand, mutates either toward virulence or toward avirulence quite readily. A virulent culture placed on culture media favoring saprophytic growth for six months at refrigerator temperature will frequently have no virulence to chickens. On the other hand, avirulent *gallinarium* passed by blind passage through a series of as many as 20 chickens normally will mutate to full virulence at some point along the way.

DR. RODERICK: Do your selected and unselected lines show any differences in symptomatic manifestations of the diseases?

DR. GOWEN: In answer to Dr. Roderick's second question, our strains of mice do show differences in the manifestiation of disease when inoculated with the same line of *Salmonella typhimurium* 11C. There are distinct differences in morbidity. The S mice show scarcely any effects of the disease even when exposed to 2,000,000 organisms. The Ba mice are obviously sick when the exposure dose is 20 to 200 organisms. The other strains act as though they were between these extremes. Clear-cut differences in the survival curves are evident when they are analyzed for their different constants: mean, mode, standard deviations, skewness, kurtosis, and type of curve. Incidentally, there are no detectable immune bodies in the bloods of any of our strains before they have contact with the pathogen. One strain has a characteristic *alpha*-globulin and the other a characteristic *beta*-globulin. These behave as though controlled by single Mendelian genes but seem to have little correlation with the pattern of disease.

Dr. Weir has mentioned the differences in leucocytes. These differences in our strains are correlated with differences in resistance to disease. However, in his selections when differences in leucocytes were established in two separate strains of mice, the differences were not accompanied by corresponding differences in resistance to disease. Macrophages also differ in their abilities to retain visible bacteria when examined microscopically during the course of infection. These differences may be correlated with digestive abilities of intracellular enzymes, since they are correlated with the resistances of the strains. It may be said that one strain or the other of our group has been found to differ significantly from the others in all of the organs yet measured. Many of these differences are correlated with the visible manifestations of the disease.

GENETICS OF SOMATIC CELLS

George Klein, M.D.

GENETICS *of* SOMATIC CELLS†

Originally, the study of heredity was mainly concerned with the ways and means by which genetic information is transmitted through the germ cells of higher organisms; and sexual crossing was the predominant, if not exclusive, method of approach on which all important concepts were based.   The first important change in this situation came with the development of microbial and, particularly, bacterial genetics.   Initially, bacteria were believed to multiply exclusively by binary fission and to lack every form of sexual reproduction.   Nevertheless, consistent attempts were made to apply the concepts derived from the genetics of higher organisms to the study of bacterial heredity, but this met with considerable skepticism in the beginning.   From the viewpoint of the geneticist, the approach often seemed too indirect and uncertain, particularly since it was impossible to distinguish between genotype and phenotype, as there was no method available by which hybridization or other forms of intercellular genetic transfer could have been accomplished.   Microbiologists who were not thoroughly acquainted with genetics were not ready to accept the view that the same basic laws may apply to bacteria as to higher organisms and both may contain essentially the same type of genetic material, organized in a similar fashion.   Variation in bacteria was often interpreted as developmental rather than genetic.   Superficial consideration of the many adaptive phenomena occurring in bacterial populations after exposure to various noxious agents, such as chemicals, antisera, bacterial viruses, and the like, has led to the view that the heredity of lower organisms may be more easily influenced by the environment than

† The work of the author and his collaborators quoted in this paper has been supported by the Swedish Cancer Society, by grants C-3700 and C-4747 from the National Cancer Institute, U.S. Public Health Service, and by the Knut and Alice Wallenberg Foundation.

the heredity of higher forms, perhaps because of the lack of an insulated germ line, and Lamarckism found its last stronghold in microbiology.

This situation has radically changed since 1940. A closer scrutiny of bacterial variation and, in particular, the analysis of clonal variance of mutations to phage resistance by Luria and Delbrück[812] have led to the understanding that most cases of adaptation find their explanation in the differential selection of spontaneous mutants. It has been increasingly realized that bacterial cultures must be regarded as populations which may have genotypically diverse components. The analogy between mutations in higher organisms and bacterial variation has been clearly emphasized by Luria.[809] He pointed out, as common features, the random, apparently spontaneous occurrence of variation at specific, generally low rates and its independence of physiologic conditions. The same agents were found capable of inducing mutation in higher and lower forms, for example, radiation and nitrogen mustards. Mutations affecting different characters occurred, as a rule, independently of one another. Often the same type of function, such as enzymatic specificity, was affected. The specific induction of mutations by environmentally induced adaptation and inheritance of acquired characters could be disproved in almost every case. The crucial proof, watertight even for the most critical, came with the achievement of indirect selection of drug-resistant bacteria by replica plating[776] or by other indirect means,[186] that is, without ever exposing the cell population to the drug.

The most important developments in the study of bacterial heredity that have led to its spectacular ascendance and present outstanding position in contemporary genetics, where it influences almost every major aspect of genetic thinking, have been undoubtedly the discoveries of various methods permitting the intercellular transfer of genetic information, such as DNA-mediated transformation, sexual recombination, and phage-mediated transduction. They have proved beyond doubt that bacteria contain linearly organized genetic material, quite analogous to the chromosomal genes of higher organisms with regard to chemical composition, structure, mutability, and function, and advanced far beyond this demonstration, revealing a wealth of new facts and principles concerning the fine structure and action of the genes.

The recently reawakened interest in somatic-cell genetics is based on the belief that genetic concepts may be helpful for the understanding of such somatic phenomena in higher organisms as differentiation, enzymatic adaptation, antibody formation, neoplasia, and some forms of viral infection. In a way, the prospects for the study of genetic phenomena in somatic cells are very similar to the situation in bacterial genetics some twenty years ago, prior to the discovery of methods for genetic transfer. This leads us immediately to the recurrent theme of this paper: although it is not certain *a priori* that analogous methods of transfer can be worked out for somatic cells, there is no more urgent need in this field than the exploration of all possibilities in this direction. Meanwhile, the genetic approach will have to remain indirect and restricted to the analysis of variation in populations of somatic cells, without any possibility of proving its intrachromosomal or extrachromosomal nature. Although even such an approach may lead to the discovery of many valuable facts, interpretation of the phenomena in terms of cellular mechanisms by analogies will be difficult, more so than during the corresponding stage of bacterial genetics, since it has to be recognized that somatic cells of higher organisms are subject to developmental processes, the cellular mechanisms of which still represent some of the most formidable unknown territories of biology.   Many attractive models are available, borrowed from various fields of genetics,[133, 417, 821, 1173, 1257] and the method of nuclear transplantation opens fascinating possibilities for direct studies on embryos,[131] but information presently available on the relationships between genotype and cellular differentiation does not offer a firm foundation on which to build the delicate superstructure of somatic-cell genetics.   The classical notion of genotypic equality in all somatic cells, based essentially on the nature of the mitotic process, has not been confirmed experimentally.   On the one hand, it is quite true that phenotypic diversity does not exclude genetic identity. That can often be proved with differentiated microörganisms with cells that can be subjected to breeding analysis and may turn out to be genetically identical, in spite of profound, permanent, and stable phenotypic differences.[935]   These differences are believed to be epigenetic[325, 934] due to mechanisms that regulate the expression of genetic potentialities, rather than truly genetic mechanisms regulating the maintenance of structural information.   On the other hand, the doctrine of genotypic identity of somatic cells is being viewed with increasing caution,[1257] particularly in the light of the nuclear transplanation experiments of Briggs and King,[131] pointing toward the existence of a true nuclear and probably chromosomal[1257] differentiation.   On the one hand, there are many workers who believe that development and differentiation are much too orderly and predetermined processes to be akin to the random changes at the genetic level known under the category of mutations, even if the term is used in a broad sense. On the other hand, there are others who disagree on this point, and genetic models of differentiation have been constructed, based on the phenomena of gene activation and the concept of controlling elements, such as dissociators, activators, and modulators.[133, 821, 1173]

Under these cirumstances, the study of somatic-cell genetics will have to be limited at the present time to the thorough experimental study of comparatively simple situations, without too many theoretical preconceptions.   Generalizations will seldom be justified, except in the form of working hypotheses.   The possible approaches can be classified as the study of appropriate phenotypic marker characteristics, detectable at the cellular level, and the direct examination of chromosomal morphology.   Since the latter subject will be discussed in the next chapter, this discussion will be mainly limited to the former type of approach and chromosomal studies will only be considered if they have direct bearing on the topic discussed.   Subsequently, an attempt will be made to review briefly various approaches toward the possible development of methods that might permit the accomplishment of genetic transfer between somatic cells. Microbial genetics will be regarded as the master discipline as far as methods and approach are concerned, while interpretations based on analogies will be viewed with caution, because of the many unknowns interposed by the developmental and organizational processes in cells of higher organisms.

STUDIES ON SOMATIC VARIATION THROUGH PHENOTYPIC MARKERS

Because of differences in methodology, in the nature of the problems involved, and in the type of information now available, this chapter will be subdivided according to the biological materials into a consideration of pertinent studies on (1) normal somatic cells *in vivo*, (2) neoplastic cells *in vivo*, and (3) tissue-culture work. Since the subject matter that might be considered is overwhelmingly large, attention will be mainly focused on certain aspects that have caught the interest of the author. No claim of completeness can be made; other reviewers would have emphasized and discussed other aspects of this huge field where so little is known and so much may be pertinent. Animal, particularly mammalian, material will be considered most and information on other groups will only be touched upon when appearing particularly relevant.

NORMAL SOMATIC CELLS *in vivo*

Somatic crossing over.—This mechanism will be considered separately, because, as pointed out below, it may permit the genetic analysis of somatic cells even in the absence of methods of genetic transfer.

The occurrence of somatic crossing over (s.c.o.) in *Drosophila* has been demonstrated by the classical work of Stern.[1275] This was based on the occurrence of twin spots, when, on *Ab/aB* individuals, the recessive phenotype *aa* appeared in an area adjacent to one with the recessive phenotype *bb*. Stern's interpretation was that s.c.o. has occurred between two of the four chromatids of the relevant homologous chromosomes. It was localized between the *B*-locus and the kinetochore, and the next mitosis led to sister cells with the genotype *aB/aB* and *Ab/Ab*. If the linked, recessive mutants *a* and *b*, originally located on opposite members of a homologous pair of chromosomes, were of a nature to produce a phenotype visible in the hypoderm when homozygous (such as *yellow* and *singed* on the X chromosome which can be identified in single bristles), then further multiplication of the crossover cells subsequently led to the formation of two adjacent patches of mutant tissue on a wildtype background. Somatic crossing over was a much rarer event than crossing over in the germinal tissues. Its frequency could be influenced by X irradiation,[68, 1386] by temperature,[138] and by maternal aging.[139, 1179] While there seems to be no basic difference between meiotic and mitotic crossing over,[1011] they appear to differ with regard to their response to modifying factors. High temperatures decreased somatic crossing over in contrast to germinal crossing over, which was increased.[1277] The agents that prevented meiotic crossing over in male *Drosophila* did not affect somatic crossing over that occurs in both sexes.[685, 1275] The *C3G* gene, which practically eliminated meiotic crossing over in female *Drosophila*, had no effect on the frequency of somatic crossing over.[771]

Particularly interesting was the finding[1275] that the various *Minute* loci increased the frequency of s.c.o. Minutes are known to occur throughout the chromosomal complement, all producing a similar phenotype, characterized by prolonged larval development, slender bristles, and lowered viability and fertility. Acting as dominants, they are thought to be small deletions, lethal when homozygous. These loci exerted varying degrees of effect on the frequency of s.c.o., the sex-linked Minutes being more effective in causing increases than the autosomal Minutes. The chromosome III Minutes exhibited a peculiar specificity in that they limited s.c.o. to the arm in which they were located. The existence of other genetic factors affecting s.c.o. was indicated by the work of Brown and Welshons[139] who found great variation in the incidence of mosaicism among the several stocks that they studied. Recently, Weaver[1360] reported a series of experiments designed to detect and characterize the genetic factors that control s.c.o. These studies led to the conclusion that there are chromosomal factors other than Minutes, not associated with any visible deficiency or aberration, which control the frequency of s.c.o. Such factors, governing the frequency of *X*-chromosomal s.c.o., occurred on all three major chromosomes. She concluded that s.c.o. is a process under the precise control of genetic factors governing the frequency of its occurrence. The genes responsible might affect either the closeness of somatic pairing, or the physical or chemical processes responsible for the exchange of chromosomal segments, or both.

Somatic crossing over has received its most important application so far in the work of Pontecorvo and his school, who demonstrated its usefulness as a tool for genetic analysis and emphasized its possible applicability to cells in the tissues of higher organisms. The latter possibility has often been dismissed previously as it was assumed that somatic pairing of chromosomes at metaphase is an absolute prerequisite for the occurrence of s.c.o. Whereas somatic pairing is a regular event in *Drosophila*, butterflies, and moths, and less regular in other lower organisms, it has only been demonstrated exceptionally in higher organisms.[119] However, the validity of this requirement may be seriously questioned. As emphasized by Stern,[1276] we are so ignorant about the mechanism of crossing over that negative cytologic evidence regarding the absence of somatic pairing can hardly be taken as a serious objection against the possible occurrence of s.c.o. Or, as Pontecorvo puts it:[1011] "The idea, hard to die, that somatic pairing as cytologically detectable at metaphase in a few organisms has something to do with mitotic crossing over is not very helpful to say the least." In five different species of fungi and in yeast, the occurrence of s.c.o. has been made unquestionable by purely genetic analysis in the complete absence of cytologic observations. Some results of this analysis will now be briefly considered.

Somatic crossing over has been most extensively studied in *Aspergillus nidulans*.[1011] Since it is a rare event, crossovers had to be selected. This was done by the use of suitable marker characteristics. Five types of selection could be applied: visual (this corresponding to the only method available for the analysis of s.c.o. in *Drosophila*); selection for recessive suppressors of a nutritional requirement; selection for recessive or semidominant resistance to harmful agents; selection by starvation (based on the fact that certain cells with two different nutritional requirements survived longer under conditions of starvation than cells with only one requirement); and selection of *cis* from *trans* heterozygotes, in cases where a strain heteroallelic in *trans* for a nutritional requirement did not grow on a medium lacking the relevant growth factor, whereas the corresponding *cis* heterozygote resulting from s.c.o. did, and could therefore be selected.

Another important requirement for the genetic analysis by s.c.o. was that the markers used for selection of the crossover homozygotes had to be located as distal from the centromere as possible, preferably quite near the tip of the chromosome. Among the homozygotes for such a distal marker, produced by s.c.o., there were some that were also homozygous at the next proximal locus (that is, nearer to the centromere), some at the next two, some at the next three, and so on. By expressing the proportion of homozygotes for any one, two, three, or more segments nearer to the centromere than the selected marker, as a fraction of all homozygotes obtained for the selected markers, it was possible to calculate the proportional incidence of exchanges in each of the corresponding intervals between the centromere and the first marker, the centromere and the second marker, and so on up to the selected marker. The centromere itself could be located in chromosomes marked on both arms, by identifying the segment where homozygosis for one of two adjacent markers was always accompanied by homozygosis for a series of other markers in one direction, and homozygosis for the other marker was always linked with homozygosis for a series of markers in the other direction.

Essentially, then, the procedure of genetic analysis by s.c.o. was to select, from a heterozygous diploid strain, segregants that were homozygous or hemizygous for certain distal markers and to analyze subsequently these segregants for their residual genotypes.

In molds, the occurrence of another important process—haploidization—complicates the s.c.o. analysis. Between 10 and 50 per cent of all phenotypically recessive patches selected from heterozygous diploid colonies turned out to be haploid. Somatic crossing over and haploidization result from two quite different processes that occur in different nuclei and do not coincide more often than expected by chance. Haploid segregants can be separated from diploid, however, by measuring the diameter of the conidia or by other methods.[1011] Haploidization turned out to be a very convenient tool for genetic analysis by itself. It yields all possible recombinations between chromosomes, but practically no recombination between linked markers. As a result, it locates any previously unlocated marker on its appropriate linkage group at once.

Genetic mapping by s.c.o. proved to be perfectly feasible in the studies on *Aspergillus*. Since the absolute incidence of s.c.o. was small, difficult to measure, and variable, these maps were not based on recombination fractions as in meiotic mapping, however. The distances had to be expressed in different units for each chromosomal arm. This was done by determining the total number of segregants for the distal selected marker and expressing the incidence of crossing over in each of the intervals between the centromere and that marker as fractions of the total. This mapping, which led to the identification of linkage groups, the order of the genes and their location, turned out to be fully dependable; indeed, from the qualitative point of view it was more dependable than meiotic mapping, due to the rarity of multiple exchanges in s.c.o. and to the occurrence of haploidization without crossing over.

It has been repeatedly emphasized by Pontecorvo and his associates[1011, 1012] that there was no reason why s.c.o. could not be used, if it occurred, for establishing the formal genetics of somatic cells in cultures of tissues and organs from higher organisms. Even processes akin to haploidization would be of great value. In *Aspergillus*, most haploids originate by nondisjunction as aneuploids, which are rapidly reduced to the monosomic condition by accidental selection in favor of the balanced haploid. No similar process is known to occur in somatic cells of higher organisms; although aneuploidy is common in established strains of cells *in vitro*, it usually leads to an increase rather than a decrease of the chromosome number and no truly or nearly haploid cells have been propagated in culture so far.

For these reasons, it appears to be of considerable importance to find out whether s.c.o. occurs in mammalian cells or not. Two phenomena have been described that might be possibly interpreted as such. One is the somatic variation of human erythrocytic antigens, the other the variation of *H-2*-isoantigens in heterozygous tumors. The former has been studied recently by Goudie[461] and by Atwood and Scheinberg.[37] The latter authors found that exceptional red cells lacking *A* agglutinogen were present in the peripheral blood of normal *A* or *AB* persons, comprising about 0.1 per cent of the erythrocytes in heterozygous young adults. Their phenotype agreed in every way with the expected consequences of the loss of the *A* allele. Some cells apparently lacking *B* agglutinogen were also found in *AB* blood and they were phenotypically *A*. Nonspecific inagglutinability was considered to have been ruled out by the fact that the exceptional cells behaved in the same way as unchanged cells when typed for *MN* components.

Considering the question whether these cells represent changes at the genetic level or phenocopies, Atwood and Scheinberg[37] suggest that this might be judged from studies on the distribution of exceptional cells among different individuals, their relation to the age of the individual, to heterozygosity, and to mutagenic agents. Evidence on these points is mostly unavailable at present, although a very recent study of Scheinberg and Reckel[1166] strengthened the genetic explanation by showing that the frequency of cells with the inagglutinable phenotype can be increased by radiation. Nevertheless, the argument remains rather hypothetical and it cannot be excluded that the exceptional cells represent phenocopies rather than true genetic changes, but, if so, they are at least accurate facsimiles of genuine mutants. While Goudie[461] considered mitotic crossing over as the most probable explanation, Atwood and Scheinberg believe that mutation is more likely, since loss of A_1 was accompanied by a great increase in the *H*-substance which they would not have expected in *B* homozygotes, such as would arise by crossing over. This argument is highly conjectural, however, since heterozygous and homozygous *B* cells have not yet been proved to be distinguishable by their reaction with anti-*H* reagents.   For this reason, the origin of the exceptional cells by mutation, somatic crossing over, or phenotypic change must be left open at the present time.   In fact, it cannot be excluded that the transitory absence of an agglutinogen might be a normal feature of the erythrocytic life cycle.   Since erythrocytes are differentiated end products, and cannot be subjected to progeny tests, it cannot be decided whether the exceptional behavior of the variant cells would breed true or not.   This difficulty can be overcome by using cells capable of division. An analogous study has, in fact, been carried out with populations of neoplastic cells, induced in F_1 hybrids of coisogenic resistant mouse strains heterozygous for the isoantigens determined by the *H*-2 locus.   The results of this study are quite compatible with the possible occurrence of somatic crossing over, although they do not conclusively prove it.   They will be described in more detail in the chapter on neoplastic cells.

Other types of genetic variation.—There is an extensive literature dealing with phenotypic variation in the somatic tissues of the same plant or animal.   Many of these are certainly nongenetic; in plants, such may arise, for example, by viral infections or by cytoplasmic changes affecting the chloroplasts.   The various genetic mechanisms that may play a rôle can be classified in different ways.   Swanson[1305] lists them as "chimeras, in which tissues of different genetic or chromosomal constitution lie adjacent to, or overlap, each other; endomitosis and somatic reduction which alters the chromosomal complement of the cell; abnormal fertilizations to give gynandromorphs or mosaics; somatic crossing over which reveals hidden heterozygosity; chromosomal elimination or fragmentation; and gene mutations."   It is obviously impossible to review all these mechanisms within the limits of this article; discussion will be restricted to a few types that have been well documented and appear to be of general significance for the analysis of somatic variation already at the present stage.   Chromosomal mechanisms as such will not be considered unless relevant to the cases discussed, since they will be dealt with in another chapter.

Chimeras have been well known in plants for several decades, but the knowledge of their existence in animals and, particularly, their experimental production are of comparatively recent date.   Most available evidence concerns the hematopoietic system, and more particularly, erythrocytic antigens.   The various possible mechanisms that can give rise to erythrocytic mosaicism have been reviewed by Cotterman[228] who classified them as natural chimeras (usually made possible through chorionic vascular anastomosis between twins during embryonic life), artificial chimeras (produced by parabiosis or by transplantation) and mutational, nonchimerical mosaics that may arise by gene mutation, somatic segregation, or crossing over, or by chromosomal aberration or variegation mechanisms of several kinds.   The first mosaics were discovered by Owen[984] who found that cattle twins were almost invariably identical in blood types, in spite of the fact that identical twins in cattle are relatively rare.   A closer examination revealed that the peripheral blood was often a mixture of two populations of cells, but never more than two.   On genetic analysis, it was found that each twin transmitted to its progeny genes corresponding to the phenotypes of one of the two cell populations. Owen concluded that an interchange of cells must have occurred between bovine embryos as a result of vascular anastomoses.   Stem cells appeared to be capable of becoming established in the hematopoietic tissues of the co-twin hosts and there continuing to provide a source of blood cells distinct from those of the host. This finding was of the greatest importance for the understanding and experimental study of immunologic tolerance.

The initial discovery of Owen has been followed by analogous findings of blood-group mosaicism, probably chimerical in origin, in sheep,[1288] man,[310] and fowl.[92] Sex-chromatin studies on neutrophilic granulocytes of human twins of different sex suggested that chimerism may involve the stem cells of granulocytes and erythrocytes as well.[117, 943]   In two pairs of human chimeras with both partners examined,[228] the ratio of the two populations of cells was distinctly different in the co-twins, the grafted population being in a minority in three of the four persons.   It was assumed that this inequality resulted from a gradual selective overgrowth of the host's cells at the expense of the co-twin's cells.

It is not always easy to distinguish between mutation and chimerism.   According to Cotterman,[228] the latter is indicated by the existence of twinning, the presence of mosaicism for two or more antigens simultaneously, and a not very unequal mixture of two populations of cells.   None of these is an infallible criterion.   Highly unequal mixtures may occur in cases of chimerism as mentioned above.   Two or more antigens could be expected to change simultaneously by some events belonging to the broad category of mutations, such as deletion, somatic reduction, and crossing over.   However, as long as the number of antigenic markers studied is relatively small and most of them are unlinked, these mechanisms generally bring about a single difference only. Most mutational mosaics are of the minute variety, with only a very small admixture of cells of the mutant type.   Cotterman lists three exceptions for which more nearly equal mixtures could arise through mutation: (a) genic instability or variegation, (b) neoplastic hemic diseases, and (c) very early genic mutation or chromosomal aberration.   The first of these is but little known in mammals and the second is discussed in the chapter on neoplasia.   The third mechanism has been brought into the foreground by the recent striking developments in the field of human cytogenetics. Several cases of mosaicism have been discovered with two different types of cells; each, with its distinctively different chromosomal equipment, could be shown to coexist in the bone marrow and in other tissues as well.   Most of these analyses concern the behavior of the sex chromosomes, and they have not been shown to involve erythrocytic antigens but are nevertheless relevant for the problem of mosaicism.   Ford et al.[388] found that the bone marrow of a patient with Klinefelter's syndrome contained two types of cells, one with 47 and one with 46 chromosomes.   There were good reasons to believe that the extra chromosome corresponded to the Y chromosome and that this was a case of an XXY/XX mosaic.   The genesis of this situation appeared to involve two different events.   It was assumed that the individual arose as a result of nondisjunction during gametogenesis, either by the fertilization of an *XX*-bearing egg originating through nondisjunction of the *X* chromosomes during meiosis, by a normal, *Y*-bearing spermatozoon, or by the fertilization of a normal, *X* egg by a spermotozoon containing *XY* due to nondisjunction during spermatogenesis.   In this *XXY* individual, cells containing the normal complement of 46 chromosomes could be expected to arise by mitotic nondisjunction.   Ford *et al.*[388] assume that such cells may exhibit a selective advantage and compete successfully with their progenitors, since the loss of the *Y* chromosome from an *XXY* cell would be a step in the direction of normality. In fact, the work of the Harwell group on serial transplantation of bone marrow through lethally irradiated mice clearly showed that different clones of cells, distinguishable by differences in their chromosomal complement, may proliferate differentially even to the extent that one clone is disseminated widely and outgrows all the rest.[53] They also obtained evidence indicating that marrow cells deficient in chromosomal segments, produced by the irradiation of normal animals, were rapidly eliminated.

Other cases of mosaicism for the sex chromosomes, reflected in the bone marrow and in other tissues, have been described by Ford[384] who found two patients with Turner's syndrome, containing a mixture of *XO* and *XX* cells.   These individuals were assumed[384] to have originated through the union of an *O* gamete, derived from nondisjunction during meiosis, with a normal *X*-bearing gamete.   Mitotic nondisjunction of an *XO* cell during development would subsequently have led to normalization and the appearance of an *XX* cell which, due to its selective advantage determined by the normal chromosomal complement, would have survived well in competition with its less normal progenitors.   Another case of mosaicism, discovered recently by Jacobs *et al.*[658] contained two cell classes, with 45 and 47 chromosomes, respectively, and detailed analysis led to the conclusion that this was an *XXX/XO* mosaic.   The proportions of the two cell types varied quite considerably between different tissues.   This case was particularly interesting since it had to be attributed to nondisjunction of the *X* chromosomes during the first cleavage division of the fertilized egg.   It could not have arisen during any of the later divisions, since a triple mosaic of the type *XO/XX/ XXX* would have resulted from such an event.

Thus aneuploidy, due to meiotic or mitotic nondisjunction, may be an important factor causing mosaicism and various developmental abnormalities in higher organisms. Of course, when considering isoantigenic substances such as blood groups, the restriction has to be made that only variant types for which the host organism is tolerant can survive and become demonstrable.   This means either that they must arise during embryonic life or that they must represent losses rather than changes to alternative antigenic states.

Somatic mutation has often been postulated as an explanation of unusual findings involving the *ABO* system to the exclusion of other blood systems.[228]   No individual had a history of twinning or transfusion in these cases.   They include two families with $O + A_2$ mosaics, with A_2 between 8 and 12 per cent in four cases.   In one woman there was clear-cut partial agglutination, heterochromia iridum partialis, and the *A* antigen was found in one small area of the decidua parietalis after the birth of a third group-O child.   The mosaicism was evidently inherited, since it was present in a younger sister.   In a second family, a similar $O + A_2$ mosaic was found in mother and son. One interesting case described by Armstrong et al.[34] was a person with true hermaphroditism (testis on right side, ovary on left), having O erythrocytes but a saliva containing B substance in nearly normal quantity.

The genetic control of antigenic specificity in higher organisms is a rather controversial field and intensified studies on somatic blood-group variation may contribute to the clarification of several issues.   At present, it appears[985] that more than one locus can affect an antigen, and a locus may affect more than one antigenic molecule.   Also, genetic variation may be concerned with haptenic substitutions at different sites of a given macromolecular species.   The development of refined techniques for the study of somatic mutations, for example, in the Rh system, may help to illuminate the relationships between genetic information and antigenic specificity and may resolve the much-debated question of multiple alleles versus closely linked genes.[228, 985] Methodologic developments, specifically aiming at the detection and study of blood-group mosaics, whether mutational or chimeric in origin, are of the greatest importance for future advancement in this field.   As one example of a promising approach, the lytic system of Hildemann[574] in mice may be mentioned, as applied to the study of artificial (radiation) chimeras by Owen.[982]   This method permits the absolute measurement in a spectrophotometer of the proportion of red cells hemolyzed by isoantibody, directed against a given specific isoantigen, in the presence of complement. Color caused by specific hemolysis can be compared with the color deriving from total osmotic hemolysis of a similar test-cell suspension in distilled water.   This proportion of cells specifically hemolyzed in chimeric animals can be compared with (a) a similar value for normal animals known to be positive to the test reagent or (b) mixtures of known positive and negative suspensions of cells.   According to Owen,[982] this permits the quantitative evaluation of intermixtures in which the proportions of positive cells range from about 5 to 100 per cent.   Care must be taken, however, that the concentration of none of the reagents becomes limiting, with reference to the total number of cells to be lysed.

In addition to isoantigenic markers and chromosomal studies, genetic differences in hemoglobin[1016] and a genetic deficiency leading to anemia[1100] have been useful in studies on bone-marrow transplantation involving artificial chimerism.   In the latter case, normal and otherwise isologous marrow has been highly successful in competing with anemic marrow.

Among the many other phenomena interpreted as somatic mutations in normal cells, straightforward genetic tests can be applied in exceptional cases where the mutated sector involves both somatic and germinal tissue.   In corn, for instance, a mutation in the tassel can give rise to a sector that produces mutant pollen.   The genetic nature of this change can be confirmed by ordinary crosses.   In animals, one particularly interesting case has been reported by Russell and Major[1117] who studied the reversion of the spontaneous color mutant *pearl* (*pe*) in mice to full color.   Such reversions are frequent on the genetic background of strain 201, while the gene appears to be stable on a C3H background.   In most cases, the spots occupied only about 0.1 per cent of the coat and the germinal tissue was not involved.   In five animals where the proportion of full-colored fur was much larger (100 per cent in two), the germinal tissue was also involved, however, and the reverted character of full color was transmitted to a certain proportion of the young.   Since even the two completely full-colored animals produced less than 40 per cent full-colored young, it was concluded that somatic rather than germinal mutation was responsible even in these individuals. These events seem to be capable of occurring early or late in development, and, when occurring early, they may involve the gonads.   Appropriate crosses have confirmed that the reversions were due to events at the *pe* locus or within 0.4 per cent crossover units from it.   Russell and Major suggested that somatic mutations of this type may be valuable for following cellular lineage in development of the mouse.

Other interesting cases where somatic mutations have included the gonads in mice have been described by Dunn[295] and Bhat.[86]   In both Dunn's and Bhat's case the mosaics (which, incidentally, were mutants for allelomorphs of the albino and the agouti series, respectively) carried the gene *s* (*piebald*), in the heterozygous condition. Dunn has pointed out that the frequent association between piebaldness and mosaicism can hardly be accidental, whatever the causal relation may be.   The majority of color mosaics in other rodents have occurred in animals with some white spotting.

Another case where both soma and gonads were involved but which turned out not to be due to simple somatic mutation upon breeding analysis has been described by Carter.[156]   A mosaic female mouse developed from a zygote of the constitution $W^v/+$, W^v being a semidominant color and macrocytic anemia mutant.   Parts of this animal showed a somatic deficiency for W^v, but she bred as though her germinal tissues were partly deficient of the wild-type allele $+^w$.   Somatic mutation could hardly have been the mechanism of formation, since two mutational steps would have to be postulated, W^v to $+$ in the soma and $+$ to W^v in the gonad.   Deletion was also improbable, since two deletions in homologous chromosomes would have to be postulated. On the other hand, nondisjunction during one of the early cleavage divisions, somatic reduction, and somatic crossing over have been regarded as possible explanations, quite compatible with the facts.   Carter[156] quotes also some other cases of mosaic animals from the literature which fit somatic reduction or somatic crossing over much better than point mutation.

The cases where both soma and gonads are involved must be regarded as highly exceptional, and most animal mosaics are purely somatic.   In plants, such somatic changes may often be subjected to further study; a "bud mutation" on a tree, for instance, can be propagated asexually and produce a new line of distinct individuals. In fact, most old plants vegetatively propagated, such as pelargoniums and potatoes, have become chimeras owing to somatic mutations at some time during their history.[240] In animals, somatic mutations have a different effect: development and life being limited, the mixtures do not survive beyond the life of the individual in which they arose. Also, since development is less simple, the mixtures are much less regular. Mutations occurring in cells which later proliferate give rise to altered spots or sectors whose size and location depends on the time in development when the mutational event has happened and the type of cell lineage in which it arose.   One interesting and highly regular type of mixture is the gynandromorph, well known in *Drosophila*, with part of the animal male and part female.   The origin of the anomaly can be attributed to a genetic difference arising between a pair of nuclei produced at an early mitosis in the embryo.   Various mixtures can arise but the commonest is the half sider.   They can arise by different types of mechanisms which can be proved by both genetic and cytologic evidence and "all the different irregularities occur which will give workable results in each organism."[240]

Another interesting example of somatic mutation in animals is the condition of mosaic fleece in sheep, described by Australian workers.[148] Sheep of a normally short-wooled breed show occasionally areas on which much longer, loosely crimped fleece is growing.   Among twenty million animals, 30 have been found that showed this condition.   The most interesting feature was the relation between the extent of skin area involved and the frequency of the condition.   The percentage area involved could be used to calculate the probable stage of segmentation when the mutation occurred. The results indicate that cells at various early stages of segmentation had about equal chances to undergo the change which occurred in frequencies between $10^{-7.15}$ and $10^{-7.3}$ per cell, thus well within the usual range for germ-cell mutations.

Somatic mutations can be produced by X rays; this has been first shown in *Drosophila*[992] but it holds true for other species as well.[1117] Flecking of pigeon feathers can be induced by X rays and these induced changes are transmitted through successive somatic generations of regenerated feathers after intermittent plucking.   A particularly informative investigation has been published recently by Russell and Major[1117] who studied the induction of somatic mutations at specific loci in mice.   Since this is the first investigation of its kind on a mammalian organism, it will be described in some detail. Their method was to irradiate embryos heterozygous for four coat-color genes and examine the adults for mosaic patches.   As controls, embryos homozygous for the wild-type alleles of the four loci studied were also irradiated, in order to differentiate between developmental effects, leading to abnormal differentiation of pigment cells, and genetic effects.   The frequency of such developmental changes was subtracted from the total mosaics in the heterozygous series in order to obtain the frequency of changes due to the expression of the four coat-color recessives.   Using this general method, 57 heterozygous first litters of the outcross used (C57BL♀ × NB♂, heterozygous for the coat-color genes p: *pink-eyed dilution*, c^{ch}: *chinchilla*, d: *dilution*, and b: *brown*, the F_1 offspring being phenotypically black) served as unirradiated controls, while the second litter of 65 females was irradiated.   A total of 60 irradiated litters was examined.   Pilot experiments indicated that $10\frac{1}{4}$-day-old embryos were particularly suitable for irradiation. This stage was chosen on basis of the consideration that it must be advantageous to irradiate embryos at a stage when the number of pigment-precursor cells was large enough to permit a sufficient number of mutations to occur, so as to be detectable in animal groups of suitable size.   On the other hand, the number of precursor cells must not be too large, since the progeny of the mutated cells will then become so small that the spot resulting in the adult fur may be too minute to be perceptible.   Irradiation on the morning of the tenth day following the observation of the vaginal plug turned out to be practical.   This stage was also very sensitive with regard to radiation-induced neonatal death.   A dose of 100 r was found to be low enough to reduce this complication and still give satisfactory mutation frequencies and was therefore chosen for most of the experiments.   The animals were examined in detail for mosaic patches at 32–58 days of age.   Nonmosaic animals were discarded at that time, while mosaics were saved and again observed 4–10 weeks later to insure that the spots were not transient.   The mosaic area was measured with calipers in fully mature mice, between 8 and 21 weeks of age. The total coat area to which the mosaic spots were related was not determined for each individual mouse but was taken to correspond to 45 cm.^{2} on an average.

In offspring of the C57BL × NB cross that had been irradiated with 100 r as 10$\frac{1}{4}$-day embryos, the incidence of mosaic animals was about 11 per cent, the unirradiated control figure being 0.5 per cent.   This figure was corrected for the spontaneous and radiation-induced frequency of animals with coat-color changes that could be attributed to developmental causes or to coat-color dominants, of somatic origin, as indicated by the frequency of spots with altered coat color in the homozygous C57BL × C57BL cross, untreated and irradiated, respectively.   Another deduction was made for the spontaneous frequency of animals with somatic expression of the recessives at the b, c, p, and d loci, as indicated by the frequency of mosaic animals in the unirradiated C57BL × NB F_1 hybrid group.   The corrected figure was still of the order of 10 per cent.   Statistical examination of the data indicated that the mosaic animals were distributed randomly among the litters.   The appearance of the mosaic spots suggested that in each mosaic animal the closely adjacent spots observed represented not more than a single event of change.   The areas, if there were more than one, were of the same color and the gap never exceeded 1 cm.   From the colors recorded, it was probable that the spots represented mutations at two loci at least, b and c.   The p locus was found to be involved by microscopic examination in at least one case.

From the size of the coat occupied by the mosaic spot the approximate number of prospective pigmented cells in the irradiated embryo could be calculated.   From this figure, it was possible to estimate the rate of mutation—and any other change leading to expression of the four coat-color recessives for which the animals were heterozygous— as being 7.0×10^{-7} per locus per r.   Of course, it could not be definitely decided whether the changes observed were really due to somatic mutations or to other causes. There was one particularly interesting case, however, that made somatic mutation more probable than some of its alternatives.   In this case the p locus, which is on the same chromosome as one of the other markers, c, was involved, the crossover distance being 16 in females and 12 in males.   Here the expression of p was changed, without a concomitant change in the expression of c that would have been expected if a whole chromosome were lost or somatic reduction occurred, giving rise to haploid cells. Somatic crossover could have been the explanation, provided that p was distal to c and the crossover occurred between them, or that p was proximal to c and there was a double crossover, or that p and c were on opposite sides of the centromere with a crossover occurring between them. Thus, while a few mechanisms by which the recessives may come to express themselves in addition to somatic mutation can be ruled out, others remain.[1117] Russell and Major consider somatic mutations or small or medium-sized category deficiencies as most probably responsible. It may be added, as still another category of phenomena, that although the occurrence of the change in heterozygotes, its absence in homozygotes, and its inducibility by X rays strongly suggest that it is determined at the genetic level, epigenetic phenomena[934] controlling genic expression rather than the genetic information itself cannot be excluded with absolute certainty. With these reservations in mind, together with the relative uncertainty of the estimate due to the difficulty of obtaining an exact measure of the mosaic spots, it is nevertheless interesting that the somatic rate of $7 \times 10^{-7}/r/$locus is of the same order of magnitude as the germinal rate for the same four loci (2.4×10^{-7}).[1134] Analogous results have been obtained in *Drosophila.*[778]

The most specific and regular kinds of somatic mutation and the most informative studies concerning the genetic control of the mutation rate can be found in reports on the genetics of maize and *Drosophila.* Rhoades has discovered that the a_1 gene of the A_1 series of alleles in maize, although ordinarily very stable, can be made to mutate at a high rate to other alleles in the series due to the presence of an entirely different gene.[821, 1055] The A_1 series is located on the third chromosome, and when the allele a_1 is present in homozygous form, no anthocyanin pigment is formed in the aleurone of the endosperm or the plant. However, in the presence of a dominant gene Dt, located on the ninth chromosome, it mutates to other alleles of the A_1 series which are dominant to a_1 and allow the production of pigment. This mutation occurs in the germ cells and the somatic tissues as well. In plants that are a_1a_1Dt-, colored spots appear in the aleurone and narrow strips of pigment in the plant parts. During development of the endosperm tissue the a_1 genes mutate to A_1, starting off centers of growth of colored tissue which become visible as spots when they get large enough; each spot is assumed to arise from a single mutation. The fact that the spots are usually of the same size indicates that the mutation takes place at a definite period of development. Since the endosperm tissue involved is triploid, the effects of several doses of the Dt gene can be compared. A single dose of the gene leads to 7.2 mutations per seed, whereas the corresponding figures are 22.2 for double and 121.9 for triple dose, respectively.

Extensive information is also available about the genetic control of somatic mutation in *Drosophila.* One example of this, the genetic control of somatic crossing over, has been already discussed. The other outstanding example is somatic variegation.[519, 786, 1171, 1172, 1173] This is similar to a mutational event, leading to somatically mosaic phenotypes, and associated with a special type of chromosomal rearrangement in which the position of the genes with respect to heterochromatic chromosomal regions determines their behavior.   The patches of mutant tissue appear in individuals of the proper constitution when a locus is abnormally placed next to a heterochromatic chromosomal region by means of the chromosomal rearrangement, and subsequently becomes variable in its expression.   The size and number of patches and the type of mutant allele appearing depend on "the distance of the locus from the point of re-arrangement, the kind of heterochromatic region, the quantitative relations of other heterochromatic regions in the chromosomal complement, and a galaxy of modifiers distributed through the chromosomes.   Moreover, the influence of these factors, at least in part, is exerted differentially at different stages of development: the effects are maximal at the later stages, but are not detectable in tissues differentiating early, such as the gonads."[1171]   The extent of the variegation can be changed considerably by a folic-acid analog such as amethopterin, and this can be reversed by thymidine.[1173]

Variegation in maize and *Drosophila* are phenomena of such complexity and subject to such precise genetic control that they have served as models in trying to bridge the gap between mutations and differentiation and between heredity and development.[133, 821, 1173]

Another field in which somatic mutations and selection of clones have been implicated recently is that of antibody formation.   Much interesting discussion has followed, but the relationship between the underlying changes and genetic phenomena *versus* differentiation is entirely a matter of conjecture at the present time.   The interested reader is referred to the thoughtful and provocative theoretical treatises of Talmage,[1308] Burnet,[149] Lederberg,[772] and Szilard.[1306]

NEOPLASTIC CELLS *in vivo*

Compared to normal tissues, neoplastic cells offer one advantage as tools for the study of somatic variation: their ability for unlimited multiplication under appropriate circumstances facilitates transplantation experiments and progeny tests of various kinds.   It must be kept in mind, however, that the findings are not necessarily applicable to normal cells.   Furthermore, transplantation *per se* will impose a selection pressure of its own on the cell population.   When purposefully used, such selective pressures can be helpful in concentrating and identifying rare variants that would have escaped detection otherwise.   On the other hand, continuous selection in the course of serial transfer will remodel the population, and variation as found in transplanted tumor cells must not be extrapolated, without further inquiry, to the original tumor.

With tumor cells, as with normal somatic cells, the main approaches presently available for the study of genetic variation can be registered as the use of phenotypic marker characteristics and the detailed morphologic examination of the chromosomes. In addition, the relationships between cellular phenotype and viral infection, particularly as regards tumor viruses, have received increasing attention recently, and the concept of infective heredity has entered the tumor field.

The choice of suitable phenotypic marker characteristics for the study of somatic variation is influenced by several considerations. It is desirable to have a fair knowledge of their genetic determination mechanism. This should be sufficiently variable in nature to permit the identification of a series of alternative forms. It is essential, furthermore, to have some kind of a selective device that permits the concentration of rare alternative forms (variants) from large populations of cells.

Among the phenotypic characteristics encountered in populations of neoplastic cells *in vivo*, some are common to most or all normal neoplastic cells of the organism while others are more or less distinctive for the neoplastic transformation itself. The characters in the former category studied best are represented by the isoantigens of the histocompatibility system in the mouse. These have been explored in considerable detail by several schools of transplantation geneticists, immunologists, and general biologists among whom Little,[794] Strong,[1292] Snell,[1248] Gorer,[456] Hoecker,[587] Amos,[18] Medawar,[862] and their co-workers may be mentioned particularly. It is neither necessary nor feasible to discuss these developments here; several excellent review articles of recent date are available on the subject.[129, 456, 528, 529, 862, 1248, 1249] Since these characteristics are common to all cells of the organism, they can be subjected to straightforward genetic analysis by crosses, and their genetic determination mechanism is fairly well known, at least in the mouse; other species are being studied to an increasing extent.[62, 348, 575, 576, 675]

The second category of properties would include the unit characters of tumor progression.[399, 401] Their mechanism of determination is unknown, but nevertheless they are of great interest because of their direct relationship to the phenomena of malignancy. Drug resistance, the ability to grow in the ascites form, and, to some extent, invasiveness and ability to metastasize are such characters that have already been studied in experiments designed to investigate the population dynamics of progression.

The selectivity of the experimental systems involved has not been studied to a very considerable extent; some reconstruction experiments with artificial mixtures of different types of cells are available for the histocompatibility system,[723] for some cases of drug resistance in murine leukemia,[720, 725, 1018] and for the convertibility of a certain solid murine sarcoma into the ascites form.[721] Among these, only the first permits the design of absolutely selective systems in which the selective force utilized (the homograft reaction) is able to inhibit the growth of one cellular phenotype completely while leaving an alternative form undamaged. In the other two systems, growth inhibition of one type is not complete, and the selective advantage of deviating variants is relative rather than absolute. This makes them less selective; in the particular cases investigated by reconstruction experiments, the minimum frequency of demonstrable variants has been of the order of 10^{-6} in the case of amethopterin resistance[720] and 4×10^{-5} with the ascites system.[726]

Studies on chromosomal morphology may yield direct information about changes at the genetic level. Correlated studies on biological changes and alterations of chromosomal morphology may lead to a true cytogenetics of populations of neoplastic cells. The number of such correlated studies is quite limited as yet, however.

The following sections will deal with a more detailed consideration of the various experimental systems in which phenotypic marker characteristics have been used for the study of variation in populations of neoplastic cells *in vivo*.

Isoantigens and transplantation characteristics.—Antigens are believed to reflect the specificity of genes more directly than other cellular characteristics.[1435] The genetic mechanism of determination for several isoantigenic systems is known, at least in the mouse, and this makes them particularly suitable as markers for studies on variation in populations of tumor cells. Owing to the impressive developments in transplantation genetics during the last decades, it is now well established that the transplantability of normal and tumor tissues depends on a fairly large number of genes, called histocompatibility genes, located at different chromosomal loci and subject to considerable genetic variation in all species of vertebrates in which they have been studied. In the mouse, one particular locus, called *H-2*, has been studied most extensively,[1249] and its allelic identity in the donor and the actual recipient appears to be one of the most important prerequisites for transplantability (for transplanted tumors, donor means the animal in which the tumor has originated). For this reason, *H-2* is often referred to as a strong locus. It is also well known that the various allelic forms of *H-2* determine the specificity of a whole complex of cellular components that can act as isoantigens in other animals devoid of the same components and are then demonstrable by serologic techniques.[18, 456, 587] The number of identified components determined by a given *H-2* allele is continuously growing, and probable crossovers within the *H-2* system indicate that this is actually a compound system of pseudoalleles.[985]

Histocompatibility characteristics have been studied in tumors ever since the earliest days of transplantation experiments on inbred mice. Early results of Tyzzer[1334] and of Little and Tyzzer[802] were interpreted by Little[791, 795] as indicating that the take and growth of a transplanted tumor in a new host depended on the simultaneous presence in the recipient of more than one gene, derived from the donor of the original tumor. Strong and Little[1299] have shown some years later that two tumors of closely similar origin and histology grew in distinctly different frequencies when inoculated simultaneously on opposite sides of the same animals belonging to a segregating hybrid generation and derived from an outcross between the strain of origin of the tumor and an unrelated strain. For a given tumor, this percentage of takes (later often referred to as "gene requirement") could change in the course of time. Strong[1295] has isolated two rapidly growing variant sublines from an adenocarcinoma of the DBA strain which differed from the original tumor with regard to their gene requirements. While the percentage of takes in the original line of a segregating DBA $\times$ A F$_2$ hybrid population were compatible with a requirement for six genes derived from the DBA strain, one subline gave a two-factor and the other a one-factor ratio. This was interpreted to mean that a genetic change has occurred in the variant lines. Similar changes, always proceeding in the direction of decreased specificity, have been described by Bittner[108] and by Cloudman.[208] Since the changes appeared to be sudden and were perpetuated from one cell generation to another, they were interpreted as probable mutations.[794]   Cloudman[208] has also found that three different adenocarcinomas of the breast, arising at essentially the same time in a single individual mouse, grew in individually different frequencies in segregating hybrids.   This was interpreted as indicative of differences in the genetic constitution of the three tumors.

Little[794] has pointed out that transplantation in known and controlled genetic material provides a more delicate test of physiologic and biologic differences between certain neoplasms than any other available test.   In his view, transplanation experiments demonstrating "somatic mutational changes in the genetic constitution of a tumor . . . afford a most helpful avenue of investigation on the nature and incidence of somatic mutation as a process of importance in cancer research."

Since different primary tumors of the same mice may differ in their transplantation characteristics, it is conceivable that different cell clones of the same tumor may also differ in this respect.   This has been demonstrated experimentally by Axelrad and Klein[42] who compared primary tumors with their metastases, the latter presumably originating from one or a few cells, and found them different in their transplantability to segregating F_2 hybrids, and by Hauschka and Levan[532] who studied the transplantation behavior of different single-cell clones isolated by micromanipulation from the Ehrlich and Krebs 2 ascites tumors, and found clear-cut differences in their homotransplantability to certain foreign genotypes.

The differences between primary tumors have been assumed to reflect differences in their genetic constitution.   It has been argued that since multiple primary tumors differ from each other, at least all but one of them must differ from the tissue of origin at the genetic level and a mutation or mutations must have been involved in their origin; in fact, this is one type of experiment on which the mutation theory of carcinogenesis has been built.

This interpretation of changes in the host range of tumors has been complicated by several findings in recent years.   In particular, it has been realized increasingly[20, 1237] that tumors may grow and kill their hosts in spite of weak histocompatibility differences.   When F_2 or backcross tests of the conventional type are being carried out with tumors arising in inbred strains, the ratios of animals killed by progressively growing tumors to survivors are usually in good agreement with Mendelian expectation, indicating a difference of only a few relevant histocompatibility genes between the two strains used for the outcross, seldom more and usually less than 4 or 5 (if n is the number of histocompatibility genes that have to be identical in tumor and host to permit progressive growth, $(3/4)^n$ expresses the Mendelian expectation for the proportion of animals killed by tumor in an F_2 and $(1/2)^n$ in a backcross test).   Skin grafting has, on the other hand, revealed a much larger number of histocompatibility differences, indicating the existence of not less and probably more than 15 loci.[52, 789]   Is this due to the antigenic simplification which, according to Gorer, occurs frequently in transplanted lines of tumor cells?[456]   Not always and not necessarily.   Preimmunization of the recipient hybrids with normal or neoplastic cells isogenic with the animal in which the tumor has originated may decrease the take percentage in segregating hybrids considerably[20] and in entirely foreign genotypes in which the tumor nevertheless grew if untreated hosts were used, prevent it more or less completely.[715, 1051, 1237] This reveals the presence of weak isoantigenic differences that can be overridden by tumor growth unless the hosts are preimmunized. It also suggests that a change in the transplantability of a tumor across weak isoantigenic barriers may occur in certain instances, not because of an antigenic loss, but rather through a change in the tumor's ability to reach irreversible size in spite of the antigenic difference.

A case in point is represented by the phenomenon discovered by Barrett and Deringer.[54, 55] These authors found a regular change in the histocompatibility requirements of a mammary carcinoma of C3H origin after passage through F_1 hybrid mice, derived by outcrossing C3H with another, resistant strain. We have studied the mechanism of this change and the findings indicated[725] that it was not due to the selection of preëxisting variant cells with different antigenicity but could be regarded as having been induced by some factor or factors present in the host environment of the F_1 mouse. It appeared that this change did not involve a loss of isoantigens but rather an increased tolerance to certain isoantibodies involved in the homograft reaction. This view was based on the observation that the difference between the original and the altered line disappeared when they were compared in preimmunized F_2 or backcross hybrids. The Barrett-Deringer change has so far been found to occur with weak antigenic systems only and could not be demonstrated for the strong, *H-2* barrier.

The question of why certain tumors grow progressively in genetically foreign hosts of the same species while others of similar origin and microscopic appearance do not, is not understood at the present time. Various mechanisms have been held responsible, such as quantitative changes in the relative amounts of certain isoantigens,[456] decreased antigenicity of the tumor cells due to chromosomal changes,[527, 529] and increased resistance against the homograft reaction, due to an increased isoantigenic content of the tumor cells.[354, 588] In a critical study, however, E. Klein[715] has compared the isoantigenicity of two sarcomas induced by methylcholanthrene. They were induced in the same genotype by the same dose of carcinogen and showed a closely similar microscopic picture but differed with regard to their homotransplantability to genetically different hosts. Inocula of the same size were compared for their isoantigenicity in certain foreign genotypes, subsequent to irradiating the cells with a lethal X-ray dose of 15,000 *r*. The cells were killed in order to prevent their subsequent multiplication which would have induced an important bias by itself, since the homotransplantable tumor can grow progressively in the foreign host while the nonhomotransplantable tumor will regress and the dose and temporal relationships of antigen emission will be quite different. Under such comparable conditions, both tumors provoked approximately the same antibody response as measured by isohemagglutinin formation and by the rejection of a second homograft. Homotransplantability could not be attributed to an increased resistance against the graft reaction either, since the cells of the nonspecific tumor were still highly sensitive to the second-set reaction initiated by one previous injection of heavily irradiated cells of their own kind.   Thus, while loss and weakening of isoantigenic components doubtless occurs, as will be discussed below, and can give rise to variant tumor sublines with specifically changed compatibilities, homotransplantability cannot be generally explained by these phenomena.   Homotransplantable tumors are still isoantigenic and can be so to the same extent as comparable non-homotransplantable tumors.   Thus, the phenomenon of tumor homotransplantability remains unexplained but perhaps more attention should be given to other characteristics of transplanted tumors than isoantigenicity or resistance to isoantibodies, as pointed out by Snell.[1239]   The speed of vascularization of the graft[868] and its subsequent growth rate are important factors that may profoundly influence the delicate balance between the time elapsing before the host can develop an efficient homograft reaction and the speed with which the tumor reaches irreversible size.   Also, the fascinating paradox of immunologic enhancement, characterized by isoantibody facilitating rather than inhibiting tumor growth in a foreign genotype, may be of considerable relevance and intensified studies are urgently needed on the cellular mechanisms involved.[674]

In suitable experimental systems it is possible to separate true and specific isoantigenic variation from nonspecific increase of homotransplantability.   We have been particularly interested in exploiting for this purpose some of the possibilities offered by the isogenic resistant (IR) lines of mice, developed by Snell.[1242,1238]   These IR lines have been bred with the intention of establishing a coisogenic background while maintaining an allelic difference at one of the histocompatibility loci.   Theoretically, they are homozygous with regard to their entire genome, except the histocompatibility gene in question and a chromosomal section of undefined length around it.   This expectation is not completely fulfilled in reality;[795,1238,1245] the existing IR lines are good approximations at best, and it can only be stated that the number of major histocompatibility differences which influence the fate of tumor transplants is usually not more than one. With a more sensitive indicator, such as skin, numerous other weak differences could be detected in the lines we have used; this was probably exceptional, however, and due to the particular history of these lines.[1245]

Tumors originating in the F_1 hybrid outcross of two IR lines are especially suitable to detect isoantigenic variation at the cellular level,[545,716,717,723,722,774,883,884] since variant cells arising by mutation or stable phenotypic changes that lead to the inactivation or loss of isoantigenic components, specifically determined by a strong histocompatibility factor, such as H-2, derived from one of the parental strains, may be expected to become selectively compatible with the other parental strain.   Model experiments with artificial mixtures of tumor cells derived from known genotypes have shown[723] that compatible cells can grow out selectively from populations of incompatible cells even if their proportion is very small (4×10^{-7} in the actual experiments) and even if compatibility is mostly a matter of a single difference in a major histocompatibility barrier.

A fairly large number of such F_1 tumors have been studied in our laboratory, including methylcholanthrene-induced sarcomas, spontaneous and estrogen-induced mammary carcinomas, spontaneous and estrogen-induced lymphomas, and estrogen-induced and dependent testicular tumors.[545, 716, 717, 722, 724]   Clear-cut isoantigenic variants could be extracted from a number of tumors by utilizing the homograft reaction of coisogenic resistant hosts as selective force.   All tumors tested were at the primary stage or in their earliest transfer generations.   They showed different types of behavior when tested in the parental strains or in foreign genotypes.   A few were nonspecific, either immediately upon their origin or after a certain number of serial passages.   They grew indiscriminately in both parental types and also in foreign strains. They grew less well or not at all in mice of the parental strains if these had been pre-immunized against the isoantigens of the opposite parent or in foreign genotypes preimmunized against the hybrid genotypes of the tumor.

The great majority of the tumors tested were highly specific in their transplantation behavior and grew regularly in the F_1 hybrid genotype of origin, but only seldom or not at all in any of the parental strains.   Occasional takes in mice of parental strains were of great potential interest for this study, and such tumors were regularly tested by further transplantation into both parental strains and into the other two coisogenic resistant lines as well, together with various F_1 combinations between these and the two parents.   Different types of behavior could be observed.   Some tumors failed to grow upon repeated testing in the same parental genotype in which they appeared. Others grew in a certain, often high, proportion of the same parental strain but not in the other parent or in foreign genotypes.   Tests in preimmunized mice gave interesting results.   Some variants were fully compatible with their new parental host even if preimmunized by one or several inoculations of normal or malignant tissues from the opposite parental strain.   Others grew only in nonimmunized mice and regressed in preimmunized animals.   When lines of the latter type were carried in nonimmunized hosts of the parental strain during a series of transfers, one of two alternatives could happen: the line died out after a number of passages, failing to take even in non-immunized hosts, or it became increasingly and sometimes fully compatible with preimmunized hosts.   The latter development was observed repeatedly with certain host-tumor combinations.[722]

Occasional takes were sometimes obtained in coisogenic, resistant mice outside the two parental strains (for example, with A × A.SW F_1 tumors in A.BY or A.CA mice or in various F_1 hybrids derived from crosses between A.BY or A.CA and one of the parental strains).   In spite of extensive testings, it was never possible to establish a subline that would breed true in these partially foreign genotypes.   Upon repeated testing, they either refused to grow at all, or turned out to be nonspecific and grew in a wide variety of strains, as the tumors did that were nonspecific from the beginning. Even in this case, growth in foreign genotypes could usually be prevented by previous immunization.

A series of similar tests were performed with tumors induced in an identical fashion in homozygous mice of the strains A, A.SW, or A.CA.   Takes were much less frequently obtained with such tumors in coisogenic resistant mice carrying foreign *H*-2 alleles than was the case when heterozygous F_1 tumors were tested in one parental strain.   Whenever such occasional takes were obtained, they failed to breed true in the new genotype and either died out after one or a few passages, or showed a highly nonspecific behavior.[716]

This experience with homozygous and heterozygous tumors indicated that we were dealing with at least two different phenomena.   One probably corresponds to what Snell describes as false positives, when the tumor grows in the presence of and in spite of a homograft reaction, and when its growth can be at least partially prevented by testing it in preimmunized hosts.[1237, 1239]   This result is not particularly frequent on the whole, and the liability of its occurrence varies from tumor to tumor.   It may occur with both homozygous and heterozygous tumors and also with the selected and originally specific sublines of the latter.   It does not seem to be limited to the parental strains or to certain specific genotypes, although it may be obtained with greater ease in one genotype than in another.   The latter variation may be attributed to variations in the strength of the homograft reaction, depending on the nature of the isoantigenic barrier between tumor and host.

The appearance of variants, selectively compatible with one of the parental strains, appears to be a different phenomenon, for the following reasons.   A certain percentage of takes was obtained in one of the parental strains.   When tested further, some of these tumors were selectively compatible with the parental strain in question but not with the opposite parent or any other strain.   They grew in preimmunized mice of their new strain to a varying extent, but in a number of cases they were compatible even with such hosts, either immediately or after one or a few passages in nonimmunized mice. From certain tumors it was possible to establish both kinds of variants, selectively compatible with the maternal or the paternal strain, respectively, but still refusing to grow in the opposite parent.   This result could only be obtained with heterozygous tumors and selective compatibility was restricted to the parental strains, excluding other coisogenic resistant mice carrying foreign alleles at the *H*-2 locus, and semi-isologous F_1 hybrids differing from the genotype of origin with regard to one substitution at the *H*-2 locus.

The question arose whether such selectively compatible variants grow in their parental hosts because they have lost the specific isoantigenicity determined by the *H*-2 factor derived from the opposite parental strain.   This was tested in a number of different ways.   The original tumor and its variant(s) were compared[64, 717, 724] for their ability to induce the formation of hemagglutinins[459] and cytotoxic antibodies[460] directed specifically against the *H*-2-determined isoantigenic products of both parental strains, for their ability to absorb preformed hemagglutinins or cytotoxic antibodies from isoimmune sera, and for their ability to provoke a second-set response subsequent to the inoculation of heavily irradiated cells into the variant-compatible type (tested by challenging the pretreated mice with another F_1-hybrid tumor derived from the same genotype and capable of temporary growth in the parental type in question).

In addition, K. E. Hellström[545] has studied the sensitivity of F_1-hybrid lymphomas and their variants towards various isoimmune sera, by the direct technique of Gorer and O'Gorman.[460]

All techniques gave essentially similar results. Variants which were able to grow in a high frequency of preimmunized mice of their new parental genotype did not seem to contain detectable amounts of isoantigens derived specifically from the oppposite parental strain. Less clear-cut results were obtained with variants that grew in non-immunized but not in preimmunized mice, but the opposite parental antigens were still detectable in several cases by the sensitive method of provoking a second-set response with preirradiated cells and by the direct cytotoxic test on the lymphoma cells. In several cases, subsequent serial transplantation in nonimmunized mice of the parental genotype led to an increased ability of such variant cells to grow in preimmunized mice. When tested again at this stage, the specific antigens of the opposite parental strain were no longer detectable with either technique.

In conclusion, specific variants can be distinguished from false positives by their selective compatibility with one of the parental strains, including the ability to grow in hosts hyperimmunized against the opposite parental type, and the apparent loss of H-2-determined isoantigens specifically derived from the opposite parent. Such variants have been found only with F_1 tumors so far and were limited to one of the parental strains (to the exclusion of other coisogenic, resistant mice) carrying a foreign H-2 factor. False positives are less discriminative in their host requirements, grow in many different genotypes to a varying extent, do not breed true to type on selective transfer, and are completely or partially inhibited upon testing in preimmunized mice. They occur with homozygous as well as with heterozygous tumors and are not limited to the parental strains. The tendency to give rise to false positives varies considerably from tumor to tumor.

Further experiments showed that the specific variants were highly stable even if returned to and carried in the F_1 hybrid genotype of origin in serial passage for prolonged periods of time. Their isoantigenic and transplantability pattern, whether fully or only partially compatible with the parental host, could be reproduced with great constancy even after prolonged F_1 passage. True variants were often more specific in their transplantation behavior than the original F_1 tumor from which they had been derived. This was particularly apparent with tumors capable of giving rise to variants in both parental strains; the original F_1 tumor took in both parental types with a certain frequency but variants selected in and compatible with P_1 refused to grow in P_2 hosts as a rule.[716] Attempts to switch variants compatible with one parental type to the opposite parental type by passage through newborn mice of the opposite parental strain were quite unsuccessful.

Since four isogenic resistant lines characterized by four different alleles at the H-2 locus were used in these studies, it was possible to study the question of whether a tumor originating in a certain F_1 hybrid (for example, H-2^aH-2^s) can give rise to variants specifically compatible with a semi-isologous F_1 genotype, produced by outcrossing one of the parental strains with a third strain. This did not appear possible. Whenever takes obtained in a semi-isologous F_1 were tested further and turned out to be specific variants, they were not specific for the F_1 in question but for the parental strain that entered the outcross. In other experiments, attempts were made to detect the possible occurrence of genetic recombination between different tumor cells by mixing two different, highly specific, F_1 tumors containing four different H-2 markers and growing them together in newborn mice of a fifth H-2 type, not yet capable of an efficient homograft reaction. The mixed tumor was tested in the two original F_1 types and the four new F_1 combinations in order to select possible genetic recombinants. Altogether 24 such experiments were carried out, all with negative results. This excludes the possibility that transfer of genetic fragments between host and tumor cells or among tumor cells might be responsible for variant formation; new, semi-isologous, F_1 variants would also be expected to appear if this were the case.

As another possibility, it might be speculated that variant formation could be due to the contact of tumor cells with antibody and subsequent immunologic enhancement. Such enhancement[674] can be induced experimentally by pretreatment of genetically foreign hosts with lyophilized tumor or with isoantiserum, and it empowers certain tumor homografts to grow in otherwise resistant hosts. Such growth may lead to a physiologic[674] change in the tumor, enabling it to grow for several passages in untreated hosts of the same strain, although it dies out eventually as a rule. The question arose whether enhancement, known to be mediated by humoral antibodies, is possibly involved in variant formation. In one series of experiments, animals of one parental strain were pretreated before inoculation by intraperitoneal injection of isoantiserum directed against the opposite parental strain or with tumor supernatant.[716] A total of 20 tumors were inoculated into parental hosts pretreated in this way, but only in 10 cases was a clear-cut enhancing effect registered. The enhanced tumors were transplanted further to untreated mice of the same parental strain and to other genotypes. Some of them turned out to be false positives, others were nonspecific, while still others were true variants, specifically compatible with preimmunized mice of the parental strain and capable of breeding true during serial transfer. When considering those neoplasms which could give rise to variants in nonenhanced mice in a certain frequency, it cannot be said that enhancement has facilitated the establishment of true variants to any detectable extent or in any specific way. More significantly, enhancement did not seem to promote the formation of any variant that did not also arise spontaneously in a given F_1 tumor-parental-host system. Neither did it permit the establishment of true variants from homozygous tumors, compatible with a foreign homozygous H-2 genotype, nor of variants from heterozygous tumors, exclusively compatible with a semi-isologous F_1 type derived from the outcross of one parental strain with a third strain.

Another attempt was made to check the possible rôle of enhancement in the establishment and maintenance of variants, by inoculating them into mice of the parental strain preimmunized against the opposite parental strain by a single dose of mixed spleen, liver, and kidney tissue, given intraperitoneally exactly 7 days prior to tumor inoculation. According to Kaliss,[674] when enhanced tumors are tested in preimmunized mice, they usually fail to grow, at least when the time between the immunizing (first) inoculum and the grafting of the enhanced tumor is short enough (for example, 7 days). The results showed that the variants grew equally well in both types of preimmunized mice.

Thus, variant formation appears to be different from immunologic enhancement. Although some mechanisms can be eliminated in this way, it is still very difficult to identify the mechanisms really at work. The ambiguities are due to the impossibility of distinguishing between genotype and phenotype. The irreversibility of the variants and the fact that they could be obtained from heterozygous but not from homozygous tumors are suggestive of a genetic phenomenon. However, stable changes of antigenic phenotype have been described at the cellular level, for example, in Paramecia and in bacteria in the absence of any genotypic change.[1257] Such changes, assumed to depend on "mechanisms that regulate the expression of genetic potentialities," as contrasted to "truly genetic mechanisms that regulate the maintenance of the structural information"[325] have been called "epigenetic" by Nanney.[934] Unless ordinary genetic tests can be performed, it is not possible to distinguish conclusively between genetic and epigenetic mechanisms and only indirect arguments can be applied. The only guide is that epigenetic changes are, as a rule, less stable and can be induced in a predictable way.[934]

The epigenetic explanation is rendered less likely by the high stability of the parent-compatible variants upon F_1 passage and the impossibility of reëstablishing the missing serologic phenotype in the course of this passage, or by antiserum treatment *in vitro* or by forcing the tumor through the opposite, incompatible parental strain by using newborn mice as recipients.[716] Furthermore, in some cases, the variants were demonstrated to have been preëxistent in the original F_1-hybrid, tumor population by chromosomal studies[64] and by isolating variants directly in preimmunized parental hosts for which the background growth of the original unchanged cell type was reduced to a minimum.[545, 716] The preëxistence of the variants in the original population, before it was exposed to the selective environmental conditions, makes an epigenetic mechanism less likely, since the latter is usually triggered by altered environmental conditions.

Turning to genetic mechanisms, point mutation is not very likely since several antigenic specificities were often lost simultaneously,[64, 545, 717] some of which are believed to be controlled by at least two different chromosomal sites (but see Owen[985] for a different interpretation). Also, all changes so far identified were due to losses rather than to new, alternative specificities which would be equally probable if point mutations were responsible. Mutagenic treatment with X rays and triethylenemelamine did not give consistent results,[265, 716, 717] although increased variant formation was evident with some tumors that already exhibited a spontaneous tendency to throw off variants of the type in question.[38, 265, 717]

Gross chromosomal changes of various kinds are conceivable as an explanation and within the realm of experimental approach. One such study has been made by comparing an heterozygous sarcoma with its 10 different variant sublines, all isolated in and compatible with the same parental strain.[64] With the exception of two variants, derived from the same original inoculum pool and that showed closely identical karyotypes and probably represented the repeated isolation of the same clone, all other variants differed from each other and from the original tumor with regard to their chromosomal complement. It is conceivable that chromosomal changes of various kinds have occurred in these lines, involving the chromosomal segment carrying *H-2*. Alternatively, the change may have a more indirect chromosomal mechanism, due to the remodeling of the genetic background that may lead to changes of modifiers, suppressors, position effects, and the like. It cannot be excluded, however, that the chromosomal differences observed merely reflect clonal variation in the tumor without being causally connected with variant formation. This suspicion is strengthened by the fact that methylcholanthrene-induced sarcomas, including the *MSWB* tumor used in this study, do have a wide range of chromosomal variability to begin with. Lymphomas are much less variable, and, in an analogous study, Hellström[545] could not demonstrate any gross chromosomal differences between F_1 lymphomas and their derived variant sublines.

Mitotic crossing over, discussed in detail in the section on normal cells, remains perhaps the most probable explanation. It would lead to apparent antigentic losses and no gain of new specificities. It would yield homozygous, rather than hemizygous variants. In this connection, it is of interest that in his work with lymphomas of F_1-hybrid origin, Hellström[545] showed that variants selected for compatibility with one parental strain (P_1) became similar to homozygous cells (of P_1 origin) with regard to their sensitivity to the cytotoxic action of anti-*H-2* isoantibodies (of the type P_2 anti-P_1, P_2 being the opposite parental strain entering the cross), and distinctly different from the heterozygous cells of the original F_1 lymphoma from which they had been derived, the F_1 cells showing a quantitatively lesser sensitivity. Also, as mentioned previously, the P_1 variants grew more specifically and were less liable to take in the opposite parental strain (P_2) than the original F_1 tumor, behavior similar to homozygous tumors. Analogous results were obtained for a number of F_1 carcinomas and sarcomas and their variants.[716] This, together with the total failure of persistent attempts to switch established variants, compatible with one parental strain, to the opposite parental type, again is in agreement with homozygous, rather than hemizygous constitution of the variant cells (unless the cell losing both *H-2* factors becomes inviable and thus escapes detection). Thus, on the whole, there is a phenotypic similarity between the variants and cells of homozygous origin. Unfortunately, such phenotypic similarity cannot be taken as evidence for homozygous genotype. A genetically hemizygous cell may mimic the homozygous phenotype, for example, if the products of the remaining allele are capable of structurally replacing those of the missing allele. Therefore it is impossible to decide between a phenocopy of the homozygote and truly homozygous cells.

Objections may be raised against the crossing-over hypothesis on the basis that the frequency of P_1 *versus* P_2 variants was highly unequal with some of the tumors studied. This was not true for all tumors. With some, particularly with those originating in the A × A.CA-F_1-hybrid genotype, nearly equal frequencies were obtained in many cases. The asymmetrical tumors may have genetic constitutions that make the expression of one variant less likely than the other, either because of differences in viability, or as a result of the histocompatibility differences in addition to *H-2* that separate the strains used, as pointed out previously. Such barriers may be of different strength for the two parental strains, making the expression of one variant type less likely than the other. In fact, a distinct general preference for one of the parental strains characteristic for each F_1 genotype that has been studied[716, 724] does appear when surveying a large number of tumors.

These interpretations have been discussed at some length to illustrate the vicissitudes of indirect reasoning and to emphasize even more the necessity of exploring all possibilities to develop methods of genetic transfer between somatic cells. As they stand, the data are best compatible with a genetic change, and, in the opinion of the author, among the various types of genetic changes that may come into consideration, somatic crossing over appears most probable. If this can be confirmed, it would be of considerable significance, since genetic mapping of somatic cells would be within reach through this process, as pointed out by Pontecorvo, analogous to his impressive studies on *Aspergillus*.[1011]

Whatever its mechanism, the formation of isoantigenic variants from F_1 tumors has a certain model relationship to some forms of carcinogenesis and tumor progression. It can be viewed as the loss of certain defined isoantigens that normally prevent growth in a given, incompatible, host genotype. Tumor progression and some forms of tumorigenesis (particularly those involving the endocrine system and certain types of chemical carcinogenesis) are characterized by the gradual loss of responsiveness to various superimposed, growth-controlling mechanisms in the autochthonous host. In a sense, variant formation may be regarded as an experimental model (even though an unnatural, synthetic one) of the deletion theory of carcinogenesis. Bearing the superficial nature of this parallel in mind, it is nevertheless interesting to note[722] certain analogies with the rules of tumor progression as formulated by Foulds.[399, 401] Progression occurs independently in different tumors and leads to independent reassortment of different cellular characters—different tumors show an individuality with regard to the occurrence and the frequency of given variants. Different unit characters undergo progression independently of each other—different isoantigenic variants can be obtained from the same tumor. Progression can occur in successive, distinct steps, and so can variant formation. Progression is essentially a one-way process; the permanence and stability of variant formation is equally irreversible. Whatever its mechanism, variant formation has demonstrated that stable, irreversible, and heritable changes can occur in somatic cells that lead to the loss of cellular components and thereby convey upon the cell a new ability to grow under circumstances in which its progenitor would have been checked by a superimposed, systemic mechanism.

When considering the question of isoantigenic variation in populations of neo-plastic cells, mention has to be made of some findings on human patients with leukemia, recently summarized and discussed by Salmon.[1146] Several patients have been des-cribed, all with acute leukemias, in whom antigens of the A group have been modified concurrently with the disease, leading to a mixture of two populations of erythrocytes, consisting either of A_1 and O, or of cells with normal and with weak A, respectively. No other blood-group anomalies were present. In one case, the patient had been grouped two years previously to the onset of leukemia as a normal A and only after the disease developed did a mixture of normal and weak A appear. In another case[445] the abnormal, weak-A-group erythrocytes diminished in frequency, parallel to a remis-sion of the leukemia. According to Salmon,[1146] chimerism could be excluded on genetic and statistical grounds. All cases hitherto observed concerned agglutinability by anti-A antibodies. Salmon attributed them to somatic mutations, occurring in stem cells that give rise to leukemic myeloblasts on the one hand and weak A mutants in the red-cell series on the other hand. However, it may be questioned whether a change in agglutinability really reflects a change in antigenicity. The work of Möller[887] on the development of certain isoantigens in newborn mice shows that the sensitivity of lymphoid cells to the cytotoxic action of isoantisera is subject to developmental changes, immature cells being less or not at all reactive. This does not necessarily mean a lack of antigens, since these can be demonstrated in absorption tests; the difference is one of reactivity, connected with maturity, rather than one of antigenic specificity. In this case, it is of interest to note that weak-A erythrocytes in the case described by Gold et al.[445] continued to absorb anti-A sera. Therefore it may be questioned whether the observed changes are really mutational rather than developmental. In the latter case, they would represent only another aspect of the immaturity of leukemic cells (all cases so far described were acute leukemias). This would also be in agreement with the decrease of the abnormal group during remission.

Before closing this discussion on tumor isoantigens, it seems appropriate to deal briefly with the question of tumor-specific antigens in experimental tumors, that is, antigens present only in the tumor cells but not in their host. For every such considera-tion it is absolutely essential to distinguish critically between the isoantigens of the histocompatibility system (always of importance in transplantation experiments in which donor and recipient are not genetically identical) and true tumor-specific antigens that are antigenic in the host of origin or in animals isogenic with it. For critical evaluation, some more rigorous test of isogenicity is required than the mere statement that an inbred strain was used. Skin grafting is probably the most reliable test presently available. In experiments carried out with the intention of studying the possible existence of host resistance against a tumor, it must be proved critically that the donor animal in which the tumor has originated and the recipients of the graft do not differ from each other with regard to their genetically determined isoantigenic composi-tion and possible positive results are not due to more or less subliminal homograft reactions. The experiments of Prehn and Main[1023] in particular have gone a long way to exclude the possibility that resistance against methylcholanthrene-induced sarcomas in mice, previously demonstrated by Foley,[383] was caused by residual heterozygosis in the inbred strains used. They have demonstrated that normal tissues from the very same mouse in which the neoplasm had been originally induced could not immunize against the tumor; neither could implants of methylcholanthrene-induced tumor tissue immunize isologous mice against skin grafts taken from the same animal in which the tumor had arisen originally. Resistance against a given methylcholanthrene-induced sarcoma was not necessarily accompanied by cross resistance against another one, induced in the same fashion and in the same strain, although there was evidence of incomplete cross reaction between certain tumors; thus the different tumors tended to be immunologically distinct.

Révész[1050] has confirmed the findings of Prehn and Main. He has also shown that no similar resistance could be induced against lymphomas of recent origin. On the other hand, with two lymphomas and one carcinoma of inbred strains that have been kept in serial transplantation through several years, a state of incompatibility could be readily demonstrated after the same pretreatment, apparently as a result of the isoantigenic differences that are known to develop between sublines of mice carried independently and therefore necessarily also between a line of inbred mice and its transplanted tumor carried in serial passage through long periods of time.

These experiments made it highly improbable that the resistance against methylcholanthrene-induced sarcomas was due to residual heterozygosis of the inbred strains, since such heterozygosis could only explain the results if it would exert a differential effect on the compatibility of methylcholanthrene-induced sarcoma cells on the one hand, and spontaneous sarcomas, carcinomas, lymphomas, and skin on the other hand. A last trace of doubt did nevertheless remain, particularly since there were data to indicate that different tissues may respond differently to weak differences in histocompatibility.[789, 1396] The final proof was obtained when it was demonstrated that resistance can also be built up in the primary autochthonous host from which the tumor had been excised.[727] The autochthonous hosts operated on, together with groups of isologous animals, were pretreated with irradiated sarcoma cells and subsequently challenged with increasing doses of viable cells. Untreated isologous controls were inoculated with the same doses of cells at the same time. Resistance could be demonstrated with 12 out of 16 tumors in the autochthonous and 19 out of 22 tumors in the isologous hosts. Resistance was relative rather than absolute, and it broke down when the dose of viable cells was progressively increased. Resistance against a given sarcoma did not lead to cross resistance against a different sarcoma induced by methylcholanthrene in the same genotype. Repeated inoculation of homologous, MC-induced sarcoma tissue prior to injection of carcinogen did not reduce the yield of primary sarcomas.

Perhaps the most unexpected among these findings was the individually different antigenicity of the MC-induced sarcomas. This cannot be explained satisfactorily at the present time, although it does indicate a cytogenetic individuality of the different neoplasms. This is also suggested by their individually different pattern of *H*-2-variant formation and by differences in the chromosomal picture.[544] At present it cannot be decided whether this individuality is related to the action of methylcholanthrene on some self-reproducing template system or to its possible ability to induce cytogenetic variation in general, with antigenic variation as one consequence and neoplastic change as another or, finally, to its presumed depressing effect on antibody formation whereby it would permit the establishment of antigenic clones of tumor cells (arising as a secondary consequence of the neoplastic change and representing just another aspect of the well-known variability of tumor-cell populations, eliminated from other tumor types by the host response, but achieving success in this case because of the depression of the immune response).   These problems must be attacked by additional investigation.   Meanwhile, it is of interest to note that the tumor-specific antigen discovered by Gorer[456] in lymphomas of mice (X factor) is now known to exist in three different forms in three different lymphomas.[457]

Drug resistance and hormonal independence.—Several excellent review articles of recent date are available dealing with drug resistance in neoplasms.[89, 527, 761, 762, 1019, 1378, 1406]   This phenomenon may be considered from various angles: as a population change, as a cytogenetic phenomenon, as a biochemical mechanism, or as an important factor in frustrating therapy.   For the present discussion, only some points will be emphasized which appear relevant for the use of markers for drug resistance in the study of somatic variation.

From the point of view of population dynamics, available experimental data are strongly suggestive in indicating that the spontaneous appearance of variant cells with a diminished responsiveness towards the drug plays an important rôle similar to resistance to toxic agents in microörganisms or insects.

This is hardly surprising if it is considered that large differences exist in the primary sensitivity of different neoplasms to the same drug. For every efficient drug so far studied, neoplasms can be found that are highly resistant from the beginning. Even if consideration is limited to a single group of tumors, for example, lymphocytic neoplasms, and a single group of drugs, such as the folic-acid analogs, there is a wide range of response from natural resistance to striking sensitivity.[761]   It is therefore hardly surprising that resistant variants may also appear in populations of sensitive cells. The ease with which resistance develops toward a certain drug in different tumors varies greatly, however.   To quote a few examples from our personal experience, it is very easy to make lymphoma *L*1210 resistant against fluorouracil, while in the Ehrlich ascites tumor, resistance develops with great difficulty and only after prolonged passage in the presence of the drug.[1048]   Amethopterin resistance developed easily and regularly (in the course of 3–4 passages) in several sublines started with small doses of cells from our ascitic line of *L*1210,[720] while Law[766] has found that sublines passed independently from the original solid form of this tumor and derived from the same small inoculum in the beginning, vary greatly with regard to the time of appearance of resistant variants.   In the experiments of Potter,[1019] prolonged passage in the presence of the drug was necessary before the lymphocytic neoplasm *P288* developed resistance against amethopterin.   Potter[1019] has also obtained data indicating that resistance to azaserine can develop in two different patterns in the plasma-cell neoplasm 70429. There were also differences between the way resistance against DON developed in two different lines of mast-cell neoplasm *P815*.

The variations among different neoplasms with regard to the development of resistance against the same drug are reminiscent of the differences in the probability that a given variant phenotype will develop from different tumor lines of closely similar origin and type,[716] and is suggestive of profound differences in the cytogenetics and/or epigenetics of the different lines of tumor cells.   The finding that resistance against the same drug can develop in different patterns in the same neoplasm has many parallels in the field of microbial resistance to drugs and is also reminiscent of the finding that the same variant phenotype can be selected in one single step or through intermediate steps from the same neoplasm.[722]

It may now be asked what evidence there is to show that resistance to drugs in neoplastic cells really has a variation-selection basis as postulated.   The following observations are relevant.   Once established, resistance is usually stable and irreversible, even in the absence of the drug.   Resistance is sometimes attained suddenly, but more often progressively, suggesting a series of changes,[1406] and in certain instances discrete, stepwise increases of resistance have been noted.[766]   Particular significance has been attributed to the fluctuation test of Luria and Delbrück,[812] modified for this problem by Law,[766] who carried out the experiments with leukemia *L1210* in DBA/2 mice, using a transplantation system *in vivo*.   Two different sets of tests were made: one in which leukemic cells were taken from fifteen independent sublines of leukemic cells (originally started from a small, common, cellular pool) and one in which ten repeated tests were made with leukemic cells derived from the same subline.   All groups received a standardized inoculum subcutaneously and were given 2.5 mg. of amethopterin per kilogram body weight every other day for 4 doses.   At 9 days postinoculation, all mice were sacrificed, and the mean weight of lymphomatous tissue was determined for each group.   It was found that this measure showed extreme variability between those groups which represented the independent sublines, while the variability was small and distributed in a uniform probability pattern in the groups inoculated with duplicate samples from the same subline.   The differences in the weight of lymphomatous tissue after amethopterin treatment was interpreted as reflecting variations in the number of resistant cells in the different sublines.   If this interpretation is correct, highly significant fluctuations in the number of resistant cells in the independent subline group, started from a common source, would strongly favor the assumption that the appearance of resistant cells is due to random intercellular variation not related to any inducing action of the drug.   However, there is one point in this experiment which is not entirely convincing.   The fifteen independent sublines were tested by inoculating fifteen different cellular suspensions prepared artificially in Locke's solution.   In contrast, aliquots of the same cellular suspension were used for making repeated tests from a single subline. Different suspensions of cells may differ greatly from each other with regard to their physiologic quality and with regard to the fraction of the population capable of continued multiplication. This may in turn result in variations in the amount of lymphomatous tissue found in different groups 9 days after inoculation. In general, the fluctuation test always presupposes that all environmental and preparative conditions are equal for the lines carried independently and for the duplicate samples of the single line. Since this does not seem to have been the case, it remains uncertain to what extent the fluctuation found reflects the actual variation in the number of preformed resistant cells.

Of great interest are the recent experiments *in vitro* of Szybalski,[1307] also based on a modified fluctuation test, and dealing with azaguanine resistance of cells *in vitro*. The strain, designated D98S and originally derived from a single-cell isolate of an established tissue-culture line originating from human bone marrow carried in tissue culture, formed well-defined colonies of uniform size, composed of epithelial-type cells and firmly attached to glass. Since such colonies contained approximately equal numbers of cells, they were regarded as analogous to the series of separate test-tube cultures used in Luria and Delbrück's fluctuation test. A standard inoculum, containing approximately 100 cells, was plated in drug-free medium. At various intervals after inoculation, four plates were selected at random. In two of these, the cells were fixed and stained while the remaining two were exposed to various concentrations of azaguanine. The two first plates were used for determining the average colony count per plate, S (reflecting the inoculum size and the plating efficiency), and the average number of cells per colony, N. The other two plates were incubated for an additional period of 10–12 days, with replacement of the azaguanine-containing medium every 3–4 days. The number of colonies arising from resistant mutants (R) were scored after fixation and staining. The expression $(S - R)/S$ was employed to calculate the proportion (P_0) of the primary colonies that did not contain resistant mutant cells after the initial incubation and were therefore sloughing off the glass after the addition of the selective medium. On the basis of these data, the average mutation rate to azaguanine resistance was calculated from the equation of Newcombe[942] as $4.9 \pm 1.8 \times 10^4$ per cell per division cycle. This figure was valid for azaguanine levels between 4 and 8 γ per ml., whereas there was an apparent twofold increase of the mutation rate when the level of drug was lowered to 2 γ per ml. The variants resistant to azaguanine were inhibited by 8-azaguanosine, but mutants resistant to this compound could be selected in an additional step. With the same method, the mutation rate from azaguanosine sensitivity to azaguanosine resistance was determined as 1.2×10^{-6} per cell per division cycle. No direct mutations from azaguanine sensitivity to azaguanosine resistance could be detected in populations as large as 10^{-7} cells.

This approach appears to be very promising and applicable to many other similar problems. With regard to the argument of preadaptive mutations, that is, appearance of mutants in the absence of the drug, it is highly suggestive, although the final proof will have to be obtained, as in microbial genetics, by replica plating or some other method of indirect selection.[186, 776]   Another approach concerned with the relationship of somatic-cell variation to drug resistance is the study of the karyotype.   It has been observed repeatedly[89, 527] that the appearance of drug-resistant lines may be accompanied by changes in modal number of chromosomes or with regard to visible chromosomal markers, but this was not a regular finding and it was not clear whether the variation exceeded what could have been expected in a series of unselected, randomly isolated clones.   In the study of Biesele *et al.*,[89] all five amethopterin-resistant sublines of leukemia *L*1210 lacked a large submetacentric chromosome that was present in the modal cellular type of the parental line and four other amethopterin-sensitive sublines.   It was pointed out that the relation between resistance and the loss of the marker had a low probability of being fortuitous.

Recently, Vogt[1349] published some interesting studies particularly stressing the problem of whether the properties of drug resistance for the various lines of cells are causally related to the differences in their karyotypes or whether they are only an additional expression of the variability between tumor cells.   She studied two clonal lines of *HeLa* cells *in vitro* and compared them to variant sublines adapted to growth on suboptimal media or in the presence of the drug, aminopterin.   Each karyotype was classified by the total number of its chromosomes and by the number of large chromosomes corresponding to the three largest pairs in the normal idiogram.   Two sublines adapted to growth under suboptimal nutritional conditions and three sublines showing an increased resistance to aminopterin all had karyotypes which had not been observed in the parental lines.   Also, clones selected for variant cellular or colonial appearance differed in their karyotypes from the parental lines and one another.   Special significance was attributed to the finding that lines selected and maintained under strong selection pressure also maintained a stable karyotype.   This was taken to indicate that the selective value of a cell depends on its karyotype.   This could be best explained on the assumption that the karyotype determined the phenotype.

While these findings support the assumption, so convincingly proved with various microörganisms, that the variation underlying the development of resistance to drugs in tumors is probably localized at the genetic (that is, DNA) level, they do not prove it. It may be well at this point to place emphasis on the truism that tumor cells are not microörganisms.   Neither the stability of the changes nor the apparently random appearance and clonal growth of the resistant variants does critically prove that changes at the genetic level are responsible.   When dealing with cells of higher organisms, even if neoplastic, it must be realized that they may be subject to developmental processes which, although still among the most obscure in biology, are regarded by the majority of workers as not being akin to mutations at the genetic level.   They are usually regarded as epigenetic changes *par excellence*, caused by modifications of genic expression rather than by a change of the genetic information itself.   A progressive limitation of genic expression may occur in the course of differentiation; this "may be of such a nature that in each cell only a special complement of loci maintains its specific activities."[1173]   Such limitations may concern, for example, the enzymes of nucleic-acid synthesis, and there is experimental evidence to indicate that different tissues in the same species may differ in their ability to utilize preformed pyrimidines.[1378] If such differences can arise in the course of differentiation, there is no reason why they could not arise in tumor cells by processes akin to those governing differentiation and not necessarily corresponding to mutations at the genic level.   If differentiation normally operates by limiting genetic potentialities, some of the enzymatic losses accompanying the development of resistance to drugs in tumors[137, 1048] might be epigenetic rather than genetic.   This question cannot be approached experimentally at present and there is little likelihood of any substantial progress in this regard until some method of genetic analysis can be worked out for somatic cells.

Another field, closely related to drug sensitivity and, in addition, of considerable relevance for the problems of tumorigenesis, cellular autonomy, and somatic variation, deals with the dependence of neoplasms originating in endocrine organs or their target tissues on the original hormonal imbalance that induced them, this being usually an overstimulation by a given hormone or the lack of a normally inhibiting hormone.   This field has been reviewed repeatedly,[87, 401, 412, 414, 413, 421, 917] and only some points relevant to the present discussion will be emphasized.   It appears that prolonged stimulation of certain target tissues with superimposed hormones often leads to neoplasms the growth of which at first depends on continued stimulation.   These forms are usually called dependent or conditioned tumors.   The same result can be achieved by the removal of an inhibiting hormone; the thyrotropic pituitary tumors in mice arising after the removal of the thyroid illustrate this.   When observed for extensive periods of time and, if necessary, through serial transplantation, it is the rule rather than the exception that such tumors change to a more autonomous state, or show a tendency to throw off autonomous variant sublines.   These no longer require the hormonal stimulus or the absence of the hormonal inhibitor that was originally needed for tumor induction and maintenance.   This progression to autonomy is random, unpredictable, although bound to happen sooner or later.   The probability of its occurrence varies from system to system, and, sometimes, from tumor to tumor within a single morphologic and etiologic group.   In certain systems, a series of successive steps can be distinguished, from full dependence, through decreasing levels of partial dependence, to full autonomy.[32, 413] There is a certain resemblance between these phenomena and resistance to drugs, and, when viewed together, it is difficult to escape the conclusion that, whenever it is possible to influence a population of tumor cells by natural or artificial, stimulating or inhibiting, chemical or hormonal factors, it will exhibit a varying degree of plasticity, tending to escape from the limiting factor.   Here, as in the case of drug resistance, it is most plausible to assume that evolutionary mechanisms of the variation-selection type are responsible.   Unfortunately, no case of hormonal dependence has yet been subjected even to the preliminary type of analysis from the point of view of population dynamics that is available for some cases of drug resistance.   It is the opinion of the reviewer that this is an important and potentially fertile field for investigation, particularly suitable for the application of fluctuation tests, model experiments with artificial mixtures, and combined *in vitro–in vivo* approaches.   Endocrine tumors do not differ from other neoplasms in principle, but they have one very important advantage: with most other tumors nothing is known about the superimposed growth-controlling devices that regulate cell division in the normal tissues of origin, although they must exist.   For this reason, it is difficult to assess the dependence or independence of the neoplastic cells in relation to the controlling forces, except in the most vague and uncertain terms.   While the full story of the control of growth is probably not known for any single endocrine tissue either, at the present stage of our knowledge it must be regarded as a tremendous advantage that something is known, and cellular multiplication can be made to depend on a single, well-characterized hormone in such systems. As pointed out repeatedly by Furth,[412, 413] endocrine tumors are therefore useful models for the study of cell dependence and autonomy in relation to growth-controlling mechanisms.

Meanwhile, it may be wise to postpone hasty conclusions about the rôle of variation and selection in the development of hormone independence.   Many attractive models can be borrowed from the field of microbial genetics, and there is no doubt that the evolutionary mechanism seems intellectually the most plausible at the present time.   Just because of this rationale, caution has to be advocated against reasoning by analogy in the absence of any experimental evidence.   Some rather preliminary information does indeed suggest that other factors, such as population size, may have a bearing on hormone dependence *versus* independence at least in special cases.[725]   A certain hormonal stimulus may be required for growth while the number of tumor cells is small, but this may no longer be necessary after the population has reached a certain critical size.   It can be speculated that tumor cells might be capable of manufacturing growth-stimulating substances that have to reach a critical concentration and can then replace the requirement for the exogenous hormonal stimulus.   Model experiments involving the mixture of metabolically active but reproductively inactive, X-irradiated cells with a small number of viable cells provide support for such a concept,[1051] suggesting that a progressional change may sometimes occur as an automatic consequence of increasing population size.   In another, as yet unpublished study, some estrogen-induced and dependent, interstitial-cell tumors of the testis were studied in our laboratory by serial transplantation into hosts with the genotype of origin.   Four tumors were strictly dependent on estrogenic stimulation.   They grew well upon transfer to estrogen-pellet-bearing males and females and more slowly but regularly in untreated females. They consistently refused to grow when inoculated into untreated males.   After they have reached a certain size in estrogen-treated males, the pellet could be removed, however, without influencing the further growth of the tumor.   The tumor continued to grow although it did not become hormone independent, because on the next transplantation the same dependence was apparent as previously.   The fact that the behavior of the large population does not breed true strongly indicates that we are dealing with a physiologic mechanism, perhaps involving self-stimulation of growth of the tumor by its own products, which would make the cells more self-sufficient and less dependent on exogenous stimulators and inhibitors. The biochemical mechanism or mechanisms involved are entirely unknown and their study would appear to be of considerable importance.

Ascites conversion, cellular adhesiveness, and related characteristics.—When 42 different murine tumors were compared with regard to their ability to grow in the ascitic form,[721] it was found that they differed in convertibility, defined as the capacity of tumor cell implanted intraperitoneally to proliferate as suspensions of free cells or clumps of cells, or both, in the peritoneal fluid and to reach there a high absolute level (of the order of 10^8 to 10^9) and relative concentration (70 to 90 per cent of all cellular types). Thirty tumors were not convertible under the experimental conditions; this group included all highly differentiated and slowly growing tumors tested. The 12 tumors that proved to be convertible were all rapidly growing and anaplastic. However, there were some rapidly growing and anaplastic tumors in the inconvertible group also, indicating that a high rate of growth or low degree of differentiation, or both, were necessary but by themselves not sufficient conditions for convertibility.

Among the 12 convertible tumors, 6 could be converted immediately, that is, a suspension of solid-tumor cells gave rise to typical ascites tumors immediately after their first intraperitoneal inoculation. Six other tumors were not convertible in the beginning but could be adapted by prolonged selective transfer of peritoneal exudate containing tumor cells. This gradual conversion was reproducible for a given tumor and has been subjected to closer analysis.[713, 714, 726] The converted line turned out to be different from the original in a stable and irreversible way. Even after having been returned to and carried in the solid form for a long series of prolonged passages (67–98 serial transfers with different tumors), the converted sublines maintained their characteristic ability to give rise to typical ascites tumors immediately upon their first intraperitoneal inoculation, while the original sublines, maintained by serial subcutaneous transfer, had to be subjected to a long series of adaptive passage by serial exudate transfer before becoming convertible. Otherwise, these biologically different pairs of sublines were closely similar in microscopic appearance and with regard to their content of nucleic acid. They differed, however, in their ability to metastasize to the lungs.[713, 1057] The adapted line metastasized earlier, more extensively, and in the form of dissociated and widely disseminated free cells, in contrast to the nonadapted line that metastasized at a late stage only, and then in the form of compact intravascular emboli. The adapted line of cells infiltrated the subcutaneous tissue more, exhibited a lower degree of cellular adhesiveness, and had a higher negative electrical charge of the surface.[1031]

As in the cases previously discussed, the question arose about whether the tumors gradually convertible underwent a process of selection in the course of the serial exudate transfer whereby a preëxistent variant fraction, possessing higher ability to become established in the ascitic form than the rest, was merely concentrated or whether this was a case of true cellular adaptation induced by prolonged exposure to the exudate environment.   Experiments carried out with the specific purpose of studying this question[713] supported the selection of preëxistent variants.   The main point was that convertible or inconvertible lines could be produced from the same original tumor at will, simply by modifying the size of the first inoculum in the beginning of the conversion experiment.   When very small inocula were used, convertible lines could be changed into inconvertible.   This can only be explained by assuming that the variants were already present in the original population of the solid tumor in a small proportion. Extrapolation of model experiments with artificial mixtures led to a tentative estimate of a probable frequency of the order of 4×10^{-7}.

These studies may be considered together with the earlier work of Coman,[218] who found that tumor cells were characterized by decreased cellular adhesiveness, and the more recent results of Ambrose *et al.*,[17] who showed that their surface carried an increased negative electrical charge when compared to homologous normal cells. Also relevant are the results of Abercrombie and Heaysman,[1] who observed that surface movements of normal fibroblasts in culture will be inhibited upon contact with other fibroblasts, leading to adherence of the cells, while such contact inhibition was largely or completely absent in cultures of sarcoma cells.   Taken together, all these data point toward a relationship between changed surface characteristics of neoplastic cells and their tendency to invade and metastasize.   Also suggesting a surface change are the findings of Wolff and his school,[1399] who showed that dissociated embryonic cells of different species associate specifically *in vitro* and build up real organ chimeras that contain a mixture of cells derived from the homologous tissues of the different animal species used.   On the other hand, malignant cells appeared to be devoid of such a surface-recognition factor since they invaded all kinds of embryonic tissues without discrimination when tested in a similar system *in vitro*.   The fact that it was possible, in the experiments on ascites conversion discussed in the previous paragraph, to select tumor sublines that differed from each other with regard to their surface characteristics as well as their ability to invade, to metastasize, and to proliferate in the ascitic form, indicates that the surface characteristic in question is not an all-or-none phenomenon but may be expressed to different degrees and is liable to undergo progression in established tumors through variation and selection.   It is to the merit of Foulds[399, 401] to have emphasized clearly that progression can proceed independently with regard to different cellular characteristics.   Changes in surface and invasive properties do not necessarily have to be correlated with changes in other qualities of the neoplastic cells, such as rate of growth, hormonal dependence, antigenic behavior, and the like. On the other hand, recent work on tumor viruses *in vitro*, particularly the associations between Rous virus and chicken fibroblast[1080] or iris epithelium[326] and the system of polyoma virus and fibroblast in hamster and mouse[1350] has demonstrated that a tumor virus may impose the entire neoplastic phenotype, including the ability to grow as a tumor upon reimplantation into an appropriate animal host, upon the recipient normal cell within a comparatively short time.   Simultaneously with this change, the surface properties and the mutual relationships of the cells are altered, as indicated by their tendency to grow in multiple layers rather than in monolayer.   Thus, it seems that the same phenotypic end result, the invasive cell, can be reached through one single step of change or through a series of multiple steps.   This is also true for some other types of progression: one of the progression rules of Foulds[401] actually stresses the existence of alternative paths of development.   In his words: "One lesion . . . may develop in several different ways by direct paths or by indirect paths traversing intermediate stages or precursor lesions.   Progression may consist of steplike advances along one path of development, or it may bring about a decisive change of path."   In genetics, it is almost a commonplace that mutants of identical or highly similar phenotype may result from changes at quite different loci; and roughly similar end effects may be caused by disturbances at any of the steps along a biosynthetic pathway.[1173]   Thus, no specific type of etiology such as serial mutations or viral oncogenesis is necessary as the exclusive cause of a certain cellular phenotype, such as the invasive cell with altered surface properties.

Other phenotypic markers studied in neoplastic cells in vivo.—Many other characteristics have been studied in tumor cells *in vivo* that are potentially useful as phenotypic markers although very few have been actually employed in experiments aimed directly at the selection of distinct sublines with different marker characteristics.   The work of Hesselbach[553] is of interest.   She showed that both pigmented and nonpigmented tumors could be derived from a melanoma in mice by selective transfer of pigmented and non-pigmented tissue, respectively.   Genetic factors controlling host melanization did not have any effect on the melanization of the tumor cells borne by the host; pigmentation appeared to depend on the inherent properties of the tumor cells themselves.

Another interesting phenotypic property is the ability of plasma-cell tumors to produce individually distinctive abnormal proteins, both in mice[332] and men.[1032] In mice, transplanted plasma-cell neoplasms maintained their individual characteristics through repeated generations of transfer.[332]   Although they had been derived from the same type of cell within the same inbred strain of mice, these lines of tumors were not equivalent, and their protein products could be clearly differentiated.   Although the nature of these proteins and their relationship to the normal serum proteins is not clear, it is tempting to seek a relationship to antibodies, presumably produced by normal plasma cells.   If there is such a relationship, there is a striking resemblance between the fact that different malignant lines of plasma cells produce individually specific substances and the postulate, inherent in the recently developed clonal selection theories of antibody formation,[149, 772, 1308] that the precursors of the antibody-forming cells should show a high degree of genetic diversity, arising from a high spontaneous mutation rate, and that each cell spontaneously produces small amounts of the antibody corresponding to its own genotype.   Mature cells are expected to proliferate extensively under antigenic stimulation and produce large clones of cells genetically preadapted to produce homologous antibody.   The analogy between antibody formation and the appearance of abnormal proteins in plasma-cell tumors must not be pursued too far at the present time, but if such an analogy exists, experimental plasma-cell tumors which can be expected to lend themselves excellently to the isolation of single-cell clones *in vivo*,[415, 532] may become important tools for the study of such problems as the ability of a single cell to produce two different specific proteins at the same time, and the mutability of product specificity in untreated cells and after exposure to X rays and various mutagenic agents. In the words of Burnet:[149] "The multiple myeloma findings provide the best possible material for displaying the salient features of the clonal selection approach to the phenomena of antibody production and malignancy. The globulins produced in multiple myelomatosis are abnormal but they remain consistently of the one physical and chemical pattern which has been, as it were, chosen from a probably unlimited number of possibilities. This constancy of pattern can hardly be interpreted in any other fashion than as a genetically controlled quality of the clone."

The findings on the plasma-cell tumors represent another example of the individual distinctiveness of different neoplasms of closely similar origin, already indicated by the previously discussed facts regarding the individual antigenicity of methycholanthrene-induced sarcomas and certain lymphomas in their autochthonous or isologous hosts. Such an individuality is also suggested by the majority of studies on chromosomes. Since the latter subject is discussed by Dr. Yerganian in another chapter, it will not be reviewed here in detail. It may be pertinent to mention some points, however, with regard to chromosomal studies that appear to be relevant for the present discussion.

Chromosomal markers are especially useful to establish the identity of cell lines and to reveal possible contaminations.[1076, 1319] They are also valuable for the study of genetic variation in populations of tumor cells and, when combined with the experimental isolation of single-cell clones,[532, 841] they can yield information about such problems as the ability of the various karyotypes to become established as clones, the relationships between biological and chromosomal variation, and the speed with which morphologically detectable chromosomal variation is being reëstablished within the originally homogeneous population of a clone. Since cells of different chromosomal constitutions can display different sensitivities to the homograft reaction within the same tumor-cell population,[531, 533, 690] it is possible, in favorable cases, to establish tumor sublines, differing in their modal chromosome number, by immunoselection from the same neoplasm.[531, 690] These can be used as tools for studies on the relationships between number of chromosomes and various biological characteristics. Such comparisons are more rigorous and reliable than conventional studies on unrelated tumors differing in numbers of chromosomes and many other additional characteristics. Comparisons of this type have been carried out with regard to parameters of growth[530] and radiosensitivity,[1052] and they have revealed information that was not available from previous analogous studies on unrelated tumor lines.

Viral tumors and infective heredity.—Recent advances in virology and particularly in bacterial and phage genetics have tended to bridge the conceptual gap between mutation and viral infection, and it is now a matter of taste whether to consider bacteriophages as genes that function at the parasitic level or as parasitism at the genetic level.[775]   These developments have also greatly influenced thinking about viral tumors, and many attractive models have been offered in which the possible mechanism of viral oncogenesis is considered in the light of the information obtained about such phenomena as transduction and lysogenic conversion.[810, 811, 1380]   It has been repeatedly emphasized that the contrast between the somatic mutation and the viral causation theory of cancer is probably more semantic than substantial.   As pointed out by Luria,[811] the concept of infective heredity is mainly based on the following three groups of findings: "the central and often exclusive rôle of viral nucleic acid in initiating virus infection; the interactions between viruses and genetic constituents of the cell; and the viral control of cellular functions through determination of the structure of specific proteins."   In Luria's view, viral infection may be considered as a class of cellular mutations, in which the primary change, the entry of the viral genome, is a genetic change.

Perhaps most suggestive of an analogy with viral tumors are the cases of lysogenic conversion in which the genome of the converting phage contains determinants that control cellular properties, such as surface antigens, which have no apparent relation to the viral function of the phage.   Cases at least superficially similar to this situation, in which viral onocogenesis has led to a new phenotypic cellular property, in addition to the characteristics recognized as related to the neoplastic change itself, have been discussed recently by Luria.[811]   He pointed out that such cases deserve particular attention in view of the fact that the amount of genetic information in many animal viruses must be quite limited and therefore any new cellular function that can be shown to be directly controlled by viral genes has "a significant chance of being *the* key function in the tumoral transformation."   One such case is the appearance of arginase in cells infected by Shope papilloma virus.[1069]   This enzyme is absent from normal rabbit epidermis but present in papilloma cells and in the cancers that originate from them.   The host animals contain a precipitin in their serum, reacting with the tumor arginase that is absent from the control animals.   Another analogous case has been described recently by Zilber[1469] who found a specific antigen in cells of the Rous sarcoma, not present in the Rous virus itself or in methylcholanthrene-induced sarcomas in chickens.   A similar situation is suggested by the recent findings of Sjögren et al.,[1215] who showed that mice preimmunized against the polyoma virus were resistant against the transplantation of polyoma-induced tumors that have arisen in isologous mice and were capable of growing in untreated isologous hosts.

This is not the suitable place to enter into lengthy hypothetical discussions about the relationship between the virus and the cellular genome in virus-induced tumors. Only one particular view will be quoted, which has been put forward recently by Vogt and Dulbecco,[1350] who studied the interactions of polyoma virus with cells of mouse and hamster *in vitro*.   They found that the virus could give rise to two types of virus-cell interaction: a cytocidal interaction, leading to extensive viral synthesis and cellular degeneration, and a moderate interaction leading to the transformation of the cells into neoplastic cells, usually unable to produce detectable virus and resistant to superinfection with the same virus.   In cultures of cells from the mouse, the cytocidal interaction was most frequent, while in cultures of cells from the hamster the moderate interaction was almost exclusive.   They explain these results by the following hypothesis: "Upon entering a cell, the virus assumes an 'uncommitted' state in both hamster and mouse cells, during which it may undergo a limited multiplication without appreciably affecting the normal properties of the cell.   From this 'uncommitted' state, the virus has the choice of entering either the state of cytocidal multiplication or the integrated state. The late transformation of the mouse cultures would be due to the selection of a few transformed cells.   In the hamster cultures, on the other hand, the choice would be almost exclusively toward the integrated state, both in the animal and in the tissue culture."[1350]   As to the state of the virus in the transformed cells, the absence or low level of production of virus and the resistance to superinfection were similar to the properties of lysogenic bacterial culture, suggesting that the integrated virus exists as a provirus.   However, as Vogt and Dulbecco pointed out, other explanations are not excluded, and it is conceivable that the virus could ultimately be lost from the transformed cells and resistance to superinfection might be a secondary consequence of transformation.

It is obviously of the greatest interest to investigate the relationship between viral and cellular genome more extensively, not only because of its significance for the understanding of relationships between virus and cell in general and viral oncogenesis in particular, but also from the point of view of somatic-cell genetics.   Provided that a truly integrated state really exists, it may permit the development of methods for genetic transfer such as already exist in microbiology in the form of transduction and lysogenic conversion.

Some general considerations.—To close this section on neoplastic cell populations *in vivo*, it has to be emphasized that the variability of such markers as isoantigens or drug resistance does not necessarily reflect any tumor-specific property.   Comparable information on normal cells *in vivo* is not available.   Neoplastic cells represent special types of somatic cells and while certain markers may display a similar variation in comparable normal and neoplastic tissues, others may not; this will have to be established from case to case.   For instance, it is not known whether normal bone-marrow cells can develop resistance against antileukemic antimetabolites in the same way as leukemic cells do.   With leukemic cells, serial transplantation in the presence of the drug is often necessary to establish a line resistant to a given compound.   Only seldom does resistance become apparent in the course of treatment in the first host, which is usually killed by the background growth of sensitive cells.[447]   Bone-marrow cells would have to be exposed under the same conditions, that is, during serial transfer, in order to investigate whether they can become resistant or not.   Such serial transfer is quite feasible through lethally irradiated mice; the marrow cells remain normal and do not become leukemic under such conditions.[385]   This system would also permit the study of isoantigenic variation in bone-marrow cells in experiments similar to those carried out with various types of neoplastic cells.   Another approach would be to compare isoantigenic variation in early dependent forms and in late autonomous forms of the same hormone-induced endocrine tumor.

Another point, already emphasized in previous papers,[719, 720] is the possibility of analyzing the various phenotypic characteristics entering into the complex phenomenon of malignancy, by selecting specific pairs of tumor sublines for comparison. The random assortment of various unit characteristics such as growth rate, degree of differentiation, invasiveness, ability to spread, dependence on superimposed hormonal and other controls, sensitivity to antimetabolites and other growth-inhibitors during tumor progression [399, 401] is strongly suggestive of independent determination mechanisms. If they are really partially or completely unrelated, it would follow that these units, rather than malignancy as a whole, should be studied separately in clear-cut systems. It is increasingly realized that conventional biochemical or other comparisons between malignant tissues (often used after long periods of serial transplantation) and their homologous, normal counterparts suffer from serious drawbacks. Representativeness is one of the problems; a given normal tissue may contain the normal ancestor cell of a given tumor in a low frequency only; pulmonary tumors and normal pulmonary tissue, or hepatic tumors of the cholangioma type and normal liver, are cases in point. Even if this source of error could be minimized, the tumor may still contain considerable amounts of irrelevant material, such as inflammatory infiltrates, fibrous stroma, and areas of necrosis. These may seriously interfere with the meaningfulness of the comparison.

Even if these risks could be avoided and homologous normal and malignant tissues were obtainable in highly purified cultures and free of necrosis, the stepwise and multiple nature of tumor progression, apparently involving different kinds of cellular changes, would make it difficult to correlate biochemical or morphologic differences with any one of the unit characteristics involved in the development of malignancy. Even such experiments may, therefore, lead to variable results and may fail to demonstrate general differences between normal and malignant cells; but are there any general differences to be expected? The tacit assumption of a common cellular pathway of carcinogenesis, implicit in many experimental designs, may be nothing but wishful thinking, as pointed out by Berenblum.[80] Perhaps it would be more reasonable to seek for cellular mechanisms that can explain, for example, the change of a conditioned tumor of an endocrine organ, caused by a certain hormonal imbalance of the host rather than by any cellular change, into its autonomous counterpart, no longer dependent on the original hormonal imbalance and caused by some change in the cells themselves.[412, 413, 414] By comparing dependent and autonomous sublines of the same original tumor, if preservable indefinitely by frozen storage, the difficulties discussed above might be largely eliminated. The populations of cells would be more comparable, and differences between them would be more liable to be correlated with the biological change than in less related populations. The biological characteristic itself is sharper and better defined than malignancy or normalcy, although clearly related to one of the most essential features of malignant behavior: the changed responsiveness to the superimposed homeostatic mechanisms of the organism.

As discussed in previous sections, it is also possible to establish, by techniques of selective transfer, parallel sublines from the same original tumor differing in chromosomal characteristics,[530, 531, 690] in isoantigenicity,[531, 717, 724] in the capacity to grow in the ascitic form, and with regard to correlated phenomena of invasiveness and ability to metastasize,[721, 1031, 1057] in hormone production,[412, 413, 414] and in drug resistance.[761, 762]   Differences between such specific pairs may reveal more about underlying mechanisms than comparisons between unrelated populations of normal and tumor cells.

TISSUE CULTURE

The recent spectacular advances in tissue culture leading to the development of cloning techniques, partially or fully defined media, suspension cultures, and highly improved procedures for chromosomal studies, bring *in-vitro* methods into the forefront of interest in every discussion on somatic-cell genetics.   It would be impossible to review these developments here in detail; consideration will be restricted to a brief survey of available marker characteristics and discussion of the relationships between systems *in vivo* and *in vitro*.

Marker characteristics.—The conscious, large-scale selection of cellular variants differing in phenotypic marker characteristics has been mainly initiated by the development of cloning techniques.   The growth of single, isolated cells *in vitro* was first achieved by Sanford, Earle, and Likely.[1150]   Convenient, large-scale procedures were developed by Puck and associates,[1025] and various new methods or modifications were introduced by others with the purpose of combining the ease and large-scale operation which is characteristic of Puck's technique with the rigorously critical assurance of single-cell origin inherent in the method of Sanford *et al.*[35, 449]   The markers studied include resistance to drugs, viruses, and radiation and nutritional, morphologic, and chromosomal characteristics.

Puck and associates[1025] have isolated clones of *HeLa* cells differing from the majority of the population with regard to their appearance, requirements for human serum, or resistance to Newcastle-disease virus.   As an interesting example, the case of lines *S*1 and *S*3 may be quoted.   When the standard medium was supplemented with 3 per cent human serum only, the plating efficiency of *S*3 was 100 per cent and that of *S*1 was zero.   The *S*1 cells displayed a cloning efficiency of 100 per cent, however, if plated on a feeder layer of irradiated *S*3 cells metabolically active but incapable of further multiplication.   The behavior of these lines was stable during continuous passage.   Other types of stable variants have been produced by X irradiation of *HeLa*-cell populations.[1025]   Some of these were morphologic variants, throwing off bizarre cells, and enlarged monstrosities, and this behavior was bred true during prolonged passage.   Other X-ray-induced variants were nutritionally different from the original type.   With regard to spontaneous or X-ray-induced variants that were recognizable by a difference in their clonal appearance, Puck stressed particularly the importance of distinguishing between reversible changes, inducible at will by certain variations in the composition of the medium, and stable differences, capable of breeding true upon repeated culture. Examination of the chromosomal complement of different clones, isolated from highly aneuploid populations, show that they may differ with regard to their karyotype and these differences tend to remain fairly stable and permanent in the course of serial passage. A semiannual recloning was found sufficient to maintain the stability of a clonal stemline.[1322]

The nutritional requirements of various standard established strains of cells *in vitro* are surprisingly similar.[314] Only a few of them exhibit some special, unusual requirements; these have been summarized recently by Eagle.[314] For this reason, it is of considerable importance to select clear-cut nutritional variants experimentally on a large scale. Their availability will facilitate the analysis of variation in populations of cultured cells, it will permit the development of a field dealing with their biochemical genetics, and it is a basic prerequisite for the exploration of the feasibility of genetic analysis through somatic crossing over, as discussed in the section on normal cells *in vivo*. Experiments in this direction are now well in progress. Hsu and Kellogg[603] reported recently the isolation of variants adapted to galactose and to xylose, respectively, from a subline of the well-known *L* strain of murine fibroblasts. While glucose supported the growth of all lines, xylose did not support any line except the xylose-adapted strain. Galactose supported only a very limited growth of the parental line while the galactose-adapted strain was growing much better. Eagle and co-workers,[312] who found that meso-inositol was an essential requirement for all strains of cells they have studied, isolated one inositol-independent line from strain *L* cultures. This line could utilize glucose for the biosynthesis of inositol, and significant amounts of the latter (of the order of 10^{-6} M) were actually released into the medium.

A method for selecting auxotrophic mutants of *HeLa* cells, that is, variants with at least one more nutritional requirement than the type of cell from which they had been derived, was recently described by De Mars and Hooper.[244] Their procedure was based on the fact that cultivated mammalian cells are usually unable to proliferate in the presence of the folic-acid analogs, aminopterin and amethopterin, this effect being reversible by a mixture of adenine, thymine, and glycine. Also, bacteria which are unable to synthesize thymine, either because of a mutational deficiency or as a result of sulfonamide treatment, die in media containing all factors necessary for growth with the exception of thymine, because of unbalanced growth.[211] On the other hand, if the thymine-free medium also lacks one of the other factors necessary for growth, the bacteria do not undergo unbalanced growth and remain viable. This preferential survival of thymine-starved cells also deprived from an additional growth factor was postulated to permit the selection of auxotrophic variants from large-cell populations made thymine deficient through aminopterin treatment. Accordingly, *HeLa* cells were treated with aminopterin and their survival in a growth-supporting, complete medium was compared with survival in a medium deficient in a single essential nutrient.

The findings were quite in accordance with the expectations.   The aminopterin-induced thymine deficiency was lethal for *HeLa* cells which ceased to proliferate but increased in size and doubled their average content of protein.   The increase in protein could be prevented by the omission of a single essential amino acid, such as arginine, from the medium.   This resulted in a 10^5-fold greater survival after 6 days when the amino acid was supplied and aminopterin was removed as compared to survival on complete medium.   Ordinary *HeLa* cells which do not require glutamine showed a very low survival in the presence of aminopterin in glutamine-free medium.   In contrast, glutamine-requiring auxotrophic mutants survived well and could be selected specifically and efficiently, even if present in low frequencies.   Small clones of auxotrophs could be detected microscopically at an early stage (48 hrs.) after exposure to aminopterin, due to their unswollen appearance, in contrast to the swollen look of the nonauxotrophic cells.   This technique appears very promising for the selection of clear-cut nutritional variants.

Drug-resistance markers have also recently entered the tissue-culture field.   Some of these studies have already been discussed in the previous section, particularly the modified fluctuation test of Szybalski,[1307] suitable for the estimation of mutation rates, and the experiments of Vogt,[1349] designed to establish the relationship between karyotype and cellular phenotype.   Viral resistance also appears useful as a marker and Vogt[1348] has reported on the chromosomal constitution of *HeLa*-cell variants resistant to poliomyelitis virus.   Increased resistance to X rays and ultraviolet may also be workable.[1058]

Chromosomal markers have frequently been applied to the study of genetic variation in tissue culture.   Most of this work has been done on established strains of cells.   No detailed consideration will be given here to this extensive field which has been reviewed and interpreted recently by Levan.[781]

One characteristic, useful as a marker and of the greatest interest by itself, is the fascinating but still ill-defined malignization of normal-cell strains *in vitro*.   Unfortunately, this subject is rather difficult to evaluate at present.   One reason is the surprising and disconcerting finding of Rothfels *et al.*[1076]   After a careful study of chromosomal markers, these authors came to the conclusion that many so-called "altered" strains of cells originally normal represent contaminations with some of the most extensively used malignant lines, such as *L* cells or *HeLa* cells.   Similar findings have been reported by other cytologists.[391, 605]   Even in cases where contamination can be critically excluded, other factors impede evaluation.   Whether because of the "anthropocentric glamor of using human tissues,"[773] or for other reasons, a large part of this work has been done with human material.   It is notoriously difficult to determine the normalcy or malignancy of a line of human cells, since malignancy is essentially a matter of host-cell relationship, and the appropriate, genetically identical host is not available for transplantation tests.   The use of an animal the heterograft response of which must first be inhibited by such means as cortisone or X rays, or both, is by no means comparable.   Failure of a line to grow may be due to residual graft reaction and not to lack of malignancy.   On the other hand, the ability of a line to grow in the completely foreign host whose own normal responses have been influenced in a rather drastic way is not necessarily indicative of true malignancy; it may be just another form of tissue culture.   The use of human volunteers does not ameliorate the situation significantly, since the homograft reaction is an equally serious complication.

Critical experiments with cells derived from inbred strains of animals leave no doubt that normal cells can change to malignant in tissue culture.   On the basis of present evidence, it is impossible, to judge, however, whether this is a frequent or a rare event and what the relationships are between the probability of its occurrence and such factors as the genetic background of the host, its age, the tissue of origin, the cultural environment, and the period of cultivation.   Opinions among tissue culturists vary from belief that malignization is a regular event, occurring invariably after prolonged cultivation of all cell strains of normal origin, to belief that it is highly exceptional.   Surprising as they are, these ambiguities can be largely attributed to the lack, in most cases, of a suitable animal recipient for the unequivocal testing of malignancy by isotransplantation.   Frequently, transplantation tests are being done with lines of cells that have not arisen in inbred strains and, if they have, there is often a gap of several years between the origin of the line and the actual test.   Absolute genetic identity between two mammalian organisms can perhaps never be achieved[515] except in the case of uniovular twins, but a good approximation is possible by the use of highly inbred strains of mice.[1129]   Even with such strains, separation of independently propagating sublines leads to genetic differentiation fairly rapidly, however.   For this reason, a line of cells serially propagated may not be fully compatible with mice belonging to the strain in which it originated, if the time interval elapsing between the first explanation and the actual test extends over several years.

As a good illustration of the dilemma posed, the admirable work of Sanford et al. can be cited.[1151, 1152]   A single cell has been isolated from an established strain of cells, originally derived from the subcutaneous tissue of an adult C3H mouse.   From the derived clonal culture, different lines of cells were established and propagated serially in vitro.   Cells of one line, designated high-sarcoma producing or high, produced sarcomas in 97 per cent of inoculated C3H mice.   Cells of another line, designated low, produced sarcomas in only 1 per cent of mice, unless the animals had been previously X irradiated; if so, almost half of the mice developed sarcomas.   This difference appeared during the first year following the separation of the two lines.   Tests were designed to investigate whether the cells were antigenically different from the recipient hosts and the X-ray effect could be attributed to the inhibition of a graft response of immunologic nature.   It was found that both lines were antigenic to C3H mice and produced cross-immunity against each other.   The question arose whether the difference between the high and the low line was really related to the elusive property of malignancy or to a difference in antigenicity.   The latter type of difference may conceivably develop in one of two ways.   Isoantigenic differences may exist between the C3H mouse from which the original cell strain had been derived several years earlier and the C3H mice in which they are actually tested.   These differences may be expressed to different degrees in the two lines, representing another case of isoantigenic variation, well demonstrated to occur in systems of transplanted tumor *in vivo*.   Alternatively, the two lines may have developed an antigenicity of their own, not yet present in the C3H mouse from which their common ancestor had been derived, in analogy with individual antigenicity of methylcholanthrene-induced sarcomas previously discussed.

Sanford *et al.*[1151, 1152] carried out critical experiments to decide the question of antigenic difference between the lines *versus* a difference in malignancy.   They found that in irradiated hosts with homograft reaction paralyzed, the two lines differed significantly with regard to the latent period elapsing before the appearance of palpable tumors.   Cells of the high line proliferated more rapidly when first implanted into mice than low-line cells.   In contrast, both lines showed the same growth rate when compared *in vitro*.   There was also a difference between the ability of the two lines to induce vascularization: high-line cells became more rapidly vascularized than equal numbers of low-line cells.   It was concluded that the difference between the two lines was not due to the fact that one was antigenic and the other was not, since both were antigenic, but to the fact that the high-line cells could produce an established tumor before resistance developed in the host.   This in turn depended on a more rapid rate of proliferation of the high-line cells *in vivo*, presumably because of a difference between the two lines in their response to some growth-controlling factor in the host.

Critical studies of this and similar type are urgently needed to clarify this field. Whenever possible, it will be highly desirable to exclude or minimize the artifact of isoantigenic divergence between cells and host strain.   This can be best insured by the use of cells derived from critically tested homozygous strains of animals.   Exchange of skin grafts is probably the best available test for isogenicity.   In addition, the time elapsing between the first explantation of the tissue and the retransplantation experiment should be restricted to the shortest possible period, preferably not more than one or two years.   If this is not feasible, the genotype of the original donor animal may be partially reproduced by combining modern techniques of frozen storage of spermatozoa with artificial insemination.   Whatever the genotype of the recipient female, if the donor male was identical or isogenic with the animal from which the strain of cells had been derived and truly homozygous, the offspring will contain all histocompatibility factors of the cell-donor mouse and thus represent a genetically fully compatible recipient, similarly to an ordinary F_1 hybrid.   The use of such hosts for critical transplantation tests would permit cleaner experimentation with old lines in tissue culture than now practiced, and all sources of erroneous interpretation now stemming from the homograft reaction would be eliminated.

Only large-scale studies of the most rigorous design will illuminate the many important problems that await answer regarding the probability of the malignant change *in vitro*, its relationship to intrinsic cellular and host factors, and to extraneous influences such as carcinogenic agents, radiation, viruses, and the cultural environment. A clear-cut test system is also essential to follow the change from normal to malignant from the biochemical or cytomorphologic and chromosomal points of view.   Frozen storage may be an invaluable help in preserving initial and intermediate stages for critical replicate experiments.

Malignization of cultured cells has been achieved unequivocally and within surprisingly short intervals by the use of certain oncogenic viruses *in vitro*, as discussed in the previous section.   This is a highly promising field which, again, may be advantageously combined with the use of genetically known material from inbred animals so as to permit a rigorous correlation between the morphologic changes observed *in vitro* and the malignant behavior that must be proven by retransplantation experiments to appropriate hosts.

Even the possibility of populations of malignant cells turning nonmalignant during prolonged cultivation has been indicated by a number of experiments.   The significance of these findings is not clear, however, since the selection on nonmalignant cells present from the beginning or the development of histoincompatibility between the cells and the strain of the recipient mouse have not been excluded.

Relationships between cells in vivo and in vitro.—The greatest problem of the field of tissue culture as related to somatic-cell genetics is the question of representativeness. Obviously, if all types of cells could be grown *in vitro* with unchanged characteristics and without limitations, nearly exclusive use of methods of tissue culture could be wholeheartedly advocated for the study of somatic-cell variation.   Unfortunately, this is by no means the case.   In the words of Eagle:[314] "Most of these dispersed cell cultures fail to carry out specialized functions.   Fibroblast cultures do not usually make collagen; melanoblast cultures do not make melanin; liver cultures are so called only because they originally derived from a bit of liver, and not because they carry out the specific metabolic activities generally associated with that organ.   The significant biochemical differences between normal and malignant tissues which are implicit in the very fact of malignancy may similarly fail to be expressed in cell culture; at least, they have not been recognized to date."   It may be added that some organ-specific and blood-group antigens disappear within surprisingly short periods of time during cultivation *in vitro*;[590, 1371] bone-marrow cells lose their ability to differentiate and to repopulate the marrow and function in lethally irradiated mice although they are perfectly capable of doing so if grown in diffusion chambers *in vivo*.[81, 82]   Many similar examples could be cited.   Since cloning techniques fail to give more than about 1 per cent clones with tissue taken directly from the body,[1024] it cannot be decided whether the few and surprisingly similar cellular phenotypes obtained from different tissues *in vitro* are due to the selection from large heterogeneous populations of certain cells preadapted for growth under the artificial conditions of modern tissue culture or to a tendency of all kinds of cells to convert to a fairly uniform common type under the pressure of the environment.   The low cloning efficiency of tissues taken directly from the organism is undoubtedly partly due to technical difficulties, but this does not have to be the whole story and it is quite conceivable that most cells are not capable of growing under the conditions of tissue culture.   The critical period of adaptation of tissue culturists, characterized by sluggish growth and loss of many cultures prior to the establishment of vigorously growing strains of cells, sounds very much like an exercise in selection.   A recent interesting paper of Sato *et al.*[1158] reports experiments specifically designed to illuminate this problem.   They have found that tissue-culture populations derived from liver of day-old rats differ markedly in antigenic structure from the liver at this age.   Furthermore, freshly prepared hepatic inocula are prevented from initiating growth in culture if pretreated with renal tissue-culture antiserum previously absorbed with liver.   On the other hand, the growth potential of hepatic inocula remained unaffected by anti-liver antiserum containing demonstrable antibodies against liver.   Also, short-term cultures derived from liver possessed sharply reduced ornithine-transcarbamylase activity and serum-albumin content; this residual activity and content could be accounted for by the microscopically demonstrable persistence of portions of the inoculum.   It was concluded that the bulk of the tissue-culture population derived from liver of day-old rats arose from a type of cell other than the parenchymal cell.   This cellular type appeared to be a small minority of the total population, since suspensions of day-old liver had a plating efficiency of about 10^{-4}.

Thus, the problem concerning the relationship between established strains of cells in tissue culture and the normal or malignant tissue from which they had been derived must be left entirely open at present.   Existing strains of cells must be regarded as interesting but completely artificial populations, useful as model systems, but by no means informative with regard to somatic-cell genetics within the mammalian organism.   Intense efforts should be directed now toward the development of media that can permit the survival and growth of many different types of cells in a state as nearly approaching and representative of their condition *in vivo* as possible.   The problem of representativeness will not be opened for serious investigation until it has become possible to clone cells taken directly from the body with high efficiency.   No cloning of cells that have already passed the critical period of adaptation in culture can replace this need.   As far as the problem of malignancy and malignization *in vitro* is concerned, a close passage-to-passage collaboration is necessary between culture *in vitro* and transplantation tests *in vivo*, with appropriate isogenic animal recipients, as discussed above.

POSSIBLE APPROACHES TO GENETIC TRANSFER BETWEEN SOMATIC CELLS

Among bacteria, three main forms of genetic transfer have been discovered: sexual recombination, DNA-mediated transformation, and phage-mediated transduction.   Lysogenic conversion by which the phage particle itself controls a certain phenotypic character of the host cell, and the field of episomes are also relevant.

Sexual recombination between somatic cells is a rather remote possibility.   It is true that fusion and segregation have been observed in somatic cells, each by itself, under particularly favorable and, as a rule, exceptional circumstances,[773, 774] but an orderly sexual cycle of nuclear fusion followed by segregation appears rather unlikely.

We have carried out extensive experiments to trace its eventual occurrence in populations of neoplastic cells by appropriate isoantigenic markers, in systems where possible recombinants were certainly demonstrable by highly selective methods.[716] The results were wholly negative.   If nuclear fusion were a regular occurrence in cellular populations and would lead to viable cells, the probability of segregation might perhaps be increased by irradiation, known to accelerate somatic variegation in various species.[773]

DNA-mediated transformation may well be within the limits of reality.   Since strongly selective methods are required to demonstrate transformants even in most cases of bacterial transformation, the H-2-isoantigenic system of mice appears presently most suitable among the markers available *in vivo*.   Selectivity can be made very high, so that 10^{-7}–10^{-8} variants can be demonstrated.   Another advantage is that the genetic determination mechanism is known and possible transformations may be readily distinguishable from spontaneous variation by selecting appropriate marker combinations in which changes to new antigenic specificities do not occur spontaneously.   To avoid the problem of incompatibility of transformants with new antigenic specificities and the enzymatic breakdown of DNA, the experiment must be carried out *in vitro*, followed by retransplantation to appropriate selective hosts.   It is hopeful that mammalian cells are known to be capable of taking up highly polymerized DNA[1214] within short periods of time after exposure.

Although their mechanism of genetic determination is unknown, drug-resistance markers may also be useful for transformation experiments, particularly in view of their successful use in bacterial transformation.   The selectivity of systems *in vivo* is usually not high enough to demonstrate the presence of minute resistant fractions since the large sensitive population may exhibit considerable residual background growth in spite of the treatment.   Experiments *in vitro* are therefore definitely preferable.   Szybalski[1307] has recently reported that the study of drug resistance *in vitro* is not only feasible but even readily amenable to quantitation.   Multiple drug-resistance markers are probably preferable to simple ones.   Again, care must be taken to distinguish transformation from spontaneous variation by appropriate controls.   Another source of error, found in some of our unpublished experiments when amethopterin-sensitive leukemic cells were treated with DNA derived from resistant variants, lies in the possibility that sensitive cells may be protected from the drug by some physiologic or chemical action of DNA that may extend into the period subsequent to the removal of the nucleic acid.   Controls should therefore always include cells treated with DNA derived from the sensitive type.

It is impossible to predict the correct conditions that may permit the possible occurrence of transformation.   Obviously, care must be taken to minimize enzymatic decomposition of DNA.   On the cellular side, competence is a highly elusive property, difficult to control even in bacterial systems and exerting a profound influence on the liability of the exposed cells to undergo transformation.   The experimental conditions necessary to secure competence are very critical and most of them have been established empirically.   For this reason, negative results should not discourage the vigorous exploration of all possibilities, since the right experimental conditions cannot be predicted and lack of success may depend on purely technical factors. As an alternative to the exposure of large populations of neoplastic or tissue culture cells to DNA derived from variants with different markers, the injection of DNA directly into eggs in cleavage may also be attempted. A possible advantage of this procedure would be the use of multiple dominant markers, controlling visible phenotypic characteristics. A mosaic individual would be the result of a successful experiment.

The feasibility of a transduction type of approach depends on the way in which a virus can be integrated with the genome of a somatic cell, its random or obligatory localization if truly integrated, the amount of genetic information it can incorporate from the host cell's genetic material before undergoing maturation, its ability to become integrated with the recipient cell without preventing it from multiplying, and the possibility of genetic exchange between the information carried by virus and the corresponding sites of the recipient cell. While this may appear to be a chain of events highly unlikely, it is not more unlikely than bacterial transduction which was hardly predictable but has now become one of the routine methods of bacterial genetics.

Since the only viruses capable of infecting higher cells without killing them are certain tumor-viruses,[1080, 1350] they are the most likely candidates for such experiments. Their possible integration with the genome of the host cell is still a matter of conjecture,[1081, 1350] although some form of integration is very probable. Nothing is known about the localization of the integrated form of the virus. In this connection it may be of interest to point out that two types of transduction are known in bacteria: generalized or nonspecific and limited or specific. In the first type, discovered by Zinder and Lederberg,[1470] certain phages are able to transmit any of the genetic characters of the donor bacteria of the recipient strain. The probability of transmission of any given character per phage is small. Phages capable of performing this type of transduction do not appear to have an obligatory chromosomal localization.[1380] In the second type of transduction, discovered by Morse, Lederberg, and Lederberg,[899] the prophage has an obligatory chromosomal localization and can transduce markers located in its immediate vicinity at a low frequency, or, under special circumstances, at a high frequency. The generalized type of transduction resembles transformation of DNA with transmission of any genetic character in suitable systems and to competent cells, while the restricted type is more similar to lysogenic conversion. The main difference between the two mechanisms apparently depends on whether or not the corresponding prophage occupies a specific position on the bacterial chromosome.[1380] It is impossible to guess the type of integration, if any, that tumor viruses may exhibit. To remain on the safe side, it will be wisest to include as many marker differences as possible in a system designed to detect the possible occurrence of transduction among somatic cells. DNA viruses such as polyoma may be more hopeful than RNA viruses from the point of view of integration, although in the absence of concrete knowledge about the rôle of RNA as possible carrier of genetic information in higher cells it cannot be excluded that RNA viruses may also become useful. Since most tumor viruses are rather small, they will incorporate very limited amounts of genetic information from the host cell at best; this also speaks for using multiply marked strains of cells. With cells of the mouse, there is an arsenal of isoantigenic markers, determined by known genetic mechanisms, that might be combined with a series of drug-resistance and nutritional markers to build up suitable, multiply marked strains of cells in tissue culture where most if not all transduction experiments will have to be done.

Possible analogies between viral tumorigenesis and lysogenic conversion have been discussed by Luria;[810, 811] this has been already considered in the section on viral tumors and infective heredity.

Tumor viruses show many analogies with the episomes of Jacob and co-workers.[653, 655, 657] These are defined as "a class of genetic elements, which are not essential constituents of the cell since they may be absent from it. When present, they may exist in two alternative states, either as autonomous units replicating independently of the bacterial chromosome, or as integrated units attached to the bacterial chromosome with which they replicate." Temperate bacteriophages, the sex factor of *E. coli* and the colicinogenic factors of certain *Enterobacteriaceae* are some examples of episomes and the recent picture given by Vogt and Dulbecco[1350] regarding the intracellular behavior of polyoma virus of mice is very reminiscent of episomic elements. It is therefore of great interest and rather hopeful that Jacob and Adelberg[654] have recently discovered that the typically episomal sexual factor *F* in *E. coli* can incorporate small segments of the bacterial chromosome and transmit them to recipient cells which become heterogenotic for the segment in question. Jacob and Adelberg suggest that each episomic element may, during its integrated phase, exchange genetic elements with a nearby chromosome segment to which it is attached and that each genetic element of the bacterial chromosome may become part of an episome.

CONCLUSIONS

In many respects, the field of somatic-cell genetics is comparable with bacterial genetics some twenty years ago. Marker characteristics are available for the study of phenotypic cellular variation. Some markers can be exploited for the selective concentration of rare variants and the semiquantitative or quantitative determination of their frequency, but it is impossible to distinguish between phenotype and genotype and genetic mechanisms cannot be separated from epigenetic changes due to altered genic expression. (Cytologically recognizable chromosomal mechanisms represent one exception, but these have been largely excluded from the subject matter of this review.) The most important need is therefore to develop some method to transfer genetic information between somatic cells. It would be surprising if none of the mechanisms now available for genetic transfer in bacteria, such as sexual recombination, DNA-mediated transformation, transduction by phage, lysogenic conversion, and episomal transduction, would be applicable in some form to somatic cells. Appropriate markers and selective experimental systems are now available for the large-scale exploration of these possibilities.

Meanwhile, existing information is often conjectural but by no means uninteresting. Somatic crossing over occurs in *Drosophila*, and the genetic and environmental factors influencing its frequency have been throroughly analyzed. In *Aspergillus*, somatic crossing over has permitted genetic mapping in no way inferior to meiotic mapping, and it has been pointed out that this mechanism could be used for the genetic analysis of somatic cells even in the absence of genetic transfer, provided that it occurs in higher organisms. Evidence on this point is not conclusive, although recent findings on the variation of erythrocytic antigens in man and tissue isoantigens in tumors derived from heterozygous, F_1, hybrid mice can be interpreted as possible mitotic crossovers. More information is urgently needed on this point.

Somatic mosaicism due to chimerism, mitotic nondisjunction, and somatic mutations is receiving increasing attention. Rare cases involving the gonads in addition to somatic tissues have been very helpful in proving the genetic nature of the underlying changes. Sometimes somatic mutations reveal new information about cellular lineage in development. They can be induced by X rays like germinal mutations and suitable experimental designs permit the estimation of their frequency, being apparently of the same order as of comparable germinal mutations. In maize and in *Drosophila*, highly informative studies are available about the genetic control of somatic variegation in various tissues. This phenomenon is so highly regulated and orderly that attempts have been made to bridge the gap between genetics and development by constructing theories of differentiation based on controlling elements at the chromosomal level, such as dissociators, activators, and modulators. Whatever the future of this theory, it is obvious that somatic variegation and its genetic control are of the greatest importance for the field of somatic-cell genetics. Changes at the genetic level have been also implicated as a basis of the clonal theory of antibody formation; this is as yet entirely at the speculative level, however.

Neoplastic cells *in vivo* are valuable tools for the study of somatic variation, although they represent a special case of somatic cells and the possible relevance of the findings for comparable normal cells will have to be established from case to case. Among the phenotypic marker characteristics that can be used for studies on population dynamics *in vivo*, isoantigens of the histocompatibility system, drug resistance, hormone dependence, ability to grow in the dissociated free-cell form of an ascites tumor and correlated surface characteristics, and, in some special cases, tumor-specific antigens and other cellular products may be mentioned. Among those tested in this respect, the isoantigenic system has the highest selectivity and represents the only marker with a known genetic determination mechanism. Isoantigenic variation can be studied particularly well in heterozygous tumors of F_1-hybrid origin, since variant sublines can be selected, compatible with one or the other of the parental strains and characterized by the loss of specific H-2-determined isoantigens. Such variants could be selected in one step or through several intermediate steps; different types of variants could be obtained from the same tumor; the isoantigenic losses were stable and irreversible after return to the F_1 genotype of origin; and different tumors of similar origin and histology were individually different with regard to variant formation. Somatic crossing over was considered as a possible explanation of this phenomenon. A certain parallel could be drawn between isoantigenic variant formation and tumor progression. The former can be viewed as the loss of isoantigens preventing growth in a foreign host genotype, while the latter is essentially a gradual loss of responsiveness to various growth-controlling forces in the autochthonous host. As a model system, variant formation demonstrated that irreversible loss of certain genetically determined cellular constituents may, under appropriate conditions, convey upon the cell a new ability to grow under circumstances in which its progenitor would have been checked by a superimposed, systemic mechanism.

At the population level, the dynamics of some changes in population have been analyzed. While a variation-selection mechanism has been demonstrated for some cases of drug resistance, certain changes in antigenic and transplantation characteristics, ascites conversion, shift of predominant karyotype, and the establishment of amelanotic melanomas, there is also one instance of a host-induced adaptive modification and an automatic change in responsiveness due to increase in population size. Specific pairs of tumor sublines, selected from a common original population, but differing with regard to well-defined unit characteristics related to progression, appear particularly suitable for the study of the cellular mechanisms determining the phenotype in question.

The recent striking developments in the field of viral tumors have introduced the concept of infective heredity into the tumor field. Some form of integration between viral and cellular genomes probably exists, at least with certain tumor viruses. In some cases, new cellular characteristics appear concurrently with viral oncogenesis, not directly related to neoplasia itself, suggesting a possible analogy with lysogenic conversion.

Tissue-culture systems provide excellent models to study somatic variation. Available markers comprise drug, virus, and radiation resistance and nutritional, morphologic, and chromosomal characteristics. Fluctuation tests and cloning techniques can be directly applied. Unfortunately, the problem of representativeness is still overwhelming and it seems as if only a limited number of cellular types could be propagated in present tissue-culture media. The development of new media that will permit a larger variety of cells to grow *in vitro* and the application of cloning techniques to cells removed directly from the body are urgently needed before tissue-culture methods will become directly useful for the study of the genetics of somatic cells as they exist in the body.

The question of malignization of tissue-culture cells of normal origin is presently rather confused due to the discovery that many of the so-called "altered" lines have been contaminated with established and highly virulent lines of malignant cells. There is no doubt that malignization *in vitro* does occur, but nothing can be said about its frequency and its relation to extrinsic and intrinsic factors. Critical experiments are yet to be done with cells derived from homozygous animals and with close passage-to-passage coordination of *in-vitro* cultivation and retransplantation to animals isogenic with the original donor. Isogenicity can be best insured by using highly inbred mouse strains, critically tested with skin grafts, and restricting the time elapsing between the origin of the line and its actual testing to the possible minimum, or, alternatively, by using some such technique as frozen storage of spermatozoa and artificial insemination to reproduce the histocompatibility factor equipment of the original donor or animals isogenic with it.

The possibilities for working out methods for transferring genetic information between somatic cells such as DNA-transformation or virus-mediated transduction have been discussed. *A priori* the feasibility of some such procedure is no less probable than it had been with bacteria.

DISCUSSION

DR. BURDETTE: Dr. Leonard A. Herzenberg will open the discussion of Dr. Klein's paper.

DR. HERZENBERG: The exciting reports of Drs. George and Eva Klein and their collaborators on isoantigenic variations in tumors stimulated us to explore the possibility of extending their genetic analysis *in vivo* to the cell-culture situation *in vitro*. In this symposium on methodology, let me indicate some of the methodologic advantages for detailed genetic analysis that cell-culture systems afford us.

First and foremost, since the demonstrations of Puck,[1029] cloning is a rapid and reproducible procedure in work utilizing cultures of cells. Populations derived from one cell are routinely available, and all the tricks of the microbiologist are applicable to the mammalian-cell system.

Cells have been cultivated from various tissues of a number of mammals in semi-defined media;[313] that is, in synthetic media to which are added a small percentage of dialyzed serum proteins. These media may contain traces of unknown materials, but the macroconstituents—amino acids, carbohydrates, purines, pyrimidines, most vitamins, and salts—are present in known concentrations. Thus, variations in nutritional needs or susceptibility to metabolic poisons can be searched for, and, if suitable ones are found, these can be employed as cellular genetic markers. One or two nutritional variants have been described.[244] However, these have not yet been too useful in genetic studies due to the difficulty in effectively selecting for these variants.

A number of drug-resistant variants of mammalian cells in culture have now been found. With some drugs, for example, 6-mercaptopurine, 8-azaguanine, and amethopterin, a resistant cell can easily be selected from a large population of sensitive cells,[366, 787] and it has been possible to answer some basic questions about the genetics of resistance to these compounds. With other drugs, for example the fluorinated pyrimidines, efficient selection of individual resistant cells from a background of sensitive cells has proved to be more difficult. Nevertheless, workers in a number of laboratories, including our own, are continuing to characterize variations in resistance to antimetabolites, with the hope that some of these may become useful genetic markers.

The main emphasis in our laboratory is an exploration of the isoantigens of cells in culture.   The *H*-2 antigens of the mouse seem particularly promising from a geneticist's point of view.   These antigens are controlled by a complex locus on the ninth chromosome of the mouse and have been the subject of extensive and detailed immunogenetic analysis in several laboratories for a number of years.[455, 1238]   At least 20 alleles have been described at this locus, and many antigenic components have been associated with most of these.   Thus if the *H*-2 antigens can be detected on cultured cells with reasonable facility, a large number of markers become immediately available for genetic studies *in vitro* by simply preparing cultures from different strains of mice. Moreover, one can have confidence that the phenotypes (serotypes) of these cells bear a direct relationship with the genotype.   One considerable advantage *a priori* is the possibility of using cultures derived from IR lines of Snell, or F_1 hybrids of these. Then one can have cultures which are genetically identical except for one gene, or at most a short region of one chromosome.

Several methods of scoring the *H*-2 antigens of cultured cells suggest themselves as possibilities.   These all involve the use of isoantisera and are potentially as specific as the sera which can be obtained with all the skills of the immunogeneticist.   We are now adapting the cytotoxic method to cultured cells.

Cells of a DBA/2 lymphoma *P*-388, whose nutritional needs *in vitro* have recently been determined,[552] are lysed when incubated with an anti-*H*-2*d* antiserum and guinea-pig complement.   This lysis has been conveniently followed with an electronic cell counter.   The viability of unlysed cells then can unambiguously be determined by plating the culture in growth medium and counting the number of clones which develop.   With these procedures, we have found that these cells in long-term cultures retain the *H*-2*d* phenotype of the strain from which they were derived.   Continued cycles of cellular killing and regrowth have selected no stable variants which have lost the *H*-2*d* antigens.   No tests have been performed for individual antigenic components.

Now that the basic selection procedure has been tried, future work will be directed toward attempts to select variants from an F_1 hybrid-derived line of cells, where loss of only one dose rather than two doses are needed and to selection of variants for some of the individual antigenic components.

Means which do not result in the death of cells exhibiting an *H*-2 phenotype must be explored.   We have demonstrated that, in the absence of complement, the isoantisera are completely without effect on cellular viability.   Thus labeled antibodies, either fluorescent or radioactive, should be tried to score cells.   The technique of mixed agglutination, where red cells which share an antigen with a cultured cell are coupled to such cells with antiserum, is another potentially useful method.

To conclude, the facility with which cells can be cloned, and the complete control of environment obtainable with methods of cellular culture is broadening the scope and increasing the analytic power of mammalian somatic-cell genetics.

Dr. Kaliss: Dr. Klein's article contains an excellent review of the literature, and the most rewarding thing that comes out of my having to discuss it is that I was able to read it and learn from it.   I think it is an excellent contribution, and certainly there would not be any point in my trying to add anything more to specific topics covered. I will, therefore, confine myself to some more general comments.

The motivating force in the study of somatic-cell genetics is the problem of the differentiation of cellular function, whether it be in the single-cell or multicellular organism.   This symposium is concerned with the genetics of the mammalian organism and, thereby, with the problem of functional differentiation as expressed in the specific assignment of functions to the various cells, tissues, organs, and organ systems.   Parenthetically, the so-called gap between genetics and embryology, about which there is often complaint, is, of course, a conceptual gap between the geneticist and the embryologist.   The fertilized ovum apparently has no difficulty in knowing in which direction it should develop.   Perhaps the original stumbling block between the two disciplines has been the controversy in genetics as to whether or not the organization of the nuclear apparatus, specifically the chromosomes, remains unaltered from generation to generation of cell division.   The prime argument offered in support of the inviolability of the chromosomes from generation to generation of the somatic cells has been the stability of the transfer of genetic information by the germ cells (barring mutation).   The point at issue is amenable to experimental investigation with nuclear transplantation into foreign cytoplasmic environments, as is being done in amphibian embryos.[130, 131, 365]

Danielli and co-workers[238] have made nuclear transfers between two species of *Amoeba* and have demonstrated that the cytoplasm of the heterospecific host left its imprint in the form of permanent and characteristic alterations in the transplanted nucleus after 500 generations of cellular division, as determined by retransfer of the nucleus to the cell of its native species.   These changes are expressed in some 10 characteristics that were studied, such as shape and locomotion of the amoebae and the rate of cellular division of a derived clone.   On the other hand, those antigenic properties presumably located in the cellular membrane were characteristic of the nuclear species, as determined by the specificities of cytotoxic antiserum.

One could postulate a number of models for organic differentiation, and I suppose it would be possible eventually to devise feedback systems picturing the way in which the DNA-RNA-protein-enzyme systems and other cellular constituents are interrelated in time and space to carry on the multitudinous functions of the whole cell.   The division of the cellular domains into nucleus and cytoplasm, in our thinking, is, of course, one of convenience; it seems to me that we should not preclude the possibility of nuclear alterations (reversible or irreversible) that constitute interlocking events in the process of differentiation.

A few words are in order as to what we may consider somatic-cell characters. Dr. Klein has covered this point in detail in his article with respect to the histocompatibility antigens, and I only wish to reëmphasize some of its aspects.   It is not clear what rôle these antigens, which are revealed by the technique of tissue transplantation, play in the economy of the cell.   Kandutsch's work[682] in this laboratory indicates that they are located in the cellular membrane.   It is possible that they may be involved in fetal-maternal isoimmunization, but aside from this they are antigens only because the experimenter uses immunologic methods for their detection; their usefulness as a genetic tool is fully documented in Dr. Klein's paper.   There are, of course, many other cellular characteristics that might be exploited.   The uniqueness of the histo-compatibility antigens is that they are amenable to study both *in vivo* and *in vitro*.   The importance of being able to utilize the two approaches is emphasized by the apparent loss of physiologic specialization, for example, by hepatic cells or hematopoietic tissue in tissue culture.   On the other hand, this very loss of function, if it is indeed a true phenomenon and does not represent selective survival of more generalized cells, such as fibroblasts, only serves to reëmphasize in another form the problems of what constitutes functional differentiation at the level of tissues and organs.

A final word on the question of neoplasia as a possible example of somatic mutation at the nuclear level; Dr. Klein discusses this in detail in his article.   It has been mentioned to some extent in the other discussions.   It has been repeatedly emphasized, as Dr. Andervont pointed out, that cancer is most probably not a single disease but a number of different diseases.   I have heard the analogy made between the entity, cancer, and the designation of fever as a disease entity during the last century, with the eventual understanding that rising temperature reflected a general physiologic expression of a large variety of infections.   If we attempt to homologize bacterial transduction with a viral etiology of cancer, we are confronted with such phenomena in the cancer process as the very long latent period between the carcinogenic stimulus and the first overt sign of the cancerous process or the progression from dependence to autonomy in the cancerous growth, as Dr. Klein mentioned.   (Incidentally, autonomy should be looked upon in broad terms, beyond those which we derive from hormone-dependent tumors alone.   Dr. Harry Greene,[495] for example, has drawn a parallel between the degree of malignancy of a given tumor, as manifested by the formation of metastases, and the ability of the tumor to survive as an heterograft.)

One could evoke genic activation to account for the latent period, but it is not clear to me what is meant by genic activation by those who use the term.   It is difficult for me to conceive how a part of a chromosome or of other constituents of the nucleus and the cytoplasm can be inactive at one time and come to life, so to speak, at another time.   I suppose one could postulate an induced alteration at some point in the cellular apparatus, perhaps the chromosome, which can be expressed as a malignancy only under certain conditions of differentiation in the animal possessing the affected cells.   The correlation between age and the increase of the incidence of cancer may depend upon continued differentiation of the host, differentiation, progressive and retrogressive, being considered the essence of life from the time of conception of the individual until death.

DR. SNELL: Dr. Klein used the *H*-2 locus in a very fascinating fashion to shed some light on somatic changes in the cell.   I would like to pick up a lead that he dropped and, reversing the coin, discuss what his studies of somatic-cell change tell us about the *H*-2 locus.

Probably most of you know that there are reports of crossing over in the *H*-2 locus. Crossovers have been observed by Dr. Sally Allen in Chicago, Dr. Gorer in London, and Dr. Hoecker in Chile.   The surprising thing in the light of what we know of compound loci in lower organisms is the incidence of crossing over, which seems to run at least 1 per cent.   If we do think of this as a single locus, it must occupy, if this evidence is reliable, a chromosomal segment of appreciable length.   The antigenic components which have been identified at the *H*-2 locus are now quite numerous.   Dr. Stimpfling can tell you better than I can what the number is; it is over 20, I believe.   These all are potential markers by which crossing over could be studied.

So far all the recombinants apparently do fall in linear order, which is, I believe, evidence that they really are due to crossing over.   We do not have visual markers on both sides of *H*-2 on the ninth chromosome.   The *T* or brachyury gene and the *Fu* or fused gene are both on that chromosome; but they are on the same side of *H*-2 and not on the opposite sides, and that has limited the efficiency of our test for actual crossing over in the locus.

The one point I would like to draw from Dr. Klein's interesting paper is this.   It does seem to me that the instance of apparent somatic crossing over, which he showed on the board, has involved a separation of *H*-2 components.   It seems to me that it is going to be difficult to interpret this in any other way than crossing over, and perhaps this does strengthen the evidence that we really do have a locus here in which crossing over of something of the order of 1 per cent can occur.

Dr. Lederberg: I would like to amplify the points that have just arisen and to answer a particular question that Dr. Klein has raised.   The latest map I have seen (you probably have a better one) indicated at least five factors in the *H*-2 region, and according to Hoecker they could be in the sequence *D, C, V, K,* and *H*.   Is there anything that would go beyond that or contradict it as far as the present data on germinal transmission go?   In many instances the decision as to order may depend on the existence of one aberrant mouse.   As things stand now, the data on the somatic segregation do not particularly reinforce the sequence.   That is, the statement that *D* and *K* fall into some relationship to the centromere cannot be checked by any data currently available in the field of murine genetics, unless there is some translocation involving the *H*-2-bearing chromosomes that might be used for an independent check. However, if one could put one more factor into the system, we could look for the mutual consistency of the type of map which is available from the isolated apparent crossover events in the germinal and somatic tissue.

About the point that Dr. Herzenberg raised, which I think is quite an important one, that of looking for more metabolic anomalies in mice, I wonder if it is necessarily true that the present inbred strains furnish the best material for it, since they have been necessarily selected for viability of homozygotes.   Perhaps we should again start inbreeding wild mice and take not the best but the worst animals that come out and take a look at them from the point of view, not of their ease of cultivation, but whether they do have noteworthy metabolic defects.

DR. OWEN: It seems to me that Dr. Klein's system should be well adapted for mapping the *H*-2 complex. In this connection, as I remember Dr. Sally Allen's early work,[12] the only study I know in which there was a genetic marker external to the *H*-2 complex, there were two instances of apparent recombination, which would have required that the *DK* components be ordered differently with reference to that marker. Dr. Snell, would you comment on this point?

DR. SNELL: One case was thoroughly substantiated. In the other the mouse died before the breeding tests were complete. If that case were a valid recombinant, then the *D* and *K* components cannot be arranged in a linear order. However, since the tests were not completed, I do not think much weight can be attached to it.

DR. BARRETT: In relation to the markers that go along with this, I have approached this subject, of course, from a different viewpoint than Drs. Snell, Hoecker, Gorer, and so on; and I have not used these antigenic components, but I take the transplantability of the tumor, or its nontransplantability, to be a quite adequate indication of its antigenic composition. As a matter of fact, I have been surprised in this discussion of methodology that, although we talked about methods involving very small parts of the animal (sometimes just the molecule), we have said very little about how to use the whole animal in these methods.

Now I do not want to give my own paper here, but in published work[54] involving very extensive observations of what Dr. Klein has called the Barrett-Deringer phenomenon, one can see that this phenomenon can be expressed by over-all changes in percentage, usually upward; but let us remember that the upward change is probably only the easier change to find. The cell that has become incompatible automatically commits biological suicide and is hard to find. The other aspect is that if one examines the over-all increases in ratio, one finds that when they are tested in a backcross population in which there are three markers and therefore eight segregants, the change in percentage among these segregants may occur not at all, it may go up a little, it may go up a lot, or it may produce tumors that grow temporarily and regress. All these characteristics can be definitely related to external markers of these animals but cannot be related to agglutination, and apparently have nothing to do with *H*-2, although I agree with you, Dr. Klein, it appears that *H*-2 is a stubborn locus in this regard. Compatibility can be related to brown, black, to agouti, and perhaps to albinism. It cannot be related to color on a statistically significant basis, but these markers are there and the change can be related to these external markers.

DR. KLEIN: I should like to add a few words concerning suitable markers. The first question regards the nature of the approach. One may simply want to study somatic-cell variation, one may want to explore the possibility of somatic crossing over, or else one may prefer to engage in attempts to achieve transfer of genetic material between somatic cells.

To study variation, one will have to choose between systems *in vivo* and *in vitro*. The approach *in vivo* will be largely confined to neoplastic cells at present, and the markers available include drug resistance, hormonal dependence, ability to grow in the ascites form, and isoantigens, to mention only a few. The only characters that have a known mechanism of genetic determination and can be localized on the chromosomal map are the isoantigens of the histocompatibility system, best known in the mouse. This system, notably the *histocompatibility*-2 system, has also the largest selectivity; and in reconstruction experiments involving artificial mixtures of cells differing at the *H*-2 locus, it was possible to isolate selectively *H*-2 compatible cells even if they only represented an extremely small fraction (10^{-7}–10^{-8}) of a large *H*-2-incompatible population. In our experience, systems of resistance to drugs or ascites convertibility have a much lower selectivity (10^{-5}–10^{-6}). Selectivity and known localization on the map gives preference to the isoantigenic markers also as far as the somatic-crossing-over approach is concerned. For genetic-transfer experiments, it is not quite necessary to have a known localization on the map, but a highly selective system appears to be very important even in this case.

When the approach *in vivo* is contrasted with the possibilities *in vitro*, it is obvious that a higher degree of precision and more controlled conditions can be achieved in tissue culture. The isoantigenic systems represent one exception in this respect, however, since one of the main advantages, the high selectivity and specificity of the homograft reaction which serves to concentrate rare variants, has so far not been duplicated *in vitro*, although it should be possible to do this, at least with types of cells sensitive to cytotoxic, humoral isoantibodies (that is, primarily with lymphatic and bone-marrow cells and their neoplastic derivatives). As far as other markers are concerned, tissue culture will usually provide a more concise method for experimentation, although it must be kept in mind that there is no reason to believe that the population of cells *in vitro* is representative of the population *in vivo*; in fact, there is a great deal of evidence to the contrary. Until large-scale, single-cell cloning of animal cells is possible *in vitro*, with tissues taken directly from the organism *in vivo*, it appears most conservative to regard *in vitro* systems merely as models of what might apply to somatic variation *in vivo*, but without attempting direct extrapolation.

As to the points raised by Drs. Snell, Lederberg, and Owen, I would hesitate to conclude, at the present stage of our experiments, that we actually deal with somatic crossing over. It is a very probable explanation, but other possibilities must still be kept in mind. It is conceivable, for instance, that *D* and *K* could be different parts of the same molecule, in line with Dr. Owen's model, and that *D* could be a precursor of *K*. Thus a loss of *D* would automatically lead to a loss of *K* also. This argument is weakened somewhat by the fact that there are strains of mice on record, characterized by *H*-2 combinations which lack *D* but contain *K*. Nevertheless, additional markers will be needed if the possibility of somatic crossing over is to be confirmed, and I certainly appreciate the suggestions of Drs. Snell and Lederberg in this regard.

George Yerganian, Ph.D.

CYTOGENETIC ANALYSIS†

The past decade has been a most decisive period for the advancement of experimental mammalian cytology. Progress today appears even more exaggerated when compared to a relatively stagnant period of some twenty-five years, resulting from negative experiences and faltering views, due primarily to limited technical facilities. Just how did this present surge of interest come about? Until 1950, the number of cytologists familiar with mammalian chromosomes was very limited and included a few who were equally active in plant cytology. The latter group of investigators reviewed animal tissues periodically to reinstate the awareness of difficulties ascribed to mammalian chromosomes by earlier workers. However, as time went on, more favorable pretreatments and sources of tissue became available and the cumulative results were most encouraging to the small but interested group of participants. A number of independent reevaluations were conducted with the squash technique, employing seminiferous tubules for meiotic bivalents,[1175] somatic chromosomes *in*

† Thanks are extended to M. Bender, J. J. Clausen, K. Fredga, T. S. Hauschka, D. Hungerford, R. Kato, I. Kline, U. Ising, G. J. Marshall, S. Makino, P. Moorhead, S. Ohno, C. G. Palmer, M. Sasaki, and J. T. Syverton for their cooperation and help in extending the coverage of this treatise to a broad variety of methods. The author wishes also to express his gratitude to Dr. Sidney Farber and other members of the staff of The Children's Cancer Research Foundation, for advice and recommendations and particularly for the facilities provided that made these investigations possible.

The investigations conducted at The Children's Cancer Research Foundation have been supported in part by grants from the National Science Foundation (G9602), the Damon Runyon Memorial Fund for Cancer Research (DRG-293), Research Grant A-4468 from the Institute of Arthritis and Metabolic Diseases, and from the National Cancer Institute, USPHS No. CY3335.

situ,[1309] tissue-culture derivatives,[600] and ascites tumor cells.[840, 783] Attention had been directed earlier by German workers to the last type of cell as a potent experimental tissue. In the interim, a number of alterations in technique, particularly those regarding hypotonicity,[607, 609] led other workers to rally to the challenge of adapting meiotic and mitotic mammalian cells to the current array of experimental approaches. Since the early 1950's, an ever-increasing number of participants have considered the prospects of utilizing chromosomal cytology as a tool to evaluate factors affecting cellular viability, mutagenicity, transplantability, and numerous physiologic and biochemical expressions. Evidence that a large number of human syndromes are associated with cytogenetic abnormalities has been developed in a remarkably brief period, and recent studies on mammals have yielded an increasing assembly of heritable patterns among rodents which provide ideal counterparts of syndromes clinically important.[190, 613, 866, 1097, 1280] An equally promising area has been opened as a result of the over-all advances in tissue culture. There is every reason to believe that the coming decade will continue to foster fertile premises for developments in the approaches that apply to genetics of somatic cells, with greater emphasis being given to controlling components of the environment.[312, 330, 895, 941, 1027, 1088, 1359] Increasing use of trials *in vitro* as compared to those *in vivo* is primarily due to the esthetic qualities of the former system with the possible simplification and/or control of environmental factors. Nevertheless, a satisfactory outcome of effectively planned studies *in vitro* will depend largely on ability to synthesize genetically controlled living sources from which to isolate the appropriate types of cells. If long-term cultures of blast-like precursors are readily obtained, the shift toward analyses *in vitro* will be hastened. Although the trend today is toward subcellular levels of study, cytogenetic methods for cells *in vivo* must be available to simplify further and to evaluate pleiotropic expressions of a given mutation when cells are placed *in vitro*.

Future immunogenetic approaches employing normal lymphoid or reticuloendothelial derivatives *in vitro* require cautious attention. Isolation of these desirable types of cells is more worthy of trial than the use of some of the present types of unknown species and sex. Invariably, continuous propagation of cells features numerous chromosomes, some or many of which may be entirely new structurally and reflect properties that would have failed to survive during the course of selective evolutionary processing of cytogenetic mechanisms.

Physical and chemical measurements of any type of cell are certain to reflect the degree of variegation that characterizes the cellular population. In the event that a type of cell is chromosomally stable, that is, displays extremely low frequency of chromosomal breakage and/or a limited variation in chromosomal distribution (aneuploidy), it is usually assumed that the frequency of genic or point mutations is equally limited. On the other hand, fluctuating chromosomal numbers and forms will lead to innumerable variants, especially when numerous sublines are distributed to other laboratories. Among a few of the prime requisites for the study of somatic cells *in vitro* are stability or orderliness of the karyotype (number and form of chromosomes), convenience of procedures for replication to isolate special types of cells from genetically known backgrounds, and minimal numbers of genetic linkage groups.

One of the more recent surveys on numbers of chromosomes in propagated tissues accrued over the past ten years has been reported by Levan.[781] Although shifts in chromosomal numbers are observed frequently the significance of such changes remains obscure when limited to near-tetraploid derivatives.   Likewise, the tendency for amplification of histocompatibility relationships among near-tetraploid sublines derived during the course of propagating near-diploids that display restrictive host specificities remains to be fully clarified, even though considerations provided by Hauschka,[527] Fox,[402] Ising[644] and Sachs and Gallily,[1145] are noteworthy.   Since the majority, if not all, of the types of cells continuously propagated undergo numerical alterations soon after displaying rapid proliferation, the cellular population eventually bears little resemblance to the original parental karyotype and, except for species-specific immunologic properties, such alterations are difficult to control without the aid of selective and repeated cloning procedures.   Klein has circumvented this difficulty during the course of experimentation *in vivo* by preserving parental immunologic differences and karyotypes of induced malignancies by appropriate methods of freezing.

The number of normal tissues and viral-induced tumor cells that retain their classic diploid (normal) karyotype when cultured is exceedingly small, and those which are reported as being diploid need further verification.   The availability of diploid cells for experiments of various designs would be ideal.   Such a development is one of the more challenging aspects of tissue culture today.   The numerous altered forms of normal cells presently utilized because of necessity and convenience in maintenance have undergone as much or more mutation as the malignant types currently employed. The recent development of a number of diploid and quasi-diploid cultured derivatives from the Chinese hamster, having 14-hour generation times or less in chemically defined media, has been an encouraging step at this laboratory.   Retention of the diploid status in cultures growing rapidly is an essential prerequisite for deriving equally stable monosomic variants by means of extensive cloning procedures.   In this way, distinct nutritional variants and mutants induced by X and ultraviolet radiation may be harvested following the use of minimal media.   On the other hand, deficient media need not be employed when specific antimetabolites are involved in conjunction with plating trials for the purpose of selecting mutants.   Specific mutations may eventually be linked with recognizable chromosomal structures (intact monosomics or structural variants involving heterochromatin).   In the event the normal chromosomal complement cannot be identified readily, suitable markers have been utilized by Hsu and Kellogg[604] and Harris and Ruddle.[522]   Simplification of cellular types and nutritional requirements must be attempted by all who are concerned with employing, with greater value, the vast genetic storehouse provided by the house mouse, the curious immunologic variegation of the Syrian hamster, differences in viral susceptibility of the primates including man, and the small number of linkage groups and readily identifiable autosomes and sex chromosomes provided by the normal and diabetic Chinese hamster.

The rudiments for mammalian somatic-cell genetics have been accumulating during this past decade; additional progress can stem from the employment of more distinctive lines of cells than those currently in use.[773]   To Lederberg's plea can be added the need for karyotypic control as a major critique when evaluating newer forms as potential replacements of the more cumbersome types of cells.

REVIEW OF TECHNIQUES

There have been extensive technical advances in preparing slides.   In addition, slide preparation may involve personal embellishments, as becomes evident to the user in trying to duplicate a particular technique.   At times, independent laboratories have similar difficulties after following accurately the step-by-step procedures described by the originator.   Generally speaking, the number of techniques is increasing, primarily because of the need to assure the maximum number of mitoses, especially for cells having limited viability or mitotic activity and cytoplasmic volume.   The increasing numbers of investigators conducting mammalian chromosomal studies are certain to evaluate the various published procedures quite thoroughly and, in turn, devise additional variations on this subject.   If this follows, it will serve to illustrate the ever widening applications of cytogenetic principles and approaches to more distant areas of research.

The procedures outlined by Ising,[643] Makino and Sasaki,[842] and Hansen-Melander[520] serve equally well as sources of earlier cytologic techniques for ascites-tumor cells.   Persons interested in chromosomes of solid tumors (primary or transplantable) will find the approaches of Bayreuther and Klein[64] and Hungerford[614] most helpful.   Methods of tissue culture have become quite varied.   Cellular types vary considerably in squashing properties, and special pretreatments and procedures have been promoted to emphasize the clarity of the individual chromosome.   These different approaches serve to illustrate the durability of cultured cells when handled properly.   For tissue-culture procedures, other than those outlined in this chapter, the reader will find variations of the general theme and approaches by Puck et al.,[1028] Hsu and Klatt,[605] Ruddle et al.,[1084] Levan and Biesele,[782] Awa et al.,[41] and Ford and Yerganian.[391]   For those who have recently joined the ranks, the air-dry schemes of Rothfels and Siminovitch,[1077] Tjio and Puck,[1322] and Moorhead et al.[893] will provide ample encouragement.   For bone marrow, the procedures of Ford and Hamerton[386] are certain to be satisfactory for many species.

At times, the minute, and occasionally extensive, listing of details of some procedures will indicate the multiplicity of trials which the individual investigator has conducted to prepare chromosomes for viewing.   Since much depends on the individual's skill during any phase of carrying out his procedures, rigid adherence to any one schedule by the novice may result in failure.   If unsuccessful after numerous attempts, he should proceed to another method and, if necessary to satisfy his own ability and requirements, select and combine those steps which appear to enhance the material in use.

Pretreatments

Mitotic arrest and hypotonicity.—Samples of solid, transplantable tumors may be readily obtained for arrested mitoses, following the procedures outlined by Bayreuther and Klein.[64]   These workers employed a small volume of 10 per cent alcohol in distilled water, given intraperitoneally as a means of dilating the vascular network about tumor implants for greater penetration of the colchicine solution subsequently administered (0.5 ml. of a 0.025 per cent solution of colchicine, followed by sacrificing after 90 minutes).   Generally, actively proliferating cells of solid implants (*in situ*) are to be found peripherally, where vascularization assures rapid colchicinization.   Similar procedures may be employed with cheek-pouch implants of both species of laboratory hamsters in which vascularization is known to be extensive.[411]   If the cheek pouch is used, colchicine may be administered intraperitoneally or locally into the cheek pouch, following light anesthesia with ether.   When ascites tumor cells are to be pretreated with colchicine, a dose of 0.17 ml. of a 0.006 per cent solution of colchicine per 10 g. body weight provides the maximum number of mitoses.   Levan presents a convenient conversion table for injections *in vivo*.[780]

To harvest carcinomas and papillomas in rabbits, Palmer[989] injected colchicine-saline (5 per cent) at a dose rate of 0.0012 mg./g. body weight intravenously for 9–12 hours.   Hungerford employed a colchicine dosage of 5×10^{-7} M for six hours to arrest primary murine fibroblasts (unpublished).   Makino and Sasaki[842] arrested mitoses in cultures from various species with a $20–50 \times 10^{-8}$ M solution of colchicine two to five hours prior to trypsinizing and centrifuging.   D. K. Ford and his collaborators[390, 391] utilized a similar schedule, except that fixation with 3:1 enabled lengthier storage of fixed cells.

The striking resistance of the Syrian or golden hamster, *Mesocricetus auratus*, to colchicine is a topic of increasing interest as this species is utilized more extensively for research.[874, 1199]   Harnois[521] has noted that the bone marrow of the Syrian hamster displays numerous c-metaphases after three hours, following treatment with 20 mg. of colchicine/100 g. body weight.   Earlier, Orsini and Pansky[970] characterized the natural resistance of this species to an endemic alkaloid.

Exposures *in vitro* of fibroblasts derived from Syrian hamsters to colchicine are proportionally resistant.   Harnois calculated an optimum number of c-metaphases among pulmonary derivatives of Syrian hamster at a colchicine concentration of 10^{-4} M.   The optimal concentration for cells from the Chinese hamster similarly derived was 10^{-6} M.   Thus, Harnois concluded that cells of the Syrian hamster were 100 times more resistant to colchicine when compared to the Chinese hamster.   This resistance is species specific and may be increased *in vitro* by means of incubating Syrian hamster cells in the presence of a relatively low dosage of colchicine added to the nutrient medium.   It should be mentioned that mitoses of the Chinese or striped-back hamster, *Cricetulus griseus*, are equally sensitive to colchicine, as are the cells of most rodents, primates, and ungulates.   Dosages prescribed for colchicine-sensitive species are equally effective for tissues of the Chinese hamster given both *in vivo* and *in vitro*. The threshold dosage of colchicine for human cultures is 5×10^{-8} M, as reported by Levan[779] and utilized effectively by Ruddle, *et al.*[1084] Renal cells of the monkey respond favorably to a colchicine concentration of $25\,\gamma$/ml. (0.0025 per cent), as

Fig. 54. BENDER'S MODIFICATION OF FORD AND HAMERTON'S[386] TECHNIQUE FOR BONE MARROW AND EMBRYONAL TISSUES.

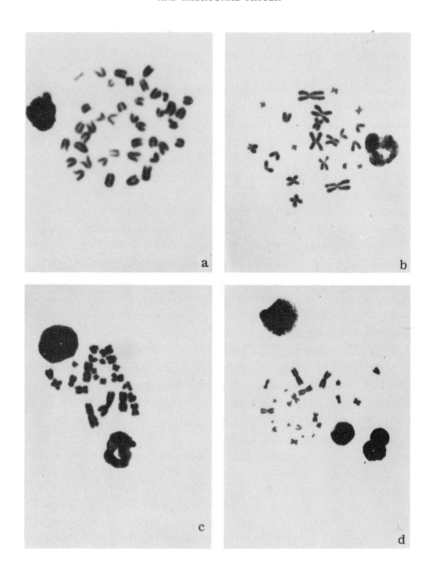

(Air-dried.   1525 X.)
TOP LEFT (a).   Embryonic cell of C3H mouse, sex unknown, aceto-orcein.
TOP RIGHT (b).   Bone marrow of Chinese hamster, male, aceto-orcein.
BOTTOM LEFT (c).   Bone marrow of Chinese hamster, male, Feulgen.
BOTTOM RIGHT (d).   Bone marrow of Chinese hamster, female, aceto-orcein.

reported by Rothfels and Siminovitch.[1077]   Tjio and Puck[1322] noted a final concentration of 0.3–0.5 γ/ml. to be effective for cells of human, opossum, and Chinese hamster in tissue culture.   The length of colchicine treatment *in vitro* will vary considerably from one cell type to another, depending on the generation time of the average cell. Nevertheless, the minimum time for colchicine should be 90 minutes and the maximum 14–18 hours for cells having generation times of 14–24 hours.

An interesting application of Vincaleukoblastine (VLB), an alkaloid of *Vinca rosea*, as a mitotic arresting agent similar to Colcemid, is reported by Palmer.[989] Exceedingly low dosages of VLB, *in vivo* or *in vitro*, lead to mitotic arrest in a number of species and strains of cells.   The J-96, human, monocytic-leukemia cell is readily arrested in metaphase at the low dosage rates of 0.01 and 0.001 gamma.   Its value may be applied whenever a cell undergoes a rate of mitosis slower than desired.

Preparations of bone marrow using the technique of Ford and Hamerton[386] are most adequate for the majority of species.   Recently, the additional step of air drying such preparations has eliminated the need for squashing delicate bone-marrow cells. Bender has followed, essentially, Ford and Hamerton's procedures up through fixation for 30 minutes.   After centrifuging and resuspending in fresh fixative, small drops of material are placed on clean, wet slides.   As the drop spreads out and the preparation dries, the cells flatten quite well, as noted in figure 54a–d.   After drying completely, several drops of aceto-orcein are added, followed by dehydration (3 : 1 and 95 per cent alcohol) and mounting in Euparal.   For similar preparations of embryonic cells of the mouse, 0.2 mg. of heparin is added to 100 ml. of the citrate solution.   This modification is apparently detrimental for bone-marrow cells, but most helpful for embryos. In the latter case, pregnant females are treated with colchicine (25×10^{-6} M/10 g. body weight for 2–6 hours).   After sacrifice, the liver and spleen of the embryo are dissected and placed in citrate solution.   By means of aspirating with a 26-gauge needle, a fairly uniform suspension of intact cells is obtained.   Thereafter, the suspended cells are treated exactly like bone-marrow cells (figure 54a).

The simplest procedure for softening surgically removed tissues for squashing is that modified further by Makino and Sasaki[842] in which they employ distilled water, followed by fixation with 50 per cent acetic acid.   The absence of ethyl alcohol, as found in Carnoy's 3 : 1 fixative, assures softness during subsequent but brief storage.   With tissue cultures, these same workers, using the method described by Awa *et al.*,[41] trypsinized and centrifuged cells previously treated with colchicine and attained hypotonicity by adding an equal volume of tap or distilled water for 5–10 minutes.   Direct fixation and staining was accomplished with acetic dahlia (0.75 gm. of dahlia violet in 30 per cent acetic acid).   Examples of their observations are given in figures 55, 57, and 61.   Ohno, Kaplan, and Kinosita[964, 966] applied distilled water to seminiferous tubules for the purpose of rendering spermatogonial chromosomes more favorable for the squash technique.   Pachytene bivalents, however, were badly affected by this treatment and, to enhance pachytene morphology, storage of freshly removed seminiferous tubules, for two hours in the refrigerator, was found satisfactory (figure 59).

Fig. 55. Culture of lung of male rat, *Rattus norvegicus*

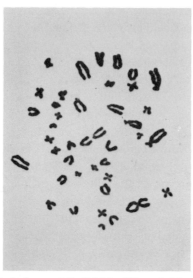

Technique of S. Makino[842] and M. Sasaki. 2000 X. (Photograph by S. Makino and M. Sasaki.)

Fig. 56. Metaphase of primary culture of ovaries and oviducts of Syrian hamster, *Mesocricetus auratus*, displaying four distinct telocentric pairs.

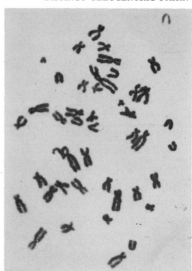

The method of Moorhead, Norwell, Melman, Batipps, and Hungerford[893] was employed following six hours of treatment with 1.3×10^{-3} M of colchicine. 2000 X. (Photograph by P. Moorhead.)

Fig. 57. Splenic culture of male guinea pig, *Cavia porcellus*.

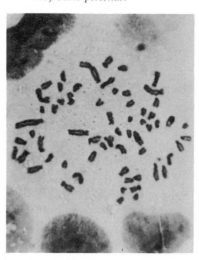

2000 X. (Photograph by S. Makino and M. Sasaki.)

Fig. 58. Primary culture of normal kidney from a cynomolgus monkey.

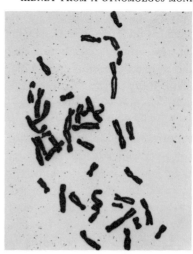

1360 X. Technique of Hsu and Klatt.[605] (Photograph provided by Dr. J. T. Syverton and Associates.)

Fetal chromosomes of the mouse were thoroughly disrupted after five minutes' pretreatment with distilled water.[615] Yerganian[1460] noted that 3:1 fixed and stored human pachytene chromosomes remained satisfactory for many months, even after detrimental effects of a lengthy necroscopy. Figure 60 illustrates Hauschka's unpublished method of pretreating testicular fragments in cold 50 per cent hypotonic Earle's solution. A useful variant of a pretreatment procedure is one employed by Tanaka and Kano.[1310] Following extirpation of the regenerated liver, small bits of tissue were placed in tap water adjusted to a pH of 7.4 for about 18 minutes. The water and tissue were heated to about 38° C. for 2–3 minutes before adding, in equal parts, a 20 per cent solution of Sudan black in propionic acid. Squashes are made

Fig. 59. LATE DIPLOTENE BIVALENTS OF HOUSE MOUSE, *Mus musculus*.

Fig. 60. FIRST MEIOTIC METAPHASE OF MALE HOUSE MOUSE, *Mus musculus*.

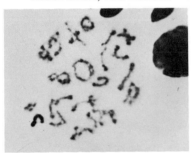

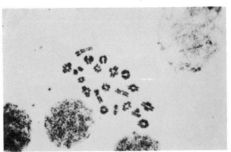

Nineteen autosomal pairs and a densely staining, X-Y pair showing end-to-end association (no chiasma is formed). 900 X. Technique described by Ohno, Kaplan, and Kinosita.[964] (Photograph provided by S. Ohno.)

Note end-to-end association of XY bivalent at top of plate. Acetic orcein squash made after 20 minutes treatment of testicular (semininiferous tubules) fragments in cold 50% hypotonic Earle's solution. 1100 X (Photograph by T. S. Hauschka.)

after one hour of storage in the latter solution. Tonomura and Yerganian[1324] treated regenerating liver by what is now considered the conventional scheme outlined herein for tissue culture. Softening was regained by transferring the tissue (stored in 3:1 solutions) into dilute, nonmordanted acetocarmine for several hours or overnight prior to squashing. This procedure helped to alleviate scattering of chromosomes in otherwise brittle hepatic cells that result after too lengthy a storage period in 3:1 solution.

The use of trypsin (10 per cent) to soften ovarian tissues for the release and squashing of oöcytes has been found to be most appropriate by Ohno, Kaplan, and Kinosita.[965] They generally employed 1- to 3-day-old females for pachytene-diplotene configurations.

Pretreatment with hot water is not applicable for cells in tissue culture. Glass-attached and suspended cultures may be handled in the manner described by Nowell[948] and Moorhead *et al.*[893] The latter procedure is most effective in spreading colchicine-resistant, Syrian hamster chromosomes (figure 56).

It appears that excessive hydration increases fuzziness of metaphase chromosomes. To offset this undesirable feature, the hypotonic pretreatment period has been reduced to five minutes when handling tissue cultures.[1077] The slow addition (drop by drop) of fixatives to nutrient medium of tissue cultures after the addition of the hypotonic agent (tap water or hypotonic salt solutions) renders the tissue more favorable both to wet and air-dried squashing procedures. This has been observed independently by several investigators.

The use of 1.0 N HCl as a softening agent[1320] is a carryover of the action noted during the Feulgen technique used in plant cytology. It dissolves the calcium pectate that binds meristematic cells and thus leads to better squashing.[239] Ruddle[1084]

Fig. 61. CULTURE (PRIMARY) OF SKIN FROM A MALE HOUSE MOUSE, *Mus musculus*.

Fig. 62. HYPOTETRAPLOID EHRLICH'S ASCITES TUMOR.

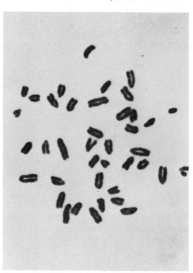

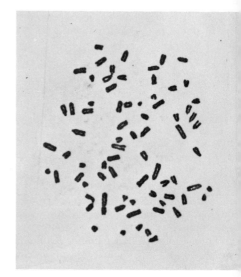

Pretreatment with colchicine and hypotonicity, fixation and staining procedures employed by Makino and Sasaki[842] and Awa, Sasaki, and Takayama.[41] (Photograph by S. Makino and M. Sasaki.) 2600 X.

Colchicine (25 × 10⁻⁶ M/10 g. body weight) administered intraperitoneally six hours prior to aspiration or removal of tumor, hypotonicity (tap water 50 per cent by volume) for 10 minutes prior to adding 3:1 fixative (50 per cent by volume) and storing. A drop of fixed cells is added to a drop of aceto carmine on a siliconed slide and squashed, following conventional procedures. 2250 X. (Phase microscopy. Photograph by R. Kato.)

employs 1.0 N HCl as a cytoplasmic clarifying agent for trypsinized, cultured cells of pigs that are directly fixed and stained with aceto-orcein. The cellular membrane becomes more elastic and thereby encourages flattening of the intact cells. The adaptation of the procedures employed by Hsu and his associates has proved most productive in the hands of Clausen and Syverton[206] (figure 58).

A variety of pretreatment schedules have appeared for ascites-tumor cells from the mouse.[781] The simplest procedure used in the writer's laboratory is illustrated by Kato's photograph (figure 62) of an Ehrlich tumor cell pretreated with colchicine *in situ* at the rate of 25×10^{-6} M/10 g. of body weight for six hours. Following removal of the exudate from the peritoneal cavity, tap water (50 per cent by volume) is added as the hypotonic agent for a period of 10 minutes. Finally, fixative is added (three parts 95 per cent ethyl alcohol: 1 part glacial acetic acid) at 50 per cent volume of hypotonic exudate, and stored until needed. Fixed cells settle to the bottom of the vial and can readily be pipetted or raised by means of an eyedropper. A drop of fixed cells is placed on a siliconed slide and a drop of acetocarmine is added. Conventional squashing procedures follow, depending on the manner of observation, that is, phase or bright-field microscopy.

Other pretreatments.—A number of simple practices have resulted in synchronizing mitoses in populations otherwise randomly dividing. For instance, Wildy and Newton[1389] subjected the *HeLa* cell to cold pretreatment (1 hour at 4° F.) and noted numerous synchronous divisions, some 18 hours later. Such a step would be useful for the evaluation of somatic pairing of homologues of noncolchicinized metaphases and for increasing the incidence of anaphases for purposes of evaluating the action of radiation and drugs. More recently, Nowell[948] noted the stimulatory effect of Bactophytohemagglutinin (Difco Laboratories) on mitoses in leucocytes of human peripheral blood cultures. It can be readily ascertained that increasing the serum concentration (1–5 per cent) of spent basal media to the normal range (10–25 per cent) may also serve to synchronize mitoses.

A masked secondary constriction on chromosome III of the Chinese hamster suddenly becomes visible following treatment with hyaluronidase for two hours.[1459] This enzyme has been quite effective for spreading human chromosomes in bone-marrow aspirates.[855] The alkylating agent, triethylenemelamine (TEM), promotes elongation and separation of chromatids within cells that are readily damaged by the agent (Yerganian, unpublished data). Minimal concentrations of the drug in the range of 1 γ/ml. of medium, or less, may readily lead to matrical anomalies that affect the spiralization diameter of major coils.

Ribonuclease has been utilized by Ohno *et al.*[964] as an agent to reduce the amount of RNA that masks the allocyclic sex bivalent of the rat and mouse during meiotic prophase. The end-to-end association of the X and Y chromosomes was readily seen after 15 minutes' incubation in 0.1 per cent solution of Armour ribonuclease in physiologic saline (figure 59). The same authors[965] noted the immediate dispersion of oöcytes of newborn rats during a 10-minute period of incubation. Hyaluronidase is reported to be quite effective in dispersing chromosomes and cells of bone marrow. Marshall *et al.*[855] placed human bone marrow in a solution containing 15 units of hyaluronidase (Wydase, or Diffusin) per 0.1 ml. of distilled water for a period of one hour. Without squashing, Wright and Giemsa smears were prepared routinely. Marshall (personal communication) states that she and her colleagues "have no supporting evidence for the hypothesis that bone-marrow cells contain 46 chromosomes. In fact, our present data (unpublished) indicate that this is much too simple a hypothesis for such a complex tissue." Could it be that the chromosome numbers of the bone-marrow complex vary purposefully as a prelude to inaugurating the various and complex processes of differentiation? Weicker and Terwey[1369] have reported an extreme variability in the chromosomal number of bone-marrow cells of the Chinese hamster. The slight aneuploidy observed in bone marrow of normal Chinese hamsters by Tonomura and Yerganian[1324] probably are similar to human derivatives in this respect.

Stains.—The use of aceto-orcein or acetocarmine, with or without phase microscopy, has eliminated the need to undertake the more time-consuming Feulgen staining for routine investigation. Although orcein and carmine are more versatile, Lacmoid, Dahlia Blue, Chlorazol Black E, and Sudan Black have been utilized occasionally. The use of nondecolorized basic fuchsin for staining of murine pachytene chromosomes has been found satisfactory by Slizinski[1217] and Jaffe.[660] A general reference recommended for staining procedures is the compact and revised report by Darlington and LaCour.[239]

The selection of a stain is regarded as a personal choice, rather than one of tried experimental comparison. Acetocarmine and propionocarmine are considered by the writer to be far more favorable for mammalian chromosomes than aceto-orcein, customarily recommended for this type of work. When viewed with phase microscopy, carmine is excellent. When employing procedures described below, carmine rarely fills the matrical portion of the chromosomes when limited to 30 seconds of staining, as contrasted to orcein which stains the chromatin as well as the pellicle or matrix deeply. The carmine-stained chromatid appears minimal in diameter, as in the case of Feulgen stain, when compared to that noted after orcein-stained trials. The finer details of chromosomal structure, that is, heterochromatic *versus* euchromatic areas, are readily visualized with light carmine stain. Excessive mordanting (see below) and lengthy periods of staining generally lead to a masking of all important structural entities. This may be illustrated by the fact that the distinct features of the mitotic sex chromosomes were not recognized readily during the past decade, although a myriad of rodent metaphases were viewed by well informed cytologists. A similar trend is now occurring in the study of human mitotic chromosomes. The majority of the cells were heavily stained with orcein and viewed either by bright-field or dark-contrast, phase microscopy. The writer considers the use of light-carmine staining and medium-to-light contrast, phase microscopy a great help in revealing heterochromatic portions and sex elements of Chinese hamster and human derivatives (figures 63, 64, and 65).

R. Lang (personal communication) has experienced no precipitation of synthetic aceto-orcein when using 85 per cent lactic and glacial acetic acids. Equal volumes of the acid solutions are warmed almost to the boiling point in a water bath prior to adding the appropriate amount of stain. The mixture is left to cool, then filtered and stored. Following this type of preparation, orcein fails to precipitate and slides resist drying.   It would be of great interest to follow this procedure, using only warm acetic acid to check the rôle which lactic acid may play in the control of precipitation.

The addition of 1 per cent lactic acid (by volume) to acetocarmine has enabled

Fig. 63. A DIPLOID, FEMALE, ADULT, FIBROBLAST DERIVATIVE (FAF 28) WITH CYCLES OF CELLULAR DIVISION OF 14 HOURS OR LESS.

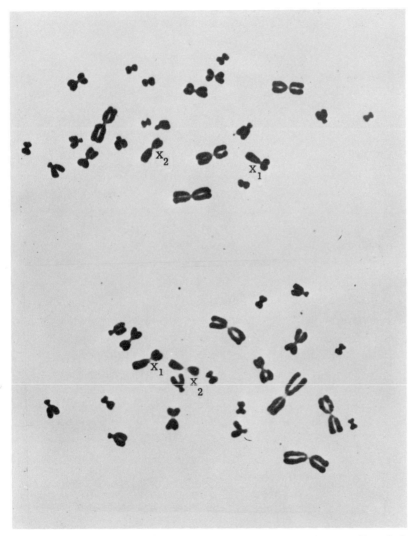

Note the duality of the paired X chromosome; the "predominant X_1" and the "retarded X_2" are different morphologically and physiologically.   1500 X approximately.

numerous plant cytologists to intensify the staining of pachytene chromosomes of *Zea mays*, without unnecessary distracting cytoplasmic staining.

When clones attached to petri dishes are to be stained, the medium is rinsed

Fig. 64. POLYOMA III.

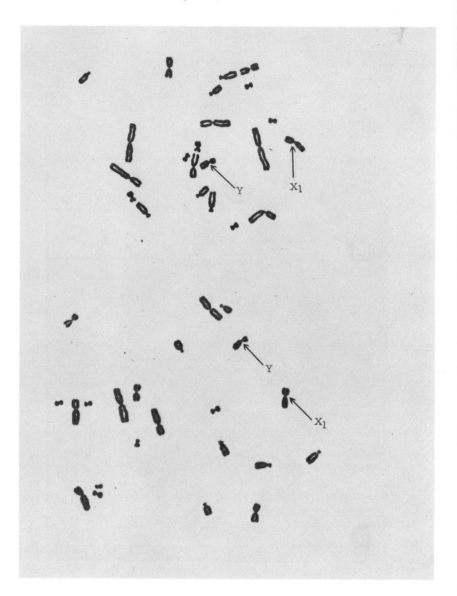

TOP   "Wet" squash preparation of *polyoma III*.   Quasidiploid state with four reciprocal translocations involving chromosomes 1, 2, 3, and 4 at the centromere.   Note intact "predominant X_1" and Y chromosomes.   1450 X.

BOTTOM   Permanent slide of *polyoma III* following Conger and Fairchild's technique.[221]   Slight fuzziness of chromosomes, due to aging of slide temporarily sealed with Kröning cement for several weeks.   Making the preparation a permanent one is best done some 72 hours after squashing to eliminate such alterations in chromosomal morphology.   1450 X.

rapidly with tap water and·the dishes turned down to drain.   Staining is easily done by bathing the clones with a 1 per cent (hematoxylin, toluidine blue, or neutral red) for 5–10 minutes, or more, if it is found desirable (figure 66).   The stain is retained for

Fig. 65. SCHEMATIC IDIOGRAM OF THE CHINESE-HAMSTER KARYOTYPE.

(colchicine pretreatment)

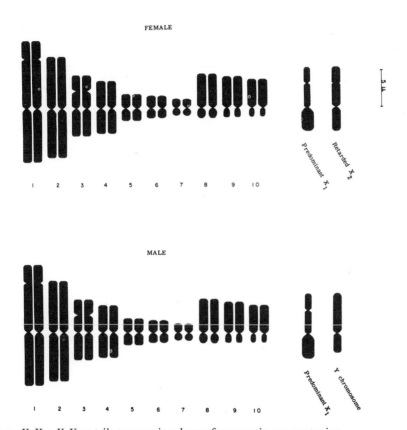

Note X_1X_2; X_1Y or triheterosomic scheme for somatic sex expression.

future use by refiltering, and the dish is rinsed several times with tap water and again turned to dry.   Visualization of stained clones is inexpensively and most effectively accomplished with the aid of a Bausch and Lomb slide projector placed in a darkened room.   Abortive clones, that is, those having 50 cells or less, can be distinguished with great assurance only when viewed with such an instrument.   Abortive clones are generally deemed to be negative when determining cloning efficiencies.

Fig. 66. CLONING EFFICIENCY OF POLYOMA DERIVATIVES OF THE CHINESE HAMSTER.

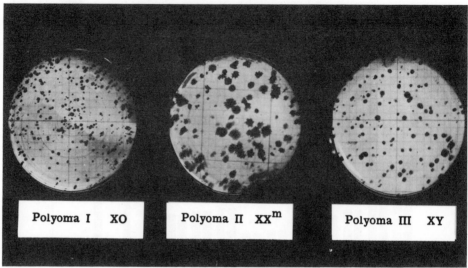

Average of 50 per cent cloning efficiency (Puck Technique) displayed by three quasi-diploid, polyoma derivatives of the Chinese hamster. The female derivatives feature variations in the sex mechanism, i.e., XO and XX^{mod}. The male derivative (XY) features four autosomal translocations and an intact sex mechanism. (See figures 64 A and B.)

Siliconed slides and procedures for permanent preparations

Invariably, the use of siliconed slides or cover glasses greatly enhances wet squashes.[1174] Silicone-treated paper for wiping eyeglasses is a good substitute, particularly when the type of cell studied has adequate cytoplasmic volume. The diploid cell, with minimal cytoplasmic volume, responds more favorably when silicone-treated slides are used. Incidentally, wet-squashed cells are flattened most satisfactorily on siliconed slides, whereas the air-drying procedures employ the principles of hypotonicity, evaporation, and resulting surface tension to spread chromosomes favorably.

In the event silicone is unavailable, a satisfactory surface may be prepared simply by touching the microscope slide to one's nose. The smear of oil is vigorously rubbed clear with a towel to provide a smooth and water-repellent surface. The use of Krönig's cement is mandatory to assure the proper seal, regardless of the manner in which the slide was wiped clean or prepared.

In the past, reports of extreme variations in chromosomal number of a given tissue led to considerable debate as to the consistency of modal numbers. The novice must be made aware of this potential pitfall when adapting techniques of chromosomal preparation to his needs. To insure against loss of chromosomes, it is urged that the principles of air-drying be employed at first.[1322, 1077] These procedures are especially helpful with normal cells having cellular membranes which are prone to rupture at the slightest pressure. Familiarity with the tissue will then encourage wet squashes as a means of determining which approach provides the better spread of chromosomes for individual needs. Delaying wet squashes for some 48–72 hours following fixation will help to harden glass-adhered mitoses for improved and intact squashing onto siliconed slides.

The procedure of Conger and Fairchild[221] for permanent preparations has become universally adopted with slight modifications, such as the use of liquid nitrogen for quicker freezing in place of dry ice (CO_2).

PROCEDURES *in vitro* FOR CHINESE-HAMSTER CULTURES

A culture vessel has been designed by the author to serve in a number of capacities in current techniques of tissue culture (figures 67, 68, and Addendum). The neck of the vessel may be bent slightly to insure against accidental spillage and contamination while the cap is left loosely turned when placed in a CO_2 incubator. The neck of the tube is reduced, as contrasted to the lengthier version of the original Leighton tube, to facilitate removal of the cover glass as well as to perform limited clonal isolation procedures.

The use of ferric hydrate or acetate mordant is relatively new for tissue cultures of mammalian cells, even though it has been extensively incorporated in procedures for the staining of plant chromosomes for well over thirty years. The use of razor blades and repeated filtration of the resulting saturated solution of ferric acetate has eliminated the adverse precipitation experienced in the past, particularly when staining botanic specimens. To reiterate, acetocarmine is preferred as a universal stain because it requires less care and rarely precipitates in the bottle after many months of exposure at room temperature. It stains the inner components of the chromosome, leaving the matrix clear. In contrast, aceto-orcein requires daily attention to remove troublesome precipitation, and it stains the entire thickness of the chromosome. Consequently, the former stain leaves the chromatids slender and distinct whereas the latter stain results in a thicker and more densely stained structure. Acetocarmine assists in revealing structural entities, such as secondary constrictions and heterochromatic segments that are consistently noted as being part of the more slender portions of chromatin, especially along the length of the sex chromosomes (figures. 63 64, and 65).

The use of petri dishes for the cultivation of cells onto flying cover glasses in a CO_2 environment is, comparatively speaking, less efficient with regard to utilization of medium and incubator space. When several cover glasses are placed in petri dishes, they generally overlap or are readily disturbed while the contents are viewed with an inverted microscope. In addition, an average of 5 ml. of medium is required for each 60-mm. petri dish. This same amount of medium, when used with the culture vessel described herein, will yield three times more surface.

In the event excessive cytoplasmic staining is experienced, propionic acid may be substituted in place of acetic acid for the preparation of stains. The chromosomes appear slimmer (less or no matrical staining) than that resulting from acetocarmine. The most intense staining results when acetocarmine and aceto-orcein are mixed (1 : 1) and further amplified with the aid of the mordanting procedure. Since aging improves stains, it is best to make large batches and store for several years, if this is practicable.

The tissue-culture preparations described below may be dried in air after being fixed primarily for wet squashing. This practice provides some versatility when appropriate attention cannot be given at the time of fixation to follow the precise, step-by-step air-drying procedures described by Rothfels and Siminovitch[1077] and Tjio and Puck.[1322] Air drying of wet-stored cultures is satisfactory only when the cellular population is minimal and the cytoplasm well spread. Following satisfactory wet squashes, duplicate fixations may be rinsed with fresh fixative prior to air drying and storing in a slide box in which the wooden length of vertical slots has been reassembled to accommodate the length of the cover glass. When excellent observations are in the making and duplicate preparations are available, it is best to proceed to complete the slide making. On many occasions, the second observation provides equal or better squashes with the help of features noted in the initial preparation.

In addition to providing adhering cells for a variety of wet and dry squashing procedures for chromosomal preparations, the cover glass provides cells for general study of clonal (phenotypic) features or responses to agents as well as intraclonal and interclonal variation of chromosomal complements. The culture vessel encourages isolation of clones by means of directing a simple, orally manipulated (latex tube) micropipette while viewing the proceedings under the low power of an inverted microscope. The cover glass can be transferred, if desired, to another vessel for pretreatments and the like, so that genetic continuity can be maintained by repopulation of the area barred by removal of the cover glass. Actively proliferating cells along the periphery of the vessel tend to migrate centrally after removal of the cover glass. By so doing, a larger sample of genetically related sublines may be retained each generation without the need for preparing and maintaining an equal number of farming bottles (without cover slips) for replicate samplings and general continuity of the stock. The size of the cover glass can vary, depending on the needs of the experiment and the ability of the strain in question to repopulate and subculture readily following minimal inoculations of cells. In general, 10 × 45 mm., 0-thickness cover glasses are employed with this vessel for purposes of sampling and maintaining sublines with minimal handling. Periodically, auxiliary subcultures are grown in large farming bottles (200-ml. serum bottles, or milk-dilution bottles) for purposes of freezing viable cells, according to the procedures described by Stulberg et al.[1301] and Hauschka et al.[534]

In our hands, cells grown in 200-ml., serum, farming bottles are prepared for freezing by simply draining off the medium and scraping with a rubber policeman. One ml. of complete medium, adjusted to *p*H 7.0 and containing 10 per cent glycerol (Merck) is added to the scrapings gathered at the base of the flask. With the aid of a Pasteur pipette two 1.5-ml., Wheaton freezing ampoules are filled with the fluid containing several million cells and restoppered with the cotton plug. The ampoules are sealed some 20 minutes later with the aid of a small propane torch and placed in a small rack or container on the top shelf of an upright deep freezer at $-4°$ C. for several hours. Thereafter, the vials are stored directly at $-79°$ C. in a cabinet designed by Stulberg et al.[1301] Practically all types of Chinese-hamster cells (over 200 genetic clones) are favorably preserved when prepared in this simple manner. Maximum periods of storage and percentages of successful regrowth have yet to be determined. It appears, however, that other laboratories have had equivalent success without the necessity of following the more elaborate approach of slow freezing as described by Hauschka et al.[534] Both approaches have been employed with tissues of the Chinese hamster, and time will prove which is the better method. In this manner, genetic continuity is assured and restoration is greatly facilitated in the event of sudden, undesirable shifts in the karyotype or outright losses due to accidents.

Wet-squash preparations of colchicine-arrested metaphases generally exhibit minimal overlapping of chromosomal arms. Some care, of course, must be taken to assure cytoplasmic integrity as contrasted to the safety of air-drying procedures. Nevertheless, various strains respond differently in regard to this vital feature.

Squashing flattens cells more effectively than drying in air, particularly when cellular membranes are exceedingly elastic or excessive clumping of chromosomes results as a response to certain treatments. In the event air-dried cells on a cover glass are preferred, the procedures outlined by Rothfels and Siminovitch[1077] and Tjio and Puck[1322] are readily combined to handle cells cultured in these vessels, along with a saving of several ml. of medium per trial.

The use of ferric acetate as a mordant for carmine stains will be discussed in detail. This adaptation, stemming from Belling's classical developments in plant cytology, has been found useful by the author for both phase-contrast and bright-field microscopy of mammalian chromosomes. Chromosomes stain intensely, along with minimal uptake of the dye in the cytoplasm.

Preliminary preparations

Several ounces of cover glasses (0 thickness, 10 × 45 mm., or less) are placed in a beaker containing 100 ml. of a 1 per cent solution of 7 X detergent and heated for a period of about one-half hour. The cover glasses are rinsed thoroughly with running tap water for several hours, with frequent agitation. Then they are rinsed and agitated thoroughly with distilled water and stored overnight in it. After draining the distilled water, 70–85 per cent ethyl alcohol is added and the jar or container covered tightly until needed. When cover glasses are placed in culture vessels, they are removed individually from the alcohol with the aid of forceps and wiped dry with lens paper to prevent fingertip smudges along the edges which are excellent adjuncts for cell proliferation. The cover glass is placed in the culture vessel and stoppered loosely for autoclaving. (CAUTION: The rubber-lined bakelite cap should be tightened only after there has been sufficient cooling following the autoclaving.)

Preparation of iron mordant

Place 6 to 10 razor blades (Gillette blue blades) in a loosely stoppered Erlenmeyer flask and add 100 ml. of 45 per cent acetic acid. Agitate occasionally during a three to six week period of brewing. When the solution is a deep amber color, filter repeatedly several times and store overnight. In the event of further precipitation, add 25–50 ml. of 45 per cent acetic acid and refilter. Store without precipitation at room temperature. A second batch of mordant should be left to age for several months; a gradual drying-out of the acetic acid and eventual erosion of the razor blades will occur. By simply adding 50 ml. of 45 per cent acetic acid to the crumbled mass and filtering, a very effective and concentrated mordant is made available immediately upon filtration. The original corroded mass is retained for future needs. Several hundred milliliters of mordant can easily be obtained in the manner described.

Culture vessels are seeded with appropriate numbers of trypsinized cells (approximately 50–100,000 maximum) and kept either in a CO_2 incubator (loosely stoppered bakelite cap) or a conventional incubator (tightly stoppered) depending on cellular performance. When cells enter the log phase of growth some 48 hours later, colchicine pretreatment, with the addition of 0.5–1.0 γ/ml. of medium, is generally initiated. Stock colchicine is prepared by dissolving 0.5 mg. of colchicine in 100 ml. of complete medium to provide a concentrate of 5 γ/ml. Five- to ten-ml. batches of filtered solution are stored at $-4°$ C. until needed. Thereafter, prolonged storage at $+4°$ C. is adequate and 0.1–0.2 ml. of this stock colchicine solution is added to provide a final concentration of 0.5–1.0 γ/ml. in the supporting medium.

The culture may be incubated for an additional two hours, or to a maximum period of 15 hours, or overnight, depending on the experimental requirements and generation time of the tissue employed. For example, some Chinese-hamster derivatives have 10- to 12-hour generation times and one must act accordingly to minimize the frequency of endoreduplications and polyploidy that can arise following prolonged colchicinization.

To reduce the degree of fuzziness displayed by the pellicle of chromosomes as a result of hypotonic pretreatment, the time of exposure has been reduced from the conventional period of 20 minutes to only 5 minutes. Hypotonicity is achieved by adding one-half ml. (10 drops) of tap water to the supporting medium in the vessel. After 5 minutes, 0.5–1.0 ml. of Carnoy's fixative (3:1) is added to the hypotonic medium. Thus, the volume of solution in the vessel is approximately doubled at this point. For best results, the culture tube should be stored at room temperature for at least 18 hours to encourage some degree of hardening of the cellular membrane for better squashes. In the event air-dried specimens are desired, several rinses with 50 per cent (one part 3:1 fixative and one part water) may be added to the schedule prior to air drying. Media containing horse serum will generally exhibit a mucoidal precipitate and should, therefore, be thoroughly rinsed and replaced with 50 per cent 3:1 fixative for delayed wet squashing. Fixed media containing fetal calf and other kinds of sera are unaffected by acetic acid and remain clear for many months.

Squash technique for chromosomal analysis

1. Transfer the cover glass upright with forceps and place upon a pedestal (30 × 60 mm. vial), as depicted in figures 67 and 68.

2. Add 3 drops of 1 to 1½ per cent acetocarmine or propionocarmine to the upper surface of the cover glass.

3. Touch the tip of an eyedropper or dip-stick retained in the iron mordant concentrate to the stain to deposit a very small amount (½ drop or less).

Fig. 67. ILLUSTRATION OF VARIOUS STEPS IN SQUASHING PROCEDURES.

A. Pretreatment(s) and Fixation

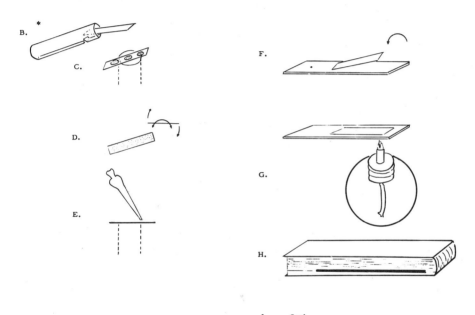

Method of G. Yerganian and associates.   Refer to figure 68 for layout of instruments and chemicals.

4. Lift cover glass with forceps and tilt back and forth to disperse the mordant throughout the stain, covering the entire surface of the cover glass.   A purple hue will be seen to spread over the entire area.   Place atop the pedestal and wait 20–30 seconds (phase-contrast microscopy), or 60–90 seconds (bright-field microscopy) before proceeding.

5. Decant (into vial) the now purple (overmordanted) stain and reset atop the pedestal.   Rinse with several drops of dilute acetocarmine or propionocarmine and decant.   Dilute stain is readily made available by simply redissolving the stain-mash that accumulates onto the filter paper (following the initial filtration) in a similar volume of 45 per cent acid and refiltering.   Straight 45 per cent acetic acid may act

too effectively as a destaining agent, thereby leaving the chromosomes only faintly stained. Whenever major spiralizations of the chromonemata become evident at low magnification after destaining, some control over this step is warranted. The use of the stain prepared secondarily will help greatly whenever the cytoplasmic staining must be corrected prior to squashing for better phase-contrast observation.

Fig. 68. LAYOUT OF CHEMICALS AND INSTRUMENTS FOR PREPARATION TO DEMONSTRATE CHROMOSOMES.

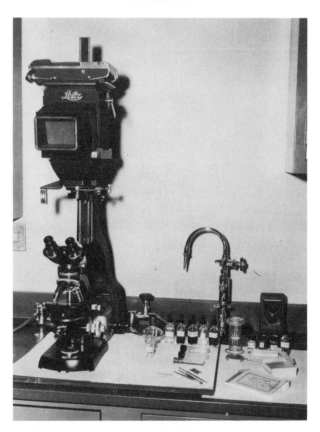

6. Invert cover glass onto siliconed microscope slide and pass over a low flame several times. Press firmly and heavily—without any side motion after placing the slide in the center of a bibulous pad. Use fingers of both hands to cover area of cover glass.

7. Observe slide under the microscope. In the event cells appear fully rounded and dense, it generally indicates: (a) the presence of too many neighboring cells or (b) insufficient pressure applied to the pad. Remove slide from the microscope and add a few drops of the dilute (0.5 per cent) acetocarmine along the edges of the cover glass, pass over the flame several times, and press again under the pad, but this time, place the

slide in the upper one-third of the pad.   Reexamine and repeat this step, if necessary. The more favorably squashed cells will appear (in the microscope field) along the upper edge in the center region.   Just why the lower edge of the cover glass responds so favorably to the squashing is not fully understood.   In the event this area of the cover glass is noted to have overlapping chromosomes and unflattened cells, added pressure is sorely needed.

    8. Seal the edges of the cover glass with Krönig's cement (see Addendum).

    9. Store for several days to several weeks.

    10. Make slides permanent after 48 hours or more of storage, following the freezing technique of Conger and Fairchild.[221]

    The above schedule is primarily for light-medium phase-contrast microscopy, using a light green filter (Zeiss Opton, Model W).   In the event dark contrast objectives are employed, such as Neofluars, the mordanting step may be eliminated or destaining may be more effectively conducted with 45 per cent acetic acid in place of 0.5 per cent acetocarmine for the final rinse.   On the other hand, bright-field microscopy may require excess mordanting and minimal destaining to enhance the differential affinity of the mordanted stain for chromosomes.   The ferric acetate mordant prepared with razor blades has eliminated the troublesome precipitate generally witnessed in the past with botanic specimens.   Excess staining of the cytoplasm may result when slides are stored for a long period.   This can be alleviated at the time the slides are to be made permanent, simply by rinsing once with 45 per cent acetic acid, or dilute stain, following separation of the cover glass and prior to initiating dehydration.

Unsquashed preparations for general morphology

    This procedure permits observations immediately after fixation and in the permanent state.   All pretreatments (hypotonic and the like) essential for chromosomal preparations must be avoided.

    1. Fix by adding 1 ml. of 3:1 fixative directly to the medium.   Store until necessary, as mentioned above, or proceed to complete the schedule.

    2. After 10–15 minutes of fixation, remove the cover glass and place atop the pedestal.

    3. Rinse with several drops of fresh 3:1 fixative.   Decant.

    4. For phase-contrast microscopy, add 3 drops of 1–1½ per cent acetocarmine (no mordant) and wait for 20 seconds before rinsing with 3:1 solution.   For bright-field microscopy, add a touch of mordant, blend into stain, and wait for 60–90 seconds. Rinse rapidly with several drops of 0.5 per cent acetocarmine and halt further destaining by adding a few drops of 3:1 fixative.

    5. Dehydrate by placing the cover glass in a vial containing 95 per cent ethyl alcohol and mount within 10 seconds, using Euparal (see Addendum) sparingly.   Pass over a low flame.   Press very gently under many layers of the bibulous pad.   Observe under low power.   Store flat until dry.

    Hematoxylin-and-eosin preparations are most satisfactory for viewing general morphology of attached and spread cells. However, the length of time required to conduct the staining, along with the many solutions needed for the schedule, tend to make such preparations a major undertaking. The procedure outlined above permits rapid and better visualization than staining with hematoxylin and eosin. The 3:1 fixative yields nuclear details with distinct clarity. Chromocentral masses, nucleoli, sex chromatin (in man), and cytoplasmic vacuoles and metabolic by-products are noticeably enhanced when using phase microscopy.

Nuclear and cytoplasmic features serve as phenotypic expressions among a large number of Chinese-hamster clones. Variations or deviations from classic diploidy, loss of sex heterochromatin, monosomaty for distinct autosomes and the like, reflect in the appearance of nuclei of 3:1-fixed cultures. The clonal cell can be utilized to record its specific phenotype-karyotype complex most conveniently with the above schedules for chromosomes and general morphology.

Routine for tissue culture

Schedules can readily dominate the attention of a cytologist beyond the desired limit. To help minimize routines, this laboratory changes the medium in flasks only twice weekly in place of the three changes customarily recommended. Also, in the process of changing the medium, only half the volume is replaced in large farming bottles, thereby conserving better than 50 per cent of the expense and labor conventionally assigned to the preparation of media. Literally thousands of cultures have been handled this way most favorably, and none have displayed deleterious effects. The media of culture tubes are never changed and thereby retain the initial $1-1\frac{1}{2}$ ml. of medium provided at the time of seeding. Stock colchicine is added to the partly spent medium at the prescribed time, or at the start of various pretreatments. This may occur as late as three days after initiating the tube culture. Chinese hamster cells proliferate rapidly in neutral or slightly acid media. In the absence of a change of medium, acidity still favors rapid metabolism and proliferation. Conservation of media is also extended to routines of tending to petri dishes placed in the CO_2 incubator for cloning trials (figure 66). In these instances, the medium is never changed during a 7- to 10-day period of incubation.

Every effort is made to insure sterility of the media and other culture solutions, to control the temperature of the incubators, and to regulate the amount of CO_2 being delivered to the incubators. Adequate pretesting of media for sterility with surplus cultures, and with thioglycollate (Difco Manual) is one of the basic steps to help retain genetic continuity of unique sublines.

Freezing and storage of pretested media is also an essential precaution, particularly if minimal inoculations of cells are to be done. Such procedures appear to be unnecessary or disregarded in laboratories where detection of genetic variations is of secondary importance. Researchers should be aware of the cytologic progression and cellular contaminations that have occurred among the various strains,[1076] despite efforts to characterize given types at various repositories.[206]

Steps to insure genetic continuity

A number of crucial reports have appeared in the literature regarding the sudden transformation of a particular cell line from its usual fibroblast-like appearance to one of epithelial features.   The extent to which such alterations have been witnessed is reviewed in part by Rothfels *et al.*[1076]   During the initial studies on Chinese-hamster cultures, Ford and Yerganian[391] noted a shift in the appearance of the six-month-old A strain toward polyploidy (70[+] chromosomes) after seven months.   The 70[+] chromosomes suggested cells of this culture to have hyperoctoploid complements, if derived from Chinese-hamster cells.   However, the chromosomes lacked any similarity to Chinese-hamster types or their immediate alterations, as noted previously among transplantable tumors.   The only plausible explanation was a technical error made when pipetting media and failing to use a separate pipette for each operation.   This measure of caution had not been fully incorporated into the routine management of the stock.   Direct cellular contamination by a line of human cells was substantiated further by employing serologic tests, using antihamster and antihuman rabbit serum.   On another occasion, a culture isolated from a normal adult Chinese hamster was sent to us for a report on the chromosomal situation on the first anniversary of its existence *in vitro*.   Immediately upon viewing the first few metaphases, it was obvious that the entire population of cells was human in origin!   Further inquiry disclosed that the culture had changed recently from a fibroblast-like form to one with epithelial features.   This, too, was another example of direct cellular contamination.   The implications of such accidents are far too complicated to suggest any remedy other than to discard the majority of the suspected forms and to replace them with equally available, and probably more serviceable cell lines of known sex and species.   Thus, adequate control and routine guarding against such events in the future would entail the rule that a pipette be used only once.   The beginner in tissue culture should either start his or her own cultures or obtain them from a reliable source that has conducted chromosomal and, if possible, immunologic comparisons.   Direct cellular contamination is far more widespread than is usually realized.[206, 1076]

Activities that help guard against accidental loss of lines of cells

1. Freeze representative samples of parental line and variant clones.[534, 1301]

2. Test filtered media for sterility with surplus cultures and by placing 1-ml. samples into sterile Bactothioglycolate while bottling media.

3. Reduce serum toxicities, particularly for trials of plating efficiency by: (a) purchasing serum from a reliable source that is aware of the problem; (b) preheat at 56° C. for 30 minutes prior to filtration (D. K. Ford, personal communication); (c) combine serum from two species, such as fetal calf serum and horse serum (Yerganian, unpublished data); (d) use dialyzed serum wherever applicable, and (e) substitute fetuin and albumin in lieu of whole or dialyzed serum, as prescribed by Fisher, Ham, and Puck.[367]   The addition of NCTC 109 (4–6 per cent) and 0.5 per cent lactalbumin

hydrolysate provides extra nutritional factors to help overcome toxicities and use of unfavorable media for general subculturing purposes (Yerganian, unpublished data).

(4) When pleuropneumonia-like organisms are suspected, substitution of 100 γ of Kanomycin sulfate (Bristol Laboratories) in place of penicillin and streptomycin for a three-week period has been demonstrated by Pollock, Kenny, and Syverton[1009] to be effective for well over six months. Retention of cultures at 41° C. for 4 hours has been found by Hayflick[536] to destroy infectious agents selectively.

5. Undertake trials of single-cell isolation for purposes of regaining the desired types of cell only when adequate supplementary inspections can be made, such as checking for chromosomal markers; viral susceptibilities of the parental stock; tumorigenic potential in homologous or heterologous hosts; growth rate or cell-generation time; and colonial features, including distribution of heterochromatic masses and shape of the nucleus.

Plating efficiency may be improved by the addition of alpha-keto acids[941] and/or the addition of 50 gamma of trypsin per ml. of medium.[322]

THE KARYOTYPE AND IDENTIFICATION OF INDIVIDUAL CHROMOSOMES

The ever-increasing accuracy with which chromosomal complements of many rodents are being identified has been most encouraging (figures 54–64). The human karyotype is but one example of the manner in which heretofore numerically complicated complements of chromosomes are being readily recognized today (figure 69). Nevertheless, the greater the proportion of telocentric forms of chromosomes, the more difficult it is to provide an accurate karyotypic analysis. Although the house mouse is favorable for genetic studies, difficulties arise when attempts are made to identify the individual chromosomes (including the sex elements). Equally disturbing is the early mitotic breakdown displayed by murine tissue cultures.[782] Renewed effort should be made by tissue culturists to meet the challenge to control the karyotype of cultured cells of the mouse and to reveal the true identity of the sex elements in mitosis. Currently only chromosomal behavior in the Chinese hamster is well understood, including its complex triheterosomic somatic sex expression. However, this phase of cytology will be in constant flux as reports on the karyotypic features of normally derived tissues continue to appear.

SEX CHROMOSOMES AND HETEROCHROMATIN

The various terms that apply to the normal, yet exceptional, behavior of sex chromosomes in the heterogametic sex (namely, the male) has recently regained prominence (figure 59). Further clarification of the phenomenon has become warranted following the interesting revelations by Ohno and his collaborators.[963, 966] The possible interrelationship between the sex chromatin of Barr and the actual sex chromosomes still is uncertain in man. Unfortunately, the presence of sex chromatin-like

Fig. 69. PERIPHERAL-BLOOD DERIVATIVES OF MAN.

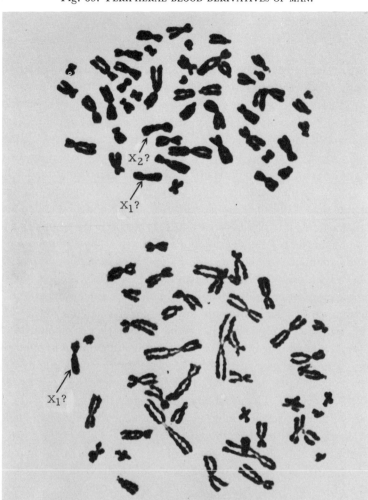

Mitotically active leucocytes prepared by the technique of Nowell[948] and stained with lactic-acetic-orcein by (Mrs.) R. Lang in the laboratory of Dr. P. Gerald, The Children's Hospital Medical Center, Boston.   (Photographs by R. Kato.)

TOP.   The author has interpreted one of the two X chromosomes from this peripheral derivative of a normal female as being the submetacentric pair exhibiting lack of chromatid separation, due probably to the retarded development of extensive segments of hetero-chromatin.   Other cells from this individual have unequivocally displayed the counterpart of the $X_1 X_2$ condition noted in the Chinese hamster.   The existence of a sex-determining (somatic) principle other than the assumed $XX:XY$ scheme in man requires application of the differential tritiated-thymidine-uptake pattern noted for sex elements by Taylor[1316] and confirmed by Yerganian and Grodzins (unpublished data).

BOTTOM.   The author has interpreted the "retarded" element as the X_1 chromosome in man.   Note the fuzziness and delayed separation of chromatids which characterize sex elements during metaphase.   2400 X.

bodies in rodent species remains to be verified, even though resting cells of female tissue appear more chromatic.[769] The distinctness of the heteropycnotic behavior of the *XY* bivalent and its respective mitotic analogs is quite striking.[966, 1459] (See figure 59.)

Ohno, Kovacs, and Kinosita[967] and Ohno and Hauschka[963] related differential heteropycnotic alterations to representing only one of the *X* chromosomes in female cells of the rat and mouse. A recent assessment and acceptable explanation of Tjio and Ostergren's[1321] observation of so-called heterochromatic stimulation by the viral agent of mammary adenocarcinoma[963] has brought the mitotic sex elements in the mouse to light for the first time in over a decade of critical observation by numerous cytologists. The great majority of reports on murine chromosomes to date have failed to be concerned with the identification of sex elements. At present, the status of sex chromosomes in practically all of the experimental tumors of the rat and mouse remains virtually unknown.[781,876] A similar situation does not exist in the Chinese hamster, since the sex chromosomes have been identified for some time. The distinctness in structure, function, and evolutionary significance of the sex chromosomes of normal and malignant cells have been reviewed by Yerganian et al.[1463] In the Chinese hamster, the *X* and *Y* chromosomes are readily identified by characteristic features and the more slender appearance of heterochromatic arms, even at magnifications as low as 200 times. The individuality of the *X* chromosome in both sexes is so readily seen that the duality of the *X* chromosome in the female is prominently brought to mind (figures 63, 64, and 65). Additional genetic and physiologic evidence for the duality (X_1 and X_2) of the *X*-chromosome pair in the female Chinese hamster has been presented elsewhere.[1316, 1459, 1461, 1464] The possibility that a triheterosomic somatic sex expression exists in many other species, including man, is most encouraging in the light of more recent morphologic evidence from this laboratory (figure 69).

Although numerous photomicrographs of normal and abnormal human complements have appeared in the literature recently, few of the authors selected a pair of metacentrics in Group 6–12, which this author and his colleagues consider either as being consistent from paper to paper, or as counterparts of the X_1 and X_2 chromosome of the Chinese hamster. Figure 69b is of a normal male, showing the X_1 quite clearly due to its partly allocyclic nature. A slight fuzziness is apparent and, at times, exaggerated when the chromatids have not separated sufficiently. Figure 69A is that of a normal (pregnant) female showing the X_1 and X_2 chromosomes. The dual nature of the *X* pair is quite apparent, even when following staining procedures outlined by others. It is most puzzling to view these structures in human cells as similar to the scheme in the Chinese hamster. A triheterosomic somatic expression or duality of the *X* chromosomes is also noted in malignant cells cultured from a Wilms' tumor and an ependymoma (Yerganian and Kato, unpublished data).

Autoradiographic procedures to reveal the course of tritiated-thymidine labeling of chromosomes have been readily adapted to known mammalian chromosomes by Taylor[1316] and Yerganian and Grodzins (unpublished data). The differential uptake of tritiated thymidine by heterochromatin as reported by Lima-de-Faria,[788] supports the writer's interpretations for the presence of two forms of the X chromosome in the Chinese hamster. Unpublished findings of Taylor and of Yerganian and Grodzins indicate that one of the small metacentrics of the Chinese hamster also is composed of heterochromatin (differential tritiated-thymidine uptake). The persistence of *Minutes*, particularly viewed among murine ascites tumors of long cultivation (figure 62) may be nothing more than residual chromosomes having a small amount of heterochromatin about the functioning kinetochore. Thus, kinetochores may persist for varying lengths of time in cultures as well as during organic evolution.

The heteropycnotic sequence noted during meiosis in the male spermatocyte of most mammals is certainly an unique circumstance. The significance of allocycly or differential stainability of heterochromatin and euchromatin remains to be unraveled. Current application of tritiated thymidine to the chromosomes of the Chinese hamster is probably the simplest system one may use to record differences in the rate of DNA (chromatid) duplication along distinct portions of known autosomes and sex chromosomes derived from both normal and malignant sources. The readily propagated lines of cells having as low as ten hour generation times, classical diploidy, quasi-diploid, and low aneuploid chromosome numbers, including monosomic (XO) sublines and variant mutant X chromosomes, provide an array of exceptional cell types to follow the uptake and/or blockage of essential and substituted purines and pyrimidines. The uptake of tagged synthetics and their distribution, whether they be specific or random, may be readily traced to given chromosomal loci (euchromatic or heterochromatic) by means of autoradiography. Such an approach may well be the basis by which one may influence the hereditary pattern of a given cell line or species. Simplicities for detecting biological and chemical transduction phenomena are currently indicated among the vast number of Chinese hamster type cultures now on hand.

DESIRABLE PROPERTIES OF SYSTEMS *in vitro*

1. Retention of the diploid or monosomic states, in primary and subsequent clonal isolations.

2. Small number of chromosome or linkage groups.

3. Recognizable autosomes, sex chromosomes, and known markers.

4. Comparatively rapid proliferation in chemically defined medium.

5. Elimination of whole serum requirements or use of dialyzed serum with little or no toxicity.

6. Relatively stable plating efficiency and general appearance; limited non-disjunction and polyploidy.

7. Derivation of type from genetically known inbred lines with known histo-compatibilities.

8. Preferable origin of type from reticuloendothelium or lymphoid tissue following appropriate induction of specific antibodies in the host.

9. Distinct susceptibilities and resistance to viral agents.

10. Viability unimpaired by routine freezing procedures.

11. Pleiotropic expressions of metabolic mutations.

Ideal types of cells may be either normal or malignant in origin. The latter form will provide the opportunity to involve additional selective pressures during the course of alternative trials *in vitro* and *in vivo*. As expected, ideal types are exceedingly rare in occurrence and certainly very limited in distribution and application at the present time. Puck, Ciecura, and Robinson[1028] reported their procedures as being most satisfactory in providing rapidly proliferating lines of human cells bearing 46 chromosomes and cells from the opossum with 22 chromosomes. Since their tissues are heterozygous and reproduction of the opossum most difficult to manage under domestication, the observed constancy of chromosomal number only partially meets requirements necessary for a system *in vitro* to be repeated effectively at the convenience of the investigator.

A rigid pattern of constancy in chromosome number may not necessarily be suited for experimental *in-vitro* genetics when new lines of cells with specific chromosomal deletions are desirable. Currently monosomaty of an autosome and nullisomaty of the X_2 or Y chromosomes (XO) are available in some strains of cells from the Chinese hamster.[1463, 1464] In the future, if disjunction can be controlled, a larger variety of viable monosomics and trisomics may be synthesized and subsequently isolated by cloning procedures. In our hands, chromosomal imbalances are characterized by alterations in the gross appearance of the cells, affinity for glass, plating efficiency, attraction or repulsion of neighboring cells, and the like. The presence or absence of visible X chromosomes also influences colonial appearance and transplantability *in vivo*.[1463, 1464] Many of the above features noted in cultures of cells from the Chinese hamster are less easily recognized in other species which are also near-tetraploids. Clonal appearance is not well correlated with chromosomal ratios after extended propagation of cells.

The influence of chromosomes on gross anatomy and on primary and secondary sex characteristics is evident in most striking fashion among disturbed human somatotypes. A bit more limited in expressiveness are the autosomal relationships that tend to result in Mongoloid-like expressions. The altered monosomic or trisomic fibroblast derivative is certain to reflect pleiotropic (biochemical) expressions in the *in-vitro* environment, even though the intact soma provides greater resources and substrate to complete defective metabolic pathways.

PROPERTIES OF THE CHINESE HAMSTER

Cytologic characteristics

1. Low chromosomal number of 10 readily recognizable autosomal pairs.

2. An X_1X_2Y or triheterosomic somatic sex expression.

3. Each of the three types of large metacentric sex chromosomes easily identifiable during meiosis and mitosis.

4. New chromosomal forms arising spontaneously, or induced with agents, are easily detected and their origin determined.

5. Single-cell cloning facilitates the isolation of monosomic and mutant forms, as well as assuring retention of the diploid or the desired parental type of lines.

6. Quasi-diploid relationships can be visualized clearly whenever the diploid number of chromosomes exists in particular types of cells.

7. Heterochromatin of sex chromosomes and autosomes is clearly correlated to variations in form and function (tritiated thymidine uptake).

8. The diploid status can be retained for long periods of time (beyond one year) in some sublines, having generation times of less than 14 hours.

9. Monosomaty of the sex chromosomes (X_1O) is quite common. The X_2 and Y chromosomes can be totally absent without imparting any deleterious effects on viability of normal or malignant cells.

Behavior of Chinese-hamster cells in tissue culture

1. Growth in various chemically defined media with whole serum (1–15 per cent) is most favorable.

2. Cellular generation times average 10–14 hours.

3. Plating efficiency is quite high among a number of lines of cells derived from normal and malignant tissue.

4. Spinner culture adaptation has been successful with certain normal hypo-tetraploid sublines.

5. The great majority of cellular types are readily preserved in the frozen state.

6. Viral susceptibility (polyoma, poliomyelitis, and measles) is moderate to high for most lines tested to date.

Current long-term derivatives in vitro

1. Adult fibroblasts from various organs.

2. Fibroblasts derived from 18-day embryos.

3. Transformed elements stemming from cells in the peritoneum following irritation resulting from the injection of distilled water 24–48 hours earlier. (Aspiration yields numerous cells without the necessity of sacrificing the animal.)

4. Spontaneous and induced malignancies from representative normal and diabetic strains.

5. *SE* polyoma-induced sarcomas (produced in collaboration with Dr. Sarah E. Stewart).

6. Sarcomas and carcinomas arising from X irradiation and treatment with carcinogenic agents.

These derivatives have been cloned repeatedly in order to provide an extensive array of karyotypes with which to conduct comparative interference microscopy, as adapted by A. C. Longwell. A large number of clonal isolates and sublines have either the X_2 or the Y chromosome missing, and apparently the loss does not interfere with cellular viability and function. Measurements such as plating efficiency appear to be affected when heterochromosomes are missing or altered. The X_2 and Y chromosomes are more similar to one another than the X_1 is to the X_2[1459, 1316] in contrast with earlier ideas regarding the origin and function of mammalian sex chromosomes.

The loss of sex heterochromatin fails to affect cellular viability, whether it be the whole or part of the long arm of the X_1 or the entire lengths of the X_2 and Y chromosomes. Similar situations are, of course, noted in human syndromes.[1026] Thus, sex heterochromatin may be considered essential for normal development of the sexes.

Malignant transformations *in vitro*, or similar changes in fibroblast-like derivatives when placed *in vivo*, have been noted to occur during or immediately following alterations in the appearance of the heterochromatic long arm of the X_1 chromosome of a line derived from a male.[1464] At no time has the short arm of the X_1 been noted to be disturbed consistently in a viable normal cell. Although evidence is still lacking, the short arm of the X_1 is probably the most vital complex of euchromatin-heterochromatin in the cell. Its integrity appears to be most essential for cell viability.

The adult animal exhibits normal antigen-antibody histocompatibility responses for grafts of skin and tumors and metabolic mutants, such as diabetes mellitus. The adults survive well over four years in the laboratory.[1462] Stress reactions following exposure to X irradiation or cortisone, or both, may be accompanied by symptoms of diabetes mellitus and diabetes insipidus depending on the sublines (Yerganian and co-workers, unpublished data). The diabetic animal exhibits the pathogenesis noted clinically.[865, 866] In addition, the biochemical and dental aspects of the disease are equally striking and compare well with human peridontal syndromes (Cohen, Shklar, and Yerganian, unpublished data). Increased alpha-2 serum proteins are currently considered useful in the early detection of potential diabetics.[494]

GENERAL REMARKS

Although the material reviewed has dealt primarily with rodent specimens, application of these procedures to cells of amphibia, reptiles, birds, and primates is also feasible and results encouraging. Additional information, particularly in the realm of pretreatments, could have been reviewed in greater detail. However, present progress makes it advisable to simplify and eliminate unnecessary technical steps, particularly when additional attention to preparation of media, routines of tissue culture, and, finally, cytology is required. There is little need for the novice to repeat the many time-consuming procedures that have led to the simpler techniques in use today.

The effectiveness of extended hypotonicity (20 minutes) has proved its value beyond any doubt when employing tissues having numerous chromosomes which are present in the majority of tissues available today.   On the other hand, colchicine alone or in conjunction with reduced hypotonicity (5 minutes or less), is most adequate for wet squashes when chromosomes are few and cytoplasm is abundant.   Examples of the latter, such as the numerous lines from the Chinese hamster, do not have the fuzziness that water-pretreated chromosomes generally display.   Clean lines delineating the chromosomal contour favor exact identification of chromosomal type and component parts, particularly centromeres, secondary constrictions, and sites of fusion involving euchromatin-heterochromatin.

The need for colchicine is also reduced when cellular generation times are rapid, and fixation or slight hypotonicity is applied during the log phase of growth.   Familiarity with the cellular type allows the reduction of pretreatments to a minimum, a procedure which promotes the clarity of details.   Overspiralization from colchicine pretreatment is helpful when chromosomes are to be counted.   However, colchicine reduces the opportunity to identify distinct chromosome types that resemble one another very closely, as in the case of the mouse and the 6–12 group of the human karyotype.   There is a real need for methods which will reveal the minute structure of the individual chromosomes (satellites and the like) in attempts to associate their appearance with metabolic and other disorders.[1323]   Other aspects to consider are the nature of very short arms, the components of the centromere, appearance of secondary constrictions following chemical pretreatments administered at nontoxic levels, and the need to start correlating component parts of mitotic chromosomes with meiotic bivalents.

It is recommended that the degree of hypotonicity be varied when attempting new trials with wet squashes.   Hypotonicity appears to be most helpful when preparing solid tissues of normal and malignant derivation for squashing.   Nevertheless, by means of colchicine or Colcemid pretreatment of the intact animal, and by using minimal hypotonicity and delaying squashing at least overnight, an adequate start can be made in obtaining satisfactory preparations.   It must be remembered that extensive hypotonicity is not always necessary and will only lead to chromatid separation and fuzzy outlines.   Storage in the fixed state, that is, 3:1 fixative added to the natural exudate of the peritoneum or to the culture medium, results in the retention of sufficient softness in practically all tissues encountered to date.

The choice of stain is a matter of the individual's preference.   Propionocarmine or acetocarmine are considered better (by the author) than aceto-orcein for chromosomes *in vitro*.   One can rely on the availability of clean stain at all times without troublesome precipitates.   Since time is an important factor, and one need not refilter the stain, as is required with the use of aceto-orcein, related stains may be preferred when they are equally applicable.   The use of iron mordant in conjunction with carmine may make up for the intensity of staining that has become associated with aceto-orcein.

The procedures of drying in air are not yet accepted by all investigators.   For best results, the log phase of growth provides sufficient numbers of mitoses in a sparse to moderately populated culture. In general, fixation before 48 hours or before cells stretch out and divide at least once, leads to unflattened, dense, rounded mitotic figures. The cell which has divided previously appears to be more satisfactorily flattened with chromosomes dispersed. This is particularly true when neighboring cells are still in the initial phases of adapting to the new surface of the present flask.

Both air-dried and wet squashes should be tried on new tissues as a means of judging their adaptability to both procedures. Elasticity of the cellular membrane may play some rôle in the degree to which the cytoplasm will respond to hypotonicity and subsequent spreading. Variations in the utilization of cations by different types of cells may be also instrumental in predicating the type of response to the technique of air drying. A most provocative observation is the excellent preparations obtained by wet squashing normal clonal derivatives and *SE* polyoma-induced sarcoma that have an extra chromosome III or involve translocations within the heterochromatin (Yerganian, unpublished data). It is quite certain from present unpublished observations that the Y chromosome of the Chinese hamster is not essential for regulating cell cytoplasmic volume or plating efficiency. However, this observation may eventually have some application in other connections.

The existence of quasi-diploidy or false diploidy in normal cells of other species remains to be fully disclosed. The frequency of quasi-diploid or subdiploid clones of the Chinese hamster, lacking the X_2 or Y chromosomes, is much higher than that expected to occur randomly and to involve the 11 chromosomes. Delayed division or formation of the sister chromatids during replication of the entire length of hetero-chromatin or the X_2 and Y chromosomes appears to be a satisfactory explanation for the outright loss of these heterochromosomes from viable cells. Likewise, extensive deletions of the heterochromatic long arm of the X_1 are often noted in viable cells and clones.[1463, 1464] Since the long arm of the X_1 is also delayed during the asynchronous DNA cycle, it too has the tendency to be deleted with greater frequency than that expected. Nevertheless, deleted heterochromatin is not lethal for rapidly proliferating cells *in vitro*.

Subsequent implantation of representative malignant sublines into cheek pouches of related animals has revealed that additional heterochromatin favors *in vivo* propagation. On the other hand, malignant transformations of male fibroblast derivatives is no greater with a modified X_1Y than with a modified X_1O sex mechanism.[1464] The rôle of sex heterochromatin during the initiation of malignancy remains obscure. However, in reviewing photographs of malignancies arising among Chinese hamsters,[387] the long arm of the X_1 chromosome was noted to exhibit increased spiralization similar to that witnessed among parental and clonal isolates of *SE* polyoma-induced sarcomas and malignant transformations maintained both *in vivo* and *in vitro*. In the absence or modifications of the X_2 or the Y chromosome, the long arm of the X_1 displays over-spiralization (reduced length) when the cellular type is capable of forming a trans-plantable tumor *in vivo*.

The number of lines propagated in culture and as ascites tumors that retain the diploid or the original parental stemline karyotype is exceedingly rare.   Consequently, the value of transformed or altered forms in somatic genetic experimentation must be reviewed with extreme caution.   The commonly noted transition toward tetraploidy during the course of proliferation may be regarded as representing an early phase of progression and adaptability of the heteroploid *in vitro*.[605, 782]   Altered types of cells rarely reflect properties of the parental stem cell *in situ* other than strain-specific immunologic features.[391, 1030]   Lines having only one or two extra normal chromosomes may be regarded as being more favorable than near-tetraploid forms for genetic studies.   Yet, the need for monosomaty and diploidy, as the more ideal forms, remains.

Studies on somatic cells from induced malignancies in the house mouse, *Mus musculus*, have been readily conducted on a short-term basis by Klein and his associates. Since long-term tissue cultures and transplantable malignancies that retain their original karyotype and immunologic properties are rare, replicable short-term cultures employing normal-cell types may serve more effectively in the future.   It is urged that investigators who are actively pursuing basic trials with the house mouse assume greater responsibilities for the development of cell lines of normal derivation that retain the diploid or near-diploid status during the course of routine propagation *in vitro*.

The normal population of somatic cells is, of course, predominantly diploid, with a small number of deviants or near diploids.   Yet innumerable experimental designs and theories regarding the malignant process are based on results with near-tetraploid variants.   Such studies must be scrutinized carefully and reevaluated in the light of new evidences accumulated by Bayreuther.[63]   The extreme and continued variation in chromosomal numbers among the numerous sublines of normal and malignant cells is evident from reports such as Levan's[781] registry of the numbers of chromosomes in various sublines from human and rodent tissues.   For the sake of economy and greater productivity of experienced cytologists, the necessity and value for continuing this determination and recording of chromosomal numbers is questioned.   Of more importance is the isolation of more appropriate lines of cells for conducting trials that are truly definitive.

TECHNICAL NOTES

REFERENCES AND HELPFUL MANUALS

Merchant, D. J., R. H. Kahn, and W. H. Murphy, Jr.: Handbook of Cell and Organ Culture. Burgess Publishing Company, Minneapolis, Minnesota.

Parker, R. C.: Methods of Tissue Culture.   Third Ed. Paul B. Hoeber, Inc., New York, 1961.

Paul, J.: Cell and Tissue Culture.   Baltimore: Williams and Wilkins; London: E. and S. Livingstone, 1959.

Puck, T. T.: Quantitative Mammalian Cell Culture, Second Preliminary Edition.   Department of Biophysics, University of Colorado Medical Center, Denver, Colorado, 1957. Address the Departmental Secretary.   Cost: $2.00.

Methods and Principles of Tissue Culture, Laboratory Manual of the Tissue Culture Association Course, University of Wisconsin, Madison, Wisconsin, June–July, 1960.

An Introduction to Cell and Tissue Culture. The Staff of the Tissue Culture Course. Cooperstown, New York. Burgess Publishing, Minneapolis, Minnesota. 1949–53.

1960 Catalogue of the Colorado Serum Company Laboratories, 4950 York Street, Boulder, Colorado.

Pamphlets on Viral Diagnostic Reagents, Microbiological Cell Culture and Repository, Media for Growth and Maintenance of Cells. Microbiological Associates, Inc., 4813 Bethesda Avenue, Bethesda 14, Maryland.

Difco Manual. Ninth Edition. Difco Laboratories, Inc., Detroit 1, Michigan.

Reagents, Media, and Cell Lines for Tissue Culture and Virus Propagation. Difco Laboratories, Inc., Detroit 1, Michigan.

Listing of Chemicals. Nutritional Biochemicals Corporation, Cleveland, 28, Ohio.

Products for the Microbiological Laboratory and Diagnostic Reagents. Baltimore Biological Laboratories, 2201 Asquith Street, Baltimore 18, Maryland.

PREPARATION OF STOCK SOLUTIONS

It should be remembered that many of the ingredients that make up the current lists of amino acids, vitamins, and salts of currently popular media are readily prepared in advance as 20–50 X concentrates, filtered in convenient aliquots, and stored either frozen or refrigerated.

When diluted proportionally, several liters of balanced salt solutions or medium are prepared at one time. Serum, antibiotics, and other supplements, if desired, can be added prior to sterilization and, in many instances, stored frozen for long periods without exhibiting detrimental effects. The use of a magnetic stirrer is helpful in dissolving stubborn mixtures, especially when additional alkaline or acid concentrates appear to have exceeded the limits of previous experience.

RECOMMENDED INGREDIENTS FOR THE PREPARATION OF SLIDES

Euparal. Made in England by Flatters and Garnett, Ltd., 309 Oxford Road, Manchester 13, England. Distributed in the United States by A. H. Thomas Company, Philadelphia, Pa.

Sealing Compound. Deckglaskitt nach Krönig fur mikroscopie (Cover glass cement according to Krönig, for microscopy). Riedel-de Haen Ag.AG., Seelze-Hanover, Germany. (Minimal order of $10.00.)

Colchicine. U.S.P. Abbott Laboratories, Chemical Sales Division, North Chicago, Illinois.

Orcein. (State natural or synthetic.) George T. Gurr, Ltd., London, S.W.6, England. Canada: ESBE Laboratory Supplies, Toronto, Canada.

Carmine. Fisher Scientific Supply Co., New York, N.Y. The Matheson Company, Inc., East Rutherford, New Jersey.

Cover Glasses. (Specify Gold Seal, thinness 0, in ½ ounce package, cut to desired sizes.) (For cytology tubes, 10 × 45 mm.)

7X Detergent. For washing tissue culture glassware. Linbro Chemical Co., Inc., New Haven, Connecticut. (Available from local laboratory supply house.)

CULTURE TUBES FOR CYTOLOGIC PREPARATIONS

For use with cover glasses up to 10 mm. wide by 45 mm. long

Require only 1–1½ ml. of medium. Useful for replicate trials and chemical dilution series. Cover glass-attached cells available for general morphologic procedures, wet or dry squash preparations for chromosomal analyses, exposure to tritiated purine and pyrimidine analogs, as well as other applications when a CO_2 environment is essential for proper growth.   The long cover glass facilitates preparation of chromosomes.   Properly squashed metaphases generally appear along the long edge of the final slide preparation.

1. Tightly stoppered type (not for CO_2 incubator)

    a. No. 14-1951 short length tissue culture tube, Leighton type.   Bellco Glass, Inc. Vineland, New Jersey.

    b. No. 14-1922 rubber stopper, silicone rubber, non-toxic.   Size No. 0.   Bellco Glass, Inc., Vineland, New Jersey.   (Silicone stoppers are manufactured by the West Company, Phoenixville, Pennsylvania.   Minimal order $10.00 for about 100 pieces.)

2. Dual model for both conventional and CO_2 incubation

    May be used in the conventional manner by turning the rubber-lined bakelite cap tightly or may be employed principally in conjunction with a CO_2 atmosphere with a loosened cap.   Autoclaving and assembly is simplified, as compared to steps necessary for preparing the above type that requires a rubber stopper.   The field is readily viewed with either a conventional or inverted microscope.   Shortened, slightly upturned neck facilitates removal of the cover glass with conventional forceps, and prevents spillage of medium.

    Bonus Laboratory Products, P.O. Box 66, North Andover, Massachusetts.

    Gold Seal cover glass, 0 thinness, is cut to any desired measurement for use in both types of tubes (local supply house).

Incubators

    An incubator design which provides excellent control of the temperature and has adequate inlets and an effective seal to retain the proper level and distribution of the desired mixtures of gases is most essential.   In this laboratory, an efficient incubator design has been found in the model 410 design by Labline, Inc.   Other laboratories have had excellent results with other designs and, therefore, the following refers to our present experience and serves purely as a guide to those who are contemplating similar needs, and is not necessarily an endorsement of the product.

    This model incubator has been found to be satisfactory, provided the speed of the built-in blower is reduced by setting the powerstat (at 60–70).   These features are presently being incorporated in the latest assemblies at the factory.   The design provides for the maximum utilization of space.   Shelves are so placed as to permit use of flasks of multiple sizes.   In our laboratory, moisture is provided by means of setting a stainless steel pan 1–2 inches deep filled with water under the bottom shelf.   In addition, the mixed gases (air and CO_2) are passed through a fritted glass tube (Corning Glass 39533) placed in a 1,000-ml. flask filled with water that is replenished directly from the faucet by means of a hose connection.   This assembly eliminates the need for opening the top of the moisturizing unit each time it becomes necessary to fill it.

    Model No. 210 CO_2 Cloning Incubator by Labline, Inc., is a smaller version of the one described above.   The model 410 is recommended as first choice because of its excellent capacity.   The smaller model 210 is useful whenever there is absolute need to refrain from disturbing the environment, particularly after ultraviolet exposures and the like.   The unit has excellent temperature control and moisture must be provided by a similar assembly described above.   A small electric heating unit placed under the evaporating flask and heated to 38–40° C. is very helpful in assuring ample moisture to the cabinet.

    An extra set of flow gauges and reducing valves needed for the proportioning of carbonic gas and air to the incubator will help to offset accidental losses, when similar items in use happen to become damaged or hampered by debris collecting along the lines.   The present model 410 in operation at this laboratory has the following CO_2 service settings: a Matheson, low-pressure, pancake regulator no. 70 is set to deliver 6 lbs. of pressure which, in turn, is reduced still further by means of a Hoke 992 flow gauge to read 1.0 liters per minute. The CO_2 is regulated by means of a Hoke Phoenix 902C reducing valve to read 6 lbs.   The needle valve is set to deliver a reading of 0.15 liters per minute on a Hoke bantam 994 flow gauge before joining, by means of a Y-shaped connecting tube, to the reduced air. The pH of media is 7.0–7.2, which is excellent for Chinese-hamster cells. The CO_2 can be reduced to 0.1 liters per minute to adjust the pH to approximately 7.3–7.4 for lines of human cells. As the interior of the incubator becomes moist less CO_2 is needed and the incoming gas mixture (about 2 per cent) is utilized more efficiently in the saturated environment.

In the event it becomes necessary to check the actual concentration of CO_2 in the incubator, especially when optimal levels are to be attained for cultures of a particular species, the Burrell Kwik-Chek gas analyzer for CO_2 may be helpful. (Burrell Corporation, Pittsburgh 19, Pa.)

Filter assemblies for sterilization of media

The following filter assemblies are listed in the order of preference based on experience in this laboratory.

Selas VFA-86-02 vacuum filtration assembly with the substitution of a heavy-wall filtration flask (Pyrex) having two tubulations (G8847, Emil Greiner Company). The upper tubulation is cotton stoppered, and this is held tightly in place with a rubber band and linked to a cotton-filled, sterile, air-filter tube (14-2340, Bellco Glass, Inc.) by a sturdy suction hose. The suction line is then linked to an intermediate catch vessel (500–1,000 c.c.) that acts as a back-suction stop when the faucet or pump is turned off. The lower tubulation is linked by means of a convenient length (11–12 inches) of latex tubing to a filling attachment (3960, Corning Glass Works, or 14-23330, Bellco Glass Co.) having an inside diameter of 20 mm. The lower end of tubing of the filling bell is cotton stoppered, the bell covered with Kraft paper, and held in position with a rubber band prior to autoclaving. The porcelain neck of the Selas-filter element is placed in the rubber stopper with the aid of Kel-F stopper grease. Extreme care must be taken to prevent accidents when assembling the glass chimney.

After autoclaving, a clamp is placed onto the latex tube and the assembly is completed. When the media has passed through the filter cone, the entire assembly is raised and held by a sturdy clamp about the upper tubulation of the filter flask to stand some 12–13 inches above the surface of the bench to facilitate filling of milk-dilution bottles with the aid of the filling bell. Repeated flaming of the base of the filling bell and necks of the milk-dilution bottle is adequate to insure continued sterility during the filling process. This filter assembly will facilitate larger volumes of media to be sterilized because of the convenience and opportunity to remove the initial liter of solution via the lower tubulation before continuing with added volumes. Larger capacity filter elements ($1\frac{1}{2}$ inches wide) provide adequate precautions for the sterilization of 2 liters of media with a 1-liter flask adapted to this scheme. The final 10 ml. of medium should be tested on surplus cultures and the batch frozen or stored in the refrigerator. On numerous occasions, the initial and final milliliter of media are added to Difco thioglycollate medium to help disclose contaminations. However, the use of surplus cultures for testing of sterility has proved to be the more sensitive system.

Fritted glass, Millipore, and Seitz filters can be substituted for the Selas-filter assembly in the order of listing, in the event adequate distilled water rinsing is not feasible. The need for a furnace to incinerate organic particles remaining in the Selas filter element may also be an initial drawback. However, every effort should be made to start with the Selas filter when genetic lines are to be maintained. Conventional stock cultures that can be replaced do not require these precautions.

Milk-dilution bottles (Corning Glass) are excellent for the storage of solutions and media. Their necks are long enough to permit setting of the bottle slant-wise in a half-pint container, along with a stainless steel graduated medicine cup (8492, Vollrath Co.) covering the opened neck. In this manner the cup is raised slightly and the media pipetted from a convenient angle to deliver to culture vessels. The stainless steel medicine cups may be comparatively expensive, but the breakage of beakers generally used for this purpose is eliminated, and the purchase price is fully warranted.

Whenever employing bakelite caps with cemented rubber linings for any kind of stopper use, limit purchases to the type distributed by Corning Glass Works, only.   Specify cemented rubber lining (grey or ochre).   It has been our experience that other kinds of bakelite caps with cemented rubber linings are inferior and always troublesome.

Heavy-duty centrifuge tubes are more serviceable.   Pipette-sterilizing boxes should be of monel metal.   The Touch-O-Matic bunsen burner (912, Microbiological Associates) is a convenient item wherever high temperature is a comfort problem.   An electric heating mantle for the base of the distilled-water flask is most helpful in reducing the heat-load in a room.   Whenever possible, the use of rubber policemen (obtainable from any supply house) to scrape cultures, rather than trypsinization, is helpful.   Use of policemen should be restricted to flasks that are to be subcultured routinely and not for cytologic use, because of the tendency to form small clumps of cells that interfere with chromosomal preparations.

DISCUSSION

Dr. Burdette: Thank you, Dr. Yerganian. Dr. T. C. Hsu, of the M. D. Anderson Hospital and Tumor Institute, will discuss Dr. Yerganian's paper.

Dr. Hsu: Dr. Yerganian's paper contains in detail the technical advances in mammalian karyology and emphasizes normalcy and uniformity of material. It is indeed of prime importance for genetic studies in somatic cells to maintain populations of cells that are primarily diploid and homogenous. At the present time, however, it is not quite feasible to grow uninterruptedly diploid human cells forever. Some investigators say they could, but extensive investigations conducted by Hayflick and Moorhead[537] show that diploid strains of cells may last as long as 50 transfers before reaching an inevitable stage of degeneration. However, large quantities of such cells can be harvested and frozen during their peak of growth so that they can be thawed when needed. This conclusion was arrived at by many workers. Nevertheless, frozen cells, after thawed and regrown, will finally degenerate as an inevitable outcome. Thus in investigations concerning somatic genetics, one faces a grave problem, namely, inability to procure a constant supply of standard and mutant materials which workers painstakingly establish.

Most of the perpetuated lines of cells available in laboratories today are aneuploid. In addition, practically all the populations of cells are polymorphic in their chromosomal constitution. Usually there is a predominating type of cell (the stemline) with its characteristic chromosomal complement, and other types with chromosome number above or below it. In essence, variability, instead of uniformity, is the rule in populations of mammalian cells *in vitro*. This variability, nevertheless, offers a unique system for genetic studies, a system that has opened a number of new avenues of research,[599] and will continue to contribute in various ways to our knowledge of differentiation, metabolism, and heredity.

How does a diploid tissue, after growth *in vitro* for a period of time, become aneuploid or heteroploid? This transformation process has been witnessed in several lines of cells from mouse and Chinese hamster.[389, 602, 782] In the mouse, the process seems to follow the diploidy-tetraploidy-heteroploidy pattern, and in the Chinese hamster, establishing stemlines around diploidy (hyperdiploidy and hypodiploidy) appears to be a popular route.   In addition to numerical changes, structural alterations of chromosomes also enter the picture to increase variability.   To students of classical genetics, great variability in chromosomal constitution is unthinkable.   In *Drosophila*, for example, a deletion of a small segment of a chromosome will result in death of the organism.   Then how could mammalian cells, with the chromosomal constitution so far away from normalcy, live and thrive *in vitro*?   The answer probably lies in the difference between intact organisms and cellular culture.   The situation mentioned in *Drosophila* is analogous to mongolism in man, in which the trisomic condition for a very small chromosome leads to serious damage of normal form and function in the individual.   It must be borne in mind that development of an organism requires both growth and differentiation, whereas in cultures of cells growth is the main concern. Thus a few missing genes or duplicated genes which affect embryonic development with great impact may exert little influence in cultures.   The well-known Grüneberg disease of the mouse, the result of a gene mutation which affects proper cartilage formation, is lethal.   Conceivably this gene, whether in its normal form, mutant form, deleted or duplicated, should not have severe influence on the growth and reproduction of cultures of epithelial cells or fibroblast cells.   By the same token, numerous genes essential to a normal organism may be useless or even detrimental to the growth of cells in culture.   The concept of genic balance must be discarded when we talk about cells *in vitro*.   In a way we must regard the cells *in vitro* as artificial species which start their evolutionary course the moment they are placed in a culture vessel.   They will have to forget the functions they are supposed to perform and live like bacteria. It is, therefore, rather surprising to see that diploid human cells can maintain diploidy for as long as a year *in vitro*.

Extending the above-mentioned concept to explain why aneuploidy is prevailing in neoplasms, cancer cells, although growing *in vivo*, tend to forget what they are supposed to do and regard the body as a giant culture vessel.

Cells in normal tissues are not always diploid.   Occasional aneuploid cells can be found if large samples of mitotic cells are analyzed.   The frequency of aneuploid cells *in situ* may even be higher than recorded, for aneuploid elements may be handicapped in normal organisms so that they do not divide as often as diploids.   In the case of cellular cultures, the situation is different.   Here the normal functions of the cells are not required.   In fact the cells are required to grow, a function normally demanded only under special circumstances such as during wound healing.   Mitotic anomalies, in an astronomical number of divisions, are bound to happen.   Cells lose and gain chromosomes or change parts.   Some of them may happen to lose genes that are detrimental to perpetual growth.   Some of them may happen to gain genes that aid perpetual growth.   Thus they may become selected by the artificial environment and emerge as the stemline.   The fact that populations of cells *in vitro* are full of chromosomal variation strongly indicates that a large number of genes are not necessary.

Variation in karyotype also offers an excellent tool to study many problems in cellular biology.   Many populations of cells contain special karyotypes with marker chromosomes. These marker chromosomes are very useful in identifying cells in mixed populations. The *D*-chromosome of strain L-P59[604] and the *E* and *F* chromosomes of strain L-M[606] have been used to advantage in population analysis. A novel attempt made by Harris and Ruddle[522] was especially interesting. A translocation between two chromosomes resulted in a very long chromosome. This element is so long that during telophase their ends are still near the equator. Thus when the daughter nuclei restitute, the long chromosome forms a finger-like projection at one side of the nulceus. In mixed populations, this type of cell can be identified with interphase nuclei.

The work by Hsu[601] suggests that different doses of a marker chromosome (the *D* chromosome of strain L-P59) will cause the cells to prefer different cultural environments. Variation in chromosomal dose (and karyotype in general) undoubtedly means variation in genetic makeup of the cells. Studies on the relationship between enzymatic activities and certain chromosomes should yield some interesting information in cellular physiology and somatic-cell genetics.

DR. BENDER: As was brought out by Dr. Snell after the last paper, there are two points of view about what we have come to call mammalian somatic-cell genetics. One is to regard it as a useful new tool for the study of classical mammalian genetics. The other is simply the converse: to consider classical mammalian genetics as a useful source of material for the study of somatic-cell genetics. There has already been a good deal of discussion of the latter point of view. Perhaps something should be said about the former.

There are already several ways in which somatic-cell techniques can help in the study of mammalian genetics. An important way is the use of techniques *in vitro* for the measurement of mammalian, and particularly human, mutation rates. We have made a beginning in this work by measuring somatic chromosome mutation rates in man, in the spider monkey, and in the Chinese hamster.[70, 71, 72, 73] We have been able to measure spontaneous and X-ray-induced aberration rates in diploid cells both *in vivo* and *in vitro*. Also, we are now making parallel measurements of human genic mutation rates at the *ABO* blood-group locus in diploid somatic cells *in vitro*, using a fluorescent antibody assay technique. It is already clear that such studies will contribute heavily to mammalian genetics in the future. It is also clear, however, that many problems of specific interest to the discipline of somatic-cell genetics will emerge from such studies.

As an example of such a problem, I will cite the difference that we have found between the rates of spontaneous chromosomal aberration found *in vivo* and *in vitro*. The epithelioid cells that have been investigated so far have a spontaneous aberration rate of about 1 per cent, while the rates for fibroblasts range up to almost 30 per cent. The mammalian somatic cells that have been studied *in vivo* have extremely low spontaneous aberration rates. To add to the puzzle, all of the somatic cells that have been investigated, *in vivo* or *in vitro*, have about the same X-ray-induced chromosomal aberration rates.

This suggests, of course, that we must be very cautious in applying *in vitro* results to whole animals. We must compare results for situations *in vitro* and *in vivo* and establish their relationships before we can substitute the quicker and easier measurement *in vitro* for those made *in vivo* on whole animals.

ADDENDUM

Preservation of a large number of normal and malignant strains and lines of Chinese hamster and human origin beyond four months storage by means of dry-ice refrigeration has been hampered because of fluctuations in temperature (-50 to $-69°C$). Recently, the adaptation to liquid nitrogen ($-179°C$) as a refrigerant for prolonged storage has greatly advanced the mechanics of handling and assuring viability of many strains currently being characterized by the Cell Culture Collection Committee of the U.S. Public Health Service, National Institutes of Health, Bethesda 14, Maryland. Information regarding the use of liquid nitrogen may be obtained from the Linde Corporation, 30 East 42nd Street, New York, New York.

The recent (June, 1961) Syverton Memorial Symposium on "Analytic Cell Culture," edited by Robert E. Stevenson, is now available as The National Cancer Institute Monograph No. 7, April, 1961, and may be obtained from the Superintendent of Documents, U.S. Government Printing Office, Washington 25, D. C. Price: $2.25. Among other topics, recent observations in the preservation and characterization of cell cultures are discussed by H. T. Meryman, C. S. Stulberg, L. L. Coriell, and E. H. Y. Chu.

Joan Staats, M.S.

APPENDIX I

*Control of the Literature on Genetics of the Mouse**

Recent decades have brought an unprecedented expansion in the amount of recorded information of importance to professional activity in all fields. This increase has been even larger in the field of genetics than in many others. Today a professionally active person must restrict his reading to an ever-narrowing area of specialization and rely on abstracts and summary reviews to be aware of trends in his own and related fields. Not only are there more papers in more journals in more languages, but also increased publication has been accompanied by greater complexity of content. To relieve this situation, many abstracting and reporting services have grown up. However, the absence of such services in many fields has forced individual organizations to set up their own information retrieval systems.

At the Jackson Memorial Laboratory, the librarian maintains one such system, the subject-strain bibliography.[1263] This is a classified, unannotated, nonselective, analytical bibliography including papers appearing in periodical literature and books in which reference is made to specific inbred strains of mice, named genes in mice, or named transplantable tumors in mice. The project was started in 1948 and was intended for use by Jackson Laboratory personnel only.

The Laboratory is a global center for genetics of the mouse, a primary trust laboratory for maintenance of murine genetic material in this country, and serves as a

* This work has been supported by the National Science Foundation under grants G5740, G11551, and G18485.

source of supply of inbred mice, tumors, and mice with mutant genes to investigators throughout the United States and abroad. Therefore, it is a natural target for queries concerning the nature and uses of this material. Gradually it became apparent that these queries should be channeled through the library and answered by means of the subject-strain bibliography. In addition to answering specific questions by mail, the librarian supplies a bibliographic supplement to *Mouse News Letter* every six months, listing selected accessions during the previous six months.

To facilitate searching the literature, more than 3,000 references were analyzed in detail to determine the fields of knowledge to which investigations using mice have contributed and the types of mice and tumors used in such work. A classification system was devised including all of the fields and types recognized in this analysis, and allowing for reasonable expansion. The key for this system lists in addition to all of the inbred strains (8 major headings, 25 minor, and other), 85 subject headings, ranging from aging and antibiotics through maternal influence and metabolism to tumor incidence and viral studies. There are 31 major subject headings and 17 indirect-sorted fields for subdivisions of fields of interest, such as 9 types of carcinogens, 27 murine diseases and human diseases studied in mice, 7 subdivisions of endocrinology, 11 types of life-history effects, 12 subdivisions of nutrition, 19 organs and systems, and 6 divisions of uses and effects of radiation. Pertinent references on all of these subjects or interactions between them can be sorted out rapidly by a small number of insertions of the stylus into the 5 by 8 inch specially printed Keysort cards on which the bibliography is classified (figure 70).

References are added to the bibliography as they appear in the literature. The Laboratory receives about 180 periodicals in the fields of biology, cancer, medicine, and psychology. These include 12 translated Russian periodicals and 14 abstract and reference tools, such as *Biological Abstracts*, *Nuclear Science Abstracts*, *Index Medicus*, and *Zoological Record—Mammalia*. In addition to these, an active reprint collection now containing over 30,000 items is maintained. The librarian scans these journals as they are received, writing a card for each pertinent reference and classifying the paper. The information on these cards is eventually transferred to the Keysort cards, which are punched according to the classification noted. A skillful operator can type 50 such cards per hour and punch 100 per hour. It is believed that through close scrutiny of the reference periodicals, scanning 166 other journals, maintaining an aggressive reprint accession policy, and utilizing bibliographies in published papers, few pertinent papers will be missed. The bibliography now contains about 16,500 references. A modification of this system is easily adapted for personal or departmental needs or other methods of sorting.

It is expedient to separate the cards according to periods of time: prior to 1930 (the earliest reference is dated 1906), 1930–1934, 1935–1939, 1940–1944, 1945–1949, and yearly thereafter. This device often helps the searcher considerably by eliminating periods of no interest. The original hand-written cards are maintained as a separate author file. The latter is indispensable for such queries as "I know John Q. Smith

Fig. 70. Card for classification of bibliography.

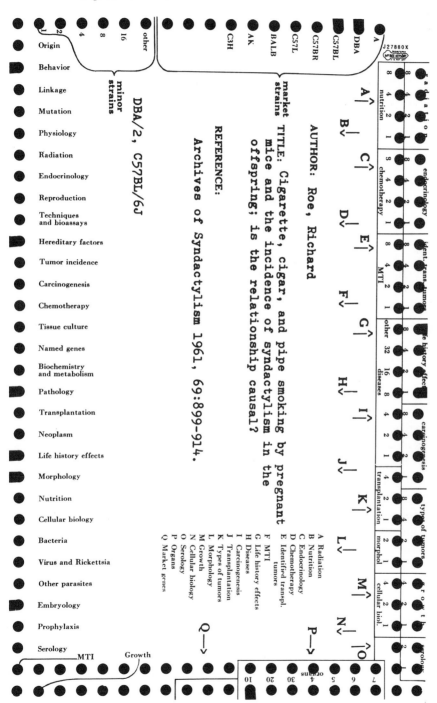

Origin
Behavior
Linkage
Mutation
Physiology
Radiation
Endocrinology
Reproduction
Techniques and bioassays
Hereditary factors
Tumor incidence
Carcinogenesis
Chemotherapy
Tissue culture
Named genes
Biochemistry and metabolism
Pathology
Transplantation
Neoplasm
Life history effects
Morphology
Nutrition
Cellular biology
Bacteria
Virus and Rickettsia
Other parasites
Embryology
Prophylaxis
Serology

DBA/2, C57BL/6J

market strains
minor strains

DBA
C57BL
C57BR
C57L
BALB
C3H
Ak

AUTHOR: Roe, Richard

TITLE: Cigarette, cigar, and pipe smoking by pregnant mice and the incidence of syndactylism in the offspring; is the relationship causal?

REFERENCE:
Archives of Syndactylism 1961, 69:899-914.

A Radiation
B Nutrition
C Endocrinology
D Chemotherapy
E Identified transpl. tumors
F MTI
G Carcinogenesis
H Diseases
I Life history effects
J Transplantation
K Types of tumors
L Morphology
M Growth
N Cellular biology
O Serology
P Organs
Q Market genes

wrote a paper some years ago, and it was in either the *Journal of Genetics, Bibliographia Genetica*, or the *Zeitschrift für Induktive Abstammungs- und Vererbungslehre.* I cannot remember the title, but it was about the agouti series." The author file makes short work of this, since the one making the request generally recognizes the reference when confronted with it. It is found, of course, that the reference in question was from the *Annual Report of the Victoria and Albert Museum.*

Separating references according to periods of time also makes possible quantitative surveys of the field and comparisons over the years (table 63). It comes as no surprise

Table 63

	Total references	Neoplasm		Radiation		Viral studies		Named genes	
		Number	%	Number	%	Number	%	Number	%
Before 1930	346	151	43.6	35	10.1	1	—	75	21.7
1930–1934	320	162	50.6	24	7.5	5	1.6	60	18.7
1940–1944	1002	593	59.2	75	7.5	10	1.0	74	7.4
1950	321	123	38.3	60	18.7	11	3.4	29	9.0
1958	1201	673	56.0	274	22.8	55	4.6	102	8.5
1959	1369	811	59.2	322	23.5	102	7.4	114	8.3

that cards with "neoplasm" punched have always accounted for about 50 per cent of the total. When Dr. C. C. Little first began inbreeding mice during the first decade of this century, his primary purpose was to have available dependable homogeneous material with which to study the genetics of tumors. Also, it comes as no surprise that studies using radiation as a tool or studies concerned somehow with viruses are increasing steadily over the years, or that the interest in basic genetics, represented by the "named genes" category, has held about the same level for the past 20 years. Morphology, chemotherapy, tumor incidence, and irradiation effects are fairly popular subjects, as are existence or characteristics of various pathologic states. A few of the more unusual requests for information are the effects of exposure to white noise, chemical composition of the mouse, development and morphology of teeth, and the surface area of the peritoneum.

Table 64, with its accompanying references, illustrates a different sort of approach to the mass of information contained in the bibliography. This type of search is done without regard to periods of time, by sorting for "origin" plus all the strains. These references may not be in all cases the original ones. An attempt has been made to list articles best describing the origin of the strain; in most cases multiple references are supplementary, offering additional information or enabling the searcher to go from a reference in which the strain is clearly named to one in which the real origin is given but the mice are not so named. The many fostered strains have not been included,

Table 64

ORIGIN OF INBRED STRAINS OF MICE

Strain	Ref. no.	Strain	Ref. no.	Strain	Ref. no.
A	1297	DE	220, 270	PBR	44
AB	220	DM	220	PET/MCV	944
AK	416,815	D103	220	PHH	1373
AKR	815	E	219, 315	PHL	1373
AU	220, 904	F	1298	PL	219
A2G	185, 220	FAKI	220	PLA	219
BALB	558, 834	FU	220	P.S.	902
BAMA	219	GFF	184	P20	915
BC	422	GM	909	RF	416
BD	220	H	220	RI	466
BDP	220, 424	HA	908	RIET	219
BL	219, 256	HR	220	R.I.L.	815
BRS	44	HR/De	258	RUS	220, 910
BRSUNT	44	I	1298	RIII	280, 282
BRVR	1361, 1362	IF	116	S	1167
BRVS	1361, 1362	IPBR	44, 220	SEA	484
BSVR	1361, 1362	JB	220	SEAC	220
BSVS	1361, 1362	JK	1298	SEAC/c	220
BRO	400	JU	912	SEC	220, 484
BR6	400	KL	200	SEC/1	220
BUA	220	L	1298	SEC/2	220
BUB	220	LCH	1377	SHC	220
BUC	220	LCL	1377	SJL	913
BUE	220	LG	450	SL	220, 625
C	1298	LGW	1167, 1467	SM	830
CBA	1298	LP	220	SMA	220
CC57BR	863, 864	L(C)	1038	ST	1269
CC57W	863, 864	L(P)	1038	STOLI	555, 834
CE	1409	LT	833	STR	219
CFCW	220	MA	927	SWR	813, 1243
CFW	220, 1243	MA/My	219	T	220
CHI	1298	MABA	219	WB	1105
CM	905	MIII	400	WC	1105
CT	220	N	1298	WH	1105
C12I	1298	NB	480, 764	WK	1105
C3H	1298	NH	1293	WLL	733
C3HeB	220, 257	NHO	44	WLO	733
C3HA	844	2NHO	1296	Y (YBR)	558
C57BL	555	NLC	1082	Y (Wilson)	220
C57BR	555	NS	404	YS	220
C57L	555, 924	NZB	546	Z	219, 466
C58	832, 836	NZC	88	ZBR	220
DA	613, 906	NZO	88, 546	IV/B	281
DB	220	NZS	911	XVII	279
DBA	928	NZY	88	101	299
DBA/2eB	220	O2O	1339	129	1135
DBA/1₀	220	P	227	1194	814

nor have most sublines, but some of the ova- and ovary-transplant lines will be found. Strains no longer in existence have not been listed, although some listed ones may also be extinct.

There is in preparation another bibliography, concerned only with genes in the mouse. This grew out of Dr. G. D. Snell's private file of references, and is in process of being put on its own form of Keysort cards. Additions to it are made from the named-genes section of the larger bibliography. Whereas the larger one is of value to all who work with mice, whether geneticists, pathologists, cancer researchers, anatomists, clinicians, or animal breeders, the gene bibliography will exist primarily to serve geneticists. Each gene and each linkage group can be sorted independently, plus a wide range of subjects covering all phases of genetic investigation.

Eventually, of course, the subject-strain bibliography will outgrow hand-sorted cards, just as it will outgrow its one-man operation. The method is described here because it is easily adapted for organization of bibliographic material in any laboratory.

Joan Staats, M.S.

APPENDIX II

International Rules of Nomenclature for Mice

On December 27, 1925, the "Mouse Club" at its meeting in New Haven agreed on certain gene symbols which "were recognized as orthodox and . . . voted into the code."[427]   This group is undoubtedly the ancestor of the present Committee on Standardized Genetic Nomenclature for Mice, which recently published the latest revisions of the rules to designate inbred strains of mice.[220]

In the intervening 35 years, many groups have had many meetings and published many papers in a continuing attempt to standardize usage of names and terms among mouse geneticists and users of mice.   The various International Genetics Congresses have served as meeting places on several occasions.   At the Seventh Congress in Edinburgh in 1939, Professors F. A. E. Crew and L. C. Dunn and Dr. George D. Snell were appointed as a Committee on Mouse Genetics Nomenclature "to deal with the nomenclature of genes in mice, the reporting of genetic progress in mouse genetics, and the preservation in a safe place of genes of value to mice geneticists."[867]   Dr. Hans Grüneberg replaced Professor Crew on the committee, because of war duties of the latter, and the first comprehensive list of rules for nomenclature was published in 1940 in the *Journal of Heredity*.[306]

Mouse Genetics News, no. 1 (1941) and 2 (1948), both carried reprintings, virtually unchanged, of the 1940 rules.[764]

In 1949, after consultation with the genetics group at the Roscoe B. Jackson Memorial Laboratory and correspondence with other geneticists, Dr. G. D. Snell contributed to *Mouse News Letter* no. 2 a list of suggested rules of nomenclature for inbred strains of mice.[903]   This list, with minor variations, was sent by Drs. Snell and T. C. Carter to geneticists in 1951, asking for a vote on two alternative systems of notation.   That giving maximum uniformity was adopted, and dba became DBA, BalbC became BALB/c, and so on.   Workers were also asked to select abbreviations of their own names for use in designating substrains.   From the point of view of this reviewer, this was the beginning of real standardization in the whole field.   Most instances of nonstandard usage are the result of unawareness or confusion regarding the rules.

The Committee on Standardized Nomenclature for Inbred Strains of Mice, as an informal working group came to be known, was responsible for issuing the first standardized nomenclature list in *Cancer Research* in 1952.[219]   This paper contained the recommended rules for symbols, a list of known inbred strains with their histories and characteristics, and a list of users of mice with abbreviations for their names or institutions.

As time went on, more substrains of existing strains were developed, by spontaneous mutations, manipulation, or merely physical separation.   New mutations and linkages were discovered, strains became more widely distributed, and the list of workers grew enormously.   Gradually it became apparent that a reappraisal of the nomenclature rules was in order.

Following the suggestion of Dr. Hans Grüneberg, the old committee on nomenclature was reactivated, reconstituted, and renamed.   During the Tenth International Genetics Congress in Montreal in 1958, 5 of the 7 members of the new group met, were named the Committee on Standardized Genetic Nomenclature for Mice, and discussed both strain and gene symbols.   This present body supersedes and represents an amalgamation of the older Mouse Genetics Nomenclature Committee (genes) and the Committee on Standardized Nomenclature for Inbred Strains of Mice (strains).

This body approved the revision every four years of the alphabetic list of inbred strains which had appeared in *Cancer Research* in 1952.   Discussion at the Montreal meeting, at a similar one in Bar Harbor a week later, and extensive correspondence among committee members, resulted in a reissue of rules for nomenclature.   These are unchanged in their essentials from the listing in 1952 but are somewhat more specific and detailed on certain points such as substrain symbols for stocks of complex origin.   The rules as they now stand are given below.

RECOMMENDED RULES FOR SYMBOLS

1. *Definition of inbred strain.*—A strain shall be regarded as inbred when it has been mated brother × sister (hereafter called b × s) for twenty or more consecutive generations.   Parent × offspring matings may be substituted for b × s matings, provided that in the case of consecutive parent × offspring matings the mating in each case is to the younger of the two parents.

2. *Symbols for inbred strains.*—Inbred strains shall be designated by a capital letter or letters in Roman type.   It is urged that anyone naming a new stock consult

Appendix 2 of Standardized Nomenclature for Inbred Strains of Mice: Second Listing[220] or *Inbred Strains of Mice*[628] to avoid duplication.   Brief symbols are preferred.

An exception is allowed in the case of stocks already widely used and known by a designation which does not conform.

3. *Definition of substrain.*—The definition of substrain presents some of the same problems as the definition of species.   In practice, the determination of whether, in published articles, substrain symbols should be added to the strain symbol, must rest with the investigators using them.   The following rules, however, may be of help.

Any strains separated after eight to nineteen generations of b × s inbreeding and maintained thereafter in the same laboratory without intercrossing for a further 12 or more generations shall be regarded as substrains.   It shall also be considered that substrains have been constituted (1) if pairs from the parental strain (or substrain) are transferred to another investigator, or (2) if detectable genetic differences become established.

4. *Designation of substrains.*—A substrain shall be known by the name of the parental strain followed by a slant line and an appropriate substrain symbol.   Substrain symbols may be of two types.

a. Abbreviated name as substrain symbol: The symbol for substrains should usually consist of an abbrevation of the name of the person or laboratory maintaining it. The initial letter of this symbol should be set in Roman capitals; all other letters should be in lower case.   Abbreviations should be brief, should as far as possible be standardized, and should be checked with published lists to avoid duplication.   Examples: A/He (Heston substrain of strain A), A/Icrc (Indian Cancer Research Centre substrain of strain A).

When a new substrain is created by transfer, the old symbol may be retained and a new one added.   Example: YBR/He, on transfer from Heston to Wilson, becomes YBR/HeWi.   The accumulation of substrain symbols in this fashion provides a history of the strain.   If the substrain symbols are not accumulated, the history of transfers should be recorded in *Inbred Strains of Mice.*

b. Numbers or lower-case letters: Numbers or lower-case letters may be used as substrain symbols in certain circumstances.   The position of these relative to other parts, if any, of the substrain symbol should be suggestive of a historic or temporal sequence.   Thus, two substrain branches, separated in and maintained by one laboratory, may be designated by terminal numbers, with or without a preceding slant line. Example: two sublines of A/HeCrgl, separated and maintained by Crgl, become A/HeCrgl/1 (or A/Crgl/1) and A/HeCrgl/2 (or A/Crgl/2).   Lower-case letters immediately following the strain symbol, with a slant line only intervening, may be employed when two substrains are separated from a common strain prior to complete inbreeding.   Example: C57BR/a and C57BR/cd.   (These were separated after nine generations of b × s.)   The use of numbers or lower-case letters immediately after the slant line, to designate lines separated after 20 or more generations b × s, is ordinarily not recommended, but may occasionally be justified for sublines widely recognized as different. Example: DBA/1 and DBA/2. Appropriate checks to avoid duplication should be made before this type of symbol is adopted.

5. *Coisogenic stocks.*—Coisogenic stocks produced by the occurrence of a single major mutation within an inbred strain, or by the introduction of a gene into an inbred background by a series of crosses, shall be designated by the strain symbol and, when appropriate (see rule 7), the substrain symbol, followed by a hyphen and the genic symbol (in italics in printed articles). Example: DBA/Ha-*D*. When the mutant or introduced gene is maintained in the heterozygous condition, this may be indicated by including a + in the symbol. Examples: A/Fa-+*c*, C3H/N-+*W^j*.

When a coisogenic strain is produced by inbreeding with forced heterozygosis, indication of the segregating locus is strictly optional. Examples: 129 or 129–*c^{ch}c* (129 is customary); SEAC–*d*+/+*se* or SEAC/Gn.

In the case of coisogenic stocks produced by repeated crosses of a dominant gene into a standard inbred strain, it may be desirable to indicate the number of backcross generations. Example: C57BL/6–+*W^v*(N8). The first hybrid or F_1 generation should be counted as generation 1, the first backcross generation as generation 2, and so forth.

6. *Substrains developed through foster nursing, ova transfer, or ovary transplant.*—Substrains developed by foster nursing shall be indicated by appending an "f" to the strain symbol. Example: C3Hf. The strain used as foster parent may be indicated if desired by the addition of its symbol or an abbrevation for the same. Example: C3HfC57BL or C3HfB (C3H fostered on C57BL). In like manner, strains developed through egg transfer or ovary transplant shall be indicated by adding an "e" or "o", respectively. Example: AeB (A ova transferred to C57BL). When the symbol for fostering or transfer might be confused with an adjoining substrain symbol, it may be used in a subscript position. Example: A/He$_f$B (Heston substrain of A fostered on C57BL).

7. *Compound substrain symbols for stocks of complex origin.*—When a stock has been produced by manipulation of a standard inbred strain, as, for example, by fostering or by introduction of a foreign gene, compound substrain symbols may be necessary. In general, the elements of such a compound symbol should be arranged in an order indicating a historic or temporal sequence. Specifically, different positions should be interpreted as follows:

a. Substrain symbol that immediately follows strain symbol: Examples: BALB/cf, DBA/2eB, C3H/Ha–*p*. In this position the substrain symbol (c, 2, or Ha in examples given) designates the substrain which was fostered or otherwise manipulated, or in which a mutation occurred.

b. Substrain symbols following symbol for manipulative process or introduced gene: Substrain symbols in this position refer either to the person performing the fostering or other manipulation, or to the person or laboratory currently maintaining the strain, or to both. The symbol or symbols may or may not be immediately preceded by a slant line. Examples: DBA/2eB/De or DBA/2eBDe (strain derived from ova of DBA/2 transferred by Deringer to C57BL, maintained by Deringer); C3H/

He$_f$/Ha (C3H/He fostered by Heston, currently maintained by Hauschka); CBA/Ca–*se*/Gn (Carter's substrain of CBA with mutation to *se*, maintained by Green). Since a single symbol in this position (for example, the De in DBA/2eBDe) may refer either to the person producing or the person maintaining the strain, the intended meaning should be clearly recorded.

8. *Indication of inbreeding.*—Where it is desired to indicate the number of generations of inbreeding b × s, this shall be done by appending, in parentheses, an F followed by the number of inbred generations. Example: A(F87). If, because of incomplete information, the number given represents only parts of the total inbreeding, this should be indicated by preceding it with a question mark and plus sign. Example: YBL (F?+10).

9. *Priority in strain symbols.*—If two inbred strains are assigned the same symbol, the symbol to be retained shall be determined by priority in publication. For this purpose, listing in *Mouse News Letter* or *Inbred Strains of Mice* shall be regarded as publication.

Margaret M. Dickie, Ph.D.

APPENDIX III

Methods of Keeping Records

Any system of keeping records must insure that searches of pedigrees may easily locate collateral relatives to the animal in question as well as its ancestors and its progeny; or to state it another way, the record system must insure that one can start with any given mouse and trace backward, forward, and laterad. Systems of keeping records should also include many biographical details of the stock in question.

The methods of keeping records for various types of mice, for example, inbred strains, genetic deviants, or experimental animals, vary widely in details but all systems have the same basic skeleton. The methods generally in use consist of (1) ledger(s), (2) pedigree card, (3) cage tag, (4) card file, and (5) pedigree chart.

Ledger and Identification System.—The ledgers in use at the Jackson Laboratory are either spiral type notebooks or bookkeeping ledgers. Two to four pages at the beginning of the ledgers are left free for notes about the animals that will be recorded in the book, about the experiments, the kinds of mice that will be recorded in the ledger, and so on. Following these note pages, on the next page the lines are numbered consecutively beginning with 01 and going through 99, then 101 through 199, 201 through 299, 301 through 399, and so forth. Note that the number 00, 100, 200 and so on are omitted since this would mean an unmarked animal in this identification system. The animals are earmarked with a common poultry punch according to the code shown in figure 71.

The ledger may be used for a single strain or mutant stock, or if the colony is small, all animals may be recorded in a single, master ledger. Whichever method is

Fig. 71. Code for marking ears.

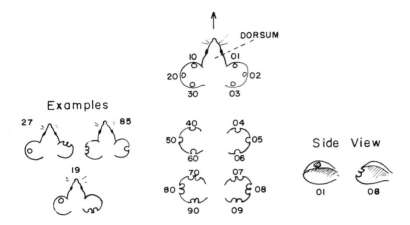

The animal is viewed from the back.   Tens are on the left ear and digits on the right ear.   Code marks at the top and bottom of the ear should be punched as close as possible to the head.   Code marks on the side of the ear should be punched close to the end of the ear fold (see side view).   If these precautions are observed, there will be little confusion about the code number.   Many times the punch pulls out the edge of the ear, but the distinction remains recognizable because the notches are very shallow in comparison to the pulled holes.

used, the information in the ledger includes the following biographical information.

1. Number of the mouse being pedigreed
2. Sex of the mouse being pedigreed
3. Phenotype, if needed
4. Fate of the animal (whether it was put in a new mating pen, set aside for a particular experiment, or classified and killed)
5. Parents of the animal
6. Generation to which it belongs
7. Date of birth

A line is drawn across the page after all animals from one litter are pedigreed.   The next line is the beginning of a new litter (figure 72).

In mutant stocks and various other types of experimental animals, it is necessary to record the phenotype of the animal.   The logical location of this information is after the sex of the animal, as previously noted.   When such notations are made, the symbol + is standard for the wild type or normal allele, especially in classification of mutant stocks (figure 73).

Some systems of keeping records employ a single ledger or two ledgers for each stock and assign each mating a double-page space for all information concerning offspring of that mating.

Fig. 72. LEDGER PAGE FOR RECORDING PEDIGREE INFORMATION.

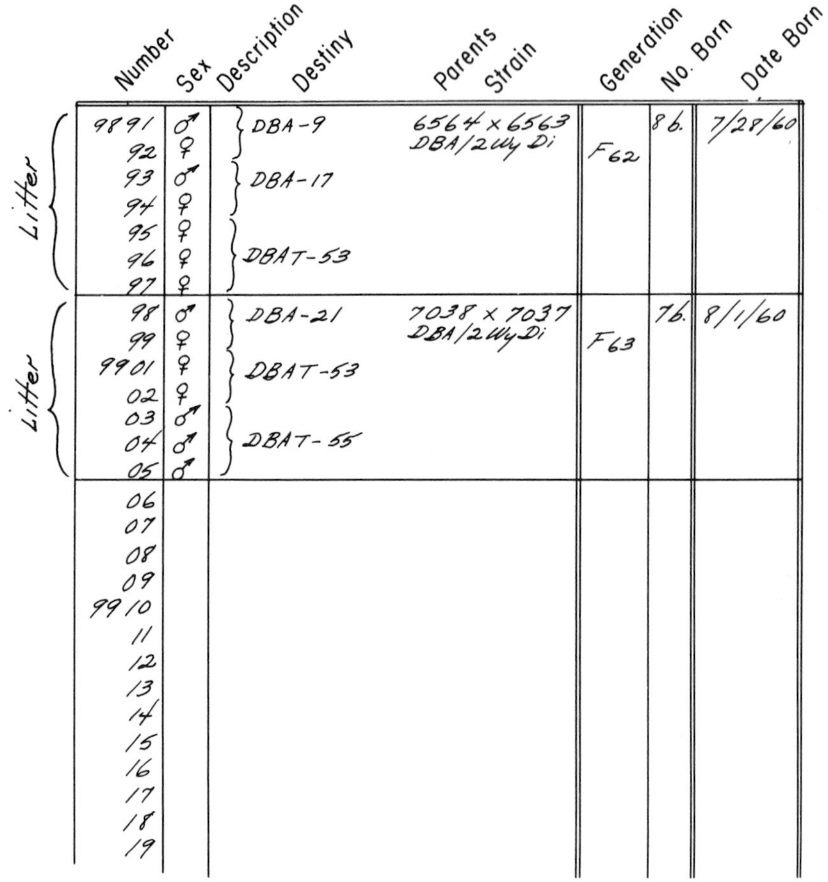

Number	Sex	Description / Destiny	Parents / Strain	Generation	No. Born	Date Born
9891	♂	} DBA-9	6564 x 6563 DBA/2Wy Di	F62	8 b.	7/28/60
92	♀					
93	♂	} DBA-17				
94	♀					
95	♀					
96	♀	} DBAT-53				
97	♀					
98	♂	} DBA-21	7038 x 7037 DBA/2WyDi	F63	76.	8/1/60
99	♀					
9901	♀	} DBAT-53				
02	♀					
03	♂					
04	♂	} DBAT-55				
05	♂					
06						
07						
08						
09						
9910						
11						
12						
13						
14						
15						
16						
17						
18						
19						

(The first litter comprises numbers 9891–97; the second litter comprises numbers 98–9905, each bracketed as "Litter.")

An example of a ledger page showing the arrangement of columns on the page and the pedigree information that has been recorded for two litters. Note in the numbering that number 9900 is omitted. Each pedigree number can be entirely written out on every line or shortened as it is in this example.

Fig. 73. LEDGER PAGE FOR A MUTANT STOCK.

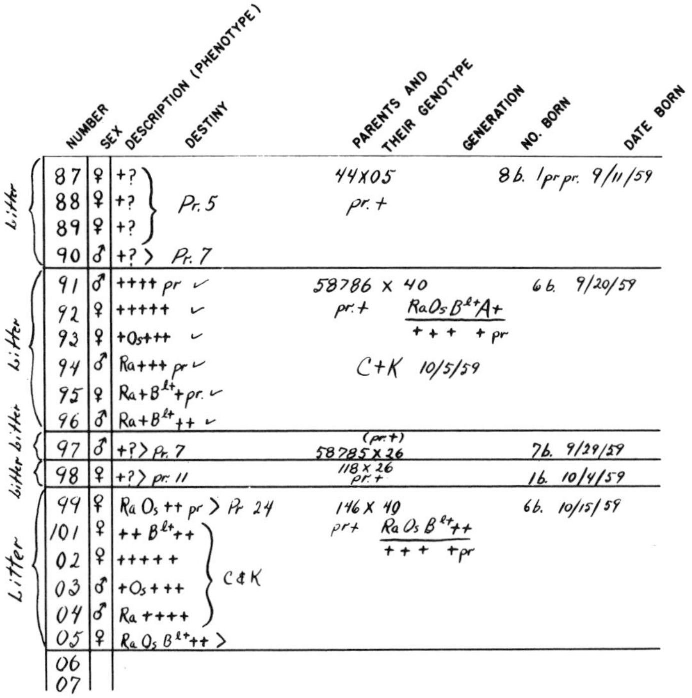

An example of a ledger page for a mutant stock showing the arrangement of columns on the page, and classification (phenotypes) of mice in several litters, their fates, and the genotypes of the parents. (Note that + is used to designate wild type or normal allele of any of these mutants.)

The Pedigree Card.—The pedigree card may be designed to suit the needs of the investigator. It should include space for the following information: date of birth, date of death, age at death, the kind of animal (strain or stock), the generation, the individual identification number, location (pen), date of birth of all litters, numbers born, number and sex of mice weaned and their fate (ledger number or a note if killed without pedigree) (figure 74A and B). In large colonies it is more efficient to have the headings on the pedigree card and in the ledger read across the card and page in the same order. It is also essential to have space on the card for the date of mating, the

Fig. 74A and B. PEDIGREE CARDS IN USE AT THE JACKSON LABORATORY.

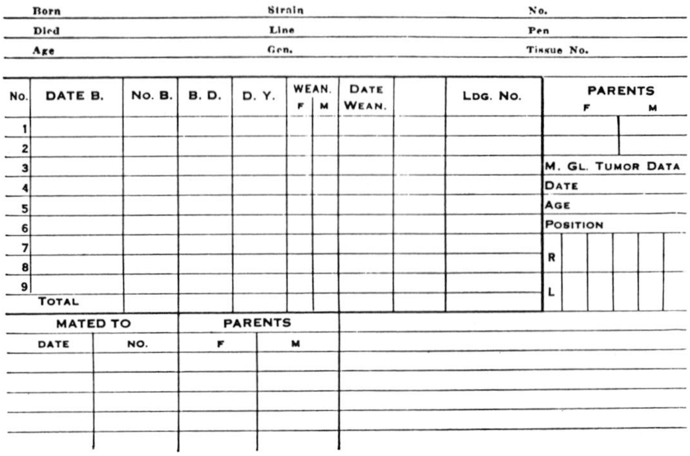

A

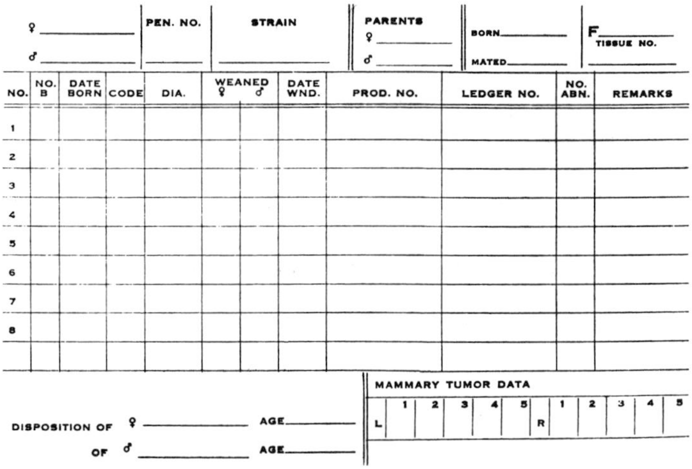

B

A. These cards provide space for the biographic details of each animal and are a permanent record of the individual.

B. Type B has been developed for use in the pedigreed expansion stocks of the laboratory. Items on the top of the cards appear in the same order as they do in the ledger. Certain columns such as b.d. (born dead) and d.y. (died young) had been converted to the uses now listed on the new card. More space has been provided for ledger numbers and remarks.

identification number of the male and kind of male, and date when he is removed. In some systems two cards are used, a mating card and a litter card (see figure 74C, D, and E).

When inbred strains are being maintained (brother × sister) it is unnecessary to provide an individual card for the male since all data for the female is applicable to the male and neither he nor she will have any other mate during their lifetime. If a male is to be used in genetic experiments, it is advisable to have a pedigree card for the male, so that one can easily obtain any information needed about him, no matter what pen he may be occupying at the time. The reverse side of the pedigree card now in use at this laboratory has been designed to allow space for comments on gross findings when animals are autopsied (figure 74F).

Fig. 74C. A PEDIGREE CARD USED IN MANY GENETIC EXPERIMENTS.

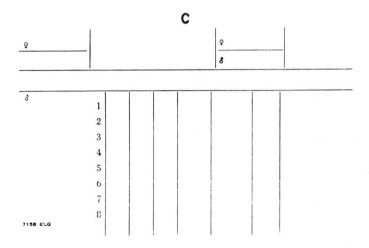

Across the top in order is found the female number, her genotype, numbers of her parents, and the cage number. On the blank line across is recorded the strain or stock and the generation and birth date. In the first column appears the number and genotype of the sire of each litter, next is his relationship to the female, the date of mating, date of birth of litter, size of litter minus number dead, the ledger number, fate of the litter, and finally, a space for remarks about the litter. The male card (not shown) is similar to the female card but does not include information about the litters. Death date is recorded in the lower right-hand corner of the card.

Fig. 74D and E. Mating card and litter card.

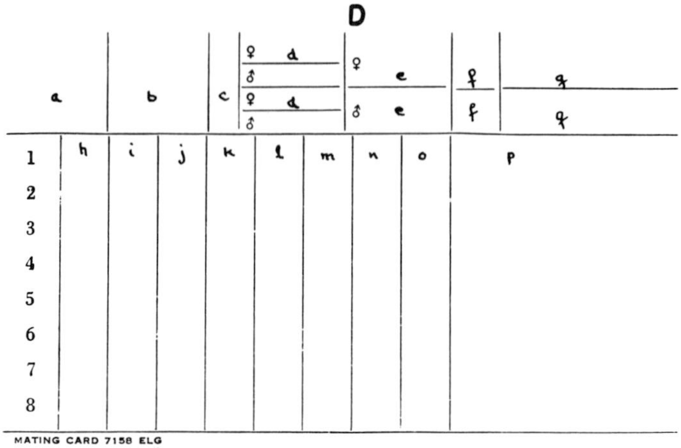

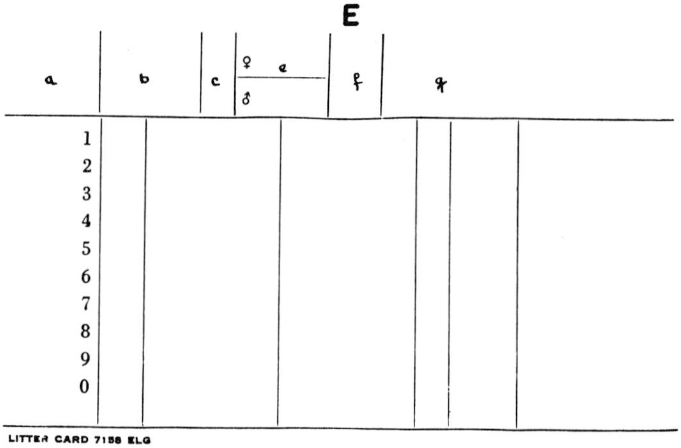

For certain types of experiments, a system has been devised which omits the use of the ledger. The system uses a mating card (D) and a litter card (E). Across the top appears the cage number[a], the strain[b], the generation[c], the parents of the pair being made up[d], the numbers of the pair[e], the dates of their birth[f], and their genotypes[g]. The columns are for date of mating[h], date of pregnancy[i], date of birth of litter[j], number born minus number dead[k], the number weaned[l], the number killed[m], the number raised[n], card number of litter[o], and remarks[p].

Card numbers for litters are stamped in serial order, and this number appears at the left of the number 1 on the litter card (E). The top of the litter card carries the same information found on the mating card, omitting only the numbers of the grandparents. The individuals in the litter are marked according to the numbers 1–0, sex recorded in the next column, then phenotype, genotype, fate of the animals, and date when their fate was determined. The last column provides space for remarks about each individual.

Fig. 74F. PEDIGREE CARD.

No.	Weight	
Skin		
Salivary Glands		
Mammary Glands		
Lymph Nodes		
Subcut.		
Abdominal		
Lumbar		
Mediastinal		
Thymus		
Liver		
Spleen		
Kidney and Ureter		
Bladder		
Stomach		
Intestine		
Pituitary		
Thyroid		
Adrenal		
Pancreas		
Gonad		
Uterus		
Vagina		
Accessory Sex		
Preputials		
Heart		
Lung		

JAX-PES-301

The reverse side of the pedigree card (fig. 74B) provides space for recording information on gross findings at autopsy of the animal.

Cage Tags.—There are two main types of cage tags, those which provide only a location number on a rack and those which provide not only the location but additional information such as the numbers of the animals in the pen, their birth and mating dates, strain and generation. On such tags pregnancy and litters are also recorded. In some genetic experiments it has been possible and efficient to use the cage tags in place of the pedigree card. When this is done, the ledger number is placed on the cage tag beside the record of the birth of the litter (figures 75 and 76).

Card File.—A card-file box is the usual repository for the pedigree cards. Several systems can be used.

1. If all the cards in one box pertain only to one strain, so indicated on the front, then it is not necessary to use any file-guide cards but merely file the pedigree cards according to date of birth (putting youngest in front) or according to cage number. The efficiency of the birth-date system is particularly noticeable in a large colony when a large number of pairs must be replaced at regular intervals. Example: It is usual in the pedigreed expansion stocks to replace breeding pairs at a stated interval which varies with the strain in use. The number to be made up varies slightly from month to month so that the number of boxes (2 pens per box) will equal a section. In C57BL/6J the replacement has been determined as 7 months. Therefore in September 1963 all pairs mated in February 1963 will be removed from the colony and replaced by the desired number of new pairs. These in turn will be replaced in 7 months (March 1964) by other new pairs. When the card file is arranged according to date of birth it is easy to see which animals should be removed in September (all those born in January, which were mated in February).

Fig. 75. CAGE TAGS.

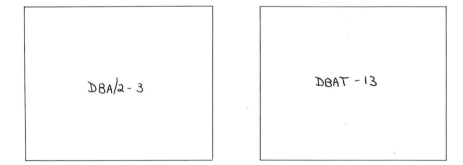

LEFT. Cage tags may show only the strain and number of the pen. RIGHT. If the cage is not a breeding cage some other notation may be used. Many times, only numbers appear on such tags.

2. When the colony is small and animals are put in and removed at irregular intervals, a file-guide system for the pens is more efficient.

3. If the system employed involves a pool of unmated females and males, as well as mated pairs and subsequent removal of pregnant females, then the card file could be handled more efficiently if it were divided into sections such as pregnant, mated, and unmated, and the cards were in numerical order within each section.

Fig. 76. CAGE TAGS.

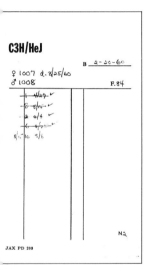 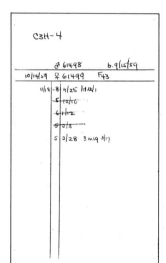 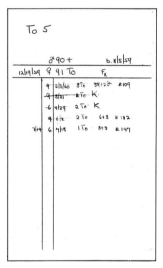

LEFT. Cage tags may provide much information about the animals housed within. In the colonies of pedigreed expansion stocks, different colors and marks present or absent across the top help insure that strains will not be mixed. The female and male number, their dates of birth and generation are recorded. The *N* in the lower corner denotes the number of generations removed from the ultimate source in the small colony of foundations stocks. Litters are recorded, crossed off at weaning, and when other types of information such as diarrhea, weights, and so on are recorded, a check mark is placed beside the litter. Mating date is not recorded in these inbred lines because all matings are set up at weaning, 25–30 days of age.

CENTER. Cage tags in a small colony may provide information similar to that listed above. The order is slightly different, and the columns may be arranged to suit individual needs. The date of mating appears before the female number.

RIGHT. Cage tags in mutant stocks can provide information usually recorded on the pedigree card, thus eliminating that card from the system. The fate of the litter is recorded here in as much or as little detail as the investigator wishes.

Pedigree Charts.—Pedigree charts are essential to insure that inbred strains of animals are being propagated through a single pair and not through series of animals that were

Fig. 77. Portion of a pedigree chart showing the source of one inbred strain.

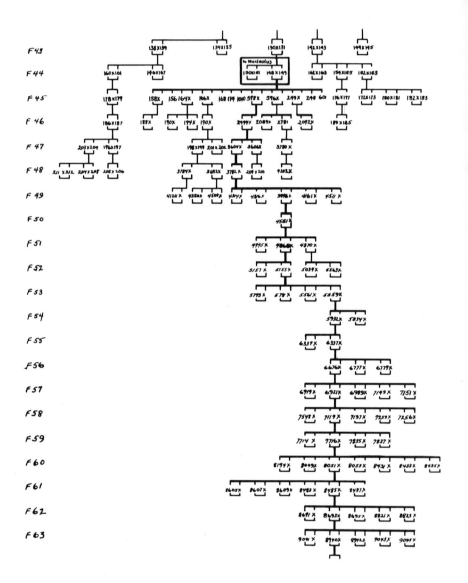

The descent of the main line through a single pair is indicated. Other siblings and cousins may be bred to produce more animals at any one time, but the pedigree history is continuous through single pairs in each generation.

descendants of sibling pairs many generations before.   This type of propagation encourages subline formation within an inbred strain.   Pedigree charts are also essential to determine relationships of deviants arising in the colony and perhaps give information about the inheritance of the deviant.   In production colonies such charts will show the portion of the pedigree that must be eliminated from the colony (figures 77, 78, and 79).

Fig. 78. Pedigree chart of the relations of a deviant that appeared in an inbred strain.

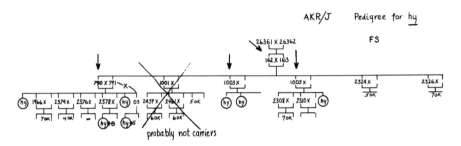

Such a chart shows (1) that the animals that should be removed from the main colony and (2) those animals that can be eliminated because they are probably not carriers of this new deviant.   Outcross tests of both the female and male to known heterozygotes would be needed to establish this conclusively.

Fig. 79. Pedigree chart of the relations of a lethal deviant that appeared in the production colony.

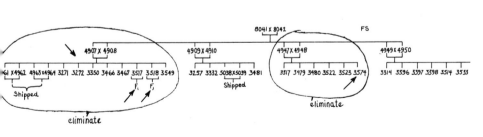

Study of the chart reveals the fate of the relatives and those animals that must be removed immediately from the colony.   All other matings were checked and found to have been discarded or not to be carriers.

Let us now work through an example of using all the parts of the system just described as they would apply to an inbred strain and to a new deviant.

A new inbred strain, NE, was procured and the supply consisted of three sibling pairs. The information given was the parentage of these siblings, their date of birth, the generation of inbreeding and the identification numbers of the individuals. Stepwise the procedure is as follows:

1. Check the ear marks of the animals against the information provided to make sure of their identity.
2. Place each female with her respective brother in a pen labeled NE1, NE2, and NE3.
3. Make out a cage tag putting on it the information desired.
4. Make out pedigree cards for each of the females and place in the card file.
5. Start a pedigree chart with the information that has been provided.

A new strain is now in business. The pens should be checked frequently to ascertain the continued good health of the animals and whether the females have become pregnant. When litters are born these are immediately recorded on the cage tag. The litters should be weaned about 25–30 days of age. If a new litter is born before that, the older litter may be set out for a few days before it is pedigreed. First litters of any females are not usually used to make up new pairs, but are set aside for experiments or for other purposes. When the second litter is ready to wean, the animals will be recorded in the ledger, new pedigree cards will be made out and new pens put into use. Those animals used as pairs will be recorded on the pedigree chart (figures 80–83).

Fig. 80. PEDIGREE CARD MADE OUT FOR ONE PAIR OF THE NEW STRAIN JUST PROCURED.

| ♀ 69025 | PEN. NO. | STRAIN | PARENTS ♀ 65647 | BORN 4/30/60 | F 319 |
| ♂ 69026 | NE 1 | NE | ♂ 65648 | MATED 6/15/60 | TISSUE NO. |

NO.	NO. B	DATE BORN	CODE	DIA.	WEANED ♀	♂	DATE WND.	PROD. NO.	LEDGER NO.	NO. ABN.	REMARKS
1	5	7/12/60		–	0	4	8/8/60		07	0	
2	7	8/6		–	3	3	8/31		18	0	
3											
4											
5											
6											
7											
8											

MAMMARY TUMOR DATA

DISPOSITION OF ♀ _____ AGE _____
OF ♂ _____ AGE _____

Fig. 81. PEDIGREE CHART THAT HAS BEEN STARTED FOR THE NEW STRAIN.

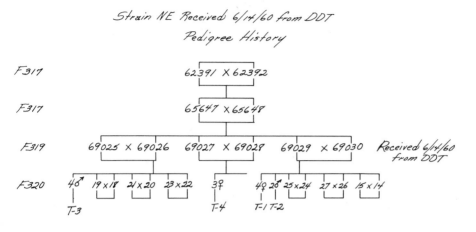

Fig. 82. LEDGER PAGE STARTED FOR THE NEW STRAIN.

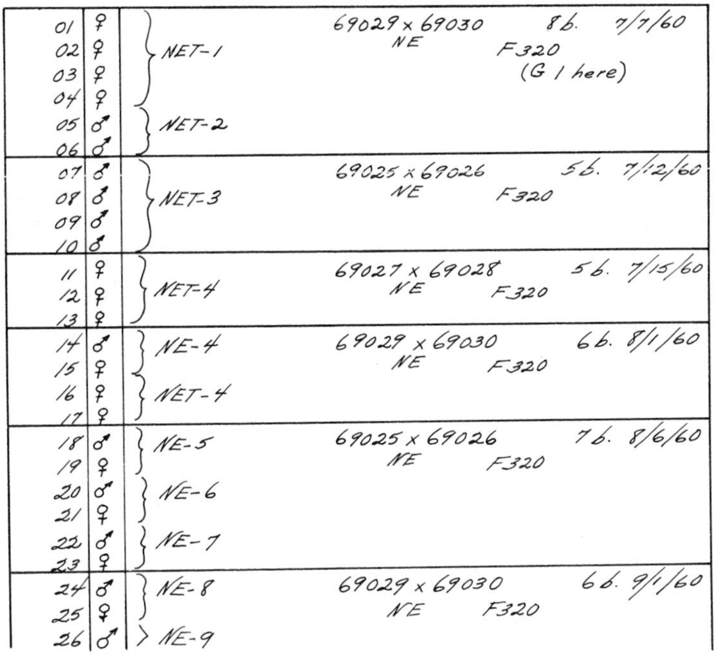

Fig. 83. Cage tag for the pair whose pedigree card is shown in figure 81.

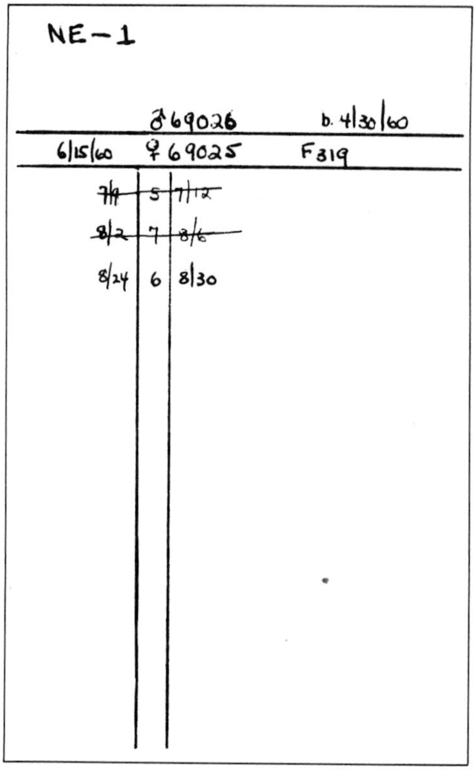

Fig. 84. Cage tag made out for new deviant.

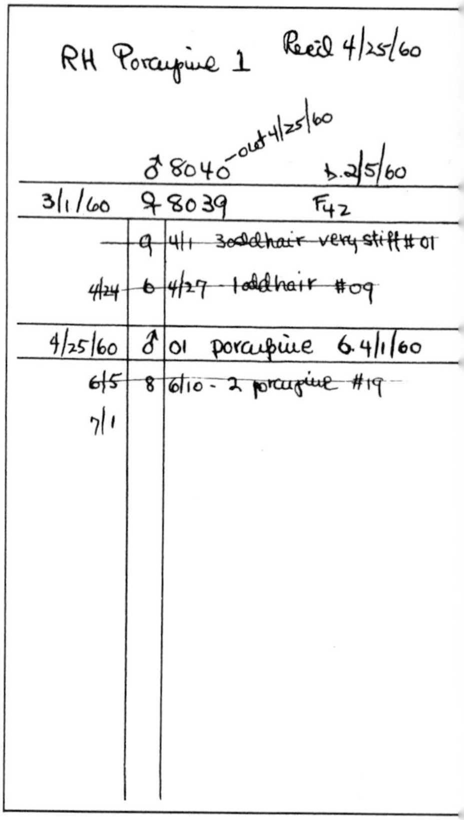

When a new deviant is found the procedure is similar but the information recorded may differ somewhat. Suppose that three mice appear that resemble porcupines. One proceeds as follows:

1. Determine the sexes of the deviants, mate father with daughters and mother with sons.
2. Earmark all individuals in the litter. If one wishes, the normal littermates may be mated *inter se*.
3. Make out cage tags with appropriate information and start a new ledger.

The pedigree of the parents of the deviant is studied and a check is made on the offspring of all collateral relatives, to see whether they have produced any abnormal offspring. Ensuing litters are classified and the stock is maintained in a manner most effective for continuation of the character (figures 84 and 85).

Fig. 85. LEDGER PAGE WHICH RECORDS INFORMATION ABOUT HISTORY AND PROGENY OF NEW DEVIANT.

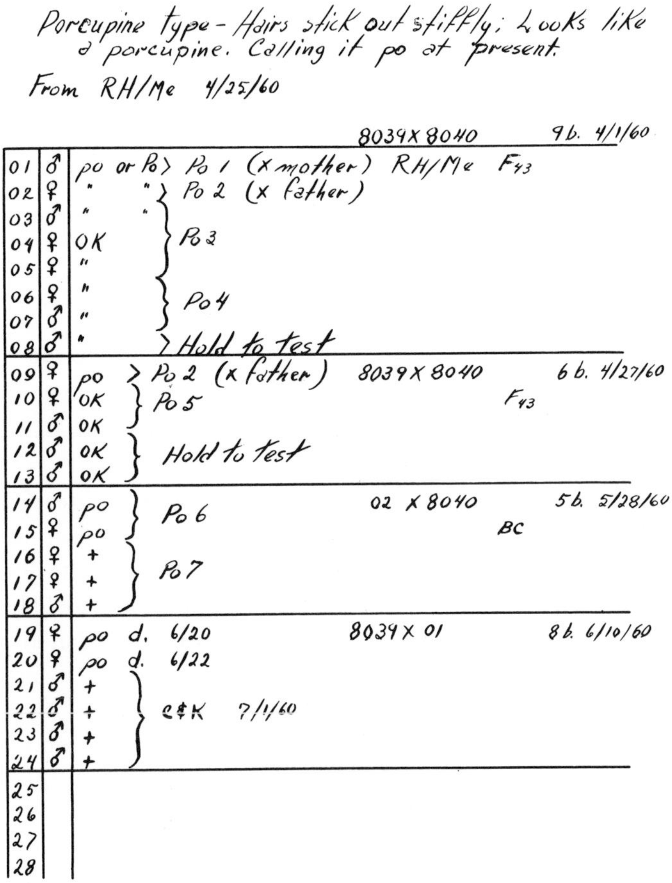

Porcupine type - Hairs stick out stiffly; Looks like
a porcupine. Calling it po at present.
From RH/Me 4/25/60

8039X 8040   9 b. 4/1/60

01 ♂	po or Po〉	Po 1 (x mother)	RH/Me F₄₃
02 ♀	" " 〉	Po 2 (x father)	
03 ♂	" "		
04 ♀ OK	〉 Po 3		
05 ♀ "			
06 ♀ "	〉 Po 4		
07 ♂ "			
08 ♂ "	〉 Hold to test		
09 ♀ po	〉 Po 2 (x father)	8039 X 8040	6 b. 4/27/60
10 ♀ OK	〉 Po 5		F₄₃
11 ♂ OK			
12 ♂ OK	〉 Hold to test		
13 ♂ OK			
14 ♂ po	〉 Po 6	02 X 8040	5 b. 5/28/60
15 ♀ po		BC	
16 ♀ +			
17 ♀ +	〉 Po 7		
18 ♂ +			
19 ♀ po d. 6/20		8034 X 01	8 b. 6/10/60
20 ♀ po d. 6/22			
21 ♂ +			
22 ♂ +	〉 ♀♀K ?/1/60		
23 ♂ +			
24 ♂ +			
25			
26			
27			
28			

The components of a system of keeping records presently in common usage have been reviewed, but it may be of interest to scan other systems that are found both in this country and abroad.[105, 166, 479, 693] Each investigator should decide what sort of information he wishes his records to provide. When this has been decided, the most efficient and least cumbersome method should be used which will provide the *raison d'être* of all such systems: easy traceability of all relatives of the animal in question.

Warren G. Hoag, D.V.M., and Edwin P. Les, Ph.D.

APPENDIX IV

Husbandry, Equipment, and Procurement of Mice†

The husbandry of laboratory mice is worth considerable attention by research workers. Too often the task of checking, maintaining, and procuring experimental mice, or other laboratory animals for that matter, is relegated to less well-trained individuals on the institutional staff. This is unfortunate since it should be remembered that in most cases the entire experimental design is based on the initial use of normal, healthy animals and on the maintenance of such animals under conditions which provide safeguards against the introduction of immeasurable variables such as fluctuations in patterns of care, environmental conditions, diet, and status of health. If experimental animals are to be used as the yardsticks or biological test tubes in research, one must make sure that such animals truly measure, either quantitatively or qualitatively, only those things intended rather than other unpredictable variables.

It is the purpose of this appendix to attempt to describe conditions and methods for the husbandry of laboratory mice which will allow the investigator to adopt those suited both to his experimental environment and to his budget. Since much descriptive literature is available concerning equipment, animal-room construction, care of mice, and the like,[547, 1233, 1333, 1366, 1410] only certain specific items will be described as examples of methods of husbandry in mice. It is important to emphasize that the

† General sources of information: Institute of Laboratory Animal Resources, 2101 Constitution Ave. N.W., Washington 25, D.C.; Universities Federation for Animal Welfare, 7a Lamb's Conduit Passage, London W.C.1; National Society for Medical Research, 920 South Michigan Boulevard, Chicago 5, Ill.

aims of proper husbandry may be attained by many methods, some expensive, some inexpensive, but one should always keep in mind these aims rather than the methods. Methodology is a constantly changing area but the philosophy of laboratory care of mice should be much less so.

The investigator using experimental mice should plan:

1. to use mice which have known lineage and uniform good health,
2. to introduce those animals into the experimental environment only after proper quarantine and observation, and
3. to maintain them in a uniform environment with a proper diet composed of known constituents under the best of sanitary conditions.

It is with these aims in mind that we describe procedures which may be used for handling, housing, and procuring mice for research purposes so that these extremely useful animals may provide their fullest value to any biological experiment.

HUSBANDRY OF MICE

Because mice are usually handled in large numbers they are often given little consideration as individual animals. The individual mouse determines in part the requirements for the whole colony and as an individual is important in its relationship to the population and the effect it has on it. In large colonies the attendant is often referred to or considers himself as a "mouse-box changer" or some similar designation which implies complete lack of consideration for the mouse as an animal deserving individual attention or even as a biological entity. It is therefore quite important to instill in the mind of the attendant the realization that he is an animal caretaker and that the animal he must care for is the mouse.

The laboratory mouse responds as do other domesticated animals to gentleness and soon becomes accustomed to routine procedures and environment. Abrupt changes in these procedures or environmental conditions are as undesirable and upsetting as to larger animals, although their manifestations or reactions are more obvious. The first rule of good husbandry is therefore that a set routine of feeding, watering, and changing of cages be established and followed faithfully.

Feeding. [120, 188, 355, 356, 444, 548, 752, 897, 898, 1390]—Food is usually made available at all times to laboratory mice. Dry feed in the form of pellets is quite satisfactory. Pellets should be a size that is easily available to the mouse through the hopper yet not easily pulled from it, and their consistency should be sufficiently hard to provide some wearing of the incisors. Hardness is determined by the formulation of the diet and the width of the pellet. The length of the pellet is in turn determined by hardness, since the usual machine extrudes compressed food which then breaks off or is scraped off.

A standardized diet for all mice or for all inbred strains of mice is desirable, but unfortunately little is known about nutritional requirements of individual strains. Many diets commercially available and specified as complete are at best subsistence formulas. Although analysis of diets may indicate a complete array of ingredients in terms of vitamins, amino acids, and other important constituents, the source material may be deficient quantitatively.   Also, even if chemical analysis indicates quantitative sufficiency of a certain item, subsequent biologic testing may indicate it is unavailable to the test animal in the form applied.   Therefore commercial diets should be evaluated critically for the source of ingredients as well as their chemical composition.   It is always advisable to examine carefully any data presented by a manufacturer as evidence for a given formulation being better than any other.   The individual using mice in his experiments should also demand that the composition of the diet for his animals be known.   Arguments are presented against this by commercial manufacturers, including the statement that standardization by chemical analysis is more important than standardization of the source of components.   This argument would be valid if more definite information were known concerning the nutritional requirements of the mouse, but unfortunately this type of information is incomplete.   Most of the information available indicates that the source of the biochemical is more important, and it is therefore more justifiable to demand a constant-ingredient formulation for mouse diets.

A good type of diet for mice has been suggested by Morris.[897, 898]   The constituents are as follows:

Skimmed milk powder	22.8	per cent
Ground whole wheat	61.5	per cent
Brewer's yeast (dried)	4.0	per cent
Cod-liver oil	2.0	per cent
Salt	1.4	per cent
Ferric citrate	.13	per cent
Corn oil	8.2	per cent

This has a calculated composition as follows:

Protein	19.6	per cent
Fat	11.7	per cent
Carbohydrate	62.7	per cent
Ash	4.8	per cent

With this as a basic formulation, various modifications may be made to suit requirements of certain inbred mouse strains.   Certain strains are inclined toward obesity and therefore an increased amount of protein and a lowered amount of fat is indicated.   This is achieved by increasing or decreasing the amounts of the major sources of protein and fat.

The feeding of mice should be directed toward meeting the physiologic demands of the specific inbred strain being used, but much work needs to be done in order to delineate appropriate parameters so that the exact nutritional requirements are known. Enough is known of the rôle nutrition plays in resistance or susceptibility to gamma radiation, agents of infectious disease, and other factors of potential hazard to the bio-logic host to emphasize that the investigator examine and evaluate the specific diet being offered to his mice as a part of the entire experimental design rather than as a casual consideration or afterthought.[291, 292, 1160]

Bedding and other environmental factors.—The bedding material selected should meet the demands of good husbandry. These include the ability to absorb the moisture of body wastes between changes of bedding. A highly absorbent material may be un-desirable for breeding mice, since such material clings to newborn and very young animals with a resultant dehydrating effect. In breeding cages, the bedding should provide material for a nest which can easily be removed from cages when they are cleaned. Some types of bedding such as baked, pulverized, or powdered clay, ground corn cobs, or ground sugar cane are rather difficult to remove from cages after use and are often found to be too absorbent. They also tend to clog drains and filters in automatic cage-washing equipment because of the difficulty in removing all of the material before washing. However, they have an advantage in being highly absorbent and hence lengthen the interval between necessary cleaning. They may therefore be suitable for certain types of experiments.

One of the most satisfactory bedding materials is dried wood shavings. These may be procured baled or in paper bags and are composed of a variety of woods. The shavings should be fine and contain no large, coarse particles. Kiln-dried soft wood such as pine makes the best shavings for bedding. Such wood shavings are absorbent and make excellent nesting material for breeding mice. Shavings from aromatic woods such as cedar have some suppressive value for ectoparasites, but on the other hand their potential for irritating the skin and possible carcinogenicity are factors to be considered. Since bedding material comes in most intimate contact with mice, such factors must be weighed carefully.

It is possible to keep mice on wire floors during experiments. These should be of the hardware cloth type with $\frac{1}{4}$- or $\frac{3}{8}$-inch mesh, depending on the size and age of the mice to be suspended. These floors should be raised at least $\frac{3}{4}$ inch above a pan containing absorbent material.

The acceptable range in temperatures of a room in which mice are maintained is $\pm 2°$ F. level. The most satisfactory temperature is usually between 70 and 75 degrees Fahrenheit. Mice withstand quite low and sometimes quite high (90° F.) temperatures but our experience indicates that for peak breeding efficiency a temperature of 72° F. is a good one. Temperatures should be maintained at a constant level the year round. Changes of air in our mouse rooms are maintained at a level of 6–7 changes per hour level; many breeders use as many as 10–12 changes of air per hour. The number of changes is of course determined by the number of animals kept in a given room (each mouse contributes an average of 0.6 B.T.U./hour/21 gram mouse from body heat) and by the desired temperature. The ventilation system should be de-signed to provide the necessary number of air changes in a manner free of drafts directly on cages. A system designed for laboratories or schoolrooms is not necessarily one satisfactory for mouse rooms. It is best to consult designers who have had experience in providing satisfactory animal room ventilation or who are competent enough to provide a design to maintain conditions selected by the investigator or breeders. Separate ventilation systems should be provided for each room to provide better containment of health problems. All air introduced should be directly from outside, although it is possible to recirculate up to 25–35 per cent of the air under certain conditions (extremely cold outside temperatures) without encountering odor problems from urine and fecal material.

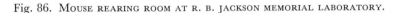

Fig. 86. MOUSE REARING ROOM AT R. B. JACKSON MEMORIAL LABORATORY.

To prevent introduction of dust particles carrying bacteria or viruses, the air intake ducts should be provided with glass-wool filters capable of removing particles as small as 0.2 microns in size or some similar system for accomplishing this. The arrangements of relative air pressures in mouse rooms and accompanying corridors, equipment-washing areas, or the like, will depend on the purpose of the unit. For breeding mice or housing normal supplies, the plan should in general provide air flow from the cleanest to the dirtiest area. This is accomplished by mounting equipment-washing machines so that a solid wall separates the discharge end of the machines (the cleanest area) from the feed-in area (the dirtiest area). The air should then flow from this clean area down corridors used solely for transportation of sanitized equipment and then into mouse rooms. The lowest air pressure should be maintained in the equipment disassembly and washing machine, feed-in area. In this way air will flow out of a mouse room when the exit doors of a room containing dirty equipment are opened and down the corridors used solely for transport of such materials. In a situation requiring containment during an experiment with a transmissible agent, some modification of this pattern must be made, depending on the particular problem being investigated.

Experimental rooms should be only large enough for use by a single principal investigator. If it is found necessary to have more investigators use a single room, one should be reponsible for supervision of the room. This procedure provides proper liaison among several investigators and animal husbandmen. In this way proper control of caretaking and procedures for control of diseases can be maintained.

The humidity requirements of mice are not well known. Practical experience indicates that a relative humidity of 30 to 50 per cent is most satisfactory. The reasons for suggesting this range are not scientific and are derived mainly from standards of human comfort. In an extremely dry atmosphere certain types of dermatoses are more prevalent in mice, although again the correlation is purely empiric.

General care and handling.[247, 1410]—Mice are as much creatures of habit as are larger animals or poultry and are as easily upset by abrupt changes in routine or unusual noises. Routines of changing cages and general caretaking should be established and strictly adhered to. Cages, water bottles, and equipment coming in direct contact with mice should be replaced weekly with sanitized or sterilized replacements. Such materials should have foreign material removed before being washed. The washing process should follow a general routine of prewashing or soaking, washing with soap or detergent, and one or more rinses with fresh water of 180° F., in the order mentioned. Disinfectants or germicidal agents are not necessary if the detergent completely removes all material from smooth surfaces and if this is followed by a thorough rinsing at 180°+ F.

Mice should be handled (figure 87) only by long forceps (except for experimental procedures). These are used by grasping the mouse firmly but gently by the base of the tail. Two or more pairs of forceps should be available, to allow the unused pairs to soak in a solution of disinfectant between use. A single pair should be used for handling only those mice in a common unit and a fresh, disinfected forceps used for the next group. The liquid disinfectant should completely immerse the lower 3 to 4 inches of the forceps and should be of a type which is neither irritating nor carcinogenic. A commonly used preparation is 0.5 per cent Wescodyne solution,† although many similar products containing bound iodine are available.

Experimental procedures should be conducted in a closely adjoinig area, and only the necessary cages of mice removed to that area. The experimental work should not be done in the aisles between cages unless it is minimal, such as visual observation, counting of newborn mice, and the like. Surfaces for experimental and other types

† West Chemical Products, Inc., 40 West Street, Long Island City, N.Y.

of work along with the hands should be disinfected between handling different groups of animals.

Handling of newborn mice in experimental procedures often results in cannibalism by the mother. To avoid this, the mother can be anesthetized lightly with diethyl ether and then replaced in her cage to recover while the newborn are being worked with. Chloroform should be avoided and should be banned from mouse rooms or adjoining areas. Certain strains of inbred mice (DBA/2 and C3H, for example) are highly susceptible to the fumes of chloroform and succumb very quickly to traces in the air.

Fig. 87. TABLE FOR CHANGING CAGES.

Mice to be culled should be removed from the room for destruction and may then be sacrificed by any suitable procedure, such as cervical dislocation or inhalation of ether or carbon dioxide.

Maintenance of health.[1333]—Diseased or latently infected mice should not be used in an experiment unless the investigator wishes to evaluate his introduced variables in terms of the agents of disease already present. It is most important therefore that (1) healthy mice be provided for an experiment and (2) the state of health be maintained throughout the work. The diseases of mice are manifold and only the groundwork has been laid in understanding latent infections. In a later section the requisites the purchaser should consider in selecting a supplier of experimental mice will be mentioned. All mice should be maintained under "pathogen-free" conditions. Simply stated, procedures of maintenance should guard against the introduction of known pathogens for mice via such items as equipment, food, and water, by personnel, and should prevent transmission of disease between cages of grouped animals or between mouse rooms.

The flow of traffic should be arranged so that cross or back traffic is avoided. This implies separate exits and entrances for used and clean equipment and materials. Personnel locks should be provided for entrance and exit of caretakers and other types of personnel to mouse rooms and to clean equipment, food, and bedding areas. Lockers and washing facilities in them should be arranged so that personnel may enter, remove street shoes and outer garments, step into a clean area or on a platform, wash and scrub hands and arms, don work clothes and footgear (only used in the mouse room to be entered), and then enter the animal area. The procedures are reversed for departure.

Mouse-room walls and floors, fixed racks, and other immovable items should be cleaned with mops, sponges, or cloths dampened with a residual type of disinfectant such as the quaternary ammonium compounds. Sweeping and dusting should be prohibited to avoid transferring agents of disease [including endoparasite eggs such as oxyurids (pinworms)]. Wet-vacuum cleaning equipment is also as useful in animal husbandry as in promoting asepsis in hospital operating suites. Different personnel should be used for each general area as follows:

1. Equipment assembly (from washing machines to autoclaves) and clean equipment, food, and material delivery.
2. Individual mouse rooms.
3. Used equipment and waste area (including the dirty phase of sanitizing equipment).

Personnel from one area should never enter another except in the above sequence or by going through all procedures of washing and changing clothes previously mentioned.

Bedding may be purchased sterilized or sanitized (ground corn cob such as San-i-cal,† clay material such as Ster-o-lit,‡ packaged shavings) or may be sterilized on the premises. Steam autoclaves may be used for sterilizing bedding and have an advantage in that the outside wrappings of feed (in plastic or metal containers) may be decontaminated by flowing steam. Ethylene-oxide autoclaving should not be used for treatment of feeds or bedding materials.

Food may also serve as a vehicle for introducing agents of disease. Treatment of food with dry heat (baking) is satisfactory provided sufficient heat-sensitive vitamins and other such ingredients are added before or after treatment to offset the loss or diminution by heat. There is a need for better techniques in the field of decontamination of food. Pasteurization, or sterilization and methods used by manufacturers, should be reviewed carefully. This should include scrutiny of actual methods of treatment, of evaluation of ingredients of feed after processing, and of methods for

† Walter F. Fisher & Sons, Inc., Bound Brook, N.J.
‡ T. C. Ashley & Co., 683 Atlantic Avenue, Boston 11, Mass.

preventing recontamination after treatment and during packaging, storage, and transportation.

Drinking water is frequently overlooked as a means of introducing pathogenic agents. Periodic and frequent examinations of tap water should be made to insure purity or, better still, water may be pasteurized by flash pasteurization of the type used in the industrial processing of milk (161° F. for 15 seconds).

The introduction of a new stock to a mouse colony always presents a potential threat to the health of the colony. New mice should be received and unpacked in quarters apart from the breeding, holding, and experimental rooms. They should be removed from shipping cartons immediately and examined carefully for physical condition. A carton of arriving animals should be transferred to the same cage when possible. Any carton containing more sick, moribund, or dead animals than usual should be noted and all these animals, including animals apparently well, should be discarded or sent to a diagnostic laboratory.

Water, food, and clean cages should be made available at once. After being unpacked, transferred, and examined in the receiving room the mice then should be moved to a quarantine room. Here they should be maintained until passing suitable checks on their state of health. A good procedure involves introduction of several mice (at least two mice that are 6–8 weeks old, preferably of the same strain) from the colony or room into which the new mice are to be admitted. These are placed in each cage with new animals and remain for four weeks. At the end of this time (or earlier if any become sick or die) these test mice are bled, killed, and necropsied. Serum samples should be tested for ectromelia (mouse pox) and organs bacteriologically examined for *Salmonella sp.* and other pathogens. The new mice may be tested upon arrival by sampling of feces for the presence of *Salmonella* or other enteric pathogens. If the test animals remain well and pass the laboratory tests (including serum tests for ectromelia) the new animals may be admitted to the animal rooms. It should always be kept in mind that sickness or death of mice recently introduced to an animal room does not imply that these animals brought in a new disease (particularly if illnesses or deaths occur one or two weeks after arrival). This could be indicative of susceptibility of new animals to diseases already present in the existing group, the latter having carried the agents in latent form.

FACILITIES AND EQUIPMENT

Literature dealing with the physical necessities for maintaining a colony of mice or other small animals should be studied thoroughly as the first step in setting up an animal colony. Several agencies for disseminating information of this type have come into existence in this country and in Great Britain. The information provided in this appendix is not intended to supplant other material on this subject, but merely to facilitate the process of establishing and maintaining a colony of mice.

Probably no two laboratories use exactly the same methods, equipment, and facilities to raise mice. A research organization contemplating construction of extensive facilities for laboratory animals would in the long run benefit by investing in the services of a qualified consultant or by sending a competent observer to visit laboratories which maintain colonies of animals.   Initial information about such colonies can be obtained from the Institute of Laboratory Animal Resources.   If a facility is to be fairly small the investment in outside consultants or designers may seem to be unwarranted.   In such cases the individuals responsible for the design of animal quarters should become acquainted with the published information on design of facilities, equipment, and husbandry.

In matters relating to the facilities and equipment necessary for maintaining a colony of mice, the methods used in any given situation should not be inflexible. Ideal circumstances for rearing mice have never been achieved by any laboratory and only experience can determine the most desirable methods.   If the arrangements described are closely followed, the experimenter will have fulfilled minimum, but adequate, maintenance requirements.   Variations from this basic pattern will be necessary to accommodate the particular type of experimentation to be pursued.   The pattern described is suitable for experiments in genetics, radiation studies, immunology, tumor biology, and for simply maintaining mice to be used elsewhere in the laboratory.

Designs for facilities for rearing mice have been worked out in considerable detail by various organizations and individuals.   Some of the published designs have been incorporated into facilities that are currently in use at the National Institutes of Health, Walter Reed Army Research Center, and University of California Medical Center. Designs such as these can be modified to suit the requirements of a particular type of research, but the essential features which make such quarters desirable should not be eliminated without compelling reasons.

Let us assume that a room is available for use and that it is suitably lighted, heated, and ventilated (or air conditioned).   The room is accessible from a laboratory and is arranged within a system of clean and dirty areas or corridors.   The floors, walls, and ceilings conform to, or exceed, the minimum standards set by the Committee on Minimum Standards for Commercial Production of Mice.   A wash basin or sink should be located in the room and equipped with a dispenser for disinfectant soap. The entrance should be a two-door vestibule.   When one door of the chamber is open the other must be closed.   This prevents air disturbances in the room and also prevents excessive loss of pressure if the room is pressurized.

Cages.—Many factors must be considered in selecting the type of cage (figure 88) to be used in a mouse colony.   The type of cage can affect the health of the mice and the operating efficiency of the colony.   Cage designs are discussed extensively in the *UFAW Handbook* and the manual *Care and Management of Laboratory Animals.*[547]   In general, the characteristics of a good mouse cage are:

1. It should confine the animals without causing harmful effects.
2. It must provide adequate space for nesting and exercise.
3. It should be made of material which permits thorough cleaning, can withstand sterilizing temperatures, and is durable.

Metal and plastic cages are used in most colonies.   Stainless steel is the most durable and most expensive metal, aluminium alloys are intermediate in cost and durability, and tinned or galvanized iron is the least expensive and most susceptible to the corrosive effects of urine.   Various kinds of plastics are being used for construction of cages. The transparent plastics are desirable because of the convenience they provide in allowing observation of the animals.   Such plastic materials, however, cannot withstand high temperatures and are quite fragile.   Molded fiberglass cages are more durable, but are not transparent.   Plastics are also easily scratched or chewed so that efficient cleaning eventually becomes a problem.

Fig. 88. Clean stainless steel cages stacked on dolly.

Two kinds of cages are in use in our colonies: (1) stackable two-compartment stainless steel cages and (2) two-compartment transparent plastic cages which are also stackable.   Only a few are of the latter type.   Each half of the two-compartment cage has a floor dimension of 4 by 11 inches.   The walls are about 6 inches high.   Each compartment provides adequate space for 5 adult mice or a mated pair with a litter.

The steel cages are made of a single, seamless, deep-drawn sheet of metal.   It has rounded corners and a polished surface.   The cage is practically indestructible, easy to clean, and can be autoclaved or washed at very high temperatures.   It is impervious to the corrosive effects of urine, feces, water, and cleansing chemicals.   Although the initial cost may be somewhat higher than for cages made of other materials the investment in terms of durability is very reasonable.

In general, a convenient size is one with a floor space 5 to 6 inches wide and 11 to 12 inches long. The wall height is essentially governed by two factors: the vertical space available and the ability of mice to jump out of an opened cage. A wall 5 to 6 inches in height is a workable compromise.

The transparent plastic cage is molded with rounded corners and reinforcing struts at points of stress. Such cages can withstand temperatures which adequately sanitize the cage provided the heat is applied when the cage is wet. The cage can be broken if it is dropped or struck against a hard surface. Care must be exercised in handling in order to maintain transparency.

Both the stainless steel and the plastic cage are manufactured in the two-compartment form. The use of such cages increases the efficiency of handling since two cages are taken to and from a shelf in a single motion. The users of two-compartment cages rapidly adapt themselves to a system of cage designation which minimizes the number of empty compartments in a given shelf space.

Fig. 89. CAGE COVERS WITH ATTACHED FOOD HOPPER SHOWING STACKING ARRANGEMENT FOR MACHINE WASHING.

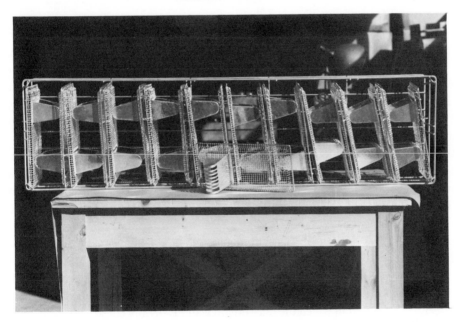

Covers.—The type of cover (figure 89) used on the cage is second in importance only to the design of the cage itself. Many types are currently in use. Some covers are attached by hinges, but in most instances they are separate. Covers fabricated of some type of metal are used almost universally. Wire mesh, expanded metal, punched sheet metal, and wire grid are perhaps the most popular. Where cages are suspended from rack shelves the cover is sometimes built into the supporting cage framework. The material may be stainless steel, tinned or galvanized iron, aluminium, or various corrosion-resistant alloys.   Brass and copper are never used because the mice may be poisoned by the corrosion products of such metals. A food hopper is usually built into or attached to the cover.

In our colony the cage cover is of hardware cloth (wire mesh) with a reinforcing rod around the edges.   The mesh is 4 squares to the inch and the wire is tinned.   A lip is formed around the edges of the cover so that when it rests on the top of the cage the depressed central portion is suspended within the walls of the cage.   The cover can be easily removed and replaced, is very light, permits adequate circulation of air into the cage, allows some observation of the contents without removal, is easy to clean, and the cost is quite low.   This cover also provides support for the water bottle.   In some other designs the water bottle may be suspended against the wall of the cage. The cover described has a heavy wire bottle support which holds the bottle at an angle of about 30 degrees.   The water delivery tube extends from the bottle stopper into the cage through a hole in the wire mesh of the cover.   Bottle supports should be adaptable for various shapes since requirements for bottles may change more frequently than cover designs.

Food hopper.—The food hopper (figure 90) should be constructed in such a way that the pellets of food are readily accessible to the mice yet cannot be removed from the hopper in excess of the amount the mice can eat.   A hardware-cloth hopper with 3 strands to the inch is tolerable for this purpose, but the mice have considerable difficulty in obtaining the food.   Wire bar grating or slotted sheet metal is much better.   Spaces between wires (or slots) should not be more than 1/4 to 5/16 inches wide.   Small mice can squeeze through a 3/8-inch slot, and some of the pellets will also fall through.

The food hopper may be an integral part of the cover or it may be removable. In some cases the food hopper is attached to the wall of the cage rather than to the cover, although the latter method is preferable.

In our colony the food hopper is attached to the cover and is located at the end toward the aisle.   The hopper is made of slotted sheet metal which is tinned after it is attached to the cover.   It is rounded on the bottom, and the sides are inclined slightly to permit stacking.   Slots and bars are equal in width; the slots extend only partway up the side.   This prevents the mice from climbing on the hopper and soiling the food with urine and feces.

Watering devices.—In most colonies bottles are used to provide water for the mice. Usually a metal or glass tube delivers the water.   The tube protrudes through a rubber stopper.   The end from which the mice obtain water is slightly constricted so that when a drop is lapped a small bubble of air passes into the bottle displacing an equal volume of water.   As long as no air leaks past the stopper and nothing breaks the surface tension of the water in the lower end of the tube the water will not run out.   Stainless steel tubing is best since water delivery tubes must be sterilized between use.   Glass tubing permits visual observation of clogging materials, but are dangerous if broken when stoppers are removed and replaced.   The water delivery tube may be bent at an angle of about 100 degrees or it may be straight, depending on how the water bottle rests on the cage cover.

Water bottles of many sizes and types are available.   In most cases the bottles were originally designed for other uses.   The size depends on the number of mice in the cages and the frequency with which clean bottles of water are provided.   Since it is most economical to have only one type of bottle it should be large enough to provide water for the maximum number of mice to be held in one cage for a one-week period.

Fig. 90. Two types of water bottles with stainless steel goosenecks.

Weekly changing of cages and water bottles is usually adequate, but all cages should be inspected daily to insure that sufficient water is available and that no bottle has spilled into a cage.

The type of water bottle we use is a one-pint, square milk bottle.   The mouth of the bottle accommodates a no. 6½ rubber stopper with a bent stainless steel tube. The bottle support on the cage cover holds the bottle at an angle of about 15 degrees from the horizontal.   The greater this angle the more water can be utilized, but the less the shelf space.

Bottle cases.—Bottles can be conveniently handled in wire cases.   Certain features are desirable regardless of bottle design.   The case should have individual compartments for each bottle to minimize breakage.   Bottoms should extend below the bottom of the bottles so that cases can be stacked even though stoppers and delivery tubes are still in place. The case should have stacking loops at the corners and should be of light, but sturdy, construction. Stainless steel wire cases are perhaps the best, but other materials, such as plated, tinned, or galvanized wire, may be suitable.

Our design of bottle cases was adapted from milk bottle carriers except for the extension of the bottom. Each case holds 25 pint bottles. The case is constructed of stainless steel wire and has stacking loops at the four corners.

Racks.—Racks for supporting mouse cages must be designed to fit the type of cage that is to be used. Spacing between cages should be large enough to permit easy removal of a single cage. Space between shelves should be at least one inch greater than the over-all height of the cage unit (with cover and water bottle attached). The top shelf of a rack should be low enough so that cages can be removed from it by a caretaker of average height. If upper shelves are high, a stepladder or similar device will be necessary.

Rack shelves may be solid or they may be made of bars (pipe). Solid shelves should be removable for cleaning. Bar shelves should be permanently fixed, with the bars close enough to support the cages firmly, but far enough apart to allow easy cleaning. Shelves of either type must be strong enough to support the load without undue bending.

Racks may be provided with casters, they may be suspended from wall or ceiling, or they may stand on the floor. Suspended racks leave the floor free of obstructions, but they also limit the flexibility of space in any room. Racks on casters must be restricted in size to permit mobility. Caution must be used in their handling since they can be easily upset if the floor is not level or they are pushed laterally instead of longitudinally. Fixation of two racks side by side by screws or clamps promotes stability and yet allows access from adjacent aisles.

Solid shelves help to prevent dust, nesting material, or feces from falling from upper cages to those below, but they also reduce the circulation of air around the cages. Bar-type shelves allow maximum circulation of air, but they do not prevent material from falling from upper cages to the cages below. If cages are constructed so that material cannot be pushed out by the mice, bar-type shelves may be adequate.

The racks used in our mouse rooms are built in sections with 7 shelves per section and room for 6 double or 12 single cages on each shelf. Thus 84 cages or 42 boxes (2 compartments per box) are in each section.

Food dispenser.—Food should be given a minimum of handling between the bag and the food hopper. Food in open boxes on top of a dispensing table or containers below the top can be easily contaminated. For this and other reasons it is best to use a container suspended above the table. The dispenser for food in our mouse room is designed to hold 50 pounds of pellets (figure 87). The container is covered by a hinged lid, and food is poured directly from the shipping bag into the dispenser. The lower part of the dispenser is tapered on all four sides to about 3 by 4 inches, and an adjustable gate attached to the body of the dispenser can be used to control the flow of the pellets into a trough located below the hole.   The shape of the trough is designed to fit the scoop used to dispense the food.   The trough can be detached and cleaned each time the container is emptied.   The entire dispenser is suspended from two upright angle iron supports fastened to the side of the table.   The bottom of the trough should be about 2 or 3 inches above the tabletop to facilitate cleaning.

Tag holders.—Various methods are used to attach cage identification tags or record cards.   If the tag holder is attached to the cover of the cage it can be moved from one cage to another at cage-changing time.   Covers need not be changed as often as the cages unless the food in the hopper becomes accidentally soiled or wet.   Tag holders for our colonies are made of sheet aluminum.   They are designed to clip onto a transverse bar about $1\frac{1}{2}$ inches from the front edge of the cover.   The edges of each holder are bent over at the bottom and the two sides to hold the card in place. Tag holders can be made in any size to accommodate the identity tags or record cards which are used.   A convenient model is designed to hold a 3 by 5 inch card.

Tables and carts.—Tables for use in mouse rooms should have smooth surfaces to prevent accumulation of dirt.   It is preferable to have separate tables for use in changing cages and for record keeping, but if both of these functions are performed on one table, the table should have space for laying out the record books or cards.   The tables should have metal surfaces (either stainless steel or galvanized iron sheets). The legs and other framework can be steel angles or pipe welded into a solid unit. Tables with large-diameter wheels (4 or 5 inches) are easiest to move.   The height of the tables should be governed by the height of the people who use them.

Tabletops should be about 42 inches long and 30 inches wide.   Two shelves under the top will make it possible to carry cages with covers and water bottles attached on each shelf.   (Although mice may seldom be removed from the room it may be necessary when experimental treatment is to be administered.)   The width of the table should be small enough to permit passage through the aisles between cage racks even when a stack of cages on a dolly is standing at one side of the aisle.

Tables for changing cages should have a food dispenser attached on the right-hand side (left side if the caretaker is left-handed).   The food dispenser previously described is quite large and overhangs the top surface of the table.   Such a food dispenser makes the table too heavy for general use.

Washing machinery.—The size of colony will determine, to a large extent, the methods to be used to sanitize the equipment used in the mouse room.   If only one person is needed to take care of the mice and to wash the cages and water bottles, extreme caution must be exercised and careful instructions must be given to be certain that the clean and dirty areas are not violated.   It is much more efficient to have two people working in the washing area, one at the dirty end and the other at the clean end.   If the colony is very small, two workers should still be used in the washing area even if on a part-time basis.

Machinery for washing cages should be adaptable to the clean-corridor system. That is, the machines themselves must function as the division point between the two areas (figure 91). For this reason a batch-type washing machine is not desirable because the cleaned equipment must be removed at the same point at which it was inserted when it was dirty. The most feasible design is one incorporating a continuous chain passing through a tunnel in which jets of water under high pressure are used to clean the equipment. In machines of this type there should be at least three stages, (1) a rinse with cold water to remove most of the soiled material, (2) a spray of hot water containing detergent or other washing compound, and (3) a very hot (sterilizing temperature, if possible) rinse at the clean end. The length of the tunnel, the speed of the conveyor, the type of cleansing chemical, water pressure from jets, number of jets, and the temperature of the water determine the effectiveness of the cleansing.

Fig. 91. CLEAN EQUIPMENT ASSEMBLY AREA.

Note washing machines mounted through walls separating clean and dirty areas.

The air pressure in the washing areas should be high on the clean side and low on the dirty side. Precautions must be taken to insure maintenance of pressure despite the loss of air through the tunnel of the washing machine. Curtains over the ends of the tunnel can be helpful in this respect.

Washing machines should be equipped with an exhaust system to remove excess water vapor, thermostatic controls to shut off the machine if the water temperature drops below the desired level, and safety devices to stop the conveyor if equipment becomes jammed in the machine.

PROCUREMENT

The usual method of obtaining mice for experimental purposes is by purchase from commercial breeders. A good list of these breeders in the United States is available from the Institute of Laboratory Animal Resources. Since this is an excellent source of such information to list them again is superfluous; it is more important to discuss methods of selecting from the sources listed.

Fig. 92. Shipping containers.

Above. Outside view of a cardboard shipping container.
Below. Inside view showing potatoes, food pellets, and bedding in place.

First, the supplier should offer stock of the genetic composition needed. The supplier should have capable geneticists on his staff or have ready access to their services. Such services should include periodic quality testing of the stocks to insure against accidental misbreeding, overlooking of mutations, and the like. The reputation of the vendor may be readily checked by communicating with other customers in the area. The health of the mice offered for sale should also be investigated. This should include information concerning ectoparasites, endoparasites, salmonellosis, ectromelia, and other diseases. The breeder should have available the services of a specialist in diseases of laboratory animals and a diagnostic laboratory.

Although delivery service by air reduces the importance of distance, it is well to remember that shipping can put undesirable stress on mice. The methods of transportation available between vendor and user should be carefully examined and schedules arranged to insure against delays or mishandling in transit. Mice should be packed in escapeproof, well-ventilated cartons with protective bedding. Water-proofed cardboard or wood cartons are usually used (figure 92). Overcrowding should be avoided, particularly in warm weather, as mice will tend to congregate around ventilators in an attempt to get cool and thus suffocate by completely shutting off the source of ventilation.

One of the best packing materials is shredded paper which should be sanitized (by dry-heat or steam pasteurization) before use. This material protects the animals against rough handling and provides insulation against heat or cold. Water is usually supplied in the form of raw potatoes or other root vegetables. These should be washed thoroughly and cut in large pieces (2–3 inches) or left whole (not over 2–3 inches in length) with ends removed. Food pellets are scattered through the litter material. Enough potatoes and feed should be added to last for the length of the trip and to guard against delays in transit. Water bottles are sometimes used instead of potatoes or other vegetables to provide water but these are unsatisfactory in most cases due to leakage during transit.

Procurement should be directly supervised by the person who is to use these animals. It is extremely foolish to spend thousands of dollars and much time on equipment and salaries and then try to save pennies and minutes on the most important parts of a biological experiment, the laboratory animal. As much care should be used in selecting, purchasing, and transporting these biological test tubes as is given to the other important features of the research project.

MANUFACTURERS AND DISTRIBUTORS OF EQUIPMENT
(Not intended as an endorsement by the authors)

Name	*Address*
EQUIPMENT WASHERS	
Better Built Machinery Corp.	78 E. 130 St., New York 37, N.Y.
Heinicke Instruments Co.	2035 Harding St., Hollywood, Fla.
Industrial Washing Machine Co.	Matawan, N.J.

Metalwash Machinery Co.	Elizabeth, N.J.
R. G. Wright Co., Inc.	2280 Niagara St., Buffalo 7, N.Y.

STERILIZING EQUIPMENT

A-C Lab Equipment Co.	673 Timson Place, New York 55, N.Y.
American Sterilizer Co.	Erie, Pa.
Atlantic Ultraviolet Co.	24–12 40th Ave., Long Island City, New York 1, N.Y.
The Hospital Supply Co., Inc.	306 E. 23rd St., New York, N.Y.
Technical Products Co.	21 N. Dunlap St., Memphis 1, Tenn.
Wilmot Castle	Box 629, Rochester 2, N.Y.

CAGES AND MISCELLANEOUS EQUIPMENT

Acme Sheet Metal Works	1121 E. 55th St., Chicago, Ill.
Aloe Scientific Co.	1831 Olive St., St. Louis 3, Mo.
American Sheet Metal Co.	1320 Ocean Ave, San Francisco, Calif.
Armstrong Metal Specialties, Inc.	No. Norwich, N.Y.
Bussey Products Co.	2700 W. 35th St., Chicago, Ill.
Consolidated Molded Products Co.	329 Cherry St., Scranton 2, Pa.
Ford Kennel Equipment, Inc.	6540 E. Westerfield Blvd., Indianapolis, Ind.
Great Falls Products Co.	P.O. Box 1706, Rochester, N.H.
Hartford Metal Products, Inc.	Box CM, Aberdeen, Md.
Hoeltge Bros., Inc.	1917 Gest St., Cincinnati 4, Ohio
Keystone Plastics Co.	701 Painter St., Media, Pa.
Lind Laboratories, Inc.	99 Commonwealth Ave., Boston, Mass.
C. H. Sommers Associates	1107 Herschel Ave., Cincinnati 8, Ohio
George H. Wahmann Mfg. Co.	1123 E. Baltimore St., Baltimore 2, Md.

Elizabeth S. Russell, Ph.D.

APPENDIX V

Techniques for the Study of Anemias in Mice

The methodology described in the body of the article on genic action in the mouse has been entirely that of research approach, with no attention to specific techniques used. For most of the analyses discussed, the author has no competence to discuss specific techniques, since they are those of a particular unfamiliar nongenetic biological discipline rather than of physiologic genetics *per se*. In the particular area of analysis of anemias in mice, which has been covered in some detail, a discussion of technical methods may be suitable and useful. No technique has been included unless the individual responsible for the description (usually, but not always, the author) has used the method extensively.

Since hematology is a rapidly developing science, new techniques are frequently described. For this reason, the most recent edition of a hematology text[1397] provides a good critique and general source of information on technique. Much of the literature cited in the body of this paper on genic action describes or refers to special modifications of techniques to make them applicable in investigations on mice. Existing surveys of the blood picture of various inbred strains of mice may provide useful baseline information,[141, 324, 1108] although many of the methods used have been superseded in more recent studies by easier and more exact techniques.

Blood collection in large amounts.—Blood may be collected from mice in several ways, choice among them depending on the size of the mouse, whether the donor is to be saved, the necessity for repeated samples, and the volume of blood required. Since the blood volume of an adult mouse is not over 2–3 ml., almost all blood-sample collections represent a considerable drain upon the erythron. The blood of mice clots exceedingly rapidly, and use of an anticoagulant, either by addition of 1 per cent heparin solution to the blood sample or by collection in dried heparinized or oxalated tubes, is very important.

For some experiments, the purpose is simply to obtain large quantities of blood or serum. A large volume (more than 1 ml.) of blood may be collected (with sacrifice of the donor on a dissecting board, under Nembutal) by allowing blood to flow from the severed brachial artery into a pocket prepared in the brachial plexus. Large quantities may also be obtained by cutting a mouse across the thorax through the heart and allowing blood to flow into a collecting tube.

Experienced operators use cardiac puncture to obtain as much as 0.6–0.8 ml. of blood from either ventricle (that is, separate samples of arterial or venous blood) of adult mice without sacrifice of the donor.[1374, 1398] A description of the procedure by H. G. Wolfe follows: For this purpose a 0.5- or 1.0-ml. syringe and no. 26 or 27 needle is used. A ring stand or other suitable device is used for attaching a short length of cord with a loop in the end through which the head of the plunger of the syringe is passed. The loop must be large enough so that the head of the plunger can be disengaged and yet snug enough so that it can be used to retract the plunger during the operation. The needle and syringe may be rinsed in an anticoagulant solution, or, if desired, no anticoagulant need be used if the blood is delivered immediately into a receptacle either with or without an anticoagulant in it, depending upon the particular experiment. The nonanesthetized mouse is held firmly in one hand with its ventral surface up. The ventral thorax should be disinfected with a germicide such as 70 per cent ethyl alcohol. This at the same time flattens the hair and enables the technician to see the exact outlines of the rib cage. A sterile needle from the mounted syringe is inserted through the rib cage near the xiphisternum and slightly to the left of it; the exact point of insertion must necessarily be determined empirically by practice and preference on the part of the technician. When the needle penetrates either the right or left ventricle, the mouse and syringe are simultaneously and gently moved away from the restrained plunger. One can always tell if the needle is properly inserted by the sudden lack of resistance between syringe and plunger. With practice and by simply altering the angle of penetration one can extract either arterial or venous blood from the left and right ventricles respectively, with a fair degree of accuracy. The blood can be drawn quite rapidly; the complete operation takes from 1 to 2 minutes once the equipment is properly set up. Age and size of the mouse are factors determining the amount of blood which can be drawn.

Collection of peripheral blood for quantitative evaluation.—When blood samples are taken for the purpose of characterizing the blood picture of the host, one of three general methods is commonly used. Blood may be collected from a slash in the lateral tail vein, especially if the whole animal has been warmed in a glass jar placed under a lamp or if the tail has been placed in hot water immediately before sample collection. Care must be taken not to massage the tail, as this often leads to erroneous counts.

A complication to bleeding from the tail is that the cut surface is frequently reopened after bleeding by contact with box or bedding, leading to unnecessary loss of blood. If a series of counts are to be made on the same animal, it is well to start near the tip of the tail and move proximally.

When repeated samples are to be taken from the same individual, and minimal loss of blood is important for maintenance of homeostasis, small samples of blood may be taken from a sinus in the inner canthus of the eye. This site has two advantages. Irrelevant loss of blood can be avoided by starting blood flow with the tip of the collecting tube. The tip of a Harshaw pipette or a capillary hematocrit tube is inserted into the network of capillaries in the medial side of the eye socket and rotated slightly. Blood is collected directly into the collecting tube, and post-treatment bleeding can be prevented by holding the eye closed for a brief period. Anesthesia is not necessary. This site can also be used for sterile collection of large volumes (0.5 + ml.) of blood for serum.[135]

Blood may be collected from newborn mice by decapitation or, for smaller quanti-ties, from a cut through the tissues of the neck and jugular vein. When applicable, the latter method is preferable, since there is less contamination with tissue fluids. This same site is suitable for collection of 1–2 lambda of blood from 15 + -day fetuses.

Quantitative determinations of the erythron.—Three basic determinations for charac-terizing the flowing blood are the number of erythrocytes per unit blood volume, the proportion of the volume of packed red cells to total blood volume (= hematocrit percentage) and the amount of hemoglobin per unit blood volume.

The classical method for determining *erythrocyte number* (erythrocytes/mm.^{3}), involving collection of an exact volume of blood in a Thoma pipette, 200 × dilution with isotonic Hayem's solution, and counting in a Neubauer counting chamber, may be used for mice. For further details of methodology the reader is referred to Wintrobe's *Clinical Hematology*.[1397] Special adaptations are required to use this method for the small volumes of blood available in fetuses and newborn mice.[1102] These include extra calibrations on the Thoma pipette, coating of the inside of the pipette with glycerin or silicone, and modification making the tip of the pipette more pointed to facilitate collecting from a small area.

A recent improvement especially valuable in murine hematology is use of the Coulter electronic cell counter[859] (Scientific Products Co., Evanston, Ill.). In this system, cell counts depend on interruptions of an electric current by cells in a known volume of fluid passing through an aperture. Very high dilutions (1/50,000) of blood in 0.9 N filtered NaCl solution are used, and the number of cells in 0.5 ml. of diluted blood is recorded electronically. (Corrections must be made for coinci-dence due to two or more cells passing through the aperture simultaneously.) Agree-ment between successive samples from the same animal is excellent, and extensive tests have been made of the reliability and accuracy of the method.[127, 859] Although initial investment is high, the savings in labor and time make possible much more extensive programs than could otherwise be undertaken. The great value of this method in studies with mice lies not only in its speed and accuracy, but also in the fact that it works very efficiently with blood samples of one lambda (0.001 ml.), making it especially suitable for newborn and fetal counts.

Hematocrit percentages can be determined with the use of Van-Allen centrifuge tubes, 3 per cent oxalate diluent, and a standard clinical centrifuge. However, recent improvement of microhematocrit methods[823, 1108, 1300] have been very useful in murine as well as human hematology. Each determination uses a very small volume of undiluted blood (± 0.05 ml.), collected in an open-ended heparinized or oxalated capillary tube (A. H. Thomas Co., Philadelphia). After heat fusion of one end in a microburner flame, these tubes are spun for six minutes at 11,000 r.p.m. in a special flat-angle hematocrit centrifuge (International Centrifuge Co., Boston). The volume of packed red cells relative to total volume of blood is then read directly in a convenient hematocrit-reader device, usually with excellent agreement between samples from the same individual. The direct readings depend on the ratio of length of column of packed cells to total blood column, so that the problem of filling a pipette to a preset level is avoided. Special oxalated pipettes collecting approximately 0.01 ml. for each reading have proved useful for newborn and fetal hematocrit determinations, but these are not read accurately in the hematocrit reader described above. The author has had no personal experience with a special reader designed to go with these tubes[1300] but has obtained fairly reliable results by reading column heights in these tubes placed on a millimeter scale in the field of a dissecting microscope.

The usual method for comparing size of erythrocytes in different animals is determination of their mean cell volumes from the ratio of hematocrit percentage to erythrocyte count:

$$\frac{\text{packed cell volume}}{\text{total blood volume}} \div \frac{\text{erythrocyte number}}{\text{total blood volume}}$$

Accuracy and repeatability of this determination depend on the variability observed in each of the component values.

A value proportional to the mean and distribution of erythrocyte volumes in a single blood sample may also be calculated directly from the relative counts for different-sized cells in the Coulter electronic counter.[477]

Determinations of hemaglobin content in blood of mice may be achieved with considerable accuracy, especially if hemoglobin is converted to the stable cyanomethemoglobin.[230] Hemoglobin standards and reagents are available (Scientific Products, Evanston, Ill.); the latter may be made up locally,[1397] and concentrations are calculated from optical density read in a photometer.

It is frequently desirable to judge the rate at which new red cells are being added to the circulation. For this, determination of the reticulocyte percentage is important. An excellent stain for the murine reticulocyte is the new methylene blue method of Brecher[124, 126] in which an approximately equal mixture of blood and dye is held in a capillary pipette for ten minutes, smeared on a cover slip, air dried, and counted under oil immersion.   The proportion of cells with a reticulum (deep blue) in a sample of at least 500 erythrocytes (pale blue-green) is determined.   Use of a Whipple ocular micrometer disk marked in a 100-square grid greatly facilitates counting.

Leucocyte counts. — It is frequently desirable in connection with analysis of anemic states to have information on levels of leucocytes and other nucleated cells.   Total leucocyte counts can be carried out in mice using the Neubauer chamber, regulation human white-counting pipettes, and gentian-violet-stained 3 per cent acetic acid diluent.   The proportion of granulocytes is much higher in mouse than in man, varying from a mean of 74 per cent to a mean of 92 per cent among 18 tested inbred strains.[1108]   Separate direct chamber counts of granulocytes and agranulocytes can be obtained by the Randolph method using propylene-glycol diluent.[141, 1260]   For many purposes differential smear counts are very satisfactory, and special modifications of staining methods have been developed for use with blood of mice.[1261]

Study of hematopoietic tissues. — Total cellularity of marrow of the mouse may be calculated from the proportion of available marrow space occupied by hematopoietic cells.[324, 1110]   A Whipple, ocular-micrometer disk marked in a 100-square grid facilitates classification, which may be made at 430 × or 970 × magnification, or preferably at both.   Proportional counts of types of hematopoietic cells may be made from touch imprints or smears stained with Wright-Giemsa.[125, 1110]   Because differences in staining properties of mouse and human marrow cells make it difficult to classify murine marrow from the appearance of the cytoplasm, the nucleus only is used in classifying members of the erythropoietic series.[125]   These methods for classifying marrow histology may be applied to all postnatal stages.

Margaret K. Deringer, Ph.D.

APPENDIX VI

Technique for the Transfer of Fertilized Ova

In studies of mammary tumors in mice, it is often desirable to have animals from which the milk agent is absent but that are susceptible to its action. One of the methods by which such animals are produced, mentioned by Dr. Heston, is the transfer of fertilized ova from a strain carrying the agent to one lacking it. A brief description of the technique of transferring ova, as employed in our laboratory, follows.

Females of donor and recipient strains, to be employed in the transfer, are mated to their brothers. Each female is allowed to produce and rear one litter before a transfer is attempted. After the litter is weaned the females are returned to the breeding cages and are observed each morning thereafter for the presence of a vaginal plug, an indication that mating has occurred. At selected periods after observation of the vaginal plug, the donor female is killed. If the transfer is made at 52 hours, the ova will be found in the oviducts. The oviducts are removed and put in a watch glass containing filtered physiologic saline at 37° C. They are minced with fine scissors allowing the ova to fall to the bottom of the watch glass where they can be observed under a dissecting microscope. If the transfer is made at 76 hours, the ova are usually in the anterior ends of the uterine horns. The uterine horns and the vagina are removed and put into saline in a watch glass. After the ova are flushed out of the horns by forcing saline through the horns by means of a hypodermic syringe inserted into the vagina, they are recovered with a finely drawn, glass pipette and are transferred in depression slides through several changes of saline.

The pregnant recipient females, at a period postplug similar to the donor females,

are anesthetized with ether.   A small midline incision is made in the ventral body
wall and one horn of the uterus exposed.   The desired number of ova to be transferred
are drawn up into a pipette with a minimal amount of saline.   After an opening has
been made in the recipient horn with a hypodermic needle, the pipette containing
the ova is inserted, pointing cephalad, and the ova are released into the horn.   It is
important to inject as little fluid as possible and to avoid injecting any air.   The opening
in the uterine horn will close without further treatment.   The incision is closed with
black silk and the female is isolated.   A daily check is made of the animal until she
produces her litter.

Various alterations in the technique have been employed by other workers.[351] [352, 827, 828]   The reader is referred to the cited articles for these.

*Staff of the Cancer Research Genetics
Laboratory, University of California*

APPENDIX VII

Current Applications of a Method of Transplantation of Tissues into Gland-free Mammary Fat Pads of Mice

A technique for the removal of host mammary gland elements before they penetrate deeply into the mammary fat pad has been developed in our laboratory. The remainder of the fat pad is then available to receive tissue transplants. The procedure, already published, is as follows: "[The inguinal (number 4) fat pads of] 3-week-old C3H/He Crgl females ... [were] cleared of host mammary gland elements by means of the following surgical procedure (figure 93). The anesthetized mouse was pinned, ventral side up, on a cork board and scrubbed with 70 per cent ethanol. An inverted Y-shaped incision was made along the abdominal midline and laterally between the fourth and fifth nipples midway down each hind leg. One at a time the resulting skin flaps, with the number 4 mammary gland attached, were carefully separated from the body wall by blunt dissection. The free edge of the skin flap was then pinned to the cork board, exposing the nipple area and most of the number 4 fat pad. In a 3-week-old female C3H mouse the mammary gland elements consist of short branching ducts which extend from the nipple area into the fat pad as is shown in figure 94. The nipple area and the large blood vessels ventral to the lymph node and between the fourth and fifth pads were cauterized (figure 94). The area containing the growing gland elements, including the adjacent portion of the fat pad and the surrounding loose connective tissue, was then excised with fine scissors (figure 94). The remaining portion of the fat pad with its circulation intact and without host mammary tissue was then ready to receive a transplant, or the skin flap was sutured and the transplantation carried out at a later date. The vascularization of the mammary fat pad in the C3H/He Crgl mouse has been investigated by Soemarwoto and Bern,[1255] whose studies indicate that the number-4 fat pad normally has three separate blood supplies and that the surgery described herein did not interfere with the circulation of the remaining fat pad.

Fig. 93. DRAWING OF A THREE-WEEK OLD FEMALE C3H/He Crgl MOUSE PREPARED FOR THE REMOVAL OF THE MAMMARY GLAND ELEMENTS FROM THE RIGHT NO.-4 (INGUINAL) FAT PAD.[225]

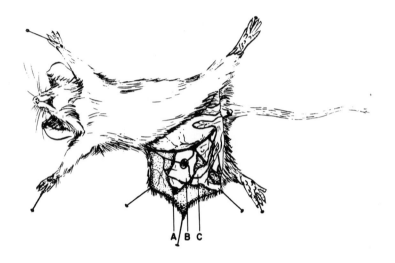

A. Nipple area; B. Right no.-4 fat pad; C. Right no.-5 fat pad.

"The transplantation site was prepared by forcing the closed points of a fine forceps into the center of the fat pad. The tissue to be transplanted was inserted into the transplantation site, with the use of the same fine forceps. The remaining number-4 fat pad was then similarly exposed and cleared and received a second transplant.... Finally, both skin flaps were returned to their normal positions, and the incision was closed with wound clips."[255]

The fate of individual tissue areas selected from the mammary gland can be followed by this transplantation method. Pieces of transplanted normal mammary tissue grow out to fill the fat pad with normal mammary gland (figure 95). Transplants of hyperplastic alveolar nodules, on the other hand, fill the fat pad with an outgrowth which is hyperactive with regard to the hormonal state of the host (figure 96). With the use of this technique, we have already reported the more frequent development of mammary tumors from transplanted hyperplastic alveolar nodules than from normal mammary gland similarly transplanted,[255] thus directly and decisively establishing the precancerous nature of the hyperplastic alveolar nodule.

The development of this method has allowed us to begin the analysis of a variety of problems.   In the following paragraphs the nature of some of our recent studies, both published and unpublished, complete and incomplete, is indicated.

The nature of the interaction of tissues cotransplanted into the same fat pad can be determined.   The transplantation of one or more pieces of mammary tissue into one fat pad has revealed the presence of a growth-regulating system in the mammary gland which determines both the extent of growth and the spacing of ductal elements.[347]

Fig. 94. Drawing of a right no.-4 fat pad from a three-week-old female mouse.

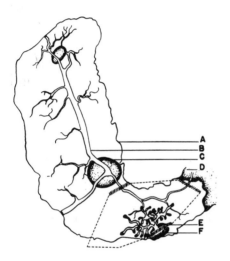

The blood vessels, fat pad, and nipple area were cauterized along the slant lines.   The fat pad and the surrounding connective tissue bounded by the broken line were removed with fine scissors.[225]   A. Boundary of no.-4 fat pad; B. Large vein; C. Inguinal lymph node; D. Portion of no.-5 fat pad; E. Branching ducts of the no.-4 mammary gland; F. Nipple area.

Figs. 95 and 96. Right no.-4 pad of a 15-week-old, virgin, C3H/Crgl female mouse.

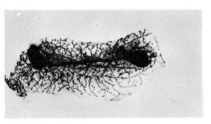

Left. The normal gland has developed from a mammary lobule (taken from a pregnant, C3H/Crgl, female mouse) which was transplanted into the pad when the host was three weeks of age.   4X.

Right. The hyperactive outgrowth has developed from an hyperplastic alveolar nodule (taken from a tumor-bearing, nonpregnant, nonlactating, multiparous, adult, C3H/Crgl female mouse) which was transplanted into the pad when the host was three weeks of age.   4X.

The tumor-forming ability of hyperplastic alveolar nodules (and of normal tissues) in a variety of hormonal milieux has been studied, and the results of endocrine manipulations both favoring and opposing tumorous transformation have been recorded.[932, 933]

Individual nodules with different morphologic or physiologic characteristics are being studied by transplantation into the fat pad, to determine their tumor-forming abilities. For example, lactating and nonlactating nodules from the same donor mouse can be transplanted contralaterally into a single host (Bern and Nandi, unpublished data). From studies such as these, it should be possible to decide whether greater hormonal sensitivity of nodules—as judged in various ways—is correlated with greater tumor-forming ability.

By serial transplantation of outgrowths from hyperplastic alveolar nodules, we have revealed the nonhomogeneity of cellular populations comprising individual nodules,[252] and we are currently searching for correlations between the tumor potential of various serially transplanted outgrowths and their morphologic (including cytologic) and physiologic characteristics. The outgrowth sublines that have been maintained have proved to be stable to a considerable degree in regard to the several properties studied.

Since tissue can be transplanted into the cleared fat pad at any age of the host, the procedure is being utilized in a study of the effect of aging upon normal and neoplastic mammary growth. Tissues that are transplanted in this fashion are potentially immortal. Such tissues can be transplanted constantly into old mice or into young mice. The effect of age of the host as distinguished from the age of the tissue can be determined, and the use of this method of transplantation in gerontologic studies seems especially promising.[253, 254]

Mammary tissues that have been subjected to various conditions in organ culture can be transplanted back into cleared fat pads and their organ-forming ability determined (Faulkin, Rivera, and DeOme, unpublished data). It seems probable that cultures of cells may also lend themselves to this kind of study.

The differentiation of embryonic mammary gland may be followed in fat-pad transplants, and some success has already been attained in this type of endeavor (Moretti and Blair, unpublished data).

Transplantation of tissues of parental strains into hybrid mice has also proved successful in our hands and in those of C. W. Bardin at Baylor University (personal communication). Normal mammary tissues appear to retain the hormonal responsiveness characteristic of their source, and this establishes further that such strain-specific responses are determined genetically at the tissue level (Nandi, unpublished data).

The possibility of transplantation of tissue into the fat pad is not limited to mammary tissues by any means. We have done very little to date with further utilization of this technique, but it is apparent that some embryonic tissues will survive to an important extent in this new environment (Blair and Moretti, unpublished data).

The use of the gland-free mammary fat pad as a site for tissue transplantation may prove to be an excellent tool not only for general investigation of the factors governing the development of mammary cancer in mice, but also for other studies where the fate of small fragments of tissues is to be followed in the living animal.   There appears to be nothing special about mammary fat *per se* as a transplantation site, other than in its great mechanical convenience.   Limited observations indicate that other accumulations of adipose tissue (perirenal, perigonadal, and the like) can also serve as sites for receipt of tissue.   To a certain extent, one can look upon transplantation into fat pads as providing a kind of organ culture *in vivo*.

BIBLIOGRAPHY

1. Abercrombie, M., and J. E. M. Heaysman: Invasiveness of Sarcoma Cells. Nature, **174:** 697, 1954.
2. Adler, F. L.: Antibody Formation after Injection of Heterologous Immune Globulin. II. Competition of Antigens. J. Immunol., **78:** 201, 1957.
3. Akabori, S., K. Ohno, and K. Narita: On the Hydrazinolysis of Proteins and Peptides: A Method for the Characterization of Carboxyl-terminal Amino Acids in Proteins. Bull. Chem. Soc. Jap., **25:** 214, 1952.
4. Alberty, R. A.: Electrochemical Properties of the Proteins and Amino Acids. *In* The Proteins. H. Neurath and K. Bailey, eds. Academic Press, New York, **1A:** 461, 1953.
5. Algire, G. H., and F. Y. Legallais: Recent Developments in the Transparent-chamber Technique as Adapted to the Mouse. J. Nat. Cancer Inst., **10:** 225, 1949.
6. Algire, G. H., J. M. Weaver, and R. T. Prehn: Growth of Cells *in vivo* in Diffusion Chambers. I. Survival of Homografts in Immunized Mice. J. Nat. Cancer Inst., **15:** 493, 1954.
7. Allard, R. W.: Formulas and Tables to Facilitate the Calculation of Recombination Values in Heredity. Hilgardia, **24:** 235, 1956.
8. Allen, D. W., W. A. Schroeder, and J. Balog: Observations on the Chromatographic Heterogeneity of Normal Adult and Fetal Human Hemoglobin: A Study of the Effects of Crystallization and Chromatography on the Heterogeneity and Isoleucine Content. J. Amer. Chem. Soc., **80:** 1628, 1958.
9. Allen, E., and W. U. Gardner: Cancer of the Cervix of the Uterus in Hybrid Mice Following Long-continued Administration of Estrogen. Cancer Res., **1:** 359, 1941.
10. Allen, J. R., B. H. Sullivan, and H. L. Dobson: Assay of Cytochrome Oxidase, Reductase and Myoglobin in Vitamin E Deficient Rabbits and Dystrophic Mice. Fed. Proc., **17:** 3, 1958.
11. Allen, S. L.: *H-2^f*, a Tenth Allele at the Histocompatibility-2 Locus in the Mouse as Determined by Tumor Transplantation. Cancer Res., **15:** 315, 1955.
12. Allen, S. L.: Linkage Relations of the Genes Histocompatibility-2 and Fused Tail, Brachyury and Kinky Tail in the Mouse as Determined by Tumor Transplantation. Genetics, **40:** 627, 1955.
13. Allison, A. C., and R. Cecil: The Thiol Groups of Normal Adult Human Haemoglobin. Biochem. J., **69:** 27, 1958.
14. Allison, A. C., and J. H. Humphrey: A Theoretical and Experimental Analysis of Double Diffusion Precipitin Reactions in Gels and its Application to Characterization of Antigens. Immunology, **3:** 95, 1960.

15. Altman, K. I.: personal communication.
16. Altman, K. I., E. S. Russell, K. Salomon, and J. K. Scott: Chemopathology of Hemoglobin Synthesis in Mice with Hereditary Anemia. Fed. Proc., **12:** 168, 1953.
17. Ambrose, E. J., A. M. James, and J. H. B. Lowick: Differences between the Electrical Charge carried by Normal and Homologous Tumour Cells. Nature, **177:** 576, 1956.
18. Amos, D. B.: Some Iso-antigenic Systems of the Mouse. Canadian Cancer Conf., **3:** 241, 1959.
19. Amos, D. B.: The Persistence of Mouse Iso-antibodies *in vivo*. Brit. J. Cancer, **9:** 216, 1955.
20. Amos, D. B., P. A. Gorer, and Z. B. Mikulska: The Antigenic Structure and Genetic Behaviour of a Transplanted Leukosis. Brit. J. Cancer, **9:** 209, 1955.
21. Amos, D. B., and J. D. Wakefield: Growth of Mouse Ascites Tumor Cells in Diffusion Chambers. II. Lysis and Growth Inhibition by Diffusible Isoantibody. J. Nat. Cancer Inst., **22:** 1077, 1959.
22. Andersen, A. C., and F. T. Shultz: Effect of Whole-body Irradiation on the Estrous Cycle and Fertility of Beagles. Radiat. Res., **12:** 417, 1960.
23. Anderson, D. E.: Genetic Aspects of Bovine Ocular Carcinoma. *In* Genetics and Cancer. R. W. Cumley, ed., Univ. of Texas Press, Austin, 364, 1959.
24. Anderson, D. E.: Studies on Bovine Ocular Squamous Carcinoma ("Cancer Eye"). X. Nutritional Effects. J. Animal Sci., **19:** 790, 1960.
25. Andervont, H. B.: Development and Genetic Characteristics of the Adenomatous Stomach Lesion in Strain I Mice. Public Health Repts., **54:** 1851, 1939.
26. Andervont, H. B.: Fate of the C3H Milk Influence in Mice of Strains C and C57 Black. J. Nat. Cancer Inst., **5:** 383, 1945.
27. Andervont, H. B.: Induction of Hemangio-endotheliomas and Sarcomas in Mice with O-Aminoazotoluene. J. Nat. Cancer Inst., **10:** 927, 1950.
28. Andervont, H. B.: Note on the Transfer of the Strain C3H Milk Influence through Successive Generations of Strain C Mice. J. Nat. Cancer Inst., **2:** 307, 1941.
29. Andervont, H. B.: Pulmonary Tumors in Mice. V. Further Studies on the Influence of Heredity upon Spontaneous and Induced Lung Tumors. Public Health Repts., **53:** 232, 1938.
30. Andervont, H. B.: Susceptibility of Mice to Spontaneous, Induced, and Transplantable Tumors. Public Health Repts., **53:** 1647, 1938.
31. Andervont, H. B., and J. H. Edgcomb: Responses of Seven Inbred Strains of Mice to Percutaneous Applications of 3-Methylcholanthrene. J. Nat. Cancer Inst., **17:** 481, 1956.
32. Andervont, H. B., M. B. Shimkin, and H. Y. Canter: Effect of Discontinued Estrogenic Stimulation upon the Development and Growth of Testicular Tumors in Mice. J. Nat. Cancer Inst., **18:** 1, 1957.
33. Anson, M. L., J. Barcroft, A. E. Mirsky, and S. Oinuma: On the Correlation between the Spectra of Various Haemoglobins and Their Relative Affinities for Oxygen and Carbon Monoxide. Proc. Roy. Soc., ser. B, **97:** 61, 1924–25.
34. Armstrong, C. N., J. E. Gray, R. R. Race, and R. B. Thompson: A Case of True Hermaphroditism; a Further Report. Brit. Med. J., **2:** 605, 1957.
35. Aronson, M., and R. W. I. Kellsel: New Method for Manipulation, Maintenance, and Cloning of Single Mammalian Cells *in vitro*. Science, **131:** 1376, 1960.
36. Atwood, K. C., and S. L. Scheinberg: Isotope Dilution Method for Assay of Inagglutinable Erythrocytes. Science, **129:** 963, 1959.
37. Atwood, K. C., and S. L. Scheinberg: Somatic Variation in Human Erythrocyte Antigens. J. Cell. Comp. Physiol., **52** Suppl. 1: 97, 1958.
38. Auerbach, C.: Chemical Mutagenesis in Animals. Abh. d. Deutsch. Adad. Wiss. Berlin, Dl. F. Medizin, **1:** 1, 1960.
39. Auerbach, C., and B. M. Slizynski: Sensitivity of the Mouse Testis to Mutagenic Action of X-rays. Nature, **177:** 376, 1956.
40. Auerbach, R.: Morphogenetic Interactions in the Development of the Mouse Thymus Gland. Develop. Biol., **2:** 271, 1960.
41. Awa, A., M. Sasaki, and S. Takayama: An *in vitro* Study of the Somatic Chromosomes in Several Mammals. Jap. J. Zool., **12:** 257, 1959.
42. Axelrad, A., and G. Klein: Differences in Histocompatibility Requirements between Primary Tumors and Their Metastases. Transplant. Bull., **3:** 100, 1956.
43. Baglioni, C.: Genetic Control of Tryptophan Peroxidase-oxidase in *Drosophila melanogaster*. Nature, **184** Suppl. 14: 1084, 1959.

44. Bagshaw, M. A., and L. C. Strong: The Occurrence of Tumors of the Forestomach in Mice after Parenteral Administration of Methylcholanthrene: A Histopathologic and Genetic Analysis. J. Nat. Cancer Inst., **11**: 141, 1950.

45. Bailey, N. T. J.: The Influence of Partial Manifestation on the Detection of Linkage. Heredity, **4**: 327, 1950.

46. Baird, D. M., A. V. Nalbandov, and H. W. Norton: Some Physiological Causes of Genetically Different Rates of Growth in Swine. J. Animal Sci., **11**: 292, 1952.

47. Baker, N., E. H. Blahd, and P. Hart: Concentrations of K and Na in Skeletal Muscle of Mice with a Hereditary Myopathy (Dyst. Musc.). Amer. J. Physiol., **193**: 530, 1958.

48. Bangham, A. D.: Distribution of Electrophoretically Different Haemoglobins among Cattle Breeds of Great Britain. Nature, **179**: 467, 1957.

49. Bangham, A. D., and B. S. Blumberg: Distribution of Electrophoretically Different Haemoglobins Among Some Cattle Breeds of Europe and Africa. Nature, **181**: 1551, 1958.

50. Bangham, A. D., and H. Lehmann: Multiple Haemoglobins in the Horse. Nature, **181**: 267, 1958.

51. Barcroft, J.: The Respiratory Function of the Blood. II. Haemoglobin. The Univ. Press, Cambridge, 1928.

52. Barnes, A. D., and P. L. Krohn: The Estimation of the Number of Histocompatibility Genes Controlling the Successful Transplantation of Normal Skin in Mice. Proc. Roy. Soc., ser. B, **146**: 505, 1957.

53. Barnes, D. W. H., C. E. Ford, S. M. Gray, and J. F. Loutit: Spontaneous and Induced Changes in Cell Populations in Heavily Irradiated Mice. Progress in Nuclear Energy, Series VI, **2**: 1, 1959.

54. Barrett, M. K., and M. K. Deringer: An Induced Adaptation in a Transplantable Tumor of Mice. J. Nat. Cancer Inst., **11**: 51, 1950.

55. Barrett, M. K., M. K. Deringer, and W. H. Hansen: Induced Adaptation in a Tumor; Specificity of the Change. J. Nat. Cancer Inst., **14**: 381, 1953.

56. Bartlett, M. S., and J. B. S. Haldane: The Theory of Inbreeding with Forced Heterozygosis. J. Genet., **31**: 327, 1935.

57. Bateman, A. J.: Mutagenic Sensitivity of Maturing Germ Cells in the Male Mouse. Heredity, **12**: 213, 1958.

58. Bateman, A. J.: Sensitivity of Immature Mouse Sperm to the Mutagenic Effects of X-rays. Nature, **178**: 1278, 1956.

59. Bateman, A. J.: The Partition of Dominant Lethals in the Mouse between Unimplanted Eggs and Deciduomata. Heredity, **12**: 467, 1958.

60. Bateman, H. J., H. A. Johnson, V. P. Bond, and H. H. Rossi: Dose Rate Studies with Monoenergetic Fast Neutrons in Mouse Testis. Radiat. Res., **12**: 420, 1960.

61. Bateson, W.: Mendel's Principles of Heredity. The Univ. Press, Cambridge, 396, 1909.

62. Bauer, J. A., Jr.: Genetics of Skin Transplantation and an Estimate of the Number of Histocompatibility Genes in Inbred Guinea Pigs. Ann. N. Y. Acad. Sci., **87**: 78, 1960.

63. Bayreuther, K.: Chromosomes in Primary Neoplastic Growth. Nature, **186**: 6, 1960.

64. Bayreuther, K., and E. Klein: Cytogenetic, Serologic, and Transplantation Studies on a Heterozygous Tumor and its Derived Variant Sublines. J. Nat. Cancer Inst., **21**: 885, 1958.

65. Beaven, G. H., and W. B. Gratzer: A Critical Review of Human Haemoglobin Variants. I. Methods for Separation and Characterization. J. Clin. Pathol., **12**: 1, 1959.

66. Beaven, G. H., and W. B. Gratzer: A Critical Review of Human Haemoglobin Variants. II. Individual Haemoglobins. J. Clin. Pathol., **12**: 101, 1959.

67. Beaven, G. H., H. Hoch, and E. R. Holiday: The Haemoglobins of the Human Foetus and Infant. Electrophoretic and Spectroscopic Differentiation of Adult and Foetal Types. Biochem. J., **49**: 374, 1951.

68. Becker, H. J.: Ueber Röntgenmosaikflecken und Defektmutationen des Auges. Zeit. ind. Abst. Vererb., **88**: 333, 1957.

69. Bell, A. E.: Physiological Factors Associated with Genetic Resistance to Fowl Typhoid. J. Infect. Diseases, **85**: 154, 1949.

70. Bender, M. A.: X-ray-induced Chromosome Aberrations in Mammalian Cells *in vivo* and *in vitro*. Symp. on "The Immediate and Low Level Effects of Ionizing Radiations." Venice, 1959. Spec. Suppl. Internat. J. Radiat. Biol., 1960.

71. Bender, M. A.: X-ray-induced Chromosome Aberrations in Normal Diploid Human Tissue Cultures. Science, **126**: 974, 1957.

72. Bender, M. A., and P. C. Gooch: Spontaneous and X-ray-induced Somatic Chromosome Aberrations in the Chinese Hamster. Internat. J. Radiat. Biol., **4**: 175, 1961.
73. Bender, M. A., and S. Wolff: X-ray-induced Chromosome Aberrations and Reproductive Death in Mammalian Cells. Amer. Nat., **95**: 39, 1961.
74. Bennett, D.: Developmental Analysis of a Mutation with Pleiotropic Effects in the Mouse. J. Morphol., **98**: 199, 1956.
75. Bennett, D.: *In vitro* Study of Cartilage Induction in T/T Mice. Nature, **181**: 1286, 1958.
76. Bennett, D., S. Badenhausen, and L. C. Dunn: The Embryological Effects of Four Late-lethal *t*-alleles in the Mouse, Which Affect the Neural Tube and Skeleton. J. Morphol., **105**: 105, 1959.
77. Bennett, D., and L. C. Dunn: Effects on Embryonic Development of a Group of Genetically Similar Lethal Alleles Derived from Different Populations of Wild House Mice. J. Morphol., **103**: 135, 1958.
78. Bennett, E. L., M. Calvin, O. Holm-Hansen, A. M. Hughes, K. K. Lonberg-Holm, V. Moses, and B. M. Tolbert: Effect of Deuterium Oxide (Heavy Water) on Biological Systems. Proc. Sec. UN Internat. Conf. on Peaceful Uses of Atomic Energy, **25**: 199, 1958.
79. Benzer, S., V. M. Ingram, and H. Lehmann: Three Varieties of Human Haemoglobin D. Nature, **182**: 852, 1958.
80. Berenblum, I.: Circumstantial Evidence Pointing to Differences Between Cancers in Terms of Etiologic Factors. Cancer Res., **16**: 675, 1956.
81. Berman, I., and H. S. Kaplan: The Cultivation of Mouse Bone Marrow *in vivo*. Blood, **14**: 1040, 1959.
82. Berman, I., and H. S. Kaplan: The Functional Capacity of Mouse Bone Marrow Cells Growing in Diffusion Chambers. Exper. Cell Res., **20**: 238, 1960.
83. Bernstein, S. E.: personal communication.
84. Bernstein, S. E., and E. S. Russell: Implantation of Normal Bloodforming Tissue in Genetically Anemic Mice, without X Irradiation of Host. Proc. Soc. Exper. Biol. Med., **101**: 769, 1959.
85. Bernstein, S. E., E. S. Russell, and F. A. Lawson: Limitations of Wholebody Irradiation for Inducing Acceptance of Homografts in Cases Involving a Genetic Defect. Transplant. Bull., **24**: 106, 1959.
86. Bhat, N. R.: A Dominant Mutant Mosaic House Mouse. Heredity, **3**: 243, 1949.
87. Bielschowsky, F., and E. S. Horning: Aspects of Endocrine Carcinogenesis. Brit. Med. Bull., **14**: 106, 1958.
88. Bielschowsky, M., F. Bielschowsky, and D. Lindsay: A New Strain of Mice with a High Incidence of Mammary Cancers and Enlargement of the Pituitary. Brit. J. Cancer, **10**: 688, 1956.
89. Biesele, J. J., J. L. Biedler, and D. J. Hutchison: The Chromosomal Status of Drug Resistant Sublines of Mouse Leukemia L1210. *In* Genetics and Cancer. R. W. Cumley, ed., Univ. of Texas Press, Austin, 295, 1959.
90. Biffen, R. H.: Mendel's Laws of Inheritance and Wheat Breeding. J. Agr. Sci., **1**: 4, 1905.
91. Billingham, R. E., and L. Brent: Acquired Tolerance of Foreign Cells in Newborn Animals. Proc. Roy. Soc., ser. B, **146**: 78, 1956.
92. Billingham, R. E., L. Brent, and P. B. Medawar: Quantitative Studies on Tissue Transplantation Immunity. III. Actively Acquired Tolerance. Phil. Trans. Roy. Soc., **239**: 357, 1956.
93. Billingham, R. E., L. Brent, P. B. Medawar, and E. M. Sparrow: Quantitative Studies on Tissue Transplantation Immunity. I. The Survival Times of Skin Homografts Exchanged between Members of Different Inbred Strains of Mice. Proc. Roy. Soc., ser. B, **143**: 43, 1954.
94. Billingham, R. E., and P. B. Medawar: A Study of the Branched Cells of the Mammalian Epidermis with Special Reference to the Fate of Their Division Products. Phil. Trans. Roy. Soc., **237**: 151, 1953.
95. Billingham, R. E., and P. B. Medawar: Pigment Spread and Cell Heredity in Guinea Pig's Skin. Heredity, **2**: 29, 1948.
96. Billingham, R. E., and P. B. Medawar: Pigment Spread in Mammalian Skin: Serial Propagation and Immunity Reactions. Heredity, **4**: 141, 1950.
97. Billingham, R. E., and P. B. Medawar: Role of Dendritic Cells in the Infective Colour Transformation of Guinea Pig's Skin. Nature, **160**: 61, 1947.
98. Billingham, R. E., and P. B. Medawar: The "Cytogenetics" of Black and White Guinea Pig Skin. Nature, **159**: 115, 1947.

98a. Billingham, R. E., and P. B. Medawar: The Technique of Free Skin Grafting in Mammals. J. Exper. Biol., **28:** 385, 1951.
99. Billingham, R. E., and W. K. Silvers: Inbred Animals and Tissue Transplantation Immunity. Transplant. Bull., **6:** 399, 1959.
100. Billingham, R. E., and W. K. Silvers: The Melanocytes of Mammals. Quart. Rev. Biol., **35:** 1, 1960.
101. Bird, G. W. G.: Haemagglutinins in Seeds. Brit. Med. Bull., **15:** 165, 1959.
102. Bishop, J., E. Allen, J. Leahy, A. Morris, and R. Schweet: Stages in Hemoglobin Synthesis in a Cell Free System. Fed. Proc., **19:** 346, 1960.
103. Bishop, J., T. Favelukes, R. Schweet, and E. Russell: Control of Specificity in Hemoglobin Synthesis. Nature, **19:** 1365, 1961.
104. Bishop, J., J. Leahy, and R. Schweet: Formation of the Peptide Chain of Hemoglobins. Proc. Nat. Acad. Sci. (Wash.), **46:** 1030, 1960.
105. Bittner, J. J.: Care and Recording. *In* Biology of the Laboratory Mouse. G. D. Snell, ed. The Blakiston Co., Philadelphia, 475, 1941.
106. Bittner, J. J.: Inciting Influences in the Etiology of Mammary Cancer in Mice. AAAS Research Conf. on Cancer, 63, 1945.
107. Bittner, J. J.: Spontaneous Lung Carcinoma in Mice. Pub. Health Repts., **53:** 2197, 1938.
108. Bittner, J. J.: The Experimental Determination of an Invisible Mutation. Papers Mich. Acad. Sci., **11:** 349, 1929.
109. Blair, P. B.: A Mutation in the Mouse Mammary Tumor Virus. Cancer Res., **20:** 635, 1960.
110. Blizzard, R. M., R. W. Chandler, B. H. Landing, M. D. Pettit, and C. D. West: Maternal Autoimmunization of Thyroid as a Probable Cause of Athyrotic Cretinism. New Eng. J. Med., **263(7):** 327, 1960.
111. Block, R. J., E. L. Durrum, and G. Zweig: A Manual of Paper Chromatography and Paper Electrophoresis. Second ed. Academic Press, New York, 1958.
112. Boardman, N. K., and S. M. Partridge: Separation of Neutral Proteins on Ion-exchange Resins. Biochem. J., **59:** 543, 1955.
113. Boche, R. D., J. H. Rust, and A. M. Budy: Studies of F_1 Generation Hybrid Mice from an Irradiated Parent. Fifth Progress Rept., USAF, **7:** 96, 1953.
114. Bock, F. C.: White Spotting in the Guinea Pig Due to a Gene (Star) Which Alters Hair Direction. Thesis, Univ. of Chicago, 54, 1950.
115. Bond, V. P., E. P. Cronkite, C. A. Sonhaus, G. Imirie, J. S. Robertson, and D. C. Borg: The Influence of Exposure Geometry on the Pattern of Radiation Dose Delivered to Large Animal Phantoms. Radiat. Res., **6:** 554, 1957.
116. Bonser, G. M.: The Hereditary Factor in Induced Skin Tumors in Mice: Establishment of a Strain Specially Sensitive to Carcinogenic Agents Applied to the Skin. J. Pathol. and Bacteriol., **46:** 581, 1938.
117. Booth, P. B., G. Plaut, J. D. James, E. W. Ikin, P. Moores, R. Sanger, and R. R. Race: Blood Chimaerism in a Pair of Twins. Brit. Med. J., **1:** 1456, 1957.
118. Borger, R.: Order of Genes in the Fifth Linkage Group of the House Mouse. Nature, **166:** 697, 1950.
119. Boss, J. M. N.: The Pairing of Somatic Chromosomes: A Survey. Texas Repts. Biol. Med., **13:** 213, 1955.
120. Bosshardt, D. K., M. Kryvokulsky, and E. E. Howe: Interrelation of Cholesterol Palmitic Acid and Unsaturated Fatty Acids in the Growing Mouse and Rat. J. Nutrition, **69:** 185, 1959.
121. Boyd, W. C.: Blood Groups and Soluble Antigens. *In* Methodology in Human Genetics, W. J. Burdette, ed. Holden-Day, Inc., San Francisco, 1962.
122. Boyd, W. C.: Fundamentals of Immunology. Third ed., Interscience Pub., New York, 1956.
123. Boyden, S. V.: The Adsorption of Proteins on Erythrocytes Treated with Tannic Acid and Subsequent Hemagglutination by Antiprotein Sera. J. Exper. Med., **93:** 107, 1951.
124. Brecher, G.: New Methylene Blue as a Reticulocyte Stain. Amer. J. Clin. Pathol., **19:** 895, 1949.
125. Brecher, G., K. M. Endicott, H. Gump, and H. P. Brawner: Effects of X-ray on Lymphoid and Hemopoietic Tissues of Albino Mice. Blood, **3:** 1259, 1948.
126. Brecher, G., and M. Schneiderman: A Time-saving Device for the Counting of Reticulocytes, Amer. J. Clin. Pathol., **20:** 1079, 1950.
127. Brecher, G., M. A. Schneiderman, and G. Z. Williams: Evaluation of Electronic Red Blood Cell Counter. Amer. J. Clin. Pathol., **26:** 1439, 1956.
128. Brenneke, H.: Strahlenschädigungen von Mäuse- und Rattensperma, Beobachtet an der Frühentwicklung der Eier. Strahlentherapie, **60:** 214, 1937.

129. Brent, L.: Tissue Transplantation Immunity. Progr. Allergy, **5:** 271, 1958.

130. Briggs, R., and T. J. King: Changes in the Nuclei of Differentiating Endoderm Cells as Revealed by Nuclear Transplantation. J. Morphol., **100:** 269, 1957.

131. Briggs, R., and T. J. King: Nuclear Transplantation Studies on the Early Gastrula (*Rana pipiens*). I. Nuclei of Presumptive Endoderm. Develop. Biol., **2:** 252, 1960.

132. Briggs, R., and T. J. King: Nucleocytoplasmic Interactions in Eggs and Embryos. *In* The Cell. J. Brachet and A. E. Mirsky, eds., Academic Press, New York, 1959.

133. Brink, R. A.: Paramutation and Chromosome Organization. Quart. Rev. Biol., **35:** 120, 1960.

134. Brinkman, R., and J. H. P. Jonxis: Alkaline Resistance and Spreading Velocity of Foetal and Adult Types of Mammalian Hemoglobin. J. Physiol., **88:** 162, 1936.

135. Briody, B. A.: Response of Mice to Ectromelia and Vaccinia Viruses. Bacteriol. Rev., **23:** 61, 1959.

136. Broadhurst, P. L.: Experiments in Psychogenetics: Applications of Biometrical Genetics to the Inheritance of Behaviour. *In* Psychogenetics and Psychopharmacology. H. J. Eysenck, ed., Routledge & Kegan Paul, London, **1:** 1, 1960.

137. Brockman, R. W., L. L. Bennett, Jr., M. S. Simpson, A. R. Wilson, J. R. Thompson, and H. E. Skipper: A Mechanism of Resistance to 8-Azaguanine. II. Studies with Experimental Neoplasms. Cancer Res., **19:** 856, 1959.

138. Brosseau, G. E.: The Environmental Modification of Somatic Crossing-over in *Drosophila melanogaster* with Special Reference to Developmental Phase. J. Exper. Zool., **136:** 567, 1957.

139. Brown, S. W., and W. Welshons: Maternal Aging and Somatic Crossing-over of Attached-*X* Chromosomes. Proc. Nat. Acad. Sci. (Wash.), **41:** 209, 1955.

140. Bruell, J. H.: Dominance and Segregation in the Inheritance of Quantitative Behavior in Mice. *In* Roots of Behavior. E. L. Bliss, ed., Hoeber, New York, 1962.

141. Budds, O. C., E. S. Russell, and G. E. Abrams: Effects of Genetics and Anaesthesia upon Granulocyte and Agranulocyte Levels in Seven Inbred Mouse Strains. Proc. Soc. Exper. Biol. Med., **84:** 176, 1953.

142. Bunker, H., and G. D. Snell: Linkage of White and Waved-1. J. Hered., **39:** 28, 1948.

143. Bunker, L. E., Jr.: Hepatic Fusion, a New Gene in Linkage Group 1 of the Mouse. J. Hered., **50:** 40, 1959.

144. Burdette, W. J.: Induced Pulmonary Tumors. J. Thoracic Surgery, **24:** 427, 1952.

145. Burdette, W. J.: The Significance of Mutation in Relation to the Origin of Tumors: A Review. Cancer Res., **15:** 201, 1955.

146. Burdette, W. J., and L. C. Strong: The Inheritance of Susceptibility to Tumors Induced in Mice. I. Tumors Induced by Methylcholanthrene in Five Inbred Strains of Mice. Cancer Res., **3:** 13, 1943.

147. Burhoe, S. O.: Blood Groups of the Rat (*Rattus norvegicus*) and Their Inheritance. Proc. Nat. Acad. Sci. (Wash.), **33:** 102, 1947.

148. Burnet, F. M.: The Biology of Cancer. Canadian Cancer Conf., **3:** 433, 1959.

149. Burnet, F. M.: The Clonal Selection Theory of Acquired Immunity. Vanderbilt Univ. Press, Nashville, 1959.

150. Bussard, A.: Description d'une Technique Combinant Simultanément l'Électrophorèse et la Précipitation Immunologique dans un Gel: l'Électrosynérèse. Biochim. Biophys. Acta, **34:** 258, 1959.

151. Bussard, A., and D. Perrin: Electrophoresis in Agar Plates. J. Lab. Clin. Med., **46:** 689, 1955.

152. Butler, J. J.: A Study of Antigens of Normal Leukocytes. J. Lab. Clin. Med., **55:** 110, 1960.

153. Cabannes, R., and Ch. Serain: Étude Électrophorétique des Hémoglobines des Mammifères Domestiques d'Algérie. C. R. Soc. Biol., **149:** 1193, 1955.

154. Cahan, A.: Fluid for Shipping Whole Blood Specimens. J. Amer. Med. Assoc., **158:** 971, 1955.

155. Carsner, R. L., and E. G. Rennels: Primary Site of Gene Action in Anterior Pituitary Dwarf Mice. Science, **131:** 829, 1960.

156. Carter, T. C.: A Mosaic Mouse with an Anomalous Segregation Ratio. J. Genet., **51:** 1, 1952.

157. Carter, T. C.: A New Linkage in the House Mouse: Undulated and Agouti. Heredity, **1:** 367, 1947.

158. Carter, T. C.: A Pilot Experiment with Mice, Using Haldane's Method for Detecting Induced Autosomal Recessive Lethal Genes. J. Genet., **56:** 353, 1959.

159. Carter, T. C.: A Search for Chromated Interference in the Male House Mouse. Zeit. ind. Abst. Vererb., **86:** 210, 1954.

160. Carter, T. C.: Genetics of the Little and Bagg X-rayed Mouse Stock. J. Genet., **54:** 311, 1956.
161. Carter, T. C.: Mouse News Letter, **16:** 15, 1957.
162. Carter, T. C.: Mouse News Letter, **21:** 40, 1959.
163. Carter, T. C.: personal communication.
164. Carter, T. C.: Radiation-induced Gene Mutation in Adult Female and Foetal Male Mice. Brit. J. Radiology, **31:** 407, 1958.
165. Carter, T. C.: Recessive Lethal Mutation Induced in the Mouse by Chronic Gamma-irradiation. Proc. Roy. Soc., ser. B, **147:** 402, 1957.
166. Carter, T. C.: Stock Recording Systems. *In* The UFAW Handbook on the Care and Management of Laboratory Animals. Second ed., A. N. Worden and W. Lane-Petter, eds., London, **10:** 151, 1957.
167. Carter, T. C.: The Genetics of Luxate Mice. II. Linkage and Independence. J. Genet., **50:** 300, 1951.
168. Carter, T. C.: The Genetics of Luxate Mice. IV. Embryology. J. Genet., **52:** 1, 1954.
169. Carter, T. C.: The Position of Fidget in Linkage Group V of the House Mouse. J. Genet., **50:** 264, 1951.
170. Carter, T. C.: The Use of Linked Marker Genes for Detecting Recessive Autosomal Lethals in the Mouse. J. Genet., **55:** 585, 1957.
171. Carter, T. C.: Wavy-coated Mice: Phenotypic Interactions and Linkage Tests Between Rex and (a) Waved-1, (b) Waved-2. J. Genet., **50:** 268, 1951.
172. Carter, T. C., and D. S. Falconer: Stocks for Detecting Linkage in the Mouse, and the Theory of Their Design. J. Genet., **50:** 307, 1951.
173. Carter, T. C., and H. Grüneberg: Linkage between Fidget and Agouti in the House Mouse. Heredity, **4:** 373, 1950.
174. Carter, T. C., M. F. Lyon, and R. J. S. Phillips: Further Genetic Studies of Eleven Trans-locations in the Mouse. J. Genet., **54:** 462, 1956.
175. Carter, T. C., and R. S. Phillips: An Experimental Attempt to Investigate the Induction of Visible Mutations in Mice by Chronic Gamma Irradiation. *In* Biological Hazards of Atomic Energy. A. Haddow, ed., Clarendon Press, Oxford, 73, 1952.
176. Carter, T. C., and R. J. S. Phillips: Ragged, a Semidominant Coat Texture Mutant in the House Mouse. J. Hered., **45:** 151, 1954.
177. Carter, T. C., and R. J. S. Phillips: The Sex Distribution of Waved-2, Shaker-2, and Rex in the House Mouse. Zeit. ind. Abst. Vererb., **85:** 564, 1953.
178. Carter, T. C., and R. J. S. Phillips: Three Recurrences of Mutants in the House Mouse. J. Hered., **41:** 252. 1950.
179. Casarett, G. W.: The Effects of Ionizing Radiations from External Sources on Gametogenesis and Fertility in Mammals. A Review, Univ. of Rochester, Atomic Energy Project Rept., UR-**441**, 1956.
180. Castle, W. E.: Studies of Heredity in Rabbits, Rats, and Mice. Carnegie Inst. Wash. Pub. No. 288, 1919.
181. Castle, W. E.: The Origin of a Polydactylous Race of Guinea Pigs. Carnegie Inst. Wash. Pub. No. 49, 1906.
182. Castle, W. E., W. H. Gates, S. C. Reed, and G. D. Snell: Identical Twins in a Mouse Cross. Science, **84:** 581, 1936.
183. Castle, W. E. and W. L. Wachter: Variations of Linkage in Rats and Mice. Genetics, **9:** 1, 1924.
184. Catalogue of Uniform Strains. Second ed. Laboratory Animals Centre, Carshalton, Surrey, 246*d*, 1958.
185. Catalogue of Uniform Strains. Second ed. Laboratory Animals Centre, Carshalton, Surrey, 606*c*, 1958.
186. Cavalli-Sforza, L. L.: Indirect Selection and Origin of Resistance. Ciba Foundation Symp. on Drug Resistance in Micro-organisms. G. E. W. Wolstenholme and C. M. O'Connor, eds., Churchill, London, 30, 1957.
187. Ceppellini, R.: L'Emoglobina Normale Lenta A_2: Suoi Rapporti con una Nuova Frazione Emoglobinica Lenta, B_2, e sua Importanza per il Reconoscimento di Varianti Talassemiche che compaiono nelle Famiglie di Portatori di *Thalassemia media* e di Emoglobinopatia H. Acta Genet. Med. et Gemellol, Suppl., **2:** 47, 1959.
188. Cerecedo, L. R. and L. Mirone: The Beneficial Effect of Folic Acid (Lactobacillus casei Factor) in Mice Maintained on Highly Purified Diets. Arch. Biochem., **12:** 154, 1947.
189. Chai, C. K.: Comparison of Two Inbred Strains of Mice and Their F_1 Hybrid in Response to Androgen. Anat. Rec., **126:** 269, 1956.

190. Chai, C. K.: Endocrine Variation. Thyroid Function in Inbred and F_1 Hybrid Mice. J. Hered., **49:** 143, 1958.
191. Chai, C. K.: Response of Inbred and F_1 Hybrid Mice to Hormones. Nature, **185:** 514, 1960.
192. Chai, C. K.: Seminal Vesicle Growth as a Function of Androgen Stimulation in Mice. Amer. J. Physiol., **186:** 463, 1956.
193. Chai, C. K., A. Amin, and E. P. Reineke: Thyroidal Iodine Metabolism in Inbred and F_1 Hybrid Mice. Amer. J. Physiol., **188:** 499, 1957.
194. Chai, C. K., and O. Mendelsohn: Response of Two Inbred Strains of Mice and Their Hybrid to Androgen, Demonstrated by Mitotic Rate in the Epithelium of the Seminal Vesicles. Growth, **20:** 53, 1956.
195. Chaplin, H., Jr., and P. L. Mollison: Preservation of Blood. *In* Preservation and Transplantation of Normal Tissues. G. E. W. Wolstenholme and M. Cameron, eds., Little, Brown & Co., Boston, 121, 1954.
196. Chapman, A. B.: personal communication.
197. Charles, D. R., and E. M. Luce-Clausen: The Kinetics of Papilloma Formation in Benzpyrene-treated Mice. Cancer Res., **2:** 261, 1942.
198. Charles, D. R., J. A. Tihen, E. M. Otis, and A. B. Grobman: Genetic Effects of Chronic X-irradiation Exposure. Univ. of Rochester, Atomic Energy Project Dept., UR-**565**, 1960.
199. Chase, H. B.: Studies on the Tricolor Pattern of the Guinea Pig. Genetics, I, **24:** 610, 1939. II, **24:** 622, 1939.
200. Chase, H. B., M. S. Gunther, J. Miller, and D. Wolffson: High Insulin Tolerance in an Inbred Strain of Mice. Science, **107:** 297, 1948.
201. Chase, H. B., and W. E. Straile: Some Effects of Accelerated Heavy Ions on Mouse Skin. Radiat. Res., **11:** 437, 1959.
202. Chernoff, A. I.: The Alkali Denaturation Procedures. Conference on Hemoglobin. National Academy of Sciences, National Research Council, Washington, D.C., **557:** 172, 1958.
203. Chernoff, A. I.: The Human Hemoglobins in Health and Disease. New Eng. J. Med., **253:** 322, 1955.
204. Chesley, P., and L. C. Dunn: The Inheritance of Taillessness (anury) in the House Mouse. Genetics, **21:** 525, 1936.
205. Cinader, B., and J. M. Dubert: Specific Inhibition of Response to Purified Protein Antigens. Proc. Roy. Soc., ser. B, **146:** 18, 1956.
206. Clausen, J. J., and J. T. Syverton: Comparison of Chromosome Number and Morphology in Twenty-eight Lines of Mammalian Cells. Fed. Proc., **19:** 388, 1960.
207. Clegg, M. D., and W. A. Schroeder: A Chromatographic Study of the Minor Components of Normal Adult Human Hemoglobin Including a Comparison of Hemoglobin from Normal and Phenylketonuric Individuals. J. Amer. Chem. Soc., **81:** 6065, 1959.
208. Cloudman, A. M.: A Genetic Analysis of Dissimilar Carcinomata from the Same Gland of an Individual Mouse. Genetics, **17:** 468, 1932.
209. Cohen, F., W. W. Zuelzer, and M. M. Evans: Identification of Blood Group Antigens and Minor Cell Populations by the Fluorescent Antibody Method. Blood, **15:** 884, 1960.
210. Cohen, J.: The Pigment Cell System in the Light Sussex Fowl. J. Embryol. Exper. Morphol., **7:** 361, 1959.
211. Cohen, S. S., and H. D. Barner: Studies on Unbalanced Growth in *Escherichia coli*. Proc. Nat. Acad. Sci. (Wash.), **40:** 885, 1954.
212. Cohn, E. J., and J. T. Edsall: The Solubility of Proteins. *In* Proteins, Amino Acids, and Peptides. Reinhold Pub. Corp. New York, 569, 1943.
213. Coleman, D. L.: Effect of Genic Substitution on the Incorporation of Tyrosine into the Melanin of Mouse Skin. Arch. Biochem. Biophys., **96:** 562, 1962.
214. Coleman, D. L.: Phenylalanine Hydroxylase Activity in Dilute and Non-dilute Strains of Mice. Arch. Biochem. and Biophys., **90:** 301, 1960.
215. Coleman, D. L., and M. E. Ashworth: Incorporation of Glycine-1-C^{14} into Nucleic Acids and Proteins of Mice with Hereditary Muscular Dystrophy. Amer. J. Physiol., **197:** 839, 1959.
216. Coleman, D. L., and M. E. Ashworth: Influence of Diet on Trans-amidinase Activity in Dystrophic Mice. Amer. J. Physiol., **199:** 927, 1960.
217. Coleman, D. L.: unpublished observations.
218. Coman, D. R.: Mechanisms Responsible for the Origin and Distribution of Blood-borne Tumor Metastases: A Review. Cancer Res., **13:** 397, 1954.
219. Committee on Standardized Genetic Nomenclature for Mice: Standardized Nomenclature for Inbred Strains of Mice. Cancer Res., **12:** 602, 1952.

220. Committee on Standardized Genetic Nomenclature for Mice: Standardized Nomenclature for Inbred Strains of Mice: Second listing. Cancer Res., **20:** 145, 1960.
221. Conger, A. D., and L. M. Fairchild: A Quick-freeze Method for Making Smear Slides Permanent. Stain Technol., **28:** 281, 1953.
222. Conger, A. D., M. L. Randolph, C. W. Sheppard, and H. J. Luippold: Quantitative Relation of RBE in Tradescantia and Average LET of Gamma-rays, X-rays, and 1.3-, 2.5-, and 14.1-Mev Fast Neutrons. Radiat. Res., **9:** 525, 1958.
223. Coombs, R. R., and D. Bedford: The A and B Antigens on Human Platelets Demonstrated by Means of Mixed Erythrocyte-platelet Agglutination. Vox Sang., **5:** 111, 1955.
224. Coombs, R. R., D. Bedford, and L. M. Rouillard: A and B Blood-group Antigens on Human Epidermal Cells Demonstrated by Mixed Agglutination. Lancet, **I:** 461, 1956.
225. Coombs, R. R., A. E. Mourant, and R. R. Race: A New Test for the Detection of Weak and "Incomplete" Rh Agglutinins. Brit. J. Exper. Pathol., **26:** 255, 1945.
226. Coombs, R. R., and F. Roberts: The Antiglobulin Reaction. Brit. Med. Bull., **15:** 113, 1959.
227. Cooper, C. B.: A Linkage between Naked and Caracul in the House Mouse. J. Hered., **30:** 212, 1939.
228. Cotterman, C. W.: Erythrocyte Antigen Mosaicism. J. Cell. Comp. Physiol., **52** Suppl. I: 69, 1958.
229. Crew, F. A. E., and P. C. Koller: The Sex Incidence of Chiasma Frequency and Genetical Crossing-over in the Mouse. J. Genet., **26:** 359, 1932.
230. Crosby, W. H., J. I. Munn, and F. W. Furth: Standardizing a Method for Clinical Hemoglobinometry. U. S. Armed Forces Med. J., **5:** 693, 1954.
231. Crow, J. F.: Possible Consequences of an Increased Mutation Rate. Eugenics Quart., **4:** 67, 1957.
232. Cuenot, L.: La Loi de Mendel et l'Hérédité de la Pigmentation chez les Souris. Arch. Zool. Exper. et Gen., Notes et Revues, **4:** 33, 1903.
233. Curry, G. A.: Genetical and Developmental Studies on Droopy-eared Mice. J. Embryol. Exper. Morphol., **7:** 39, 1959.
234. Curtis, H. J.: unpublished observations.
235. Czimas, L.: Preparation of Formalinized Erythrocytes. Proc. Soc. Exper. Biol. Med., **103:** 157, 1960.
236. Dagg, C. P.: Sensitive Stages for the Production of Developmental Abnormalities in Mice with 5-Flourouracil. Amer. J. Anat., **106:** 89, 1960.
237. D'Amour, F. E., and F. R. Blood: Manual for Laboratory Work in Mammalian Physiology. Univ. of Chicago Press, 1948.
238. Danielli, J. F., I. J. Lorch, M. J. Ord, and E. G. Wilson: Nucleus and Cytoplasm in Cellular Inheritance. Nature, **176:** 1114, 1955.
239. Darlington, C. D., and L. F. LaCour: The Handling of Chromosomes. George Allen and Unwin, Ltd., London, 1960.
240. Darlington, C. D., and K. Mather: The Elements of Genetics. Macmillan, New York, 1949.
241. Dausset, J.: Immuno-hematologic Biologique et Clinique. Editions médicales Flammarion. (22 Rue de Vaugirarad), Paris, 1956.
242. Dausset, J.: Presence of A and B Antigens in Leukocytes Disclosed by Agglutination Tests. C. R. Soc. Biol., **148:** 1607, 1954.
243. DeBruyn, W. M.: Discussion of article (The Future of Tissue Culture in Cancer Research, by J. H. Hanks). J. Nat. Cancer Inst., **19:** 835, 1957.
244. DeMars, R., and J. L. Hooper: A Method of Selecting for Auxotrophic Mutants of *HeLa* Cells. J. Exper. Med., **111:** 559, 1960.
245. Dempster, E. R., and I. M. Lerner: Heritability of Threshold Characters. Genetics, **35:** 212. 1950.
246. Denenberg, V. H.: Learning Differences in Two Separate Lines of Mice. Science, **130:** 451, 1959.
247. Denenberg, V. H., and G. C. Karas: Effects of Differential Infantile Handling upon Weight Gain and Mortality in the Rat and Mouse. Science, **130:** 629, 1959.
248. Deol, M. S.: The Anatomy and Development of the Mutants Pirouette, Shaker-1, and Waltzer in the Mouse. Proc. Roy. Soc., ser. B, **145:** 206, 1956.
249. Deol, M. S.: The Anomalies of the Labyrinth of the Mutants Varitint-waddler, Shaker-2, and Jerker in the Mouse. J. Genet., **52:** 562, 1954.
250. Deol, M.S.: Genetical Studies on the Skeleton of the Mouse. 28. Tail-short. Proc. Roy. Soc. (Biol.), **155:** 78, 1961.

251. DeOme, K. B., H. A. Bern, S. Nandi, D. R. Pitelka, and L. J. Faulkin, Jr.: The Precancerous Nature of the Hyperplastic Alveolar Nodules Found in the Mammary Glands of Old Female C3H/Crgl Mice. *In* Genetics and Cancer. R. W. Cumley, *et al.*, eds., Univ. of Texas Press, Austin, 327, 1959.

252. DeOme, K. B., P. B. Blair, and L. J. Faulkin, Jr.: Serial Transplantation of Hyperactive Outgrowths Derived from Hyperplastic Alveolar Nodules Transplanted into Gland-free Mammary Fat Pads of C3H/Crgl Mice. Proc. Amer. Assoc. Cancer Res., **3:** 106, 1960.

253. DeOme, K. B., P. B. Blair, and L. J. Faulkin, Jr.: Some Characteristics of the Preneoplastic Hyperplastic Alveolar Nodules of C3H/Crgl Mice. Acta Un. Int. Cancr., **17:** 973, 1961.

254. DeOme, K. B., P. B. Blair, and L. J. Faulkin, Jr.: The Growth and Tumor-producing Potential of Transplanted Mouse Mammary Tissue. Program, Fifth Congr. Internat. Assoc. Gerontol. San Francisco, 44, 1960.

255. DeOme, K. B., L. J. Faulkin, Jr., H. A. Bern, and P. B. Blair: Development of Mammary Tumors from Hyperplastic Alveolar Nodules Transplanted into Gland-free Mammary Fat Pads of Female C3H Mice. Cancer Res., **19:** 515, 1959.

256. Deringer, M. K.: Necrotizing Arteritis in Strain BL/De Mice. Lab. Invest., **8:** 1461, 1959.

257. Deringer, M. K.: Occurrence of Tumors, Particularly Mammary Tumors, in Agent-free Strain C3HeB Mice. J. Nat. Cancer Inst., **22:** 995, 1959.

258. Deringer, M. K.: Spontaneous and Induced Tumors in Haired and Hairless Strain HR Mice. J. Nat. Cancer Inst., **12:** 437, 1951.

259. Deringer, M. K.: The Effect of Subcutaneous Inoculation of 4-*o*-Tolylazo-*o*-Toluidine in Strain HR Mice. J. Nat. Cancer Inst., **17:** 533, 1956.

260. Deringer, M. K., and W. E. Heston: Development of Pulmonary Tumors in Mice Segregated with Respect to the Three Genes: Dominant Spotting, Caracul, and Fused. J. Nat. Cancer Inst., **16:** 763, 1955.

261. Derrien, Y.: Studies on Proteins by Means of Salting-out Curves. I. Method of Establishment of Salting-out Curves of Proteins. Biochem. Biophys. Acta, **8:** 631, 1952.

262. Derrien, Y.: Studies on the Heterogeneity of Adult and Fetal Hemoglobins by Salting-out, Alkali Denaturation and Moving Boundary Electrophoresis. Conf. on Hemoglobin. National Academy of Sciences, National Research Council, **557:** 183, 1958.

263. Detlefsen, J. A.: The Linkage of Dark-eye and Color in Mice. Genetics. **10:** 17, 1925.

264. Detlefsen, J. A., and L. S. Clementes: Linkage of a Dilute Color Factor and Dark-eye in Mice. Genetics, **9:** 247, 1924.

265. Dhaliwal, S. S.: Mutation Studies on Tumor Viruses and Mammalian Tumor Cells. I. The Study of Histocompatibility Mutations in Mouse Tumor Cells Using Isogenic Strains, Academic Thesis, Univ. of Edinburgh, 1959.

266. Dickie, M. M.: Alopecia, a Dominant Mutation in the House Mouse. J. Hered., **46:** 31, 1955.

267. Dickie, M. M.: personal communication, includes Strain MA/MyjHu—listed by K. P. Hummel *in* Inbred Strains of Mice, No. 2, 1961.

268. Dickie, M. M.: personal communication.

269. Dickie, M. M.: The Tortoise Shell House Mouse. J. Hered., **45:** 158, 1954.

270. Dickie, M. M.: The Use of F_1 Hybrid and Backcross Generations to Reveal New and/or Uncommon Tumor Types. J. Nat. Cancer Inst., **15:** 791, 1954.

271. Dickie, M. M., and P. W. Lane: Adrenal Tumors, Pituitary Tumors and Other Pathological Changes in F_1 Hybrids of Strain DE and Strain DBA. Cancer Res., **16:** 48, 1956.

272. Dickie, M. M., and P. W. Lane: Mouse News Letter, **17:** 52, 1957.

273. Dickie, M. M., J. Schneider, and P. J. Harman: A Juvenile Wabbler-lethal in the House Mouse. J. Hered., **43:** 283, 1952.

274. Dickie, M. M., and G. W. Woolley: Fuzzy Mice. J. Hered., **41:** 193, 1950.

275. Dickie, M. M., and G. W. Woolley: Linkage Studies with the Pirouette Gene in the Mouse. J. Hered., **37:** 335, 1946.

276. Dickie, M. M., and G. W. Woolley: Location of the Pirouette Gene in *Mus musculus.* J. Hered., **39:** 288, 1948.

277. Dietrich, L. S., I. M. Friedland, and L. A. Kaplan: Pyridine Nucleotide Metabolism: Mechanism of Action of the Niacin Antagonist, 6-Amino-nicotinamide. J. Biol. Chem., **233:** 964, 1958.

278. Dixon, F. J., D. W. Talmage, and P. H. Maurer: Radiosensitive and Radioresistant Phases in the Antibody Response. J. Immunol., **68:** 693, 1952.

279. Dobrovolskaïa-Zavadskaïa, N.: Efficacité de la Sélection en Vue de l'Élimination des Facteurs Héréditaires Responsables du Cancer Spontané dans une Lignée de Souris (lignée XVII n. c.). C. R. Soc. Biol., **126**: 287, 1937.

280. Dobrovolskaïa-Zavadskaïa, N.: Sur l'Hérédité de la Prédisposition au Cancer Spontané chez la Souris. C. R. Soc. Biol., **101**: 518, 1929.

281. Dobrovolskaïa-Zavadskaïa, N.: Sur une Lignée de Souris, Pauvre en Adénocarcinome de la Mammelle. C. R. Soc. Biol., **104**: 1193, 1930.

282. Dobrovolskaïa-Zavadskaïa, N.: Sur une Lignée de Souris, Riche en Adénocarcinome de la Mammelle. C. R. Soc. Biol., **104**: 1191, 1930.

283. Dobzhansky, T.: Variation and Evolution. Proc. Amer. Philos. Soc., **103**, 252, 1959.

284. Dossel, W. E.: New Method of Intracoelomic Grafting. Science, **120**: 262, 1954.

285. Douglas, C. G., J. S. Haldane, and J. B. S. Haldane: The Laws of Combination of Hemoglobin with Carbon Monoxide and Oxygen. J. Physiol., **44**: 275, 1912.

286. Drabkin, D. L.: A Simplified Technique for a Large Scale Crystallization of Human Oxyhemoglobin: Isomorphous Transformations of Hemoglobin and Myoglobin in the Crystalline State. Arch. Biochem., **21**: 224, 1949.

287. Drabkin, D. L.: Spectrophotometric Studies. XIV. The Crystallographic and Optical Properties of the Hemoglobin of Man in Comparison with Those of Other Species. J. Biol. Chem., **164**: 703, 1946.

288. Drasher, M. L.: Strain Differences in the Response of the Mouse Uterus to Estrogens. J. Hered., **46**: 190, 1955.

289. Dray, S.: Three γ-Globulins in Normal Human Serum Revealed by Monkey Precipitins. Science, **132**: 1313, 1960.

290. Dray, S., and G. O. Young: Genetic Control of Two γ-Globulin Isoantigenic Sites in Domestic Rabbits. Science, **131**: 738, 1960.

291. Dubos, R. J., and R. W. Schaedler: Effect of Nutrition on the Resistance of Mice to Endotoxin and on the Bacterial Power of Their Tissues. J. Exper. Med., **110**: 935, 1959.

292. Dubos, R. J., and R. W. Schaedler: Nutrition and Infection. J. Pediat., **55**: 1, 1959.

293. Duncan, D. B.: Multiple Range and Multiple F Tests. Biometrics, **11**: 1, 1955.

294. Duncan, D. B.: Multiple Range Tests for Correlated and Heteroscedastic Means. Biometrics **13**: 164, 1957.

295. Dunn, L. C.: Analysis of a Case of Mosaicism in the House Mouse. J. Genet., **29**: 317, 1934.

296. Dunn, L. C.: Independent Genes in Mice. Genetics, **5**: 344, 1920.

297. Dunn, L. C.: Linkage in Mice and Rats. Genetics, **5**: 325, 1920.

298. Dunn, L. C.: Mouse News Letter, **18**: 24, 1958.

299. Dunn, L. C.: Studies of Multiple Allelomorphic Series in the House Mouse. I. Description of Agouti and Albino Series of Allelomorphs. J. Genet., **33**: 443, 1936.

300. Dunn, L. C.: The Sable Varieties of Mice. Amer. Nat., **54**: 247, 1920.

301. Dunn, L. C.: Types of White Spotting in Mice. Amer. Nat., **54**: 465, 1920.

302. Dunn, L. C., and E. Caspari: A Case of Neighboring Loci with Similar Effects. Genetics, **30**: 543, 1945.

303. Dunn, L. C., and E. Caspari: Close Linkage between Mutations with Similar Effects. Proc. Nat. Acad. Sci. (Wash.), **28**: 205, 1942.

304. Dunn, L. C., and S. Gluecksohn-Waelsch: A Genetical Study of the Mutation "Fused" in the House Mouse, with Evidence Concerning its Allelism with a Similar Mutation "Kink." J. Genet., **52**: 383, 1954.

305. Dunn, L. C., and S. Gluecksohn-Waelsch: The Failure of a t- allele (t^3) to Suppress Crossing-over in the Mouse. Genetics, **38**: 512, 1953.

306. Dunn, L. C., H. Grüneberg, and G. D. Snell: Report of the Committee on Mouse Genetics Nomenclature. J. Hered., **31**: 505, 1940.

307. Dunn, L. C., E. C. Macdowell, and G. A. Lebedeff: Studies on Spotting Patterns. III. Interaction between Genes Affecting White Spotting and Those Affecting Color in the House Mouse. Genetics, **22**: 307, 1937.

308. Dunn, T. B., W. E. Heston, and M. K. Deringer: Subcutaneous Fibrosarcomas in Strains C3H and C57BL Female Mice, and F_1 and Backcross Hybrids of These Strains. J. Nat. Cancer Inst., **17**: 639, 1956.

309. Dunning, W. F., and M. R. Curtis: Multiple Peritoneal Sarcoma in Rats from Intraperitoneal Injection of Washed, Ground *Taenia* Larvae. Cancer Res., **6**: 668, 1946.

310. Dunsford, I., C. C. Bowley, A. M. Hutchinson, J. S. Thompson, R. Sanger, and R. R. Race: A Human Blood Group Chimera. Brit. Med. J., **2**: 81, 1953.

311. Durrum, E. L.: A Microelectrophoretic and Microionophoretic Technique. J. Amer. Chem. Soc., **72:** 2943, 1950.
312. Eagle, H.: Amino Acid Metabolism in Mammalian Cell Cultures. Science, **130:** 432, 1959.
313. Eagle, H.: Animal Cells and Microbiology. Bacteriol. Rev., **22:** 217, 1958.
314. Eagle, H.: Metabolic Studies with Normal and Malignant Human Cells in Culture. Harvey Lectures, 1958–1959, Academic Press, New York, 156, 1960.
315. Eaton, O. N.: Crosses between Inbred Strains of Mice. J. Hered., **32:** 393, 1941.
316. Ebert, J. D.: Immunochemical Analysis of Development. Symp. Chemical Basis of Development. W. D. McElroy and B. Glass, eds., Johns Hopkins Univ. Press, 339, 1958.
317. Edington, C. W., and M. L. Randolph: A Comparison of the Relative Effectiveness of Radiations of Different Average Linear Energy Transfer on the Induction of Dominant and Recessive Lethals in *Drosophila*. Genetics, **43:** 715, 1958.
318. Edman, P.: Method for Determination of the Amino Acid Sequence in Peptides. Acta Chem. Scand., **4:** 283, 1950.
319. Eichwald, E. J., E. C. Lustgraaf, and M. Strainer: Genetic Factors in Parabiosis. J. Nat. Cancer Inst., **23:** 1193, 1959.
320. Eldjarn, L., P. Alexander, and B. Shapiro: Cysteamine-cystamine: On the Mechanism for the Protective Action Against Ionizing Radiation. Proc. Internat. Conf. on Peaceful Uses of Atomic Energy, **11:** 335, 1956.
321. Elftmann, H., and O. Wegelius: Anterior Pituitary Cytology of the Dwarf Mouse. Anat. Rec., **135:** 43, 1959.
322. Elkind, M. M., and H. Sutton: X-ray Damage and Recovery in Mammalian Cells in Culture. Nature, **184:** 1293, 1959.
323. Emmens, C. W., ed.: Hormone Assay. Academic Press, New York, 1950.
324. Endicott, K. M., and H. Gump: Hemograms and Myelograms of Healthy Female Mice of the C-57 Brown and CFW Strains. Blood, Special Issue No. **1:** 60, 1947.
325. Ephrussi, B.: The Cytoplasm and Somatic Cell Variation. J. Cell. Comp. Physiol., **52** Suppl. 1: 35, 1958.
326. Ephrussi, B., and H. M. Temin: Infection of Chick Iris Epithelium with the Rous Sarcoma Virus *in vitro*. Virology, **11:** 547, 1960.
327. Eriksson, K.: Hereditary Forms of Sterility in Cattle. Lund. Hakan Ohlssons Boktrycheri, 1943.
328. Eschenbrenner, A. B., and E. Miller: Effects of Long-continued Total-body Gamma Irradiation on Mice, Guinea Pigs, and Rabbits. V. Pathological Observations, Biological Effects of External X and Gamma Radiation. R. E. Zirkle, ed., National Nuclear Energy Series, McGraw-Hill, New York, **22B:** 169, 1954.
329. Evans, V. J., J. C. Bryant, M. C. Fioramonti, W. T. McQuilkin, K. K. Sanford, and W. R. Earle: Studies of Nutrient Media for Tissue Cells *in vitro*. I. A Protein-free Chemically Defined Medium for Cultivation of Strain L Cells. Cancer Res., **16:** 77, 1956.
330. Evans, V. J., J. C. Bryant, W. T. McQuilkin, M. C. Fioramonti, K. K. Sanford, B. B. Westfall, and W. R. Earle: Studies of Nutrient Media for Tissue Cells *in vitro*. II. An Improved Protein-free Chemically Defined Medium for Long-term Cultivation of Strain L-929 Cells. Cancer Res., **16:** 87, 1956.
331. Evans, V. J., J. W. B. King, B. L. Cohen, H. Harris, and F. L. Warren: Genetics of Haemoglobin and Blood Potassium Differences in Sheep. Nature, **178:** 849, 1956.
332. Fahey, J. L., M. Potter, F. J. Gutter, and T. B. Dunn: Distinctive Myeloma Globulins Associated with a New Plasma Cell Neoplasm of Strain C3H Mice. Blood, **15:** 103, 1960.
333. Failla, G., and P. McClement: The Shortening of Life by Chronic Whole Body Irradiation. Amer. J. Roentgenol., **78:** 946, 1957.
334. Falconer, D. S.: A Totally Sex-linked Gene in the House Mouse. Nature, **169:** 664, 1952.
335. Falconer, D. S.: Introduction to Quantitative Genetics. (Oliver and Boyd, Edinburgh.) Ronald Press, New York, 1960.
336. Falconer, D. S.: Linkage in the Mouse: The Sex-linked Genes and "Rough." Zeit. ind. Abst. Vererb., **86:** 263, 1954.
337. Falconer, D. S.: Linkage of Rex with Shaker-2 in the House Mouse. Heredity, **1:** 133, 1947.
338. Falconer, D. S.: Location of "reeler" in Linkage Group III in the Mouse. Heredity, **6:** 255, 1952.
339. Falconer, D. S.: Mouse News Letter, **15:** 24, 1956.
340. Falconer, D. S.: Mouse News Letter, **17:** 40, 1957.
341. Falconer, D. S.: Patterns of Response in Selection Experiments with Mice. Cold Spring Harbor Symp., Quant. Biol., **20:** 178, 1955.

342. Falconer, D. S.: Total Sex-linkage in the House Mouse.   Zeit. ind. Abst. Vererb., **85:** 210, 1953.

343. Falconer, D. S., and J. W. B. King: Mouse News Letter, **9** suppl.: 7, 1953.

344. Falconer, D. S., and G. D. Snell: Two New Hair Mutants, Rough and Frizzy, in the House Mouse.   J. Hered., **43:** 53, 1952.

345. Falconer, D. S., and W. R. Sobey: The Location of Trembler in Linkage Group VII.   J. Hered., **44:** 159, 1953.

346. Fano, U.: Principles of Radiological Physics.   *In* Radiation Biology.   A. Hollaender, ed., McGraw-Hill, New York, **1:** 1, 1954.

347. Faulkin, L. J., Jr., and K. B. DeOme: Regulation of Growth and Spacing of Gland Elements in the Mammary Fat Pad of the C3H Mouse.   J. Nat. Cancer Inst., **24:** 953, 1960.

348. Favour, C. B.: Comparative Immunology and the Phylogeny of Homo-transplantation.   Ann. N. Y. Acad. Sci., **73:** 590, 1958.

349. Feinberg, J. G.: Quantitative Immuno-analysis in Gel Plates.   Nature, **185:** 555, 1960.

350. Fekete, E.: Differences in the Effect of Uterine Environment upon Development in the DBA and C57 Black Strains of Mice.   Anat. Rec., **98:** 409, 1947.

351. Fekete, E., and C. C. Little: Observations on the Mammary Tumor Incidence of Mice Born from Transferred Ova.   Cancer Res., **2:** 525, 1942.

352. Fekete, E., and H. K. Otis: Observations on Leukemia in AKR Mice Born from Transferred Ova and Nursed by Low Leukemic Mothers.   Cancer Res., **14:** 445, 1954.

353. Feldman, H. W.: Linkage of Albino Allelomorphs in Rats and Mice.   Genetics, **9:** 487, 1924.

354. Feldman, M., and L. Sachs: Immunogenetic Properties of Tumors that Have Acquired Homo-transplantability.   J. Nat. Cancer Inst., **20:** 513, 1958.

355. Fenton, P. F., and G. R. Cowgill: Reproduction and Lactation in Highly Inbred Strains of Mice on Synthetic Diets.   J. Nutrition, **33:** 703, 1947.

356. Fenton, P. F., G. R. Cowgill, M. A. Stone, and D. H. Justice: The Nutrition of the Mouse. VIII. Studies on Panthothenic Acid, Biotin, Inositol and P-Aminobenzoic Acid.   J. Nutrition, **42:** 257, 1950.

357. Ferm, V. H.: Placental Absorption of Toxins.   Twenty-third Ross Pediatric Conf., S. J. Onesti, Jr., ed., 64, 1957.

358. Field, E. O., and J. R. P. O'Brien: Dissociation of Human Hemoglobin at Low pH.   Biochem. J., **60:** 656, 1955.

359. Fielder, J. H.: The *Taupe* Mouse, A New Coat-color Mutation.   J. Hered., **43:** 75, 1952.

360. Fine, J. M., J. Uriel, and J. Faure: Étude Électrophorétique en Gélose de Diverses Hémoglobines Animales.   Bull Soc. Chem. Biol., **38:** 649, 1956.

361. Fine, J. M., and E. Waszczenko-Zacharczenko: Preparation of Transparent Starch Gels After Electrophoresis.   Nature, **181:** 269, 1958.

362. Finerty, J. C.: Parabiosis in Physiological Studies.   Physiol. Rev., **32:** 277, 1952.

363. Finney, D. J.: The Detection of Linkage.   Ann. Eugen., **10:** 171, 1940.

364. Finney, D. J.: The Estimation of the Frequency of Recombinations. I. Matings of Known Phase.   J. Genet., **49:** 159, 1949.

365. Fischberg, M., J. B. Gurdon, and T. R. Elsdale: Nuclear Transfer in Amphibia and the Problem of the Potentialities of the Nuclei of Differentiating Tissues.   Exper. Cell Res., **6** suppl.: 161, 1959.

366. Fischer, G. A.: Nutritional and Amethopterin-resistant Characteristics of Leukemic Clones.   Cancer Res., **19:** 372, 1959.

367. Fisher, H. W., R. G. Ham, and T. T. Puck: Macromolecular Growth Requirements of Single, Diploid, Mammalian Cells.   Fed. Proc., **19:** 387, 1960.

368. Fisher, R. A.: A Preliminary Linkage Test With *Agouti* and *Undulated* Mice.   Heredity, **3:** 229, 1949.

369. Fisher, R. A.: A System of Scoring Linkage Data with Special Reference to the Pied Factors in Mice.   Amer. Nat., **80:** 568, 1946.

370. Fisher, R. A.: On the Mathematical Foundations of Theoretical Statistics.   Philos. Trans. Roy. Soc., ser. A, **222:** 309, 1922.

371. Fisher, R. A.: Statistical Methods for Research Workers.   Eleventh ed.   Oliver and Boyd, Edinburgh, 354, 1950.

372. Fisher, R. A.: The Detection of Linkage with "Dominant" Abnormalities.   Ann. Eugen., **6:** 187, 1935.

373. Fisher, R. A.: The Genetical Theory of Natural Selection.   The Clarendon Press, Oxford, 1930.

374. Fisher, R. A.: The Linkage of *Polydactyly* with *Leaden* in the House Mouse.   Heredity, **7:** 91, 1953.

375. Fisher, R. A.: The Theory of Inbreeding. Oliver & Boyd, Edinburgh, 1949.
376. Fisher, R. A., and N. T. J. Bailey: The Estimation of Linkage with Differential Viability. Parts I, II, III. Heredity, **3**: 215, 1949.
377. Fisher, R. A., and W. Landauer: Sex Differences of Crossing-over in Close Linkage. Amer. Nat., **87**: 116, 1953.
378. Fisher, R. A., M. F. Lyon, and A. R. G. Owen: The Sex Chromosome in the House Mouse. Heredity, **1**: 355, 1947.
379. Fisher, R. A., and G. D. Snell: A Twelfth Linkage Group of the House Mouse. Heredity, **2**: 271, 1948.
380. Fisk, R. T., and C. A. McGee: The Use of Gelatin in Rh Testing and Antibody Determination. Amer. J. Clin. Pathol., **17**: 737, 1947.
381. Fitzpatrick, T. B., P. Brunet, and A. Kukita: The Nature of Hair Pigment. *In* The Biology of Hair Growth. W. Montagna and R. A. Ellis, eds., Academic Press, New York, 255, 1958.
382. Fitzpatrick, T. B., and A. Kukita: Tyrosinase Activity in Vertebrate Melanocytes. *In* Pigment Cell Biology. M. Gordon, ed., Academic Press, New York, 489, 1959.
383. Foley, E. J.: Antigenic Properties of Methylcholanthrene-induced Tumors in Mice of the Strain of Origin. Cancer Res., **13**: 835, 1953.
384. Ford, C. E.: Human Cytogenetics: Its Present Place and Future Possibilities. Amer. J. Human Genet., **12**: 104, 1960.
385. Ford, C. E.: personal communication.
386. Ford, C. E., and J. L. Hamerton: A Colchicine Hypotonic Citrate, Squash Sequence for Mammalian Chromosomes. Stain Technol., **31**: 247, 1956.
387. Ford, C. E., and R. H. Mole: Chromosomes and Carcinogenesis: Observations on Radiation-induced Leukemias. Progress in Nuclear Energy, Series VI. Biological Sciences. J. G. Buger, J. Coursaget, and J. F. Loutit, eds., Pergamon Press, Inc., London, **2**: 11, 1959.
388. Ford, C. E., P. E. Polani, J. H. Briggs, and P. M. F. Bishop: A Presumptive Human *XXY/XX* Mosaic. Nature, **183**: 1030, 1959.
389. Ford, D. K.: Chromosomal Changes Occurring in Chinese Hamster Cells during Prolonged Culture *in vitro*. Canadian Cancer Conf., **3**: 171, 1959.
390. Ford, D. K., R. Wakonig, and G. Yerganian: Further Observations on the Chromosomes of Chinese Hamster Cells in Tissue Culture. J. Nat. Cancer Inst., **22**: 765, 1959.
391. Ford, D. K., and G. Yerganian: Observations on the Chromosomes of Chinese Hamster Cells in Tissue Culture. J. Nat. Cancer Inst., **21**: 393, 1958.
392. Foreman, C.: Electromigration Properties of Mammalian Hemoglobins: Taxonomic Criteria. Amer. Midland Nat., **64**: 177, 1960.
393. Fortuine, R., G. Mathe, and H. Baxter: Bibliography of Hematopoietic Tissue Transplantation. Transplant. Bull., **7**: 434, 1960.
394. Foster, M.: Enzymatic Studies of the Physiological Genetics of Guinea Pig Coat Coloration. I. Oxygen Consumption Studies. Genetics, **41**: 396, 1956.
395. Foster, M.: Enzymatic Studies of Pigment-forming Abilities in Mouse Skin. J. Exper. Zool., **117**: 211, 1951.
396. Foster, M.: Physiological Studies of Melanogenesis. *In* Pigment Cell Biology. M. Gordon, ed., Academic Press, New York, 301, 1959.
397. Foster, M., and S. R. Brown: The Production of DOPA by Normal Pigmented Mammalian Skin. J. Biol. Chem., **225**: 247, 1957.
398. Foster, M., and L. Thomson: Allelic Relations at the Brown Locus of the House Mouse. Genetics **46**: 865, 1961.
399. Foulds, L.: The Experimental Study of Tumor Progression: A Review. Cancer Res., **14**: 327, 1954.
400. Foulds, L.: The Histologic Analysis of Mammary Tumors of Mice. I. Scope of Investigations and General Principles of Analysis. J. Nat. Cancer Inst., **17**: 701, 1956.
401. Foulds, L.: The Natural History of Cancer. J. Chronic Diseases, **8**: 2, 1958.
402. Fox, A. S.: Genetics of Tissue Specificity. Ann. N. Y. Acad. Sci., **73**: 611, 1958.
403. Francis, T.: The Development of the Pituitary and Hereditary Anterior Pituitary Dwarfism in Mice. Munksgaard, Copenhagen, Denmark, 1944.
404. Fraser, F. C., T. D. Fainstat, and H. Kalter: The Experimental Production of Congenital Defects with Particular Reference to Cleft Palate. Études néo-natales **2**: 43, 1953.
405. Fraser, F. C., B. E. Walker, and D. G. Trasler: Experimental Production of Congenital Cleft Palate: Genetic and Environmental Factors. Pediatrics, **19**: 782, 1957.
406. Freund, J.: The Mode of Action of Immunologic Adjuvants. Advances Tuberc., **10**: 130, 1956.

407. Frings, H., M. Frings, and M. Hamilton: Experiments with Albino Mice from Stocks Selected for Predictable Susceptibility to Audiogenic Seizures. Behaviour, **9**: 44, 1956.
408. Fuller, J. L., L. D. Clark, and M. Waller: The Effects of Chlorpromanine upon Psychological Development in the Puppy. Psychopharmacologia, **1**: 393, 1960.
409. Fuller, J. L., C. Easler, and M. E. Smith: Inheritance of Audiogenic Seizure Susceptibility in the Mouse. Genetics, **35**: 622, 1950.
410. Fuller, J. L., and R. W. Thompson: Behavior Genetics. Wiley, New York, 1960.
411. Fulton, G. P., B. R. Lutz, D. I. Patt, and G. Yerganian: The Cheek Pouch of the Chinese Hamster (*Cricetulus griseus*) for Cinephotomicroscopy of Blood Circulation and Tumor Growth. J. Lab. Clin. Med., **44**: 145, 1954.
412. Furth, J.: Conditioned and Autonomous Neoplasms: A Review. Cancer Res., **13**: 477, 1953.
413. Furth, J.: Discussion of Problems Related to Hormonal Factors in Initiating and Maintaining Tumor Growth. Cancer Res., **17**: 454, 1957.
414. Furth, J.: Experimental Pituitary Tumors. Recent Progress Hormone Res., **11**: 221, 1955.
415. Furth, J., and Kahn, M. C.: The Transmission of Leukemia of Mice with a Single Cell. Amer. J. Cancer, **31**: 276, 1937.
416. Furth, J., H. R. Siebold, and R. R. Rathbone: Experimental Studies on Lymphomatosis of Mice. Amer. J. Cancer, **19**: 521, 1933.
417. Gall, J. G.: Chromosomal Differentiation. *In* The Chemical Basis of Development. W. D. McElroy, and B. Glass, eds., Johns Hopkins Univ. Press, 103, 1958.
418. Garber, E. D.: Bent-tail, a Dominant, Sex-linked Mutation in the Mouse. Proc. Nat. Acad. Sci. (Wash.), **38**: 876, 1952.
419. Gardner, W. U.: Estrogenic Effects of Adrenal Tumors of Ovariectomized Mice. Cancer Res., **1**: 632, 1941.
420. Gardner, W. U.: Hormonal Aspects of Experimental Tumorigenesis. Advances in Cancer Res., **1**: 173, 1953.
421. Gardner, W. U.: Hormones and Carcinogenesis. Canadian Cancer Conf., **2**: 207, 1957.
422. Gardner, W. U., and J. Rygaard: Further Studies on the Incidence of Lymphomas in Mice Exposed to X-rays and Given Sex Hormones. Cancer Res., **14**: 205, 1954.
423. Gardner, W. U., and L. C. Strong: Strain-limited Development of Tumors of the Pituitary Gland in Mice Receiving Estrogens. Yale J. Biol. and Med., **12**: 543, 1940.
424. Gates, W. H.: A Case of Non-disjunction in the Mouse. Genetics, **12**: 295, 1927.
425. Gates, W. H.: Linkage of the Factor Shaker with Albinism and Pink-eye in the House Mouse. Zeit. ind. Abst. Vererb., **59**: 220, 1931.
426. Gates, W. H.: Linkage of the Factors for Short-ear and Density in the House Mouse (*Mus musculus*). Genetics, **13**: 170, 1928.
427. Gates, W. H.: Symbols for Mutations in Mice. Science, **64**: 328, 1926.
428. Gates, W. H., and T. Pullig: The Linkage of Dominant White Spotting with Hairless in the House Mouse (*Mus musculus*). Genetics, **30**: 4 (abstr.), 1945.
429. Gerald, P. S., and L. K. Diamond: The Diagnosis of Thalassemia Trait by Starch Block Electrophoresis of the Hemoglobin. Blood, **13**: 61, 1958.
430. Ginsburg, B. E.: Genetic Control of the Ontogeny of Stress Behavior. APA Symp., 1960, in press.
431. Ginsburg, B. E.: Genetics and the Physiology of the Nervous System. Research Pub. Assoc. Nervous Mental Diseases, **33**: 39, 1954.
432. Ginsburg, B. E.: Genetics as a Tool in the Study of Behavior. Perspectives in Biol. Med., **1**: 397, 1958.
433. Ginsburg, B.: The Effects of the Major Genes Controlling Coat Color in the Guinea Pig on the Dopa Oxidase Activity of Skin Extracts. Genetics, **29**: 176, 1944.
434. Giroud, A., H. Gounelle, and M. Martinet: Données Quantitatives sur le Taux de la Vitamine A chez le Rat Lors d'Expériences de Tératogénèse par Hypervitaminose A. Bull. Soc. Chimie Biol., **39**: 331, 1957.
435. Giroud, A., G. Lévy, and J. Lefebvres-Boisselot: Taux de la Riboflavine chez le Foetus de Rat Présentant des Malformations dues à la Déficience B₂. Internat. Zeit. Vitaminforsch., **22**: 308, 1950.
436. Giroud, A., and M. Martinet: Morphogénèse de l'Anencéphalie. Arch. Anat. Micro. Morphol. Exper., **46**: 247, 1957.
437. Gladner, J. A., and J. E. Folk: Carboxypeptidase B. II. Mode of Action on Protein Substrates and Its Application to Carboxyl Terminal Group Analysis. J. Biol. Chem., **231**: 393, 1958.
438. Glasstone, S.: Source Book on Atomic Energy. D. Van Nostrand Co., Inc., Princeton, N. J., 1958.

439. Gluecksohn-Schoenheimer, S.: The Effect of an Early Lethal (t°) in the House Mouse. Genetics, **25:** 391, 1940.

440. Gluecksohn-Schoenheimer, S.: The Embryonic Development of Mutants of the Sd Strain in Mice. Genetics, **30:** 29, 1945.

441. Gluecksohn-Waelsch, S.: The Effect of Maternal Immunization Against Organ Tissues on Embryonic Differentiation in the Mouse. J. Embryol. Exper. Morphol., **5:** 83, 1957.

442. Gluecksohn-Waelsch, S.: The Inheritance of Hemoglobin Types and of Other Biochemical Traits in Mammals. J. Cell. Comp. Physiol., **56** Suppl. 1: 89, 1960.

443. Gluecksohn-Waelsch, S., H. M. Ranney, and B. F. Sisken: The Hereditary Transmission of Hemoglobin Differences in Mice. J. Clin. Invest., **36:** 753, 1957.

444. Goettsch, M.: Comparative Protein Requirement of the Rat and Mouse for Growth, Reproduction and Lactation Using Casein Diets. J. Nutrition, **70:** 307, 1960.

445. Gold, E. R., G. H. Tovey, W. E. Benney, and F. J. W. Lewis: Changes in the Group A Antigen in a Case of Leukaemia. Nature, **183:** 892, 1959.

446. Goldberg, C. A.: Identification of Human Hemoglobins. Clin. Chem., **3:** 1, 1957.

447. Goldin, A., J. M. Venditti, S. R. Humphreys, D. Dennis, and N. Mantel: Studies on the Management of Mouse Leukemia (L1210) with Antagonists of Folic Acid. Cancer Res., **15:** 742, 1955.

448. Goldstein, M. B., and M. B. Feiner: unpublished.

449. Goldstein, M. N.: A Method for More Critical Isolation of Clones Derived from Three Human Cell Strains *in vitro*. Cancer Res., **17:** 357, 1957.

450. Goodale, H. D.: A Study of the Inheritance of Body Weight in the Albino Mouse by Selection. J. Hered., **29:** 101, 1938.

451. Goodman, J. W., and C. C. Congdon: Effect of Blood Injection on Bone Marrow Transplantation in Lethally Irradiated Mice. Arch. Pathol., **72:** 18, 1961.

452. Goodman, J. W., and R. D. Owen: Red Cell Repopulation in Irradiated Mice Treated with Plethoric Homologous Bone Marrow. Amer. J. Physiol., **198:** 309, 1960.

453. Goodman, J. W., and L. H. Smith: Erythrocyte Life Span in Normal Mice and in Radiation Bone Marrow Chimeras. Amer. J. Physiol. **200:** 764, 1961.

454. Goodwins, I. R., and M. A. C. Vincent: Further Data on Linkage between Short-ear and Maltese Dilution in the House Mouse. Heredity, **9:** 413, 1955.

455. Gorer, P. A.: Some Recent Data on the H-2 System of Mice. *In* Biological Problems of Grafting. F. Albert, and P. B. Medawar, eds., Blackwell, Oxford, 25, 1959.

456. Gorer, P. A.: Some Recent Work on Tumor Immunity. Advances Cancer Res., **4:** 149, 1956.

457. Gorer, P. A.: The Isoantigens of Malignant Cells. UNESCO Symp. on Biological Aspects of Cancer Chemotherapy. Louvain, 1960. Academic Press, New York, in press.

458. Gorer, P. A., S. Lyman, and G. D. Snell: Studies on the Genetic and Antigenic Basis of Tumor Transplantation. Linkage between a Histocompatibility Gene and "Fused" in Mice. Proc. Roy. Soc., ser. B, **135:** 499, 1948.

459. Gorer, P. A., and Z. B. Mikulska: The Antibody Response to Tumor Inoculation: Improved Methods of Antibody Detection. Cancer Res., **14:** 651, 1954.

460. Gorer, P. A., and P. O'Gorman: The Cytoxic Activity of Isoantibodies in Mice. Transplant. Bull., **3:** 142, 1956.

461. Goudie, R. B.: Somatic Segregation of "Inagglutinable" Erythrocytes. Lancet, **1:** 1333, 1957.

462. Gowen, J. W.: Genetic Effects in Nonspecific Resistance to Infectious Disease. Bacteriol. Rev., **24:** 192, 1960.

463. Gowen, J. W.: Humoral and Cellular Elements in Natural and Acquired Resistance to Typhoid. Amer. J. Human Genet., **4:** 255, 1952.

464. Gowen, J. W.: Inheritance of Immunity in Animals. Ann. Rev. Microbiol., **2:** 215, 1948.

465. Gowen, J. W.: Significance and Utilization of Animal Individuality in Disease Research. J. Nat. Cancer Inst., **15:** 555, 1954.

466. Gowen, J. W., and M. L. Calhoun: Factors Affecting Genetic Resistance of Mice to Mouse Typhoid. J. Infect. Diseases, **73:** 40, 1943.

467. Gowen, J. W., and R. G. Schott: A Genetic Technique for Differentiating between Acquired and Genetic Immunity. Amer. J. Hyg., **18:** 688, 1933.

468. Gowen, J. W., J. Stadler, H. H. Plough, and H. N. Miller: Virulence and Immunizing Capacity of *Salmonella typhimurium* as Related to Mutation in Metabolic Requirements. Genetics, **38:** 531, 1953.

469. Gowen, J. W., and M. R. Zelle: Irradiation Effects on Genetic Resistance of Mice to Mouse Typhoid. J. Infect. Diseases, **77:** 85, 1945.

470. Grabar, P.: Immunoelectrophoretic Analysis. *In* Methods of Biochemical Analysis. D. Glick, ed., Interscience Publishers Inc., New York, **7**: 1, 1959.

471. Grabar, P., and C. A. Williams, Jr.: Méthode Immuno-électrophorétique d'Analyse de Mélanges de Substances Antigéniques. Biochim. Biophys. Acta, **17**: 67, 1955.

472. Grabar, P., and C. A. Williams Jr.: Méthode Permettant l'Étude Conjuguée des Propriétés Électrophorétiques et Immunochimiques d'un Mélange de Protéines. Application au Sérum Sanguin. Biochim. Biophys. Acta, **10**: 193, 1953.

473. Grahn, D.: Acute Radiation Response of Mice from a Cross between Radiosensitive and Radio-resistant Strains. Genetics, **43**: 835, 1958.

474. Grahn, D.: The Genetic Control of Physiological Processes: The Genetics of Radiation Toxicity in Animals. *In* Radioisotopes in the Biosphere. R. S. Caldecott, ed., Center for Continuation Study, Minneapolis, Minn., 181, 1960.

475. Grahn, D., and K. F. Hamilton: Genetic Variation in the Acute Lethal Response of Four Inbred Mouse Strains to Whole Body X Irradiation. Genetics, **42**: 189, 1957.

476. Grahn, D., G. A. Sacher, and H. Walton, Jr.: Comparative Effectiveness of Several X-ray Qualities for Acute Lethality in Mice and Rabbits. Radiat. Res., **4**: 228, 1956.

477. Grant, J. L., M. C. Britton, and T. E. Kurtz: Measurement of Red Blood Cell Volume with the Electronic Cell Counter. Amer. J. Clin. Pathol., **33**: 138, 1960.

478. Green, A. A., E. J. Cohn, and M. H. Blanchard: Studies on the Physical Chemistry of the Proteins. XII. The Solubility of Human Hemoglobin in Concentrated Salt Solutions. J. Biol. Chem., **109**: 631, 1935.

479. Green, E. L.: A System for Keeping Records in Mouse Breeding Experiments. Unpublished manuscript, 1950.

480. Green, E. L.: Genetic and Non-genetic Factors which Influence the Type of the Skeleton in an Inbred Strain of Mice. Genetics, **26**: 192, 1941.

481. Green, E. L.: Radioisotopes and the Genetic Mechanism: The Problem of Genetic Effects of Low-level External Radioactivity. *In* Radioisotopes in the Biosphere, R. S. Caldecott, ed., Center for Continuation Study, Minneapolis, Minn., 146, 1960.

482. Green, E. L.: The Genetics of a Difference in Skeletal Type between Two Inbred Strains of Mice (BALB/c and C57B). Genetics, **36**: 391, 1951.

483. Green, E. L.: Theoretical Consequences of Systems of Mating Used in Mammalian Genetics. *In* Methodology in Mammalian Genetics, W. J. Burdette, ed., Holden-Day, Inc., San Francisco, 1962.

484. Green, E. L., and M. C. Green: Effect of the Short-ear Gene on Number of Ribs and Presacral Vertebrae in the House Mouse. Amer. Nat., **80**: 619, 1946.

485. Green, E. L., and M. C. Green: Transplantation of Ova in Mice. An Attempt to Modify the Number of Presacral Vertebrae. J. Hered., **50**: 109, 1959.

486. Green, M. C.: Effects of the Short-ear Gene in the Mouse on Cartilage Formation in Healing Bone Fractures. J. Exper. Zool., **137**: 75, 1958.

487. Green, M. C.: Himalayan, a New Allele of Albino in the Mouse. J. Hered., **52**: 73, 1961.

488. Green, M. C.: Mouse News Letter, **22**: 34, 1960.

489. Green, M. C.: Mouse News Letter, **23**: 34, 1960.

490. Green, M. C.: Mouse News Letter, **23**: 34, 1960, and unpublished.

491. Green, M. C.: The Position of Luxoid in Linkage Group II of the Mouse. J. Hered. **52**: 297, 1961.

492. Green, M. C., and M. M. Dickie: Linkage Map of the Mouse. J. Hered., **50**: 3, 1959.

493. Green, M. C., and E. F. Woodworth: unpublished.

494. Green, M. N., G. Yerganian, and H. Meier: Elevated alpha-2-serum Proteins as a Possible Genetic Marker in Spontaneous Hereditary Diabetes Mellitus of the Chinese Hamster (*Cricetulus griseus*). Experientia, **16**: 503, 1960.

495. Greene, H. S. N.: The Significance of the Heterologous Transplantability of Human Cancer. Cancer, **5**: 24, 1952.

496. Greene, H. S. N.: Uterine Adenomata in the Rabbit. III. Susceptibility as a Function of Constitutional Factors. J. Exper. Med., **73**: 273, 1941.

497. Griffen, A. B.: Occurrence of Chromosomal Aberrations in Pre-spermatocytic Cells of Irradiated Male Mice. Proc. Nat. Acad. Sci. (Wash.), **44**: 691, 1958.

498. Grüneberg, H.: A Three-factor Linkage Experiment in the Mouse. J. Genetics, **31**: 157, 1935.

499. Grüneberg, H.: An Annotated Catalogue of the Mutant Genes of the House Mouse. Med. Research Council Mem., **33**: 28, 1956.

500. Grüneberg, H.: Further Linkage Data on the Albino Chromosome of the House Mouse. J. Genetics, **33**: 255, 1936.

501. Grüneberg, H.: Genetical Studies on the Skeleton of the Mouse.   IV. Quasicontinuous Variations. J. Genet., **51:** 95, 1952.
502. Grüneberg, H.: Genetical Studies on the Skeleton of the Mouse.   XXIII. The Development of
Brachyury and Anury. J. Embryol. Exper. Morphol., **6:** 424, 1958.
503. Grüneberg, H.: Inherited Macrocytic Anemias in the House Mouse. Genetics, **24:** 777, 1939.
504. Grüneberg, H.: Inherited Macrocytic Anemias in the House Mouse.   II. Dominance Relationships. J. Genet., **43:** 285, 1942.
505. Grüneberg, H.: The Anemia of Flexed-tailed Mice (*Mus musculus L.*).   I. Static and Dynamic
Hematology. J. Genet., **43:** 45, 1942.
506. Grüneberg, H.: The Anemia of Flexed-tailed Mice (*Mus musculus L.*).   II. Siderocytes.
J. Genet., **44:** 246, 1942.
507. Grüneberg, H.: The Genetics of the Mouse. Second ed.   M. Nijhoff, ed., The Hague, 1952.
508. Grüneberg, H.: The Relations of Microphthalmia and White in the Mouse. J. Genet., **51:**
359, 1953.
509. Grüneberg, H., and G. M. Truslove: Two Closely Linked Genes in the Mouse. Genet.
Res., **1:** 69, 1960.
510. Grunt, J. A., and W. C. Young: Consistency of Sexual Behavior Patterns in Individual Male
Guinea Pigs Following Castration and Androgen Therapy. J. Comp. Physiol. Psychol.,
46: 138, 1953.
511. Gullbring, B.: Investigation on the Occurrence of Blood Group Antigens in Spermatozoa from
Man, and Serological Demonstration of the Segregation of Character. Acta Med. Scand.,
159: 169, 1957.
512. Gutter, F. J., H. A. Sober, and E. A. Peterson: The Effect of Mercaptoethanol and Urea on the
Molecular Weight of Hemoglobin. Arch. Biochem. Biophys., **62:** 427, 1956.
513. Guyer, M. F., and E. A. Smith: Further Studies on Inheritance of Eye Defects Induced in
Rabbits. J. Exper. Zool., **38:** 449, 1924.
514. Haas, F. L., and C. O. Doudney: Interrelations of Nucleic Acid and Protein Syntheses in
Radiation Induced Mutation Induction in Bacteria. Proc. Second UN Internat. Conf.
on Peaceful Uses of Atomic Energy, **22:** 336, 1958.
515. Haldane, J. B. S.: The Amount of Heterozygosis to be Expected in an Approximately Pure Line.
J. Genet., **32:** 375, 1936.
516. Haldane, J. B. S.: The Causes of Evolution. Harper and Bros., New York, 1932.
517. Haldane, J. B. S.: The Detection of Autosomal Lethals in Mice Induced by Mutagenic Agents.
J. Genet., **54:** 327, 1956.
518. Haldane, J. B. S., A. D. Sprunt, and N. M. Haldane: Reduplication in Mice. J. Genet.,
5: 133, 1915.
519. Hannah, A.: Localization and Function of Heterochromatin in *Drosophila melanogaster*. Advances in Genet., **4:** 87, 1951.
520. Hansen-Melander, E.: Accelerated Evolution of Cancer Stemlines Following Environmental
Changes. Hereditas, **44:** 471, 1958.
521. Harnois, M. C.: The Effect of Colchicine on Hamster Cells Grown *in vitro*. M.S. Thesis,
Univ. of Pittsburgh, 1960.
522. Harris, M., and F. H. Ruddle: Growth and Chromosome Studies on Drug Resistant Lines of
Cells in Tissue Culture. *In* Cell Physiology of Neoplasia. Univ. of Texas Press, Austin,
524, 1960.
523. Haruna, I., and S. Akabori: Studies on the C-Terminal and Acidic Amino Acid Containing
Peptides Obtained from Ovalbumin by Partial Hydrazinolysis. J. Biochem. (Tokyo), **47:**
513, 1960.
524. Hasserodt, U., and J. Vinograd: Dissociation of Human Carbonmonoxy-hemoglobin at High
*p*H. Proc. Nat. Acad. Sci. (Wash.), **45:** 12, 1959.
525. Haugaard, G., and T. D. Kroner: Partition Chromatography of Amino Acids with Applied
Voltage. J. Amer. Chem. Soc., **70:** 2135, 1948.
526. Haurowitz, F., R. L. Hardin, and M. Dicks: Denaturation of Hemoglobins by Alkali. J. Phys.
Chem., **58:** 103, 1954.
527. Hauschka, T. S.: Correlation of Chromosomal and Physiologic Changes in Tumors. J. Cell.
Comp. Physiol., **52** Suppl 1: 197, 1958.
528. Hauschka, T. S.: Immunologic Aspects of Cancer: A Review. Cancer Res., **12:** 615,
1952.
529. Hauschka, T. S.: Tissue Genetics of Neoplastic Cell Populations. Canadian Cancer Conf.,
2: 305, 1957.

530. Hauschka, T. S., S. T. Grinnell, L. Revesz, and G. Klein: Quantitative Studies on the Multiplication of Neoplastic Cells *in vivo*. IV. Influence of Doubled Chromosome Number on Growth Rate and Final Population Size. J. Nat. Cancer Inst., **19:** 13, 1957.

531. Hauschka, T. S., B. J. Kvedar, S. T. Grinnell, and D. B. Amos: Immunoselection of Polyploids from Predominantly Diploid Cell Populations. Ann. N. Y. Acad. Sci., **63:** 683, 1956.

532. Hauschka, T. S., and A. Levan: Cytologic and Functional Chacterization of Single Cell Clones Isolated from the Krebs-2 and Ehrlich Ascites Tumors. J. Nat. Cancer Inst., **21:** 77, 1958.

533. Hauschka, T. S., and A. Levan: Inverse Relationship Between Chromosome Ploidy and Host-specificity in Sixteen Transplantable Tumors. Exper. Cell. Res., **4:** 457, 1953.

534. Hauschka, T. S., J. T. Mitchell, and D. J. Niederpruem: A Reliable Frozen Tissue Bank: Viability and Stability of 82 Neoplastic and Normal Cell Types after Prolonged Storage at −78° C. J. Nat. Cancer Inst., **19:** 643, 1959.

535. Hawk, H. W., J. N. Wiltbank, H. E. Kidder, and L. E. Casida: Embryonic Mortality between 16 and 34 Days Post-breeding in Cows of Low Fertility. J. Dairy Sci., **38:** 673, 1955.

536. Hayflick, L.: Decontaminating Tissue Cultures Infected with Pleuropneumonia-like Organisms. Nature, **185:** 783, 1960.

537. Hayflick, L., and P. S. Moorhead: The Serial Cultivation of Human Diploid Cell Strains. Exper. Cell Res., in press.

538. Hayman, B. J.: The Theory and Analysis of Diallel Crosses. Genetics, **45:** 155, 1960.

539. Hays, F. A.: The Application of Genetic Principles in Breeding Poultry for Egg Production. Poultry Sci., **4:** 43, 1925.

540. Hays, F. A.: The Inheritance of Persistency and its Relation to Fecundity. Proc. Third World's Poultry Congress, 92, 1927.

541. Hazzard, W. R., and S. L. Leonard: Phosphoglucomutase Activity in Hereditary Muscular Dystrophy in Mice. Proc. Soc. Exper. Biol., **102:** 720, 1960.

542. Heidenthal, G.: A Colorimetric Study of Genic Effect on Guinea Pig Coat Color. Genetics, **25:** 197, 1940.

543. Hekker, A. C., W. Klomp-Magnee, H. W. Krijnen, and J. J. Van Loghem: A Papain Slide Test for Rh Mass Typing. Vox Sang., **2:** 128, 1957.

544. Hellström, K. E.: Chromosomal Studies on Primary Methylcholanthrene-induced Sarcomas in the Mouse. J. Nat. Cancer Inst., **23:** 1019, 1959.

545. Hellström, K. E.: Studies on Isoantigenic Variation in Mouse Lymphomas. J. Nat. Cancer Inst., **25:** 237, 1960.

546. Helyer, B. J., and J. B. Howie: Spontaneous Haemolytic Anaemia in NZB/Bl Mice. 36th Ann. Rept. Brit. Empire Cancer Campaign, 458, 1958.

547. Henthorne, R. D.: Care and Management of Laboratory Animals. Walter Reed Army Inst. of Research, Walter Reed Army Med. Center, Washington, D.C., 1954.

548. Hershgold, E. J., and M. B. Riley: Diet Induced Variations in Tolerance to Altitude Hypoxia in the Mouse (24792). Proc. Soc. Exper. Biol. Med., **100:** 831, 1959.

549. Hertwig, P.: Neue Mutationen und Koppelungs-gruppen bei der Hausmaus. Zeit. ind. Abst. Vererb., **80:** 220, 1942.

550. Hertwig, P.: Vererbare Semisterilität bei Mäusen Nach Röntgenbestrahlung, Verursacht Durch Reziproke Chromosomentranslokationen, Zeit. ind. Abst. Vererb., **79:** 27, 1940.

551. Hertwig, P.: Zwei Subletale Rezessive Mutationen in der Nachkommenschaft von Röntgenbestrahlten Mäusen. Erbarzt, **4:** 41, 1939.

552. Herzenberg, L. A., and R. A. Roosa: Nutritional Requirements for Growth of a Mouse Lymphoma in Cell Culture. Exper. Cell Res., **21:** 430, 1960.

553. Hesselbach, M. L.: Control of Melanization of S91 Tumors by Selective Transfer, and Biochemical Studies of the Tumors Produced. *In* Pigment Cell Growth. M. Gordon, ed., Academic Press, New York, 189, 1953.

554. Heston, W. E.: Development of Inbred Strains in the Mouse and Their Use in Cancer Research. Roscoe B. Jackson Memorial Laboratory 20th Commemoration Lectures on Genetics, Cancer, Growth, and Social Behavior. Bar Harbor Times, Bar Harbor, Maine, 1, 1949.

555. Heston, W. E.: Development of Inbred Strains in the Mouse and Their Use in Cancer Research. Roscoe B. Jackson Memorial Laboratory 20th Commemoration Lectures on Genetics, Cancer, Growth, and Social Behavior. Bar Harbor Times, Bar Harbor, Maine, 9, 1949.

556. Heston, W. E.: Effects of Genes Located on Chromosomes III, V, VII, and XIV on the Occurrence of Pulmonary Tumors in the Mouse. Proc. Internat. Genetics Symp., Cytologia, Suppl. Vol., 219, 1957.

557. Heston, W. E.: Genetic Analysis of Susceptibility to Induced Pulmonary Tumors in Mice. J. Nat. Cancer Inst., **3:** 69, 1942.
558. Heston, W. E.: Genetics of Mammary Tumors in Mice. *In* Mammary Tumors in Mice. AAAS Pub., **22:** 55, 1945.
559. Heston, W. E.: Inheritance of Susceptibility to Spontaneous Pulmonary Tumors in Mice. J. Nat. Cancer Inst., **3:** 79, 1942.
560. Heston, W. E.: Mammary Tumors in Agent-free Mice. Ann. N. Y. Acad. Sci., **71:** 931, 1958.
561. Heston, W. E.: Relationship between Susceptibility to Induced Pulmonary Tumors and Certain Known Genes in Mice. J. Nat. Cancer Inst., **2:** 127, 1941.
562. Heston, W. E., and M. K. Deringer: Relationship Between the Agouti Gene and Mammary Tumor Development in Mice. Acta Unio Internat. contre le Cancer, **6:** 262, 1948.
563. Heston, W. E., M. K. Deringer, and H. B. Andervont: Gene-milk Agent Relationship in Mammary-tumor Development. J. Nat. Cancer Inst., **5:** 289, 1945.
564. Heston, W. E., M. K. Deringer, and T. B. Dunn: Further Studies on the Relationship between the Genotype and the Mammary Tumor Agent in Mice. J. Nat. Cancer Inst., **16:** 1309, 1956.
565. Heston, W. E., M. K. Deringer, T. B. Dunn, and W. D. Levillain: Factors in the Development of Spontaneous Mammary Gland Tumors in Agent-free Strain C3Hb Mice. J. Nat. Cancer Inst., **10:** 1139, 1950.
566. Heston, W. E., M. K. Deringer, I. R. Hughes, and J. Cornfield: Interrelation of Specific Genes, Body Weight, and Development of Tumors in Mice. J. Nat. Cancer Inst., **12:** 1141, 1952.
567. Heston, W. E., and T. B. Dunn: Tumor Development in Susceptible Strain A and Resistant Strain L Lung Transplants in LAF$_1$ Hosts. J. Nat. Cancer Inst., **11:** 1057, 1951.
568. Heston, W. E., and M. A. Schneiderman: Analysis of Dose-response in Relation to Mechanism of Pulmonary Tumor Induction in Mice. Science, **117:** 109, 1953.
569. Heston, W. E., and C. H. Steffee: Development of Tumors in Fetal and Adult Lung Transplants. J. Nat. Cancer Inst., **18:** 779, 1957.
570. Heston, W. E., and G. Vlahakis: Influence of the *Ay* Gene on Mammary-gland Tumors, Hepatomas, and Normal Growth in Mice. J. Nat. Cancer Inst., **26:** 969, 1961.
571. Heston, W. E., G. Vlahakis, and M. K. Deringer: Delayed Effect of Genetic Segregation on the Transmission of the Mammary Tumor Agent in Mice. J. Nat. Cancer Inst., **24:** 721, 1960.
572. Heston, W. E., G. Vlahakis, and M. K. Deringer: High Incidence of Spontaneous Hepatomas and the Increase of this Incidence with Urethan in C3H, C3Hf, and C3He Male Mice. J. Nat. Cancer Inst., **24:** 425, 1960.
573. Hetzer, H. O.: The Genetic Basis for Resistance and Susceptibility to *Salmonella aertrycke* in Mice. Genetics, **22:** 264, 1937.
574. Hildemann, W. H.: A Method for Detecting Hemolysins in Mouse Isoimmune Serums. Transplant. Bull., **4:** 148, 1957.
575. Hildemann, W. H.: Scale Homotransplantation in Goldfish (*Carassius auratus*). Ann. N.Y. Acad. Sci., **64:** 775, 1957.
576. Hildemann, W. H., and R. L. Walford: Chronic Skin Homograft Rejection in the Syrian Hamster. Ann. N. Y. Acad. Sci., **87:** 56, 1960.
577. Hill, A. B., J. M. Hatswell, and W. W. C. Topley: The Inheritance of Resistance, Demonstrated by the Development of a Strain of Mice Resistant to Experimental Inoculation with a Bacterial Endotoxin. J. Hyg., **40:** 538, 1940.
578. Hill, R. J., and W. Konigsberg: The Isolation of Peptides from Tryptic Digests of the α-Chain from Human Hemoglobin. J. Biol. Chem., **235:** PC21, 1960.
579. Hill, R. L., and H. C. Schwartz: A Chemical Abnormality in Haemoglobin G. Nature, **184** Suppl. 9: 641, 1959.
580. Hill, R. L., and E. L. Smith: Leucine Amino Peptidase. VII. Action on Long-chain Polypeptide and Proteins. J. Biol. Chem., **228:** 577, 1957.
581. Hilschmann, N., and G. Braunitzer: Über die N-terminale Sequenz der β-Kette des Menschlichen Häemoglobins. Zeit. Physiol. Chem., Hoppe-Seyler's, **317:** 285, 1959.
582. Hine, G. J., and G. L. Brownell: Radiation Dosimetry. Academic Press, New York, 1956.
583. Hirs, C. H. W.: Chromatography of Enzymes on Ion Exchange Resins. *In* Methods in Enzymology. S. P. Colowick, and N. O. Kaplan, eds., Academic Press, New York, **1:** 113, 1955.
584. Hirsch, J., and J. C. Boudreau: The Heritability of Phototaxis in a Population of *Drosophila melanogaster*. J. Comp. Physiol. Psychol., **51:** 647, 1958.
585. Hirsch, W., P. Moores, R. Sanger, and R. R. Race: Notes on Some Reactions of Human Anti-M and Anti-N Sera. Brit. J. Haematol., **3:** 134, 1957.

586. Hitzig, W. H., and R. Gitzelman: Transplacental Transfer of Leukocyte Agglutinins. Vox Sang., **4:** 445, 1959.

587. Hoecker, G.: Genetic Mechanisms in Tissue Transplantation in the Mouse. Cold Spring Harbor Symp. Quant. Biol., **21:** 355, 1956.

588. Hoecker, G.: On the So-called Unspecificity of Long Transplanted Tumors. Transplant. Bull., **1:** 201, 1954.

589. Hoecker, G., A. Martinez, S. Markovic, and O. Pizzaro: Agitans, a New Mutation in the House Mouse with Neurological Effects. J. Hered., **45:** 10, 1954.

590. Högman, C. F.: Blood Group Antigens A and B Determined by Means of Mixed Agglutination on Cultured Cells of Human Fetal Kidney, Liver, Spleen, Lung, Heart, and Skin, unpublished.

591. Hollaender, A., C. C. Congdon, D. G. Doherty, T. Makinodan, and A. C. Upton: New Developments in Radiation Protection and Recovery. Proc. Second Internat. Conf. on Peaceful Uses of Atomic Energy, **23:** 3, 1958.

592. Hollander, W. F.: Mouse News Letter, **15:** 29, 1956.

593. Hollander, W. F.: Mouse News Letter, **20:** 34, 1959.

594. Hollander, W. F., J. H. D. Bryan, and J. W. Gowen: Pleiotropic Effects of a Mutant at the p Locus from X-irradiated Mice. Genetics, **45:** 413, 1960.

595. Hollander, W. F., and J. W. Gowen: An Extreme Non-agouti Mutant in the Mouse. J. Hered., **47:** 221, 1956.

596. Hollander, W. F., and L. C. Strong: Pintail, a Dominant Mutation Linked with Brown in the House Mouse. J. Hered., **42:** 179, 1951.

597. Hommes, F. A., J. Santema-Drinkwaard, and T. H. J. Huisman: The Sulfhydryl Groups of Four Different Human Haemoglobins. Biochim. Biophys. Acta, **20:** 564, 1956.

598. Horowitz, N. H., and M. Fling: Genetic Determination of Tyrosinase Thermostability in Neurospora. Genetics, **38:** 360, 1953.

599. Hsu, T. C.: Chromosomal Evolution in Cell Populations. Internat. Rev. Cytol., in press.

600. Hsu, T. C.: Mammalian Chromosomes *in vitro*. I. The Karyotype of Man. J. Hered., **43:** 167, 1952.

601. Hsu, T. C.: Mammalian Chromosomes *in vitro*. XIII. Cyclic and Directional Changes of Population Structure. J. Nat. Cancer Inst., **25:** 1339, 1960.

602. Hsu, T. C., D. Billen, and A. Levan: Mammalian Chromosomes *in vitro*. XV. Patterns of Transformation. J. Nat. Cancer Inst., **27:** 515, 1961.

603. Hsu, T. C., and D. S. Kellogg: Genetics of *in vitro* Cells. *In* Genetics and Cancer. R. W. Cumley, *et al.*, eds., Univ. of Texas Press, Austin, 183, 1959.

604. Hsu, T. C., and D. S. Kellogg, Jr.: Mammalian Chromosomes *in vitro*. XII. Experimental Evolution of Cell Populations. J. Nat. Cancer Inst., **24:** 1067, 1960.

605. Hsu, T. C., and O. Klatt: Mammalian Chromosomes *in vitro*. IX. On Genetic Polymorphism in Cell Populations. J. Nat. Cancer Inst., **21:** 437, 1958.

606. Hsu, T. C., and D. J. Merchant: Mammalian Chromosomes *in vitro*. XIV. Genotypic Replacement in Cell Populations. J. Nat. Cancer Inst., **26:** 1075, 1961.

607. Hsu, T. C., and C. M. Pomerat: Mammalian Chromosomes *in vitro*. II. A Method for Spreading the Chromosomes of Cells in Tissue Culture. J. Hered., **44:** 23, 1953.

608. Huff, C. G.: Natural Immunity and Susceptibility of Culicine Mosquitoes to Avian Malaria. Amer. J. Trop. Med., **15:** 427, 1935.

609. Hughes, A.: Some Effects of Abnormal Tonicity on Dividing Cells in Chick Tissue Cultures. Quart. J. Micro. Sci., **93:** 207, 1952.

610. Huisman, T. H. J.: A Method for the Characterization of Abnormal Human Hemoglobins Based upon Differences in Chromatographic Behavior on Amberlite IRC 50. Conference on Hemoglobin. National Academy of Sciences, National Research Council, Pub. No. **557:** 165, 1958.

611. Huisman, T. H. J., E. A. Martis, and A. Dozy: Chromatography of Hemoglobin Types on Carboxymethylcellulose. J. Lab. Clin. Med., **52:** 312, 1958.

612. Huisman, T. H. J., and H. K. Prins: Chromatographic Estimation of Four Different Human Hemoglobins. J. Lab. Clin. Med., **46:** 255, 1955.

613. Hummel, K. P.: The Inheritance and Expression of Disorganization, an Unusual Mutation in the Mouse. J. Exper. Zool., **137:** 389, 1958.

614. Hungerford, D. A.: Chromosome Number of Ten-day Fetal Mouse Cells. J. Morphol., **97:** 497, 1955.

615. Hungerford, D. A., and M. A. Di Berardino: Cytological Effects of Prefixation Treatment. J. Biophys. Biochem. Cytol., **4:** 391, 1958.

616. Hunt, J. A.: Identity of the α-Chains of Adult and Foetal Human Haemoglobins. Nature, **183:** 1373, 1959.

617. Hunt, J. A., and V. M. Ingram: Abnormal Human Haemoglobins. II. The Chymotryptic Digestion of the Trypsin-resistant "Core" of Haemoglobins A and S. Biochim. Biophys. Acta, **28:** 546, 1958.

618. Hunt, J. A., and V. M. Ingram: A Terminal Peptide Sequence of Human Haemoglobin. Nature, **184** Suppl. 9: 640, 1959.

619. Hunt, J. A., and V. M. Ingram: The Genetical Control of Protein Structure: The Abnormal Human Haemoglobins. *In* Biochemistry of Human Genetics, Ciba Foundation Symp., G. E. W. Wolstenholme, and C. M. O'Connor, eds., Little, Brown & Co., 114, 1959.

620. Hunt, J. A., and H. Lehmann: Haemoglobin "Bart's": A Foetal Haemoglobin Without α-Chains. Nature, **184:** 872, 1959.

621. Hurlbut, H. S., and J. I. Thomas: The Experimental Host Range of the Arthropod-borne Animal Viruses in Arthropods. Virology, **12:** 391, 1960.

622. Huseby, R. A., and J. J. Bittner: Differences in Adrenal Responsiveness to Post-castrational Alteration as Evidenced by Transplanted Adrenal Tissue. Cancer Res., **11:** 954, 1951.

623. Huseby, R. A., and J. J. Bittner: Studies on the Inherited Hormonal Influence, Acta Unio Internat. contra Cancer, **6:** 197, 1948.

624. Ibsen, H. L., and B. L. Goertzen: Whitish, a Modifier of Chocolate and Black Hairs in Guinea Pigs. J. Hered., **42:** 231, 1951.

625. Ichikawa, Y., and S. Amano: A New Type of Virus Found in a Spontaneous Mammary Tumor of SL Mice and its Proliferating Modus Observed in Ultra-thin Sections under the Electron Microscope. Gann, **49:** 57, 1958.

626. Iljin, N. A.: Ruby Eye in Animals and its Heredity. Trans. Lab. Exper. Biol., Zoo Park, Moscow, **1:** 1, 1926.

627. Iljin, N. A.: The Distribution and Inheritance of White Spots in Guinea Pigs. Trans. Lab. Exper. Biol., Zoo Park, Moscow, **4:** 255, 1928.

628. Inbred Strains of Mice. (An informal biennial mimeographed document listing laboratories maintaining inbred strains of mice, with histories and characteristics of those strains.) J. Staats, ed., Jackson Mem. Lab., Bar Harbor, Maine.

629. Ingalls, T. H., P. R. Avis, F. J. Curley, and H. M. Temin: Genetic Determinants of Hypoxia-induced Congenital Anomalies. J. Hered., **44:** 185, 1953.

630. Ingbar, S. H., and E. H. Kass: Sulfhydryl Content of Normal Hemoglobin and Hemoglobin in Sickle-cell Anemia. Proc. Soc. Exper. Biol. Med., **77:** 74, 1951.

631. Ingraham, J. S.: The Preparation and Use of Formalized Erythrocytes with Attached Antigens or Haptens to Titrate Antibodies. Proc. Soc. Exper. Biol. Med., **99:** 452, 1958.

632. Ingram, V. M.: Abnormal Human Haemoglobins. I. The Comparison of Normal Human and Sickle-cell Haemoglobins by "Fingerprinting." Biochim. Biophys. Acta, **28:** 539, 1958.

633. Ingram, V. M.: A Specific Chemical Difference between the Globins of Normal and Sickle-Cell Anaemia Haemoglobin. Nature, **178:** 792, 1956.

634. Ingram, V. M.: Chemistry of the Abnormal Human Haemoglobins. Brit. Med. Bull., **15:** 27, 1959.

635. Ingram, V. M.: Separation of the Peptide Chains of Human Globin. Nature, **183:** 1795, 1959.

636. Ingram, V. M.: How do Genes Act? Sci. Amer., **198:** 68, 1958.

637. Ingram, V. M.: Sulphydryl Groups in Haemoglobins. Biochem. J., **59:** 653, 1955.

638. Ingram, V. M.: The Stepwise Reductive Cleavage of DNP-Peptides. Biochim. Biophys. Acta, **20:** 577, 1956.

639. Institute of Laboratory Animal Resources, National Research Council: Animals for Research, A Catalogue of Commercial Sources. June, 1959.

640. Irako, Y.: Effect of X Radiation on the Ascites Tumors: Difference in Radiosensitivity Among Various Transplant-strains of the Ascites Hepatoma of the Rat. Gann, **51:** 33, 1960.

641. Irwin, M. R.: Immunogenetics. *In* Advances in Genetics. Academic Press, New York, **1:** 113, 1947.

642. Irwin, M. R.: The Inheritance of Resistance to the *Danysz* Bacillus in the Rat. Genetics, **14:** 337, 1929.

643. Ising, U.: Chromosome Studies in Ehrlich Mouse Ascites Cancer after Heterologous Transplantation Through Hamsters. Brit. J. Cancer, **9:** 592, 1955.

644. Ising, U.: Effect of Heterologous Transplantation on Chromosomes of Ascites Tumors. Acta Pathol. et Microbiol. Scand., Suppl. **127**, 1958.

645. Itano, H. A.: A Third Abnormal Hemoglobin Associated with Hereditary Hemolytic Anemia. Proc. Nat. Acad. Sci. (Wash.), **37:** 775, 1951.

646. Itano, H. A.: Electrophoretic Analyses of the Abnormal Human Hemoglobins. Conference on Hemoglobin. National Academy of Sciences, National Research Council, **557:** 144, 1958.

647. Itano, H. A.: The Human Hemoglobins: Their Properties and Genetic Control. Advances in Protein Chemistry. C. B. Anfinsen, Jr., K. Bailey, M. L. Anson, and J. T. Edsall, eds., Academic Press, New York, **12:** 215, 1957.

648. Itano, H. A., W. R. Bergren, and P. Sturgeon: The Abnormal Human Hemoglobins. Medicine, **35:** 121, 1956.

649. Itano, H. A., and E. Robinson: Demonstration of Intermediate Forms of Carbonmonoxy- and Ferrihemoglobin by Moving Boundary Electrophoresis. J. Amer. Chem. Soc., **78:** 6415, 1956.

650. Itano, H. A., and E. Robinson: Formation of Normal and Doubly Abnormal Haemoglobins by Recombination of Haemoglobin I with S and C. Nature, **183:** 1799, 1959.

651. Itano, H. A., and E. Robinson: Properties and Inheritances of Haemoglobin by Asymmetric Recombination. Nature, **184:** 1468, 1959.

652. Itano, H. A., and S. J. Singer: On Dissociation and Recombination of Human Adult Hemoglobins A, S, and C. Proc. Nat. Acad. Sci. (Wash.), **44:** 522, 1958.

653. Jacob, F.: Genetic Control of Viral Functions. Harvey Lectures, 1958–1959, p. 1, 1960.

654. Jacob, F., and E. A. Adelberg: Transfer de Caractères Génétiques par Incorporation au Facteur Sexuel d'*Escherichia coli*. C. R. Acad. Sci., **249:** 189, 1959.

655. Jacob, F., and E. L. Wollman: Les Épisomes, Éléments Génétiques Ajoutés. C. R. Acad. Sci., **247:** 154, 1958.

656. Jacob, G. F., and N. C. Tappen: Abnormal Haemoglobins in Monkeys. Nature, **180:** 241, 1957.

657. Jacob, R., P. Schaeffer, and E. L. Wollman: Episomic Elements of Bacteria. Symp. Soc. Gen., Microbiol., in press.

658. Jacobs, P. A., D. G. Harnden, W. M. Brown-Court, J. Goldstein, H. G. Close, T. N. McGregor, N. McLean, and J. A. Strong: Abnormalities Involving the *X* Chromosome in Women. Lancet, **1:** 1213, 1960.

659. Jacquot-Armond, Y., M. Theoleyre, and S. Filitti-Wurmser: Sur L'Absorption des Anti-B par les Hematies de Lapin et d'Opossum. Rev. d'Hematol., **11:** 63, 1956.

660. Jaffe, J. J.: Cytological Observations Concerning Inversion and Translocation in the House Mouse. Amer. Nat., **86:** 101, 1952.

661. Jennings, H. S.: Behavior of the Lower Organisms. Columbia Univ. Press, New York, 1906.

662. Jennings, H. S.: The Numerical Results of Diverse Systems of Breeding. Genetics, **1:** 53, 1916.

663. Jerne, N. K.: The Presence in Normal Serum of Specific Antibody against Bacteriophage *T*4 and its Increase During the Earliest Stages of Immunization. J. Immunol., **76:** 209, 1956.

664. Johnson, H. A., and E. P. Cronkite: The Effect of Tritiated Thymidine on Mouse Spermatogonia. Radiat. Res., **11:** 825, 1959.

665. Jones, A. R.: Dextran as a Diluent for Univalent Antibodies. Nature, **165:** 118, 1950.

666. Jones, A. R., and S. Silver: The Detection of Minor Erythrocyte Populations by Mixed Agglutinates. Blood, **13:** 763, 1958.

667. Jones, R. T., W. A. Schroeder, and J. R. Vinograd: Identity of the α-Chains of Hemoglobins A and F. J. Amer. Chem. Soc., **81:** 4749, 1959.

668. Jones, R. T., W. A. Schroeder, J. E. Balog, and J. R. Vinograd: Gross Structure of Hemoglobin H. J. Amer. Chem. Soc., **81:** 3161, 1959.

669. Jonxis, J. H. P.: Foetal Haemoglobin and *Rh* Antagonism. *In* Haemoglobin. F. J. W. Roughton, and J. C. Kendrew, eds., Interscience Publishers Inc., New York, 261, 1949.

670. Jope, H. M., and J. R. P. O'Brien: Crystallization and Solubility Studies on Human Adult and Foetal Haemoglobins. *In* Haemoglobin. F. J. W. Roughton, and J. C. Kendrew, eds., Interscience Publishers Inc., New York, 269, 1949.

671. Josephson, A. M., M. S. Masri, L. Singer, D. Dworkin, and K. Singer: Starch Block Electrophoretic Studies of Human Hemoglobin Solutions. II. Results in Cord Blood, Thalassemia and Other Hematologic Disorders: Comparison with Tiselius Electrophoresis. Blood, **13:** 543, 1958.

672. Kabat, E. A.: Size and Heterogeneity of the Combining Sites on an Antibody Molecule. J. Cell. Comp. Physiol., **50** Suppl. 1: 79, 1957.

673. Kabat, E. A., and M. M. Mayer: Experimental Immunochemistry, Second ed. C. C Thomas, Springfield, Ill., 1961.

674. Kaliss, N.: Immunological Enhancement of Tumor Homografts in Mice: A Review. Cancer Res., **18:** 992, 1958.

675. Kallman, K. D., and M. Gordon: Genetics of Fin Transplantation in Xiphophorin Fishes. Ann. N.Y. Acad. Sci., **73:** 599, 1958.

676. Kalmus, H., J. D. Metrakos, and M. Silverberg: Sex Ratio of Offspring from Irradiated Male Mice. Science, **116:** 274, 1952.

677. Kalter, H.: Factors Influencing the Frequency of Cortisone-induced Cleft Palate in Mice. J. Exper. Zool., **134:** 449, 1957.

678. Kalter, H.: Teratogenic Action of a Hypocaloric Diet and Small Doses of Cortisone. Proc. Soc. Exper. Biol., **104:** 518, 1960.

679. Kalter, H.: The Inheritance of Susceptibility to the Teratogenic Action of Cortisone in Mice. Genetics, **39:** 185, 1954.

680. Kalter, H., and J. Warkany: Congenital Malformations in Inbred Strains of Mice Induced by Riboflavin-deficient, Galactoflavin-containing Diets. J. Exper. Zool., **136:** 531, 1957.

681. Kalter, H., and J. Warkany: Experimental Production of Congenital Malformations in Mammals by Metabolic Procedures. Physiol. Rev., **39:** 69, 1959.

682. Kandutsch, A. A.: Intracellular Distribution and Extraction of Tumor Homograft-enhancing Antigens. Cancer Res., **20:** 264, 1960.

683. Kandutsch, A. A., and A. E. Russell: Creatine and Creatinine in Tissues and Urine of Mice with Hereditary Muscular Dystrophy. Amer. J. Physiol., **194:** 553, 1958.

684. Kaplan, H. S., B. B. Hirsch, and M. B. Brown: Indirect Induction of Lymphomas in Irradiated Mice. IV. Genetic Evidence of the Origin of the Tumor Cells from the Thymic Grafts. Cancer Res., **16:** 434, 1956.

685. Kaplan, W. D.: The Influence of Minutes upon Somatic Crossing-over in *Drosophila melanogaster*. Genetics, **38:** 630, 1953.

686. Kaplan, W. D., and M. F. Lyon: Failure of Mercaptoethylamine to Protect Against the Mutagenic Effects of Radiation. II. Experiments with Mice. Science, **118:** 777, 1953.

687. Karvonen, M. J.: A Solubility Study of Foetal and Adult Sheep Hemoglobin. *In* Hemoglobin. F. J. W. Roughton, and J. C. Kendrew, eds., Interscience Publishers Inc., New York, 279, 1949.

688. Katz, A. M., and A. I. Chernoff: Structural Similarities between Hemoglobins *A* and *F*. Science, **130:** 1574, 1959.

689. Katz, J. J., H. L. Crespi, A. J. Finkel, R. J. Hasterlik, J. F. Thomson, W. Lester, Jr., W. Chorney, N. Scully, R. L. Shaffer, and Sung Huang Sun: The Biology of Deuterium. Proc. Second UN Internat. Conf. on Peaceful Uses of Atomic Energy, **25:** 173, 1958.

690. Kaziwara, K.: Derivation of Stable Polyploid Sublines from a Hyperdiploid Ehrlich Ascites Carcinoma. Cancer Res., **14:** 795, 1954.

691. Keeler, C. E.: Hereditary Blindness in the House Mouse with Special Reference to its Linkage Relationships. Howe Lab. Ophthalmol. Bull., **3:** 11, 1930.

692. Keeler, C. E.: Rodless Retina, an Ophthalmic Mutation in the House Mouse (*Mus musculus*). J. Exper. Zool., **46:** 355, 1927.

693. Keeler, C. E.: The Breeding of Mice in Laboratories. *In* The Laboratory Mouse. Harvard Univ. Press, Cambridge, **7:** 47, 1931.

694. Keeler, C. E.: The Independence of Dominant Spotting and Recessive Spotting (Piebald), in the House Mouse. Proc. Nat. Acad. Sci. (Wash.), **17:** 101, 1931.

695. Keeler, C. E., and H. C. Trimble: Inheritance of Position Preference in Coach Dogs. J. Hered., **31:** 50, 1940.

696. Keighley, G., P. M. Lowy, H. Borsook, E. Goldwasser, A. S. Gordon, T. C. Prentice, W. A. Rambach, F. Stohlman, Jr., and D. C. Van Dyke: A Cooperative Assay of a Sample with Erythropoietic Stimulating Activity. Blood, **16:** 1424, 1960.

697. Keighley, G., E. S. Russell, and P. M. Lowy: Response of Normal and Genetically Anemic Mice to Erythropoietic Stimuli. Brit. J. Haemat., in press.

698. Keil, B.: On Proteins. XLIII. Isolation of Dinitrophenyl Derivatives of Amino Acids and Peptides by Ion-exchange. Coll. Czechoslov. Chem. Communs., **23:** 740, 1958.

699. Kekwick, R. A., and H. Lehmann: Sedimentation Characteristics of the γ-Chain Haemoglobin (Haemoglobin "Bart's"). Nature, **187:** 158, 1960.

700. Kempthorne, O.: An Introduction to Genetic Statistics. J. Wiley and Sons, New York, 1957.

701. Kempthorne, O.: International Symposium on Biometrical Genetics, Pergamon Press, New York, 1960.

702. Kempthorne, O., and O. B. Tandon: The Estimation of Heritability by Regression of Offspring on Parent. Biometrics, **9:** 90, 1953.

703. Kendrew, J. C.: Foetal Haemoglobin. Endeavour, **8:** 80, 1949.

704. Kety, S. S.: Biochemical Theories of Schizophrenia. Science, **129** (Part I): 1528, 1959.
705. Kety, S. S.: Biochemical Theories of Schizophrenia. Science, **129** (Part II): 1590, 1959.
706. Kiddy, C. A., W. H. Stone, and L. E. Casida: Immunologic Studies on Fertility and Sterility. II. Effects of Treatment of Semen with Antibodies on Fertility in Rabbits. J. Immunol., **82:** 125, 1959.
707. Kidwell, J. F.: Mouse News Letter, **24:** 39, 1961.
708. Kimball, R. F., N. Gaither, and S. M. Wilson: Recovery in Stationary-phase Paramecia from Radiation Effects Leading to Mutation. Proc. Nat. Acad. Sci. (Wash.), **45:** 833, 1959.
709. King, J. W. B.: Linkage Group XIV of the House Mouse. Nature, **178:** 1126, 1956.
710. King, S. C., and C. R. Henderson: Variance Components Analysis in Heritability Studies. Poultry Sci., **33:** 147, 1954.
711. Kirschbaum, A., M. Frantz, and W. L. Williams: Neoplasms of the Adrenal Cortex in Noncastrate Mice. Cancer Res., **6:** 707, 1946.
712. Klatskin, G., O. M. Reinmuth, and W. Barnes: A Study of the Densitometric Method of Analyzing Filter Paper. Electrophoretic Patterns of Serum. J. Lab. Clin. Med., **48:** 476, 1956.
713. Klein, E.: Gradual Transformation of Solid into Ascites Tumors. Evidence Favoring the Mutation-selection Theory. Exper. Cell Res., **8:** 188, 1955.
714. Klein, E.: Gradual Transformation of Solid into Ascites Tumors. Permanent Difference between the Original and the Transformed Sublines. Cancer Res., **14:** 482, 1954.
715. Klein, E.: Isoantigenicity of X-ray-inactivated Implants of a Homotransplantable and Nonhomotransplantable Mouse Sarcoma. Transplant. Bull., **6:** 420, 1959.
716. Klein, E., G. Klein, and K. E. Hellström: Further Studies on Isoantigenic Variation in Mouse Carcinomas and Sarcomas. J. Nat. Cancer Inst., **25:** 271, 1960.
717. Klein, E., G. Klein, and L. Revesz: Permanent Modification (Mutation?) of a Histocompatibility Gene in a Heterozygous Tumor. J. Nat. Cancer Inst., **19:** 95, 1957.
718. Klein, G.: Cancer Studies in Fields Collateral to Tissue Culture. J. Nat. Cancer Inst., **19:** 795, 1957.
719. Klein, G.: The Usefulness and Limitations of Tumor Transplantation in Cancer Research: A Review. Cancer Res., **19:** 343, 1959.
720. Klein, G.: Variation and Selection in Tumor Cell Populations. Canadian Cancer Conf. **3:** 215, 1959.
721. Klein, G., and E. Klein: Conversion of Solid Neoplasms into Ascites Tumors. Ann. N.Y. Acad. Sci., **63:** 640, 1956.
722. Klein, G., and E. Klein: Cytogenetics of Experimental Tumors. *In* Genetics and Cancer. R. W. Cumley, *et al.*, eds., Univ. of Texas Press, Austin, 241, 1959.
723. Klein, G., and E. Klein: Detection of an Allelic Difference at a Single Locus in a Small Fraction of a Large-tumour-cell Population. Nature, **178:** 1389, 1956.
724. Klein, G., and E. Klein: Histocompatibility Changes in Tumors. J. Cell. Comp. Physiol., **52** Suppl. 1: 125, 1958.
725. Klein, G., and E. Klein: The Evolution of Independence from Specific Growth Stimulation and Inhibition in Mammalian Tumor-cell Populations. Symp. Soc. Exper. Biol., **11:** 305, 1957.
726. Klein, G., and E. Klein: Variation in Cell Populations of Transplanted Tumors as Indicated by Studies on the Ascites Transformation. Exper. Cell Res., Suppl., **3:** 218, 1955.
727. Klein, G., H. O. Sjögren, E. Klein, and K. E. Hellström: Demonstration of Resistance Against Methylcholanthrene Induced Sarcomas in the Primary Autochthonous Host. Cancer Res., **20:** 1561, 1960.
728. Klontz, G. W., G. J. Ridgway, and G. P. Wilson: An Illuminator for Observing and Photographing Precipitin Reactions in Agar. J. Biol. Photog. Assoc., **28:** 11, 1960.
729. Kohn, H. I.: The Effect of Paternal X-ray Exposure on the Secondary Sex Ratio in Mice (F_1 Generation). Genetics, **45:** 771, 1960.
730. Kokorina, E. P.: Higher Nervous Activity and Milk Production in Cattle. Zhurnal Obsch'chei Biol., **19:** 148, 1958.
731. Koller, P. C.: Segmental Interchange in Mice. Genetics, **29:** 247, 1944.
732. Körber, E.: Über Differenzen des Blutfarbstoffes. Inaug. Dissertation, Dorpat. 1866, cited by F. von Krüger: Vergleichende Untersuchungen über die Resistenz des Hämoglobins verschiedener Tiere. Zeit. Vergl. Physiol., **2:** 254, 1925.
733. Kreyberg, L.: The Origin and Development of the "White Label" Mouse Strains. Brit. J. Cancer, **6:** 140, 1952.

734. Krohn, P. L.: Litters from C3H and CBA Ovaries Orthotopically Transplanted into Tolerant A Strain Mice.   Nature, **181**: 1671, 1958.
735. Kröning, F.: Die Dopareaktion bei Verschiedenen Farbenrassen des Meerschweinchens und des Kaninchens.   Wilhelm Roux' Arch. Entwicklungsmechanik der Organismen, **121**: 470, 1930.
736. Kunkel, H. G.: Zone Electrophoresis and the Minor Hemoglobin Components of Normal Human Blood.   Conf. on Hemoglobin.   National Academy of Sciences, National Research Council, Pub. **557**: 157, 1958.
737. Kunkel, H. G.: Zone Electrophoresis.   *In* Methods of Biochemical Analysis.   D. Glick, ed., Interscience Pub., Inc., New York, **1**: 141, 1954.
738. Kunkel, H. G., R. Ceppellini, U. Muller-Eberhard, and J. Wolf: Observations on the Minor Basic Hemoglobin Component in the Blood of Normal Individuals and Patients with Thalassemia.   J. Clin. Invest., **36**: 1615, 1957.
739. Kunkel, H. G., and R. J. Slater: Zone Electrophoresis in a Starch Supporting Medium.   Proc. Soc. Exper. Biol. Med., **80**: 42, 1952.
740. Kunkel, H. G., and G. Wallenius: New Hemoglobin in Human Adult Blood.   Science, **122**: 288, 1955.
741. Kunze, H.: Die Erythropoese bei einer erblichen Anämie röntgenmutierter Mause.   Folia Haematol., **72**: 392, 1954.
742. Laboratory Animals Centre, Medical Research Council: Catalogue of Uniform Strains of Laboratory Animals Maintained in Great Britain, Second ed., 1958.
743. Lacassagne, A., and G. Gricouroff: Action of Radiation on Tissues: An Introduction to Radio- therapy.   Grune and Stratton, Inc., New York, 1958.
744. Lalezari, P., M. Nussbaum, S. Gelman, and T. H. Spaet: Neonatal Neutropenia due to Maternal Isoimmunization.   Blood, **15**: 236, 1960.
745. Lambert, W. V.: Silver Guinea Pigs.   A Recessive Color Variation.   J. Hered., **26**: 279, 1935.
746. Lambert, W. V., and C. W. Knox: The Inheritance of Resistance to Fowl Typhoid in Chickens.   Iowa State Coll. J. Sci., **2**: 179, 1928.
747. Landauer, W.: On the Chemical Production of Developmental Abnormalities and of Pheno- copies in Chicken Embryos.   J. Cell. Comp. Physiol., **43** Suppl. 1: 261, 1954.
748. Landauer, W.: Phenocopies and Genotype with Special Reference to Sporadically-occurring Developmental Variants.   Amer. Nat., **81**: 79, 1957.
749. Landsteiner, K.: The Specificity of Serological Reactions.   Second ed.   Harvard Univ. Press, Cambridge, Mass., 1945.
750. Landsteiner, K., and P. Levine: On Group Specific Substances in Human Spermatozoa.   J. Immunol., **12**: 415, 1926.
751. Lane, P. W.: Lists of Mutant Genes and Mutant-bearing Stocks of the Mouse.   Mimeographed, Prod. Dept., Jackson Mem. Lab., 1962.
752. Lane, P. W.: Modern Ways of Feeding Laboratory Animals.   Proc. Nutrition Soc., **16**: 59, 1957.
753. Lane, P. W., and M. M. Dickie: Linkage of Warbler-lethal and Hairless in the Mouse.   J. Hered., **52**: 159, 1961.
754. Lane, P. W.: Mouse News Letter, **23**: 36, 1960.
755. Lane, P. W.: personal communication.
756. Lane, P. W.: The Pituitary-gonad Response of Genetically Obese Mice in Parabiosis with Thin and Obese Siblings.   Endocrinology, **65**: 863, 1959.
757. Lane, P. W., and M. C. Green: Mahogany, a Recessive Color Mutation in Linkage Group V of the Mouse.   J. Hered., **51**: 228, 1960.
758. Lapp, R. E., and H. L. Andrews: Nuclear Radiation Physics.   Second ed., Prentice-Hall, New York, 1954.
759. Larson, C. D., and W. E. Heston: Effects of Cystine and Calorie Restriction on the Incidence of Spontaneous Pulmonary Tumors in Strain A Mice.   J. Nat. Cancer Inst., **6**: 31, 1945.
760. Larson, D. L., and H. M. Ranney: Filter Paper Electrophoresis of Human Hemoglobin.   J. Clin. Invest., **32**: 1070, 1953.
761. Law, L. W.: Differences between Cancers in Terms of Evolution of Drug Resistance.   Cancer Res., **16**: 698, 1956.
762. Law, L. W.: Genetic Studies in Experimental Cancer.   Advances Cancer Res., **2**: 281, 1954.
763. Law, L. W.: Maternal Influence in Experimental Leukemia in Mice.   Ann. N.Y. Acad. Sci., **57**: 575, 1954.

764. Law, L. W.: Mouse Genetics News No. 2.   J. Hered., **39**: 300, 1948.

765. Law, L. W.: Present Status of Nonviral Factors in the Etiology of Reticular Neoplasms of the Mouse.   Ann. N.Y. Acad. Sci., **68**: 616, 1957.

766. Law, L. W.: Origin of the Resistance of Leukaemic Cells to Folic Acid Antagonists.   Nature, **169**: 628, 1952.

767. Law, L. W.: Some Aspects of the Etiology of Leukemia.   Proc. Third Canadian Cancer Conf., Academic Press, New York, 145, 1959.

768. Law, L. W.: The Flexed-tail-anemia Gene (*f*) and Induced Leukemia in Mice.   J. Nat. Cancer Inst., **12**: 1119, 1952.

769. Law, L. W., and M. Potter: The Behavior in Transplant of Lymphocytic Neoplasms Arising from Parental Thymic Grafts in Irradiated, Thymectomized, Hybrid Mice.   Proc. Nat. Acad. Sci. (Wash.), **42**: 160, 1956.

770. Lawrence, H. S., ed.: Cellular and Humoral Aspects of the Hypersensitive States.   Hoeber-Harper, New York, 1959.

771. LeClerc, G.: Occurrence of Mitotic Crossing-over without Meiotic Crossing-over.   Science, **103**: 553, 1946.

772. Lederberg, J.: Genes and Antibodies.   Science, **129**: 1649, 1959.

773. Lederberg, J.: Genetic Approaches to Somatic Cell Variation: Summary Comment.   J. Cell. Comp. Physiol., **52** Suppl. 1: 383, 1958.

774. Lederberg, J.: Prospects for a Genetics of Somatic and Tumor Cells.   Ann. N.Y. Acad. Sci., **63**: 662, 1956.

775. Lederberg, J.: Viruses, Genes and Cells.   Bacteriol. Rev., **21**: 133, 1957.

776. Lederberg, J., and E. M. Lederberg: Replica Plating and Indirect Selection of Bacterial Mutants.   J. Bacteriol., **63**: 399, 1952.

777. Lederer, E., and M. Lederer: Chromatography.   Second ed.   Elsevier Pub. Co., New York, 1957.

778. Lefevre, G., Jr.: X-ray Induced Genetic Effects in Germinal and Somatic Tissue of *Drosophila melanogaster*.   Amer. Nat., **84**: 341, 1950.

779. Levan, A.: Chromosome Studies on Some Human Tumors and Tissues of Normal Origin Grown *in vivo* and *in vitro* at the Sloan-Kettering Institute.   Cancer, **9**: 648, 1956.

780. Levan, A.: Colchicine-induced C-Mitosis in Two Mouse Ascites Tumors.   Hereditas, **40**: 1, 1954.

781. Levan, A.: Relation of Chromosome Status to the Origin and Progression of Tumors: The Evidence of Chromosome Numbers.   *In* Genetics and Cancer.   R. W. Cumley, *et al.*, eds., Univ. of Texas Press, Austin, 151, 1959.

782. Levan, A., and J. J. Biesele: Role of Chromosomes in Carcinogenesis, as Studied in Serial Tissue Culture of Mammalian Cells.   Ann. N.Y. Acad. Sci., **71**: 1022, 1958.

783. Levan, A., and T. S. Hauschka: Chromosome Numbers of Three Mouse Ascites Tumors.   Hereditas, **38**: 251, 1952.

784. Levine, P., F. Ottensooser, M. J. Celano, and W. Pollitzer: On Reactions of Plant Anti-*N* with Red Cells of Chimpanzees and Other Animals.   Amer. J. Phys. Anthropol., **13**: 29, 1955.

785. Levitt, M., and H. S. Rhinesmith: A Quantitative Study of the Number of N-terminal Amino Acid Residues and the Number and Kind of N-terminal Peptides in Horse Hemoglobin.   J. Amer. Chem. Soc., **82**: 975, 1960.

786. Lewis, E. B.: The Phenomenon of Position Effect.   Advances in Genet., **3**: 73, 1950.

787. Lieberman, I., and P. Ove: Estimation of Mutation Rates with Mammalian Cells in Culture.   Proc. Nat. Acad. Sci. (Wash.), **45**: 872, 1959.

788. Lima-de-Faria, A.: Incorporation of Tritiated Thymidine into Meiotic Chromosomes.   Science, **130**: 503, 1959.

789. Linder, O., and E. Klein: Skin and Tumor Grafting in Coisogenic Resistant Lines of Mice and Their Hybrids.   J. Nat. Cancer Inst., **24**: 707, 1960.

790. Lippincott, S. W., J. E. Edwards, H. G. Grady, and H. L. Stewart: A Review of Spontaneous Neoplasms in Mice.   J. Nat. Cancer Inst., **3**: 199, 1942.

791. Little, C. C.: A Possible Mendelian Explanation for a Type of Inheritance Apparently Non-Mendelian in Nature.   Science, **40**: 904, 1914.

792. Little, C. C.: Coat Color Genes in Rodents and Carnivores.   Quart. Rev. Biol., **33**: 103, 1958.

793. Little, C. C.: The Genetics of Spontaneous Tumor Formations.   *In* The Biology of the Laboratory Mouse.   G. D. Snell, ed., The Blakiston Co., New York, 248, 1956.

794. Little, C. C.: The Genetics of Tumor Transplantation.   *In* The Biology of the Laboratory Mouse.   G. D. Snell, ed., The Blakiston Co., New York, 279, 1956.

795. Little, C. C.: The Heredity of Susceptibility to a Transplantable Sarcoma (J. W. B.) of the Japanese Waltzing Mouse.   Science, **51**: 467, 1920.
796. Little, C. C.: The Inheritance of Black-eyed White Spotting in Mice.   Amer. Nat., **49**: 727, 1915.
797. Little, C. C.: The Inheritance of Coat Color in Dogs.   Comstock Publishing Associates, Ithaca, N.Y., 1957.
798. Little, C. C.: The Relation of Coat Color to the Spontaneous Incidence of Mammary Tumors in Mice.   J. Exper. Med., **59**: 229, 1934.
799. Little, C. C.: The Relation of Yellow Coat Color and Black-eyed White Spotting of Mice in Inheritance.   Genetics, **2**: 433, 1917.
800. Little, C. C., and H. J. Bagg: The Occurrence of Four Heritable Morphological Variations in Mice and Their Possible Relation to Treatment with X-rays.   J. Exper. Zool., **41**: 45, 1924.
801. Little, C. C., and H. J. Bagg: The Occurrence of Two Heritable Types of Abnormality among the Descendants of X-rayed Mice.   Amer. J. Roentgenol., **10**: 975, 1923.
802. Little, C. C., and E. E. Tyzzer: Further Studies on Inheritance of Susceptibility to a Transplantable Tumor of Japanese Waltzing Mice.   J. Med. Res., **33**: 393, 1916.
803. Longsworth, L. G.: Electrophoresis.   *In* Methods in Medical Research.   A. C. Corcoran, ed., The Year Book Pub., Inc., Chicago, **5**: 63, 1952.
804. Longsworth, L. G.: Recent Advances in the Study of Proteins by Electrophoresis.   Chem. Rev., **30**: 323, 1942.
805. Loosli, R., E. S. Russell, W. K. Silvers, and J. L. Southard: Variability of Incidence and Clinical Manifestation of Mouse Hereditary Muscular Dystrophy on Heterogeneous Genetic Backgrounds.   Genetics, **46**: 347, 1961.
806. Löw, B.: A Practical Method Using Papain and Incomplete Rh-Antibodies in Routine Rh Blood-Grouping.   Vox Sang., **5**: 94, 1955.
807. Lowy, P. H., G. Keighley, H. Borsook, and A. Graybiel: On the Erythropoietic Principle in the Blood of Rabbits Made Severely Anemic with Phenylhydrazine.   Blood, **14**: 262, 1959.
808. Lucas, D. R.: Inherited Retinal Dystrophy in the Mouse: Its Appearance in Eyes and Retinae Cultured *in vitro*.   J. Embryol. Exper. Morphol., **6**: 589, 1958.
809. Luria, S. E.: Recent Advances in Bacterial Genetics.   Bacteriol. Rev., **11**: 1, 1947.
810. Luria, S. E.: Viruses as Determinants of Cellular Functions.   Canadian Cancer Conf., **3**: 261, 1959.
811. Luria, S. E.: Viruses, Cancer Cells, and the Genetic Concept of Virus Infection.   Symp. on Possible Role of Viruses in Cancer.   Cancer Res., **20**: 677, 1960.
812. Luria, S. E., and M. Delbruck: Mutations of Bacteria from Virus Sensitivity to Virus Resistance.   Genetics, **28**: 491, 1943.
813. Lynch, C. J.: Influence of Heredity and Environment upon Number of Tumor Nodules Occurring in Lungs of Mice.   Proc. Soc. Exper. Biol. Med., **43**: 186, 1940.
814. Lynch, C. J.: Studies on the Relation Between Tumor Susceptibility and Heredity. III. Spontaneous Tumors of the Lung in Mice.   J. Exper. Med., **43**: 339, 1926.
815. Lynch, C. J.: The RIL Strain of Mice: Its Relation to the Leukemic AK Strain and AKR Substrains.   J. Nat. Cancer Inst., **15**: 161, 1954.
816. Lyon, M. F.: Hereditary Hair Loss in the Tufted Mutant of the House Mouse.   J. Hered., **47**: 101, 1956.
817. Lyon, M. F.: Linkage Relations and Some Pleiotropic Effects of the Dreher Mutant in the House Mouse.   Genet. Res., **2**: 122, 1961.
818. Lyon, M. F.: Twirler: A Mutant Affecting the Inner Ear of the House Mouse.   J. Embryol. Exper. Morphol., **6**: 105, 1958.
819. Lyon, M. F., and R. J. S. Phillips: Crossing-over in Mice Heterozygous for *t*-Alleles.   Heredity, **13**: 23, 1959.
820. McClearn, G. E., and D. A. Rodgers: Genetic Factors in Alcohol Preference of Laboratory Mice. J. Comp. Physiol. Psychol., **54**: 116, 1961.
821. McClintock, B.: Controlling Elements and the Gene.   Cold Spring Harbor Symp. Quant. Biol., **21**: 197, 1956.
822. McDonald, H. J.: Ionography; Electrophoresis in Stabilized Media.   Year Book Pub., Chicago, 1955.
823. McGovern, J. J., A. R. Jones, and A. G. Steinberg: The Hematocrit of Capillary Blood.   New Eng. J. Med., **253**: 308, 1955.
824. McGregor, J. F., A. P. James, and H. B. Newcombe: Mutation as a Cause of Death in Offspring of Irradiated Rats.   Radiat. Res., **12**: 61, 1960.

825. McLaren, A. and D. Michie: Factors Affecting Vertebral Variation in Mice. II. Further Evidence on Intra-strain Variation. J. Embryol. Exper. Morphol., **3**: 366, 1955.
826. McLaren, A., and D. Michie: Factors Affecting Vertebral Variation in Mice. IV. Experimental Proof of the Uterine Basis of a Maternal Effect. J. Embryol. Exper. Morphol., **6**: 645, 1958.
827. McLaren, A., and D. Michie: Studies on the Transfer of Fertilized Mouse Eggs to Uterine Foster-mothers. I. Factors Affecting the Implantation and Survival of Native and Transferred Eggs. J. Exper. Biol., **33**: 394, 1956.
828. McLaren, A., and D. Michie: Studies on the Transfer of Fertilized Mouse Eggs to Uterine Foster-mothers. II. The Effect of Transferring Large Numbers of Eggs. J. Exper. Biol., **36**: 40, 1959.
829. McNeil, C., E. F. Trentelman, N. P. Sullivan, and C. I. Argall: A New Rapid *Rh* Tube Test Using Polyvinylpyrrolidone (PVP). Amer. J. Clin. Pathol., **22**: 1216, 1952.
830. MacArthur, J. W.: Genetics of Body Size and Related Characters. I. Selecting Small and Large Races of the Laboratory Mouse. Amer. Nat., **78**: 142, 1944.
831. MacArthur, J. W.: Selection for Small and Large Body Size in the House Mouse. Genetics, **34**: 194, 1949.
832. MacDowell, E. C.: Genetic Aspects of Mouse Leukemia. Amer. J. Cancer, **26**: 85, 1936.
833. MacDowell, E. C.: "Light"—a New Mouse Color. J. Hered., **41**: 35, 1950.
834. MacDowell, E. C., E. Allen, and C. G. MacDowell: The Prenatal Growth of the Mouse. J. Gen. Physiol., **11**: 57, 1927.
835. MacDowell, E. C., J. S. Potter, and M. J. Taylor: Mouse Leukemia. XII. The Role of Genes in Spontaneous Cases. Cancer Res., **5**: 65, 1945.
836. MacDowell, E. C., and M. N. Richter: Mouse Leukemia. IX. The Role of Heredity in Spontaneous Cases. Arch. Pathol., **20**: 709, 1935.
837. MacDowell, E. C., and M. J. Taylor: Mouse Leukemia. XIII. A Maternal Influence that Lowers the Incidence of Spontaneous Cases. Proc. Soc. Exper. Biol. Med., **68**: 571, 1948.
838. Mackensen, J. A.: Mouse News Letter, **24**: 41, 1961.
839. Maisin, J., H. Maisin, A. Dunjic, and P. Maldague: Cellular and Histological Radiolesions. Their Consequences and Repair. Proc. Internat. Conf. on Peaceful Uses of Atomic Energy, **11**: 315, 1956.
840. Makino, S.: A Cytological Study of the Yoshida Sarcoma, an Ascites Tumor of White Rats. Chromosoma, **4**: 649, 1952.
841. Makino, S.: The Chromosome Cytology of the Ascites Tumors of Rats, with Special Reference to the Concept of the Stemline Cell. Internat. Rev. Cytol., **6**: 25, 1957.
842. Makino, S., and M. Sasaki: Cytological Studies of Tumors. XXI. A Comparative Ideogram Study of the Yoshida Sarcoma and its Subline Derivatives. J. Nat. Cancer Inst., **20**: 465, 1958.
843. Makinodan, T., and N. G. Anderson: Physicochemical Properties of Circulating Red Blood Cells of Lethally X-irradiated Mice Treated with Rat Bone Marrow. Blood, **12**: 984, 1957.
844. Maliugina, L. L., and O. G. Prokof'eva: Oncological Characteristics of Mice of Strain C3HA. Prob. Oncolog., (Eng. translation) **3**: 201, 1957.
845. Mallyon, S. A.: A Pronounced Sex Difference in Recombination Values in the Sixth Chromosome of the House Mouse. Nature, **168**: 118, 1951.
846. Mandl, A. M.: The Effect of Beta-mercaptoethylamine on the Sensitivity of Oocytes to X Irradiation. Proc. Roy. Soc., ser. B, **150**: 72, 1959.
847. Mandl, A. M.: The Effect of Cysteamine on the Survival of Spermatogonia after X Irradiation. Internat. J. Radiat. Biol., **1**: 131, 1959.
848. Mange, A. P., and W. H. Stone: A Spectrophotometric Technic for Measuring Erythrocyte Chimerism in Cattle. Proc. Soc. Exper. Biol. Med., **102**: 107, 1959.
849. Marglis, D. S.: Studies on the Bar Series of *Drosophila*. V. The Effects of Reduced Atmospheric Pressure and Oxygen on Facet Number in Bar-eyed *Drosophila*. Genetics, **24**: 15, 1939.
850. Marinelli, L. D., and L. S. Taylor: The Measurement of Ionizing Radiations for Biological Purposes. *In* Radiation Biology. A. Hollaender, ed., McGraw-Hill, New York, **1**: 145, 1954.
851. Market, C. L.: Biochemical Embryology and Genetics. Symp. on Normal and Abnormal Differentiation and Development. Nat. Cancer Inst. Monogr., **2**: 3, 1960.
852. Markert, C. L.: Substrate Utilization in Cell Differentiation. Ann. N.Y. Acad. Sci., **60**: 1003, 1955.

853. Markert, C. L., and W. K. Silvers: Effects of Genotype and Cellular Environment on Melano-cyte Morphology. *In* Pigment Cell Biology. M. Gordon, ed., Academic Press, New York, 241, 1959.

854. Markert, C. L., and W. K. Silvers: The Effects of Genotype and Cell Environment on the Melanoblast Differentiation in the House Mouse. Genetics, **41**: 429, 1956.

855. Marshall, G. J., E. M. Wood, L. Gingras, and H. R. Bierman: Hyaluronidase as an Effective Agent for Separating Chromosomes in Normal and Neoplastic Tissues. Proc. Amer. Assoc. Cancer Res., **3**: 131, 1960.

856. Marsi, M. S., A. M. Josephson, and K. Singer: Starch-block Electrophoretic Studies of Human Hemoglobin Solutions. I. Technic and Results in the Normal Adult. Blood, **13**: 533, 1958.

857. Masouredis, S. P.: Rh_0 (D) Genotype and Red Cell Rh_0 (D) Antigen Content. Science, **131**: 1442, 1960.

858. Mather, K.: The Measurement of Linkage in Heredity, Second ed. J. Wiley and Sons, New York, 1951.

859. Mattern, C. F. T., F. S. Brackett, and B. J. Olson: Determination of Number and Size of Particles by Electrical Gating: Blood Cells. J. Appl. Physiol., **10**: 56, 1957.

860. Mayr, E.: Where Are We? Cold Spring Harbor Symp. Quant. Biol., **24**: 1, 1959.

861. Medawar, P. B.: Reactions to Homologous Tissue Antigens in Relation to Hypersensitivity. *In* Cellular and Humoral Aspects of the Hypersensitive States. H. S. Lawrence, ed., Hoeber-Harper, New York, 504, 1959.

862. Medawar, P. B.: The Immunology of Transplantation. Harvey Lectures, **52**: 144, 1957.

863. Medvedev, N. N.: K Onkologicheskoi Kharakteristike Nizkorakovykh Myshei CC-57-Korichnevye. (On the Oncological Characteristic of the Low-tumorous CC-57-Brown Stock Mice.) Bull. Moskov. Obshchestva Ispytatelei Prirody, **62**: 63, 1957.

864. Medvedev, N. N.: O lineinykh myshakh CC-57-belye. (The CC-57-White Mice Lines.) Doklady Akad. Nauk USSR, **119**: 369, 1958.

865. Meier, H., and G. Yerganian: Spontaneous Hereditary Diabetes Mellitus in Chinese Hamster (*Cricetulus griseus*). I. Pathological Findings. Proc. Soc. Exper. Biol. Med., **100**: 810, 1959.

866. Meier, H., and G. Yerganian: Spontaneous Diabetes Mellitus in the Chinese Hamster (*Cricetulus griseus*). II. Findings in the Offspring of Diabetic Parents. Diabetes, **10**: 12, 1960.

867. Men and Mice at Edinburgh: Reports from the Genetics Congress. J. Hered., **30**: 371, 1939.

868. Merwin, R. M., and E. L. Hill: Fate of Vascularized and Nonvascularized Subcutaneous Homografts in Mice. J. Nat. Cancer Inst., **14**: 819, 1954.

869. Michelson, A. M., E. S. Russell, and P. S. Harmon: *Dystrophia muscularis*: A Hereditary Primary Myopathy in the House Mouse. Proc. Nat. Acad. Sci. (Wash.), **41**: 1079, 1955.

870. Michie, D.: A New Linkage in the House Mouse: Vestigial and Rex. Nature, **170**: 585, 1952.

871. Michie, D.: Genetical Studies with "Vestigial Tail" Mice. I. The Sex Difference in Crossing-over Between Vestigial and Rex. J. Genet., **53**: 270, 1955.

872. Michie, D.: Genetical Studies with "Vestigial Tail" Mice. II. The Position of Vestigial in the Seventh Linkage Group. J. Genet., **53**: 280, 1955.

873. Michie, D., and A. McLaren: Control of Pre-natal Growth in Mammals. Nature, **187**: 363, 1960.

874. Midgley, A. R., B. Pierce, and F. J. Dixon: Nature of Colchicine Resistance in Golden Hamster. Science, **130**: 40, 1959.

875. Miquel, J., B. Horvath, and I. Klatzo: A Chromatographic Technique for Quantitative Study of the Preciptin Reaction. J. Immunol., **84**: 545, 1960.

876. Miles, C. P., and A. S. Koons: Sexual Dimorphism of Rat Cells *in vitro*. Science, **131**: 740, 1960.

877. Miller, D. S., and M. Z. Potas: Cordovan, a New Allele of Black and Brown Color in the Mouse. J. Hered., **46**: 293, 1955.

878. Miller, J. R.: Clinical and Experimental Studies on the Etiology of Skull, Vertebra, Rib and Palate Malformations. Ph.D. thesis, McGill Univ., May 1959.

879. Mintz, B.: Embryological Development of Primordial Germ-cells in the Mouse: Influence of a New Mutation W^j. J. Embryol. Exper. Morphol., **5**: 396, 1957.

880. Mintz, B., and E. S. Russell: Gene-induced Embryological Modifications of Primordial Germ Cells in the Mouse. J. Exper. Zool., **134**: 207, 1957.

881. Mintz, B., and E. Wolff: The Development of Embryonic Chick Ovarian Medulla and Its Feminizing Action in Intracoelomic Grafts. J. Exper. Zool., **126**: 511, 1954.

882. Mitchell, H. K., and L. A. Herzenberg: Zone Electrophoresis on Sponge Rubber. Anal. Chem., **29**: 1229, 1957.

883. Mitchison, N. A.: Antigens of Heterozygous Tumours as Material for the Study of Cell Heredity. Proc. Roy. Phys. Soc., **250:** 45, 1956.
884. Mitchison, N. A.: Tissue Transplantation and Cellular Heredity. Symp. Soc. Exper. Biol., **12:** 255, 1958.
885. Mixter, R., and H. R. Hunt: Anemia in the Flexed Tailed Mouse, *Mus musculus.* Genetics, **18:** 367, 1933.
886. Mole, R. H.: Impairment of Fertility by Whole Body Irradiation of Female Mice. Internat. J. Radiat. Biol., **1:** 107, 1959.
887. Möller, G.: Studies on the Development of the Isoantigens of the H-2 System in Newborn Mice. J. Immunol., **86:** 56, 1961.
888. Mollison, P. C.: Measurement of Survival and Destruction of Red Cells in Haemolytic Syndromes. Brit. Med. Bull., **15:** 59, 1959.
889. Monie, I. W., M. M. Nelson, and H. M. Evans: Abnormalities of the Urinary System of Rat Embryos Resulting from Transitory Deficiency of Pteroylglutamic Acid during Gestation. Anat. Rec., **127:** 711, 1957.
890. Moore, S., and W. H. Stein: Chromatography of Amino Acids on Sulfonated Polystyrene Resins. J. Biol. Chem., **192:** 663, 1951.
891. Moore, S., and W. H. Stein: Column Chromatography of Peptides and Proteins. *In* Advances in Protein Chemistry. M. L. Anson, K. Bailey, and J. T. Edsall, eds., Academic Press, New York, **11:** 191, 1956.
892. Moore, S., and W. H. Stein: Procedures for the Chromatographic Determination of Amino Acids on Four Per Cent Cross-linked Sulfonated Polystyrene Resins. J. Biol. Chem., **211:** 893, 1954.
893. Moorhead, P., P. C. Nowell, W. J. Mellman, D. M. Batipps, and D. A. Hungerford: Chromosome Preparations of Leukocytes Cultured from Human Peripheral Blood. Exper. Cell Res., **20:** 613, 1960.
894. Mordkoff, A. M., and J. L. Fuller: Variability in Activity within Inbred and Crossbred Mice: a Study in Behavior Genetics. J. Hered., **50:** 6, 1959.
895. Morgan, J. F., H. J. Morton, and R. C. Parker: Nutrition of Animal Cells in Tissue Culture. Initial Studies on a Synthetic Medium. Proc. Soc. Exper. Biol. Med., **73:** 1, 1950.
896. Morgan, T. H.: The Scientific Basis of Evolution. W. W. Norton and Co., New York, 1932.
897. Morris, H. P.: Diet and Some Other Environmental Influences in the Genesis and Growth of Mammary Tumors in Mice. AAAS Pub. **22:** 140, 1945.
898. Morris, H. P.: Review of the Nutritive Requirements of Normal Mice for Growth, Maintenance, Production, and Lactation. J. Nat. Cancer Inst., **5:** 115, 1944.
899. Morse, M. L., E. M. Lederberg, and J. Lederberg: Transduction in *Escherichia coli* K-12. Genetics, **41:** 142, 1956.
900. Morton, J. A., and M. M. Pickles: The Proteolytic Enzyme Test for Detecting Incomplete Antibodies. J. Clin. Pathol., **4:** 189, 1951.
901. Morton, N. E., J. F. Crow, and H. J. Muller: An Estimate of the Mutational Damage in Man from Data on Consanguineous Marriages. Proc. Nat. Acad. Sci. (Wash.), **42:** 855, 1956.
902. Mouriquand, J., C. Mouriquand, and J. Petat: Premières Observations à propos d'une Nouvelle Souche de Souris Hautement Cancérigene. C. R. Soc. Biol., **154:** 632, 1960.
903. Mouse News Letter. (An informal semi-annual mimeographed publication distributed by the Laboratory Animals Centre, Woodmansterne Road, Carshalton, Surrey, England, ed., M. C. Lyon.)
904. Mouse News Letter, **5:** 12, 1951.
905. Mouse News Letter, **9:** 1, 1953.
906. Mouse News Letter, **12:** 29, 1955.
907. Mouse News Letter, **15,** 1956.
908. Mouse News Letter, **17:** 54, 1957.
909. Mouse News Letter, **17:** 56, 1957.
910. Mouse News Letter, **17:** 84, 1957.
911. Mouse News Letter, **17:** 92, 1957.
912. Mouse News Letter, **18:** 5, 1958.
913. Mouse News Letter, **23:** 36, 1960.
914. Moyer, F.: Some Effects of Pigment Mutations on the Fine Structure of Mouse Melanin Granules. J. Exper. Zool., **138:** 372, 1960.
915. Mühlbock, O.: On the Susceptibility of Different Inbred Strains of Mice for Estrone. Acta Brev. Neerl., **15:** 18, 1947.

916. Mühlbock, O., and L. M. Boot: Induction of Mammary Cancer in Mice without the Mammary Tumor Agent by Isografts of Hypophyses. Cancer Res., **19**: 402, 1959.

917. Mühlbock, O., and L. M. Boot: The Mechanisms of Hormonal Carcinogenesis. Ciba Foundation Symp. on Carcinogenesis. G. E. W. Wolstenholme, and M. O'Connor, eds., Little, Brown & Co., Boston, Mass., and Churchill, London, 83, 1959.

918. Muller, C. J.: Separation of the α- and β-Chains of Globins by Means of Starch-gel Electrophoresis. Nature, **186**: 643, 1960.

919. Munoz, J.: Production in Mice of Large Volumes of Ascites Fluid Containing Antibodies. Proc. Soc. Exper. Biol. Med., **95**: 757, 1957.

920. Munro, S. S., I. L. Kosin, and E. L. Macartney: Quantitative Genic-hormone Inter-actions in the Fowl. I. Relative Sensitivity of Five Breeds to an Anterior Pituitary Extract Possessing both Thyrotropic and Gonadotrophic Properties. Amer. Nat., **77**: 256, 1943.

921. Murayama, M.: Titratable Sulfhydryl Groups of Horse, Sheep, Dog, and Cow Hemoglobins at 0° and 38°. J. Biol. Chem., **233**: 594, 1958.

922. Murayama, M., and V. M. Ingram: Comparison of Normal Adult Human Haemoglobin with Haemoglobin I by "Fingerprinting." Nature, **183**: 1798, 1959.

923. Murphy, M. L.: Teratogenic Effects of Tumor-inhibiting Chemicals in the Foetal Rat. Ciba Foundation Symp. on Congenital Malformations. G. E. W. Wolstenholme, and C. M. O'Connor, eds., Little, Brown & Co., Boston, Mass., 78, 1960.

924. Murray, J. M.: "Leaden", a Recent Coat Color Mutation in the House Mouse. Amer. Nat., **67**: 278, 1933.

925. Murray, J. M., and C. V. Green: Inheritance of Ventral Spotting in Mice. Genetics, **18**: 481, 1933.

926. Murray, J. M., and G. D. Snell: Belted, a New Sixth Chromosome Mutation in the Mouse. J. Hered., **36**: 266, 1945.

927. Murray, W. S.: Genetic Segregation Mammary Cancer to No Mammary Cancer in the Mouse. Amer. J. Cancer, **34**: 434, 1938.

928. Murray, W. S.: The Breeding Behavior of the Dilute Brown Stock of Mice (Little dba). Amer. J. Cancer, **20**: 573, 1934.

929. Nachtsheim, H.: Krampfbereitschaft und Genotypus; die Epilipsie der Seissen Wiener-Kaninchen. Zeit. f. Menschl. Vererb. u. Konst. Lehre, **22**: 791, 1939.

930. Nalbandov, A. V.: Reproductive Physiology. W. H. Freeman and Co., San Francisco, 1958.

931. Nalbandov, A. V., and L. E. Card: Endocrine Identification of the Broody Genotype of Cocks. J. Hered., **36**: 34, 1945.

932. Nandi, S., H. A. Bern, and K. B. DeOme: Attempts to Delineate the Hormone Requirements for Tumorigenesis from Hyperplastic Mammary Nodules Transplanted into Mammary Gland-Free Fat Pads in C3H/Crgl Mice. Mem. Soc. Endocrinol., **10**: 129, 1961.

933. Nandi, S., H. A. Bern, and K. B. DeOme: Effect of Hormones on Growth and Neoplastic Development of Transplanted Hyperplastic Alveolar Nodules of the Mammary Gland of C3H/Crgl Mice. J. Nat. Cancer Inst., **24**: 883, 1960.

934. Nanney, D. L.: Epigenetic Control Systems. Proc. Nat. Acad. Sci. (Wash.), **44**: 712, 1958.

935. Nanney, D. L.: Microbiology, Developmental Genetics and Evolution. Amer. Nat., **94**: 167, 1960.

936. Nasrat, G. E.: Estimation of the Recombination Fraction between the Two Linked Genes *Re* and *sh*-2 in the House Mouse When the Female is the Heterozygous Parent. Proc. Zool. Soc. (Bengal), **9**: 85, 1956.

937. Neary, G. J., R. J. Munson, and R. H. Mole: Chronic Radiation Hazards: An Experimental Study with Fast Neutrons. Pergamon Press, New York, 1957.

938. Neel, J. V.: Genetic Aspects of Abnormal Hemoglobins. Conf. on Hemoglobins. National Academy of Sciences, **557**: 253, 1958.

939. Neel, J. V., and W. J. Schull: The Effect of Exposure to the Atomic Bombs on Pregnancy Termination in Hiroshima and Nagasaki. National Academy of Sciences, National Research Council, **461,** 1956.

940. Nelson, M. M.: Teratogenic Effects of Pteroylglutamic Acid Deficiency in the Rat. Ciba Foundation Symp. on Congenital Malformations. G. E. Wolstenholme, and C. M. O'Connor, eds., Little, Brown & Co., Boston, Mass, 134, 1960.

941. Neuman, R. E., and T. A. McCoy: Growth-promoting Properties of Pyruvate Oxalacetate and Alpha-ketoglutarate for Isolated Walker Carcinosarcoma 256 Cells. Proc. Soc. Exper. Biol. Med., **98**: 303, 1958.

942. Newcombe, H. B.: Delayed Phenotypic Expression of Spontaneous Mutations in *Escherichia coli*. Genetics, **33**: 447, 1948.

943. Nicholas, J. W., W. J. Jenkins, and W. L. Marsh: Human Blood Chimeras: A Study of Surviving Twins. Brit. Med. J., **1**: 1458, 1957.

944. Nichols, S. E., Jr., and W. M. Reams, Jr.: The Occurrence and Morphogenesis of Melanocytes in the Connective Tissues of the PET/MCV Mouse Strain. J. Embryol. Exper. Morphol., **8**: 24, 1960.

945. Niu, C., and H. Fraenkel-Conrat: C-Terminal Amino-acid Sequence of Tobacco Mosaic Virus Protein. Biochim. Biophys. Acta, **16**: 597, 1955.

946. Niu, C., and H. Fraenkel-Conrat: Determination of C-Terminal Amino Acids and Peptides by Hydrazinolysis. J. Amer. Chem. Soc., **77**: 5882, 1955.

947. Novitski, E., and L. Sandler: Further Notes on the Nature of Non-random Disjunction in *Drosophila melanogaster.* Genetics, **41**: 194, 1955.

948. Nowell, P. C.: Phytohemagglutinin: An Initiator of Mitosis in Cultures of Normal Human Leukocytes. Cancer Res., **20**: 462, 1960.

949. Oakberg, E. F.: Constitution of Liver and Spleen as a Physical Basis for Genetic Resistance to Mouse Typhoid. J. Infect. Diseases, **78**: 79, 1946.

950. Oakberg, E. F.: Effects of Radiation on the Gonads of Mice. Oklahoma Conference Radioisotopes in Agriculture. TID-7578, USAEC, p. 157, 1959.

951. Oakberg, E. F.: Gamma-ray Sensitivity of Oocytes of Young Mice. Anat. Rec., **137**: 385, 1960.

952. Oakberg, E. F.: Gamma-ray Sensitivity of Spermatogonia of the Mouse. J. Exper. Zool., **134**: 343, 1957.

953. Oakberg, E. F.: Initial Depletion and Subsequent Recovery of Spermatogonia of the Mouse after 20 r of Gamma Rays and 100, 300, and 600 r of X-rays. Radiat. Res., **11**: 700, 1959.

954. Oakberg, E. F.: manuscript in preparation.

955. Oakberg, E. F.: Sensitivity and Time of Degeneration of Spermatogenic Cells Irradiated in Various Stages of Maturation in the Mouse. Radiat. Res., **2**: 369, 1955.

956. Oakberg, E. F.: The Effect of X-rays on the Mouse Ovary. Proc. Tenth Internat. Congr. Genetics, **2**: 207, 1958.

957. Oakberg, E. F., and R. L. DiMinno: X-ray Sensitivity of Primary Spermatocytes of the Mouse. Internat. J. Radiat. Biol., **2**: 196, 1960.

958. Ochinskaia, G. K.: On the Difference in Biological Action of Roentgen Rays and Gamma Rays of Radioactive Cobalt on Mammals. Med. Radiol. (Moscow), **4**: 29, 1959.

959. Oettlé, A. G.: Spontaneous Carcinoma of the Glandular Stomach in *Rattus (Mastomys) natalensis,* an African Rodent. Brit. J. Cancer, **11**: 415, 1957.

960. Ohno, K.: On the Structure of Lysozyme. I. Quantitative Estimation of Carboxyl-terminal Amino Acid by Improved Hydrazinolysis Method. J. Biochem. (Tokyo), **40**: 621, 1953.

961. Ohno, K.: On the Structure of Lysozyme. II. Characterization of Aspartyl, Asparaginyl, and Glutaminyl Residues in Lysozyme. J. Biochem. (Tokyo), **41**: 345, 1954.

962. Ohno, K.: On the Structure of Lysozyme. III. On the Carboxyl-terminal Peptide. J. Biochem. (Tokyo), **42**: 615, 1955.

963. Ohno, S., and T. S. Hauschka: Allocycly of the *X*-chromosome in Tumors and Normal Tissues. Cancer Res., **20**: 541, 1960.

964. Ohno, S., W. D. Kaplan, and R. Kinosita: Heterochromatic Regions and Nucleolus Organizers in Chromosomes of the Mouse, *Mus musculus.* Exper. Cell Res., **13**: 358, 1957.

965. Ohno, S., W. D. Kaplan, and R. Kinosita: On Isopycnotic Behavior of the XX-bivalent in Oocytes of *Rattus norvegicus.* Exper. Cell Res., **19**: 637, 1960.

966. Ohno, S., W. D. Kaplan, and R. Kinosita: On the End-to-end Association of the *X* and *Y* Chromosomes of *Mus musculus.* Exper. Cell Res., **18**: 282, 1959.

967. Ohno, S., E. T. Kovacs, and R. Kinosita: On the *X*-chromosomes of Mouse Mammary Carcinoma Cells. Exper. Cell Res., **16**: 462, 1959.

968. Oppenheimer, H., A. DeLuca, and A. T. Milhorat: Serum and Plasma Proteins, Lipeproteins and Glycoproteins. III. Hereditary Muscular Dystrophy in Mice. Proc. Soc. Exper. Biol. Med., **100**: 568, 1959.

969. Oppenheimer, H., H. R. Terry, E. Forsyth, and A. T. Milhorat: Muscle Proteins in Mice with Hereditary Muscular Dystrophy. Fed. Proc., **18**: 116, 1959.

970. Orsini, M. W., and B. Pansky: The Natural Resistance of the Golden Hamster to Colchicine. Science, **115**: 88, 1952.

971. Osborne, R., and W. S. B. Paterson: On the Sampling Variance of Heritability Estimates Derived from Variance Analyses. Proc. Roy. Soc., ser. B, **64**: 456, 1952.

972. Oster, I. I., S. Zimmering, and H. J. Muller: Evidence of the Lower Mutagenicity of Chronic Than Intense Radiation in *Drosophila* Gonia. Science, **130:** 1423, 1959.

973. Otis, E.: Prenatal Mortality Rates of Seventeen Radiation-induced Translocations in Mice. Univ. of Rochester, Atomic Energy Project Rept., UR-**291**, 1953.

974. Oudin, J.: Allotypy of Rabbit Serum Proteins. I. Immunochemical Analysis Leading to the Individualization of Seven Main Allotypes. J. Exper. Med., **112:** 107, 1960.

975. Oudin, J.: Allotypy of Rabbit Serum Proteins. II. Relationships between the Various Allotypes: Their Common Antigenic Specificity, Their Distribution in a Sample Population; Genetic Implication. J. Exper. Med., **112:** 125, 1960.

976. Oudin, J.: Méthode d'Analyse Immunochimique par Précipitation Spécifique en Milieu Gélifié. C. R. Acad. Sci., **222:** 115, 1946.

977. Oudin, J.: Réaction de Précipitation Spécifique entre des Sérums d'Animaux de Même Espèce. C. R. Acad. Sci., **242:** 2489, 1956.

978. Owen, A. R. G.: The Analysis of Multiple Linkage Data. Heredity, **7:** 247, 1953.

979. Owen, J. A., H. J. Silberman, and C. Got: Detection of Haemoglobin, Haemoglobin-Haptoglobin Complexes and Other Substances with Peroxidase Activity after Zone Electrophoresis. Nature, **182:** 1373, 1958.

980. Owen, R. D.: Antigenic Characteristics of Rat Erythrocytes and Their Use as Markers for Parabiotic Exchange. Genetics, **33:** 623, 1948.

981. Owen, R. D.: Erythrocyte Antigens as Markers for Repopulation by Homologous Erythropoietic Tissues in Irradiated Mice. Radiat. Res., **9:** 164, 1958.

982. Owen, R. D.: Erythrocyte Repopulation after Transplantation of Homologous Erythropoietic Tissues into Irradiated Mice. Bull. Soc. Intern. Chirurgie, **28:** 289, 1959.

983. Owen, R. D.: Genetic Aspects of Tissue Transplantation and Tolerance. J. Med. Educ., **34:** 366, 1959.

984. Owen, R. D.: Immunogenetic Consequences of Vascular Anastomoses between Bovine Twins. Science, **102:** 400, 1945.

985. Owen, R. D.: Immunogenetics. Proc. Tenth Internat. Cong. Genetics, **1:** 364, 1959.

986. Owen, R. D.: Immunological Tolerance. Fed. Proc., **16:** 581, 1957.

987. Ozawa, H., and K. Satake: On the Species Difference of N-Terminal Amino Acid Sequence in Hemoglobin I. J. Biochem. (Tokyo), **42:** 641, 1955.

988. Ozawa, H., K. Tamai, and K. Satake: Studies on Hemoglobin. III. Arginine-peptides in the Tryptic and Chymotryptic Hydrolysates of Bovine and Horse Globins. J. Biochem. (Tokyo), **47:** 244, 1960.

989. Palmer, C. G.: The Cytology of Rabbit Papillomas and Derived Carcinomas. J. Nat. Cancer Inst., **23:** 241, 1959.

990. Parsons, P. A.: A Balanced Four-point Linkage Experiment for Linkage Group XIII in the House Mouse. Heredity, **12:** 77, 1958.

991. Parsons, P. A.: Additional Three-point Data for Linkage Group V of the Mouse. Heredity, **12:** 357, 1958.

992. Patterson, J. T.: The Production of Mutations in Somatic Cells of *Drosophila melanogaster* by Means of X-rays. J. Exper. Zool., **53:** 327, 1929.

993. Pauling, L.: Genetic and Somatic Effects of Carbon-14. Science, **128:** 1183, 1958.

994. Pauling, L., H. A. Itano, S. J. Singer, and I. C. Wells: Sickle-cell Anemia, a Molecular Disease. Science, **110:** 543, 1949.

995. Pearl, R.: Introduction to Medical Biometry and Statistics. W. B. Saunders Co., Philadelphia, 1940.

996. Perkoff, G. T., and F. H. Tyler: Creatine Metabolism in the Bar Harbor 129 Strain Dystrophic Mouse. Metabolism, **7:** 745, 1958.

997. Perutz, M. F., A. M. Liquori, and F. Eirich: X-ray and Solubility Studies of the Haemoglobin of Sickle-cell Anaemia Patients. Nature, **167:** 929, 1951.

998. Perutz, M. F., M. G. Rossmann, A. F. Cullis, H. Muirhead, G. Will, and A. C. T. North: Structure of Haemoglobin. A Three-dimensional Fourier Synthesis at 5.5-A° Resolution, Obtained by X-Ray Analysis. Nature, **185:** 416, 1960.

999. Peterson, E. A., and H. A. Sober: Chromatography of Proteins. I. Cellulose Ion-exchange Absorbents. J. Amer. Chem. Soc., **78:** 751, 1956.

1000. Peterson, R. R.: Electron Microscope Observations on the Pituitary Gland of the Dwarf Mouse. Anat. Rec., **133:** 322, 1959.

1001. Phillips, R. J. S.: *Jimpy*, a New Totally Sex-linked Gene in the House Mouse. Zeit. ind. Abst. Vererb., **86:** 322, 1954.

1002. Phillips, R. J. S.: Lurcher, a New Gene in Linkage Group XI of the House Mouse.   J. Genet, **57**: 35, 1960.

1003. Phillips, R. J. S.: Mouse News Letter, **15**: 28, 1956.

1004. Phillips, R. J. S.: The Linkages of Congenital Hydrocephalus in the House Mouse.   J. Hered., **47**: 302, 1956.

1005. Pickels, E. G.: Ultracentrifugation.   *In* Methods in Medical Research.   A. C. Corcoran, ed., Year Book Pub., Inc., Chicago, **5**: 107, 1952.

1006. Pillemer, L., L. Blum, L. Lepow, I. H. Wurz, and E. W. Todd: The Properdin System and Immunity.   III. The Zymosan Assay of Properdin.   J. Exper. Med., **103**: 1, 1956.

1007. Pincus, G., and K. V. Thimann, eds.: The Hormones.   Academic Press, New York, vol. I, 1948; vol. II, 1950; vol. III, 1955.

1008. Pinsky, L., and F. C. Fraser: Congenital Malformations After a Two-hour Inactivation of Nicotimamide in Pregnant Mice.   Brit. Med. J., **2**: 195, 1960.

1009. Pollock, M. E., G. E. Kenny, and J. T. Syverton: Isolation and Elimination of Pleuro-Pneumonia-like Organisms from Mammalian Cultures.   Proc. Soc. Exper. Biol. Med., **105**: 10, 1960

1010. Polson, A.: An Electrophoresis Cell for Analysing Four Samples Simultaneously.   Biochim. Biophys. Acta, **13**: 451, 1954.

1011. Pontecorvo, G.: Trends in Genetic Analysis.   Columbia Univ. Press, New York, 1958.

1012. Pontecorvo, G., and E. Käfer: Genetic Analysis Based on Mitotic Recombination.   Advances Genet., **9**: 71, 1958.

1013. Pope, R. S., and E. D. Murphy: Survival of Strain 129 Dystrophic Mice in Parabiosis.   Amer. J. Physiol., **199**: 1097, 1960.

1014. Popp, R. A.: Regression of Grafted Bone Marrow in Homologous Irradiated Mouse Chimeras, J. Nat. Cancer Inst., **26**: 629, 1961.

1015. Popp, R. A., and G. E. Cosgrove: Solubility of Hemoglobins as Red Cell Marker in Irradiated Mouse Chimeras.   Proc. Soc. Exper. Biol. Med., **101**: 754, 1959.

1016. Popp, R. A., G. E. Cosgrove, Jr., and R. D. Owen: Genetic Differences in Hemoglobin as Markers for Bone Marrow Transplantation in Mice.   Proc. Soc. Exper. Biol. Med., **99**: 692, 1958.

1017. Popp, R. A., and W. St. Amand: Studies on the Mouse Hemoglobin Locus.   I. Identification of Hemoglobin Types and Linkage of Hemoglobin with Albinism.   J. Hered., **51**: 141, 1960.

1018. Potter, M.: Biologic Studies on the Development of DON Resistance in a Mast-cell Neoplasm of the Mouse.   J. Nat. Cancer Inst., **23**: 163, 1959.

1019. Potter, M.: Variation in Resistance Patterns in Different Neoplasms.   Ann. N.Y. Acad. Sci., **76**: 630, 1958.

1020. Poulik, M. D., and O. Smithies: Comparison and Combination of the Starch-gel and Filter-paper Electrophoretic Methods Applied to Human Sera: Two-dimensional Electrophoresis. Biochem. J., **68**: 636, 1958.

1021. Preer, J. R., Jr.: A Quantitative Study of a Technique of Double Diffusion in Agar.   J. Immunol., **77**: 52, 1956.

1022. Prehn, R. T.: Tumors and Hyperplastic Nodules in Transplanted Mammary Gland.   J. Nat. Cancer Inst., **13**: 859, 1953.

1023. Prehn, R. T., and J. M. Main: Immunity to Methylcholanthrene-induced Sarcomas.   J. Nat. Cancer Inst., **18**: 769, 1957.

1024. Puck, T. T.: Growth and Genetics of Somatic Mammalian Cells *in vitro*.   J. Cell. Comp. Physiol., **52** Suppl. 1: 287, 1958.

1025. Puck, T. T.: The Genetics of Somatic Mammalian Cells.   Advances Biol. Med. Phys., **5**: 751957.

1026. Puck, T. T.: Tissue Culture.   *In* Methodology in Human Genetics, W. J. Burdette, ed., Holden-Day, Inc., San Francisco, 1962.

1027. Puck, T. T., S. J. Cieciura, and H. W. Fisher: Clonal Growth *in vitro* of Human Cells with Fibroblastic Morphology; Comparison of Growth and Genetic Characteristics of Single Epithelioid and Fibroblast-like Cells From a Variety of Human Organs.   J. Exper. Med., **106**: 145, 1957.

1028. Puck, T. T., S. J. Cieciura, and A. Robinson: Genetics of Somatic Mammalian Cells.   III. Long-term Cultivation of Euploid Cells from Human and Animal Subjects.   J. Exper. Med., **108**: 945, 1958.

1029. Puck, T. T., and P. I. Marcus: A Rapid Method for Viable Cell Titration and Clone Production with *HeLa* Cells in Tissue Culture: The Use of X-irradiated Cells to Supply Conditioning Factors.   Proc. Nat. Acad. Sci. (Wash.), **41**: 432, 1955.

1030. Puck, T. T., and M. Oda: Antibody Specificity of Mammalian Cells Grown *in vitro*. Fed. Proc., **19:** 206, 1960.
1031. Purdom, L., E. F. Ambrose, and G. Klein: A Correlation between Electrical Surface Charge and Some Biological Characteristics during the Stepwise Progression of a Mouse Sarcoma. Nature, **181:** 1586, 1958.
1032. Putnam, F. W.: Aberrations of Protein Metabolism in Multiple Myeloma. I. Interrelationships of Abnormal Serum Globulins and Bence-Jones Proteins. Physiol. Rev., **37:** 512, 1957.
1033. Quevedo, W. C., Jr.: Effect of Biotin Deficiency on Follicular Melanocytes of Mice. Proc. Soc. Exper. Biol. Med., **93:** 260, 1956.
1034. Rabinowitz, J. L.: Glucorolactone Decarboxylase in the Dystrophic Mouse. Atompraxis, **4:** 317, 1958.
1035. Rabinowitz, J. L.: Studies on Dystrophic Mice and Their Littermates. Fed. Proc., **18:** 306, 1959.
1036. Race, R. R., and R. Sanger: Blood Groups in Man. Second ed. Blackwell, Oxford, 1954, Third ed. C. C Thomas, Springfield, Ill., 1958.
1037. Rajam, P. C., and A. L. Jackson: Labelling of Antibody Against the Ehrlich Mouse Ascites Carcinoma with Tritium (H^3). J. Lab. Clin. Med., **55:** 46, 1960.
1038. Ranadive, K. J., and S. A. Hakim: A Biological Study of Strain L(P) and its Response to 20-Methylcholanthrene Treatment. Brit. J. Cancer, **12:** 44, 1958.
1039. Ranney, H. M., and S. Gluecksohn-Waelsch: Filter-paper Electrophoresis of Mouse Haemoglobin: Preliminary Note. Ann. Hum. Genet., **19:** 269, 1955.
1040. Rawles, M. E.: Origin of Pigment Cells from the Neural Crest in the Mouse Embryo. Physiol. Zool., **20:** 248, 1947.
1041. Rawles, M. E.: The Development of Melanophores From Embryonic Mouse Tissues Grown in the Coelom of Chick Embryos. Proc. Nat. Acad. Sci. (Wash.), **26:** 673, 1940.
1042. Reams, W. M., S. E. Nichols, and H. G. Hager: Chemical Evocation of Melanocyte Branching in the Chick Embryo. Anat. Rec., **134:** 667, 1959.
1043. Reed, S. C.: The Inheritance and Expression of Fused, a New Mutation in the House Mouse. Genetics, **22:** 1, 1937.
1044. Reed, S. C., and J. M. Henderson: Pigment Cell Migration in Mouse Epidermis. J. Exper. Zool., **85:** 409, 1940.
1045. Reed, T. E.: Variable Specificity of an Anti-A Serum Labelled in Three Different Ways. Acta Haematol. (Basel), **25:** 355, 1961.
1046. Reeve, E. C. R.: The Variance of the Genetic Correlation Coefficient. Biometrics, **11:** 357, 1955.
1047. Reichert, E. T., and A. P. Brown: The Differentiation and Specificity of Corresponding Proteins and Other Vital Substances in Relation to Biological Classification and Organic Evolution: The Crystallography of Hemoglobins. Carnegie Inst. Wash. Pub. No. **116,** 1909.
1048. Reichard, P., O. Sköld, and G. Klein: Possible Enzymic Mechanism for the Development of Resistance Against Fluorouracil in Ascites Tumors. Nature, **183:** 939, 1959.
1049. Rennels, E. G., and W. McNutt: The Fine Structure of the Anterior Pituitary Cells of the Dwarf Mouse. Anat. Rec., **131:** 591, 1958.
1050. Révész, L.: Detection of Antigenic Differences in Isologous Host-tumor Systems by Pretreatment with Heavily Irradiated Tumor Cells. Cancer Res., **20:** 443, 1960.
1051. Révész, L.: Effect of Lethally Damaged Tumor Cells upon the Development of Admixed Viable Cells. J. Nat. Cancer Inst., **20:** 1157, 1958.
1052. Révész, L., and U. Norman: Relationship between Chromosome Ploidy and Radiosensitivity in Selected Tumor Sublines of Common Origin. J. Nat. Cancer Inst., **25:** 1041, 1960.
1053. Rhinesmith, H. S., W. A. Schroeder, and N. Martin: The N-Terminal Sequence of the β-Chains of Normal Adult Human Hemoglobin. J. Amer. Chem. Soc., **80:** 3358, 1958.
1054. Rhinesmith, H. S., W. A. Schroeder, and L. Pauling: A Quantitative Study of the Hydrolysis of Human Dinitrophenyl (DNP) globin: The Number and Kind of Polypeptide Chains in Normal Adult Human Hemoglobin. J. Amer. Chem. Soc., **79:** 4682, 1957.
1055. Rhoades, M. M.: The Genetic Control of Mutability in Maize. Cold Spring Harbor Symp. Quant. Biol., **9:** 138, 1941.
1056. Riggs, A.: The Nature and Significance of the Bohr Effect in Mammalian Hemoglobins. J. Gen. Physiol., **43:** 737, 1960.
1057. Ringertz, N., E. Klein, and G. Klein: Histopathologic Studies of Peritoneal Implantation and Lung Metastasis at Different Stages of the Gradual Transformation of the MC1M Mouse Sarcoma into Ascites Form. J. Nat. Cancer Inst., **18:** 173, 1957.

1058. Rixon, R. H., and J. F. Whitfield: Comparison of the Effects of Ultraviolet Light on Multiplication of Normal and X-ray Resistant Mouse Cells. Exper. Cell Res., **20:** 220, 1960.

1059. Roberts, E., and L. E. Card: The Inheritance of Resistance to Bacillary White Diarrhea. Poultry Sci., **6:** 18, 1926.

1060. Roberts, E., and J. H. Quisenberry: Linkage of the Genes for Non-yellow (y) and Pink-eye ($p2$) in the House Mouse (*Mus musculus*). Amer. Nat., **69:** 181, 1935.

1061. Robertson, A.: Experimental Design in the Evaluation of Genetic Parameters. Biometrics, **15:** 219, 1959.

1062. Robertson, A., and I. M. Lerner: The Heritability of All-or-none Traits: Viability of Poultry. Genetics, **34:** 395, 1949.

1063. Robertson, G. G.: Ovarian Transplantation in the House Mouse. Proc. Soc. Exper. Biol. Med., **44:** 302, 1940.

1064. Robinson, A. R., M. Robson, A. P. Harrison, and W. W. Zuelzer: A New Technique for Differentiation of Hemoglobin. J. Lab. Clin. Med., **50:** 745, 1957.

1065. Robinson, E., and H. A. Itano: Asymmetrical Recombination of Alkali-dissociated Haemoglobin Mixtures. Nature, **185:** 547, 1960.

1066. Robinson, G. E., Jr.: Variations in Gonadotrophic and Thyrotrophic Potency of Pituitaries from Pregnant and Non-pregnant Swine. Ph.D. Thesis, Univ. of Illinois, 1950.

1067. Robinson, R.: Sable and Umbrous Mice. Genetics, **29:** 319, 1959.

1068. Roche, J., Y. Derrien, and M. Moutte: Sur la Spécificité des Hémoglobines et sur l'Existence Probable de Deux Hémoglobines dans le Sang de Divers Mammifères. C.R. Soc. Biol., **135:** 1235, 1941.

1069. Rogers, S.: Induction of Arginase in Rabbit Epithelium by the Shope Rabbit Papilloma Virus. Nature, **183:** 1815, 1959.

1070. Rogers, S.: Studies of the Mechanism of Action of Ultraviolet Irradiation in Initiating Tumors in the Lung Tissue of Mice. *In* Radiation Biology and Cancer. R. W. Cumley, *et al.,* eds., Univ. of Texas Press, Austin, 1959.

1071. Rood, J. J. van, A. van Leeuwen, and J. G. Eernisse: Leucocyte Antibodies in Sera of Pregnant Women. Vox Sang., **4:** 427, 1959.

1072. Rosa, J., G. Schapira, J. S. Dreyfus, J. de Grouchy, G. Mathe, and J. Bernard: Different Heterogeneities of Mouse Haemoglobin According to Strains. Nature, **182:** 947, 1958.

1073. Roscoe B. Jackson Memorial Laboratory, unpublished List of Incidences of Tumors in Inbred Strains of Mice.

1074. Rosenkrants, H.: Dehydrogenase Levels in Mice with Muscularis Dystrophiea. Fed. Proc., **18:** 312, 1959.

1075. Rossi, H. H., J. L. Bateman, V. P. Bond, L. J. Goodman, and F. E. Stickley: The Dependence of RBE upon the Energy of Fast Neutrons. I. Physical Design and Measurement of Absorbed Dose. Radiat. Res., **13:** 503, 1960.

1076. Rothfels, K. H., A. A. Axelrad, L. Siminovitch, E. A. McCulloch, and R. C. Parker: The Origin of Altered Cell Lines from Mouse, Monkey, and Man, as Indicated by Chromosome and Transplantation Studies. Canadian Cancer Conf., **3:** 189, 1959.

1077. Rothfels, K. H., and L. Siminovitch: An Air-drying Technique for Flattening Chromosomes in Mammalian Cells Grown *in vitro*. Stain Technol., **33:** 73, 1958.

1078. Rothman, S., H. F. Krysa, and A. M. Smiljanic: Inhibitory Action of Human Epidermis on Melanin Formation. Proc. Soc. Exper. Biol. Med., **62:** 208, 1946.

1079. Rubin, B. A.: Comment. Ann. N.Y. Acad. Sci., **87:** 130, 1960.

1080. Rubin, H.: An Analysis of the Assay of Rous Sarcoma Cells *in vitro* by the Infective Center Technique. Virology, **10:** 29, 1960.

1081. Rubin, H., and H. M. Temin: Radiation Studies on Lysogeny and Tumor Viruses. *In* Radiation Biology and Cancer. R. W. Cumley, *et al.,* eds., Univ. of Texas Press, Austin, 359, 1959.

1082. Rudali, G., N. Yourkovski, L. Juliard, and M. Fautrel: Sur Quelques Caractères des Souris Appartenant à la Nouvelle Lignée Cancéreuse. Lignee NLC de la Fondation Curie. Bull. Assoc. Franc. Cancer, **43:** 364, 1956.

1083. Ruddle, F. H.: Chromosome Variation in Clonal Lines of Tissue Culture Cells Derived From Pig Kidney. Proc. Tissue Culture Assoc., 22, 1960.

1084. Ruddle, F. H., L. Berman, and C. S. Stulberg: Chromosome Analysis of Five Long-term Cell Culture Populations Derived from Nonleukemic Human Peripheral Blood. (Detroit Strains.) Cancer Res., **18:** 1048, 1958.

1085. Rugh, R.: Chronic Low-level Exposures of Young Mice to Ionizing Radiation and the Effect on Fertility. J. Pediat., **44:** 248, 1954.

1086. Rugh, R., and J. Wolff: Evidence of Some Chemical Protection of the Mouse Ovary Against X-irradiation Sterilization. Radiat. Res., **7:** 184, 1957.

1087. Rugh, R., and J. Wolff: X-irradiation Sterilization of the Female Mouse. Fertility and Sterility, **7:** 546, 1956.

1088. Runner, M. N.: Inheritance of Susceptibility to Congenital Deformity. Metabolic Clues Provided by Experiments with Teratogenic Agents. Pediatrics, **23:** 245, 1959.

1089. Runner, M. N.: Linkage of Brachypodism. A New Member of Linkage Group V of the House Mouse. J. Hered., **50:** 81, 1959.

1090. Runner, M. N., and C. P. Dagg: Metabolic Mechanisms of Teratogenic Agents During Morphogenesis. Symp. Normal and Abnormal Differentiation and Development. Nat. Cancer Inst., Monogr. **2:** 41, 1960.

1091. Russell, E. S.: Analysis of Pleiotropism at the W-Locus in the Mouse: Relationship between the Effects of W and W^v Substitution on Hair Pigmentation and on Erythrocytes. Genetics, **34:** 708, 1949.

1092. Russell, E. S.: A Quantitative Histological Study of the Pigment Found in the Coat-color Mutants of the House Mouse. I. Variable Attributes of the Pigment Granules. Genetics, **31:** 327, 1946.

1093. Russell, E. S.: A Quantitative Histological Study of the Pigment Found in the Coat-color Mutants of the House Mouse. II. Estimates of the Total Volume of Pigment. Genetics, **33:** 228, 1948.

1094. Russell, E. S.: A Quantitative Histological Study of the Pigment Found in the Coat-color Mutants of the House Mouse. III. Interdependence Among the Variable Granule Attributes. Genetics, **34:** 133, 1949.

1095. Russell, E. S.: A Quantitative Histological Study of the Pigment Found in the Coat-color Mutants of the House Mouse. IV. The Nature of the Effects of Genic Substitution in Five Major Allelic Series. Genetics, **34:** 146, 1949.

1096. Russell, E. S.: A Quantitative Study of Genic Effects on Guinea Pig Coat Colors. Genetics, **24:** 332, 1939.

1097. Russell, E. S.: Review of the Pleiotropic Effects of W-Series Genes on Growth and Differentiation. Aspects of Synthesis and Order of Growth. D. Rudnick, ed., Princeton Univ. Press, Princeton, N.J., **13:** 113, 1954.

1098. Russell, E. S.: unpublished results.

1099. Russell, E. S.: Genetic Aspects of Implantation of Blood-forming Tissue. Fed. Proc. **19:** 573, 1960.

1100. Russell, E. S., S. E. Bernstein, F. A. Lawson, and L. J. Smith: Long-continued Function of Normal Blood-forming Tissue Transplanted into Genetically Anemic Hosts. J. Nat. Cancer Inst., **23:** 557, 1959.

1101. Russell, E. S., and E. Fekete: Analysis of W-Series Pleiotropism in the Mouse; Effect of W^vW^v Substitution on Definitive Germ Cells and on Ovarian Tumorigenesis. J. Nat. Cancer Inst., **21:** 365, 1958.

1102. Russell, E. S., and E. L. Fondal: Quantitative Analysis of the Normal and Four Alternative Degrees of an Inherited Macrocytic Anemia in the House Mouse. I. Number and Size of Erythrocytes. Blood, **6:** 892, 1951.

1103. Russell, E. S., and P. S. Gerald: Inherited Electrophoretic Hemoglobin Patterns Among 20 Inbred Strains of Mice. Science, **128:** 1569, 1958.

1104. Russell, E. S., G. Keighley, H. Borsook, and P. Lowy: Effects of Erythropoietic Stimulating Factor on Inherited Anemia in Mice. Physiologist, **2:** 3, 1959.

1105. Russell, E. S., and F. A. Lawson: Selection and Inbreeding for Longevity of a Lethal Type. J. Hered., **50:** 19, 1959.

1106. Russell, E. S., F. Lawson, and G. Schabtach: Evidence for a New Allele at the W-Locus of the Mouse. J. Hered., **48:** 119, 1957.

1107. Russell, E. S., L. M. Murray, E. M. Small, and W. K. Silvers: Development of Embryonic Mouse Gonads Transferred to the Spleen: Effects of Transplantation Combined with Genotypic Autonomy. J. Embryol. Exper. Morphol., **4:** 347, 1956.

1108. Russell, E. S., E. F. Neufeld, and C. T. Higgins: Comparison of Normal Blood Picture of Young Adults from 18 Inbred Strains of Mice. Proc. Soc. Exper. Biol. Med., **78:** 761, 1951.

1109. Russell, E. S., L. J. Smith, and F. A. Lawson: Implantation of Normal Blood Forming Tissue in Radiated Genetically Anemic Hosts. Science, **124:** 1076, 1956.

1110. Russell, E. S., C. M. Snow, L. M. Murray, and J. P. Cormier: The Bone-marrow in Inherited Macrocytic Anemia in the House Mouse. Acta Haematol., **10:** 247, 1953.

1111. Russell, L. B.: Dominant Lethals Induced at a Highly Sensitive Stage in Mouse Oogenesis. Anat. Rec., **125:** 647, 1956.

1112. Russell, L. B.: Mouse News Letter, **23:** 59, 1960.

1113. Russell, L. B.: Radiation Hazards during Embryonic Development. IX. Internat. Congr. Radiology. B. Rajewsky, ed., Georg Thieme-Verlag und Verlag Urban & Schwarzenberg, Munich, 1959.

1114. Russell, L. B., and M. K. Freeman: Comparison of the Effects of Acute and Fractionated Irradiation Fertility of the Female Mouse. Anat. Rec., **128:** 615, 1957.

1115. Russell, L. B., and M. H. Major: A High Rate of Somatic Reversion in the Mouse. Genetics, **41:** 658, 1956.

1116. Russell, L. B., and M. H. Major: Dominant Lethals in Mouse Oocytes Inducted by X-rays in Air and in 5% Oxygen. Genetics, **38:** 687, 1953.

1117. Russell, L. B., and M. H. Major: Radiation-induced Presumed Somatic Mutations in the House Mouse. Genetics, **42:** 161, 1957.

1118. Russell, L. B., and W. L. Russell: An Analysis of the Changing Radiation Response of the Developing Mouse Embryo. J. Cell. Comp. Physiol., **43:** 103, 1954.

1119. Russell, L. B., and W. L. Russell: A Study of the Physiological Genetics of Coat Color in the Mouse by Means of the Dopa Reaction in Frozen Sections of Skin. Genetics, **33:** 237, 1948.

1120. Russell, L. B., and W. L. Russell: Genetic Effects of Radiation in the Female Mouse. Proc. Tenth Internat. Cong. Genetics, **2:** 245, 1958.

1121. Russell, L. B., and W. L. Russell: Hazards to the Embryo and Fetus from Ionizing Radiation. Proc. UN Internat. Conf., Peaceful Uses of Atomic Energy, 1955, **11:** 175, 1956.

1122. Russell, L. B., and W. L. Russell: Radiation Hazards to the Embryo and Fetus. Radiology, **58:** 369, 1952.

1123. Russell, L. B., and W. L. Russell: The Sensitivity of Different Stages in Oogenesis to the Radiation Induction of Dominant Lethals and Other Changes in the Mouse. *In* Progress in Radiobiology. J. S. Mitchell, B. E. Holmes, and C. L. Smith, eds., Oliver and Boyd, Ltd., Edinburgh, 187, 1955.

1124. Russell, L. B., and R. J. Spear: Relation between Dominant Lethal Incidence and Stage in Oogenesis Irradiated. Radiat. Res., **3:** 342, 1955.

1125. Russell, L. B., and R. J. Spear: X-ray Induced Dominant Lethals in Mouse Oocytes and Their Relation to Irradiation-to-ovulation Interval. Genetics, **39:** 991, 1954.

1126. Russell, L. B., K. F. Stelzner, and W. L. Russell: Influence of Dose Rate on Radiation Effect on Fertility of Female Mice. Proc. Soc. Exper. Biol. Med., **102:** 471, 1959.

1127. Russell, L. B., and L. Wickham: The Incidence of Disturbed Fertility among Male Mice Conceived at Various Intervals after Irradiation of the Mother. Genetics, **42:** 392, 1957.

1128. Russell, W. L.: Genetic Effects of Radiation in Mammals. *In* Radiation Biology. A. Hollaender, ed., McGraw-Hill Book Co., New York, vol. **1,** 1954.

1129. Russell, W. L.: Inbred and Hybrid Animals and Their Value in Research. *In* Biology of the Laboratory Mouse. G. D. Snell, ed., The Blakiston Co., New York, 325, 1941.

1130. Russell, W. L.: Investigation of Physiological Genetics of Hair and Skin Color in the Guinea Pig by Means of the Dopa Reaction. Genetics, **24:** 645, 1939.

1131. Russell, W. L.: Lack of Linearity between Mutation Rate and Dose for X-ray Induced Mutations in Mice. Genetics, **41:** 658, 1956.

1132. Russell, W. L.: Mammalian Radiation Genetics. *In* Symposium on Radiobiology. J. J. Nickson, ed., J. Wiley and Sons, New York, 427, 1952.

1133. Russell, W. L.: Shortening of Life in the Offspring of Male Mice Exposed to Neutron Radiation from an Atomic Bomb. Proc. Nat. Acad. Sci. (Wash.), **43:** 324, 1957.

1134. Russell, W. L.: X-Ray-induced Mutations in Mice. Cold Spring Harbor Symp. Quant. Biol., **16:** 327, 1951.

1135. Russell, W. L., and J. G. Hurst: Pure Strain Mice Born to Hybrid Mothers Following Ovarian Transplantation. Proc. Nat. Acad. Sci. (Wash.), **31:** 267, 1945.

1136. Russell, W. L., J. C. Kile, Jr., and L. B. Russell: Failure of Hypoxia to Protect Against the Radiation Induction of Dominant Lethals in Mice. Genetics, **36:** 574, 1951.

1137. Russell, W. L., and L. B. Russell: Radiation-induced Genetic Damage in Mice. Proc. Second UN Internat. Conf. on Peaceful Uses of Atomic Energy, **22:** 360, 1958.

1138. Russell, W. L., and L. B. Russell: The Genetic and Phenotypic Characteristics of Radiation-induced Mutations in Mice. Radiat. Res., Suppl. **1:** 296, 1959.

1139. Russell, W. L., L. B. Russell, and M. B. Cupp: Dependence of Mutation Frequency on Radiation Dose Rate in Female Mice. Proc. Nat. Acad. Sci. (Wash.), **45:** 18, 1959.

1140. Russell, W. L., L. B. Russell, and E. M. Kelly: Radiation Dose Rate and Mutation Frequency. Science, **128:** 1546, 1958.

1141. Russell, W. L., L. B. Russell, and A. W. Kimball: The Relative Effectiveness of Neutrons from a Nuclear Detonation and from a Cyclotron in Inducing Dominant Lethals in the Mouse. Amer. Nat., **88:** 269, 1954.

1142. Russell, W. L., L. B. Russell, and E. F. Oakberg: Radiation Genetics of Mammals. *In* Radiation Biology and Medicine. W. D. Claus, ed., Addison-Wesley Pub. Co. Inc., Reading, Mass., 189, 1958.

1143. Russell, W. L., L. B. Russell, M. H. Steele, and E. L. Phipps: Extreme Sensitivity of an Immature Stage of the Mouse Ovary to Sterilization by Irradiation. Science, **129:** 1288, 1959.

1144. Sacher, G. A.: On the Relation of Radiation Lethality to Radiation Injury, and its Relevance for the Prediction Problem. Proc. Ninth Internat. Cong. Radiol., Urban and Schwarzenberg, Munich, 1223, 1960.

1145. Sachs, L., and R. Gallily: The Chromosomes and Transplantability of Tumors. II. Chromosome Duplication and the Loss of Strain Specificity in Solid Tumors. J. Nat. Cancer Inst., **16:** 803, 1956.

1146. Salmon, C.: Leucemie Aigue et Mutations Somatiques des Substances de Groupe Sanguin. Rev. Hematol., **14:** 205, 1959.

1147. Sanderson, M. H., and S. P. Stearner: The Effect of Ionizing Radiation on Fertility in the Female Mouse. Argonne National Laboratory, Quart. Rept. Biol. Med. Res., ANL-**5576:** 57, 1956.

1148. Sandow, A., and M. Brust: Contractility of Dystrophic Mouse Muscle. Amer. J. Physiol., **194:** 557, 1958.

1149. Sanford, K. K.: Clonal Studies on Normal Cells and on Their Neoplastic Transformation *in vitro.* Cancer Res., **18:** 747, 1958.

1150. Sanford, K. K., W. R. Earle, and G. D. Likely: The Growth *in vitro* of Single Isolated Tissue Cells. J. Nat. Cancer Inst., **9:** 229, 1948.

1151. Sanford, K. K., R. M. Merwin, G. L. Hobbs, M. C. Fioramonti, and W. R. Earle: Studies on the Difference in Sarcoma-producing Capacity of Two Lines of Mouse Cells Derived *in vitro* from One Cell. J. Nat. Cancer Inst., **20:** 121, 1958.

1152. Sanford, K. K., R. M. Merwin, G. L. Hobbs, J. M. Young, and W. R. Earle: Clonal Analysis of Variant Cell Lines Transformed to Malignant Cells in Tissue Culture. J. Nat. Cancer Inst., **23:** 1035, 1959.

1153. Sanger, F.: The Free Amino Groups of Insulin. Biochem. J., **39:** 507, 1945.

1154. Sanger, F., and E. O. P. Thompson: The Amino-acid Sequence in the Glycyl Chain of Insulin. Biochem. J., **53:** 353, 1953.

1155. Sarvella, P. A., and L. B. Russell: Steel, a New Dominant Gene in the House Mouse. J. Hered., **47:** 123, 1956.

1156. Sasakawa, S., and K. Satake: Studies on Hemoglobin. II. Arginine-containing Peptides from the Hydrolysate by *Streptomyces griseus* Proteinase. J. Biochem. (Tokyo), **45:** 867, 1958.

1157. Sasakawa, S., and K. Satake: Studies on Hemoglobin. IV. Histidine Peptides Derived from Horse Globin with *Streptomyces griseus* Proteinase. J. Biochem. (Tokyo), **47:** 672, 1960.

1158. Sato, G., L. Zaroff, and S. E. Mills: Tissue Culture Populations and Their Relation to the Tissue of Origin. Proc. Nat. Acad. Sci. (Wash.), **46:** 963, 1960.

1159. Schachman, H. K.: Ultracentrifugation in Biochemistry. Academic Press, New York, 1959.

1160. Schaedler, R. W., and R. J. Dubos: Effect of Dietary Proteins and Amino Acids on the Susceptibility of Mice to Bacterial Infections. J. Exper. Med., **110:** 921, 1959.

1161. Schaefer, H. J.: Radiation and Man in Space. *In* Advances in Space Science, F. I. Ordway, ed., Academic Press, New York, **1:** 267, 1959.

1162. Schaefer, T.: The Effects of Early Experience: Infant Handling and Later Behavior in the White Rat. Doctoral dissertation, Univ. of Chicago, 1957.

1163. Schaible, R. H.: Mouse News Letter, **22:** 32, 1960.

1164. Scheinberg, I. H.: The Structural Basis of Differences in Electrophoretic Behavior of Human Hemoglobins. Conf. on Hemoglobin. National Academy of Sciences, National Research Council, **557:** 227, 1958.

1165. Scheinberg, I. H., R. S. Harris, and J. L. Spitzer: Differential Titration by Means of Paper Electrophoresis and the Structure of Human Hemoglobins. Proc. Nat. Acad. Sci. (Wash.), **40:** 777, 1954.

1166. Scheinberg, S. L., and R. P. Reckel: Induced Somatic Mutations Affecting Erythrocyte Antigens. Science, **131:** 1887, 1960.

1167. Schott, R. G.: The Inheritance of Resistance to *Salmonella aertrycke* in Various Strains of Mice. Genetics, **17**: 203, 1932.
1168. Schroeder, W. A., and G. Matsuda: N-Terminal Residues of Human Fetal Hemoglobin. J. Amer. Chem. Soc., **80**: 1521, 1958.
1169. Schull, W. J.: Empirical Risks in Consanguineous Marriages: Sex Ratio, Malformation and Viability. Amer. J. Hum. Genet., **10**: 294, 1958.
1170. Schull, W. J., and J. V. Neel: Radiation and the Sex Ratio in Man. Science, **128**: 343, 1958.
1171. Schultz, J.: Malignancy and the Genetics of the Somatic Cell. Ann. N.Y. Acad. Sci., **71**: 994, 1958.
1172. Schultz, J.: The Function of Heterochromatin. Proc. Seventh Internat. Cong. Genetics, 257, 1939.
1173. Schultz, J.: The Role of Somatic Mutation in Neoplastic Growth. *In* Genetics and Cancer. R. W. Cumley, *et al.*, eds., Univ. of Texas Press, Austin, 25, 1959.
1174. Schultz, J., and D. A. Hungerford: Characteristics of Pairing in the Salivary Gland Chromosomes of *Drosophila melanogaster*. Genetics, **38**: 689, 1953.
1175. Schultz, J., and P. St. Lawrence: A Cytological Basis for a Map of the Nucleolar Chromosome in Man. J. Hered., **40**: 30, 1949.
1176. Schultz, W.: Erzeugung der Winterschwartz. Willkürliche Schwarzung gelber Haare. Arch. Entwickl. mech. Organ., **51**: 338, 1918.
1177. Schultz, W.: Kälteschwärzung eines Säugetieres und Ihre Allgemeinbiologischen Hinweise. Arch. Entwickl. mech. Organ., **47**: 43, 1924.
1117a. Schumann, H.: Die Entstehung bei Scheckung bei Mäusen mit weissen Blasse. Devel. Biol. **2**: 501, 1960.
1178. Schwartz, D.: Electrophoretic and Immunochemical Studies with Endosperm Proteins of Maize Mutants. Genetics, **45**: 1419, 1960.
1179. Schwartz, D.: Studies on the Mechanism of Crossing Over. Genetics, **39**: 692, 1954.
1180. Schwartz, H. C., T. H. Spaet, W. W. Zuelzer, J. V. Neel, A. R. Robinson, and S. F. Kaufman: Combinations of Hemoglobin G, Hemoglobin S, and Thalassemia Occurring in One Family. Blood, **12**: 238, 1957.
1181. Schweet, R., H. Lamfrom, and E. Allen: The Synthesis of Hemoglobin in a Cell-free System. Proc. Nat. Acad. Sci. (Wash.), **44**: 1029, 1958.
1182. Scott, J. P.: Effects of Single Genes on the Behavior of *Drosophila*. Amer. Nat., **77**: 184, 1943.
1183. Scott, J. P.: The Effects of Selection and Domestication upon the Behavior of the Dog. J. Nat. Cancer Inst., **15**: 739, 1954.
1184. Scott, J. P.: The Embryology of the Guinea Pig. II. The Polydactylous Monster. A New Teras Produced by the Genes PxPx. J. Morphol., **62**: 299, 1938.
1185. Scott, J. P.: The Embryology of the Guinea Pig. III. The Development of the Polydactylous Monster. A Case of Growth Accelerated at a Particular Period by a Semi-dominant Lethal Gene. J. Exper. Zool., **77**: 123, 1937.
1186. Scott, J. P.: The Genetic and Environmental Differentiation of Behavior. *In* The Concept of Development. D. B. Harris, ed., Univ. of Minnesota Press, Minneapolis, 1957.
1187. Scott, J. P., and M. S. Charles: Genetic Differences in the Behavior of Dogs: A Case of Magnification by Thresholds and by Habit Formation. J. Genet. Psychol., **84**: 175, 1954.
1188. Scott, J. P., J. L. Fuller, and J. A. King: The Inheritance of Annual Breeding Cycles in Hybrid Basenji-Cocker Spaniel Dogs. J. Hered., **50**: 254, 1959.
1189. Searle, A. G.: A Lethal Allele of Dilute in the House Mouse. Heredity, **6**: 395, 1952.
1190. Searle, A. G.: Genetical Studies on the Skeleton of the Mouse. XI. The Influence of Diet on Variation Within Pure Lines. J. Genet., **52**: 413, 1954.
1190a. Searle, A. G.: Tipsy, a New Mutant in Linkage Group VII of the Mouse. Genet. Res., **2**: 122, 1961.
1191. Sen, N. N., K. C. Das, and B. K. Aikat: Foetal Haemoglobin in the Monkey. Nature, **186**: 977, 1960.
1192. Serra, J. A.: Constitution of Hair Melanins. Nature, **157**: 771, 1946.
1193. Shapiro, J. R., and A. Kirschbaum: Intrinsic Tissue Response to Induction of Pulmonary Tumors. Cancer Res., **11**: 644, 1951.
1194. Shaver, S. L.: X-irradiation Injury and Repair in the Germinal Epithelium of Male Rats. I. Injury and Repair in Adult Rats. Amer. J. Anat., **92**: 391, 1953.
1195. Shavit, N., R. G. Wolfe, and R. A. Alberty: The Electrophoresis and Titration of Fumarase. J. Biol. Chem., **233**: 1382, 1958.

1196. Shelton, J. R., and W. A. Schroeder: Further N-Terminal Sequences in Human Hemoglobins A, S, and F by Edman's Phenylthiohydantoin Method. J. Amer. Chem. Soc., **82**: 3342, 1960.

1197. Shimkin, M. B., H. G. Grady, and H. B. Andervont: Induction of Testicular Tumors and Other Effects of Stilbestrol-cholesterol Pellets in Strain C Mice. J. Nat. Cancer Inst., **2**: 65, 1941.

1198. Shimkin, M. B., and G. B. Mider: Induction of Tumors in Guinea Pigs with Subcutaneously Injected Methylcholanthrene. J. Nat. Cancer Inst., **1**: 707, 1941.

1199. Shope, R. E.: Summary of Informal Discussions. Symp. on Possible Role of Viruses in Cancer. Cancer Res., **20**: 784, 1960.

1200. Shull, R., and R. B. Alfin-Slater: Tissue Lipids of Dystrophia Muscularis, a Mouse with Inherited Muscular Dystrophy. Proc. Soc. Exper. Biol. Med., **97**: 403, 1958.

1201. Shultz, F. T., and A. C. Andersen: Reproductive Fitness in Female Dogs Exposed to Whole-Body X-radiation. Radiat. Res., **9**: 182, 1958.

1202. Silvers, W. K.: An Experimental Approach to Action of Genes at the Agouti Locus in the Mouse. II. Transplants of Newborn *aa* Ventral skin to $a^t a$, $A^w a$ and *aa* Hosts. J. Exper. Zool., **137**: 181, 1958.

1203. Silvers, W. K.: An Experimental Approach to Action of Genes at the Agouti Locus in the Mouse. III. Transplants of Newborn A^w-, *A*-, and a^t- skin to A^y-, A^w-, *A*- and *aa* Hosts. J. Exper. Zool., **137**: 189, 1958.

1204. Silvers, W. K.: A Histological and Experimental Approach to Determine the Relationship between Gold-impregnated Dendritic Cells and Melanocytes. Amer. J. Anat., **100**: 225, 1957.

1205. Silvers, W. K.: Melanoblast Differentiation Secured from Different Mouse Genotypes after Transplantation to Adult Mouse Spleen or to Chick Embryo Coelom. J. Exper. Zool., **135**: 221, 1957.

1206. Silvers, W. K.: Origin and Identity of Clear Cells Found in Hair Bulbs of Albino Mice. Anat. Rec., **130**: 135, 1958.

1207. Silvers, W. K.: Pigment Cells: Occurrence in Hair Follicles. J. Morphol., **99**: 41, 1956.

1208. Silvers, W. K., and E. S. Russell: An Experimental Approach to Action of Genes at the Agouti Locus in the Mouse. J. Exper. Zool., **130**: 199, 1955.

1209. Simmonds, D. H.: Automatic Equipment for Determination of Amino Acids Separated on Columns of Ion Exchange Resins. Anal. Chem., **30**: 1043, 1958.

1210. Singer, S. J., and H. A. Itano: On the Asymmetrical Dissociation of Human Hemoglobin. Proc. Nat. Acad. Sci. (Wash.), **45**: 174, 1959.

1211. Singer, S. J., and E. S. Russell: Some Physical Chemical Experiments with Hemoglobins from Normal and Genetically Anemic Mice. Proc. Nat. Acad. Sci. (Wash.), **40**: 6, 1954.

1212. Sirlin, J. L.: Location of Vacillans in Linkage Group VIII of the House Mouse. Heredity, **11**: 259, 1957.

1213. Sirlin, J. L.: Vacillans, a Neurological Mutant in the House Mouse Linked with Brown. J. Genet., **54**: 42, 1956.

1214. Sirotnak, F. M., and D. J. Hutchison: Absorption of Deoxyribonucleic Acid by Mouse Lymphoma Cells. Biochim. Biophys. Acta, **36**: 246, 1959.

1215. Sjögren, H. O., I. Hellstrom, and G. Klein: Transplantation of Polyoma Virus-induced Tumors in Mice. Cancer Res. **21**: 329, 1961.

1216. Slatis, H. M., R. H. Reis, and R. E. Hoene: Consanguineous Marriages in the Chicago Region. Amer. J. Hum. Genet., **10**: 446, 1958.

1217. Slizynski, B. M.: A Preliminary Pachytene Chromosome Map of the House Mouse. J. Genet., **49**: 242, 1949.

1218. Slizynski, B. M.: Cytological Analysis of Translocations in the Mouse. J. Genet., **55**: 122, 1957.

1219. Slizynski, B. M.: Partial Sex Linkage in the Mouse. Nature, **174**: 309, 1954.

1220. Smith, E. W., and J. V. Torbert: Study of Two Abnormal Hemoglobins with Evidence for a New Genetic Locus for Hemoglobin Formation. Bull. Johns Hopkins Hosp., **102**: 38, 1958.

1221. Smith, L. J.: A Morphological and Histochemical Investigation of a Preimplantation Lethal (t^{12}) in the House Mouse. J. Exper. Zool., **132**: 51, 1956.

1222. Smith, P. E., and E. C. MacDowell: An Hereditary Anterior-pituitary Deficiency in the Mouse. Anat. Rec., **46**: 249, 1930.

1223. Smith, P. E., and E. C. MacDowell: The Differential Effect of Hereditary Mouse Dwarfism on the Anterior-pituitary Hormones. Anat. Rec., **50**: 85, 1931.

1224. Smith, W. E.: The Neoplastic Potentialities of Mouse Embryo Tissues. V. The Tumors Elicited with Methylcholanthrene from Pulmonary Epithelium. J. Exper. Med., **91:** 87, 1950.

1225. Smith, W. E.: The Tissue Transplant Technic as a Means of Testing Materials for Carcinogenic Action. Cancer Res., **9:** 712, 1949.

1226. Smithberg, M., and M. N. Runner: Pregnancy Induced in Genetically Sterile Mice. J. Hered., **48:** 97, 1957.

1227. Smithies, O.: An Improved Procedure for Starch-gel Electrophoresis: Further Variations in the Serum Proteins of Normal Individuals. Biochem. J., **71:** 585, 1959.

1228. Smithies, O.: Zone Electrophoresis in Starch Gels and its Application to Studies of Serum Proteins. *In* Advances in Protein Chemistry. C. B. Anfinson, Jr., *et al.*, eds., Academic Press, New York, **14:** 65, 1959.

1229. Smithies, O.: Zone Electrophoresis in Starch Gels: Group Variations in the Serum Proteins of Normal Human Adults. Biochem. J., **61:** 629, 1955.

1230. Snedecor, G. W.: Statistical Methods. Fifth ed., Iowa State Coll. Press, Ames, 1956.

1231. Snell, G. D.: A Cross-over Between the Genes for Short-ear and Density in the House Mouse. Proc. Nat. Acad. Sci. (Wash.), **14:** 926, 1928.

1232. Snell, G. D.: An Analysis of Translocations in the Mouse. Genetics, **31:** 157, 1946.

1233. Snell, G. D. ed.: Biology of the Laboratory Mouse. The Blakiston Co., New York, 1941.

1234. Snell, G. D.: Ducky, a New Second Chromosome Mutation in the Mouse. J. Hered., **46:** 27, 1955.

1235. Snell, G. D.: Dwarf, a New Mendelian Recessive Character of the House Mouse. Proc. Nat. Acad. Sci. (Wash.), **15:** 733, 1929.

1236. Snell, G. D.: Effect of Injection of Anterior-pituitary Extract on the Thyroids of Mice with Hereditary Dwarfism. Anat. Rec., **47:** 316, 1930.

1237. Snell, G. D.: Histocompatibility Genes of the Mouse. I. Demonstration of Weak Histocompatibility Differences by Immunization and Controlled Tumor Dosage. J. Nat. Cancer Inst., **20:** 787, 1958.

1238. Snell, G. D.: Histocompatibility Genes of the Mouse. II. Production and Analysis of Isogenic Resistant Lines. J. Nat. Cancer Inst., **21:** 843, 1958.

1239. Snell, G. D.: Incompatibility Reactions to Tumor Homotransplants with Particular Reference to the Role of the Tumor: A Review. Cancer Res., **17:** 2, 1957.

1240. Snell, G. D.: Inheritance in the House Mouse, the Linkage Relations of Short Ear, Hairless, and Naked. Genetics, **16:** 42, 1931.

1241. Snell, G. D.: Linkage of Jittery and Waltzing in the Mouse. J. Hered., **36:** 279, 1945.

1242. Snell, G. D.: Methods for the Study of Histocompatibility Genes. J. Genet., **49:** 87, 1948.

1243. Snell, G. D.: Mouse Genetic News, R. B. Jackson Mem. Lab., Bar Harbor, Maine, **1:** 7, 1941.

1244. Snell, G. D., and L. C. Stevens: Histocompatibility Genes of the Mouse. III. H-1 and H-4, Two Histocompatibility Loci in the First Linkage Group. Immunology, **4:** 366, 1961.

1245. Snell, G. D.: Note on Results of Linder and Klein with Coisogenic Resistant Lines of Mice. J. Nat. Cancer Inst., **25:** 1191, 1960.

1246. Snell, G. D.: Preliminary Data on Crossing-over Between *H-2, Fu, Ki,* and *T* in the Mouse. Heredity, **6:** 247, 1952.

1247. Snell, G. D.: The Early Embryology of the Mouse. *In* Biology of the Laboratory Mouse. G. D. Snell, ed., The Blakiston Co., New York, 1, 1941.

1248. Snell, G. D.: The Genetics of Transplantation. J. Nat. Cancer Inst., **14:** 691, 1953.

1249. Snell, G. D.: The Homograft Reaction. Ann. Rev. Microbiol., **11:** 439, 1957.

1250. Snell, G. D.: The Induction by X-rays of Hereditary Changes in Mice. Genetics, **20:** 545, 1935.

1251. Snell, G. D., S. Counce, P. Smith, L. Dube, and D. E. Kelton: A 5th Chromosome Histocompatibility Locus Identified in the Mouse by Tumor Transplantation. Proc. Amer. Assoc. Cancer Res., **2:** 46, 1955.

1252. Snell, G. D., M. M. Dickie, P. Smith, and D. E. Kelton: Linkage of Loop-tail, Leaden, Splotch and Fuzzy in the Mouse. Heredity, **8:** 271, 1954.

1253. Snell, G. D., and L. W. Law: A Linkage between Shaker-2 and Wavy-2 in the House Mouse. J. Hered., **30:** 447, 1939.

1254. Snell, G. D., and J. Staats: Inbred Strains of Mice, No. 1. Companion Issue to Mouse News Letter No. 21, 1959.

1255. Soemarwoto, I. N., and H. A. Bern: The Effect of Hormones on the Vascular Pattern of the Mouse Mammary Gland. Amer. J. Anat., **103:** 403, 1958.

1256. Sonneborn, T. M.: Gene and Cytoplasm. I. The Determination and Inheritance of the Killer Character in Variety 4 of *Paremecium aurelia*. II. The Bearing and the Determination and Inheritance of Characters in *Paramecium aurelia* on the Problems of Cytoplasmic Inheritance, Penumococcus Transformations, Mutations and Development. Proc. Nat. Acad. Sci. (Wash.), **29:** 329, 1943.

1257. Sonneborn, T. M.: The Gene and Cell Differentiation. Proc. Nat. Acad. Sci. (Wash.), **46:** 149, 1960.

1258. Sorsby, A., P. C. Keller, M. Attfield, J. B. Davey, and D. R. Lucas: Retinal Dystrophy in the Mouse: Histological and Genetic Aspects. J. Exper. Zool., **125:** 171, 1954.

1259. Spackman, D. H., W. H. Stein, and S. Moore: Automatic Recording Apparatus for Use in Chromatography of Amino Acids. Anal. Chem., **30:** 1190, 1958.

1260. Speirs, R. S.: Principles of Eosinophil Diluents. Blood, **7:** 550, 1952.

1261. Speirs, R. S., and M. E. Dreisbach: Quantitative Studies of the Cellular Responses to Antigen Injections in Normal Mice. Technic for Determining Cells in the Peritoneal Fluid. Blood, **11:** 44, 1956.

1262. Sproul, J. A., Jr.: Estimates of Recovery Rate in Mice Exposed to Neutrons and Gamma Rays. Radiat. Res., **9:** 187, 1958.

1263. Staats, J.: A Classified Bibliography of Inbred Strains of Mice. Science, **119:** 295, 1954.

1264. Stadler, J., and J. W. Gowen: Contributions to Survival Made by Body Cells of Genetically Differentiated Strains of Mice Following X Irradiations. Biol. Bull., **112:** 400, 1957.

1265. Stadler, J., and J. W. Gowen: Radiological Effects on Resistance Mechanisms of Genetically Differentiated Strains of Mice Exposed to *Salmonella typhimurium*. J. Infect. Diseases, **100:** 284, 1957.

1265a. Stadler, J., and J. W. Gowen: Radiation Effects on Active Acquired Immunity to *Salmonella typhimurium* in Mice. J. Infect. Diseases, **100:** 300, 1957.

1266. Staff of the Roscoe B. Jackson Memorial Laboratory: The Existence of Non-chromosomal Influence in the Incidence of Mammary Tumors in Mice. Science, **78:** 465, 1933.

1267. St. Amand, W., and M. B. Cupp: Mouse News Letter, **17:** 88, 1957.

1268. St. Amand, W., and M. B. Cupp: Mouse News Letter, **19:** 38, 1958.

1269. Stamer, S.: Effect of a Carcinogenic Hydrocarbon on Manifest Malignant Tumors in Mice. Copenhagen, Munksgaard, 1943.

1270. Stavitsky, A. B.: Micromethods for the Study of Proteins and Antibodies. I. Procedure and General Applications of Hemagglutination and Hemagglutination-inhibition Reactions with Tannic Acid and Protein-treated Red Blood Cells. J. Immunol., **72:** 360, 1954.

1271. Stein, W. H.: Observations of the Amino Acid Composition of Human Hemoglobins. Conference on Hemoglobin. National Academy of Sciences, National Research Council, **557:** 220, 1958.

1272. Stein, W. H., H. G. Kunkel, R. D. Cole, D. H. Spackman, and S. Moore: Observations on the Amino Acid Composition of Human Hemoglobins. Biochim. Biophys. Acta, **24:** 640, 1957.

1273. Steinberg, A. G.: A Reconsideration of the Mode of Development of the Bar Eye of *Drosophila melanogaster*. Genetics, **26:** 325, 1941.

1274. Steinberg, D., M. Vaughn, C. B. Anfinsen, and J. Gorry: Preparation of Tritiated Proteins by the Wilzbach Method. Science, **126:** 447, 1957.

1275. Stern, C.: Somatic Crossing Over and Segregation in *Drosophila melanogaster*. Genetics, **21:** 625, 1936.

1276. Stern, C.: The Nucleus and Somatic Cell Variation. J. Cell. Comp. Physiol., **52** Suppl. 1: 1, 1958.

1277. Stern, C., and V. Rentschler: The Effect of Temperature on the Frequency of Somatic Crossing-Over in *Drosophila melanogaster*. Proc. Nat. Acad. Sci. (Wash.), **22:** 451, 1936.

1278. Stevens, L. C., and K. P. Hummel: A Description of Spontaneous Congenital Testicular Teratomas in Strain 129 Mice. J. Nat. Cancer Inst., **18:** 719, 1957.

1279. Stevens, L. C., and J. A. Mackenson: The Inheritance and Expression of a Mutation in the Mouse Affecting Blood Formation, the Axial Skeleton, and Body Size. J. Hered., **49:** 153, 1958.

1280. Stevens, L. C., J. Mackenson, and S. E. Bernstein: A Mutation Causing Neonatal Jaundice in the House Mouse. J. Hered., **50:** 35, 1959.

1281. Stevens, L. C., E. S. Russell, and J. L. Southard: Evidence on Inheritance of Muscular Dystrophy in an Inbred Strain of Mice Using Ovarian Transplantation. Proc. Soc. Exper. Biol. Med., **95:** 161, 1957.

1282. Stevenson, A. C.: The Load of Hereditary Defects in Human Populations. Radiat. Res., Suppl. **1:** 306, 1959.

1283. Stone, S. H.: Method for Obtaining Venous Blood From the Orbital Sinus of the Rat or Mouse. Science, **119:** 100, 1954.
1284. Stone, W. H., and J. Beckstrom: A Technique for Preserving Cattle Whole Blood in Shipment. Immunogen. Letter, **1:** 12, 1960.
1285. Stone, W. H., W. J. Tyler, and M. R. Irwin: A Technique of Freezing and Storing Cattle Erythrocytes for Use in Blood Typing. J. Animal Sci., **17:** 1218, 1958.
1286. Stormont, C.: Linked Genes, Pseudoalleles and Blood Groups. Amer. Nat., **89:** 105, 1955.
1287. Stormont, C., Y. Suzuki, and W. J. Miller: Bacterially Mediated False-positive Reactions in Lytic Blood-typing Tests. Nature, **186:** 247, 1960.
1288. Stormont, C., W. C. Weir, and L. L. Lane: Erythrocyte Mosaicism in a Pair of Sheep Twins. Science, **118:** 695, 1953.
1289. Strandskov, H. H.: Effects of X Rays in an Inbred Strain of Guinea Pigs. J. Exper. Zool., **63:** 175, 1932.
1290. Stratton, F., and P. H. Renton: Practical Blood Grouping. C. C Thomas, Springfield, Ill., 1958.
1291. Stretton, A. O. W., and V. M. Ingram: An Amino Acid Difference between Human Hemoglobins A and A_2. Fed. Proc., **19:** 343, 1960.
1292. Strong, L. C.: A Genetic Analysis of the Factors Underlying Susceptibility to Transplantable Tumors. J. Exper. Zool., **36:** 67, 1922.
1293. Strong, L. C.: A Genetic Analysis of the Induction of Tumors by Methylcholanthrene with a Note on the Origin of the NH Strain of Mice. Amer. J. Cancer, **39:** 347, 1940.
1294. Strong, L. C.: Genetic Analysis of the Induction of Tumors by Methylcholanthrene. IX. Induced and Spontaneous Adenocarcinomas of the Stomach in Mice. J. Nat. Cancer Inst., **5:** 339, 1945.
1295. Strong, L. C.: On the Occurrence of Mutations within Transplantable Neoplasms. Genetics, **11:** 294, 1926.
1296. Strong, L. C.: Susceptibility to Fibrosarcomas in 2NHO Mice. Yale J. Biol. Med., **24:** 109, 1952.
1297. Strong, L. C.: The Establishment of the A Strain of Inbred Mice. J. Hered., **27:** 21, 1936.
1298. Strong, L. C.: The Origin of Some Inbred Mice. Cancer Res., **2:** 531, 1942.
1299. Strong, L. C., and C. C. Little: Tests for Physiological Differences in Transplantable Tumors. Proc. Soc. Exper. Biol. Med., **18:** 45, 1920.
1300. Strumia, M. M., A. B. Sample, and E. D. Hart: An Improved Micro-hematocrit Method. Amer. J. Clin. Pathol., **24:** 1016, 1954.
1301. Stulberg, C. S., H. D. Soule, and L. Berman: Preservation of Human Epithelial-like and Fibroblast-like Cell Strains at Low Temperature. Proc. Soc. Exper. Biol. Med., **98:** 428, 1958.
1302. Sumner, F. B.: Continuous and Discontinuous Variations and Their Inheritance in *Peromyscus*. Amer. Nat., I, **52:** 177; II, **52:** 250; III, **52:** 439, 1918.
1303. Sumner, F. B.: Genetic, Distributional, and Evolutionary Studies of the Subspecies of Deer Mice (*Peromyscus*). Bibliog. Genet., **9:** 1, 1932.
1304. Svedberg, T., and K. O. Pederson: Theory of Sedimentation: Elementary Theory. *In* The Ultracentrifuge. R. H. Fowler and P. Kapitza, eds., Clarendon Press, Oxford, 5, 1940.
1305. Swanson, C. P.: Cytology and Cytogenetics. Prentice-Hall, Inc., Englewood Cliffs, N.J., 1957.
1306. Szilard, L.: The Molecular Basis of Antibody Formation. Proc. Nat. Acad. Sci. (Wash.), **46:** 293, 1960.
1307. Szybalski, W.: Genetics of Human Cell Lines. II. Method for Determination of Mutation Rate to Drug Resistance. Exper. Cell Res., **18:** 588, 1959.
1308. Talmage, D. W.: Immunological Specificity. Science, **129:** 1643, 1959.
1309. Tanaka, T.: A Study of the Somatic Chromosomes in Various Organs of the White Rat (*Rattus norvegicus*) Especially with Regard to the Number and its Variations. Res. Genet, **2:** 39, 1951.
1310. Tanaka, T., and K. Kano: On the Somatic Chromosomes of Rats. Proc. Internat. Genetics Symp., Cytologia Suppl. 196, 1956.
1311. Tannebaum, A., and H. Silverstone: Genesis and Growth of Tumors, Effects of Varying the Proportion of Protein (Casein) in the Diet. Cancer Res., **9:** 162, 1949.
1312. Tannenbaum, A., and H. Silverstone: The Influence of the Degree of Caloric Restriction on the Formation of Skin Tumors and Hepatomas in Mice. Cancer Res., **9:** 724,1949.
1313. Tansley, K.: Hereditary Degeneration of the Mouse Retina. Brit. J. Ophthalmol., **35:** 573, 1951.
1314. Tassoni, J. P., R. L. Curtis, and M. B. Hollingshead: Progessive Biochemical and Histochemical Changes in Muscular Dystrophy. Anat. Rec., **133:** 342, 1959.

1315. Taylor, A. C.: Survival of Rat Skin and Changes in Hair Pigmentation Following Freezing. J. Exper. Zool., **110:** 77, 1949.

1316. Taylor, J. H.: Asynchronous Duplication of Chromosomes in Cultured Cells of Chinese Hamster. J. Biophys. Biochem. Cytol., **7:** 455, 1960.

1317. Terasaki, P. I., J. A. Cannon, and W. P. Longmire, Jr.: Antibody Response to Homografts. I. Technic of Lymphoagglutination and Detection of Lympho Agglutinins Upon Spleen Injection. Proc. Soc. Exper. Biol. Med., **102:** 280, 1959.

1318. Tiselius, A.: A New Apparatus for Electrophoretic Analysis of Colloidal Mixtures. Trans. Faraday Soc., **33:** 524, 1937.

1319. Tjio, J. H., and A. Levan: Chromosome Analysis of Three Hyperdiploid Ascites Tumors of the Mouse. Kgl. Fysiografiska Sällskapets Handlingar, Lund, **65:** 1, 1954.

1320. Tjio, J. H., and A. Levan: Some Experiences with Aceto Orcein in Animal Chromosomes. Anal. Est. Exper. Aula Dei, **3:** 225, 1954.

1321. Tjio, J. H., and G. Ostergren: The Chromosomes of Primary Mammary Carcinomas in Milk Virus Strains of the Mouse. Hereditas, **44:** 451, 1958.

1322. Tjio, J. H., and T. T. Puck: Genetics of Somatic Mammalian Cells. II. Chromosomal Constitution of Cells in Tissue Culture. J. Exper. Med., **108:** 259, 1958.

1323. Tjio, J. H., T. T. Puck, and A. Robinson: The Human Chromosomal Satellites in Normal Persons and in Two Patients with Marfan's Syndrome. Proc. Nat. Acad. Sci. (Wash.), **46:** 532, 1960.

1324. Tonomura, A., and G. Yerganian: Aneuploidy in the Bone Marrow of the Chinese Hamster. Anat. Rec., **127:** 377, 1957.

1325. Topley, W. W. C.: Spread of Bacterial Infection. Lancet, **2:** 1, 45, 91, 1919.

1326. Totter, J. R., M. R. Zelle, and H. Hollister: Hazard to Man of Carbon-14. Science, **128:** 1490, 1958.

1327. Trasler, D. G.: Genetic and Other Factors Influencing the Pathogenesis of Cleft Palate in Mice. Ph. D. thesis, McGill Univ., 1958.

1328. Trasler, D. G.: Influence of Uterine Site on Occurrence of Spontaneous Cleft Lip in Mice. Science, **132:** 420, 1960.

1329. Trasler, D. G., and F. C. Fraser: Factors Underlying Strain, Reciprocal Cross, and Maternal Weight Differences in Embryo Susceptibility to Cortisone-induced Cleft Palate in Mice. Proc. Tenth Internat. Congr. Genetics, **2:** 296, 1958.

1330. Trasler, D. G., B. E. Walker, and F. C. Fraser: Congenital Malformations Produced by Amniotic-sac Puncture. Science, **124:** 439, 1956.

1331. Trentin, J. J., and W. U. Gardner: Site of Gene Action in Susceptibility to Estrogen-induced Testicular Interstitial-cell Tumors of Mice. Cancer Res., **18:** 110, 1958.

1332. Tuchmann-Duplessis, H., and L. Mercier-Parot: Sur l'Action Tératogène de l'Acide *x*-Méthylfolique chez la Souris. C. R. Acad. Sci., **245:** 1963, 1957.

1333. Tuffery, A. A.: The Health of Laboratory Mice. A Comparison of General Health in Two Breeding Units Where Different Systems are Employed. J. Hyg., **57:** 386, 1959.

1334. Tyzzer, E. E.: A Study of Inheritance in Mice with Reference to Their Susceptibility to Transplantable Tumors. J. Med. Res., **21:** 519, 1909.

1335. Uetake, H., S. E. Luria, and J. W. Burrous: Conversion of Somatic Antigens in *Salmonella* by Phage Infection Leading to Lysis or Lysogeny. Virology, **5:** 68, 1958.

1336. Upton, A. C., J. A. Sproul, Jr., and M. L. Randolph: Survival of Mice under Chronic Exposure to Neutrons and Gamma Rays. Radiat. Res., **9:** 197, 1958.

1337. Valmet, E., and H. Svensson: Some Problems Inherent in Paper Electrophoresis: The LKB Paper Electrophoresis Apparatus. Science Tools, **1:** 3, 1954.

1338. van der Helm, H. J., G. van Vliet, and T. H. J. Huisman: Investigations on Two Different Hemoglobins of the Sheep. Arch. Biochem. Biophys., **72:** 331, 1957.

1339. Van Gulick, P. J., and R. Korteweg: Susceptibility to Follicular Hormone and Disposition to Mammary Cancer in Female Mice. Amer. J. Cancer, **38:** 506, 1940.

1340. Vella, F.: Haemoglobin Types in Ox and Buffalo. Nature, **181:** 564, 1958.

1341. Venge, O.: Studies of the Maternal Influence on the Birth Weight in Rabbits. Acta Zool., **31:** 1, 1950.

1342. Vesselinovitch, S. D.: A Simple Method for Making Starch-gel Electrophoretic Strips Transparent. Nature, **182:** 665, 1958.

1343. Vicari, E. M.: Establishment of Differences in Susceptibility to Sound-induced Seizures in Various Endocrinic Types of Mice. Anat. Rec., **97:** 3, 1947.

1344. Vicari, E. M.: Fatal Convulsive Seisures in the DBA Mouse Strain. J. Psychol., **32:** 79, 1951.

1345. Vinograd, J., and W. D. Hutchinson: Carbon-14 Labelled Hybrids of Haemoglobin.   Nature, **187**: 216, 1960.

1346. Vinograd, J., W. D. Hutchinson, and W. A. Schroeder: C^{14}-Hybrids of Human Hemoglobins.   II. The Identification of the Aberrant Chain in Human Hemoglobin S.   J. Amer. Chem. Soc., **81**: 3168, 1959.

1347. Vlahakis, G., and W. E. Heston: Relationship between Recessive Obesity and Induced Pulmonary Tumors in Mice.   J. Hered., **50**: 99, 1959.

1348. Vogt, M.: A Genetic Change in a Tissue Culture Line of Neoplastic Cells.   J. Cell. Comp. Physiol. 52 Suppl. 1: 271, 1958.

1349. Vogt, M.: A Study of the Relationship between Karotype and Phenotype in Cloned Lines of Strain HeLa.   Genetics, **44**: 1257, 1959.

1350. Vogt, M., and R. Dulbecco: Virus-cell Interaction with a Tumor Producing Virus.   Proc. Nat. Acad. Sci. (Wash.), **46**: 365, 1960.

1351. Vos, O., J. W. Goodman, and C. C. Congdon: Transplantation of Chimeric Bone Marrow.   Internat. J. Radiat. Biol., **3**: 29, 1961.

1352. Waddington, C. H.: The Strategy of the Genes.   The Macmillan Co., New York, 1957.

1352a. Wagener, G.: Die Entstehung der Scheckung bei dem Haubenratte.   Biol. Zbl., **78**: 451, 1959.

1353. Wagner, R. P., and H. K. Mitchell: Genetics and Metabolism.   J. Wiley and Sons, New York, 1955.

1354. Walker, B. E., and F. C. Fraser: The Embryology of Cortisone-induced Cleft Palate.   J. Embryol. Exper. Morphol., **5**: 201, 1957.

1355. Wallace, M. E.: A Balanced Three-point Experiment for Linkage Group V of the House Mouse.   Heredity, **11**: 223, 1957.

1356. Wallace, M. E.: Locus of the Gene "Fidget" in the House Mouse.   Nature, **166**: 407, 1950.

1356a. Wallace, M. E.: New Linkage and Independence Data for Ruby and Jerker in the Mouse.   Heredity, **12**: 453, 1958.

1357. Warkany, J.: Congenital Malformations Induced by Maternal Nutritional Deficiency.   J. Pediat., **25**: 476, 1944.

1358. Warkany, J., J. G. Wilson, and J. F. Geiger: Myeloschisis and Myelomeningocele Produced Experimentally in the Rat.   J. Comp. Neurol., **109**: 35, 1958.

1359. Waymouth, C.: Rapid Proliferation of Sublines of NCTC Clone 929 (Strain L) Mouse Cells in a Simple Chemically Defined Medium (MB752/1).   J. Nat. Cancer Inst., **22**: 1003, 1959.

1360. Weaver, E. C.: Somatic Crossing Over and its Genetic Control in *Drosophila*.   Genetics, **45**: 345, 1960.

1361. Webster, L. T.: Inheritance of Resistance of Mice to Enteric Bacterial and Neurotropic Virus Infections.   J. Exper. Med., **65**: 261, 1937.

1362. Webster, L. T.: Inherited and Acquired Factors in Resistance to Infection.   I. Development of Resistant and Susceptible Lines of Mice Through Selective Breeding.   J. Exper. Med., **57**: 793, 1933.

1363. Webster, L. T.: Inherited and Acquired Factors in Resistance to Infection.   II. A Comparison of Mice Inherently Resistant or Susceptible to *Bacillus enteritidis* Infection with Respect to Fertility, Weight and Susceptibility to Various Routes and Types of Infection.   J. Exper. Med., **57**: 819, 1933.

1364. Webster, L. T.: Microbe Virulence and Host Susceptibility in Mouse Typhoid Infection.   J. Exper. Med., **37**: 231, 1923.

1365. Webster, L. T.: The Application of Experimental Methods of Epidemiology.   Amer. J. Hyg., **4**: 134, 1924.

1366. Wedum, A. G., E. Hanel, G. B. Phillips, and O. T. Miller: Laboratory Design for Study of Infectious Disease.   Amer. J. Pub. Health, **46**: 1102, 1956.

1367. Wegelius, O.: The Dwarf Mouse—An Animal with Secondary Myxedema.   Proc. Soc. Exper. Biol. Med., **101**: 225, 1959.

1368. Wegelius, O., and H. Elftmann: Physiological and Cytological Evidence for Deficiency of Thyrotropic Hormone in the Dwarf Mouse.   Acta Pathol. Micro. Scand., **47**: 209, 1959.

1369. Weicker, H., and K. H. Terwey: Die Chromosomenzahl der Erythroblasten.   Klin. Wchnschr., **36**: 1132, 1958.

1370. Weil, L., and T. S. Seibles: Specificity of Protaminase.   Fed. Proc., **17**: 332, 1958.

1371. Weiler, E.: A Cellular Change in Hamster Kidney Cultures: Loss of Tissue-specific Antigens.   Exper. Cell Res., Suppl. **7**: 244, 1959.

1372. Weinstock, I. M., S. Epstein, and A. T. Milhorat: Enzyme Studies in Muscular Dystrophy.   III. In Hereditary Muscular Dystrophy in Mice. Proc. Soc. Exper. Biol. Med., **99**: 272, 1958.

1373. Weir, J. A.: Association of Blood-*p*H with Sex Ratio in Mice. J. Hered., **44:** 133, 1953.
1374. Weir, J. A.: Blood-*p*H as a Factor in Genetic Resistance to Mouse Typhoid. J. Infect. Diseases, **84:** 252, 1949.
1375. Weir, J. A.: Genetics and Laboratory Animal Diseases. Proc. Animal Care Panel, **10:** 177, 1960.
1376. Weir, J. A.: Genetic Resistance of Mice to Live and Heat-killed *Salmonella typhimurium.* M.S. Thesis. Iowa State Coll., 1942.
1377. Weir, J. A., R. H. Cooper, and R. D. Clark: The Nature of Genetic Resistance to Infection in Mice. Science, **117:** 328, 1953.
1378. Welch, A. D.: The Problem of Drug Resistance in Cancer Chemotherapy. Cancer Res., **19:** 359, 1959.
1379. Welling, W., and D. W. van Bekkum: Different Types of Haemoglobin in Two Strains of Mice. Nature, **182:** 946, 1958.
1380. Wellman, E. L., and F. Jacob: Lysogeny, Transduction and Cancer Genesis. *In* Genetics and Cancer. R. W. Cumley, *et al.*, Univ. of Texas Press, Austin, 43, 1959.
1381. Wells, I. C., and H. A. Itano: Ratio of Sickle-cell Anemia Hemoglobin to Normal Hemoglobin in Sicklemics. J. Biol. Chem., **188:** 65, 1951.
1382. Welsohns, W. J., and L. B. Russell: The *Y*-Chromosome as the Bearer of Male-determining Factors in the Mouse. Proc. Nat. Acad. Sci. (Wash.), **45:** 560, 1959.
1383. West, W. T., and E. D. Murphy: Histopathology of Hereditary Progressive Muscular Dystrophy in Inbred Strain 129 Mice. Anat. Rec., **137:** 279, 1960.
1384. White, J. C., and G. H. Beaven: A Review of the Varieties of Human Haemoglobin in Health and Disease. J. Clin. Pathol., **7:** 175, 1954.
1385. White, J. C., and G. H. Beaven: Foetal Haemoglobin. Brit. Med. Bull., **15:** 33, 1959.
1386. Whittinghill, M.: Some Effects of Gamma Rays on Recombination and on Crossing Over in *Drosophila melanogaster.* Genetics, **36:** 332, 1951.
1387. Wieland, T., and E. Fischer: Über Electrophorese auf Filtrierpapier. Naturwissenschaften, **35:** 29, 1948.
1388. Wiener, A. S., and L. Katz: Studies on the Use of Enzyme-treated Red Cells in Tests for *Rh* Sensitization. J. Immunol., **66:** 51, 1951.
1389. Wildy, P., and A. A. Newton: The Synchronous Division of HeLa Cells. Proc. Biochem. Soc. (Biochem. J.), **68:** 14, 1958.
1390. Williams, W. L., J. B. Cardle, and R. D. Meader: The Nature of Dietary Fat and the Pattern of Hepatic Liposis in Choline-deficient Mice. Yale J. Biol. Med., **31:** 263, 1959.
1391. Wilson, J. G.: Differentiation and the Reaction of Rat Embryos to Radiation. J. Cell. Comp. Physiol., **43:** 11, 1954.
1392. Wilson, J. G.: Factors Involved in Causing Congenital Malformations. Bull. N.Y. Acad. Med., **36:** 145, 1960.
1393. Wilson, J. G., A. R. Beaudoin, and H. J. Free: Studies on the Mechanism of Teratogenic Action of Trypan Blue. Anat. Rec., **133:** 115, 1959.
1394. Wilson, J. G., C. B. Roth, and J. Warkany: An Analysis of the Syndrome of Malformations Induced by Maternal Vitamin A Deficiency. Effects of Restoration of Vitamin A at Various Times During Gestation. Amer. J. Anat., **92:** 189, 1953.
1395. Wilson, S., and D. B. Smith: Separation of the Valyl-leucyl- and Valyl-glutamyl-polypeptide Chains of Horse Globin by Fractional Precipitation and Column Chromatography. Canadian J. Biochem. Physiol., **37:** 405, 1959.
1396. Winn, H. J., L. C. Stevens, and G. D. Snell: Tests of Alternative Methods for Demonstrating the Histocompatibility-1 Isoantigen in Mice. Transplant. Bull., **5:** 18, 1958.
1397. Wintrobe, M. M.: Clinical Hematology. Fifth ed. Lea and Febiger, Philadelphia, 1961.
1398. Wolfe, H. G.: Blood-*p*H Differences in Two Inbred Strains of Mice. J. Hered., **50:** 155, 1959.
1399. Wolff, E., and E. Wolff: Les Résultats d'une Nouvelle Méthode de Culture de Cellules Cancéreuses *in vitro.* Rev. Franc. Etud. Clin. Biol., **3:** 945, 1958.
1400. Wolff, G. L.: A Sex Difference in the Coat Color Change of a Specific Guinea Pig Genotype. Amer. Nat., **88:** 381, 1954.
1401. Wolff, G. L.: The Effects of Environmental Temperature on Coat Color in Diverse Genotypes of the Guinea Pig. Genetics, **40:** 90, 1955.
1402. Wood, D. C.: A Preliminary Study of Some Effects of Antibody on Eye Differentiation in Utero. S. J. Onesti, ed., Twenty-third Ross Pediatric Conf., 77, 1957.
1403. Woodruff, M. F. A., and M. Sparrow: Induction of Tolerance to Homografts of Thyroid and Adrenal in Rats. Quart. J. Exper. Physiol., **43:** 91, 1958.

1404. Woolley, G. W.: Misty Dilution in the Mouse. J. Hered., **36:** 269, 1945.
1405. Woolley, G. W.: The Adrenal Cortex and its Tumors. Ann. N.Y. Acad. Sci., **50:** 616, 1949.
1406. Woolley, G. W.: Tumor Cell Resistance to Antimetabolites and Possible Genetic Implications. *In* Genetics and Cancer. R. W. Cumley, *et al.*, eds., Univ. of Texas Press, Austin, 349, 1959.
1407. Woolley, G. W., M. M. Dickie, and C. C. Little: Adrenal Tumors and Other Pathological Changes in Reciprocal Crosses in Mice. I. Strain DBA and Strain CE and the Reciprocal. Cancer Res., **12:** 142, 1952.
1408. Woolley, G. W., M. M. Dickie, and C. C. Little: Adrenal Tumors and Other Pathological Changes in Reciprocal Crosses in Mice. II. An Introduction to Results of Four Reciprocal Crosses. Cancer Res., **13:** 231, 1953.
1409. Woolley, G. W., and C. C. Little: The Incidence of Adrenal Cortical Carcinoma in Gonadectomized Female Mice of the Extreme Dilution Strain. I. Observations on the Adrenal Cortex. Cancer Res., **5:** 193, 1945.
1410. Worden, A. N., and W. Lane-Petter: The UFAW Handbook on the Care and Management of Laboratory Animals. Universities Fed. for Animal Welfare, London, 1957.
1411. World Health Organization Technical Report Series No. 166, Effect of Radiation on Human Heredity: Investigations of Areas of High Natural Radiation. Geneva, 1959.
1412. Wright, M. E.: Two Sex-linkages in the House Mouse with Unusual Recombination Values. Heredity, **1:** 349, 1947.
1413. Wright, S.: A Mutation of the Guinea Pig, Tending to Restore the Pentadactyl Foot When Heterozygous, Producing a Monstrosity When Homozygous. Genetics, **20:** 84, 1935.
1414. Wright, S.: An Analysis of Variability in Number of Digits in an Inbred Strain of Guinea Pigs. Genetics, **19:** 506, 1934.
1415. Wright, S.: A Quantitative Study of Variations in Intensity of Genotypes of the Guinea Pig at Birth. Genetics, **44:** 1001, 1959.
1416. Wright, S.: An Intensive Study of the Inheritance of Color and Other Coat Characters in Guinea Pigs with Especial Reference to Graded Variation. Carnegie Inst. Wash. Pub. No. 241, Part II, 1916.
1417. Wright, S.: Classification of the Factors of Evolution. Cold Spring Harbor Symp. Quant. Biol., **20:** 16, 1956.
1418. Wright, S.: Coefficients of Inbreeding and Relationship. Amer. Nat., **56:** 330, 1922.
1419. Wright, S.: Color Inheritance in Mammals. J. Hered., **8:** 224, 1917.
1420. Wright, S.: Effects of Age of Parents on Characteristics of the Guinea Pig. Amer. Nat., **60:** 552, 1926.
1421. Wright, S.: Estimates of the Amounts of Melanin in the Hair of Diverse Genotypes of the Guinea Pig from Transformation of Empirical Grades. Genetics, **34:** 245, 1949.
1422. Wright, S.: Evolution in Mendelian Populations. Genetics, **16:** 97, 1931.
1423. Wright, S.: Evolution in Populations in Approximate Equilibrium. J. Genet., **30:** 257, 1935.
1424. Wright, S.: Genetics of Abnormal Growth in the Guinea Pig. Cold Spring Harbor Symp. Quant. Biol., **2:** 137, 1934.
1425. Wright, S.: Genetics and the Hierarchy of Biological Sciences. Science, **130:** 959, 1959.
1426. Wright, S.: Inbreeding and Homozygosis. Proc. Nat. Acad. Sci. (Wash.), **19:** 411, 1933.
1427. Wright, S.: Modes of Selection. Amer. Nat., **90:** 5, 1956.
1428. Wright, S.: On the Genetics of Hair Direction in the Guinea Pig. I. Variability in the Patterns Found in Combinations of the R and M Loci. J. Exper. Zool., **112:** 303, 1949.
1429. Wright, S.: On the Genetics of Hair Direction in the Guinea Pig. II. Evidences for a New Dominant Gene Star and Tests for Linkage with Eleven Other Loci. J. Exper. Zool., **112:** 325, 1949.
1430. Wright, S.: On the Genetics of Hair Direction in the Guinea Pig. III. Interactions between the Processes Due to Loci R and ST. J. Exper. Zool., **113:** 33, 1950.
1431. Wright, S.: On the Genetics of Several Types of Silvering in the Guinea Pig. Genetics, **32:** 115, 1947.
1432. Wright, S.: On the Genetics of Silvering in the Guinea Pig with Especial Reference to Interaction and Linkage. Genetics, **44:** 387, 1959.
1433. Wright, S.: On the Genetics of Subnormal Development of the Head (Otocephaly) in the Guinea Pig. Genetics, **19:** 471, 1934.
1434. Wright, S.: Physiological and Evolutionary Theories of Dominance. Amer. Nat., **68:** 25, 1934.
1435. Wright, S.: Physiological Aspects of Genetics. Ann. Rev. Physiol., **7:** 75, 1945.

1436. Wright, S.: Physiological Genetics, Ecology of Populations and Natural Selection. Evolution After Darwin. I. The Evolution of Life. S. Tax, ed., The Univ. of Chicago Press, Chicago, 429, 1959.

1437. Wright, S.: Postnatal Changes in Intensity of Coat Color in Diverse Genotypes of the Guinea Pig. Genetics, **45**: 1503, 1960.

1438. Wright, S.: Qualitative Differences among Colors of the Guinea Pig Due to Diverse Genotypes. J. Exper. Zool., **142**: 75, 1959.

1439. Wright, S.: Residual Variability in Intensity of Coat Color at Birth in a Guinea Pig Colony. Genetics, **45**: 583, 1960.

1440. Wright, S.: Silvering (si) and Diminution (dm) of Coat Color of the Guinea Pig, and Male Sterility of the White or Near-white Combination of These. Genetics, **44**: 563, 1959.

1441. Wright, S.: Summary of Patterns of Mammalian Gene Action. J. Nat. Cancer Inst., **15**: 837, 1954.

1442. Wright, S.: Systems of Mating. Genetics, **6**: 117, 1921.

1443. Wright, S.: Systems of Mating. I. The Biometric Relations between Parent and Offspring. Genetics, **6**: 111, 1921.

1444. Wright, S.: The Analysis of Variance and the Correlations between Relatives with Respect to Deviations from an Optimum. J. Genet., **30**: 243, 1935.

1445. Wright, S.: The Effects of Inbreeding and Crossbreeding on Guinea Pigs. I. Decline in Vigor. U.S. Dept. Agric., Bull. **1090**: 1, 1922.

1446. Wright, S.: The Effects of Inbreeding and Crossbreeding on Guinea Pigs. II. Differentiation among Inbred Families. U.S. Dept. Agric., Bull. **1090**: 37, 1922.

1447. Wright, S.: The Effects of Inbreeding and Crossbreeding on Guinea Pigs. III. Crosses Between Highly Inbred Families. U.S. Dept. Agric., Bull. **1121**: 1, 1922.

1448. Wright, S.: The Genetics of Quantitative Variability. Agric. Research Council: Quantitative Inheritance. Her Majesty's Stationery Office, London, 5, 1952.

1449. Wright, S.: The Physiological Genetics of Coat Color of the Guinea Pig. Biol. Symp., **6**: 337, 1942.

1450. Wright, S.: The Physiology of the Gene. Physiol. Rev., **21**: 487, 1941.

1451. Wright, S.: The Results of Crosses between Inbred Strains of Guinea Pigs Differing in Numbers of Digits. Genetics, **19**: 537, 1934.

1452. Wright, S.: The Roles of Mutation, Inbreeding, Crossbreeding and Selection in Evolution. Proc. Sixth Internat. Cong. Genet., **1**: 356, 1932.

1453. Wright, S., and Z. I. Braddock: Colorimetric Determination of the Amounts of Melanin in the Hair of Diverse Genotypes of the Guinea Pig. Genetics, **34**: 223, 1949.

1454. Wright, S., and H. B. Chase: On the Genetics of the Spotted Pattern of the Guinea Pig. Genetics, **21**: 758, 1936.

1455. Wright, S., and O. N. Eaton: Factors Which Determine Otocephaly in Guinea Pigs. J. Agric. Res., **26**: 161, 1923.

1456. Wright, S., and O. N. Eaton: Mutational Mosaic Color Patterns of the Guinea Pig. Genetics, **11**: 333, 1926.

1457. Wright, S., and P. A. Lewis: Factors in the Resistance of Guinea Pigs to Tuberculosis with Especial Regard to Inbreeding and Heredity. Amer. Nat., **55**: 20, 1921.

1458. Wright, S., and K. Wagner: Types of Subnormal Development of the Head from Inbred Strains of Guinea Pigs and Their Bearing on the Classification and Interpretation of Vertebrate Monsters. Amer. J. Anat., **54**: 383, 1934.

1459. Yerganian, G.: Chromosomes of the Chinese Hamster (*Cricetulus griseus*). I. The Normal Complement and Identification of the Sex Chromosomes. Cytologia, **24**: 66, 1959.

1460. Yerganian, G.: Cytological Maps of Some Isolated Human Pachytene Chromosomes. Amer. J. Hum. Genet., **9**: 42, 1957.

1461. Yerganian, G.: Radiation Effects on Mammalian Sex Chromosomes. Radiat. Res., **12**: 485, 1960.

1462. Yerganian, G.: The Striped-back or Chinese Hamster (*Cricetulus griseus*). J. Nat. Cancer Inst., **20**: 705, 1958.

1463. Yerganian, G., R. Kato, M. J. Leonard, H. Gagnon, and L. A. Grodzins: Sex Chromosomes in Malignancy, Transplantability of Growths, and Aberrant Sex Determination. Cell Physiology of Neoplasia. The Univ. of Texas M.D. Anderson Hospital and Tumor Institute, 14th Ann. Symp. Fundamental Cancer Research. Univ. of Texas Press, Austin, 49, 1960.

1464. Yerganian, G., M. J. Leonard, and H. J. Gagnon: Chromosomes of the Chinese Hamster, (*Cricetulus griseus*). II. Onset of "Malignant Transformation" *in vitro* and the Appearance of the X_1 Chromosome. Pathol. Biol., **9**: 533, 1961.

1465. Yoon, C. H.: Homeostasis Associated with Heterozygosity in the Genetics of Time of Vaginal Opening in the House Mouse. Genetics, **40**: 297, 1955.

1466. Yoon, C. H., and E. P. Les: Quivering, a New First Chromosome Mutation in Mice. J. Hered., **48**: 176, 1957.

1467. Zelle, M. R.: Genetic Constitutions of Host and Pathogen in Mouse Typhoid. J. Infect. Diseases, **71**: 131, 1942.

1468. Zierler, K. L.: Aldolase Leak from Muscle of Mice with Hereditary Muscular Dystrophy, Bull. Johns Hopkins Hosp., **102**: 17, 1958.

1469. Zilber, L. A.: An Immunological Approach to Tumor Growth Control. UNESCO Symp. "Biological Aspects of Cancer Chemotherapy," Louvain, Academic Press, New York, in press.

1470. Zinder, N. D., and J. Lederberg: Genetic Exchange in *Salmonella*. J. Bacteriol., **64**: 679, 1952.

AUTHOR INDEX

Abercrombie, M., 444
Adelberg, E. A., 459
Akabou, S., 316, 317
Allard, R. W., 56, 73, 74
Allen, D. W., 311–314
Allen, S., 466, 467
Ambrose, E. J., 444
Amos, D. B., 366, 423
Anderson, D. E., 248
Andervont, H. B., 254, 264, 266, 465
Andrews, H. L., 128, 129
Anson, M. L., 322
Armstrong, C. N., 417
Atwood, K. C., 151, 413
Awa, A., 472, 475
Axelrad, A., 425

Badenhausen, S., 223
Bagg, H. J., 132
Bailey, N. T. J., 61
Bardin, C. W., 568
Barr, M. L., 494
Barrett, M. K., 340, 379, 426, 467
Bartlett, M. S., 3, 42, 48, 49, 51, 53
Bateman, A. J., 136
Batipps, D. M., 476
Bayreuther, K., 472, 473, 503
Beaver, G. H., 301
Beckstrom, J., 364
Bedford, D., 365
Bender, M. A., 475, 509
Bennett, E. L., 130, 222, 223
Benzer, S., 318
Berenblum, I., 449
Bern, H. A., 565–569, 566, 568
Bernstein, S. E., 322
Bhat, N. R., 418
Biesele, J. J., 440, 472
Biffin, R. H., 400

Billingham, R. E., 104, 327, 335
Bittner, J. J., 254, 256, 260, 261, 424
Blair, P. B., 568
Blizzard, R. M., 240
Boardman, N. K., 307, 308
Boche, R. D., 145, 147
Bock, F. C., 175
Bond, V. P., 132
Boudreau, J. C., 292
Boyd, W. C., 369, 371
Boyden, S. V., 367
Brennecke, H., 133
Brent, L., 327
Briggs, R., 409
Broadhurst, P. L., 291, 292
Brown, A. P., 311
Brown, S. R., 340
Brown, S. W., 411
Brownell, G. L., 128
Bruell, J. H., 291, 292
Burdette, W. J., v, 41, 54, 152, 188, 191, 229,
    231, 257, 268, 280, 293, 320, 336, 374,
    399, 462, 507
Burnet, F. M., 422, 446
Bussard, A., 370

Cahan, A., 364
Calhoun, M. L., 401
Card, L. E., 393
Carter, T. C., 36, 61, 63, 64, 70, 71, 76, 89, 137,
    139, 140, 141, 152, 153, 418, 518
Cepellini, R., 305
Chai, C. K., 228, 267
Chapman, A. B., 150
Charles, D. R., 135, 140, 264
Chase, H. B., 150, 188, 325
Ciecura, S. J., 498
Clausen, J. J., 478
Clegg, M. D., 311

Cloudman, A. M., 424, 425
Cohen, C., 379
Cohen, J., 333
Coleman, D. L., 189, 229, 232, 342
Coman, D. R., 444
Conger, A. D., 485
Coombs, R. R., 361, 363, 365
Cooper, C. B., 76
Cotterman, C. W., 364, 414, 415
Crew, F. A E., 76, 517
Cronkite, E. P., 133
Crow, J. F., 55, 144, 152
Czimas, L., 368

Dagg, C. P., 190
Danielli, J. F., 464
Darlington, C. D., 480
Degenhardt, K. H., 151, 190
Delbrück, M., 408, 438
DeMars, R., 451
Dempster, E. R., 248
Denenberg, V. H., 291
DeOme, K. B., 256, 261, 568
Deringer, M. K., 426, 563–564
Dickie, M. M., 522–537
Dobzhansky, T., 185
Doolittle, D. P., 3–41, 42, 46, 49, 51, 53, 54, 55, 249
Douglas, C. G., 322
Drabkin, D. L., 311
Drasher, M. L., 276, 277
Dray, S., 376, 379
Dulbecco, R., 447, 459
Duncan, D. B., 339
Dunn, L. C. 222, 223, 326, 400, 418, 517

Eagle, H., 451, 455
Earle, W. R., 450
Eaton, O., 90
Edman, P., 316
Eichwald, E. J., 373
Eldjarn, L., 149

Fairchild, L. M., 485
Falconer, D. W., 56, 63, 64, 75, 76, 193–216, 247
Fano, U., 128
Faulkin, L. J., Jr., 568
Fedorov, V. K., 266
Fekete, E., 275
Field, E. O., 310
Finerty, J. C., 373
Finney, D. J., 56, 63, 67, 68, 69, 70
Fisher, H. W., 493
Fisher, R. A., 3, 5, 8, 61, 62, 63, 64, 66, 70, 87, 185
Fitzpatrick, T. B., 340
Foley, E. J., 436
Ford, C. E., 143, 267, 415, 416, 472, 475
Ford, D. K., 473, 492, 493
Foster, M., 171, 336, 340, 342

Foulds, L., 434, 444, 445
Fox, A. S., 471
Fraenkel-Conrat, H., 317
Fraser, F. C., 233–246, 244
Freund, J., 353, 377
Fuller, J. L., 283–293, 295
Furth, J., 442

Gallily, R., 471
Gardner, W. U., 260
Geroud, A., 238, 241
Ginsburg, B. E., 54, 171, 231, 232, 284, 292, 293
Glasstone, S., 128
Gluecksohn-Waelsch, S., 240–309
Gold, E. R., 435
Goldstein, A., 155
Goodman, J. W., 363
Gordon, F. B., 402
Gorer, P. A., 358, 375, 423, 425, 430, 437, 466, 467
Goudie, R. B., 413
Gowen, J. W., 54, 55, 88, 151, 154, 228, 281, 379, 383–399, 400, 401, 403, 404
Grabar, P., 370
Grahn, D., 127–151, 152, 154
Green, C. V., 325
Green, E. L., 3–41, 42, 46, 49, 51, 53, 54, 55, 148, 151, 248, 249
Green, M. C., 56–82, 256
Greene, H. S. N., 465
Griffen, A. B., 135
Grodzins, L. A., 496, 497
Grüneberg, H., 79, 133, 179, 247, 508, 517, 518
Guyer, M. F., 240

Hagar, H. G., 333
Haldane, J. B. S., 3, 42, 48, 49, 51, 53, 139, 152
Ham, R. G., 493
Hamerton, J. L., 472, 475
Hansen-Melander, E., 472
Harnois, M. C., 473
Harris, M., 471, 509
Haruna, I., 317
Hasserodt, U., 310
Haugaard, G., 303
Hauschka, T. S., 425, 471, 477, 486, 496
Hawke, H. W., 276
Hayflick, L., 494, 507
Hazel, L. N., 150
Heaysman, J. E. M., 444
Hellström, K. E., 430, 433
Henderson, J. M., 172, 206
Hertwig, P., 133, 137
Herzenberg, L. A., 232, 342, 378, 462
Hesselbach, M. L., 445
Heston, W. E., 151, 228, 247–264, 266, 267, 268, 281, 294, 342
Hetzer, H. A., 386, 387, 390, 393, 395, 400
Hildemann, W. H., 417
Hill, A. B., 401

Hill, R. L., 316, 318
Hine, G. J., 128
Hirs, C. H. W., 308
Hirsch, J., 292
Hoag, W. G., 538–557
Hoecker, G., 423, 466, 467
Hooper, J. L., 451
Hsu, T. C., 451, 471, 472, 476, 478, 507, 509
Huff, C. G., 403
Huisman, T. H. J., 308
Hungerford, D. A., 472, 473, 476
Hunt, J. A., 316, 318, 320
Hurlbut, H. S., 402, 403
Huseby, R. A., 260, 261
Hutchinson, W. D., 310

Ingraham, J. S., 368
Ingram, V. M., 316, 317, 318, 320
Irwin, M. R., 160, 393, 400
Ising, U., 471, 472
Itano, H. A., 301, 310, 312

Jacob, F., 459
Jacobs, P. A., 416
Jacquot-Armand, Y., 356
Jaffe, J. J., 480
Jay, G. E., Jr., 83–123, 250
Jennings, H. S., 285
Johnson, H. A., 133
Jones, R. T., 320

Kabat, E. A., 369–371
Kaliss, N., 432, 464
Kalmus, H., 144
Kalter, H., 235
Kandutsch, A. A., 465
Kano, K., 477
Kaplan, W. D., 149, 260, 475, 477
Kato, R., 479, 496
Katz, J. J., 130
Keil, B., 316
Kellogg, D. S., Jr., 451, 471
Kempthorne, O., 9, 203, 204
Kenny, G. E., 494
Kiddy, C. A., 365
King, J. A., 291
King, S. C., 206
King, T. J., 409
Kinosita, R., 475, 477, 496
Kirschbaum, A., 259
Klatt, O., 472, 476
Klein, E., 426, 462, 473, 503
Klein, G., 143, 266, 267, 377, 407–462, 465, 467, 471, 472
Klontz, G. W., 370
Knox, C. W., 393
Koller, P. C., 76
Kovacs, E. T., 496
Krohn, P. L., 251
Kroner, T. D., 303
Kröning, F., 171

Kukita, A., 340
Kunkel, H. G., 304

LaCour, L. F., 480
Lambert, W. V., 393, 400
Landauer, W., 244
Landsteiner, K., 377
Lane, P. W., 27, 36
Lang, R., 480
Lapp, R. E., 128, 129
Law, L. W., 257, 260, 437, 438
Lawson, F., 326
Lebedeff, G. A., 326
Lederberg, E. M., 458
Lederberg, J., 154, 422, 458, 466, 468, 472
Lerner, I. N., 248
Les, E. P., 538–557
Levan, A., 425, 471, 472, 473, 474
Likely, G. D., 450
Lima-de-Faria, A., 497
Little, C. C., 132, 248, 256, 326, 335, 423, 424, 425, 514
Longwell, A. C., 500
Löw, B., 359
Luce-Clausen, E. N., 264
Luria, S. E., 408, 438, 447, 459
Lush, J. L., 150
Lynch, C. J., 268
Lyon, M. F., 149

Macdowell, E. C., 254, 257, 326–400
Main, J. M., 435
Major, M. H., 417, 418, 419, 421
Makino, S., 472, 473, 475, 476
Mandl, A. N., 149
Marinelli, L. D., 128
Marshall, G. J., 479
Martinet, M., 238
Masouredis, S. P., 378, 379
Mather, K., 56, 61
Mayer, K., 56, 61
Mayer, M. M., 369–371
Medawar, P. B., 327, 335, 423
Mellman, W. J., 476
Michie, D., 248
Miguel, J., 368
Mikulska, Z. B., 358
Möller, G., 435
Mollison, P. C., 363
Monie, I. W., 238
Moore, S., 308–314
Moorhead, P., 472, 476, 477, 507, 537
Moretti, J., 568
Morris, H. P., 540
Morse, M. L., 458
Morton, N. E., 144, 359
Mourant, A. E., 361
Moyer, F., 189
Mühlbock, O., 89, 111
Muller, H. J., 133, 144
Munoz, J., 363

Munro, S. S., 277
Murayama, M., 315
Murray, J. M., 325
McIntosh, W. B., 54
McLaren, A., 248

Nachtsheim, H., 231
Nalbandov, A. V., 266, 269–280, 281, 282, 341
Nandi, S., 568
Nanney, D. L., 432
Neel, J. V., 144
Nelson, M. M., 239
Newton, A. A., 479
Nichols, S. E., 333
Nowell, P. C., 476, 477, 479

Oakberg, E. F., 133, 151
O'Brien, J. R. P., 310
O'Gorman, P., 430
Ohno, K., 316, 317
Ohno, S., 475, 477, 494, 496
Orsini, M. W., 473
Ostergren, G., 496
Ouchterlony, Ö., 369, 370
Oudin, J., 369, 370, 376
Owen, R. D., 150, 342, 347–374, 378, 414, 415, 417, 432, 467, 468

Palmer, C. G., 473, 475
Pansky, B., 473
Partridge, S. M., 307, 308
Pauling, L., 130
Pearl, R., 145
Peterson, E. A., 377
Phillips, R. J. S., 137
Pickels, E. G., 309
Pickles, M. M., 359
Pilgrim, H. I., 53, 54, 231, 266
Pillemer, L., 361
Pollock, M. E., 494
Pontecorvo, G., 411, 413, 434
Popp, R. A., 189, 218, 299–320, 322
Potter, M., 260, 437, 438
Preer, J. R., Jr., 370
Prehn, R. T., 261, 435
Prins, H. K., 308
Puck, T. T., 143, 450, 462, 472, 475, 486, 487, 493, 498

Quevedo, W. C., Jr., 331

Race, R. R., 356, 361, 363, 365
Raffel, S., 363
Rawles, M. E., 323
Reams, W. M., 333
Reed, S. C., 172
Reed, T. E., 379
Reeve, E. C. R., 203
Reichart, E. T., 311
Révész, L., 436
Rhinesmith, H. S., 319
Ridgway, G. J., 369

Riggs, A., 315
Roberts, E., 393
Roberts, F., 363
Robertson, A., 248
Robertson, G. G., 373
Robinson, A., 498
Robinson, E., 301
Roderick, T. H., 403, 404
Rogers, S., 260
Rothfels, K. H., 452, 472, 475, 486, 487, 493
Rothman, S., 232
Rubin, B. A., 373
Ruddle, F. H., 471, 472, 474, 478, 509
Rugh, R., 149
Runner, M. N., 241
Russell, E. S., 150, 189, 217–229, 231, 232, 294, 320, 322, 326, 327, 337, 338, 340, 558–562
Russell, L. B., 136, 138, 144, 146, 148, 151, 152, 417, 418, 419, 421
Russell, W. L., 133, 136, 137, 138, 140, 141, 144, 145, 146, 148, 151, 152, 171

Sachs, L., 471
St. Amand, W., 309
Salmon, C., 435
Sanford, K. K., 450, 454
Sanger, F., 315, 318
Sanger, R., 356, 363, 365
Sasaki, M., 472, 473, 475, 476
Sato, G., 456
Schabtach, G., 326
Schachman, H. K., 309
Schaefer, H. J., 129
Schaible, R. H., 342
Scheinberg, I. H., 317
Scheinberg, S. L., 151, 413
Schott, R. G., 388, 389, 390, 393, 395, 400
Schroeder, W. A., 311
Schull, W. J., 143, 144
Schwartz, D., 370
Schwartz, H. C., 316, 318, 320
Schweet, R., 321
Scott, J. P., 178, 190, 283–293, 296
Searle, A. G., 137
Shapiro, J. R., 259
Shultz, W., 231
Silvers, W. K., 104, 323–336, 340, 341, 342
Siminovitch, L., 472, 475, 486, 487
Singer, S. J., 310
Slater, R. J., 304
Slatis, H. M., 55, 143, 341
Slizynski, B. M., 76, 77, 480
Smith, E. A., 240
Smith, E. W., 320
Smith, L. H., 363
Smith, W. E., 260
Smithies, O., 370
Snell, G. D., 3, 17, 27, 32, 58, 64, 66, 75, 87, 133, 138, 279, 326, 374, 376, 378, 423, 427, 463, 465, 467, 468, 516, 517, 518

Sober, H. A., 377
Sobey, W. R., 75
Soemarwoto, I. N., 566
Sparrow, M., 373
Sproul, J. A., Jr., 142
Staats, J., 17, 511–516, 517–521
Stadler, J., 386, 396
Stavitski, A. B., 367
Stein, W. H., 308–314
Steinberg, A. G., 191
Stern, C., 410, 411
Stewart, S. E., 499
Stimpfling, J. H., 359, 466
Stone, W. H., 364
Strong, L. C., 257, 400, 423, 424
Stulberg, C. S., 486, 487
Sumner, F. B., 185
Svedberg, T., 309
Swanson, C. P., 414
Syverton, J. T., 478, 494
Szilard, L., 422
Szybalski, W., 452, 457

Talmage, D. W., 422
Tanaka, T., 477
Tandom, O. B., 203, 204
Taylor, J. H., 496, 497
Taylor, L. S., 128
Terasaki, P. I., 365
Terwey, K. H., 480
Thompson, E. O. P., 318
Tischer, E., 303
Tiselius, A., 301
Tjio, J. H., 472, 475, 486, 487, 496
Tonomura, A., 477, 480
Topley, W. W. C., 400
Torbert, J. V., 320
Totter, J. R., 130
Trasler, D. G., 244

Trentin, J. J., 260
Tyzzer, E. E., 424

Uphoff, D., 151
Upton, A. C., 142

Venge, O., 273
Vinograd, J., 310–319
Vogt, M., 440, 447, 452, 459
Vos, O., 365

Wachtel, I. J., 174
Waddington, C. H., 176
Wakefield, J. D., 366
Warkany, J., 235, 237, 238
Weaver, E. C., 411
Webster, L. T., 393, 395, 400, 401
Weicker, H., 480
Weir, J. A., 386, 399
Welshons, W., 411
Whitten, W. K., 280
Wieland, T., 303
Wildy, P., 479
Winn, H. J., 375, 378
Wolfe, H. G., 559
Wolff, J., 149, 231
Wood, D. C., 240
Woodruff, M. F. A., 373
Wright, S., 3, 42, 55, 111, 159–188, 189, 190, 191, 231, 247, 254, 281, 293, 296, 325, 335, 340, 400

Yerganian, G., 143, 154, 189, 267, 446, 469–507
Yoon, C. H., 195

Zelle, M. R., 386, 398, 403
Zilber, L. A., 447
Zinder, N. D., 458

SUBJECT INDEX

$AB/ab \times AB/ab$ matings, 68
$Ab/aB \times Ab/aB$ matings, 68
$AB/ab \times AB/ab$ matings, with A and B semi-dominant, 69
$AB/ab \times AB/ab$ matings, with A semidominant, 69
$Aa/Bb \times aa/bb$ matings, 67
Absorption
    analysis leading to recognition of two specificities, 355
    ion-exchange, 307
    of antibodies *in vivo*, 376
Actinomycin D, 152
Action, genic, in the mouse, 217
Adaptation through behavior, 285
Additive variance, 198
Adenoma
    of pituitary, 250
    of stomach, 250
Adenosine, 154
Adhesiveness, cellular, 443
Adjuvants
    Freund's, 353, 377
    paraffin-oil, 377
    useful in producing immunization, 353
Adrenal
    carcinoma, 260
    cortical tumors, 250
    transplantation, 373
Agar-gel
    electrophoresis, 307
    immunochemical analysis, 376
Age
    development of pigment with, 341
    effect of, on coat color in guinea pigs, 169
Agglutination
    saline, 349, 357
    systems for, 358
Agouti locus, analysis of action of genes, 328

Albinism, 330
Albino, 60
    series of genes, 218
Alkali denaturation of hemoglobin, 311
Alleles
    cumulative action of multiple, 182
    selective values of, 182
Allocycly, 497
Allotypes, gamma-globulin, 379
Allotypy, 376
Alpha-2-serums, 190
Amino acid
    analysis, 314
    N-terminal, 321
Amoeba, 464
Analysis
    amino acid, 314
    C-terminal, 316
    cytogenetic, 469
    end-group, 315
    N-terminal, 315
    of gametic pairs with respect to linked locus, 47, 50
    of hormonal defects by substitution therapy, 221
    of pleiotropisms, 226
    of variance for tyrosinase activity, 339
    resin paper, 322
    retrograde, 222, 227
    starch-gel, 322
    sulphydril, 315
Anaphylactic response, 353
Anemias
    hereditary in mice, 223, 231
    radiation response of mice with, 225
    response of mice with, to erythropoietic stimuli, 226
    techniques for study of in mice, 558
    tissue localization of action of W-gene, 225

Anemias—*continued*
  W-series, 224
Anencephaly, 191, 238
Aneuploidy, 416
  of cells in culture, 507
Animal-breeding rooms, reverse-lighting patterns in, 154
Animal mutations chemically produced, 154
Anomalies
  metabolic in mice, 466
  of urinary tract, 238
Anophthalmia, 189
Antibodies, 348
  absorption of *in vivo*, 376
  cellular, 375
  cytotoxic, 429
  fluorescent, 379, 463
  labeled, 378, 463
  radioactive, 463
  reagents, preparations of, 354
  tumor, 269
Antibody
  formation, 422
  production, differences in between strains, 379
  titer, 193
Antigen-antibody reactions, 347
Antigenic
  differences, 193
    and malignancy, 453
  simplification in transplanted tumors, 425
  specificity
    genetic control of, 417
    loss of, 432
Antigens, 348
  clearance of labeled, 363
  coupled with red cells, 367
  histocompatibility, 465
  macromolecular, 348
  molecular, 348
  soluble, 366
  tumor-specific, 460
  univalent, 349
Antiglobulin tests, 361
Antimetabolites, resistance of cells *in vitro* to, 423
Antisera, procurement of, 350
Array, frequency, 42
Arthropods, susceptibility to viruses, 402
Ascites
  conversion, 433
  form of tumor cells, 423, 468
Aspergillus, 161, 434, 460
Assortment, random, 48
Atavistic digits, 176
Atomic recoil, 130
Attitude, genetic variance, 216
Audiogenic seizures, 230, 285, 287
Autonomy of tumors, 441
Autosomal linkage groups, 76
Auxotrophic mutants, selection of, 451
Average rate fixation of loci, 55

Avian
  embryos, use of for heterologous graft, 327
  malaria, 403
Avidity, 350

Backcrosses, 4, 5, 58
  double, 57
  genotypes among offspring, 61
  method of testing mutations, 137
  single, 57
Backcross system, 18
  generation matrix of, 20
  generations required to obtain given percentage of incrosses, 21
  mating types and their probabilities, 19
  of mating, 53
  probability of
    heterozygosity in, 70
    incrosses for, 20
Bacteria, selection of drug resistant, 408
Bacteriophages, temperate, 459
Balance, genic, 508
Barking, variations in, 286
Barrett-Deringer
  change in transplanted tumors, 426
  phenomenon, 467
Barrier, uterine, 241
Bedding for mice, 541, 545
Behavior
  development of, 286
  influences of environment on, 287
  measurement of, 288
Behavioral
  differences, 283
    pattern, 284
    relationship to genetic variation, 285
    unit of, 284
  dominance, 402
  genetics, 295
    developmental threshold in, 296
    methodology of, 293
  pattern
    frequency of, 288
    genotypic *versus* phenotypic mutation in, 290
    intensity of, 288
    latency of, 288
    mechanism of, tests for, 292
  phenotypes, 292, 295
  traits, 148
  variation, relationship to genetic variation, 285
Behavioral differences, 283–293
Belts of radiation, 129
Bibliography, 571–621
  subject strain, 511
Biochemical
  effects of teratogens on development, 240
  pathways
    genetics of, 297
    relation of genes to behavior, 292

Biochemical—*continued*
    properties of embryo at various developmental
        stages, 238
Blood, collection of, in mice, 558
Blood-group genes, 161
B-mercaptoethylamine, 149
Body weight
    and hormonal secretion, 273
    and rank order in cattle, 274
Bone marrow
    post-irradiation injection of, 148
    transplantation of, 374
Bottle cases, 551
Bovine ocular carcinoma, 248
Brachyury, 222
Breeding
    choice of system, 39
    methods, 1
    performance, 54
    systems (see Systems of breeding)
Breed sensitivity of chickens to exogenous
        hormones, 277
Broody instinct in chickens, 278
Brother-sister inbreeding, 5, 12, 53
    generation matrix for, 13
    probabilities of incrosses for, 15
    with heterozygosis forced by backcrossing, 5,
        32
    with heterozygosis forced by intercrossing, 5,
        36
Brother-sister mating, 48, 49, 53
    gamete diagram, 45
    matrices for, 16
    zygote diagram, 44

Cages
    for mice, 547
        bedding for, 541
        covers for, 547
        food hopper for, 550
    racks for, 552
Cancer (see also Tumors)
    adrenal, 260
    bovine ocular, 248
    estimation of number of susceptibility genes,
        254
    lysogeny, relationship to cancer, 268
    mammary gland, 265
    of skin, 249
    of stomach in *Mastomys*, 254
    pancreatic, in hamsters, 267
    transduction, relationship to cancer, 268
Cannibalism in mice, 544
Carbon-14, genetic hazards of in man, 130
Carbonmonoxyhemoglobins
    of mice
        analyzed by crystallography, 312
        analyzed by salting-out methods, 313
        moving-boundary electrophoresis of, 302
Carcinogen, method of producing pulmonary
        tumors with, 255

Carcinogenesis
    dose response curves, 264
    relationship to mutagenesis, 263
Carcinoma (see also Cancers and Tumors)
    adrenal, 260
    bovine ocular, 248
    estimation of number of susceptibility genes,
        254
    mammary gland, 265
    of skin, 249
    of stomach in *Mastomys*, 254
Carts and tables for care of mice, 553
Cases, bottle, 551
Cattle
    correlation of twinning, lactation, and growth,
        273
    pigment-cell studies in, 333
    rank order in twinning, lactation, and body
        weight, 274
Cells
    frozen, 507
    genetic continuity of lines of, 493
    in culture
        aneuploidy of, 507
        drug-resistant variants, 462
        freezing of, 507
        isoantigens of, 463
    in suspension, use of immunogenetics, 365
    malignization of, *in vitro*, 452, 461
    neoplastic
        in vivo, 422
        phenotypic markers in, 448
    preservation of, by freezing, 471
    relationships *in vivo* and *in vitro*, 455
    single, solution of, for tissue culture, 450
    somatic, genetic transfer between, 456
Cellular
    adhesiveness, 443
    antibodies, 375
    contamination of tissue cultures, 493
    function, differentiation of, 464
Celom of chick embryo, transplantation of
        tissue to, 330
Cellulose, ion-exchange chromatography, 377
Cesarean section, use of in procuring sublines
        free of milk agent, 251
Cesium-137, 128
Chance in random selection, 54
Character
    quantitative, 193
    threshold, 247
Chemical methods for analysis of hemoglobins, 34
Chiasmata counts, 76
Chickens
    breed sensitivity to exogenous hormones, 277
    broody instinct in, 278
    inbred lines, 342
    size of comb in, 271
Chimeras, 414
    organ, 444
    radiation, 417

Chinese hamsters, 189, 471, 473, 493, 498, 507, 509
    cytologic characteristics of, 498
    diabetes mellitus in, 267
    idiogram of, 483
    pancreatic tumors in, 267
    pituitary tumors in, 267
    sex chromosomes of, 496
    strains of, in development, 116
    tissue cultures, 485
Chi-square, 58, 59
    for linkage for three kinds of matings, 59
    for segregation at individual loci, 59
    heterogeneity, 73
Choice of suitable breeding system, 39
Chromatography
    cellulose ion-exchange, 377
    column, 307, 321
    paper, 340
Chromosomal
    breaks
        and abnormal embryonic development, 152
        and restitution, kinetics of, 131
    changes in transplanted tumors, 433
    markers, 452
        in tumors, 446
    morphology, 423
Chromosomes
    identification of individual chromosomes, 494
    layout for preparation of, 490
    mean length of heterozygous, 22
    pattern of alteration of numbers *in vitro*, 507
    preparation of, utilizing vinkaleukoblastine, 475
    sex, 494
        Chinese hamster, 496
    staining, 480
    techniques
        for permanent preparations of, 484
        for preparation, 472
    unsquashed preparations of, 491
    use of
        colchicine in preparation of, 473, 501
        hyaluronidase in preparing, 479
        ribonuclease in preparing, 479
Chronic exposures to radiation, 130
Cleft
    lip, 243
        palate, 238, 241, 243, 244, 245
Clones, 450
Cloning, 462
    single-cell, 468
    techniques of, 461
Clustering of malformations, 243
Coat color
    environmental effects on, 169
    of guinea pigs, factor interactions in, 161
Cobalt-60, 128
Code for marking ears, 523
Coefficient of inbreeding, 44, 55, 194

Coisogenic color lines
    development of, 325
    utilization in studying genic interactions, 326
Colchicine, use of in chromosomal preparations, 473, 501
Colicinogenic factors, 459
Collection of blood in mice, 558
Color
    factors and activity of tyrosinase, 170
    interactions, 172
    lines, coisogenic
        development of, 325
        utilization in studying genic interaction, 326
    stocks, development of inbred, 324
Column chromatography, 307, 321
Comb and size of, in chickens, 271
Combinations
    of locus control and relationship, 5
    of teratogens, effect of, 241
Committee on Standardized Genetic Nomenclature for Inbred Strains of Mice, 83
Comparison of mortality of mice and man, 147
Complement fixation, 370
    test, 376
Concept of thresholds, 293
Congenital deformities produced by acute X irradiation, 153
Constant effects, 179
Containers, shipping, for mice, 555
Contamination, cellular, of tissue cultures, 493
Continuously distributed traits, 296
Control, karyotypic, 472
Control of the literature on genetics of the mouse, 511–516
Conversion
    ascites, 443
    lysogenic, 447, 456, 459
Correlated changes in rates of hormonal secretion, 273
Correlation
    between weight and size of litter in guinea pigs, 281
    phenotypic, between members of families, 203
Corti's organ, differentiation of, 219
Cosmic
    particles, 129
    radiation, 129
Counts of chiasmata, 76
Coupling phase of linkage, 57
Covers for cages for mice, 549
Crest, neural, 34
Cross-backcross-intercross system, 27, 28, 53
    cross-generation matrix for, 30
    cycle matrix for, 30
    cycle matrix of, 31
    generation matrix of, 31
    mating, 3, 5
        intercross generation matrix of, 31
        types and their probabilities, 27

Cross-backcross-intercross system—*continued*
  probability
    of heterozygosity, 30
    of incrosses for, 29
  probabilities of mating types, 29
Crosses, 4
Crossing over, 77
  in *H-2* locus, 466
  somatic, 410, 460, 467, 468
    in transplanted tumors, 433
Cross-intercross system, 5, 22, 23, 53
  cycle matrix of, 25
  mating types, 22
  matrices for, 26
  probabilities of various mating types, 24
  probability of heterozygotes, 24
Crystallography and solubility of hemoglobin, 311
C-terminal analysis, 316
Cultures, tissue
  cellular contamination, 493
  Chinese hamster, 485
Culture tubes, 504
  cytologic preparations for, 504
Cumulative
  action of multiple alleles, 182
  effects, 179
Current applications of a method of transplantation of tissues into gland-free mammary fat pads of mice, 565–569
Cystamine, 149
Cysteamine, 149
Cysticerus fasciolaris, relationship to sarcomas in rats, 253
Cytocidal interaction of tumor viruses, 447
Cytogenic analysis, 469–507
Cytogenic analysis of radiation effects, 143
Cytogenetics, references and manuals useful in, 503
Cytologic
  characteristics of Chinese hamster, 498
  preparations, culture tubes for, 504
Cytotoxic
  antibodies, 429
  technique, 375
  tests, 378
    in non-*H-2* systems, 378

Deformities, congenital, produced by acute X irradiation, 153
Degree of genetic determination, 194, 216
Demes, 185
Denaturation, alkali, 311
Deoxyribonucleic acid, 217
Dependence, hormonal, 441, 468
Deringer-Barrett change in transplanted tumors, 426, 467
Detection of linkage between *ru* and *je*, 66
Determination, genetic, 194, 195
  experimental design, 196

Deuterium, introduction into biological systems, 130
Deuterons, 129
Development, immunologic aspects of, 240
Developmental
  method for studying behavior, 286
  threshold in behavioral genetics, 296
  stage at which teratogen is used, 236
Diabetes mellitus, 189
  in Chinese hamsters, 267
Diagram
  gamete, 45
  zygote, 44
Dibenzanthracene, 255
Differences
  antigenic, 193
  behavioral, 283
  in stability of hemoglobin, 322
Differential selection, 211, 213
Differentiation
  alteration of, by genes, 222
  of cellular function, 464
Digits, atavistic, 176
Dilute, lethal, 230
Diploid karyotype, retention of, 471
Diploidy, 507
Discrepancy between action of teratogenic agents in humans and laboratory animals, 237
Disease resistance, transmission of, 386
Dispenser, food, 552
Disturbed segregations, 61
Dominance, behavioral, 402
Dominant
  lethal effect of hypoxia, 148
  lethals from radiation, 135
  mutations, 136
  spotting (*Wᵛ*), 70, 71
  visible mutation, 140
Dopa reaction, 171
Dose
  effect of on mutation rate, 141
  rate of radiation, 142
Dosimetry, 128
Double backcross, 57, 58
Double heterozygote, 57
Double matings, 388
  technique of, 402
Double recombinant, 75
Drift
  genetic, 267
  random, 187, 192
Drosophila, 48, 55, 129, 131, 137, 141, 142, 144, 152, 154, 185, 290, 292, 341, 400, 460, 508
  gynandromorphs, 419
  somatic crossing over in, 410
  somatic mutation in, 421
Drug resistance, 460, 468
  and hormonal independence, 437
  and karyotype, 440

Drug resistance—*continued*
  markers, 452, 457
  of tumor cells, 423
Drug resistant variance of cells in culture, 462

Ears, code for marking, 523
E. coli, sex factor of, 459
Effect of
  genes on specific tissues, 218
  temperature on genic action, 231
Electrons, low energy, 129
Electrophoresis
  agar-gel, 307
  immune, 307
  moving-boundary, 301, 302
  paper, 303
  starch-block, 304
  starch-gel, 305, 370
  zone, 302
Electrophoretic analysis of hemoglobins, 300
Elementary genic action, 160
Elimination rates of labeled cells, 363
Embryo, biochemical properties at various stages
  of development, 238
Embryonic
  development affected by genes, 222
  stages, effect of radiation on, 151
Embryos, avian, use of for heterologous grafts, 327
End-group analyses, 315
Endocrine defects, 221
Endocrinology, basic concepts of, 269
Endomitosis, 414
Energy
  electron, 129
  spectrum, 129
Enhancement, tolerance, and paralysis, 373
  immunologic, 431
Environment, influence of on behavior, 287
Environmental
  effects on coat color, 169
  variance, 195
Enzymatic inhibition, 367
Epigenetic, mechanisms in transplanted tumors, 432
Episomal transduction, 459
Episomes, 459
Epistasis, 199
Equipment
  and facilities of mouse colony, 546
  husbandry, and procurement of mice, 538
Erythrocytic mosaicism, 414
Erythron
  in mutants, 223
  quantitative determinations of, 560
Erythropoietic stimuli, response of anemic mice
  to, 226
Escherichia coli, 161
Established strains of animals (see Strains)
Estimate of hormonal secretion, indirect, 271

Estimation
  of heterogeneity, 70
  of number of genes, 216
  of recombination, 72, 78
    frequencies, 56
Estrogen, response to, by different inbred strains, 276
Eumelanin, 162, 324
  in guinea pigs, 163, 164, 165
Evolution
  genic interaction in, 183
  in homogenous and subdivided populations,
    comparison of, 188
Exposure variables of radiation, 131
Extrachromosomal factor in resistance to leukemia, 253
Eye of mammal, transplantation to, 330

F_1 hybrids, 253
  change in histocompatibility requirements
    after passage through, 426
  use of in localizing genic action, 258
F_1 tumors
  behavior of, 427
  isoantigenic variance from, 434
Facilities and equipment of mouse colony, 546
Factor interactions in coat color of guinea pig, 161
Fast neutrons, 128
Fat pad, mammary gland, transplantation of, 261
Feeding of mice, 539
Fertility and gonadotrophic hormone, 270
Fertilized ova, technique for transfer of, 563
Fetal lung, transplants of, 259
Fetal-maternal isoimmunization, 465
Fibonacci series, 45
Filter assemblies for sterilization of media, 506
Filtration of air in animal rooms for mice, 542
"Fingerprinting," 318, 321
Fitness, 185
  natural, 214
Fixation, complement, 370
Flexed tail, pleiotropic effect of, 257
Fluctuation test, 438, 439, 442, 452, 461
Fluorescent antibodies, 379, 463
5-fluorouracil, 190
Food
  dispenser, 552
  for mice, introduction of pathogens in, 545
  hopper for cages for mice, 550
Forced heterozygosis, 257
Forehead spot, 175
Foster nursing, 251
Fractionated exposure to radiation, 130, 142
Freezing
  of cells in culture, 507
  preservation of cells, 471
Frequencies
  genic, 42
  genotypes, 6, 7, 43
  of mating types, 4, 11

Frequency
  array, 42
  of behavior, 288
Freund's adjuvant, 353, 363, 377
Frozen cells, 507
Full-time energies, 128

Gametic
  diagram, 44
  pairs, analysis of with respect to linked locus,
    47, 50
Gamma globulin
  allotypes, genetic control of, 379
  groups in rabbits, 379
Gamma radiation, 128
Gastrointestinal cancer, 251
Gel-diffusion serology, 369, 370
Gene-frequency systems
  mean selective values of, 187
  trajectories of, 186
Generation matrix, 8
  for brother-sister inbreeding system, 13
  for the backcross system, 20
Generation sequence relative to type of matrix, 5
Generations required to obtain a given percent-
    age of incrosses, 21
Genes
  and rate of hormonal secretion, 279
    at the *Agouti* locus, 328
  blood-group, 161
  effect of, on specific tissues, 219
  embryonic development and differentiation,
    222
  histocompatibility, 375
  hormones, and pattern of response, 221
  interaction, with teratogens, 244
  in the mouse, 79
  number of, 214
    susceptibility, in cancer, 254
  proteins, and enzymatic activity, 218
Genetic
  analysis in mammals, 132
  characters, threshold, 267
  continuity of lines of cells, 493
  control of differences in sensitivity to hor-
    mones, 272
  damage, protection against, 148
  determination, 194, 195, 216
    degree of, 194
    experimental design, 196
    heritability, 198, 216
  differences
    in reactions to water imbibition, 232
    in response to loading, 232
  drift, 267
  effects of radiation
    qualitative, 134
    quantitative, 143
  mechanisms in transplanted tumors, 432
  radiation, 125
  research, radiation as a tool for, 150

Genetic—*continued*
  sterility, 279
  strains and stocks, 83
  transfer
    between somatic cells, 456
    experiments, 468
  variance, 185
  variation, relationship to behavioral variation,
    285
Genetic strains and stocks, 83–123
Genetics
  behavioral, 295, 297
  of the mouse, control of literature on, 511
  physiologic, 157
Genetics of infectious diseases, 383–399
Genetics of neoplasia, 247–264
Genetics of reproductive physiology, 269–280
Genetics of somatic cells, 407–462
Genic
  action, 160
    in the mouse, 217
    localization of in organs and tissues, 258
    paths of, 258
    temperature effect on, 231
  balance, 508
  frequencies, 42
  interaction, 159
    absent effect of, in some combinations, 180
    history of, 159
    in evolution, 183
    opposite direction of effect in different
      combinations, 180
    selective, value of, 181
    utilization of coisogenic lines in studying,
      326
  mutations, 136
  substitution, 188
Genic interactions, 159–188
Genotype frequencies, 6, 7, 43
Genotypes in offspring of a single backcross,
  61
Genotypic *versus* phenotypic orientation in
  behavioral patterns, 290
Geotactic maze, 292
Gestation, hormonal secretion and length of, 273
Globulin, serum, 348
G matrix, 11, 20
Gonadotrophic hormone, 270
  amount secreted in different strains of swine,
    281
Graft, black and white in guinea pig, 334
Grafting of skin, 371
Groups, linkage, 76
Growth
  and maturation, 147
  in cattle, 273
Guinea pig
  atavistic digits, 176
  black and white graft in, 334
  coat color of, 162
    and activity of tyrosinase, 170

Guinea pig—*continued*
  correlation between weight and size of litter, 281
  environmental effects on coat color, 169
  established strains of, 112
  eumelanin
    and phaeomelanin in, 163, 164
    in at birth and six months later, 165
  factor interactions
    between hair direction and white forehead spot, 173
    in coat color, 161
  forehead spot in, 175
  hair direction in, 172
  homology, 178
  lymphosarcomas in, 254
  mean intensity index for various genotypes of coat color, 166
  mean quality index for various genotypes of coat color, 166
  opposite direction of effect in different combinations, 180
  pigment-cell studies in, 333
  polydactyly in, 176, 293
  replicative homologs, 178
  selective value, 181
  spotting, 162
  stocks of, of genetic interest, 113
  strains of, in development, 112
  symbols for designating substrains of, 113
  types of genic interaction in, 179
Gynandromorphs, 414
  in *Drosophila*, 419

H-2 locus, crossing over in, 466
Hagedorn effect, 186
Hair direction
  factor interactions in, 173
  in guinea pigs, 172
  pleiotropic effect of, on white forehead spot, 173
Half molecules, of hemoglobin dissociation, 310
Hamsters
  cancer of pancreas in, 267
  Chinese, 189, 471, 493, 498, 507, 509
    cytologic characteristics of, 498
    diabetes mellitus in, 267
    idiogram of, 483
    sex chromosomes, 496
    strains of, in development, 116
    tissue cultures, 485
  pituitary tumors in, 267
  symbols for designating substrains, 117
  Syrian, 471, 473
    established strains of, 114
    stocks of, of genetic interest, 116
    strains of, in development, 115
Handling of mice, 543
Haploidization in molds and somatic crossing over, 412
Harderian-gland tumors, 250

Hare lip, 178
Hemagglutinating materials, diverse, 351
Hemangioendothelioma, 250
Hematocrit determination, 561
Hematopoietic tissues, 373
  study of, 562
Hemes, 322
Hemoglobins
  chemical methods for analysis of, 314
  differences
    in segregation patterns, 321
    in stability, 322
  dissociation into half molecules, 310
  in mice, analyzed by "fingerprinting," 319
  mammalian, 299
  physical methods for analysis of, 300
  physical methods for study, 300
    alkali denaturation, 311
    column chromatography, 307
    crystallography and solubility, 311
    electrophoretic analysis of, 300
    moving-boundary electrophoresis, 301
    ultracentrifugation, 309
    zone electrophoresis, 302
  solubility, 314
  structure of, 218, 300
Hemoglobinopathies, human, 224
Hemolysis, 349, 360
  test systems, 360
Hepatomas, 249, 265
Hereditary sterility from radiation, 134
Heredity, infective, 422, 461
  and viral tumors, 446
Heritability, 194, 198, 212, 216
  offspring-parent regression, 201
  sib analyses, 205
  standard error of, 209
Heterochromatin, 494
Heterogeneity
  chi-square, 72
  estimation of, 70
Heterogenous strains, 194
Heterosis, 214
Heterozygosis, 32, 44
  forced, 257
  limiting ratio of, 49, 53
  ratio of, 48
  relative, recurrence equation for, 49
  residual, and resistance to induced sarcomas, 436
Heterozygosity, probability of, 4
  cross-backcross-intercross systems, 30
  cross-intercross system, 24
  with heterozygosis forced by backcrossing, 34, 35
  with heterozygosis forced by intercrossing, 37, 38
Heterozygotes
  double, 57
  high selective value for, 53
Heterozygous tumors, behavior of, 429

High-energy protons, 129
Hiroshima, 144
Histocompatibility, 32, 229, 231, 457
 antigens, 465
 in chickens, 342
 loci, sequence of, 466
 of genes, 375
 system, 460, 468
  in the mouse, isoantigens of, 423
History of genic interaction, 159
Holders, tag, 553
Homallelic populations, 183
 with six modifiers, 184
Homeostatic control, 176
Homology, 178
Homotransplantability, increase of, 427
Homozygous tumors, behavior of, 428
Hormonal
 content of adenohypophysis in different strains
  of swine, 274
 dependence, 468
  of tumors, 441
 independence, 468
  and drug resistance, 437
 secretion
  and body weight, 273
  and length of gestation, 273
  and litter size, 273
  and rate of ovulation, 273
  correlated changes in rates of, 273
  indirect estimate of, 271
  rate of and genes, 279
  rates of, 271
Hormone
 dependent tumors, 465
 exogenous sensitivity of chickens to, 277
 genic control of differences in sensitivity to, 272
 gonadotrophic, 270
  amount secreted in swine, 281
 target-specific, 280
Host-parasite relationship, 381
 in cancer, 262, 264
 symbiotic relationship of parasite to, 403
 vector, 403
Humidity of animal rooms for mice, 543
Husbandry, equipment, and procurement of
  mice, 538–557
Husbandry of mice, 539
Hyaluronidase, use of in preparing chromosomes, 479
Hybrid
 extrachromosomal factors, 253
 recipient, preimmunization of, 425
 substances, 160
 use of, in revealing extrachromosomal factors,
  253
Hybrids
 F$_1$, 253
  change in histocompatibility requirements
   after passage through, 426
  use of in localizing genic action, 258

Hydrocephalus, 178
Hypotonicity and mitotic arrest, 473
Hypoxia, effect of, on radiation, 148

Identification of chromosomes, 494
Idiogram of Chinese hamster, 483
Immigration, 187
 pressure, 185
Immune electrophoresis, 307
Immunization, 353
Immunochemical analysis, agar-gel, 376
Immunoelectrophoresis, 370
Immunogenetic(s), 345
 mammalian, 347
 tests, using mixtures of cells, 363
Immunologic
 aspects of development, 240
 enhancement, 431
 tolerance, 327
Inbred color stocks, development of, 324
Inbred strains
 differences between physiologic characteristics
  in, 294
 for cancer research, 248
 for use in studies in neoplasms, 249
 of mice, resistance to murine typhoid, 384
Inbreeding, 54, 194
 and crossing, 214
 brother-sister, 5
  generation matrix for, 13
  probability of incrosses for, 15
  with heterozygosis forced by backcrossing,
   32, 35
  with heterozygosis forced by intercrossing, 36
 coefficients, 44, 55, 194
 depression, 214
 parent-offspring, 4
Increase of homotransplantability, 427
Incrosses, 4
 probabilities of, for six mating systems, 39
 probability of, with heterozygosis forced by
  backcrossing, 33, 35
 probability of, with heterozygosis forced by
  intercrossing, 37
Incubators, 505
Index
  mean intensity, 166
  mean quality, 166
Indices, panmictic, 45
Indirect estimate of hormonal secretion, 271
Induced sterility, 133
Induced tumors, isoantigenicity of, 426
Inductive failure, 222
Infant-mortality data in mice, 146
Infection, resistance to and selection, 401
Infectious diseases, genetics of, 383
Infective heredity, 422, 461
 and viral tumors, 446
Inheritance
 quantitative, 193
 polygenic, 193

Inhibition
  enzymatic, 367
  systems with soluble antigens, 366
Instinct, broody, 278
Intensity of behavioral pattern, 288
Intensity-stimulus, relationship between, in two
    populations, 289
Interaction
  effects, nonadditive, 161
  genic, in evolution, 183
  of genes and teratogens, 244
  types of, 179
Intercross generation matrix, 31
Intercrosses, 4, 57, 58
  in repulsion, 59
Interdemic selection, 192
Intermediate optima, 191
Internal emitters, 129
International cooperation in genetic studies, 151
International rules of nomenclature for mice,
    517–521
Interstitial cell tumors of testis, 250
Intrademic selection, 185
Intrauterine variables related to malformations,
    243
Ion-exchange absorption, 307
Ionization, 130
  density, 141
Iron mordant, 488
Isoantigenic
  divergence, 454
  markers, 417, 468
  variants from F_1 tumors, 434
  variation, 460
Isoantigenicity of induced tumors, 426
Isoantigens, 468
  and transplantation characteristics, 424
  of cells in culture, 463
Isoimmunization, fetal-maternal, 465

Kappa particles in *Paramecium*, 262
Karyotype, 494
  and drug resistance, 440
  diploid, retention of, 471
  variation, 508
Karyotypic control, 472
Kidney, absence of and *Short-Danforth* gene, 222
Klinefelter's syndrome, 415

Labeled
  antibodies, 378, 463
  antigens, clearance of, 363
  cells, elimination rates of, 363
Labor of measurement, 202
Lactation in cattle, 273
  rank order and, 274
Latency of behavior, 288
Layout for preparation of chromosomes, 490
Leaky enzyme, 230
Lethal(s), 55
  dominant, from radiation, 135

Lethal(s)—*continued*
  linked procedure, 138
  mutations, 136
  sex-linked, 140
Leucocyte counts, 562
Leukemia, 249, 260, 268
  genetics of in mice, 254
  genetic studies on, 253
  murine, 423
  resistance to, 253
Liesegang phenomenon, 376
Limiting ratio of heterozygosis, 49, 53
Linear accelerator, 129
Linear order of loci, 73
Line breeding, 4
Linkage
  chi-square, for three kinds of mating, 57
  coupling phase, 57
  crosses, proportion of offspring in, 62
  detection of, between *ru* and *je*, 66
  groups, 76
    autosomal, 76
  map, 56, 76
    distances between loci, 77
    of the mouse, 77, 78
  methods for testing, 56
  nonrandom associations, 194
  of *ru* and *je*, 70
  repulsion phase, 57
  sex, 77
  stocks for testing, 75, 76
  testing segregations and detecting, 58
  types of matings useful in determining, 56
Linked genes, order of, 56
Linked lethal procedure, 138, 153
Linked locus, analysis of gametic pairs with
    respect to, 47, 50
Literature, control of, on genetics of the mouse,
    511
Litter
  correlation between weight and size in guinea
    pig, 281
  reduction of size following radiation, 139
  size of, 271, 281
    and hormonal secretion, 273
  size, variance of in mice, 208
Liquid nitrogen, use of as refrigerant for cells *in
    vitro*, 510
Loci
  average rate of fixation of, 55
  linear order of, 73
Locus, *piebald*, mutation rate of, 138
Lung
  fetal transplants of, 259
  tumors of, 249, 254, 258, 259, 264, 266
    estimating susceptibility genes for, 248
    estimation of number of genes, 254
    in mice, 251
    method of induction with carcinogens, 255
    production of
      with nitrogen mustard, 255

Lung—*continued*
  tumors of—*continued*
    with sulphur mustard, 255
    urethan-induced, 200, 255
Luxate, 70, 71
  use of in retrograde analysis, 227
Lymphosarcoma in guinea pig, 254
Lysogeny, relationship to cancer, 268
Lysogenic conversion, 447, 456, 459

Machinery, washing for sanitizing equipment,
    553
Macromolecular antigens, 348
Maize
  somatic mutation in, 421
  variegation in, 422
Malaria, avian, 403
Maldevelopment, analysis of, 237
Malformations
  clustering of, 243
  intrauterine variables related to, 243
  mother-fetal relationship in, 241
  pathogenic mechanisms underlying, 238
Malignancy and antigenic differences, 453
Malignization of cells *in vitro*, 452, 461
Mammalian
  hemoglobins, 299
  immunogenetics, 347
  radiation genetics, 127
Mammalian hemoglobins, 299–320
Mammalian radiation genetics, 127–151
Mammals, radiation genetic analysis of, 132
Mammary fat pads of mice, transplantation of
    tissues into, 565
Mammary gland
  cancer, 265
  effect of pituitary hormone, 261
  transplantation into fat pad of, 261
  tumor, 249, 256, 260
    agent, loss of, by dilution, 267
    virus, 262
Manuals and references useful in cytogenetics,
    503
Map, linkage, 56, 76
  distances between loci, 77
  of the mouse, 77, 78
Markers, 461, 467, 468
  chromosomal, 452
    in tumors, 446
  drug resistance, 452, 457
  isoantigenic, 417, 468
  phenotypic, 459
    and studies of somatic variation, 410
    in neoplastic cells, 448
    in tumors, 445
    in vivo, 460
Mass selection, 192
Mastomys, carcinoma of glandular stomach in,
    254
Maternal
  behavior in cocks, 278

Maternal—*continued*
  fetal
    isoimmunization, 465
    relationship resulting in malformations, 241
  weight and teratogenesis, 242
Matrices
  for backcross system, 20, 21
  for brother-sister mating, 16
  for cross-intercross system, 26
Matrix
  cross-backcross-intercross system, 30
  cycle, cross-intercross system, 25
  G, 11, 20
  generation, 8
    for cross-backcross-intercross system, 30, 31
  method, 54
  sequence relative to type of, 5
Mating
  brother-sister, 48, 49, 53
  double, 388
    technique of, 402
  frequencies of types, 4, 6, 7, 11
  gamete diagram, brother-sister, 45
  matrix of types, 49
  probabilities of types, 4
    in backcross system, 18
    in cross-backcross-intercross system, 27, 29
    in cross-intercross system, 22
  random, 4, 6
  systems (see Systems of breeding)
  type frequencies, 4, 6, 7, 11
    vector of, 8
Maximum-likelihood
  estimate of probability, 62
  methods, 61
    for estimation of, 61
  scores, 63, 65, 73, 74
Maze, geotactic, 292
Mean
  intensity index for various genotypes of coat
      color in guinea pigs, 166
  length of heterozygous chromosomes, 22
  quality index for genotypes of coat color in
      guinea pigs, 166
  selection values of gene-frequency systems, 187
  selection values of homallelic populations, 183
    with six modifiers, 184
  superiority, 211
Measurement
  labor of, 202
  of behavior, 288
Mechanization of tests for behavioral patterns,
    292
Media
  sterilization, filter assemblies for, 506
  synthetic, for tissue culture, 462
Melanin pigmentation in mammals, biochem-
    istry of, 336
  turbidimetric estimates of, 338
Melanocytes, 162
  origin of, in black and white graft, 334

Mercaptoethylamine, 149
Metabolic anomalies in mice, 466
Method for analyzing regular mating systems, 8
Methodology of behavioral genetics, 293
Methodology of experimental mammalian teratology, 233–246
Methods
   in mammalian immunogenetics, 347
   for measuring behavioral differences
      choice of conditions for testing, 289
      controlling environment, 287
      developmental, 286
      genotypic *versus* phenotypic orientation, 290
      measurement of behavior, 288
      statistical analysis, 291
   for pigment cell-research, 323
   for testing linkage, 56
   of breeding, 1
   of keeping records, 522
   of procurement of blood, 351
   of tissue culture, 492
   of tissue transplantation, 371
   useful in physiologic genetics, 227
Methods for testing linkage, 56–82
Methods in mammalian immunogenetics, 347–374
Methods of keeping records, 522–537
Methylcholanthrene, 255
Mice, 55
   animal rooms
      filtration of air in, 541
      humidity in, 543
      temperature, 541
      traffic flow in, 545
   bedding in cages for, 541, 545
   cages for, 547
   cannibalism in, 544
   collection of blood in, 558
   comparison of rate of ovulation and reproductive efficiency of, 276
   covers for cages, 549
   established strains of, 86
   feeding of, 539
   food for, introduction of pathogens in, 545
   food hoppers for cages of, 550
   handling of, 543
   husbandry, 539
      equipment, and procurement of, 538
   international rules of nomenclature for, 517
   procurement of, 554
   shipping containers for, 555
   symbols for designating substrains of, 101
   tables and carts for care of, 553
   technique for study of anemia in, 558
   transplantation of tissues into mammary fat pads of, 565
   vendors of equipment useful in care of, 556
   watering devices for, 550
Microbial genetics, relationship to somatic cell genetics, 407
Microphthalmia, 178, 191, 243

Mitotic arrest
   and hypotonicity, 473
   and vincaleukoblastine, 475
Mixed cross, 57
Molds, haploidization in, and somatic crossing over, 412
Molecular antigens, 348
Mongolism, 505
Morbidity, relationship of, to genic differences in susceptibility, 404
Mordant, iron, 488
Morphology, chromosomal, 423
Mortality
   infant, in mice, 146
   neonatal, infant, and childhood, of mice compared to man, 147
   of mice from *Salmonella*, 392
Mosaicism, 460
Mosaics, 414
Mouse
   colony, facilities and equipment, 546
   linkage map of, 77, 78
   names of genes in, 79
Moving-boundary electrophoresis, 301
   of carbonmonoxyhemoglobins, 302
Multiple-gene stocks, 75
Mutation(s)
   backcross test of, 137
   chemically-induced, 154
   dominant, 136
      lethals, effect of hypoxia on, 148
      visible, 140
   effect of purine and adenosines on, 154
   following radiation, 136
   lethal, 136
      from radiation, 138
   microbial, 142
   probability of, 54
   random, 192
   rate
      affected by actinomycin D, 152
      for sex-linked lethals, 140
      in somatic cells, 509
      relationship to dose of radiation, 141
      spermatogonial, 141
   recessive, 136
      visible, 141
   recovery process, 138
   somatic, 416, 417, 460, 465
      in *Drosophila*, 421
      in maize, 421
      in translated tumors, 432
      X rays in, 419
   specific locus test of, 137
   visible, 136
Murine leukemia, 423
Murine typhoid, 384
Muscular dystrophy in mouse, 220
Mutagenesis, relationship to carcinogenesis, 263
Mutants
   auxotrophic, selection of, 451

Mutants—*continued*
  of *Peromyscus maniculatus*, 121
Myelomeningocele, 238
Myoepitheliomas, 250

Nagasaki, 144
Natural fitness, 214
Natural selection, 185
Net selection coefficients, 187
Neoplasia (see also Cancer, Carcinoma, Tumors)
  estimation of number of susceptibility genes, 254
  genetics of, 247
  hepatomas, 265
  host-parasite relationship, 262
  in species other than mouse, 253
  leukemia, 268
  linkage between genic loci, 256
  localization of genic action in organs and tissues, 258
  multiple-factor inheritance, 247
  nongenetic factors, 247
  pancreatic, in hamsters, 267
  paths of genic action in, 258
  pituitary, in hamsters, 267
  pulmonary, 264, 266
Neoplastic cells
  in vivo, 422
    phenotypic characteristics of, 423
    phenotypic markers in, 448
Neoplasms, inbred strains for studies on, 249
Neural crest, 341
Neurospora, 161, 218, 231
Neutrons, 128
  fast, 128
  somatic lethal effects, 142
  thermal, 128
Nitrogen, liquid for refrigeration of cells *in vitro*, 510
Nitrogen mustard, production of pulmonary tumors with, 255
No effect of some genic combinations, 180
Nomenclature
  for inbred strains, 84
  of mice, international rules for, 517
  standardized for inbred strains, 84
Nonadditive variance, 198
Nondisjunction, 416, 460
Nongenetic aspects of mammalian pigmentation, 325
Nonrandom linkage associations, 194
Nonuniformity of absorbed dose, 128
Nuclear transfers, 464
Nuclear transplantation, 409
Number of genes, 214
Nutritional requirements of cells *in vitro*, 451
N-terminal analysis, 315, 321

Observed characters, relationship between genome, external environment, and, 160
Ocular carcinoma in cattle, 248

Official characters, 248
Offspring
  from recombinant and nonrecombinant gametes in three kinds of mating, 57
  parent regression, 201
Oligodactyly, 60
Order of linked genes, 56
Organ chimeras, 444
Origin of pigment cells, 341
Otocephaly, 247
Ova, fertilized, technique for transfer of, 563
Ovarian transplants, 373
Ovaries
  transplantation of, 252
  tumors of, 250, 261
Ovulation, rate of, 271
  and hormonal secretion, 273
  and reproductive efficiency, 276
  in mice, 276
Oxygen consumption of skin in different genotypes, 337
Oxyhemoglobins of mice
  analyzed by starch-gel electrophoresis, 306
  analyzed by ultracentrifugation, 310

Paired-dose method of radiation, 131
Panmictic indices, 45
Paper chromatography, 340
Paper electrophoresis, 303
Papilloma of skin, 249
Parabiosis, 373, 414
Paraffin-oil adjuvants, 377
Paralysis, tolerance, and enhancement, 373
Paramecium, kappa particles in, 262
Parameters, radiation, 128
Parasite-host relationship, 381
  in cancer, 262, 264
  symbiotic relationships to host, 403
Parental gamete, 58
Parent-offspring inbreeding, 4
Partial sterility, 135
Path analysis, 42
Path-coefficient analysis, 54
Pathogenic mechanisms underlying malformations, 238
Pathogens
  effect of numbers on survival, 389
  introduction in food for mice, 545
  relationship between survival and selection for resistance to, 391
Pattern of alteration of chromosomal numbers, 507
Peptide analysis, 317
Percentages, recombination, 77
Permanent preparations of chromosomes, techniques for, 484
Peromyscus maniculatus, 185
  mutant stocks of, 121
  stocks of possible genetic interest, 122
Phaeomelanin, 162, 324, 340
  in guinea pigs, 163, 164

Phage resistance, clonal variance of mutations to, 408
Phenomenon, Liesegang, 376
Phenotypic
  characteristics in neoplastic cells *in vivo*, 423
  correlation between members of families, 203
  markers, 459
    and studies of somatic variation, 410
    in neoplastic cells, 448
    in tumors, 445
    in vivo, 460
  value of character, 194
  variance, 216
  versus genotypic orientation in behavioral patterns, 290
Phenotypes, behavioral, 292, 295
Photoreactivation, 152
Physical methods for analyzing hemoglobins, 300
Physiologic characteristics, differences between inbred strains, 294
Physiologic genetics, 157
Piebald locus, 138, 147
Pigment
  development of with age, 341
  spread, 333
Pigmentation
  inherent in tumor cells, 445
  in mammals, nongenetic aspects of, 325
  melanin biochemistry of, 336
Pigment cell(s)
  origin of, 341
  research, 323
  studies in
    cattle, 333
    guinea pig, 333
    rabbits, 333
Pituitary adenoma, 250
Pituitary tumors in hamsters, 267
Plasma-cell tumors, abnormal proteins in, 445
Plating, replica, 408
Pleiotropic
  effect of
    flexed tail, 257
    genes causing anemia, 224
  effects, 231
Pleiotropism, 226, 230
  W-series and *steel*, 226
Pleiotropy
  effect of on hair direction on white forehead spot, 173
  secondary, 161
Polydactylism and irradiation, 190
Polydactyly, 189, 190, 191, 247, 254
  in guinea pigs, 176, 293
Polydipsia in mice, genetic differences in, 232
Polygenic inheritance, 193
Polyoma virus, cytocidal interaction, 447
Population
  comparison of evolutionary processes in, 188
  homallelic, 183
    with six modifiers, 184

Population—*continued*
  of limited size, 46
Postimplantation losses, 152
Precipitation system, 368
Preimmunization of recipient hydrids, 425
Preimplantation losses, 151
Premutational damage, 142
Preparation of slides, 504
Preservation of red cells, 364
Pressure immigration, 185
Prevention of loss of lines of cells, 493
Probabilities of mating types, 4
  in backcross system, 18
  in cross-backcross-intercross system, 27, 29
  in cross-intercross system, 22
Probability of heterozygosity, 4
  in backcross system, 20
  in cross-backcross-intercross system, 30
  in cross-intercross system, 24
  with heterozygosis forced by intercrossing, 37, 38
Probability of incrosses
  for backcross system, 20
  for brother-sister inbreeding system, 15
  for cross-backcross-intercross system, 29
  for six mating systems, 39
  with heterozygosis forced by backcrossing, 33
Probability of mutation, 54
Probability of various mating types in cross-intercross system, 24
Problems and potentialities in the study of genic action in the mouse, 217–229
Procurement
  husbandry, and equipment of mice, 538
  of blood, methods for, 351
  of mice, 554
Progression, tumor, 423, 434, 444
Prolactin, effect on cocks and phenotype, 278
Proportions
  of offspring in linkage crosses, 62
  recombinations, 75
Protection against induced genetic damage, 148
Proteins
  abnormal, in plasma-cell tumors, 445
  alpha-2, 190
Protons, high-energy, 129
Pulmonary tumors, 249, 254, 258, 259, 264, 266
  estimating susceptibility genes for, 248
  estimation of number of genes, 254
  fetal transplants of, 259
  in mice, 251
  method of induction with carcinogens, 255
  production of
    with nitrogen mustard, 255
    with sulphur mustard, 255
  urethan-induced, 200, 255
Purines, 154

Qualitative genetic effects of radiation, 134

Quantitative
  character, 193
    number of genes, 214
  degree of genetic variation, 216
  determinations of the erythron, 560
  inheritance, 193
  variability, 191
Quantitative inheritance, 193–216
Quasicontinuous variation, 241–247

Rabbits
  gamma-globulin groups in, 379
  pigment cells, 333
  stocks of, of genetic interest, 120
  strains of, in development, 118
  symbols for designating strains of, 120
  uterine tumors in, 254
Racks for cages, 552
Radioactive
  antibodies, 463
  tyrosine, 342
Radiation
  chimeras, 417
  chronic exposure, 130
  congenital deformities produced by, 153
  cosmic, 129
  dominant lethals, 135
  dominant visible mutations, 140
  dose-rate effect, 142
  dosimetry, 128
  effect, cytogenetic analysis of, 143
  effect of actinomycin D on, 152
  effect of dose rate of, on mutation rate, 141
  effect of hypoxia on, 148
  effect of, on embryonic stages, 151
  exposure variables, 131
  exposure, techniques of, 131
  fractionated exposure to, 130, 142
  gamma, 128
  genetic analysis in mammals, 132
  genetics, 125
  hereditary partial sterility, 134
  induced sterility, 133
  internal emitters, 129
  international cooperation in studies, 151
  mutations, 136
    dominant, 136
    lethal, 136, 138
    recessive, 136, 138
    visible, 136
  neutrons, 128
  nonuniformity of absorbed dose, 128
  paired-dose method, 131
  parameters, 128
  premutational damage, 142
  pressures to reduce hazard of, 152
  protection against induced genetic damage,
    148
  qualitative, genetic effects of, 134
  quantitative, genetic effects of, 143
  recessive visible mutation, 141

Radiation—*continued*
  reduction of litter size following, 139
  reproductive performance following, 133
  response of anemic mice, 225
  selection of species and systems of mating,
    149
  sex ratio following, 144
  single exposure to, 130
  space, 129
  sources, 128
  temporal factors in, 130
  tool for genetic research, 150
  viability following, 144
  X, 128
Radius, swept, 76
Random
  assortment, 48
  drift, 187, 191, 192
  mating, 4, 6
  mutation, 192
  selection, chance in, 54
Rates
  of hormonal secretion, 271
    correlated changes in, 273
  of ovulation, 271
    and hormonal secretion, 273
    and reproductive efficiency in mice, 276
Ratio of heterozygosis, 48
Rats
  established strains of, 105
  lists of symbols for designating substrains, 110
  stocks of genetic interest, 110
RBE, 128, 141, 142, 145
  nonlinear relation between, and energy, 129
  values, 129
Recessive mutations, 136
  lethal, from radiation, 138
  visible, 141
Recombinant
  double, 75
  gamete, 58
Recombination
  between *ru* and *je*, estimation of, 72
  differences between sexes, 78, 79
  estimate of, 78
  estimation of, between *lx* and W^v, 71
  frequencies, estimation of, 56
  percentages, 77
  probability of, 61
  proportion, 75
  sexual, 408, 459
Records, methods of keeping, 522
Recovery from mutation, 138
Recurrence equation for relative heterozygosis,
    49
Red blood cells
  antigens coupled to, 367
  immunogenetics of, 350
  preservation of, 364
  test systems for immunogenetic studies, 357
Reflection-meter readings, 162

References, 571
  and manuals useful in cytogenetics, 503
Refrigeration of cells *in vitro* with liquid nitrogen, 510
Regression
  offspring-parent, 201
  of litter size following radiation, 139
Regular mating systems, 8
Relationship between genome, external environment, and observed characters, 160
Relative biological effectiveness, 128
Relative heterozygosis, recurrence equation for, 49
Replica plating, 408
Replicative homologs, 178
Reproductive efficiency and rate of ovulation, 276
Reproductive performance, 138
  following radiation, 133
Reproductive physiology, genetics of, 269
Repulsion
  intercross in, 59
  phase of linkage, 57
Requirements, nutritional, of cells *in vitro*, 451
Residual heterozygosis and resistance to induced sarcomas, 436
Resin-paper analysis, 322
Resistance
  disease, transmission of, 386
  drug, 460, 468
    and hormonal independence, 437
    and karyotype, 440
    of tumor cells, 423
  mechanisms, relationship to X-ray irradiation, 396
  of cells *in vitro* to antimetabolites, 423
  phage, clonal variance of mutations to, 408
  to disease, historical sketch, 400
  to infection, selection for, 401
  to murine typhoid, 384
Response, second-set, 429
Retinal degeneration, 219
Retrograde analysis, 222, 227
  of metabolic deviation, 220
Response to selection, 211
Reverse-lighting patterns in animal breeding rooms, 154
Rex, 75
Ribonuclease, use of in preparing chromosomes, 479
Ricin, route of entry and survival of mice inoculated with, 395
Rodless retina (r), 77
Route of entry and survival of mice inoculated with ricin, 395
Route of infection, relation to severity of disease, 393
ru, 72
Ruby, 64, 77
Ruby eye, 58

Rules, international for nomenclature, 517

Saline agglutination, 349, 357
Salmonella gallinarum, 403
Salmonella, mortality of mice from, 392
Salmonella typhimurium, 389
  11C, 384, 403
  survival of mice related to dose of, 390
Sarcoma, subcutaneous, 250
Scores
  and information per individual, table of,
    $Aa/Bb \times aa/bb$ mating, 67
    $AB/ab \times AB/ab$ mating, 68
    $Ab/aB \times Ab/aB$ mating, 68
    $AB/ab \times AB/ab$ mating, with A semi-dominant, 69
    $AB/ab \times AB/ab$ mating, with A and B semidominant, 69
  methods, 70
  maximum likelihood, 63, 65, 70, 73, 74
Second-set response, 429
Segregation(s)
  disturbed, 61
  in individual loci, chi-square, 59
  pattern, differences in for hemoglobin, 321
  testing and detecting linkage, 58
Seizures
  audiogenic, 230, 285, 287
  in mouse, 232
  in rabbits, 231
  phenylalanine hydroxylase level, 232
  relationship between pigment and audiogenic, 232
Selection
  coefficient, 187
  differential, 211, 213
  final level reached, 211
  for resistance to infection, 401
    relation to survival, 391
  in neoplastic cells *in vivo*, 423
  intrademic, 185, 192
  mass, 192
  natural, 185
  of auxotrophic mutant, 451
  of drug-resistant bacteria, 48
  of species and systems of mating for radiation genetic studies, 149
  pressure on transplanted cells, 422
  random chance in, 54
  response to, 211
  time to reach limit, 211
  two-way, 212
Selection-variation mechanism, 461
Selective peaks, 191
Selective value
  in genic interaction, 181
  mean of homallelic populations, 183
  of alleles, 182
Semilethals, 55
Sequence, of histocompatibility loci, 466
Sera, methods for procurement of, 351

Series, Fibonacci, 45
Serology, gel-diffusion, 369, 370
Serum
    alpha-2, 190
    globulin, 348
Severity of disease, related to route of infection, 393
Sex
    Chinese hamster, 496
    chromosomes, 494
    factor of *E. coli*, 459
    linkage, 77
        lethals, 140
        ratio, 144
Sexual recombination, 408, 459
Shaker-2, 75
Shape determining effects of genes, 223
Shipping containers for mice, 555
Sib analyses, 205
    with unequal numbers in families, 207
Silver (si), 77
Simplification, antigenic, in transplanted tumors, 425
Single backcross, 57
Single cells
    cloning, 468
    isolation of, for tissue culture, 450
Single exposure to radiation, 130
Size, population of limited, 46
Skin
    carcinoma of, 249
    grafting, 371
Slides, preparation of, 504
Solubility
    and crystallography of hemoglobin, 311
    of hemoglobin, 314
Soluble antigens, 366
Solutions, stock, 504
Somatic
    cell
        genetic relationship to microbial genetics, 407
        genetic transfer between, 456
        genetics of, 405
        mutation rates in, 509
        variation, 467
    crossing over, 410, 460, 467, 468
        in transplanted tumors, 433
    lethal effects of neutrons, 142
    mutation, 416, 417, 460, 465
        hypothesis of oncogenesis, 268
        in *Drosophila*, 421
        in maize, 421
    variation, 423, 468
        studies through phenotypic markers, 410
Sources
    of gamma ray, 128
    of radiation, 128
Space radiation, 129
Special sublines for study of mammary tumor agent, 251
Specificity, antigenic, loss of, 432

Specific locus tests, 153
    for mutation, 137
Spermatogonial mutations, 141
Spleen, transplantation of tissue to, 330
Spotting, 326
    in guinea pigs, 169
    white, 330
Spread of pigment, 333
Squashing procedures, illustration of, 489
Staining of chromosomes, 480
Standard error of heritability, 209
Starch-block electrophoresis, 304
Starch-gel
    analysis, 322
    electrophoresis, 305, 370
Stemline, 507
Sterility, 133
    and gonadotrophic hormone, 270
    genetic, 279
    hereditary, from radiation, 134
    induced by radiation, 133
    partial, 135
    reciprocal translocation association with, 134
Sterilization of media, filter assemblies for, 506
Stimulus-intensity, relationship between, in two populations, 289
Stock(s) (see also Strains)
    for testing linkage, 75, 76
    genetic, 83
    multiple gene, 75
    solutions, 504
Stomach
    adenoma of, 250
    carcinoma of, in *Mastomys*, 254
Strains (see also Stocks)
    established, in mice, 86
    genetic, 83
    of Chinese hamsters in development, 116
    of guinea pigs
        established, 112
        in development, 112
    of rabbits in development, 118
    of rats, established, 103, 114
    of Syrian hamsters
        established, 114
        in development, 115
Stripped nuclei, 129
Subject-strain bibliography, 511
Sublines of tumors, comparison of pairs, 449
Substrains of mice, symbols for, 101
Sulphydryl analysis, 315
Sulphur mustard, production of pulmonary tumors with, 255
Summary of systems of mating, 41
Superiority, mean, 211
Survival
    effect of numbers of pathogens on, 389
    of mice related to dose of *Salmonella typhimurium*, 390
Susceptibility
    of arthropods to viruses, 402

Susceptibility—*continued*
relationship of genic differences to morbidity, 404
to disease, historical sketch, 400
Suspended cells, use of in immunogenetics, 365
Swept radius, 76
Swine
amount of gonadotrophic hormone secreted, 281
hormonal content of adenohypophyses in different strains of, 274
Symbiotic relationship of parasite to host, 403
Symbols
for designating substrains of guinea pigs, 113
for designating substrains of hamsters, 117
for designating substrains of mice, 101
for designating substrains of rabbits, 120
for designating substrains of rats, 110
Syndrome
Klinefelter's, 415
Turners, 416
Synergism between teratogens, 241
Syrian hamsters, 471, 473
established strains of, 114
stocks of, of genetic interest, 116
strains of, in development, 115
Systems based on relationship, 4
histocompatibility, 423, 468
Systems of breeding, 3, 4
agglutination, 358
based on controlling locus of interest, 5
brother-sister inbreeding, 5, 12
backcross, 18, 53, 57
with heterozygosis forced by backcrossing, 5, 32, 33
with heterozygosis forced by intercrossing, 5, 36
choice of, 39
combination of locus control and relationship, 5
cross-backcross-intercross, 3, 27, 28, 53
cross-intercross, 22, 23
double backcross, 57
generation matrix in the backcross system, 20
generations required to obtain a given percentage of incrosses, 21
intercross, 57
matrices for brother-sister, 16
mixed cross, 57
probability of heterozygosity in the backcross system, 20
probability of incrosses for the backcross system, 20
random mating, 5, 6
selection of species and for radiation genetic studies, 149
single backcross, 57
summary, 41
theory of, 3

Systems of breeding—*continued*
used in cancer research, 249
used in mammalian genetics, 3
Systems of mating used in mammalian genetics, 3–41

Tables
and carts for care of mice, 553
Finney's, 67, 68, 69, 70
Tactics in pigment-cell research, 323–336
Tag holders, 553
Target-specific hormones, 280
Techniques
cloning, 461
cytotoxic, 375
for preparing chromosomes, 472
for the study of anemias in mice, 558
for the transfer of fertilized ova, 563
of radiation exposure, 131
Techniques for the study of anemias in mice, 558–562
Techniques for the transfer of fertilized ova, 563–564
Temperate bacteriophages, 459
Temperature
effect of, on coat color in guinea pigs, 169
effect of, on genic action, 231
effective period, 191
of animal rooms for mice, 541
Temporal factors in radiation, 130
Teratogenesis, pathogenetic mechanisms underlying malformations, 238
Teratogens, 188
biochemical effects on development, 240
effect of combinations of, 241
interaction with genes, 244
synergism between, 241
Teratomas, testicular, 250
Teratology
agents used, 235
clustering, 243
developmental stages at which agents used, 236
dosage of agents used, 237
experimental, 233
influence of maternal weight on, 242
interaction of genes and teratogens, 244
mammalian, 233
proper use of controls, 234
relationship between mother and fetus, 241
summary of methodology, 245
types of animals used for study, 233
use of inbred lines, 234
Tester stock, 137
Testing segregations and detecting linkage, 58
Test
complement-fixation, 376
fluctuation, 438, 439, 442, 452, 461
Tests, cytotoxic, 378
in non-*H*-2 systems, 378

Testis
  interstitial-cell tumors of, 250
  teratoma, 250
  tumors of, 260
Thalassemia, 304
Therapy, X-ray, 128
Thermal neutrons, 128
Threshold
  characters, 247, 248
  concept of, 285, 293
  developmental in behavioral genetics, 296
  genetic characters, 267
Timing of developmental stages in teratogenic
    studies, 236, 237
Tissue culture, 450, 492
  behavior of Chinese hamster cells in, 499
  cellular contamination of, 493
  Chinese hamster, 485
  methods of, 492
  synthetic media, 462
Tissue localization of action of *W* gene, 225
Tissue transplantation, methods for, 371
Titer, antibody, 193
To (*tortoise*), 140
Toe, small, factors controlling development of
    in guinea pig, 177
Tolerance
  immunologic, 327
  paralysis and enhancement, 373
Tortoise (*To*), 140
Total phenotypic, variance, 198
Tradescantia, 129
Traffic, flow of, in animal rooms for mice, 545
Traits, continuously distributed, 296
Trajectories of gene-frequency systems, 186
Transduction, 408, 447, 458, 459, 465
  episomal, 459
  relationship to cancer, 268
Transfer
  genetic, between somatic cells, 456
  genetic, experiments, 468
  nuclear, 464
  of fertilized ova, 251, 563
Transformation, 408, 457, 459
Translocation, reciprocal in sterility, 134
Transmission of disease resistance, 386
Transmutation, 130
Transplantation
  adrenal, 373
  characteristics and isoantigens, 424
  hematopoietic tissues, 373
  nuclear, 409
  of bone marrow, 374
  ovarian, 373
Transplantation of tissues, 371
  for studying mammalian pigmentation, 327
  into mammary fat pads of mice, 565
  to mammalian eye, spleen, and celom of chick
    embryo, 330

Transplanted pulmonary tissue, 259
Transplanted tumors
  antigenetic simplification of, 425
  Barrett-Deringer change in, 426
  chromosomal changes in, 433
  crossing over in, 433
  genetic and epigenetic mechanisms of, 432
  point mutations in, 432
Trembler, 75
Trypsin, use of tissues, 477
Tubes, cultural for cytologic preparations, 504
Tumors
  adenoma of stomach, 250
  adrenal, 260
    cortical, 250
  antibodies, 266
  ascites, 468
  autonomy of, 441
  cells, ascites, form of, 423
  chromosomal markers in, 446
  F_1
    behavior of, 427
    isoantigenic variants from, 434
  genetic, viral and environmental influences on
    etiology, 264
  harderian gland, 250
  hemangioendothelioma, 250
  hepatomas, 265
  heterozygous, behavior of, 429
  homozygous, behavior of, 428
  hormonal dependence of, 441, 465
  host-parasite relationship in, 262
  induced, isoantigenicity of, 426
  interstitial cell of testis, 250
  leukemia, 254, 260, 268
  linkage between genic loci, 256
  localization of genic action in organs and
    tissues, 258
  mammary gland, 249, 256, 260
  myoepithelioma, 250
  ovarian, 250, 261
  pancreatic in hamsters, 267
  paths of genic action in, 258
  pigmentation inherent in, 445
  pituitary adenoma, 250
  pituitary in Chinese hamster, 267
  pulmonary, 249, 254, 258, 259, 264, 266
    estimating susceptibility genes for, 248
    estimation of number of genes, 254
    in mice, 251
    method of induction with carcinogen,
      255
    production of, 200, 255
  progression, 423, 434, 444
  relation of growth curve to incidence, 266
  relationship of stress to susceptibility, 266
  subcutaneous sarcoma, 250
  sublines, comparison of pairs, 449
  teratomas, testicular, 250
  testicular, 260

Tumors—*continued*
transplanted
antigenic simplification of, 425
Barrett-Deringer change in, 426
chromosomal changes in, 433
crossing over in, 433
genetic and epigenetic mechanisms in, 432
point mutations in, 432
uterine, in rabbits, 254
viral, 461
and infective heredity, 446
viruses
cytocidal, 447
in relation to, 265
Tumor-specific antigens, 460
Turbidimetric estimates of natural melanin content, 338
Turner's syndrome, 416
Twinning in cattle, 273
rank order and, 273
Types of interaction, 179
Types of mating useful in determining linkage, 56
Typhoid, murine, 384
Tyrosinase, 164, 189
activity, analysis of variance, 339
color factors and activity of, 170
molecule, 218
Tyrosine, radioactive, 342

Ultracentrifugation, 309
Univalent
antibodies, 349
antigens, 349
Unsquashed preparations of chromosomes, 491
Uranium, 128, 235
Urethan, pulmonary tumors induced by, 255
Urinary tract, anomalies of, 238
Uterine
carrier for teratogens, 241
tumors in rabbits, 254

Van Allen belt, 129
Variance
additive, 198
analysis of, for tyrosinase activity, 339
environmental, 195
genetic, 185
in litter size in mice, 208
nonadditive, 198
phenotypic, 198, 216
Variants
analysis of, 291

Variants—*continued*
assignment of, to genetic and environmental components, 291
drug-resistant, of cells in culture, 462
isoantigenic, from F_1 tumors, 434
Variation, 195
in barking, 286
in karyotype, 508
isoantigenic, 460
quantitative, degree of genetic, 216
quasicontinuous, 241, 247
relationship between genetic and behavioral, 285
somatic, 423, 467, 468
studies through phenotypic marker, 410
Variation-selection mechanism, 461
Variegation in maize, 422
Vector
host, 403
of mating-type frequency, 8
Vendors of equipment useful in care of mice, 556
Viability following radiation, 144
Vincaleukoblastine
and mitotic arrest, 475
and preparation of chromosomes, 475
Viral
hypothesis of oncogenesis, 268
tumors and infective heredity, 446
tumors, 461
Viruses
relation to etiology of tumors, 265
susceptibility of arthropods to, 402
tumor, cytocidal interaction, 447
Visible mutations, 136
dominant, 140
recessive, 141

wabbler-lethal mice, 219
Washing machinery for sanitizing equipment, 553
Watering devices for mice, 550
Weight, correlation of, with size of litter in guinea pigs, 281
White spotting, 330

X factor, 437
X irradiation, 128
and somatic mutations, 419
polydactylism, 190
relationship to resistance mechanisms, 396
therapy, 128

Zone electrophoresis, 302
Zygote diagram of brother-sister mating, 44